Bioconjugate
Techniques

Bioconjugate
Techniques

Greg T. Hermanson

Pierce Chemical Company
Rockford, Illinois

Academic Press

An Imprint of Elsevier

San Diego   New York   Boston   London   Sydney   Tokyo   Toronto

Front Cover Illustration: A DNA double helix chemically modified at the N2 of a guanine residue to possess a γ-aminobutyric acid (GABA) group. The molecular model was kindly provided by Dr. George Pack of the University of Illinois College of Medicine at Rockford.

This book is printed on acid-free paper.

Copyright ©1996, Elsevier (USA).

Academic Press
An imprint of Elsevier
525 B Street, Suite 1900, San Diego, California 92101-4495, USA
http://www.academicpress.com

Academic Press
84 Theobald's Road, London WC1X 8RR, UK
http://www.academicpress.com

Library of Congress Cataloging-in-Publication Data

Hermanson, Greg T.
   Bioconjugate Techniques / Greg T. Hermanson.
      p.   cm.
   Includes bibliographical references and index.
   International Standard Book Number: 0-12-342335-X (case)
   International Standard Book Number: 0-12-342336-8 (paperback)
1.   Bioconjugates–Synthesis. I. Title
[DNLM: 1. Biochemistrry–methods. 2. Proteins-chemistry.
3.   Peptides-chemistry. 4. Nucleic Acids–chemistry. QV 25 H552b 1995]
QP517.B49H47   1995
574.19'296–dc20
DNLM/DLC
For Library of Congress   95-16511
                                                 CIP
PRINTED IN THE UNITED STATES OF AMERICA
            05   06   07   9   8

For Amy and Meghan

Contents Overview

Detailed Contents

Preface

Bioconjugation involves the linking of two or more molecules to form a novel complex having the combined properties of its individual components. Natural or synthetic compounds with their individual activities can be chemically combined to create unique substances possessing carefully engineered characteristics. Thus, a protein able to bind discretely to a target molecule within a complex mixture may be cross-linked with another molecule capable of being detected to form a traceable conjugate. The detection component provides visibility for the targeting component, producing a complex that can be localized, followed through various processes, or used for measurement.

The technology of bioconjugation has affected nearly every discipline in the life sciences. The application of the available cross-linking reactions and reagent systems for creating novel conjugates with peculiar activities has made possible the assay of minute quantities of substances, the *in vivo* targeting of molecules, and the modulation of specific biological processes. Modified or conjugated molecules have been used for purification, for detection or localization of specific cellular components, and in the treatment of disease.

The ability to chemically attach one molecule to another has caused the birth of billion-dollar industries serving research, diagnostics, and therapeutic markets. A significant portion of all biological assays, including clinical testing, is now done using unique conjugates that have the ability to interact with particular analytes in solutions, cells, or tissues. Cross-linking and modifying agents can be applied to alter the native state and function of peptides and proteins, sugars and polysaccharides, nucleic acids and oligonucleotides, lipids, and almost any other imaginable molecule that can be chemically derivatized. Through careful modification or conjugation strategies, the structure and function of proteins can be investigated, active site conformation discovered, or receptor–ligand interactions revealed. Without the development of bioconjugate chemistry to produce the associated labeled, modified, or conjugated molecules, much of life science research as we know it today would be impossible.

Bioconjugate Techniques attempts to capture the essence of this field through three main sections: its chemistry, reagent systems, and principal applications. Although the scope of bioconjugate technology is enormous, this book provides for the first time a practical overview that condenses this breadth into a single volume. Part I, Bioconjugate Chemistry, begins with a review of the major chemical groups on target molecules that can be used in modification or cross-linking reactions. The chemical reactivities and native properties of proteins, carbohydrates, and nucleic acids are examined in separate chapters, with a view toward designing conjugation strategies that work. Next is a discussion on how to create particular functional groups on these molecules where none exist, or how to transform one chemical group into another. Blocking agents also are examined in this section. The last chapter in Part I summarizes all the major reactions used in bioconjugate chemistry in brief, easy-to-follow descriptions, with liberal references to the literature and to other parts of the book where the reactions are put to use.

Part II, Bioconjugate Reagents, provides a detailed overview organized both by reagent type and by chemical reactivity to present all the major modification and conjugation chemicals commonly used today. The first section in this part examines true cross-linking agents. Zero-length cross-linkers, homobifunctional and heterobifunctional cross-linking agents, and the new trifunctional reagents are discussed with regard to their reactivities, physical properties, and commercial availability. In many cases, conjugation strategies and suggested protocols are presented to illustrate how the reagents may be used in real applications. The next section, Tags and Probes, discusses modification reagents capable of adding fluorescent, radioactive, or biotin labels to molecules. Major fluorophores, including fluorescein, rhodamine, and coumarin derivatives as well as many others, are presented with modification protocols for attaching them to proteins and other molecules. In addition, procedures and compounds for adding radiolabels to molecules, including iodination reagents for ^{125}I-labeling and bifunctional chelating agents to facilitate labeling with other radioisotopes, are discussed. Finally, numerous biotinylation reagents are presented along with protocols for adding a biotin handle to macromolecules for subsequent detection using avidin or streptavidin conjugates.

Part III is by far the largest portion of the book. Bioconjugate Applications discusses how to prepare unique conjugates and labeled molecules for use in particular application areas. This includes: (1) preparing hapten–carrier conjugates for immunization, antibody production, or vaccine research; (2) manufacturing antibody–enzyme conjugates for use in enzyme immunoassay systems; (3) preparing antibody–toxin conjugates for use as targeted therapeutic agents; (4) making lipid and liposome conjugates and derivatives; (5) producing conjugates of avidin or streptavidin for use in avidin–biotin assays; (6) labeling molecules with colloidal gold for sensitive detection purposes; (7) producing polymer conjugates with PEG or dextran to modulate bioactivity or stability of macromolecules; (8) enzyme modification and conjugation strategies; and (9) nucleic acid and oligonucleotide conjugation techniques.

Each of these application areas involves cutting-edge technologies that rely heavily on bioconjugate techniques. In many cases, without the basic ability to attach one molecule to another much of the research progress in these fields would grind to a halt. Bioconjugation thus is not the end but the means to providing the reagent tools necessary to do other research or to produce assays, detection systems, or therapeutic agents.

The purpose of this book is to capture this field in an understandable and practical way, providing the foundation and techniques required to design and synthesize any bioconjugate desired. To aid in this process, over 1100 pertinent references are cited and over 650 illustrations depicting reactions and chemical compounds are presented. Hundreds of bioconjugate reagents are examined for use in dozens and dozens of potential applications.

The choices available for producing any one conjugate can be overwhelming. I have attempted to identify the best reagents for use in particular application areas, but the presentation is by no means exhaustive. In addition, most of the protocols included in the book are generalized or based on personal experience or literature citations directed at particular applications. Occasionally, applying a bioconjugate protocol that works well in one instance to another application may not work as expected. One or more of the components of the conjugate may lose activity, the conjugate may precipitate, or yields may not be acceptable. In almost every case, some optimization of reaction conditions or reagent choices will have to be done to produce the best possible conjugate or modified molecule for use in a new application. Even protocols as common as antibody–enzyme conjugation techniques may need to be altered somewhat for each new antibody complex produced. The best strategy is to use the suggested protocols, literature citations, and insights gained from this book as starting points to create a bioconjugate that will work well in your own unique application.

Greg T. Hermanson

Acknowledgments

I acknowledge the many individuals who made valuable contributions to the field of bioconjugation in general and to the contents of this book in particular. First, thanks to the thousands of researchers, some of whose names appear in the reference section, who developed and optimized the hundreds of reagents and applications related to modification and conjugation of biomolecules. Then specifically, many thanks to Ed Fujimoto for making available many of the cross-linkers we have come to rely on and introducing me to the technology "o so many years ago." Thanks also to Krishna Mallia for pioneering cross-linking applications at Pierce; Laura Sykaluk for her optimization of the heterobifunctional cross-linking of antibodies and enzymes, including their analysis by HPLC gel filtration and capillary electrophoresis to determine conjugate yield and purity; Juli Zanocco for her contributions to oligonucleotide conjugation using phosphoramidate methods and hapten–carrier conjugation techniques; Barb Olson for developing solubility and purity data on many cross-linkers, including her expert application study on the trifunctional reagent sulfo-SBED; Keld Sorensen for showing us how to remove excess enzyme from antibody–enzyme conjugates; Randy Krohn for making them all work more than once; Paul Davis for his constant encouragement and many gifts of reprints of pertinent articles; Patti Domen for her contributions to cationized carriers for immunogen production; Dennis Klenk for his review of the manuscript; and especially Robb Anderson for allowing it all to happen. Additional special thanks go to Barb Tanaglia and Sally Etheridge for their expert help in obtaining journal and literature references. I also owe a great debt of gratitude to Gloria Mattson, Dean Savage, Surbhi Desai, Ed Conklin, Sallie Morgensen, and George Nielander for writing so many fine instruction manuals and contributing much technical data.

Finally, I acknowledge and thank Pierce Chemical Company for its support. They have provided much-needed computer and library resources, without which this book would have been much more difficult to write. However, Pierce Chemical Company has not reviewed the text of this book for accuracy or completeness and does not endorse its technical content nor its references to other companies as sources for reagents and materials.

It is my sincere hope that the techniques covered in this book will be useful to scientists in many fields. I welcome any suggestions or references to new techniques that may make future volumes even more useful. Please feel free to call or write me at Pierce Chemical Company or by e-mail at GregTH@AOL.com.

Greg T. Hermanson

PART I

Bioconjugate Chemistry

Functional Targets

Modification and conjugation techniques are dependent on two interrelated chemical reactions: the reactive functional groups present on the various cross-linking or derivatizing reagents and the functional groups present on the target macromolecules to be modified. Without both types of functional groups being available and chemically compatible, the process of derivatization would be impossible. Reactive functional groups on cross-linking reagents, tags, and probes provide the means to label specifically certain target groups on ligands, peptides, proteins, carbohydrates, lipids, synthetic polymers, nucleic acids, and oligonucleotides. Knowledge of the basic mechanisms by which the reactive groups couple to target functional groups provides the means to design intelligently a modification or conjugation strategy. Choosing the correct reagent systems that can react with the chemical groups available on target molecules forms the basis for successful chemical modification.

The process of designing a derivatization scheme that works well in a given application is not as difficult as it may seem at first glance. A basic understanding of about a dozen reactive functional groups that are commonly present on modification and cross-linking reagents combined with knowledge of about half that many functional target groups can provide the minimum skills necessary to plan a successful experiment.

Fortunately, the principal reactive functional groups commonly encountered on bioconjugate reagents are now present on scores of commercially obtainable compounds. The resource that this arsenal of reagents provides can assist in solving almost any conceivable modification or conjugation problem. The following sections describe the predominant targets for these reagent systems. The functional groups discussed are found on virtually every conceivable biological molecule, including amino acids, peptides, proteins, sugars, carbohydrates, polysaccharides, nucleic acids, oligonucleotides, lipids, and complex organic compounds. A careful understanding of target molecule structure and reactivity provides the foundation for the successful use of all of the modification and conjugation techniques discussed in this book.

1. Modification of Amino Acids, Peptides, and Proteins

Protein molecules are perhaps the most common targets for modification or conjugation techniques. As the mediators of specific activities and functions within living organisms, proteins can be used *in vitro* and *in vivo* to effect certain tasks. Having enough of a protein that can bind a particular target molecule can result in a way to detect or assay the target, providing the protein can be followed or measured. If such a protein does not possess an easily detectable component, it often can be modified to contain a chemical or biological tracer to allow detectability. This type of protein complex can be designed to retain its ability to bind its natural target, while the tracer portion can provide the means to find and measure the location and amount of target molecules.

Detection, assay, tracking, or targeting of biological molecules by using the appropriately modified proteins are the main areas of application for modification and conjugation systems. The ability to produce a labeled protein having specificity for another molecule provides the key component for much of biological research, clinical diagnostics, and human therapeutics.

In this section, the structure, function, and reactivity of amino acids, peptides, and proteins will be discussed with the goal of providing a foundation of successful derivatization. The interplay of amino acid functional groups and the three-dimensional folding of polypeptide chains will be seen as forming the basis for protein activity. Understanding how the attachment of foreign molecules can affect this tenuous relationship, and thus alter protein function, ultimately will create a rational approach to protein chemistry and modification.

1.1. Protein Structure Reactivity

Amino Acids

Peptides and proteins are composed of amino acids polymerized together through the formation of peptide (amide) bonds. The peptide bonded polymer that forms the backbone of polypeptide structure is called the α-chain. The peptide bonds of the α-chain are rigid planar units formed by the reaction of the α-amino group of one amino acid with the α-carboxyl group of another (Fig. 1). The peptide bond possesses no rotational freedom due to the partial double bond character of the carbonyl-amino amide bond. The bonds around the α-carbon atom, however, are true single bonds with considerable freedom of movement.

The sequence and properties of the amino acid constituents determine protein structure, reactivity, and function. Each amino acid is composed of an amino group

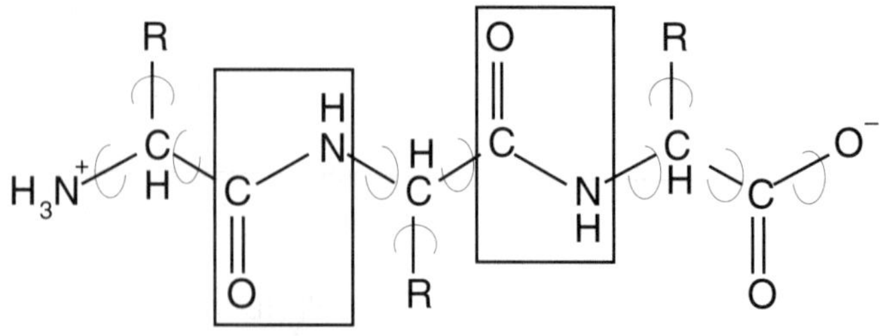

Figure 1   Rigid peptide bonds link amino acid residues together to form proteins. Other bonds within the polypeptide structure may exhibit considerable freedom of rotation.

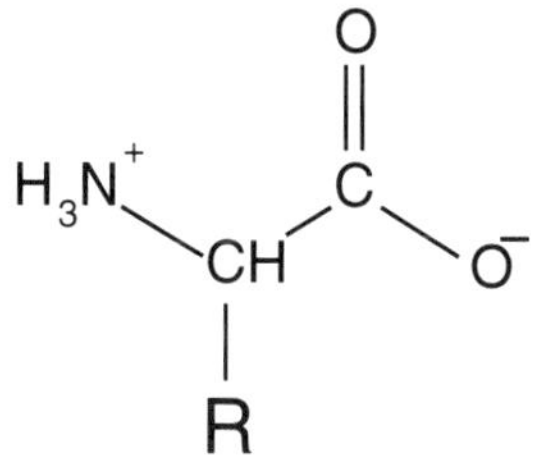

Figure 2   Individual amino acids consist of a primary (α) amine, a carboxylic acid group, and a unique side chain structure (R). At physiological pH the amine is protonated and bears a positive charge, while the carboxylate is ionized and possesses a negative charge.

and a carboxyl group bound to a central carbon, termed the α-carbon. Also bound to the α-carbon is a hydrogen atom and a side chain unique to each amino acid (Fig. 2). There are 20 common amino acids found throughout nature, each containing an identifying side chain of particular chemical structure, charge, hydrogen bonding capability, hydrophilicity (or hydrophobicity), and reactivity. The side chains do not participate in polypeptide formation and are thus free to interact and react with their environment.

Amino acids may be grouped by type depending on the characteristics of their side chains. There are seven amino acids that contain aliphatic side chains that are relatively nonpolar and hydrophobic: glycine, alanine, valine, leucine, isoleucine, methionine, and proline (Fig. 3). Glycine is the simplest amino acid—its side chain consisting of only a hydrogen atom. Alanine is next in line, possessing just a single methyl group for its side chain. Valine, leucine, and isoleucine are slightly more complex with three or four carbon branched-chain constituents. Methionine is unique in that it is the only reactive aliphatic amino acid, containing a thioether group at the terminus of its

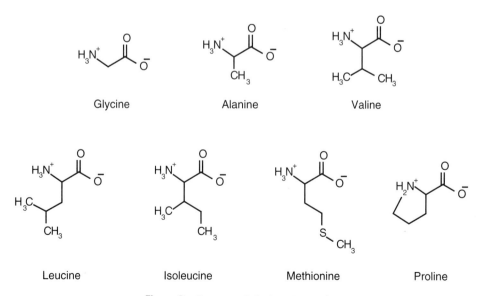

Figure 3   Common aliphatic amino acids.

hydrocarbon chain. Proline is actually the only *imino* acid. Its side chain forms a pyrrolidine ring structure with its α-amino group. Thus, it is the only amino acid containing a secondary α-amine. Due to its unique structure, proline often causes severe turns in a polypeptide chain. Proteins rich in proline, such as collagen, have tightly formed structures of high density. Collagen also contains a rare derivative of proline, 4-hydroxyproline, found in only a few other proteins. Proline, however, cannot be accommodated in normal α-helical structures, except at the ends where it may create the turning point for the chain. Poly-proline α-helical structures have been formed, but the structural characteristics of these artificial polypeptides are quite different from native protein helices.

Phenylalanine and tryptophan contain aromatic side chains that, like the aliphatic amino acids, are also relatively nonpolar and hydrophobic (Fig. 4). Phenylalanine is unreactive toward common derivatizing reagents, whereas the indolyl ring of tryptophan is quite reactive, if accessible. The presence of tryptophan in a protein contributes more to its total absorption at 275–280 nm on a mole-per-mole basis than any other amino acid. The phenylalanine content, however, adds very little to the overall absorbance in this range.

All of the aliphatic and aromatic hydrophobic residues often are located at the interior of protein molecules or in areas that interact with other nonpolar structures such as lipids. They usually form the hydrophobic core of proteins and are not readily accessible to water or other hydrophilic molecules.

There is another group of amino acids that contains relatively polar constituents and is thus hydrophilic in character. Asparagine, glutamine, threonine, and serine (Fig. 5) are usually found in hydrophilic regions of a protein molecule, especially at or near the surface where they can be hydrated with the surrounding aqueous environment. Asparagine, threonine, and serine often are found post-translationally modified with carbohydrate in N-glycosidic (Asp) and O-glycosidic linkages (Thr and Ser). Although these side chains are enzymatically derivatized in nature, the hydroxyl and amide portions have relatively the same nucleophilicity as that of water and are therefore difficult to modify with common reagent systems under aqueous conditions.

The most significant amino acids for modification and conjugation purposes are the ones containing ionizable side chains: aspartic acid, glutamic acid, lysine, arginine, cysteine, histidine, and tyrosine (Fig. 6). In their unprotonated state, each of these side chains can be a potent nucleophile to engage in addition reactions (see the following discussion on nucleophilicity).

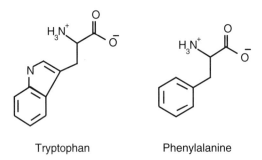

Tryptophan   Phenylalanine

Figure 4   The two nonpolar aromatic amino acids.

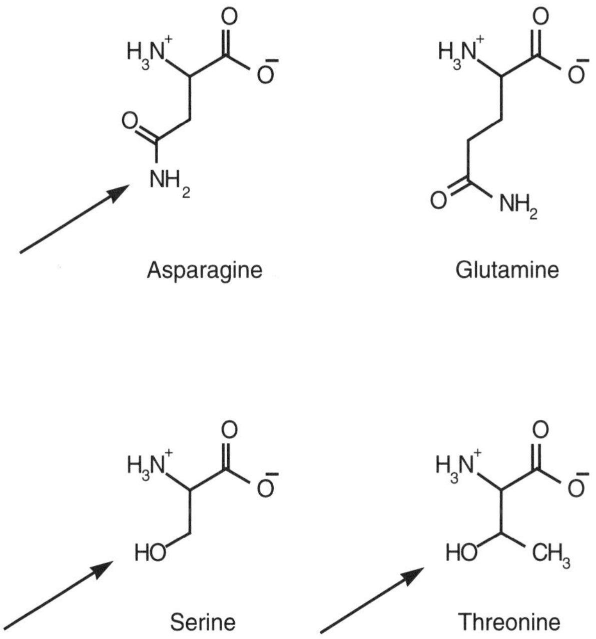

Figure 5 The four polar amino acids. The arrows show the attachment points for carbohydrate residues on glycoproteins.

Both aspartic and glutamic acids contain carboxylate groups that have ionization properties similar to those of the C-terminal α-carboxylate. The theoretical pK$_a$ of the β-carboxyl of aspartic acid (3.7–4.0) and the γ-carboxyl of glutamic acid (4.2–4.5) are somewhat higher than the α-carboxyl groups at the C-terminal of a polypeptide chain (2.1–2.4). At pH values above their pK$_a$, these groups are generally ionized to negatively charged carboxylates. Thus at physiological pH, they contribute to the overall negative charge contribution of an intact protein (see following section).

Carboxylate groups in proteins may be derivatized through the use of amide bond forming agents or through active ester or reactive carbonyl intermediates (Fig. 7). The carboxylate actually becomes the acylating agent to the modifying group. Amine-containing nucleophiles can couple to an activated carboxylate to give amide derivatives. Hydrazide compounds react in a manner similar to that of amines. Sulfhydryls, while reactive and resulting in a thioester linkage, form unstable derivatives that hydrolyze in aqueous solutions.

Lysine, arginine, and histidine have ionizable amine containing side chains that, along with the N-terminal α-amine, contribute to a protein's overall net positive charge. Lysine contains a straight four-carbon chain terminating in a primary amine group. The ε-amine of lysine differs in pK$_a$ from the primary α-amines by having a slightly higher ionization point (pK$_a$ of 9.3–9.5 for lysine versus pK$_a$ of 7.6–8.0 for α-amines). At pH values lower than the pK$_a$ of these groups, the amines are generally protonated and possess a positive charge. At pH values greater than the pK$_a$, the amines are unprotonated and contribute no net charge. Arginine contains a strongly basic chemical constituent on its side chain called a guanidino group. The ionization point of this residue is so high (pK$_a$ >12.0) that it is virtually always protonated and

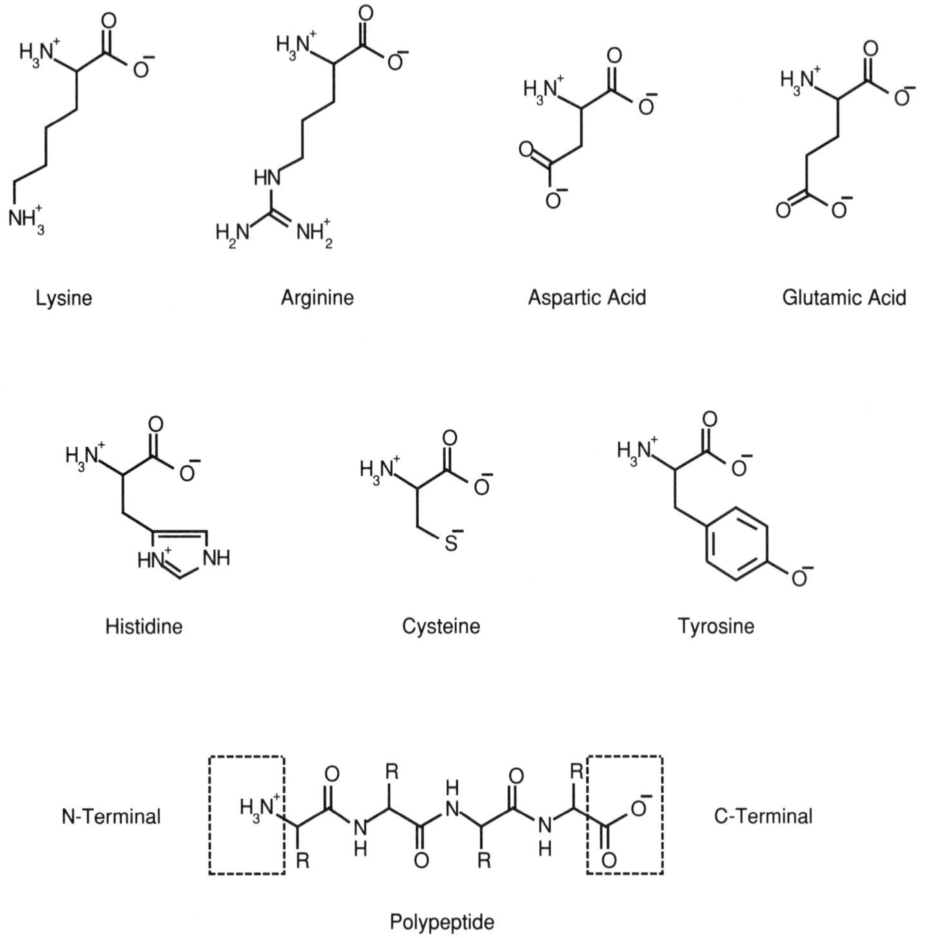

Lysine   Arginine   Aspartic Acid   Glutamic Acid

Histidine   Cysteine   Tyrosine

N-Terminal   C-Terminal

Polypeptide

Figure 6   The ionizable amino acids possess some of the most important side-chain functional groups for bioconjugate applications. The C- and N-terminal of each polypeptide chain also is included in this group.

carries a positive charge. Histidine's side chain is an imidazole ring that is potentially protonated at slightly acidic pH values (pK_a 6.7–7.1). Thus, at physiological pH, these residues contribute to the overall net positive charge of an intact protein molecule.

The amine-containing side chains in lysine, arginine, and histidine typically are exposed on the surface of proteins and can be derivatized with ease. The most important reactions that can occur with these residues are alkylation and acylation (Fig. 8). In alkylation, an active alkyl group is transferred to the amine nucleophile with loss of one hydrogen. In acylation, an active carbonyl group undergoes addition to the amine. Alkylating reagents are highly varied and the reaction with an amine nucleophile is difficult to generalize. Acylating reagents, however, usually proceed through a carbonyl addition mechanism, as shown in Fig. 9. The imidazole ring of histidine also is an important reactive species in electrophilic reactions, such as in iodination using radioactive ^{125}I or [131]I (Chapter 8, Section 4).

Cysteine is the only amino acid containing a sulfhydryl group. At physiological pH,

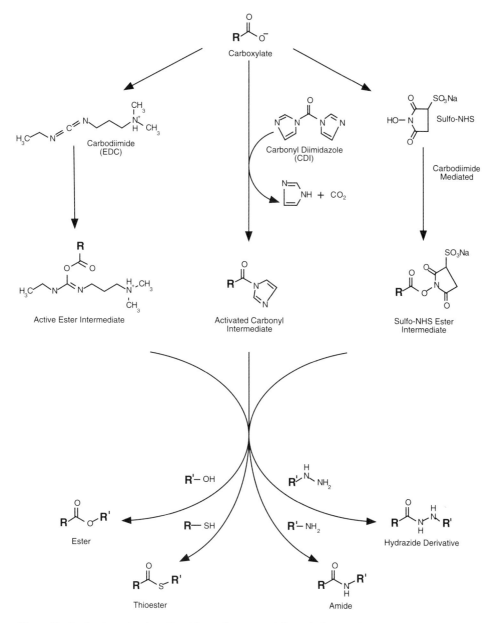

Figure 7 Derivatives of carboxylic acids can be prepared through the use of active intermediates, which react with target functional groups to give acylated products.

this residue is normally protonated and possesses no charge. Ionization only occurs at high pH (pK_a 8.8–9.1) and results in a negatively charged thiolate residue. The most important reaction of cysteine groups in proteins is the formation of disulfide cross-links with another cysteine molecule. Cysteine disulfides (called cystine residues) often are key points in stabilizing protein structure and conformation. They frequently occur between polypeptide subunits, creating a covalent linkage to hold two chains

Derivatization of the side chain sulfhydryl of cysteine is one of the most important reactions of modification and conjugation techniques for proteins.

Tyrosine contains a phenolic side chain with a pK_a of about 9.7–10.1. Due to its aromatic character, tyrosine is second only to tryptophan in contributing to a protein's overall absorptivity at 275–280 nm. Although the amino acid is only sparingly soluble in water, the ionizable nature of the phenolic group makes it often appear in hydrophilic regions of a protein—usually at or near the surface. Thus tyrosine derivatization proceeds without much need for deforming agents to further open protein structure.

Tyrosine may be targeted specifically for modification through its phenolate anion by acylation, through electrophilic reactions such as the addition of iodine or diazonium ions, and by Mannich condensation reactions. The electrophilic substitution reactions on tyrosine's ring all occur at the *ortho* position to the —OH group (Fig. 11). Most of these reactions proceed effectively only when tyrosine's ring is ionized to the phenolate anion form.

In summary, protein molecules may contain up to nine amino acids that are readily derivatizable at their side chains: aspartic acid, glutamic acid, lysine, arginine, cysteine, histidine, tyrosine, methionine, and tryptophan. These nine residues contain eight principal functional groups with sufficient reactivity for modification reactions: primary amines, carboxylates, sulfhydryls (or disulfides), thioethers, imidazolyls, guanidinyl groups, and phenolic and indolyl rings. All of these side chain functional groups in addition to the N-terminal α-amino and the C-terminal α-carboxylate form the full complement of polypeptide reactivity within proteins (Fig. 12).

Nucleophilic Reactions and the pI of Amino Acid Side Chains

Ionizable groups within proteins can exist in one of two forms: protonated or unprotonated. Carboxylate groups below their pK_a values exist in the protonated state and are therefore in the conjugate acid form and carry no charge. However, at pH values above the pK_a of the carboxylic group, the acid is ionized and therefore unprotonated to a negative charge. This same relationship is true of the —OH group on the phenol ring of tyrosine. At pH values below its pK_a, tyrosine's side chain is uncharged. Above the pK_a, however, the hydrogen ionizes off, leaving a negatively charged phenolate. Conversely, amine nucleophiles below their pK_a values are in a protonated state and possess a positive charge. At pH values above the pK_a of the amino group, it is then ionized and unprotonated to neutrality.

Each type of ionizable group in proteins will have a unique pK_a based upon the theoretical value for the amino acid and modulated from that value by its own surrounding microenvironment. Minute environmental changes will cause amine containing residues at different structural locations to have different ionization potentials, even if the groups are otherwise chemically identical.

Thus, the actual pK_a of each ionizable group within protein molecules may range considerably lower or higher than the theoretical values as the microenvironment of individual groups changes. Identical side chains in differing parts of a protein molecule may have widely varying pK_a values depending on the immediate chemical milieu. Such factors as the presence of other amino acid side chains in the vicinity, salts, buffers, temperature, ionic strength, and other effects of the solvent medium all play crucial roles in creating microenvironmental changes that affect the ionization potential of these groups (Tanford and Hauenstein, 1956; Schewale and Brew, 1982).

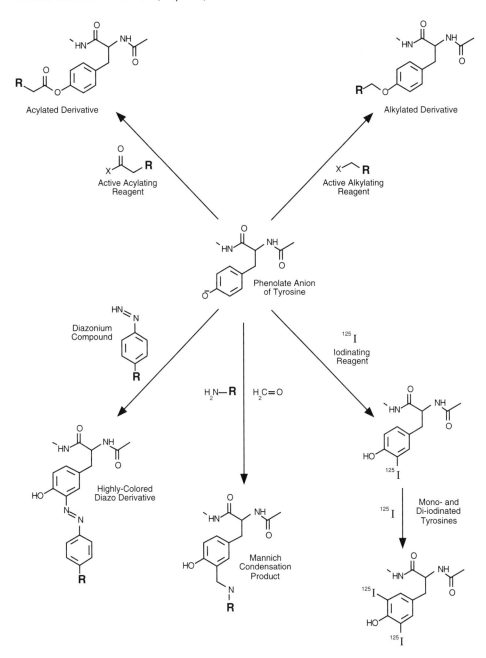

Figure 11 Tyrosine residues are subject to nucleophilic and electrophilic reactions. The phenolate ion may be alkylated or acylated using a variety of bioconjugate reagents. Its aromatic ring also may undergo electrophilic addition using diazonium chemistry or Mannich condensation, or be halogenated with radioactive isotopes such as [125]I.

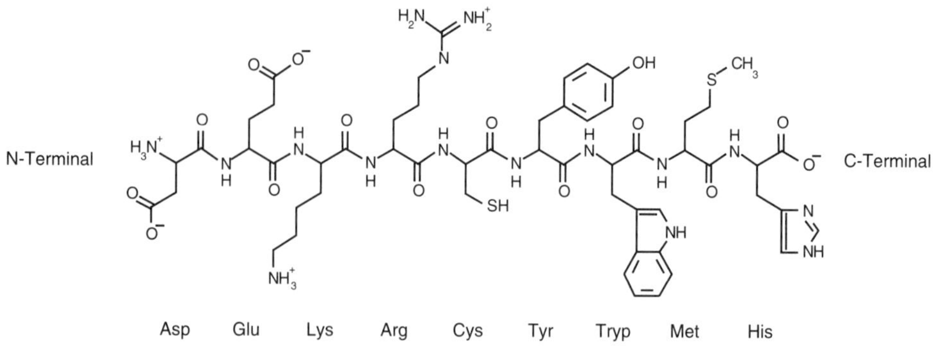

Figure 12 The full complement of polypeptide functional groups is represented by these nine amino acids. Bioconjugate chemistry may occur through the C- and N-terminals of each polypeptide chain, the carboxylate groups of aspartic and glutamic acids, the ε-amine of lysine, the guanidino group of arginine, the sulfhydryl group of cysteine, the phenolate ring of tyrosine, the indol ring of tryptophan, the thioether of methionine, and the imidazole ring of histidine.

The Henderson–Hesselbalch equation (1) explains the relationship of pH and pK_a to the relative ratios of protonated (acid) and unprotonated (base) forms of an ionizable group. Note that the ionized form of such a group does not have to possess a negative charge, as in the case of unprotonated primary amines. Indeed, in that instance it is the protonated amine that bears a charge of positive one. According to the mathematical implications of this equation, an ionizable group at its pK_a value is exactly 50 percent ionized. This means that aspartic acid side chains placed in a medium with a pH equal to its pK_a should have half of its carboxylates ionized to a negative charge and half of them unionized with no charge:

$$pH = pK_a + \log\{[\text{base}]/[\text{acid}]\}. \tag{1}$$

Further implications of this equation are that at one pH unit below or above the pK_a, an ionizable group will be 91% unionized (protonated) or 91% ionized (unprotonated), respectively. Two pH units below or above translate to a 99% unionized or 99% ionized state.

The absolute ratio of protonated-to-unprotonated forms will change from this theoretical approach based upon the microenvironment each group experiences. The reactivity of amino acid side chains is directly related to their being in an unprotonated or ionized state. Many reactions of modification and conjugation occur efficiently only when the nucleophilic species is in an ionized form. As the unprotonated form increases in concentration, the relative nucleophilicity of the ionizable group increases. Many of the reactive groups commonly used for protein modification will couple in greater yield as the pH of the reaction is raised closer to the pK_a of the ionizable target. However, continuing to increase the pH beyond the pK_a may not be necessary for increased yield, and may even be detrimental, because many reactive groups will begin to lose activity through hydrolysis at high pH values.

A nucleophile is any atom containing an unshared pair of electrons or an excess of electrons able to participate in covalent bond formation. Nucleophilic attack at an atomic center of electron deficiency or positive charge is the basis for many of the coupling reactions that occur in chemical modifications. Thus, an uncharged amine group is a more powerful nucleophile than the protonated form bearing a positive charge. Likewise, a negatively charged carboxylate has greater nucleophilicity than its uncharged, protonated conjugate acid form. In addition, an unprotonated thiolate, bearing a negative charge (RS^-), is a much more powerful nucleophile than its protonated, uncharged sulfhydryl form.

According to the theory of nucleophilicity (Bunnett, 1963; Edwards and Pearson, 1962; Pearson *et al.*, 1968), the relative order of nucelophilicity relative to the major groups in biological molecules can be summarized as follows:

$$R—S^- > R—SH$$

$$R—NH_2 > R—NH_3^+$$

$$R—COO^- > R—COOH$$

$$R—O^- > R—OH$$

$$R—OH = H—OH$$

and finally,

$$R—S^- > R—NH_2 > R—COO^- = R—O^-$$

Using these relationships, it is obvious that the strongest nucleophile in protein molecules is the sulfhydryl group of cysteine, particularly in the ionized, thiolate form. Next in line are the amine groups in their uncharged, unprotonated forms, including the α-amines at the N-terminals, the ϵ-amines of lysine side chains, the secondary amines of histidine imidazolyl groups and tryptophan indole rings, and the guanidino amines of arginine residues. Finally, the least potent nucleophiles are the oxygen-containing ionizable groups including the α-carboxylate at the C-terminal, the β-carboxyl of aspartic acid, the γ-carboxyl of glutamic acid, and the phenolate of tyrosine residues.

According to the theoretical pK_a values for the ionizable side chains of amino acids, nucleophilic substitution reactions involving primary amines or sulfhydryl groups on proteins should not be efficient below a pH of about 8.5 (Table 1). In practice, however, reactions can be done with these groups in high yield at pH values not much higher than neutrality. This discrepancy relates to the changes in pK_a due to microenvironmental effects experienced by the residues within the three-dimensional structure of the protein molecule. In reality, the ϵ-amine groups on lysine side chains with proteins, having theoretical pK_a values over 10, nonetheless exist in sufficient quantity in an unprotonated form even at a pH of 7.2 that modification easily occurs.

One important point should be noted, however. The changes that occur in the pK_a of ionizable groups in protein molecules due to microenvironmental effects make it nearly impossible to select exclusively certain residues for modification simply by careful modulation of reaction conditions. For instance, overlap of the pK_a range for sulfhydryls and amine-containing residues eliminates any chance of directing a reaction solely toward —SH groups by adjusting the pH of the reaction medium. Thus, in practice, for many modification reagents, pH alone cannot be used as an effective modulator of reaction targeting. To site-direct a modification reaction, the proper choice of chemical reactions and reactive groups is far more important than slight changes in pH.

Table 1 pK$_a$ of Ionizable Amino Acids

Group location	Functionality	pK$_a$ range
α-Amine; N-Terminus	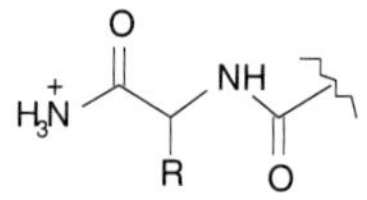	7.6–8
Lysine's ε-amine	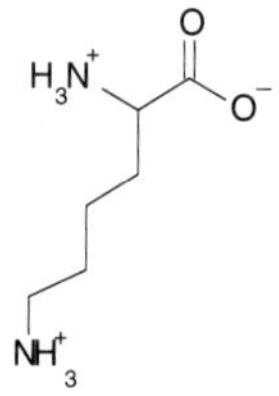	9.3–9.5
Histidine's imidazolyl nitrogen	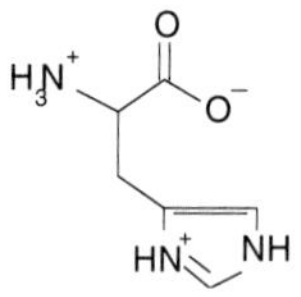	6.7–7.1
Arginine's guanidinyl group	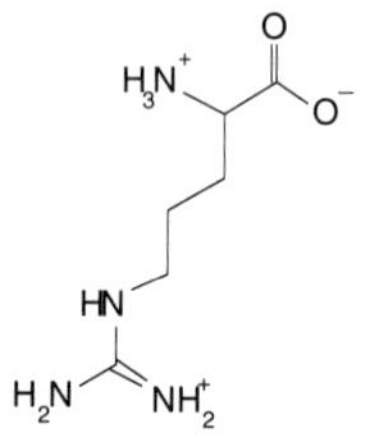	>12
Tyrosine's phenolic hydroxyl	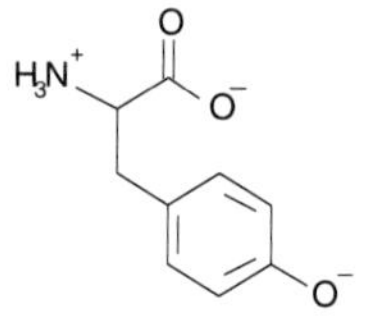	9.7–10.1
α-Carboxyl; C-terminus	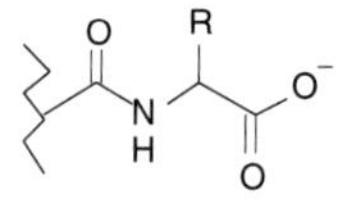	2.1–2.4
Aspartic acid's γ-carboxyl	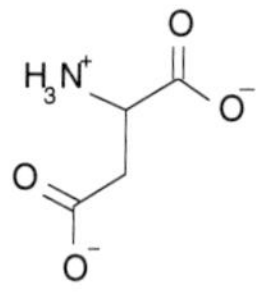	3.7–4
Gutamic acid's γ-carboxyl	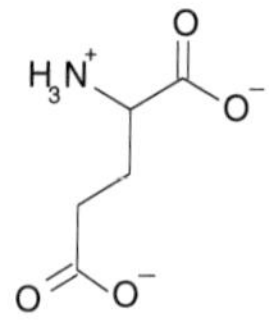	4.2–4.5
Cysteine's sulfhydryl	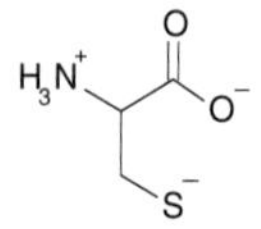	8.8–9.1

Secondary, Tertiary, and Quaternary Structure

Amino acids are linked through peptide bonds to form long polypeptide chains. The *primary* structure of protein molecules is simply the linear sequence of each residue along the α-chain. Each amino acid in the chain interacts with surrounding groups through various weak, noncovalent interactions and through its unique side chain functional groups. Noncovalent forces such as hydrogen bonding and ionic and hydrophobic interactions combine to create each protein's unique organization.

It is the sequence and types of amino acids and the way that they are folded that provide protein molecules with specific structure, activity, and function. Ionic charge, hydrogen bonding capability, and hydrophobicity are the major determinants for the resultant three-dimensional structure of protein molecules. The α-chain is twisted, folded, and formed into globules, α-helices, and β-sheets based upon the side chain amino acid sequence and weak intramolecular interactions such as hydrogen bonding between different parts of the peptide backbone (Fig. 13). Major secondary structures of proteins such as α-helices and β-sheets are held together solely by massive hydrogen bonding created through the carbonyl oxygens of peptide bonds interacting with the hydrogen atoms of other peptide bonds (Fig. 14).

In addition, negatively charged residues may become bonded to positively charged groups through ionic interactions. Nonpolar side chains may attract other nonpolar residues and form regions of hydrophobicity to the exclusion of water and other ionic groups. Occasionally, disulfide bonds also are found holding different regions of the polypeptide chain together. All of these forces combine to create the *secondary* structure of proteins, which is the way the polypeptide chain folds in local areas to form larger, sometimes periodic, structures.

On a larger scale, the unique folding and structure of one complete polypeptide chain is termed the *tertiary* structure of protein molecules. The difference between

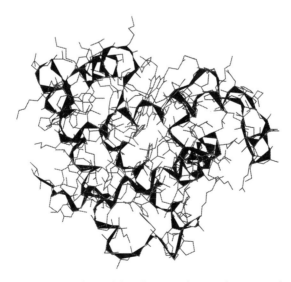

Figure 13 The α-chain structure of myoglobin illustrates the complex nature of polypeptide structure within proteins.

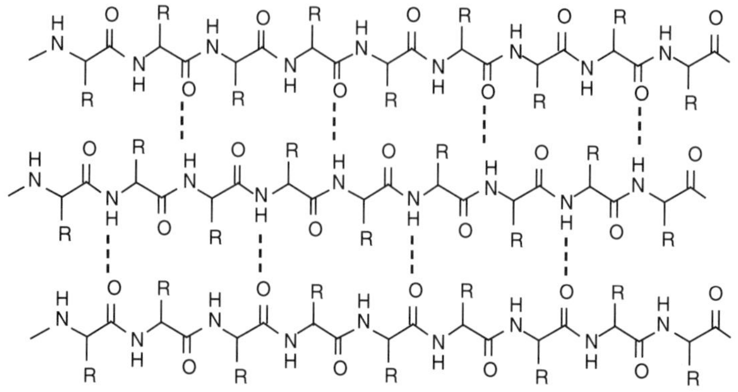

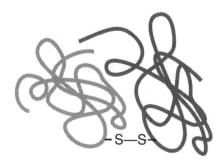

Figure 14 Secondary structure within proteins may be stabilized through hydrogen bonding between adjacent α-chains, forming β-sheet conformations.

local secondary structure and complete polypeptide tertiary structure is arbitrary and sometimes of little practical difference.

Larger proteins often contain more than one polypeptide chain. These multi-subunit proteins have a more complex shape, but are still formed from the same forces that twist and fold the local polypeptide. The unique three-dimensional interaction between different polypeptides in multi-subunit proteins is called the *quaternary* structure. Subunits may be held together by noncovalent contacts, such as hydrophobic or ionic interactions, or by covalent disulfide bonds formed from the cysteine residue of one polypeptide chain being cross-linked to a cysteine sulfhydryl of another chain (Fig. 15).

Thus, aside from the covalently polymerized α-chain itself, the majority of protein structure is determined by weak, noncovalent interactions that potentially can be disturbed by environmental changes. It is for this reason that protein structure can be easily disrupted or denatured by fluctuations in pH or temperature or by substances that can alter the structure of water, such as detergents or chaotropes.

Not surprisingly, chemical modification to the amino acid constituents of a polypeptide chain also may cause significant disruption in the overall three-dimensional structure of a protein. If amino acid residues critical to folding near functionally important regions are modified with chemical groups that change the charge, hydrophilicity, or hydrogen bonding character of the polypeptide chain, protein structure

Figure 15 Polypeptide chains may be bound together through disulfide linkages occurring between cysteine residues within each subunit.

may be altered and activity may be compromised. This concept will be discussed further in subsequent sections.

Prosthetic Groups, Cofactors, and Post-translational Modifications

Proteins may contain structures other than polypeptide chains that are important for biological function. Prosthetic groups and cofactors are small organic compounds that are sometimes tightly bound to a protein and aid in forming the active center. A prosthetic group is usually carried within the three-dimensional protein structure in a firm-fitting pocket or even attached through a covalent bond, such as the heme ring associated with cytochrome *c* molecules which is bonded through thioether linkages with adjacent cysteine residues (Fig. 16). Cofactors, by contrast, may be only transiently bound to proteins during periods of activity. Enzymes often require cofactors to act as donors or acceptors of chemical groups that are added to or cleaved from a substrate molecule. Some common cofactors are ATP, ascorbic acid, coenzyme A, NAD, NADP, FAD, FMN, and biotin. Sometimes, the enzyme cofactor also is an energy source for the catalytic reaction, as in the case of ATP-dependent reactions.

Frequently, metal ions are associated with the prosthetic group or cofactor. Heme rings usually contain a chelated iron atom. Occasionally, however, these metals are merely bound within folded polypeptide regions with no additional organic constituents required. Many metal ions are known to participate in enzymatic activity. One or more of the ions of Na, K, Ca, Zn, Cu, Mg, and Mn, as well as Co and Mo are often required by enzymes to maintain activity.

Prosthetic groups and cofactors, whether organic or metallic, may be removed from a protein to create an inactive *apo* protein or enzyme. Loss of these groups may occur through environmental changes, such as removing metal ions from solution or adding denaturants to unfold protein structure. In many cases, full activity can be restored simply by reintroducing the needed group into the surrounding medium.

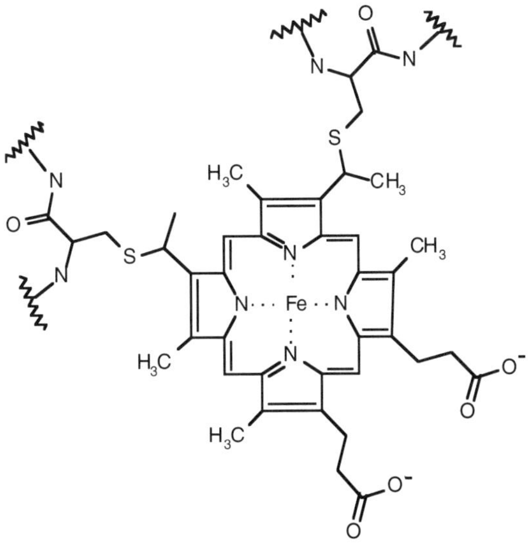

Figure 16 The heme ring of cytochrome *c* is a non-amino acid, prosthetic group bound to the protein through two cysteine residues.

In addition to small organic molecules or metal ions, proteins may have other components tightly associated with them. Nucleoproteins, for instance, contain non-covalently bound DNA or RNA, as in some of the structural proteins of viruses. Lipoproteins contain associated lipids or fatty acids and may also carry cholesterol, as in the high-density and low-density lipoproteins in serum.

During modification or conjugation reactions, prosthetic groups and other associated molecules may be lost or damaged. Metal ions temporarily may be removed by the inclusion of a chelating agent added to maintain sulfhydryl stability during coupling through the —SH groups of a protein. To restore activity after conjugation, it is necessary to remove the chelator and add the required metal salts. Other changes to the prosthetic carriers may not be so easily corrected. For instance, heme-containing molecules are sensitive to the presence of agents that can form a coordination complex with or modify the oxidation state of the chelated metal ion. Some reagent systems may permanently inactivate the heme-containing protein.

Thus, loss of activity can occur not only through changes to the amino acid constituents of a protein, but through prosthetic group or cofactor loss or damage as well. Most of these potential difficulties can be overcome through careful selection of the reaction conditions and through knowledge of the cofactor dependencies that are critical to the activity of the protein being modified.

Post-translational modifications to protein structure are covalent changes that occur as the result of controlled enzymatic reactions or due to chemical reactions not under enzymatic regulation. The most common cellular modification performed on proteins after ribosomal synthesis is glycosylation. Proteins newly synthesized on ribosomes may be transported to the Golgi apparatus where specific glycosyl transferases catalyze the coupling of carbohydrate residues to the polypeptide chains. Glycoproteins and mucoproteins are formed by the coupling of polysaccharides through O-glycosidic linkages to serine, threonine, or hydroxylysine and through N-glycosidic linkages with the amide side chain group of asparagine.

The structure of most glycoprotein carbohydrate is branched with the sugars mannose, N-acetyl glucosamine, sialic acid, glactose, and L-fucose being prevalent. Asparagine-linked polysaccharides are well characterized and are known to be constructed of a core unit consisting of three mannose residues and 2 N-acetyl glucosamine (GlcNAc) residues. The GlcNAc residues are bound to the Asp side chain amide nitrogen through a β1 linkage (Kornfield and Kornfield, 1985). The three mannose groups then usually form the first branch point in the oligosaccharide chain (Section 2).

The content by weight of carbohydrate in glycoproteins may vary from only a few percent to over 50% in some proteins in mucous secretions. Although the function of the polysaccharide in most glycoproteins is unknown, in some case it may provide hydrophilicity, recognition, and points of noncovalent interaction with other proteins through lectin-like affinity binding.

The presence of carbohydrate on protein or peptide molecules provides important points of attachment for modification or conjugation reactions. Coupling only through polysaccharide chains often can direct the reaction away from active centers or critical points in the polypeptide chain, thus preserving activity. Polysaccharides can be specifically targeted on glycoproteins through mild sodium periodate oxidation. Periodate cleaves adjacent hydroxyl groups in sugar residues to create highly reactive aldehyde functionalities (Section 4.4). The level of periodate addition can be adjusted to cleave selectively only certain sugars in the polysaccharide chain. For instance, a concentration of 1mM sodium periodate specifically oxidizes sialic acid residues to aldehydes, leaving all other monosaccharides untouched. Increasing the concentration to 10 mM, however, will cause oxidation of other sugars in the carbohydrate chain, including galactose and mannose. The generated aldehydes then can be used in coupling reactions with amine- or hydrazide-containing molecules to form covalent linkages. Amines can react with formyl groups under reductive amination conditions using a suitable reducing agent such as sodium cyanoborohydride. The result of this reaction is a stable secondary amine linkage (Chapter 2, Section 5.3). Alternatively, hydrazides spontaneously react with aldehydes to form hydrazone linkages, although the addition of a reducing agent increases the efficiency of the reaction (Chapter 2, Section 5.1).

Another form of post-transitional modification that may add carbohydrate to a polypeptide is nonenzymatic glycation. This reaction occurs between the reducing ends of sugar molecules and the amino groups of proteins and peptides. See Section 2.1 for further details and the reaction sequence behind this modification.

Protecting the Native Conformation and Activity of Proteins

The goal of most protein modification or conjugation procedures is to create a stable product with good retention of the native state and activity. Ideally, any derivatization should result in a protein that performs exactly as it would in its unmodified form, but with the added functionality imparted by whatever is conjugated to it. Thus, an antibody molecule tagged with a fluorophore should retain its ability to bind to antigen and also have the added function of fluorescence.

One of the best ways to ensure retention of activity in protein molecules is to avoid chemical reaction at the active center. The active center is that portion of the protein where ligand, antigen, or substrate binding occurs. In simpler terms, the active center (or active site) is that part that has specific interaction with another substance (Means and Feeney, 1971). For the preparation of enzyme derivatives, it is important to protect the site of catalysis where conversion of substrate to product occurs. When working with antibody molecules, it is crucial to stay away from the two antigen-binding sites.

The best chemical procedures avoid the active site by selecting functional groups away from that area or by protecting the site through the incorporation of additives. In some cases, the inclusion of substrates, cofactors, ligands, inhibitors, or antigens in the modification reaction will protect the active site. Addition of the appropriate substance can bind the active site and mask it from modification by cross-linking agents. In enzyme derivatization procedures, this is often just a matter of adding a reversible inhibitor or substrate analog. For instance, when working with alkaline phosphatase merely doing the reaction in phosphate buffer protects the active center from chemical modification, since phosphate ions bind in the catalytic site. With trypsin, the incorporation of benzamidine similarly masks and protects the active site.

However, protecting the antigen-binding sites on an antibody molecule by using this method is often more difficult. Inclusion of antigen to mask the binding sites is effective in blocking these areas, but it also may cause irreversible cross-linking of the antigen to the antibody. This is especially true when the antigen is a peptide or a protein having the same chemical reactive groups as the antibody. Any modification reactions that are directed at the antibody may modify the antigen as well. Therefore, only use this method if the antigen is lacking in the chemical targets that are going to be used on the antibody. For instance, if the polysaccharide chains on the antibody are to be modified, then using a protein antigen that does not contain carbohydrate to block the antigen binding sites may work well.

An equally effective method of protecting the activity of a protein is by using site-directed reactions that result in modifications away from the active center. In some cases, specific functionalities are known to be present only at restricted sites within the three-dimensional structure of a protein. If these functional groups are not present close to the active site, then using them exclusively for modification reactions should ensure good retention of activity. For instance, sulfhydryl groups or carbohydrate chains are often present in limited quantity and in specific regions on a protein. Selecting reagent systems that target these groups ensures derivatization only at restricted sites within the protein molecule, thus potentially avoiding the active center.

Surprisingly, the goal of some protein modification or conjugation schemes is to somewhat alter the native presentation of the product. This is especially true in hapten–carrier conjugation, as used for immunogen or vaccine production. In this case, the main objective is to modify the environment of the hapten to create an immunological response *in vivo*. A hapten is usually a small molecule that is not able to generate an immune response on its own, but can react with the products of such a response once generated. Most often these products are antibodies having specificity for the hapten.

The complexities involved in achieving a successful conjugation strategy are best illustrated in the problems and concerns dealing with hapten–carrier conjugation. In order to produce the initial immune response to a small molecule, the hapten is typically coupled to a larger protein that can generate a response on its own. In simple terms, the larger carrier protein confers immunogenicity to the smaller hapten. The native presentation of the hapten is altered toward the immune system, thus creating the immune response.

The site of attachment of the hapten to the carrier and the nature of the cross-linker are both important to the specificity of the resultant antibodies generated against it. For proper recognition, the hapten must be coupled to the carrier with the appropriate orientation. For an antibody to recognize subsequently the free hapten without carrier, the hapten–carrier conjugate must present the hapten in an exposed and accessible form. Optimal orientation is often achieved by directing the cross-linking reaction to specific sites on the hapten molecule. With peptide haptens, this is typically done by attaching a terminal cysteine residue during synthesis. This provides a sulfhydryl group on one end of the peptide for conjugation to the carrier. Cross-linking through this group provides hapten attachment only at one end, therefore ensuring consistent orientation.

In hapten–carrier conjugation, the goal is not to maintain the native state or stability of the carrier, but to present the hapten in the best possible way to the immune system. In reaching this goal, the choice of conjugation chemistry may control the resultant titer, affinity, and specificity of the antibodies generated against the hapten. It may be important in some cases to choose a cross-linking agent containing a spacer arm long enough to present the antigen in an unrestricted fashion. It also may be important to control the density of the hapten on the surface of the carrier. Too little hapten substitution may result in little or no response. A hapten density too high actually may cause immunological suppression and decrease the response. In addition, the cross-linker itself may generate an undesired immune response. Fortunately, for the majority of hapten–carrier conjugation problems, a few main cross-linking techniques provide a workable compromise to solving all these concerns and ultimately generating an effective immune response (Chapter 9).

1.2. Protein Cross-linking Methods

The cross-linking of two proteins using a simple homobifunctional reagent (Chapter 4) potentially can result in a broad range of conjugates being produced (Avrameas, 1969). The reagent initially may react with either one of the proteins, forming an active intermediate. This activated protein may then form cross-links with the other protein or with another molecule of the same protein. The activated protein also may react intramolecularly with other functional groups on part of its own polypeptide chain. Other cross-linking molecules may continue to react with these conjugated species to form various mixed products, including severely polymerized proteins that may fall out of solution (Fig. 17).

The problems of indeterminate conjugation products are amplified in single-step reaction procedures using homobifunctional reagents (Chapter 4). Single-step procedures involve the addition of all reagents at the same time to the reaction mixture. This technique provides the least control over the cross-linking process and invariably leads

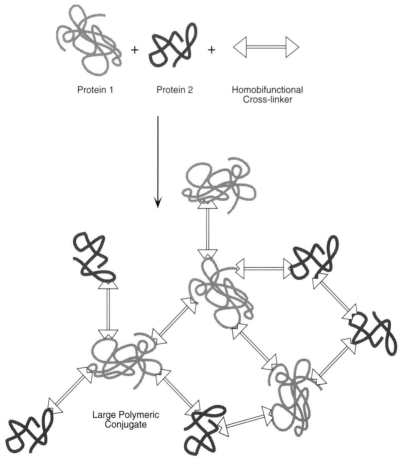

Protein 1   Protein 2   Homobifunctional
Cross-linker

Large Polymeric
Conjugate

Figure 17   Protein cross-linking reactions done using homobifunctional reagents can result in large polymeric complexes of multiple sizes and indefinite structure.

to a multitude of products, only a small percentage of which represent the desired or optimal conjugate. Excessive conjugation may cause the formation of insoluble complexes that consist of very high molecular weight polymers. For example, one-step glutaraldehyde conjugation of antibodies and enzymes (Chapter 10, Section 1.2) often results in significant oligomers and precipitated conjugates. To overcome this shortcoming, multistep reaction procedures have been developed using both homobifunctional and heterobifunctional reagents (Chapter 5). Controlled, multistep conjugation protocols alleviate the polymerization problem and form relatively low molecular weight, soluble antibody–enzyme complexes (Chapter 10, Section 1.1).

In two-step protocols, one of the proteins to be conjugated is reacted or "activated" with a cross-linking agent and excess reagent and by-products are removed. In the second stage, the activated protein is mixed with the other protein or molecule to be conjugated, and the final conjugation process occurs (Fig. 18).

The use of homobifunctional reagents in two-step protocols still creates many of the problems associated with single-step procedures, because the first protein can cross-link and polymerize with itself long before the second protein is added. Homobifunctional reagents by definition have the same reactive group on either end of the cross-linking molecule. Since the protein to be activated has target functional groups on every molecule that can couple with the reactive groups on the cross-linker, both ends of the reagent potentially can react. This inherent potential to polymerize uncontrollably unfortunately is characteristic of all homobifunctional reagents, even in multistep protocols.

The greatest degree of control in cross-linking procedures is afforded using heterobifunctional reagents (Chapter 5). Since a heterobifunctional cross-linker has different reactive groups on either end of the molecule, each side can be specifically directed toward different functional groups on proteins. Using a multistep conjugation protocol with a heterobifunctional reagent can allow one macromolecule to be activated, excess cross-linker removed, and then a second macromolecule added to induce the final linkage. Directed conjugation will occur as long as the first protein activated does not have groups able to couple with the second end of the cross-linker, whereas the second molecule does possess the correct functionalities.

Occasionally, the second protein does not naturally have the target groups necessary to couple with the second end of the cross-linker. In such cases, a specific functional group usually can be created to make the conjugation successful (Section 4). In such three-step systems, the first protein is activated with the heterobifunctional reagent and purified away from excess cross-linker. The second protein is then modified to contain the specific target groups required for the second stage of the conjugation. Finally in step three, the two modified proteins are mixed to cause the coupling reaction to happen (Fig. 19).

Two- and three-step protocols using heterobifunctional cross-linkers often are designed around amine-reactive and sulfhydryl-reactive chemical reactions. Many of these reagents utilize NHS esters on one end for coupling to amine groups on the first protein and maleimide groups on the other end that can react with sulfhydryls on the second protein. The NHS ester end is reacted with the first protein to be conjugated, forming an activated intermediate containing reactive maleimide groups. Fortunately, the maleimide end of such cross-linkers is relatively stable to degradation; thus the activated protein can be isolated without loss of sulfhydryl coupling ability. Additionally, if the second protein does not contain indigenous sulfhydryls, these can be created by an abundance of methods (Section 4.1). After mixing the maleimide-

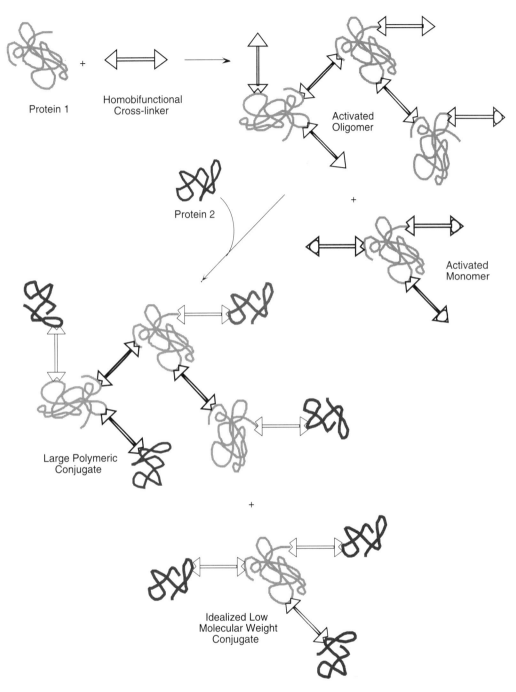

Figure 18 A two-step protocol using a homobifunctional cross-linking agent offers more control than single step methods, but still may result in oligomer formation.

activated protein with the sulfhydryl-containing protein, conjugation can occur only in one direction.

Control of the products of conjugation increases as the protocols progress from

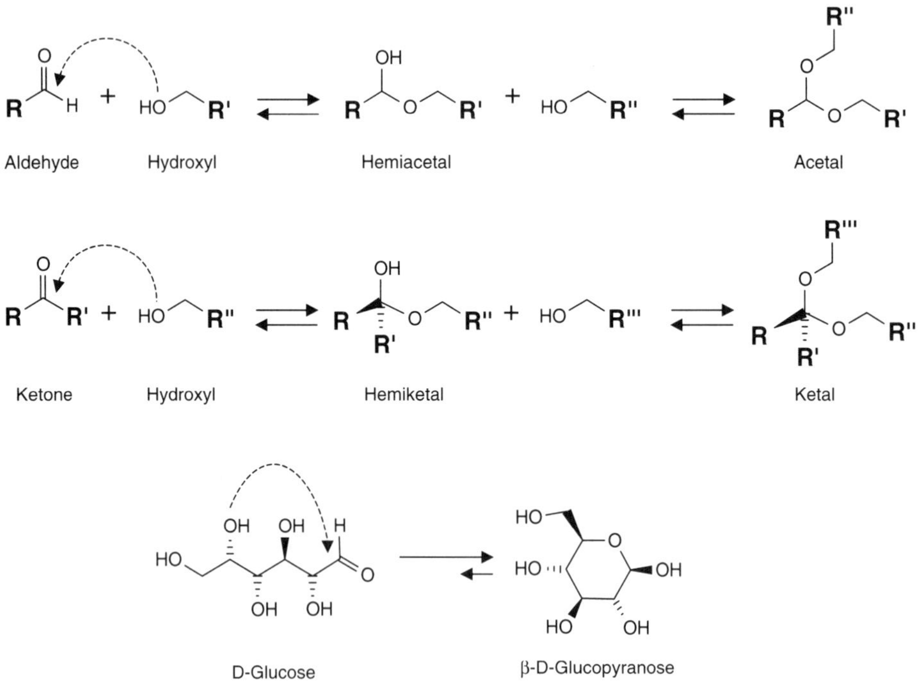

Figure 20 Carbonyl groups and hydroxyls may react to form acetal or ketal products. Sugars naturally undergo these reactions to form ring structures in aqueous solution.

reagent. However, due to this predominance of the cyclic structure of monosaccharides, they do not have the capability of reacting with bisulfite or Schiff reagent as do normal unblocked aldehydes and ketones. Thus, the carbonyl functional groups of sugars have reduced reactivity, because of hemiacetal and hemiketal formation.

Figure 21 shows the structures of some of the most common monosaccharide molecules: D-glyceraldehyde, D-erythrose, D-ribose, D-arabinose, D-xylose, D-glucose, D-glucosamine, N-acetyl-D-glucosamine, D-mannose, D-galactose, D-galactosamine, N-acetyl-D-galactosamine of the aldose family and dihydroxyacetone, D-ribulose, D-fructose, D-N-acetylneuraminic acid of the ketose family. Formation of the cyclic structure of each of these sugars can result in one of two stereoisomers, designated α and β, depending on the orientation of the aldehyde group or ketone group during hemiacetal formation. For aldoses, the α form is drawn in the standard Haworth projection with the number 1 carbon hydroxyl pointing down. For ketoses, the α form consists of the No. 2 carbon hydroxyl pointing down. All the common monosaccharide structures shown in Fig. 21 are in the β-stereoisomer form.

Since in aqueous solutions the cyclic form of monosaccharides is in equilibrium with their corresponding open forms, the α and β structures continually interconvert. At equilibrium, one form usually predominates. For instance, glucose dissolved in water consists of about a 2:1 ratio of β-D-glucose to α-D-glucose. Although their chemical constituents are identical, the biochemical properties between the α and the β forms can be quite different. Monosaccharides linked together to form disaccharides and polysaccharides cannot continue to interconvert and are therefore frozen in the α or β forms. Changing one monosaccharide in a complex carbohydrate to its opposite

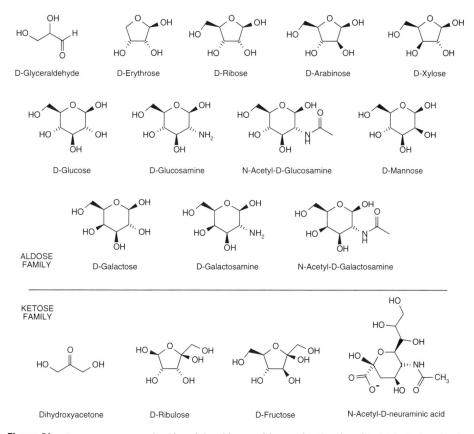

Figure 21 Common monosaccharides of the aldose and ketose families found in biological molecules.

stereoisomer form can produce radical structural changes in the polysaccharide chain and significantly alter its biochemical properties.

Sugar Functional Groups

Monosaccharide functional groups consist of either a ketone or an aldehyde, several hydroxyls, and the possibility of amine, carboxylate, sulfate, or phosphate groups as additional constituents. Amine-containing sugars may possess a free primary amine, but often are modified to the *N*-acetyl derivative, such as the *N*-acetylglucosamine residue of chitin. Sulfate-containing monosaccharides frequently are found in certain mucopolysaccharides, including chondroitin sulfate, dermatan sulfate, heparin sulfate, and keratin sulfate (Fig. 22). Carboxylate-containing sugars include sialic acid as well as many aldonic, uronic, oxoaldonic, and ascorbic acid derivatives (Fig. 23). Phosphate-containing monosaccharides are almost exclusively created in metabolic processes involving energy utilization, such as in the production of glucose 1-phosphate formed during glycogen breakdown and glucose 6-phosphate produced during glycolysis. Perhaps the most common phosphate sugar derivative, however, is the 5′-phosphate of D-ribose or D-2-deoxyribose found as a repeating component of RNA and DNA, respectively.

Modification and conjugation reactions can be designed to target many of these

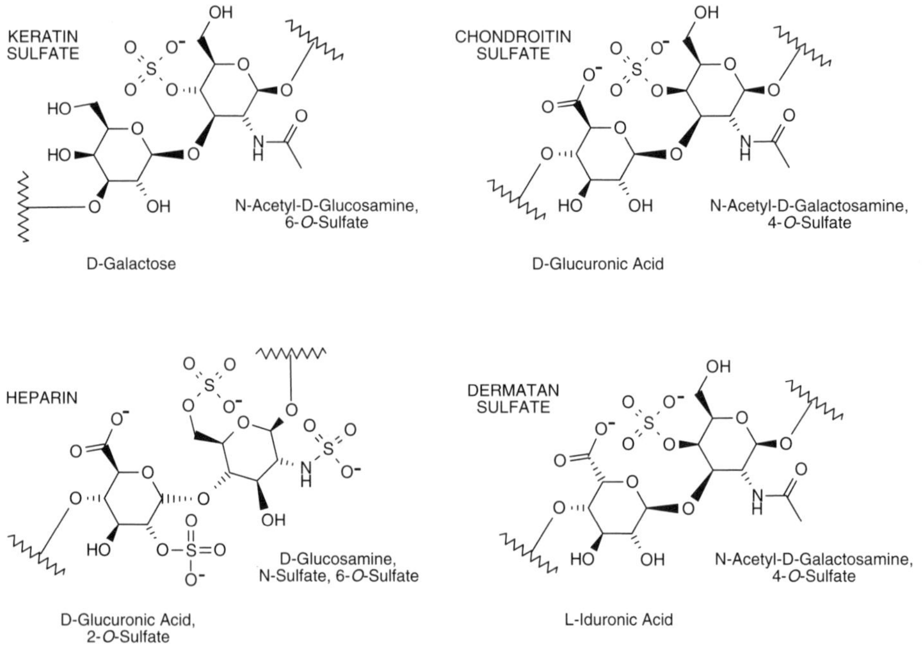

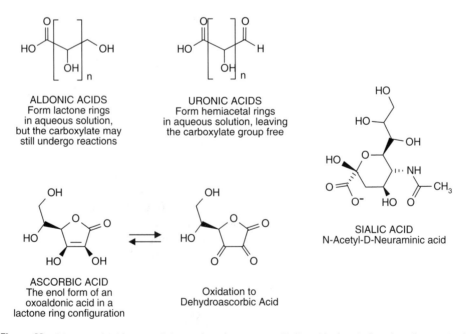

KERATIN SULFATE

N-Acetyl-D-Glucosamine, 6-O-Sulfate

D-Galactose

CHONDROITIN SULFATE

N-Acetyl-D-Galactosamine, 4-O-Sulfate

D-Glucuronic Acid

HEPARIN

D-Glucosamine, N-Sulfate, 6-O-Sulfate

D-Glucuronic Acid, 2-O-Sulfate

DERMATAN SULFATE

N-Acetyl-D-Galactosamine, 4-O-Sulfate

L-Iduronic Acid

Figure 22   Common sulfonated polysaccharides of biological origin.

ALDONIC ACIDS
Form lactone rings
in aqueous solution,
but the carboxylate may
still undergo reactions

URONIC ACIDS
Form hemiacetal rings
in aqueous solution, leaving
the carboxylate group free

SIALIC ACID
N-Acetyl-D-Neuraminic acid

ASCORBIC ACID
The enol form of an
oxoaldonic acid in a
lactone ring configuration

Oxidation to
Dehydroascorbic Acid

Figure 23   Monosaccharides containing carboxylate groups. Sialic acid often is found at the terminal residues of polysaccharides within glycoproteins.

functional groups. Sugar hydroxyl groups, for example, may be derivatized by acylating or alkylating reagents, similar to the principal reactions of primary amines (Section 1). However, acylation of a hydroxyl group usually creates an unstable ester derivative that is subject to hydrolysis in aqueous solution. An exception to this is acylation by a carbonylating reagent such as CDI (Chapter 2, Section 4.2) or DSC (Chapter 2, Section 4.3), which can produce stable carbamate linkages after subsequent conjugation with an amine containing molecule. By contrast, alkylating reagents, such as alkyl halogen compounds (Chapter 2, Section 4.6) typically form more stable ether bonds after reaction with hydroxyls. Figure 24 shows the reactions associated with alkylation and acylation of hydroxyl residues.

Carbohydrates containing hydroxyl groups on adjacent carbon atoms may be treated with sodium periodate (Section 4.4) to cleave the associated carbon-carbon bond and oxidize the hydroxyls to reactive formyl groups (Bobbitt, 1956). Modulating the concentration of sodium periodate can direct this oxidation to modify exclusively sialic acid groups (using 1 mM concentration) or to convert all available diols to aldehydes (using 10 mM or greater concentrations). Specific monosaccharide residues may be targeted with selective sugar oxidases to generate similar aldehyde functions only on discrete points of a polysaccharide chain (Section 4.4) (Avigad et al., 1962; Gahmberg, 1978). The creation of formyl groups in this manner may be done on purified polysaccharide molecules, as in the case of soluble dextrans (Chapter 15, Section 2.1), or may be selectively performed on carbohydrate constituents of glycoproteins and other glycoconjugates. Once formed, aldehyde groups may be covalently coupled with amine-containing molecules by reductive amination using sodium cyanoborohydride (Chapter 3, Section 4) (Dottavio-Martin and Ravel, 1978; Cabacungan et al., 1982).

The native reducing ends of carbohydrates also may be conjugated to amine-containing molecules by reductive amination. The reaction, however, typically is less efficient than using periodate-created aldehydes, since the open structure is in low concentration in aqueous solutions compared to the cyclic hemiacetal form. The

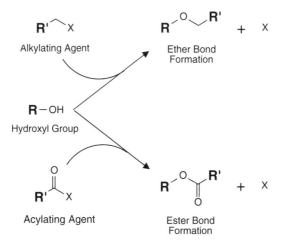

Figure 24   Hydroxyl groups within sugar residues may undergo alkylation or acylation reactions, forming ether or ester linkages.

reaction is usually allowed to continue for a week or more to reach good yields of coupling. Proteins may be modified to contain carbohydrate using this procedure (Gray, 1974, 1978; Baues and Gray, 1977; Schwartz and Gray, 1977).

The reducing ends of oligosaccharides can be modified with β-(p-amino-phenyl)ethylamine to yield terminal arylamine derivatives (Zopf *et al.,* 1978a; Jeffrey *et al.,* 1975). The aromatic armines then can be diazotized for coupling to active-hydrogen-containing molecules, such as the tyrosine phenolic residues in proteins (Zopf *et al.,* 1978b). Alternatively, the arylamines may be transformed into isothiocya-nate derivatives for coupling to amine-containing molecules, such as proteins (D. F. Smith *et al.,* 1978). The aromatic amine also may be used to conjugate the modified oligosaccharide directly with amine-reactive cross-linking agents or probes.

Another potential reaction of created or native aldehyde groups on carbohydrates is with hydrazide functionalities to form hydrazone linkages. Hydrazide-containing probes or cross-linking reagents may be conjugated with periodate-oxidized polysac-charides or with the reducing ends of sugars. The hydrazone bonds may be reduced with sodium cyanoborohydride to more stable linkages (Chapter 2, Section 5.1). The reduction step is recommended for long-term stability of cross-linked molecules. An example of this modification strategy is the use of biotin-hydrazide (Chapter 8, Section 3.3) to label specifically glycoproteins at their carbohydrate locations.

Reducing sugars can be detected by reaction with phenylhydrazine to yield a hy-drazone product, except that the result of the reaction is not what one might imagine giving the structure of aldoses and ketoses. Glucose, for example, can react with phenylhydrazine to yield the anticipated 1-phenylhydrazone derivative. In an excess of phenylhydrazine, however, the reaction continues to yield a 1,2-phenylhydrazone product, called an osazone, with concomitant production of aniline and ammonia (Fig. 25). Exactly how the No. 2 hydroxyl group gets oxidized to react with another molecule of phenylhydrazine is not entirely clear, but it probably proceeds through an enol intermediate. This reaction is typical of all α-hydroxy aldehydes and α-hydroxy ketones, not just those occurring in carbohydrate molecules. Thus, glucose, mannose, and fructose all yield the same osazone product upon reaction with phenylhydrazine, since the stereochemical differences about carbons 1 and 2 are eliminated. Reversal of the phenylhydrazone linkage with an excess of benzaldehyde yields an osone, a 1-aldehyde-2-keto- derivative of the sugar. Many simple hydrazide-containing re-agents probably are capable of forming similar 1,2-hydrazone derivatives with reduc-ing sugars, provided their size does not cause steric difficulties.

Polysaccharides, glycoproteins, and other glycoconjugates therefore may be specifi-cally labeled on their carbohydrate by creating aldehyde functional groups and subse-quently derivatizing them with another molecule containing an amine or a hydrazide group. This route of derivatization is probably the most common way of modifying carbohydrates.

The hydroxyl residues of polysaccharides also may be activated by certain com-pounds that form intermediate reactive derivatives containing good leaving groups for nucleophilic substitution. Reaction of these activated hydroxyls with nucleophiles such as amines results in stable covalent bonds between the carbohydrate and the amine-containing molecule. Activating agents that can be employed for this purpose include carbonyl diimidazole (Chapter 2, Section 4.2 and Chapter 3, Section 3), certain chloroformate derivatives (Chapter 2, Section 4.3), tresyl- and tosyl chloride, cyanogen bromide, divinylsulfone, cyanuric chloride (Chapter 15, Section 1.1), disuc-cinimidyl carbonate (Chapter 4, Section 1.7), and various *bis*-epoxide compounds

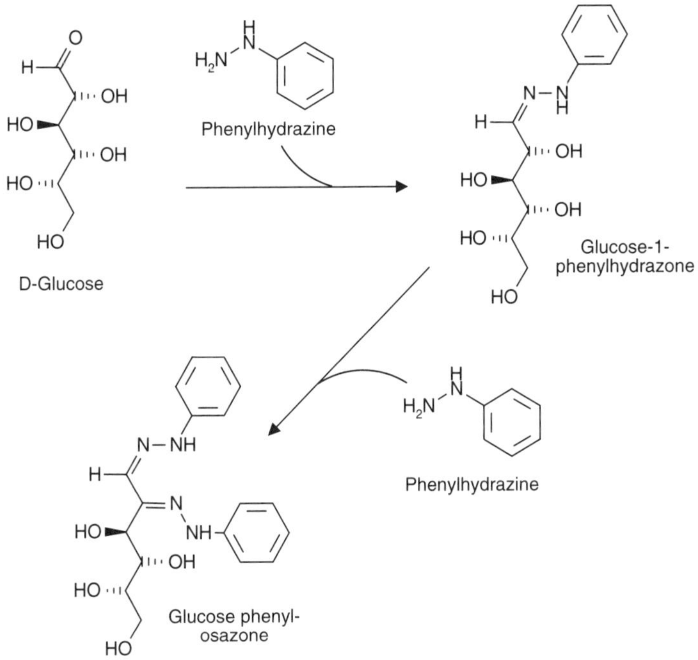

Figure 25 Phenylhydrazine can react with aldehyde or ketone groups within carbohydrates to give detectable products.

(Chapter 2, Section 1.7). Such activation steps are frequently done in nonaqueous solutions (i.e., dry dioxane, acetone, DMF, or DMSO) to prevent hydrolysis of the active species. Although many pure polysaccharides can tolerate these organic environments, many biological glycoconjugates cannot. Thus, these methods are suitable for activating pure polysaccharides such as dextran, cellulose, agarose, and other carbohydrates, but are not appropriate for modifying sugar residues on glycoproteins. Many of these hydroxyl-activating reagents also can be used to activate polysaccharide chromatography supports and other hydroxyl-containing synthetic polymers such as polyethylene glycol. For a complete treatment of polysaccharide chromatographic support activation through hydroxyl groups, see Hermanson *et al.* (1992). For a description of the activation of soluble polysaccharides and synthetic polymers, see Chapter 15.

The hydroxyl groups of carbohydrate molecules are only mildly nucleophilic—approximately equal to water in their relative nucleophilicity. Since the majority of reactive functional groups on bioconjugation reagents are dependent on nucleophilic reactions to initiate covalent bond formation, specific hydroxyl group modification is usually not possible in aqueous solution. Hydrolysis of the active groups on cross-linking reagents occurs faster than hydroxyl group modification, due to the relative high abundance of water molecules compared to the amount of carbohydrate hydroxyls present. In some cases, even if modification does occur, the resultant bond may be unstable. For instance, NHS esters (Chapter 2, Section 1.4) can react with hydroxyls to form ester linkages, which are themselves unstable to hydrolysis.

Anhydrides, such as acetic anhydride (Sections 4.2 and 5.1), may react with carbohydrate hydroxyls even in aqueous environments to form acyl derivatives. The reaction, however, is reversible by incubation with hydroxylamine at pH 10–11.

Epoxide-containing reagents, such as the homobifunctional 1,4-(butanediol) di-glycidyl ether (Chapter 4, Section 7.1), can react with polysaccharide hydroxyl groups to form stable ether bonds. Bis-epoxy compounds have been used to couple sugars and polysaccharides to insoluble matrices for affinity chromatography (Sundberg and Porath, 1974). The reaction of epoxides, however, is not specific for hydroxyl groups and will cross-react with amine and sulfhydryl functional groups, if present.

Hydroxyl groups on carbohydrates may be modified with chloroacetic acid to produce a carboxylate functional group for further conjugation purposes (Plotz and Rifai, 1982). In addition, indigenous carboxylate groups, such as those in sialic acid residues and aldonic or uronic acid-containing polysaccharides, may be targeted for modification using typical carboxylate modification reactions (Chapter 2, Section 3). However, when these polysaccharides are part of macromolecules containing other carboxylic acid groups such as glycoproteins, the targeting will not be specific for the carbohydrate alone. Pure polysaccharides containing carboxylate groups may be coupled to amine-containing molecules by use of the carbodiimide reaction (chapter 3, Section 1). The carboxylate is activated to an O-acylisourea intermediate which is in turn attacked by the amine compound. The result is the formation of a stable amide linkage with loss of one molecule of isourea.

Carbohydrate molecules containing amine groups, such as D-glucosamine, may be easily conjugated to other macromolecules using a number of amine-reactive chemical reactions and cross-linkers (Chapter 2, Sections 1 and 2). Some polysaccharides containing acetylated amine residues, such as chitin which contains N-acetyl-glucosamine, may be deacetylated under alkaline conditions (Jeanloz, 1963) to free the amines (forming chitosan in this case).

Amine functional groups also may be created on polysaccharides (Section 4.3). The reducing ends of carbohydrate molecules (or generated aldehydes) may be reacted with small diamine compounds to yield short alkylamine spacers that can be used for subsequent conjugation reactions. Hydrazide groups may be similarly created using bis-hydrazide compounds (Section 4.5).

Phosphate-containing carbohydrates that are stable, such as the 5′-phosphate of the ribose derivatives of oligonucleotides, may be targeted for modification using a carbodiimide-facilitated reaction (Section 4.3). The water-soluble carbodiimide EDC can react with the phosphate groups to form highly reactive phospho-ester intermediates. These intermediates can react with amine- or hydrazide-containing molecules to form stable phosphoramidate bonds.

Polysaccharide and Glycoconjugate Structure

Aldose monosaccharide units are frequently bound together through the No. 1 carbon hydroxyl group of one sugar to another sugar's No. 4 or 6 hydroxyl group, forming a complete acetal linkage. Two monosaccharides coupled in this fashion are termed a disaccharide. Numerous monosaccharides bound together to form a chain are called a polysaccharide. The most abundant polysaccharides in nature, starch and cellulose, consist of glucose bound together in α-1,4, β-1,4, and, to a lesser extent, α-1,6 acetal linkages (Fig. 26). Although the hemiacetal, cyclic structure of individual sugars shows some reversibility under equilibrium conditions, the acetal linkage between two monosaccharides is quite stable, only hydrolyzing under severe pH extremes.

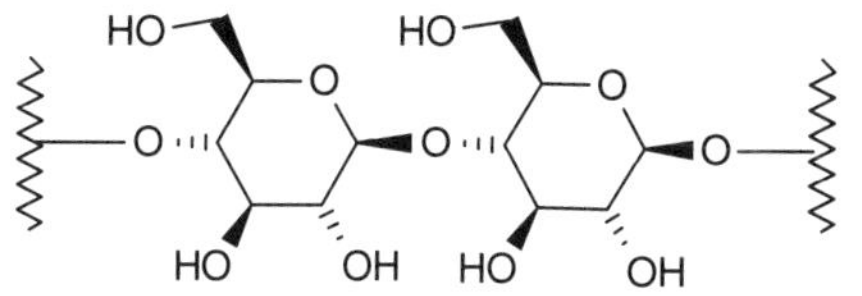

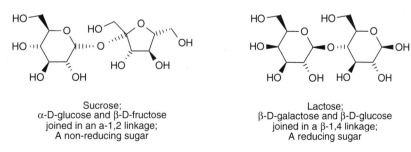

Cellobiose repeating unit
of cellulose;
D-Glucose joined in
β-1,4-linkages

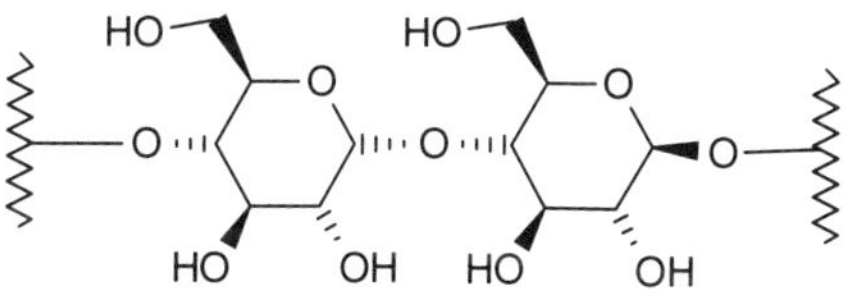

Maltose repeating unit
of starch (amylose);
D-Glucose joined in
α-1,4-linkages

Figure 26   The repeating units of cellulose and starch, two of the most common polysaccharides in nature.

Similarly, ketose sugars participate in polysaccharide formation by reaction of their anomeric carbon with a hydroxyl of another monosaccharide to create a ketal linkage. The acetal and ketal bonds within polysaccharides are termed O-glycosidic linkages.

Hemiacetal hydroxyl groups of carbohydrate molecules also may be coupled to amine-containing molecules to form N-glycosidic linkages, such as those in nucleic acids and ogligonucleotides.

Polysaccharides may or may not have reducing power, depending on the way they are linked together and whether the terminal, potentially reducing end is available. The structure of simple disaccharides can illustrate this point. Of the most common disaccharides, sucrose and lactose, sucrose is a nonreducing sugar since β-D-fructose is linked through its reducing C-2 hydroxyl, and lactose remains a reducing sugar, since the terminal glucose is linked to β-D-galactose through its C-4 hydroxyl, leaving its reducing end free (Fig. 27).

Polysaccharide synthesis is under enzymatic control, but does not occur from a

Sucrose;
α-D-glucose and β-D-fructose
joined in an a-1,2 linkage;
A non-reducing sugar

Lactose;
β-D-galactose and β-D-glucose
joined in a β-1,4 linkage;
A reducing sugar

Figure 27   Comparison of a reducing and a nonreducing disaccharide.

template as in protein synthesis. For this reason, each molecule of a particular polysaccharide will have its own unique molecular weight. The molecular weight of a carbohydrate polymer is usually expressed as an average. Starch or cellulose chains, for example, may vary by several hundred thousand in their molecular weights between individual molecules. For an excellent review of carbohydrate chemistry, see Binkley (1988).

Due to their polyhydroxylic structures, all carbohydrates are polar and will possess associated water molecules in aqueous solution, but they may not be fully water-soluble. Large polysaccharides such as cellulose form intricate matrices created from extensive hydrogen bonding. Neighboring monosaccharide units hydrogen bond within the same chain, whereas neighboring polymers form interchain hydrogen bonds between hydroxyls. The three-dimensional structure of a carbohydrate to a large extent is determined by these hydrogen bonds—sometimes resulting in sheeted or helical structures, as in the triple helix of agarose polysaccharide chains. Water will be intimately associated in this internal arrangement, but the overall multipolymer structure often is too large to allow for complete water solubility. For a review, see Preis (1980).

Polysaccharide solubility in aqueous solutions usually is dependent on polymer size and its allied three-dimensional structure. Even water-insoluble carbohydrates may be solubilized by controlled hydrolysis of O-glycosidic linkages to create smaller polysaccharide molecules. Thus, cellulose may be solubilized by heating in an alkaline solution until the polymers are broken up sufficiently to reduce their average molecular weight. Many such soluble forms of common polysaccharides are available commercially.

Carbohydrate also is an important constituent of many biological molecules. Polysaccharides may be found covalently conjugated to proteins and lipids, forming glycoproteins, proteoglycans, glycolipids, and lipopolysaccharides. Such glycoconjugates are produced in the cell through controlled, enzymatic processes. With proteins, the modification occurs after translational synthesis of the polypeptide chain at the ribosome.

Proteins newly synthesized on ribosomes may be transported to the Golgi apparatus where specific glycosyl transferases catalyze the coupling of monosaccharides to the polypeptide chains. Glycoproteins and mucoproteins are formed by the coupling of polysaccharides through O-glycosidic linkages to serine, threonine, or hydroxylysine in addition to N-glycosidic linkages with the amide side chain group of asparagine (Fig. 28). For reviews of glycoconjugate structure and function, see Hynes (1987); Lennarz (1980); Jentoft (1990); and Steer and Ashwell (1986).

The structure of most glycoprotein carbohydrate consists of a complex, branched heteropolysaccharide with the sugars mannose, N-acetyl glucosamine, sialic acid, galactose, and L-fucose being prevalent. Asparagine-linked polysaccharides are well characterized and are known to be constructed of a core unit consisting of three mannose residues and two N-acetyl glucosamine (GlcNAc) residues. The GlcNAc residues are bound to the Asp side chain amide nitrogen through a β1 linkage (Kornfield and Kornfield, 1985). The three mannose groups then usually form the first branch point in the oligosaccharide chain (Fig. 29). Much of the detailed structural knowledge of glycoconjugates is developed using controlled chemical or enzymatic degradation of the polysaccharides followed by analysis by gas chromatography and mass spectrometry (Biermann and McGinnis, 1989; McCleary and Matheson, 1986; Sweeley and Nunez, 1985; Vliegenthart et al., 1983).

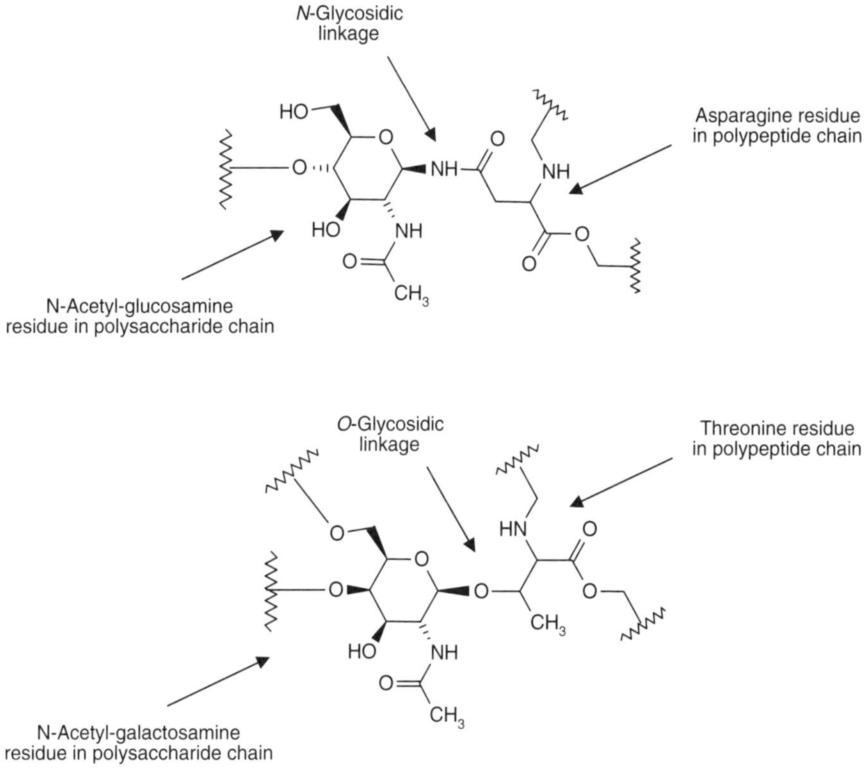

Figure 28   Common attachment points for polysaccharide chains on glycoproteins.

The content by weight of carbohydrate in glycoproteins may vary from only a few percent to as much as 70% in some proteins in mucous secretions. Although the exact function of the polysaccharide in most glycoproteins is unknown, in some cases it may provide hydrophilicity, recognition, and points of noncovalent interaction with other proteins through lectin-like affinity binding. In addition, extensive polysaccharide modification is helpful in preventing proteolytic digestion of the underlying polypeptide chain.

Another form of post-translational modification that may add carbohydrate to a polypeptide is nonenzymatic glycation. This reaction occurs between the reducing ends of sugar molecules and the amino groups of proteins and peptides. The aldehyde group of a reducing sugar first forms a reversible Schiff's base linkage with the α-amino or ε-amino groups of the protein. This bond then can undergo an Amadori rearrangement to form a stable ketoamine derivative (Fig. 30). The result is a blocked amine containing a sugar derivative with available hydroxyl residues. This reaction commonly occurs with proteins continually exposed to reducing sugars, such as glucose in blood. The measurement of glycated hemoglobin is a clinically important parameter in the management of diabetes mellitus. Increases in the blood sugar level in diabetes cause concomitant increases in the level of nonenzymatic glycation of blood proteins. Measuring the relative amount of glycated hemoglobin provides the physician with information concerning a diabetic patient's blood glucose control.

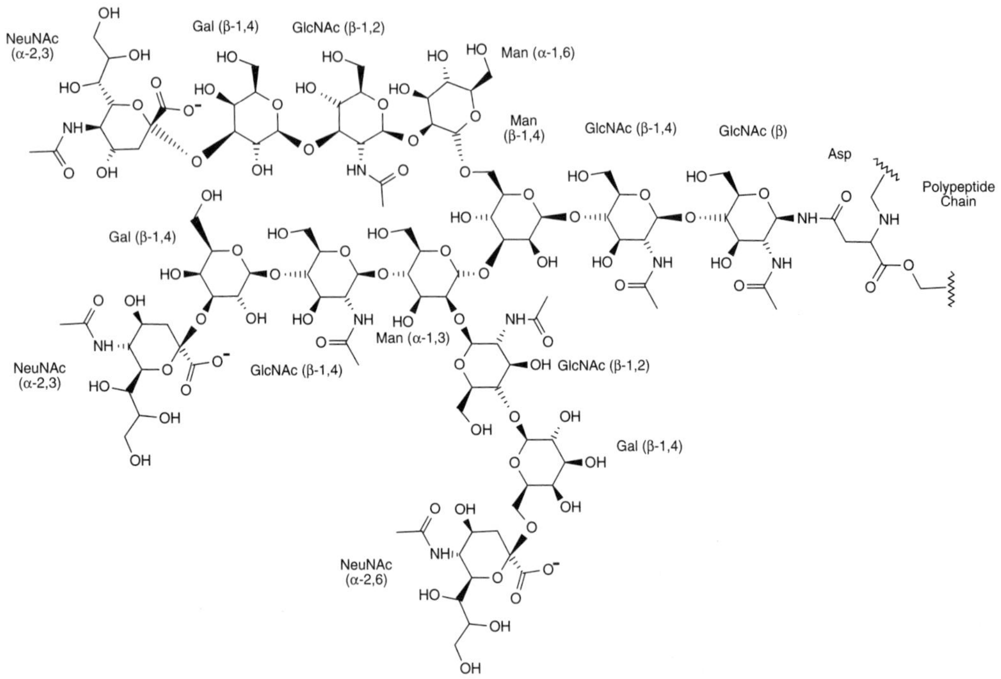

Figure 29 The complex structure of an asparagine-linked polysaccharide. Note the branched nature of the polymer with terminal sialic acid residues on each chain.

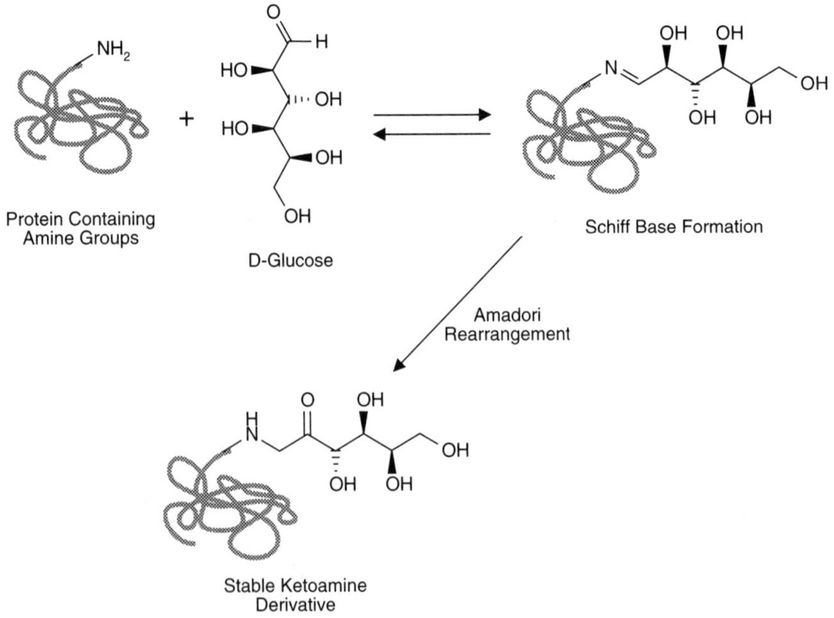

Figure 30 A reducing sugar may modify protein amine groups through Schiff base formation followed by an Amadori rearrangement to give a stable ketoamine product. Glucose is a common *in vivo* modifier of blood proteins through this process.

2.2. Carbohydrate Cross-linking Methods

The presence of carbohydrate on biomolecules provides important points of attachment for modification and conjugation reactions. Coupling only through polysaccharide chains often can direct the reaction away from active centers or critical points in protein molecules, thus preserving activity. Cross-linking strategies involving polysaccharides or glycoconjugates usually involve a two- or three-step reaction sequence. If no reactive functional groups other than hydroxyl groups are present on the carbohydrate, then the first step is to create sufficiently reactive groups to couple with the functional groups of a second molecule.

Perhaps the easiest way to target specifically polysaccharides on glycoproteins is through mild sodium periodate oxidation. Periodate cleaves adjacent hydroxyl groups in sugar residues to create highly reactive aldehyde functional groups (section 4.4). It is an aqueous reaction that is tolerated by most biological glycoconjugates and pure polysaccharide molecules. Particularly convenient is that the level of periodate addition can be adjusted to cleave selectively only certain sugars in the polysaccharide chain. A concentration of 1 mM sodium periodate specifically oxidizes sialic acid residues to aldehydes, leaving all other monosaccharides untouched. Increasing the concentration to 10 mM, however, will cause oxidation of other sugars in the carbohydrate chain, including galactose and mannose residues on glycoproteins. The generated aldehydes then can be used in coupling reactions with amine- or hydrazide-containing molecules to form covalent linkages. Amines can react with formyl groups under reductive amination conditions using a suitable reducing agent such as sodium cyanoborohydride. The result of this reaction is a stable secondary amine linkage (Chapter 2, Section 5.3). Hydrazides spontaneously react with aldehydes to form hydrazone linkages, although the addition of a reducing agent greatly increases the efficiency of the reaction and the stability of the bond (Chapter 2, Section 5.1).

Oxidized glycoconjugates usually are stable enough to be stored freeze-dried without loss of activity prior to a subsequent conjugation reaction. Storage in solution, however, may cause slow polymerization if the molecule also contains amine groups, as in glycoproteins. Sometimes the protein can be treated to block its amines prior to periodate oxidation, as in the procedure often used with the enzyme horseradish peroxidase (HRP) (Chapter 16, Section 1), thus eliminating the potential for self-conjugation.

If the second molecule to be coupled to the oxidized glycoconjugate already has the requisite amines or hydrazide groups, then directly mixing the two components together in the presence of a reductant is all that is needed. This is an example of a two-step procedure. However, if the second molecule possesses none of the appropriate functional groups for coupling, then modifying it to contain them must be done prior to the conjugation reaction (see Sections 4.3 and 4.5). Thus, a three-step protocol results. The use of other functional groups (either indigenous or created) on polysaccharide molecules to effect a cross-linking reaction can be done in similar two- or three-step strategies.

Occasionally, it is important to conjugate a polysaccharide-containing molecule to another molecule while retaining, as much as possible, the carbohydrate's original chemical and three-dimensional structure. For instance, in the preparation of immunogen conjugates by coupling a polysaccharide molecule to a carrier, care should be taken to preserve the structure of the carbohydrate to ensure antibody recognition of the native molecule. In this case, periodate oxidative techniques may not be best choice to effect cross-linking due to the potential for extensive ring opening throughout the chain. Under controlled conditions, however, where periodate is carefully used in limiting quantities, this method has proved successful in creating oligosaccharide–carrier conjugates (Anderson *et al.*, 1989).

Retention of native carbohydrate structure also is important in applications that utilize the conjugated polysaccharide in binding studies with receptors or lectins. In these cases, the carbohydrate should be modified at limited sites, preferentially only at its reducing end.

3. Modification of Nucleic Acids and Oligonucleotides

The nucleic acid polymers DNA and RNA form the most basic units of information storage within cells. The conversion of their unique information code into proteins and enzymes is the fundamental step in controlling all cellular processes. Targeting segments of this encoded data with labeled probes that are able to bind to specific genetic regions allows detection, localization or quantification of discrete oligonucleotides. This targeting capability is made possible by the predictable nature of nucleic acid interactions. Despite the complexity of the genetic code, the base-pairing process, which causes one oligonucleotide to bind to its complementary sequence, is rather simplistic. Nucleic acids are the only type of complex biological molecule wherein their binding properties can be fully anticipated and incorporated into synthetic oligonucleotide probes. Thus, a short DNA segment can be synthetically designed and used to target and hybridize to a complementary DNA strand within a much larger chromosome. If the small oligonucleotide is labeled with a detectable component that does not interfere in the base-pairing process, then the targeted DNA can be assayed.

Bioconjugate techniques involving nucleic acids are becoming one of the most important application areas of cross-linking and modification chemistry. As the secrets of the genetic code are broken by such mammoth efforts as the Human Genome Project, knowledge of the DNA sequence that governs specific protein synthesis is leading to diagnostic tests able to assess the presence of critical genetic markers associated with certain disease states. To test for particular target sequences, complementary oligonucleotide probes are used that possess conjugated enzymes, fluorophores, haptens, radiolabels, or other such groups that can be used to detect a hybridization signal. Such oligonucleotide conjugates can be used to discover target sequences in blots, electrophoresis gels, tissues, or cells, or immobilized to surfaces or in solution.

The power and advantages of assessing cellular processes at their most fundamental level is propelling the science of oligonucleotide probe detection into the most prominent position in bioconjugate chemistry. Some are predicting that in the not-too-distant future hundreds of tests will be done routinely in the physician's office—each monitoring different aspects of genetic information—all with the use of specific oligonucleotide probes.

In this section, the chemistry and structure of nucleic acids and oligonucleotides is discussed with a view to creating functional conjugates with detectable molecules. The corresponding strategies and protocols associated with DNA or RNA modification and conjugation can be found in Chapter 17.

3.1. Polynucleotide Structure and Function

Polymers of nucleic acids are characterized by the types of base residues present and the structure of their sugar backbone. The bases are nitrogenous ring compounds consisting of either purine or pyrimidine derivatives. A purine is a fused-ring compound containing one six-member ring attached to a five-member ring, whereas a pyrimidine consists of a single six-member ring (Fig. 31).

Nucleic acids can contain of any one of three kinds of pyrimidine ring systems (uracil, cytosine, or thymine) or two types of purine derivatives (adenine or guanine). Adenine, guanine, thymine, and cytosine are the four main base substituents found in DNA. In RNA molecules, three of these four bases are present, but with thymine replaced by uracil to make up the fourth. Some additional minor derivatives are found in messenger RNA (mRNA), transfer RNA (tRNA), and ribosomal RNA (rRNA), particularly the N^4, N^4-dimethyladenine and N^7-methylguanine varieties.

Nucleic acid sugar residues are attached to the associated base units in an N-glycosidic bond, involving the No. 1 nitrogen of pyrimidine bases or the No. 9 nitrogen of purines directly linked to the No. 1 carbon of the monosaccharide derivative (Fig. 32). The sugar group consists of either a β-D-ribose unit (found in RNA) or a β-D-2-deoxyribose unit (in DNA) (Fig. 33). In mRNA and rRNA, a minor sugar derivative, a 2′-O-methylribosyl group, also is found.

The nomenclature of nucleic acid chemistry further characterizes the structure of the associated groups. A *nucleoside* contains only a base group and an attached sugar. A *nucelotide* consists of a base and a sugar plus a phosphate group. At this point, the naming system gets somewhat confusing due to the fact that the nucleoside name is a derivative of the base name. Table 2 shows this relationship and their associated abbreviations (which are simpler to remember).

In each nucleotide monomer of DNA or RNA molecules, a phosphate group is attached to the C-5 hydroxyl of each sugar residue in an ester (anhydride) linkage. These phosphate groups in turn are linked in diester bonds to neighboring sugar groups of adjacent nucleotides through their 3′-ribosyl hydroxyl to create the oligonucleotide polymer backbone (Fig. 34). Thus, the phosphate–sugar repeating unit produces the linear sequence within the DNA or RNA structure, while the four types of base units protrude out from this backbone, creating the unique code making up the genetic information.

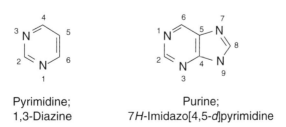

Pyrimidine;
1,3-Diazine

Purine;
7*H*-Imidazo[4,5-*d*]pyrimidine

Figure 31 The pyrimidine and purine ring structures common to nucleic acids.

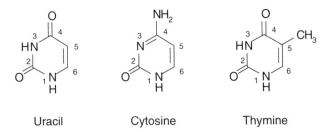

Uracil   Cytosine   Thymine

Figure 35 The three pyrimidine bases common to nucleic acid construction. Cytosine and thymine are found in DNA, while in RNA, uracil residues replace thymine. The associated sugar groups are bound in N-glycosidic linkages to the N-1 nitrogen.

Addition of a nucleophile to the C-6 position of cytosine often results in fascile displacement reactions occurring at the N_4 location. With hydroxylamine attack, nucleophilic displacement causes the formation of an N_4-hydroxy derivative. A particularly important reaction for bioconjugate chemistry, however, is that of nucleophilic bisulfite addition to the C-6 position. Sulfonation of cytosine can lead to two distinct reaction products. At acid pH wherein the N-3 nitrogen is protonated, bisulfite reaction results in the 6-sulfonate product followed by spontaneous hydrolysis. Raising the pH to alkaline conditions causes effective formation of uracil. If bisulfite addition is done in the presence of a nucleophile, such as a primary amine or hydrazide compound, then transamination at the N_4 position can take place instead of hydrolysis (Fig. 38). This is an important mechanism for adding spacer arm functionalities and other small molecules to cytosine-containing oligonucleotides (see Chapter 17, Section 2.1).

Electrophilic reagents also can modify the pyrimidine rings of nucleic acids. Alkylation and acylation reactions can take place at several sites on all three bases. Figure 39 illustrates the principal locations where electrophilic attack can occur. In particular, the heteroatoms (oxygen and nitrogen) are the best positions of high electron density, therefore functioning as nucleophiles in reaction processes. Of the pyrimidine residues, however, it is the N-3 position of cytosine derivatives that is the most susceptible to alkylation. Reactions can occur with ethylenimine compounds (Section 4.3), alkyl halogens (Chapter 2, Section 2.1), epoxides (Chapter 2, Section 1.7), and many other strong alkylating agents [for review, see Brown (1974)].

Acylation reactions can be done at the nucleophilic sites on pyrimidines using

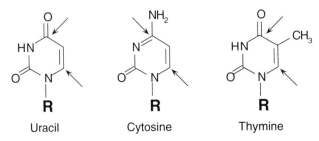

Uracil   Cytosine   Thymine

Figure 36 Pyrimidine bases are subject to nucleophilic displacement reactions primarily at the C-4 and C-6 positions.

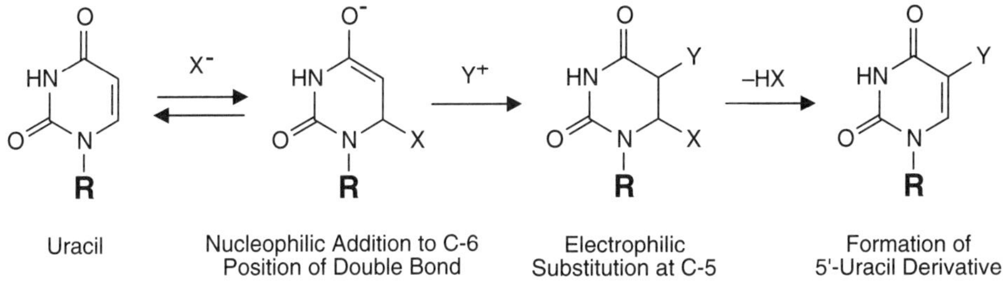

Figure 37 Nucleophilic addition at C-6 of the pyrimidine double bond can cause electrophilic substitution to occur at the C-5 position.

activated forms of carboxylic acids. Acylation of functional groups in nucleotides typically is used for protection during synthesis (Reese, 1973). However, for bioconjugate applications, the reactivity of native groups on pyrimidines is not as great as that obtained using an amine-terminal spacer derivative, such as those described in Chapter 17, Section 2.1. Yields and reaction rates are typically low for direct acylation or alkylation of pyrimidine bases, especially in aqueous environments.

The N-3 position of uracil also can be modified with carbodiimide reagents. In

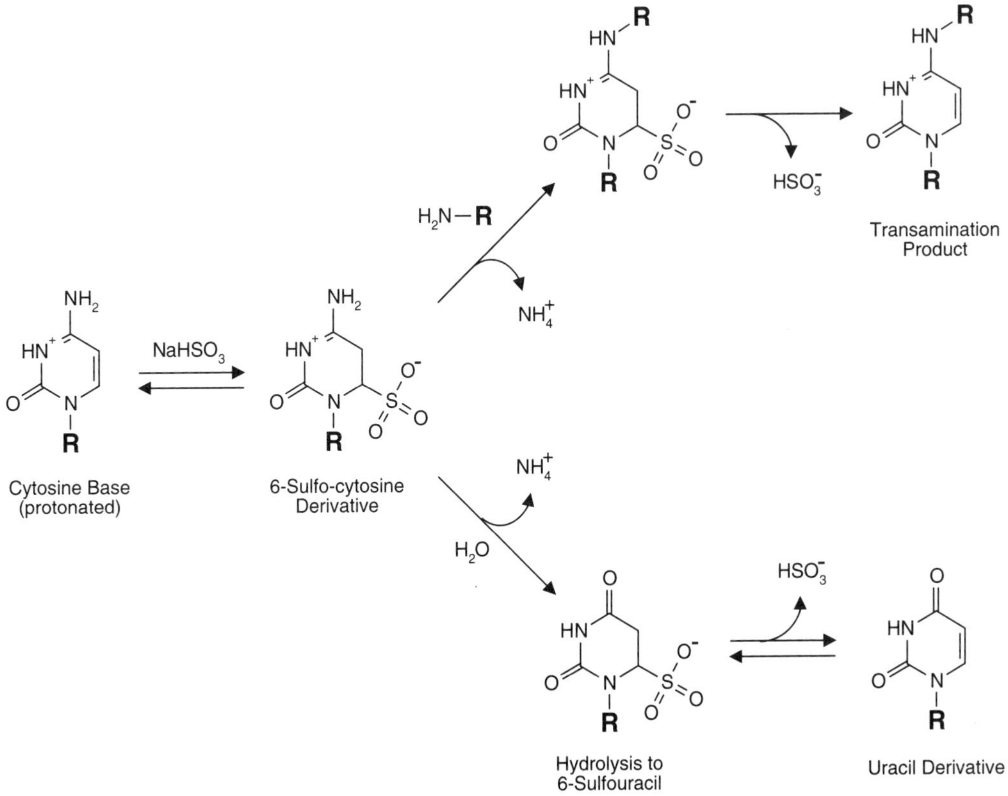

Figure 38 Reaction of bisulfite with cytosine bases is an important route of derivatization. It can lead to uracil formation or, in the presence of an amine (or hydrazide)-containing compound, transamination can occur, resulting in covalent modification.

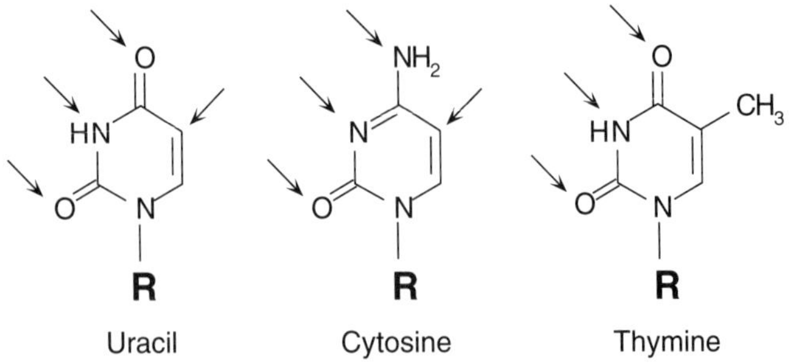

Figure 39   Potential sites of electrophilic attack on pyrimidine bases.

particular, the water-soluble carbodiimide CMC [1-cyclohexyl-3-(2-morpholino-ethyl) carbodiimide, as the metho *p*-toluene sulfonate salt] can react with the N-3 nitrogen at pH 8 to give an unstable, charged adduct. The derivative is reversible at pH 10.5, regenerating the original nucleic acid base (Fig. 40). Cytosine is unreactive in this process.

Halogenation of pyrimidine bases may be done with bromine or iodine. Bromination occurs at the C-5 of cytosine, yielding a reactive derivative that can be used to couple diamine spacer molecules by nucleophilic substitution (Fig. 41) (Traincard *et al.*, 1983; Sakamoto *et al.*, 1987; Keller *et al.*, 1988). Other pyrimidine derivatives also are reactive to bromine compounds at the C-5 position. Either an aqueous solution of bromine or the compound N- bromosuccinimide can be used for this reaction. The brominated derivatives can be used to couple amine-containing compounds to the pyrimidine ring structure (chapter 17, Section 2.1).

Other reactions characterized for pyrimidine residues include mercuration at C-5 of cytosine or uracil (Hopman *et al.*, 1986), cyloaddition to the 5,6 double bond of thymine and uracil (Cimino *et al.*, 1985), and thiolation at the C-4 amino group of cytosine (Malcolm and Nicolas, 1984).

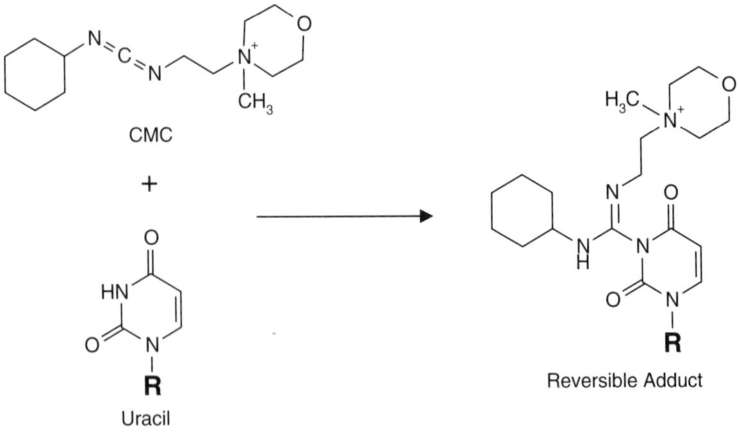

Figure 40   The carbodiimide CMC can react with the N-3 nitrogen to yield a reversible product.

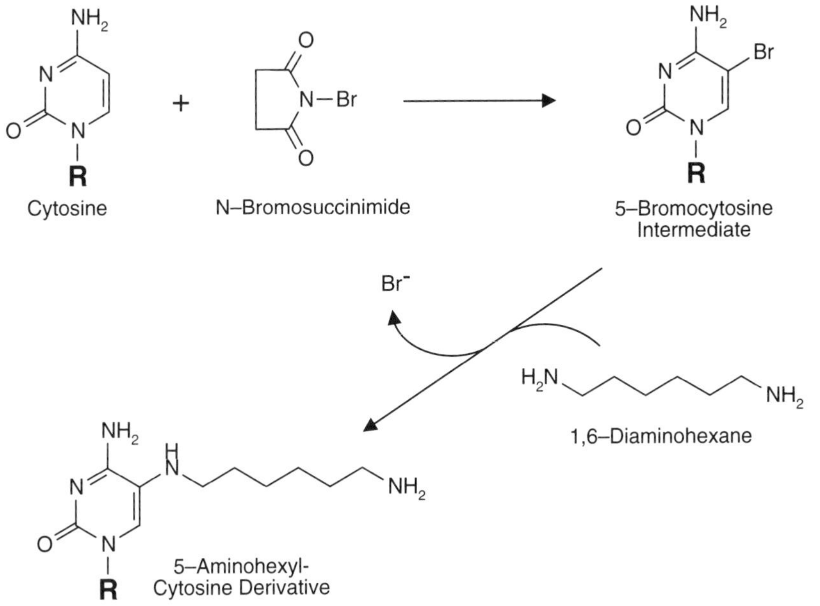

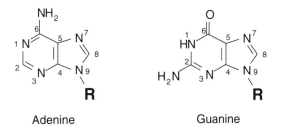

Figure 41 Cytosine bases are susceptible to bromination at the C-5 double bond position, resulting in active intermediates capable of reacting with amine nucleophiles.

Adenine and Guanine Residues

The purine bases of nucleic acids are constructed of a two-ring system made from a pyrimidine-type, six-member ring fused with a five-member imidazole ring. Adenine and guanine are present in both RNA and DNA. They differ in their six-member ring structures by an additional point of unsaturation between C-6 and N-1 (in adenine) and by the presence of amine or ketone groups attached to C-2 or C-6 (Fig. 42). Attachment to ribose or deoxyribose in nucleosides is made through an N-glycosidic linkage at N-9 of the imidazole ring on either purine.

As in the case of pyrimidine bases discussed previously, adenine and guanine are subject to nucleophilic displacement reactions at particular sites on their ring structures (Fig. 43). Both compounds are reactive with nucleophiles at C-2, C-6, and C-8, with C-8 being the most common target for modification. However, the purines are

Figure 42 The structures of the common purine bases of RNA and DNA. The associated sugar groups are bound in N-glycosidic linkages to the N-9 position.

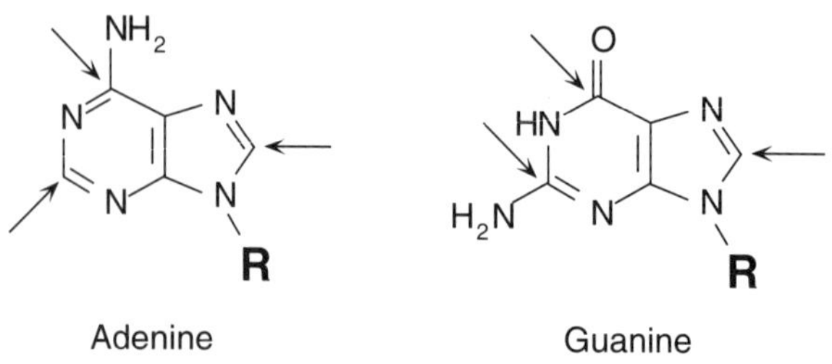

NH₂ ... Adenine

Guanine

Figure 43 Primary nucleophilic displacement sites on purine bases.

much less reactive to nucleophiles than the pyrimidines. Hydrazine, hydroxylamine, and bisulfite—all important reactive species with cytosine, thymine, and uracil—are almost unreactive with guanine and adenine.

With purines, reaction with electrophilic species is the most important route to derivatization. Figure 44 identifies the major sites of electrophilic attack on adenine and guanine. On both bases it is the heteroatoms that make up the majority of sites. Alkylation reactions thus can occur at N-1, N-3, and N-7 in adenine or N-3 and N-7 in guanine. However, the greatest location of electron density (nucleophilicity) occurs at N-7 on the imidazole ring of guanine, followed by N-1 of adenine. According to Brown (1974), the order of reactivity of nucleosides toward alkylation by esters of strong acids is guanine > adenosine > cytidine >> uridine (nearly unreactive).

As with pyrimidines, the water-soluble carbodiimide CMC may react with guanine derivatives to give a reversible adduct at N-1 (Fig. 45). Raising the pH to highly alkaline conditions regenerates the purine group. Adenine residues, however, display no reactivity in this process.

One of the most important reactions of purines is the bromination of guanine or adenine at the C-8 position. It is the site that is the most common point of modification for biconjugate techniques using purine bases (Fig. 46). Either an aqueous solution of bromine or the compound N-bromosuccinimide can be used for this reaction. The brominated derivatives then can be used to couple amine-containing compounds to the pyrimidine ring structure by nucleophilic substitution (Chapter 17, Section 2.1).

Adenine also may undergo an additional reaction at its C-6 amine group using a Fischer–Dimroth rearrangement mechanism. Alkylation at N-1 can result in a rearrangement to give the C-6 alkylated product. The reaction at N-1 usually requires

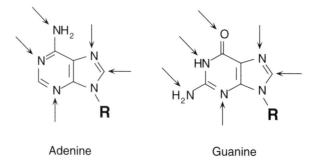

Adenine   Guanine

Figure 44 Electrophilic attack can occur at a number of sites on both purine bases.

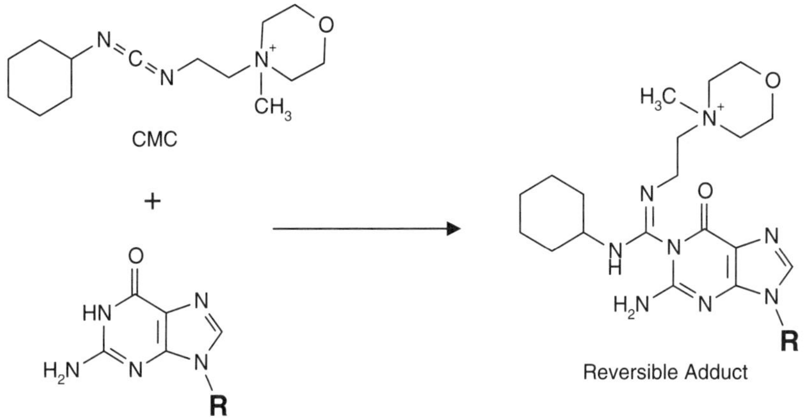

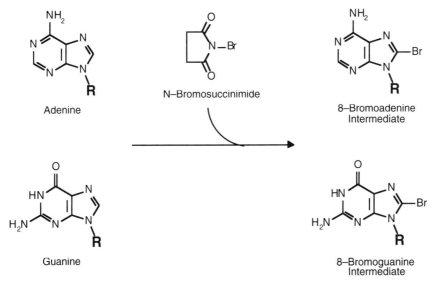

Figure 45 The carbodiimide CMC can react with guanine at the N-1 position to form a reversible complex.

extended time to obtain good yields. For instance, alkylation with iodoacetic acid takes 5–10 days at pH 6.5. Under alkaline conditions and elevated temperatures, the six-member ring then is broken and reformed, resulting the 6-aminoalkylated product containing a terminal carboxylate group (Fig. 47). The resultant acid can be used in further derivatization reactions to facilitate conjugate formation (Lowe, 1979).

An additional reaction reported for adenine involves the coupling of glutaraldehyde to the 6-amino group (Matthews and Kricka, 1988). However, reaction at this group with electrophilic reagents such as those discussed in Part II proceeds more slowly than

Adenine

N–Bromosuccinimide

8–Bromoadenine
Intermediate

Guanine

8–Bromoguanine
Intermediate

Figure 46 The purine bases are subject to bromination reactions at the C-8 position, forming an important reactive intermediate for derivatization purposes.

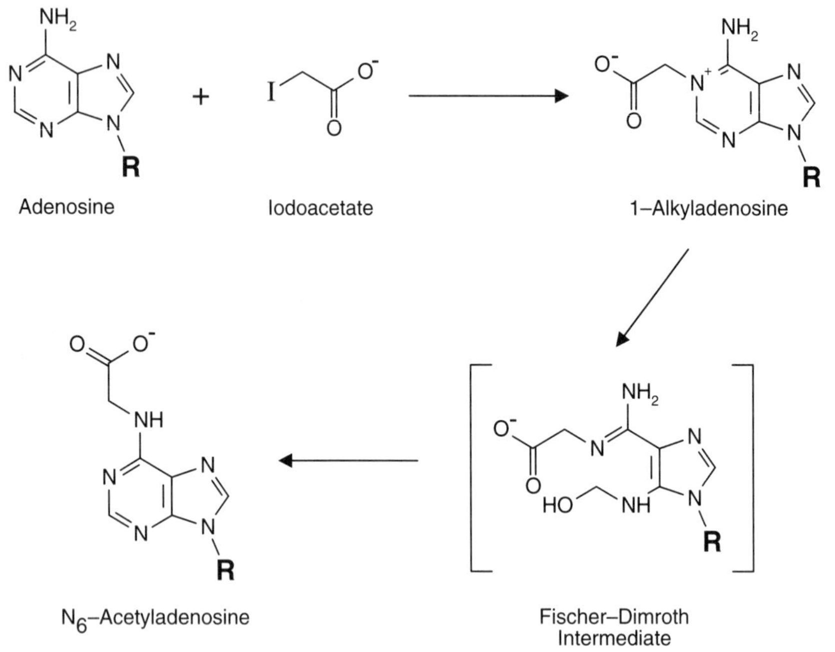

Figure 47   Alkylation reactions can occur at the N-1 position of adenosine, resulting in a Fischer–Dimroth rearrangement to yield an N_6 derivative.

that possible using a primary aliphatic amine. In general, bioconjugate chemistry done with nucleic acid bases involves the formation of an intermediate derivative containing a spacer arm terminating in an amine, sulfhydryl, or carboxylate to obtain acceptable reactivity and yields.

Sugar Groups

The sugar portion of oligonucleotides is a 5-carbon pentose occurring in one of two forms. In RNA, it is β-D-ribose in a ring structure. In DNA, the monosaccharide is β-D-2-deoxyribose, wherein the number 2′ carbon of the ring lacks an hydroxyl group. An individual nucleotide will have its 1′ hydroxyl group of the ribose unit tied up in an N-glycosidic bond with the associated base and its C-5 hydroxyl group bound to phosphate in an ester linkage. If the nucleotide is of the deoxy form, then the only remaining hydroxyl is on the 3′ carbon of the sugar unit. Ribonucleic acids, by contrast, contain a diol group formed from the two hydroxyls on the 2′ and 3′ carbons of ribose (Fig. 48). Polymers of nucleic acids are created through diester phosphate bonds, mainly connected between the 5′ hydroxyl of one sugar group and the 3′ hydroxyl of the next adjacent sugar. Thus, DNA contains no hydroxyl groups except the single one at the 3′ terminal of each strand. RNA has one hydroxyl at each nucleotide sugar unit and a diol group at the 3′ end.

Conjugation or modification reactions may be done through the 3′ hydroxyl group of deoxyribonucleic acids or the 2′,3′-diol of ribonucleic acids. Hydroxyls may be targeted for coupling using strong alkylating agents under alkaline conditions. Epox-

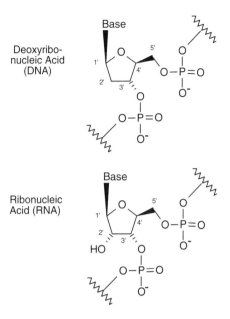

Figure 48　The similar structures of DNA and RNA basic units.

ide compounds (Chapter 4, Section 7) are particularly effective at modifying hydroxyl groups. The most common method of conjugation through nucleotide sugar units, however, is periodate oxidation of the adjacent hydroxyls of ribonucleic acids. Treatment with periodate breaks the carbon—carbon bond between the two hydroxyl residues and creates two aldehyde groups (Seela and Waldeck, 1975). A procedure for oxidizing carbohydrates with sodium periodate can be found in Section 4.4. This method can be used to create RNA conjugates through directed coupling only at the 3' end or to immobilize ribonucleic acids such as ATP to insoluble supports for affinity chromatography (Lowe, 1979).

Phosphate Groups

The phosphate groups of nucleotides are joined to the 5' hydroxyl group of the sugar component in an ester or anhydride linkage. Several forms of nucleoside phosphate compounds are possible, containing up to three esterified phosphate groups polymerized off the ribose or deoxyribose unit. The presence of these groups contributes an overall negative charge to the nucleotide—minus two for the terminal phosphate group and minus one for each internal phosphate under alkaline conditions. Multiple esterified phosphates contain considerable potential energy from their easily hydrolyzed anhydride bonds. This energy is the basis for many biochemical transformations in biological systems. It is the triphosphate form of nucleosides that is utilized in DNA and RNA synthesis *in vivo*. However, nucleoside triphosphates and diphosphates such as ATP and ADP have numerous contributions to cellular metabolism beyond just oligonucleotide construction. Controlled hydrolysis of their multiple phosphate ester bonds releases energy for many biological operations. Other derivatives of nucleoside phosphate compounds provide cofactors for enzymes (such as coenzyme A) or are involved in signal transduction processes [such as cyclic AMP (cAMP)]. Figure 49 shows some of these common nucleoside phosphate derivatives.

The phosphate groups of nucleotides may be targeted for modification reactions using condensation agents such as carbodiimides. In aqueous environments, EDC (Chapter 3, Section 1.1) may be used to couple amine-containing compounds to the terminal phosphate group of an oligonucleotide, forming a phosphoramidate linkage. In DNA or RNA chains, the internal phosphate groups do not react under the pH conditions of the modification. In this way, the 5′ phosphate group may be specifically targeted for modification or conjugation, thus avoiding potential interference with hydrogen bonding interactions with complementary polynucleotide strands. Chapter 17, Sections 2.1 and 2.2 describe the use of this reaction in bioconjugate applications.

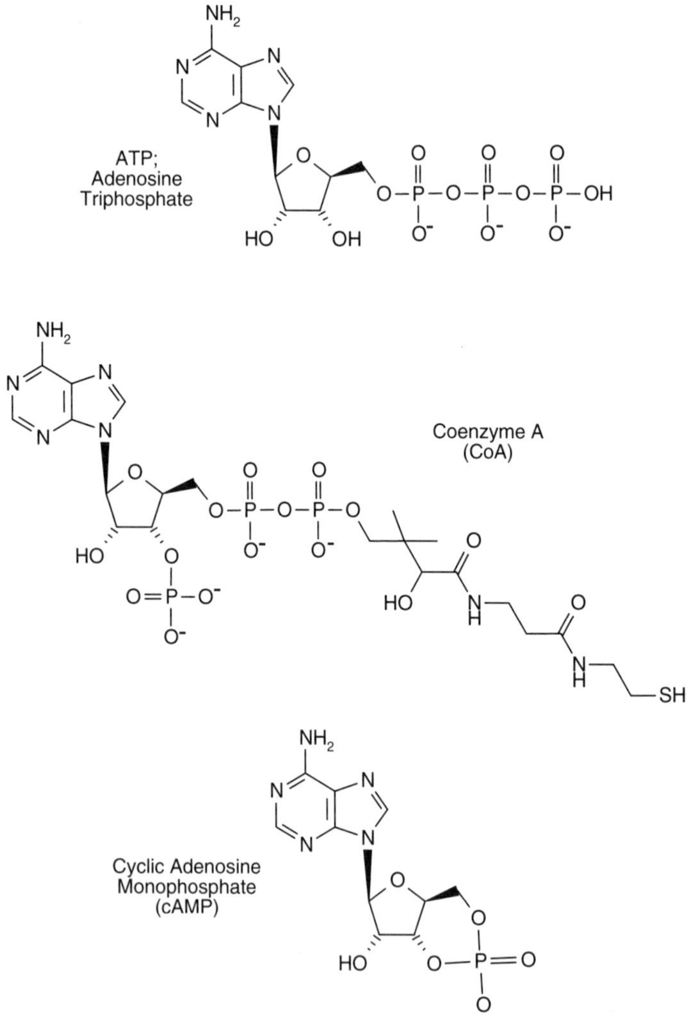

Figure 49 Nucleotide derivatives have additional functions *in vivo* beyond their role in oligonucleotide construction.

Another phosphate modification procedure that is effective at adding detectable components to oligonucleotide probes is to use a phosphoramidite derivative. The common method of automated oligonucleotide synthesis is to use phosphoramidite chemistry to add nucleotides to the growing sequence. A functionalized phosphoramidite nucleotide derivative can be added at particular points in the synthetic process to create labeled probes of known structure. Nonnucleotide phosphoramidites also may be used to produce modified probes containing fluorescent molecules, biotin, chelating groups, or spacer groups with amines for further derivatization. Most of these techniques require an automated DNA synthesizer. The methods of DNA modification during synthesis have been reviewed and are beyond the scope of this book (Beaucage and Iyer, 1993).

RNA and DNA Structure

The nucleotides forming RNA or DNA molecules are linked together in phosphodiester bonds with sugar–phosphate repeating units. The esters are directionally linked between the 3′ hydroxyl of one ribosyl group and the 5′ hydroxyl of the next. The fundamental step in cellular DNA synthesis involves the reaction of a deoxynucleoside triphosphate group with the 3′ end of an existing chain. The nucleotide sequence of a new strand is enzymatically controlled by use of a complementary chain as a template. Each new nucleotide addition is facilitated by the energy released through hydrolysis of two phosphates from the triphosphate group of the incoming nucleoside. The resulting succession of nucelotides encodes the message for protein synthesis, with each three-base code signaling a particular amino acid in a polypeptide sequence.

Nucelotide bases projecting from the sugar–phosphate backbone of a polynucleotide are able to interact with other strands through hydrogen bonding. Hydrogen bonding can occur between cytosine and guanine base units in different strands of DNA through interaction of the C-2 ketone oxygen, the N-3 nitrogen, and C-4 amine groups of cytosine with the C-2 amine, N-1 nitrogen, and the C-6 ketone oxygen of guanine. In a similar fashion, thymine (or uracil) residues can hydrogen bond with adenine groups through the N-3 nitrogen and C-4 ketone oxygen of thymine interacting with the N-1 nitrogen and C-6 amine of adenine (Fig. 50).

This specific base-pairing capability of oligonucleotides defines the structure of

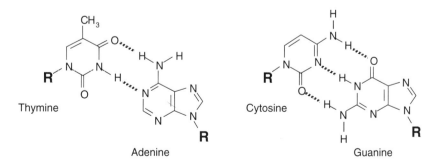

Figure 50 Base-pairing can occur between complementary bases in opposing oligonucleotide strands. These predictable interactions form the basis for using synthetic oligonucleotide probes to target particular DNA sequences.

complementary DNA molecules. In the classic Watson–Crick model, two complementary DNA strands interact in an antiparallel fashion to form a right-handed double helix. Thus, one chain runs in the 3′ to 5′ direction while the complementary chain runs in the 5′ to 3′ direction through the helical structure. This standard double helix, now called the B form, occurs often in aqueous solution and is the most stable structure under physiological conditions (Fig. 51). However, there are several other forms that double-stranded DNA can take in solution. Another right-handed helical construction, the A form, can occur under nonaqueous conditions and is more compact than the B form. A completely different DNA structure, the Z form, is a left-handed helix that can occur in some segments containing an abundance of alternating pyrimidines and purines. Short segments of Z structure have been found in some cells. Finally, some rare DNA sequences can form triple-helical regions through normal and non-Watson–Crick base-pairing.

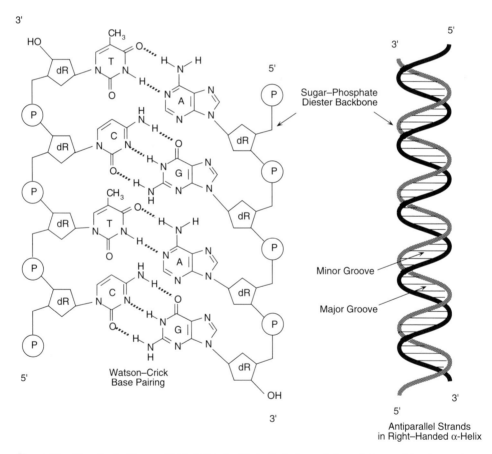

Figure 51   The classic Watson–Crick DNA double helix is formed through base-pairing interactions between two antiparallel strands. In physiological conditions, the two strands take on an α-helical shape with about 10 bp per turn of the helix. The phosphate–sugar backbone of the helix faces outward, while hydrogen bonding between opposing bases occurs in the middle of the wrapped strands. This configuration creates minor and major grooves between the phosphate–sugar backbones, potentially exposing the internal bases to interactions with other molecules.

Unlike the double-stranded nature of DNA, RNA molecules usually occur as single strands. This does not mean they are unable to base-pair as DNA can. Complementary regions within an RNA molecule often base-pair and form complex tertiary structures, even approaching the three-dimensional nature of proteins. Some RNA molecules, such as transfer RNA (tRNA) possess several helical areas and loops as the strand interacts with itself in complementary sections. Other hybrid molecules such as the enzyme RNase P contain protein and RNA portions. The RNA part is highly complex with many circles, loops, and helical regions creating a convoluted structure.

The predictable nature of DNA and RNA base-pairing make their interactions the most defined of any biological system. The specific affinity of one strand for its complementary sequence makes it possible to target genetic markers with extreme accuracy. Synthetic segments of RNA or DNA can be used to detect or quantify their complementary targets, even in highly dilute environments containing many other oligonucleotide molecules. If the oligonucleotide probe is labeled with a highly detectable component, then specific base-pairing interactions can be assayed. This ability has created an extensive utilization of labeled probes in molecular biology. Detection of target DNA or RNA can be done in cells, tissue sections, blots, or electrophoresis gels, or after amplification by PCR techniques, or in solution. The ability to detect single-copy genes through the use of labeled oligonucleotide probes will make this field one of the leading application areas for bioconjugate techniques.

3.2. Polynucleotide Cross-linking Methods

The unique properties of oligonucleotides create cross-linking options that are far different from those of any other biological molecule. Nucleic acids are the only major class of macromolecule that can be specifically synthesized *in vitro* by enzymatic means. The addition of modified nucleoside triphosphates to an existing DNA strand by the action of polymerases or transferases allows addition of spacer arms or detection components at random or discrete sites along the chain. Alternatively, chemical methods that modify nucleotides at selected functional groups can be used to produce spacer arm derivatives or activated intermediates for subsequent coupling to other molecules.

Thus, both chemical and enzymatic derivatization techniques can be used to form oligonucleotide probes of high activity in hybridization assays. The main consideration for successful polynucleotide cross-linking, as in other bioconjugate applications, is to avoid probe inactivation during the modification or conjugation process. Since the purpose in constructing a DNA or RNA probe is to hybridize and detect a complementary oligonucleotide through hydrogen bond interactions, any derivatization procedure that significantly interferes with Watson–Crick base-pairing should be avoided. This means that large amounts of base derivatization along a polynucleotide chain has potential for causing obstructions in the hybridization process, sometimes dramatically reducing or eliminating base-pairing efficiency. In general, base modifications within an oligonucleotide probe should be limited to no more than about 30–40 sites per 1000 bases to maintain hybridization ability.

By contrast, derivatization at the ends of an oligo or at the sugar–phosphate backbone usually produces little interference in base-pairing. Conjugates may be created by enzymatic polymerization of functionalized nucleoside triphosphates off the 3' end or by chemical modification of the 5' phosphate group with minimal to no interference in hybridization potential. The application of these strategies to creating labeled oligonucleotide probes is discussed in Chapter 17.

4. Creating Specific Functional Groups

It is often desirable to alter the native structure of a macromolecule to provide functional targets for modification or conjugation. The use of most reagent systems requires the presence of particular chemical groups to effect coupling. For instance, heterobifunctional cross-linkers may contain two different reactive species that are directed against different functional groups. One target molecule must contain chemical groups able to react with one end of the cross-linker, while the other target molecule must contain groups able to react with the other end. Occasionally, the required chemical groups are not present on one of the target molecules and must be created. This usually can be done by reacting an existing chemical group with a modification reagent that contains or produces the desired functional group upon coupling. Thus, an amine can be "changed" into a sulfhydryl or a carboxylate can be altered to yield an amine simply by using the appropriate reagent.

This same type of modification strategy also can be used to create highly reactive groups from functional groups of rather low reactivity. For instance, carbohydrate chains on glycoproteins can be modified with sodium periodate to transform their rather unreactive hydroxyl groups into highly reactive aldehydes. Similarly, cystine or disulfide residues in proteins can be selectively reduced to form active sulfhydryls, or 5' phosphate groups of DNA can be transformed to yield modifiable amines.

Alternatively, spacer arms can be introduced into a macromolecule to extend a reactive group away from its surface. The extra length of a spacer can provide less steric hindrance to conjugation and often yields more active complexes.

The use of modification reagents to create specific functional groups is an important technique to master. In one sense, the process is like using building blocks to construct on a target molecule any desired functional groups necessary for reactivity. The success of many conjugation schemes depends on the presence of the correct chemical groups. Care should be taken in choosing a modification strategy, however, since some chemical changes will radically affect the native structure and activity of a macromolecule. A protein may lose its capacity to bind a specific ligand. An enzyme may lose the ability to act upon its substrate. A DNA probe may no longer be able to hybridize to its complementary target. In many cases, the potential for inactivation relates to changing conformational structures, blocking active sites, or modifying critical functional groups. Trial and error and careful literature searches are often necessary to optimize any modification tactic.

4.1. Introduction of Sulfhydryl Residues (Thiolation)

The sulfhydryl group is a popular target in many modification strategies. Cross-linking agents that have more than one reactive group often employ a sulfhydryl-reactive functional group at one end to direct the conjugation reaction to a particular part of a target macromolecule. The frequency of sulfhydryl occurrence in proteins or other molecules is usually low (or nonexistent) compared to that of other groups like amines or carboxylates. The use of sulfhydryl-reactive chemical reactions thus can restrict modification to only a limited number of sites within a target molecule. Limiting modification greatly increases the chances of retaining activity after conjugation, especially in sensitive proteins like some enzymes. Unfortunately, sulfhydryl groups often need to be generated (from reduction of indigenous disulfides) or created (from use of the appropriate thiolation reagent systems). The following sections describe the most popular techniques of creating these functional groups. Some of these reagent systems are specifically designed to form —SH groups, while others are cross-linkers that also can serve the dual purpose of sulfhydryl-generating agents.

Sulfhydryl groups are susceptible to oxidation and formation of disulfide cross-links. To prevent disulfide bond formation, remove oxygen from all buffers by degassing under vacuum and bubbling an inert gas (i.e., nitrogen) through the solution. In addition, EDTA ($0.01-0.1$ M) may be added to buffers to chelate metal ions, preventing metal-catalyzed oxidation of sulfhydryls. Some proteins of serum origin (particularly BSA) contain so many contaminating metal ions (presumably iron from hemolyzed blood) that 0.1 M EDTA is required to prevent this type of oxidation.

Modification of Amines with 2-Iminothiolane (Traut's Reagent)

Perham and Thomas (1971) originally prepared an imidoester compound containing a sulfhydryl group, methyl 3-mercaptopropionimidate hydrochloride. The imidoester group can react with amines to form a stable, charged linkage (Chapter 2, Section 1.10), while leaving a sulfhydryl group available for further coupling (Fig. 52). Traut *et al.* (1973) subsequently synthesized an analogous reagent containing one additional carbon, methyl 4-mercaptobutyrimidate. Later, this compound was found to cyclize as a result of the sulfhydryl group reacting with the intrachain imidoester, forming 2-iminothiolane (Jue *et al.*, 1978). The cyclic imidothioester still can react with primary amines in a ring-opening reaction that regenerates the free sulfhydryl (Fig. 53).

Traut's Reagent;
2-Iminothiolane
MW 137.6
8.1 Å

Traut's reagent is fully water-soluble and reacts with primary amines in the pH range $7-10$. The cyclic imidothioester is stable to hydrolysis at acid pH values, but its half-life in solution decreases as the pH increases beyond neutrality. However, even at pH 8 in 25 mM triethanolamine the rate of sulfhydryl formation without added primary amine was found to be negligible. On addition of dipeptide amine, the reagent reacted quickly as evidenced by the production of Ellman's reagent color. The rate of reaction also can be followed by 2-iminothiolane's absorbance at 248 nm (λ_{max}; $\varepsilon = 8840$ $M^{-1}cm^{-1}$). As the cyclic imidate reacts with amines, its absorbance at

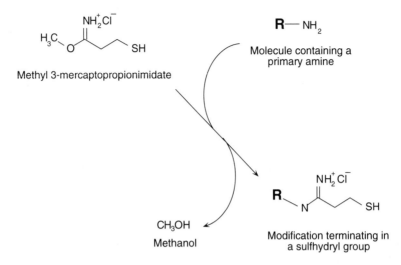

Figure 52   Thiolation of an amine-containing compound with methyl 3-mercaptopropionimidate. The modification preserves the positive charge on the primary amine.

this wavelength decreases. With addition of the dipeptide glycylglycine, the starting absorbance of a solution of Traut's reagent decreased over 80% within 20 min (Jue *et al.*, 1978). Thus, protein modification with 2-iminothiolane is very efficient and proceeds rapidly at slightly basic pH.

At high pH (10), Traut's reagent also is reactive with aliphatic and aromatic hydroxyl groups, although the rate of reaction with these groups is only about 0.01 that of primary amines. In the absence of amines, however, carbohydrates such as agarose or cellulose membranes can be modified to contain sulfhydryl residues (Alagon and

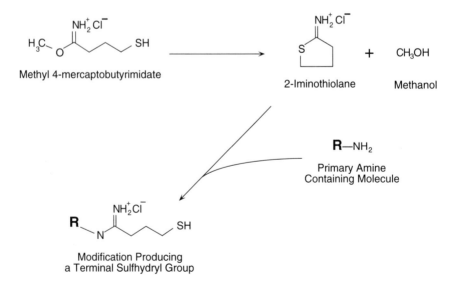

Figure 53   Methyl 4-mercaptobutyrimidate forms 2-iminothiolane, which can react with a primary amine to create a sulfhydryl group. The modification preserves the positive charge of the original amine.

King, 1980). Polysaccharides modified in this manner are effective in covalently cross-linking antibodies for use in immunoassay procedures.

Proteins modified with 2-iminothiolane are subject to disulfide formation on sulfhydryl oxidation. This can cause unwanted conjugation, potentially precipitating the protein. The addition of a metal chelating agent such as EDTA (0.01–0.1 M) will prevent metal-catalyzed oxidation and maintain sulfhydryl stability. In the presence of some serum proteins (i.e., BSA) a 0.1 M concentration of EDTA may be necessary to prevent metal-catalyzed oxidation, presumably due to the high contamination of iron from hemolyzed blood.

Traut's reagent has been used successfully in the investigation of ribosomal proteins (Sun *et al.*, 1974; Jue *et al.*, 1978; Kenny *et al.*, 1979; Blattler *et al.*, 1985 a, b; Lambert *et al.*, 1983), RNA polymerase (Hillel and Wu, 1977), and progesterone receptor subunits (Birnbaumer *et al.*, 1979), and in the synthesis of enzyme-labeled DNA hybridization probes (Ghosh, *et al.*, 1990). It is an excellent thiolation reagent for use in the preparation of immunotoxins (Chapter 11). Recently, it has been used to modify and introduce sulfhydryls into oligosaccharides from asparagine-linked glycans (Tarentino *et al.*, 1993).

Protocol

1. Prepare the protein or macromolecule to be thiolated in a non-amine-containing buffer at pH 8.0. For the modification of ribosomal proteins (often cited in the literature) use 50 mM triethanolamine hydrochloride, 1 mM $MgCl_2$, 50 mM KCl, pH 8. The magnesium and potassium salts are for stabilization of some ribosomal proteins. If other proteins are to be thiolated, the same buffer may be used without added salts for stabilization. Alternatively, 50 mM sodium phosphate, 0.15 M NaCl, pH 8, or 0.1 M sodium borate, pH 8.0 may be used. For the modification of polysaccharides, use 20 mM sodium borax, pH 10, to produce reactivity toward carbohydrate hydroxyl residues. Dissolve the protein to be modified at a concentration of 10 mg/ml in the reaction buffer of choice. Lower concentrations also may be used with a proportional scaling back of added 2-iminothiolane.

2. Dissolve the Traut's reagent (Pierce) in water at a concentration of 2 mg/ml (makes a 14.5 mM stock solution). The solution should be used immediately. For the modification of IgG at a concentration of 10 mg/ml using a 10-fold molar excess of Traut's reagent, add 45.8 µl of the stock solution to each milliliter of the protein solution.

3. React for 1 h at room temperature (a 4°C reaction temperature may be used successfully as well).

4. Purify the thiolated protein from unreacted Traut's reagent by dialysis or gel filtration using your buffer of choice (i.e., 20 mM sodium phosphate, 0.15 M NaCl, 1 mM EDTA, pH 7.2). The addition of EDTA to this buffer helps to prevent oxidation of the sulfhydryl groups and the resultant disulfide formation.

5. The degree of —SH modification may be determined using the Ellman's assay (Section 4.1).

When 2-iminothiolane is used to modify proteins in tandem with 4,4'-dipyridyl disulfide, a protected sulfhydryl can be introduced in a single step (King *et al.*, 1978). The simultaneous reaction between a protein, 2-iminothiolane, and 4,4'-dipridyl disulfide yields a modification containing pyridyl disulfide groups. The pyridyl disulfide may be subsequently reduced with DTT to yield a free sulfhydryl. Pyridyl disulfides also are highly reactive toward sulfhydryls through disulfide interchange (Section 5.2). The protocol is a modification of the method of King *et al.* (1978).

Protocol

1. Dissolve 1–10 mg of a protein to be modified in 1.0 ml of 0.025 *M* sodium borate, pH 9.
2. Dissolve 2-iminothiolane in 0.025 *M* sodium borate to a concentration of 0.02 *M*.
3. Dissolve 4,4'-dipyridyl disulfide at a concentration of 2 mg/ml in acetonitrile.
4. Add 0.2 ml of (3) and 1.0 ml of (2) to the protein solution.
5. React for 2 h at room temperature.
6. Purify the modified protein by gel filtration or dialysis.

Occasionally, a protein modified in this manner will begin to precipitate as the reaction proceeds. Stopping the reaction earlier or adding a smaller quantity of modifying reagents may limit this effect.

Modification of Amines with SATA

A versatile reagent for introducing sulfhydryl groups into proteins is SATA, *N*-succinimidyl *S*-acetylthioacetate (Duncan *et al.*, 1983). The active NHS ester end of SATA reacts with amino groups in proteins and other molecules to form a stable amide linkage (Fig. 54) (Chapter 2, Section 1.4). The modified protein then contains a

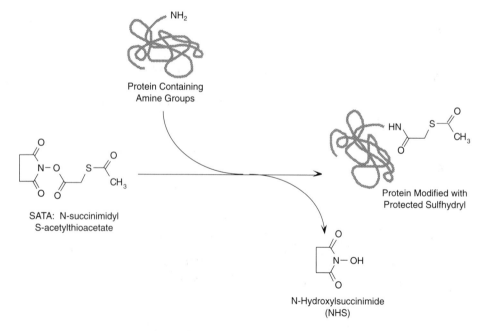

Figure 54   SATA can react with available amine groups in proteins and other molecules via its NHS ester end to form protected sulfhydryl derivatives.

protected sulfhydryl that can be stored without degradation and subsequently de-protected as needed with an excess of hydroxylamine (Fig. 55). Since the protecting group can be removed without adding disulfide reducing agents like DTT, disulfides indigenous to the native protein will not be affected. This is an important consideration if disulfides are vital to activity, such as in the case of some protein toxins.

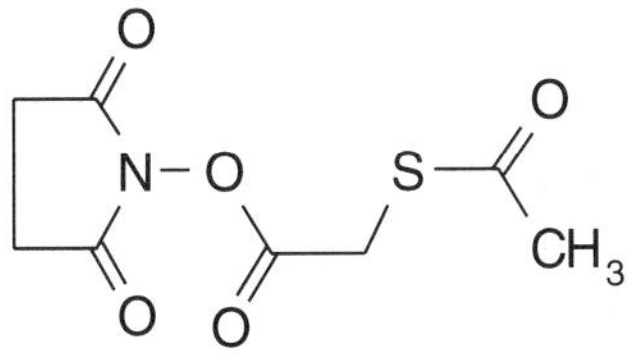

SATA;
N-succinimidyl
S-acetylthioacetate
MW 231.2

SATA is often used to form antibody–enzyme conjugates utilizing maleimide-containing heterobifunctional cross-linking agents. Most polyclonal antibody molecules may be modified to contain up to about six SATA molecules per immunoglobulin with minimal effect on antigen binding activity. Some sensitive monoclonal antibodies, however, may be susceptible to modification and should be tested on a case-by-case basis. The modified antibody then may be deprotected and reacted with a maleimide-activated enzyme to form a conjugate useful in immunoassays (Chapter 10, Section 1.1). Conjugates formed using SATA are usually of low molecular weight with very few high-molecular-weight oligomers. They also maintain a bivalent antibody, ensuring a conjugate containing two antigen binding sites. This is an advantage over reduction schemes that break the antibody molecule into two heavy–light chain pairs to create sulfhydryls, since disulfide cleavage yields antibody fragments with only one antigen binding site.

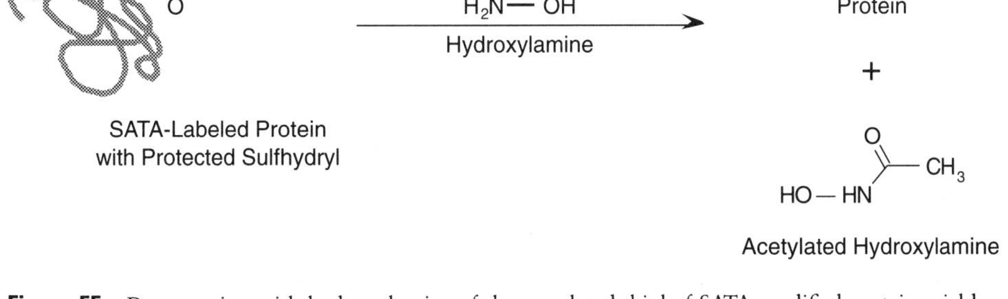

Figure 55 Deprotection with hydroxylamine of the acetylated thiol of SATA-modified proteins yields a free sulfhydryl group.

SATA has been used to form conjugates with avidin or steptavidin with excellent retention of activity (Chapter 13, Section 3.1). It has been used in the formation of a therapeutically useful toxin conjugate with recombinant CD4 (Ghetie *et al.*, 1990A).

SATA is freely soluble in many organic solvents. In use, it is typically dissolved as a stock solution in DMSO, DMF, or methylene chloride, and then an aliquot of this solution is added to an aqueous reaction mixture containing the protein to be modified.

The thiolation method described below is generally applicable for the modification of proteins with SATA, particularly for subsequent conjugation with a maleimide-activated secondary protein. The degree of modification described usually yields 3–4 mol of —SH groups per mole protein when thiolating immunoglobulins. Other macromolecules containing primary amines may be modified using a similar procedure. The degree of modification observed with other molecules may vary depending on the number of available primary amines and their relative reactivity. For comparison purposes, the molar ratio of SATA to immunoglobulin added to a reaction for the modification of rabbit polyclonal IgG versus the degree of sulfhydryl incorporation is illustrated in Fig. 56 (Sykaluk,, 1994).

The following protocol represents a generalized method for protein thiolation using SATA. For comparison purposes, contrast the variation of this SATA modification method as outlined in Chapter 10, Section 1.1 for use in the preparation of antibody–enzyme conjugates.

Protocol

1. Dissolve the protein to be thiolated at a concentration of 1–5 mg/ml in 50 mM sodium phosphate, pH 7.5, containing 1–10 mM EDTA. Other non-amine-containing buffers such as borate, Hepes, and bicarbonate also may be used as the reaction medium. The effective pH for the NHS ester modification reaction is in the range 7 to 9.

2. Dissolve the SATA reagent (Pierce) in DMSO at a concentration of 65 mM (15 mg/ml). Note: DMSO should be handled in a fume hood.

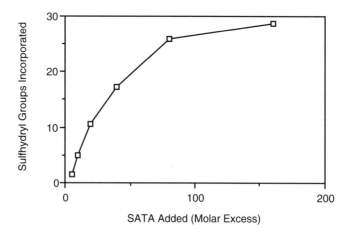

Figure 56   SATA modification of rabbit polyclonal IgG with the resultant sulfhydryl incorporation level.

3. Add 10 μl of the SATA solution to each milliliter of protein solution.
4. Mix and react for 30 min at room temperature.
5. Separate modified protein from unreacted SATA and reaction by-products by dialysis against 50 mM sodium phosphate, pH 7.5, containing 1 mM EDTA or by gel filtration on a Sephadex G-25 column (Pharmacia) using the same buffer.
6. Deprotect the acetylated —SH groups as needed by adding 100 μl of 0.5 M hydroxylamine hydrochloride in 50 mM sodium phosphate, 25 mM EDTA, pH 7.5, to each milliliter of the SATA-modified protein solution.
7. Mix and react for 2 h at room temperature.
8. Purify the sulfhydryl-modified protein by dialysis against 50 mM sodium phosphate, 1 mM EDTA, pH 7.5, or by gel filtration on a Sephadex G-25 column using the same buffer.

The deacetylated protein should be used immediately to prevent loss of sulfhydryl content through disulfide formation. The degree of —SH modification may be determined by performing an Ellman's assay (Section 4.1).

Modification of Amines with SATP

SATP, succinimidyl acetylthiopropionate, is an analog of SATA (Section 4.1) containing one additional carbon atom in length (Fuji *et al.*, 1985). The compound retains all the advantages of a protected sulfhydryl, including stability of the modified protein and selective release of the protecting group with hydroxylamine to free the sulfhydryl as needed (Fig. 57). SATP is soluble in DMF and methylene chloride. It is usually first solubilized in organic solvent and an aliquot added to an aqueous solution containing the macromolecule to be modified. It is particularly useful in adding an N-terminal —SH group at the completion of peptide synthesis.

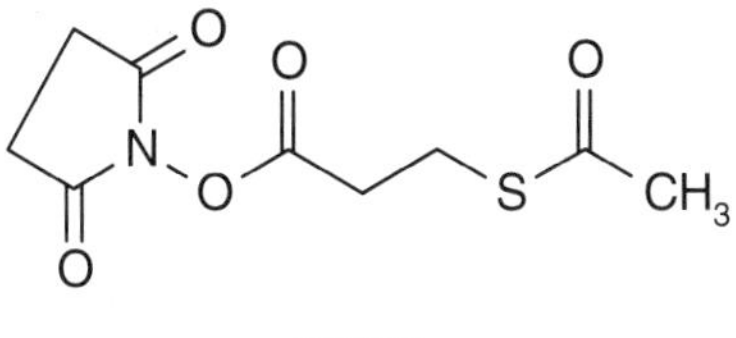

SATP
Succinimidyl acetyl-
thiopropionate
MW 245

Protocol

1. Dissolve the protein or peptide to be thiolated at a concentration of 10 mg/ml in 50 mM sodium phosphate, pH 7.5, containing 1 mM EDTA. Other non-amine-containing buffers such as borate, Hepes, and bicarbonate also may be used as the reaction medium. The effective pH for the NHS ester modification reaction is in the range 7 to 9.
2. Dissolve the SATP reagent (Molecular Probes) in DMF at a concentration of 65 mM (16 mg/ml). Note: DMF should be handled in a fume hood.
3. Add 10 μl of the SATP solution to each milliliter of protein or peptide solution.
4. Mix and react for 8 h (or overnight) at room temperature.

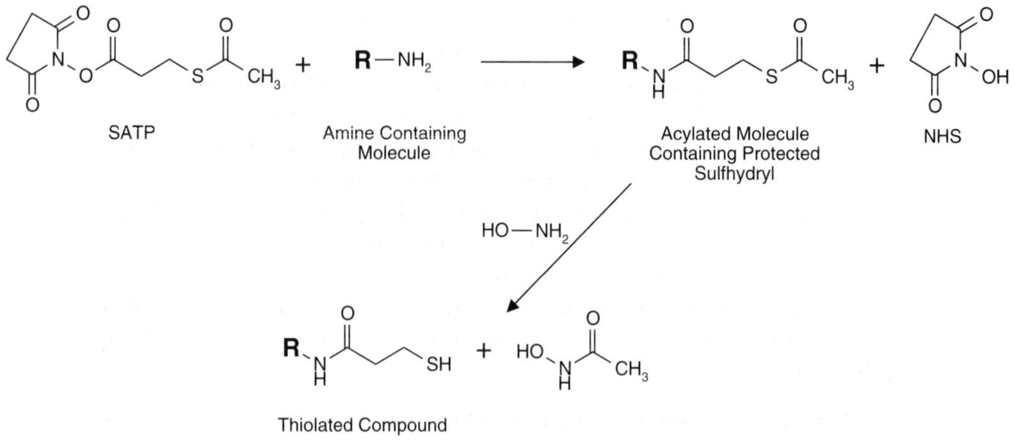

Figure 57   SATP reacts with amine-containing proteins or other molecules via its NHS ester end to create protected sulfhydryl derivatives in a manner similar to that of SATA. Deprotection can be done with hydroxylamine to free the thiol.

5. Separate modified protein from unreacted SATP and reaction by-products by dialysis against 50 mM sodium phosphate, pH 7.5, containing 1 mM EDTA or by gel filtration on a Sephadex G-25 column (Pharmacia) using the same buffer. If a peptide of low molecular weight is being modified, careful gel filtration using a matrix having a low exclusion limit will separate the peptide from the reaction by-products. In this case, use either Sephadex G-25 or Sephadex G-10 for the chromatography.

6. Deprotect the acetylated —SH groups as needed by adding 100 µl of 0.5 M hydroxylamine hydrochloride in 50 mM sodium phosphate, 25 mM EDTA, pH 7.5, to each milliliter of the SATP-modified protein solution.

7. Mix and react for 2 h at room temperature.

8. Purify the sulfhydryl-modified protein by dialysis against 50 mM sodium phosphate, 1 mM EDTA, pH 7.5, or by gel filtration on a Sephadex G-25 column using the same buffer. Again, if a peptide of low molecular weight is being modified, use gel filtration for purification.

The deacetylated protein should be used immediately to prevent loss of sulfhydryl content through disulfide formation. The degree of —SH modification may be determined by performing an Ellman's assay (Section 4.1).

Modification of Amines with SPDP

SPDP, N-succinimidyl 3-(2-pyridyldithio)propionate, is one of the most popular heterobifunctional cross-linking agents (Chapter 5, Section 1.1). The NHS ester end of SPDP reacts with amine groups to form an amide linkage, while the 2-pyridyldithiol group at the other end can react with sulfhydryl residues to form a disulfide linkage (Carlsson et al., 1978). The cross-linker is used extensively to form immunotoxin conjugates for in vivo administration (Chapter 11, Section 2.1). The reagent is also useful in creating sulfhydryls in proteins and other molecules. Once modified with

SPDP, a protein can be treated with DTT (or other disulfide reducing agents, see Section 4.1) to release the pyridine-2 thione leaving group and form the free sulfhydryl (Fig. 58). The terminal —SH group then can be used to conjugate with any cross-linking agents containing sulfhydryl-reactive groups, such as maleimide or iodoacetyl (for covalent conjugation) or 2-pyridyldithiol groups (for reversible conjugation).

There are three forms of SPDP analogs currently commercially available (Pierce Chemical): the standard SPDP, a long-chain version designated LC-SPDP, and a water-soluble, sulfo-NHS form also containing a extended chain, called Sulfo-LC-SPDP (Chapter 5, Section 1.1). The main disadvantage to using SPDP to create sulfhydryls is the necessity of using a reducing agent to remove the pyridine-2-thione group. Reducing agents will also affect indigenous disulfides within a protein molecule, cleaving and reducing them. This method therefore works well for proteins containing no sulf-hydryls or no disulfides that are critical to function, but it may cause loss of activity or subunit breakdown in proteins containing essential disulfides.

The following procedure is similar to the method of Cumber et al. (1985), but with some modifications.

Protocol

1. Dissolve the protein or macromolecule to be thiolated at a concentration of 10 mg/ml in 50 mM sodium phosphate, 0.15 M NaCl, pH 7.2. Other non-amine-containing buffers such as borate, Hepes, and bicarbonate also may be used in this reaction. The effective pH for the NHS ester modification reaction is in the range 7.0 to 9.0.

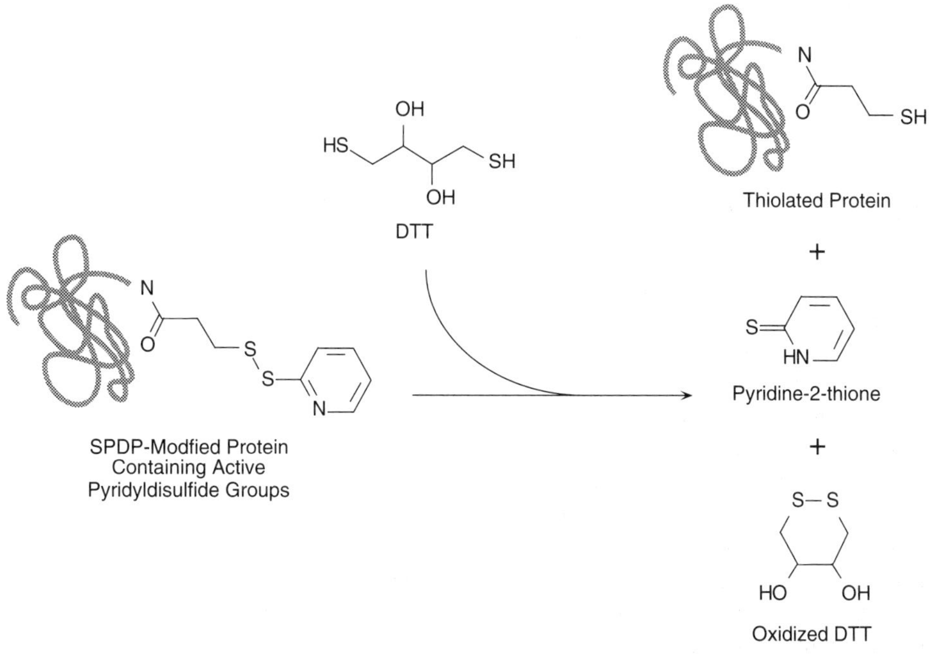

Figure 58 SPDP-modified proteins can be reduced with DTT to yield free sulfhydryl groups for conjugation.

2. Dissolve SPDP (Pierce) at a concentration of 6.2 mg/ml in DMSO (makes a 20 mM stock solution). Alternatively, LC-SPDP may be used and dissolved at a concentration of 8.5 mg/ml in DMSO (also makes a 20 mM solution). If the water-soluble Sulfo-LC-SPDP is used, a stock solution in water may be prepared just prior to adding an aliquot to the thiolation reaction. In this case, prepare a 10 mM solution of Sulfo-LC-SPDP by dissolving 5.2 mg/ml in water. Since an aqueous solution of the cross-linker will degrade by hydrolysis of the sulfo-NHS ester, it should be used quickly to prevent significant loss of activity. If a sufficiently large amount of protein will be modified to allow accurate weighing of Sulfo-LC-SPDP, the solid may be added directly to the reaction mixture without preparing a stock solution in water.

3. Add 25 μl of the stock solution of either SPDP or LC-SPDP in DMSO to each milliliter of the protein to be modified. If Sulfo-LC-SPDP is used, add 50 μl of the stock solution in water to each milliliter of protein solution.

4. Mix and react for at least 30 min at room temperature. Longer reaction times, even overnight, will not adversely affect the modification.

5. Purify the modified protein from reaction by-products by dialysis or gel filtration using 50 mM sodium phosphate, 0.15 M NaCl, pH 7.2.

6. To release the pyridine-2-thione leaving group and form the free sulfhydryl, add DTT (Pierce) at a concentration of 0.5 mg DTT per milligram of modified protein. A stock solution of DTT may be prepared to make it easier to add it to a small amount of protein solution. In this case, dissolve 20 mg of DTT per milliliter of 0.1 M sodium acetate, 0.1 M NaCl, pH 4.5. Add 25 μl of this solution per milligram of modified protein. Release of pyridine-2-thione can be followed by its characteristic absorbance at 343 nm ($\varepsilon = 8.08 \times 10^3 \, M^{-1}cm^{-1}$).

7. Mix and react at room temperature for 30 min.

8. Purify the thiolated protein from excess DTT by dialysis or gel filtration using 50 mM sodium phosphate, 0.15 M NaCl, 1 mM EDTA, pH 7.2. The modified protein should be used immediately in a conjugation reaction to prevent sulfhydryl oxidation and formation of disulfide cross-links.

Modification of Amines with SMPT

SMPT, succinimidyloxycarbonyl-α-methyl-α-(2-pyridyldithio)toluene, contains an NHS ester end and a pyridyldisulfide end similar to those of SPDP, but its hindered disulfide makes conjugates formed with this reagent more stable (Thorpe *et al.*, 1987) (Chapter 5, Section 1.2). The reagent is especially useful in forming immunotoxin conjugates for *in vivo* administration (Chapter 11, Section 2.1). A water-soluble analog of this cross-linker containing an extended spacer arm is also commercially available as Sulfo-LC-SMPT (Pierce).

SMPT or Sulfo-LC-SMPT may be used as thiolation reagents by first reacting its NHS ester end with an amine-containing molecule and then releasing the pyridine-2-thione leaving group with DTT to free the sulfhydryl (Fig. 59). The disadvantage of this approach is the necessity of using a reducing agent to create the —SH group modification. This method of thiolation only should be used if there are no disulfides in the target molecule that are critical to function. If a reductant cannot be used, choose a thiolation method that does not need DTT treatment, such as the use of Traut's reagent or SATA (Section 4.1).

Since SMPT is not soluble in aqueous solutions it must be first dissolved in organic

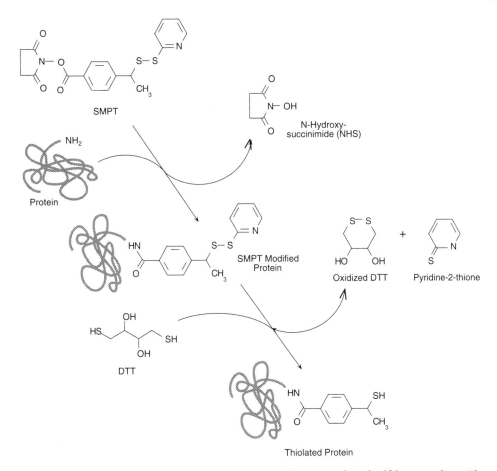

Figure 59 SMPT can be used to modify the amine groups of proteins to form disulfide intermediates. The disulfides can be reduced with DTT to create free thiols for subsequent conjugation purposes.

solvent and an aliquote of this stock solution transferred to the reaction solution. The reagent is soluble in DMF and DMSO, but is much more stable in solutions of acetonitrile. A stock solution of SMPT in acetonitrile may be kept frozen without loss of activity. The NHS ester of SMPT also is extraordinarily stable to hydrolysis in water. Even when an SMPT/acetonitrile aliquot is added to an aqueous solution and stored at room temperature, SMPT will only lose about 5% of its activity after 16 h. By contrast, other NHS esters usually have half-lives of only 2–6 h in aqueous environments.

Sulfo-LC-SMPT is not as stable as SMPT. The sulfo-NHS ester is more susceptible to hydrolysis in aqueous solutions and the pyridyldisulfide group is more easily reduced to the free sulfhydryl. Stock solutions of Sulfo-LC-SMPT may be prepared in water, but should be used immediately to prevent loss of amine coupling ability.

Protocol

1. Dissolve the protein or macromolecule to be thiolated at a concentration of 10 mg/ml in 50 mM sodium phosphate, 0.15 M NaCl, pH 7.2. Other non-amine-containing buffers such as borate, Hepes, and bicarbonate also may be used as

3. For the bicarbonate reaction, gently mix for 20 h at 4°C. For the silver-catalyzed reaction, continue for 1 h or until the silver complex has fully dissolved.
4. To remove the silver mercaptide formed from the facilitated protein thiolation reaction, add an excess of thiourea to convert all the silver into a soluble Ag(thiourea)$_2^+$ complex and free the sulfhydryl modifications.
5. Remove unreacted N-acetylhomocysteinethiolactone and reaction by-products by gel filtration or dialysis against 10 mM sodium phosphate, 0.15 M NaCl, 10 mM EDTA, pH 7.2. Other buffers suitable for individual protein stability may be used as desired. For the silver nitrate-containing reaction, removal of the silver–thiourea complex may be done by adsorption onto Dowex 50, and the protein subsequently eluted from the resin by 1 M thiourea. Removal of the thiourea then may be done by gel filtration or dialysis.

Including EDTA in the final preparation inhibits metal-catalyzed oxidation of the sulfhydryl groups to disulfides. The modified peptide or protein should be used immediately to ensure full sulfhydryl reactivity.

Modification of Amines with SAMSA

S-Acetylmercaptosuccinic anhydride, or SAMSA, is an amine-reactive reagent containing a protected sulfhydryl much like SATA described previously. The anhydride portion opens in response to the attack of an amine nucleophile, yielding an amide linkage (Weston et al., 1980; Klotz and Heiney, 1962; Klots and Keresztes-Nagy, 1962). The ring-opening reaction, however, does produce a free carboxylate group that lends a negative charge to the modified molecule where once there was a positive charge (Fig. 61). This charge reversal may affect the conformation and activity of some sensitive proteins. After the initial modification step, the thiolated derivative is formed by releasing the acetylated sulfhydryl protecting group with hydroxylamine.

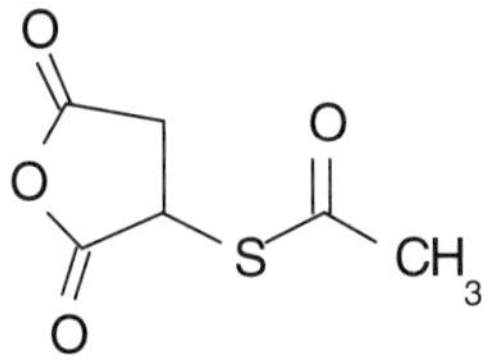

SAMSA;
S-Acetylmercaptosuccinic
Anhydride
MW 174

Protocol

1. Dissolve the protein or other amine-containing macromolecule in 0.1 M sodium phosphate, 0.15 M NaCl, pH 7.5, at a concentration of 5 mg/ml.
2. Dissolve SAMSA in DMF at a concentration of 25 mg/ml.
3. Add 20 μl of the stock SAMSA solution to each milliliter of the protein solution, with mixing.
4. React at room temperature for 30 min.
5. Remove excess reagent and reaction by-products by dialysis or gel filtration

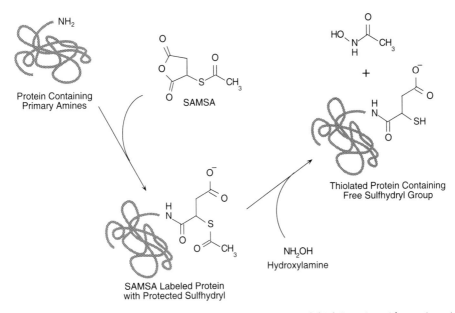

Figure 61   SAMSA is an anhydride compound containing a protected thiol. Reaction with protein amine groups yields amide bond linkages. Deprotection of the acetylated thiol produces free sulfhydryl groups for conjugation.

using 0.1 *M* sodium phosphate, 0.15 *M* NaCl, 10 m*M* EDTA, pH 7.5. For chromatographic separation use Sephadex G-25 (Pharmacia) or the equivalent. The SAMSA-modified protein may be stored at -20°C until needed.

6. To deprotect the acetylated sulfhydryl group of SAMSA-modified proteins, add 100 μl of 0.5 *M* hydroxylamine hydrochloride in 50 m*M* sodium phosphate, 25 m*M* EDTA, pH 7.5, to each milliliter protein solution.

7. Mix and react for 2 h at room temperature.

8. Purify the sulfhydryl-modified protein by dialysis against 50 m*M* sodium phosphate, 1 m*M* EDTA, pH 7.5, or by gel filtration on a Sephadex G-25 column using the same buffer.

The deacetylated protein should be used immediately to prevent loss of sulfhydryl content through disulfide formation. The degree of —SH modification may be determined by performing an Ellman's assay (Section 4.1).

Modification of Aldehydes or Ketones with AMBH

AMBH (2-acetamido-4-mercaptobutyric acid hydrazide) is a unique hydrazide derivative that can thiolate aldehydes and ketones to form reactive sulfhydryl groups (Taylor and Wu, 1980). It is particularly useful in converting oxidized carbohydrates. In this respect, glycoproteins or other carbohydrate- and diol-containing molecules may be treated with sodium periodate under mild conditions to form aldehyde residues (see Section 4.4). The aldehydes readily react with the hydrazide groups of AMBH to form hydrazone linkages, leaving a free terminal sulfhydryl residue to use in further conjugation reactions (Fig. 62).

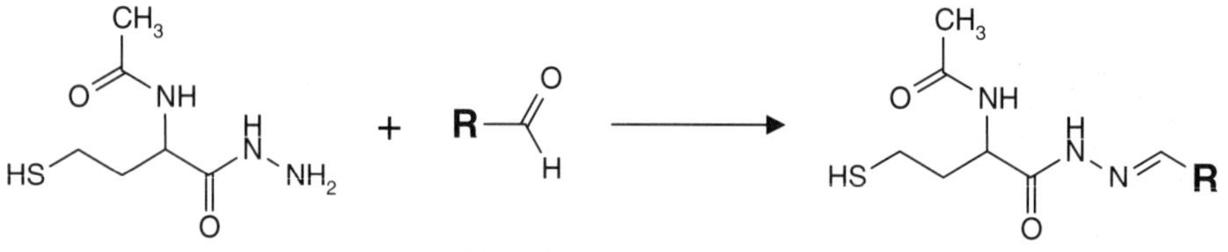

Figure 62 AMBH is a hydrazide-containing compound that reacts with carbonyl groups to form hydrazone bonds. The free thiol can be used for subsequent conjugation reactions.

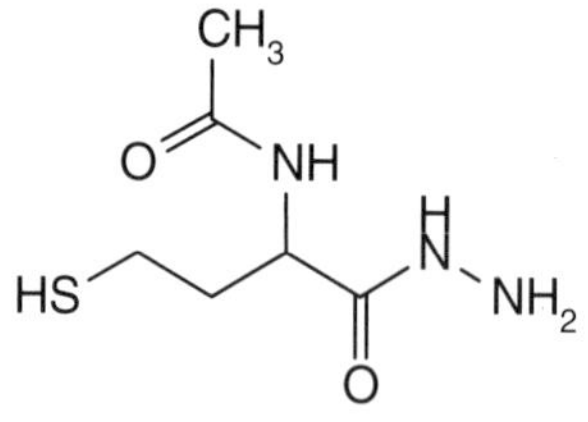

AMBH
2-Acetamido-4-mercaptobutyric
acid hydrazide
MW 191

Protocol

1. Dissolve an aldehyde-containing macromolecule to be modified (i.e., a periodate-oxidized glycoprotein) in 0.01 M sodium phosphate, 0.15 M NaCl, pH 7.4, containing 1 mM EDTA. A suitable concentration range for a protein is 1–10 mg/ml.
2. Add a 10–fold molar excess of AMBH (predissolved in ethanol) (Molecular Probes) over the expected amounts of aldehydes to be modified.
3. React for 2 h at room temperature.
4. Purify the modified protein by gel filtration

Modification of Carboxylates or Phosphates with Cystamine

Cystamine is decarboxylated cystine [or 2,2'-dithiobis(ethylamine)], a small disulfide-containing molecule with primary amines at both ends. This versatile reagent can be used in several conjugation techniques. Cystamine may be used to introduce sulfhydryl residues in proteins, nucleic acids, and other molecules, or as the active species in disulfide exchange cross-linking reactions, or in reversible conjugation procedures. The reagent can be used to create sulfhydryl groups in proteins or other molecules by first conjugating its terminal amino groups with the carboxylates on a target molecule using the carbodiimide reaction (Chapter 2, Section 1.11 and Chapter 3, Section 1). Subsequent reduction of the disulfide group liberates the free sulfhydryl (Section 4.1) (Fig. 63). This same modification procedure also can be used to introduce sulfhydryl residues at the 5' phosphate group of DNA (Chu et al., 1986; Ghosh et al., 1990). The

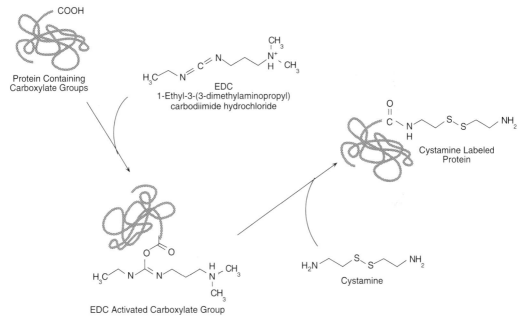

Figure 63 Cystamine may be used to label protein carboxylate groups using the water-soluble carbodiimide EDC.

carbodiimide activates the phosphate and the amines of cystamine may then react with this active species to form a phosphoramidate bond (Chapter 17, Section 2.2) (Fig. 64). Specific labeling of DNA probes only at the 5′ end is possible using this technique.

Cystamine;

2,2′-dithiobis(ethylamine)
MW 152

The carbodiimide of choice used to couple cystamine to carboxylate- or phosphate-containing molecules is most often the water-soluble carbodiimide EDC (1-ethyl-3-(3-dimethylaminopropyl)carbodiimide hydrochloride; Chapter 3, Section 1.1). This reagent rapidly reacts with carboxylates or phosphates to form an active complex highly reactive toward primary amines. The reaction is efficient from pH 4.7 to 7.5, and a variety of buffers may be used, providing they do not contain competing groups.

Cystamine also is used as an activating reagent for disulfide exchange reactions. In this procedure, the reagent is used to modify one of two proteins to be conjugated. The cystamine-modified protein is then mixed with the other protein that contains, or is thiolated to contain, a sulfhydryl group. By disulfide exchange, the sulfhydryl-containing molecule cleaves the disulfide of the cystamine-modified protein, releasing 2-mercaptoethylamine and forming a disulfide cross-link (Fig. 65).

Using this approach, EGF has been successfully conjugated by disulfide exchange to

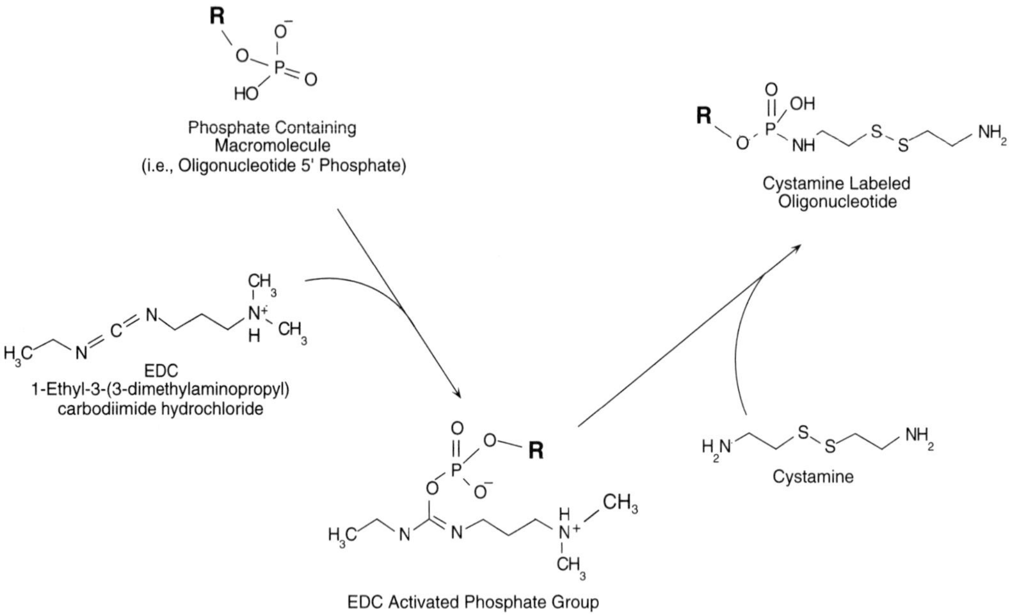

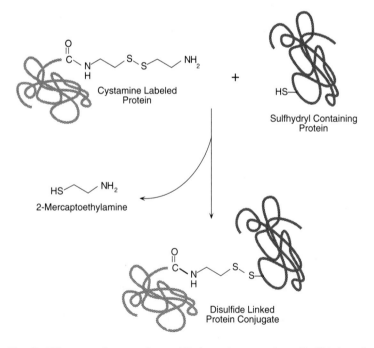

Figure 64 Cystamine may be used to label phosphate groups, such as on nucleic acids, via a carbodiimide reaction using EDC. The resultant phosphoramidate linkage is a common way to modify oligonucleotides at the 5′ end.

Figure 65 The disulfide group of a cystamine-modified protein may undergo disulfide interchange reactions with another sulfhydryl-containing protein to yield a disulfide-linked conjugate.

the A chain of diphtheria toxin (Shimisu *et al.*, 1980). A cystaminyl derivative of insulin also could be conjugated to the A chain of diphtheria toxin by this method (Miskimins and Shimizu, 1979). Other references to disulfide exchange using cystamine include Oeltmann and Forbes (1981) and Bacha *et al.* (1983), who prepared antibody–toxin and peptide–toxin conjugates, respectively.

Finally, cystamine may be used to conjugate two macromolecules through its terminal amine groups. In this case, the internal disulfide bridge remains intact, forming a reversible conjugate of the two molecules through reduction of the disulfide bond. Using this approach, the first molecule is modified with cystamine by use of the EDC reaction. A second molecule is then reacted with the free amines of cystamine on the first molecule by use of an amine-reactive chemistry. Typically, this reaction scheme is used if the first molecule initially contains no reactive amines and the second molecule is often an amine-reactive fluorescent tag or other probe. For instance, DNA probes may be cystamine-modified through their 5' phosphate group using this method and amine-reactive biotin labels subsequently attached. The biotin label is then reversible by virtue of the cystamine cross-bridge through simple disulfide reduction.

Modification of Proteins with Cystamine

The following protocol is useful for the modification of proteins with cystamine with subsequent reduction to create the free sulfhydryl.

Protocol

1. Dissolve the protein to be modified at a concentration of 10 mg/ml in a buffer having a pH between 4.7 and 7.5. Avoid buffers or other components containing groups competing with the carbodiimide reaction (i.e., carboxylates or amines). For the lower pH conditions, 0.1 M MES, pH 4.7, works best. For a physiological pH environment, 0.1 M sodium phosphate, 0.15 M NaCl, pH 7.2, also will give good incorporation of cystamine. For other concentrations of protein in solution, proportionally adjust the amount of reagents added.
2. Dissolve cystamine (Aldrich) in the reaction buffer at a concentration of 2.25 mg/ml (10 mM). Add an aliquot of this solution to the protein solution to be modified. Use about a 10- to 20-fold molar excess of cystamine over the amount of protein present. For a protein of MW 100,000 at a concentration of 10 mg/ml, add 10 μl of the stock cystamine solution to each milliliter of protein solution to obtain a 10-fold molar excess.
3. Add EDC (Pierce) to the solution prepared in (2) to obtain at least a five-fold molar excess over the amount of cystamine present. React for 2 h at room temperature.
4. Separate excess cystamine and EDC (and reaction by-products) from the modified protein by dialysis or gel filtration using 10 mM sodium phosphate, 0.15 M NaCl, pH 7.2. A desalting column may be used for the gel filtration procedure (i.e., Sephadex G-25 from Pharmacia).
5. To reduce the disulfide groups, add DTT (Pierce) at a concentration of 0.5 mg DTT per milligram of modified protein. A stock solution of DTT may be prepared to make it easier to add it to a small amount of protein solution. In this case, dissolve 20 mg of DTT per milliliter of 0.1 M sodium acetate, 0.1 M NaCl, pH 4.5. Add 25 μl of this solution per milligram of modified protein.

6. Mix and react at room temperature for 30 min.
7. Purify the thiolated protein from excess DTT by dialysis or gel filtration using 50 mM sodium phosphate, 0.15 M NaCl, 1 mM EDTA, pH 7.2. The modified protein should be used immediately in a conjugation reaction to prevent sulfhydryl oxidation and formation of disulfide cross-links.

Modification of Nucleic Acids and Oligonucleotides with Cystamine

DNA or RNA also may be modified with cystamine at the 5′ phosphate group using a carbodiimide reaction. See Chapter 17, Section 2.2 for a complete discussion of the labeling protocol.

Use of Disulfide Reductants

One of the most convenient ways of generating sulfhydryl groups is by reduction of indigenous disulfides. Many proteins contain cystine disulfides that are not critical to structure or activity. In some cases, mild reducing conditions can free one or more —SH groups for conjugation or modification purposes. The creation of free sulfhydryls in this manner allows for site-directed modification at a limited number of locations within the protein molecule.

This method of creating sulfhydryls for conjugation purposes should be avoided, however, if the indigenous disulfides are important for maintaining native structure and activity. Disulfides are often the point of attachment for subunits within a protein molecule. The cystine bonds may be crucial for maintaining quaternary integrity. Reduction may cause a protein to break up into two or more subunits with little or no remaining activity. Disulfides also may be critical for retention of ligand binding activity. Deformation of an active site may occur if important disulfides are reduced. In these cases, the best mode of thiolation is through the use of a reagent system that does not require a disulfide reducing agent, such as 2-iminothiolane or SATA (Section 4.1).

Occasionally, even a protein containing critical disulfides can be partially reduced to yield a useful thiolated derivative. IgG molecules contain disulfide groups that hold together the two heavy chains as well as disulfides holding the light chain–heavy chain pairs together. Selective reduction of only the hinge region disulfides between the heavy chains can result in a monovalent antibody molecule that still maintains its antigen binding capability. Reductants such as DTT or 2-mercaptoethylamine in a nondenaturing environment can be used at moderate concentrations to perform this type of partial cleavage. The thiolated "half" antibody so generated then can be successfully conjugated with enzymes or other molecules through the sulfhydryl residue(s) in the exposed hinge region (Chapter 10, Section 1.1).

Disulfide reductants are also used to investigate protein structural properties. In this case, retention of activity is not the critical issue, but complete reduction of all disulfides is paramount. The standard method of doing protein subunit molecular weight determinations by SDS polyacrylamide gel electrophoresis often depends on complete disulfide reduction. When total reduction needs to be ensured, the reductants must contain a deforming agent to unfold protein tertiary structure. This is typically done by including high concentrations of denaturants such as 8 M urea or guanidine or detergents such as SDS. Under severely deforming conditions, proteins unfold exposing internal disulfides to the reducing agent. Without these added reagents to deform native protein structure, many buried disulfides would remain unaffected by the reductants.

The following reducing agents represent the most popular options for cleaving disulfide bonds. Their properties and use vary widely. The decision of which reagent is best often is governed by the molecule being reduced and the potential application. Careful review of these properties may sway the success or failure of a conjugation protocol.

Cleland's Reagent: DTT and DTE

Dithiothreitol (DTT) and dithioerythritol (DTE) are the *trans* and *cis* isomers of the compound 2,3-dihydroxy-1,4-dithiolbutane. The reducing potential of these versatile reagents was first described by Cleland in 1964. Due to their low redox potential (-0.33 V) they are able to reduce virtually all accessible biological disulfides and maintain free thiols in solution despite the presence of oxygen. The compounds are fully water-soluble with very little of the offensive odor of the 2-mercaptoethanol they were meant to replace. Since Cleland's original report, literally hundreds of references have appeared citing the use of mainly DTT for the reduction of cystine and other forms of disulfides.

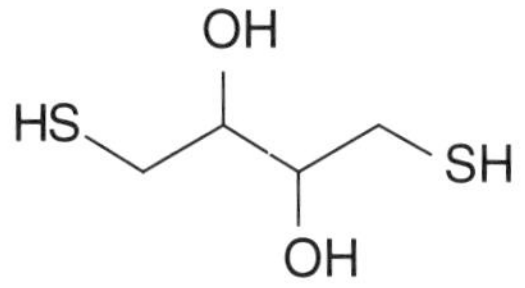

DTT;
Dithiothreitol
MW 154.25

The unique characteristics of DTT and DTE are mainly reflected in their ability to form intramolecular ring structures upon oxidation. Disulfide reductants such as 2-mercaptoethanol, 2-mercaptoethylamine, glutathione, thioglycolate, and 2,3-dimercaptopropanol cleave disulfide bonds in a two-step reaction that involves the formation of a mixed disulfide (Fig. 66). In the second stage of the reducing process, the mixed disulfide is cleaved by another molecule of reductant, freeing the sulfhydryl and forming a dimer of the reducing agent through the formation of a intermolecular disulfide bond. For simple reductants containing only one thiol, the equilibrium for disulfide exchange is nearly equivalent for the reductant and target protein. Thus, monothiol compounds are usually required in extreme excess to drive the reaction to completion.

The presence of two sulfhydryl groups in DTT and DTE, however, allows the formation of a favored cyclic disulfide during the course of target protein reduction (Fig. 67). This drives the equilibrium toward the reduction of target disulfides. Therefore, complete reduction is possible with much lower concentrations of DTT or DTE than when using monothiol systems.

As with all reductants, DTT and DTE will reduce disulfides only if they are accessible. The three-dimensional structure of a protein molecule often contains disulfides buried deep in the inner structure of the polypeptide chains. A protein retaining its native conformation is frequently protected from complete reduction. In the absence

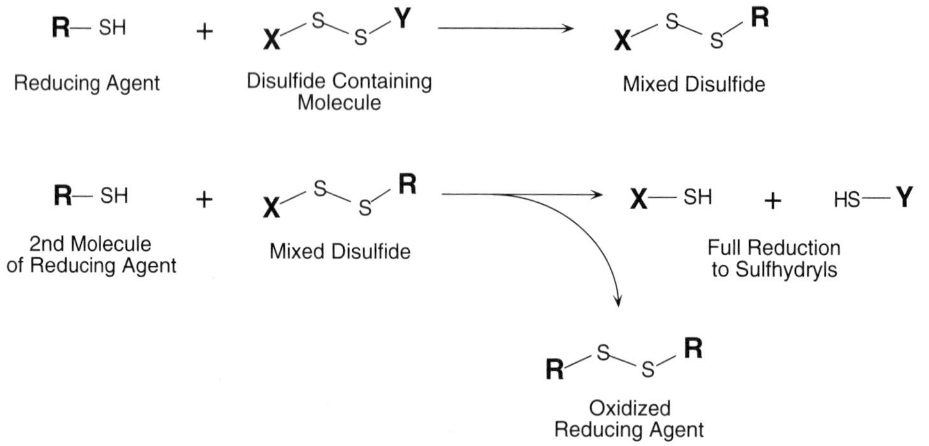

Figure 66 Thiol-containing disulfide reductants reduce disulfide groups through a multistep process producing a mixed disulfide intermediate.

of denaturants such as urea, guanidine, or SDS, DTT is not capable of reducing all available disulfides within some proteins (Bewley *et al.*, 1968; Bewley and Li, 1969). For instance, at moderate concentrations of DTT and no denaturants, limited cleavage of disulfides in antibody molecules can result in reducing only the bonds between the heavy chains of the immunoglobulin. This produces two half-antibody molecules, each containing one antigen binding site and free sulfhydryls in the hinge region. This limited reduction process can be used to site-direct sulfhydryl-reactive conjugation reagents away from the antigen binding sites, thus preserving activity (de Rosario *et al.*, 1990). However, using an appropriate concentration of deforming agents, DTT

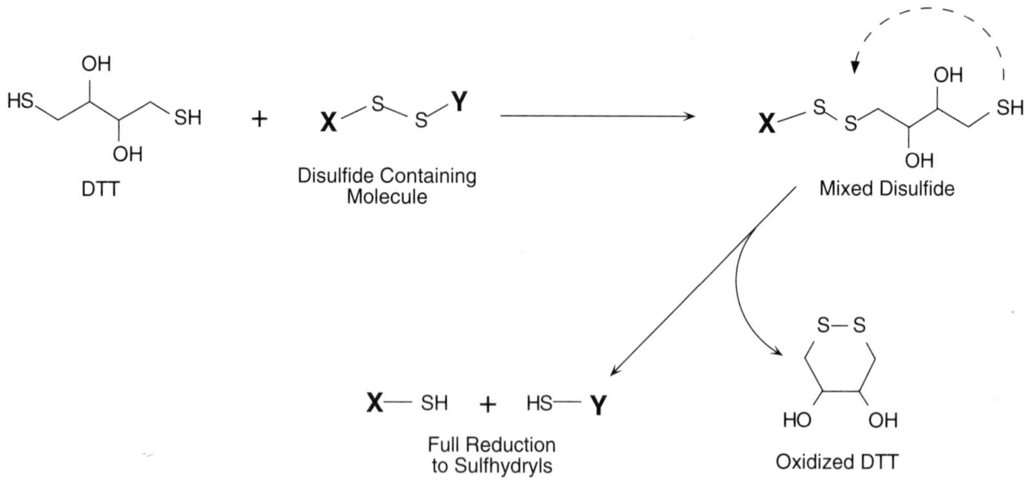

Figure 67 DTT is highly efficient at reducing disulfides, since a single molecule can reduce the intermediate mixed disulfide by forming a ring structure.

efficiently reduces all protein disulfides in the antibody and allows subunit separation for analysis (Konigsberg, 1972).

DTT also may be used to cleave disulfide-containing modification and cross-linking reagents. For thiolation procedures, DTT may be used to remove a dithiopyridyl group or cleave other disulfides to produce a free sulfhydryl. In this case, the presence of a denaturant usually is not required to access and reduce the disulfide of the modification reagent. Similarly, disulfides of cross-linking agents may be reduced after two macromolecules have been conjugated to release them as desired. This technique is often used to analyze receptor–ligand interactions or to discover how two proteins associate *in vivo*.

Complete Reduction of Disulfides in Protein Molecules Using DTT

Protocol

1. Dissolve a disulfide-containing protein or peptide at a concentration of 1–10 mg/ml in 6 M guanidine hydrochloride, 0.01 M sodium phosphate, 0.15 M NaCl, pH 7.4. Alternative denaturant conditions may be used [i.e., 8 M urea or 2.3% (w/w) SDS] along with any other buffer salts and pH values desired. A pH between 7.0 and 8.1 usually works best.
2. Add DTT (Pierce) to a final concentration of 10–100 mM.
3. Incubate for 2 h at room temperature. For some buried disulfides to become exposed and fully reduced, it may be necessary to heat the solution (in a capped test tube) at 50°C for 30 min. Some procedures use a 2-min incubation in a boiling water bath to completely denature the protein.
4. For removal of excess DTT, a protein of molecular weight greater than 5000 may be isolated by gel filtration using Sephadex G-25. To maintain the stability of the exposed sulfhydryl groups, include 1–10 mM EDTA in the chromatography buffer. The presence of oxidized DTT can be monitored during elution by measuring the absorbance at 280 nm. The protein should elute in the first peak and the DTT reaction products in the second peak.

Use of DTT to Cleave Disulfide-Containing Cross-linking Agents

The following method may be used to reduce the disulfide bonds of some cross-linking agents, thus cleaving conjugated proteins. This procedure will reduce the pyridyl disulfide group of SPDP (Section 4.1) to create a thiolated species. It also may be used to reduce partially the indigenous disulfides in some protein molecules. In this regard, DTT under nondenaturing conditions has been used to reduce selectively the disulfides between the heavy chains of immunoglobulin G (Edelman *et al.*, 1968). Without an added denaturant to open the polypeptide chain, internally buried disulfides typically will remain unreduced.

Protocol

1. Dissolve a cross-linked protein or peptide that has been conjugated with the use of a disulfide containing cross-linker at a concentration of 1–10 mg/ml in 0.01 M sodium phosphate, 0.15 M NaCl, pH 7.4. Alternative buffer conditions and pH levels may be used; however, a pH between 7.0 and 8.1 usually works best.

2. Add DTT to a final concentration of 1–10 mM.
3. Incubate for 2 h at room temperature.
4. For removal of excess DTT, a protein of molecular weight greater than 5000 may be isolated by gel filtration using Sephadex G-25. To maintain the stability of the exposed sulfhydryl groups, include 10 mM EDTA in the chromatography buffer. The presence of oxidized DTT can be monitored during elution by measuring the absorbance at 280 nm. The protein should elute in the first peak and the DTT reaction products in the second peak.

2-Mercaptoethanol

2-Mercaptoethanol is one of the most common agents used for disulfide reduction. Sometimes referred to as β-mercaptoethanol, it is a clear, colorless liquid with an extremely strong odor. All operations with this chemical should be performed in a well-ventilated fume hood. The reduction of protein disulfides with 2-mercaptoethanol proceeds rapidly via a two-step process involving an intermediate mixed disulfide (Fig. 68). Due to its strong reducing properties, the reagent is used most often when complete disulfide reduction is required. It also can be used to cleave disulfide-containing cross-linking agents. Usually a concentration of 0.1 M 2-mercaptoethanol will cleave a disulfide-containing cross-linker and liberate conjugated proteins (Chapter 7, Section 1).

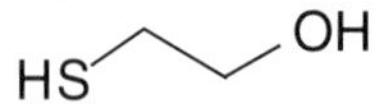

HS⌇⌇OH

2-ME;
2-Mercaptoethanol
MW 78.13

2-Mercaptoethanol is used as a reducing additive in a number of biochemical reagents. It is used as a reductant for a Gram-negative bacteria lysis buffer (Schwinghamer, 1980; Scopes, 1982), as the second-dimensional equilibration buffer for 2D electrophoresis (Dunbar, 1987), as the sample reducing buffer for SDS–polyacrylamide gel electrophoresis (Laemmli, 1970), and as a participant in the o-phthalaldehyde reaction for the detection of primary amines (Jones and Gilligan, 1983).

Protocol for Preparation and Use of a Gram-Negative Bacteria Lysis Buffer

1. Prepare a solution consisting of 2.5 ml glycerol, 100 µl of 10% Triton X-100 (Pierce Surface-Amps X-100), and 10 µl 2-mercaptoethanol.
2. Add 10 g of wet packed cells to the lysis buffer and stir vigorously for 30 min.
3. Add 30 ml of an extraction buffer consisting of 20 mM potassium phosphate, pH 7, 1 mM EDTA, 0.2 mg/ml lysozyme, and 10 µg/ml DNase I.
4. Add 5 mg PMSF dissolved in 0.5 ml acetone and 0.1 mg pepstatin A.
5. Centrifuge for 20 min at 15,000 g. Recover the extracted, solubilized material in the supernatant.

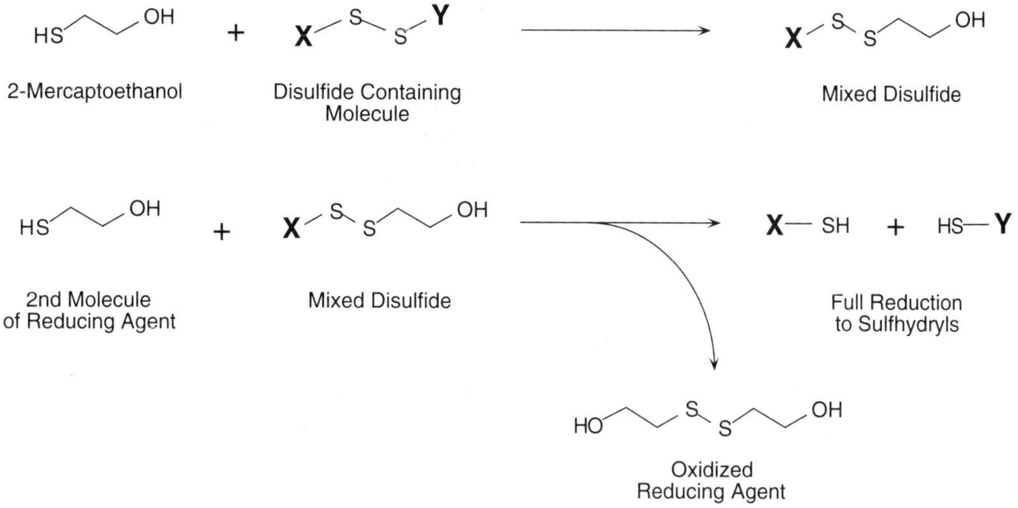

Figure 68 The reduction of disulfides by 2-mercaptoethanol proceeds through a mixed disulfide inter-mediate.

Protocol for Preparation and Use of the Second-Dimension Equilibration Buffer for 2D Gels

The following procedure relates to electrophoretic protocols where the first dimension is developed by isoelectric focusing (in tube gels) and the second dimension is a size exclusion separation by SDS–polyacrylamide electrophoresis in a slab gel.

1. Add 4 g SDS and 20 ml of 10% glycerol to 150 ml of 0.125 M Tris, pH 6.8, and adjust the final volume to 200 ml. Once dissolved, add a few crystals of bromophenol blue, mix, and pass the solution through a 0.2-μm filter. For storage, freeze in 10- to 15-ml aliquots.
2. Immediately before use, add 2-mercaptoethanol to a final concentration of 0.5–0.8%.
3. Incubate the first dimensional electrophoresis tube gel in this reducing buffer for 15 min. Drain off excess buffer and electrophorese in the second dimension.

SDS Sample Buffer for Running Electrophoresis Size Separations under Reducing Conditions

1. Dissolve 2 g of SDS, 0.75 g Tris base, and 10 ml of glycerol in 90 ml of water. Adjust the pH to 6.8 and bring the final volume to 100 ml.
2. To a small aliquot of the above buffer, add 2-mercaptoethanol to obtain a final concentration of 2–5%. Only 200 μl of this buffer typically is required to treat and reduce about 10–500 μg of protein. Solubilize the protein sample in this buffer.
3. Incubate a sealed tube at 95°C for 5–10 min or in a boiling bath for 1–2 min. Electrophorese immediately.

o-Phthalaldehyde Solution for the Fluorescent Detection of Primary Amines (see Section 4.3, OPA)

1. Add 3 ml of the detergent Brij-35 (as a 30% solution) and 2 ml of 2-mercaptoethanol to 950 ml of fluoraldehyde reagent diluent (all reagents from Pierce).
2. Dissolve 0.5–0.8 g of *o*-phthalaldehyde crystals in about 10 ml of methanol.
3. Mix the OPA solution with the solution from (1) and store under nitrogen in sealed glass bottles at 4°C. The addition of an aliquot of this solution to a sample containing primary amines will yield an intense blue fluorescence.

2-Mercaptoethylamine

2-Mercaptoethylamine is a disulfide reducing agent that has found widespread application in the partial reduction of immunoglobulin molecules. The reagent is supplied as a solid in the hydrochloride form (Pierce) and possesses very little of the sulfhydryl odor of 2-mercaptoethanol. When used under nondenaturing conditions, 2-mercaptoethylamine can cleave, almost selectively, the disulfide bonds between the heavy chains of IgG. This directed reduction is important for generating sulfhydryls while preserving antigen binding activity.

$$HS\diagup\diagdown\diagup NH_3^+Cl^-$$

2-MEA;
2-Mercaptoethylamine
Hydrochloride
MW 113.62

The complex structure of an antibody molecule creates two antigen binding sites from the interaction of the hypervariable regions on both the heavy and the light chains. For this reason, heavy–light chain pairing must remain intact during any modification procedure to ensure that antigen binding activity is retained. In addition, it is important that any chemistry take place away from the antigen binding sites so they are not sterically blocked by modification reagents or by subsequent conjugation steps. 2-Mercaptoethylamine can be used to cleave disulfides primarily in the hinge region of IgG—away from the antigen binding sites—thus preserving the disulfides that hold the heavy and light chains together (Yoshitake *et al.*, 1979). It also can be used to reduce F(ab')$_2$ fragments, because they still retain the hinge region disulfides of intact IgG (Fig. 69).

Once reduced with 2-mercaptoethylamine, immunoglobulins will be cleaved in half, forming two heavy chain–light chain molecules of MW 75,000–80,000 and each containing one antigen binding site. These half molecules of IgG will possess reactive sulfhydryls in the hinge region that can be used in conjugation protocols with sulfhydryl-reactive cross-linking reagents. For instance, a reduced antibody may be used to make a conjugate with a maleimide-activated enzyme, forming a reagent useful in immunoassays (Chapter 10, Section 1.1). Similarly, F(ab')$_2$ fragments may be reduced to yield two molecules, each containing an antigen binding site. Making conjugates with this low-molecular-weight fragment can dramatically reduce background in assay systems or provide access to antigens restricted to higher-molecular-weight con-

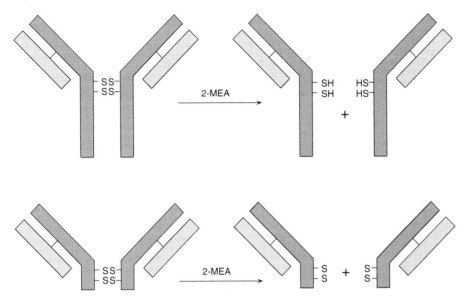

Figure 69 Disulfide reducing agents such as 2-mercaptoethylamine can be used to cleave the disulfide bonds in the hinge region of antibody molecules. Either intact IgG molecules or F(ab′)₂ fragments may be reduced in this manner to yield monofunctional antigen binding fragments.

jugates made with intact antibody (such as in immunohistochemical staining techniques).

Protocol

1. Dissolve the antibody to be reduced at a concentration of 10 mg/ml in 20 m*M* sodium phosphate, 0.15 *M* NaCl, pH 7.4, containing 1–10 m*M* EDTA.
2. To each milliliter of the antibody solution, add 6 mg of 2-mercaptoethylamine hydrochloride (final concentration is 50 m*M*). Mix to dissolve.
3. Incubate the solution in a sealed tube for 90 min at 37°C.
4. Purify the reduced IgG from excess 2-mercaptoethylamine and reaction by-products by dialysis or gel filtration using Sephadex G-25. All buffers should contain 1–10 m*M* EDTA to preserve the free sulfhydryls from metal-catalyzed oxidation. The sulfhydryl containing half-antibody may now be used in conjugation protocols that use —SH-reactive heterobifunctional cross-linkers (Chapter 10, Section 1.1).

TCEP

The reduction of disulfide bonds with trivalent phosphines has been known for some time (Ruegg and Rudingder, 1977; Kirley, 1989; Levison *et al.*, 1969). Unfortunately, trialkylphosphines generally are water-insoluble, undergo autoxidation, and are extremely odious.

The water-soluble tris(2-carboxyethyl)phosphine (TCEP) was synthesized and used to cleave rapidly organic disulfides to sulfhydryls in water (Burns *et al.*, 1991). The advantage of using this phosphine derivative in disulfide reduction as opposed to previous ones is its excellent stability in aqueous solution, its lack of reactivity with other common functionalities, and its freedom from odor.

The reaction of TCEP with biological disulfides proceeds with initial cleavage of the S—S bond followed by oxidation of the phosphine (Fig. 70). The stability of the phosphine oxide bond that is formed in this process is great enough to prevent reversal of the reaction. Since this reaction is performed without any added —SH compounds, subsequent conjugation with the generated sulfhydryl groups can be done without removal of excess TCEP or reaction by-products (provided the conjugation step does not involve disulfide exchange reactions, such as with the active disulfide-containing reagent SPDP; Chapter 5, Section 1.1).

Although TCEP is capable of rapidly and quantitatively reducing simple organic disulfides in solution, it requires the presence of a deforming agent to reduce fully all disulfides in proteins. Without opening up the internal disulfides in many protein molecules, TCEP will not be able to reduce them. For complete reduction of IgG, it was found that 20 mM TCEP and 5 min of boiling was needed (Hines, 1992). Partial reduction, however, is possible of some more accessible disulfides in aqueous buffers at room temperature.

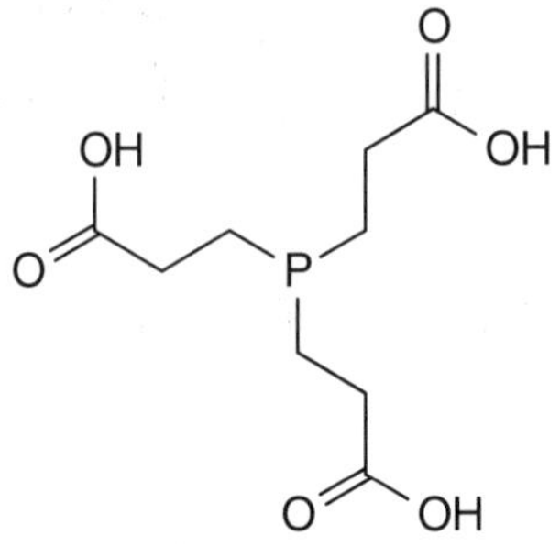

TCEP
Tris(2-carboxyethyl)phosphine
(hydrochloride)
MW 250.19

Protocol for the Complete Reduction of Disulfide Bonds within Protein Molecules

1. Dissolve the protein to be reduced at a concentration of 1–10 mg/ml in 20 mM sodium phosphate, 0.15 M NaCl, pH 7.4. Other buffers and pH values also may be used. A strong denaturant may be added (6 M guanidine or 8 M urea) to this solution to promote protein unfolding and make buried disulfides more accessible.
2. Add TCEP to a final concentration of 20 mM.
3. Place in a sealed tube and incubate in a boiling water bath for 5 min. If a denaturant was included in the buffer from (1), then high temperature may not be necessary. Alternatively, incubate the sample at 50°C for 30 min.
4. To remove excess TCEP and reaction by-products, dialyze the solution or purify by gel filtration using a buffer containing 1–10 mM EDTA.

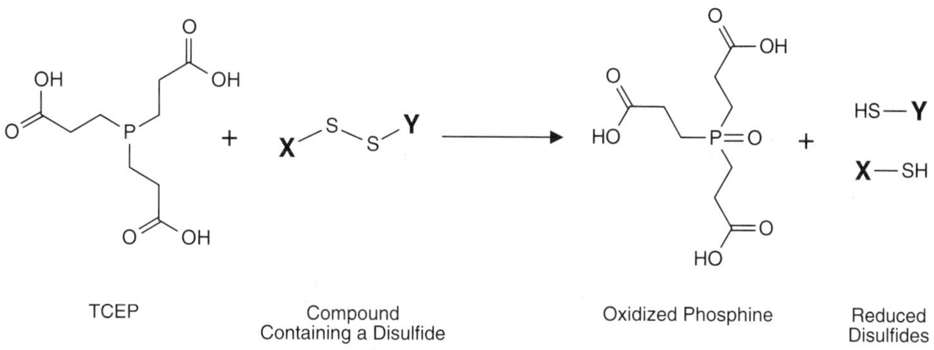

Figure 70   TCEP reduction of disulfides proceeds without the use of thiol compounds.

Protocol for Partial Reduction of Protein Disulfides or for Cleaving Disulfide-Containing Modification Reagents

1. Dissolve the protein to be reduced at a concentration of 1–10 mg/ml in 20 mM sodium phosphate, 0.15 M NaCl, pH 7.4. Other buffers and pH values also may be used. Do not add a denaturant to unfold protein structure.
2. Add TCEP to a final concentration of 20 mM.
3. Incubate for 2 h at room temperature.
4. To remove excess TCEP and reaction by-products, dialyze the solution or purify the protein by gel filtration using a buffer containing 1–10 mM EDTA.

Immobilized Disulfide Reductants

Many extracellular proteins like immunoglobulins, protein hormones, serum albumin, pepsin, trypsin, ribonuclease, and others contain one or more indigenous disulfide bonds. For functional and structural studies of proteins, it is often necessary to cleave these disulfide bridges. Disulfide bonds in proteins are commonly reduced with small, soluble mercaptans, such as 2-mercaptoethanol, thioglycolic acid, and cysteine. High concentrations of mercaptans (molar excess of 20- to 1000-fold) are usually required to drive the reduction to completion.

Cleland (1964) showed that dithiothreitol (DTT) and dithioerythritol (DTE) are superior reagents in reducing disulfide bonds in proteins (Section 4.1). DTT and DTE have low oxidation–reduction potential and are capable of reducing protein disulfides at concentrations far below that required with 2-mercaptoethanol. However, even these reagents must be used in an approximately 20-fold molar excess in order to get close to 100% reduction of a protein.

An immobilized disulfide reductant usually consists of an insoluble beaded support material such as agarose that has been modified with a small ligand containing a terminal sulfhydryl group. The presence of densely coupled sulfhydryl groups on the matrix creates enormous disulfide reducing potential. Simply mixing a solution of a disulfide-containing peptide or protein with the immobilized reductant efficiently breaks any disulfide linkages and creates free sulfhydryls. This is done without extraneous sulfhydryl contamination by the reductant, as in the case of soluble reductants.

The use of immobilized disulfide reductants thus have the following advantages over solution phase agents:

1. Immobilized disulfide reductants can be used to reduce all types of biological disulfides without liberating product or by-product contaminants.
2. Soluble components that interfere with the assay of free thiol groups are not present if immobilized disulfide reductants are used.
3. Small molecules containing disulfide bonds (such as cystine-containing peptides) may be reduced and isolated simply by removing the immobilized reductant. Separation of reduced molecules from reductant is much more difficult if a soluble reducing agent is used with low-molecular-weight disulfides.
4. Immobilized disulfide reductants can be easily regenerated and reused many times.

Immobilized dihydrolipoamide (thioctic acid) (Gorecki and Patchornick, 1973, 1975) and immobilized N-acetyl-homocysteine thiolactone (Eldjarn and Jellum, 1963; Jellum, 1964) are the two most commonly used immobilized disulfide reductants. The author, along with Krishna Mallia of Pierce Chemical, has successfully used immobilized reductants to reduce many types of biological disulfides, including small molecules like oxidized glutathione and bovine insulin.

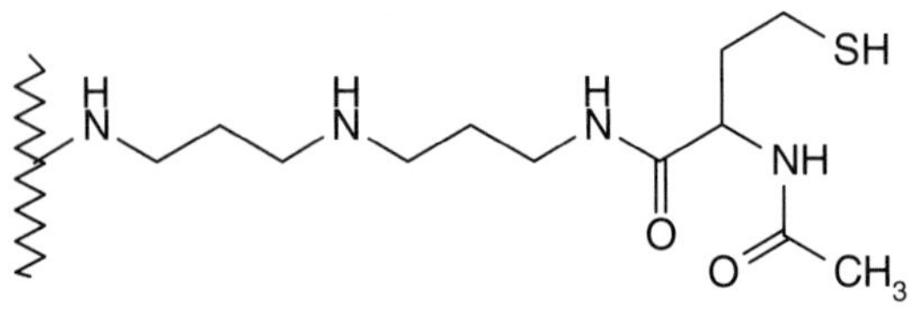

Immobilized N-Acetyl homocysteine
attached to a diaminodipropylamine spacer

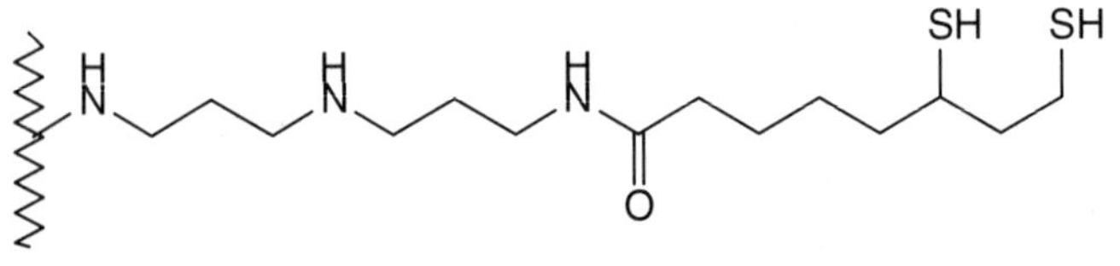

Immobilized Dihydrolipoamide
attached to a diaminodipropylamine spacer

Immobilized disulfide reductants may be synthesized as described in Hermanson *et al.* (1992) or obtained commercially (Pierce).

Reduction of Peptides Using Immobilized Reductants

Note: For optimal reduction of peptides, the following steps should be performed at room temperature.

1. Pack an immobilized reductant gel (2 ml settled gel) in a disposable polypropylene column and wash with 5 ml of 0.1 M sodium phosphate buffer, pH 8, containing 1 mM EDTA (equilibration buffer).

2. Prepare the sulfhydryl column by washing with a disulfide reducing agent. Apply 10 ml of freshly made 10 mM DTT solution (15.4 mg of DTT dissolved in 10 ml of equilibration buffer). This treatment converts the immobilized ligands into a fully reduced form (free —SH groups).
3. Wash the column with 20 ml of equilibration buffer 1 to remove free DTT.
4. Apply to the column 1 ml of peptide solution (dissolved in equilibration buffer) to be reduced. Normally, small peptides (molecular weight less than or equal to that of insulin) require no deforming agent (denaturant) such as guanidine to be completely reduced.
5. After the sample has completely entered into the gel bed, wash the column with 9 ml of equilibration buffer, while collecting 1-ml fractions.
6. Monitor the elution of reduced peptide from the column by measuring the absorbance at 280 nm (if peptide absorbs at this wavelength) as well as by performing an Ellman's assay (Section 4.1) for sulfhydryl groups using a small aliquot (10–20 μl) of each collected fraction.
7. Regenerate the sulfhydryl containing support by following steps 2 and 3 above. Such columns can be regenerated and reused at least 10 times without any significant decrease in the reductive capacity.
8. Store the column in 0.02% sodium azide at 4°C.

Reduction of Proteins Using Immobilized Reductants

Note: For optimal reduction of proteins, the following steps must be performed at room temperature.

1. Pack an immobilized reductant gel (2 ml) in a disposable polypropylene column and wash with 5 ml of 0.1 M sodium phosphate buffer, pH 8, containing 1 mM EDTA (equilibration buffer 1).
2. Prepare the sulfhydryl column by washing with a disulfide reducing agent. Apply 10 ml of freshly made 10 mM DTT solution (15.4 mg of DTT dissolved in 10 ml of equilibration buffer 1).
3. Wash the column with 10 ml of equilibration buffer 1 and 10 ml of 0.1 M sodium phosphate buffer, pH 8, containing 1 mM EDTA and 6 M guanidine hydrochloride (equilibration buffer 2) to remove free DTT.
4. Apply to the column 1 ml of protein solution (dissolved in equilibration buffer 2) to be reduced. The inclusion of a denaturant in the solution deforms the protein structure so that inner disulfides are available to the immobilized reductant. Without the presence of guanidine or another deforming agent (i.e., urea, SDS), only partial reduction of the protein is possible.
5. After the sample has completely entered the gel bed, incubate the column at room temperature for 1 h.
6. Wash the column with 9 ml of equilibration buffer 2 while 2-ml fractions are collected.
7. Monitor elution of reduced protein from the column by measuring the absorbance at 280 nm as well as by performing an Ellman's assay for sulfhydryl groups (Section 4.1) using a small aliquot (50–100 μl) of each collected fraction.
8. Regenerate the sulfhydryl containing column by following steps 2 and 3 above.

Such columns can be regenerated and reused at least 10 times without any significant decrease in the reductive capacity.

9. Store the column in 0.02% sodium azide at 4°C.

Sodium Borohydride

Perhaps the simplest route to the reduction of disulfide groups in peptides is the use of sodium borohydride ($NaBH_4$). This common reducing agent often used in organic synthesis is able to reduce specifically disulfides to free thiols without affecting any of the other major functional groups in proteins. Gailit (1993) developed a protocol for borohydride reduction that avoids any purification steps to remove the reducing agent after the reaction. Thus, peptides reduced by this protocol can be used immediately in bioconjugate applications without additional steps.

Protocol

1. Dissolve the peptide to be reduced in a buffer at pH 8–10. Sodium phosphate or sodium bicarbonate at 0.1 M work well.
2. Add sodium borohydride (Aldrich) to the peptide solution to obtain a final concentration of 0.1 M. Generation of hydrogen bubbles will occur as the borohydride is dissolved.
3. Incubate at room temperature for 30–60 min.
4. Adjust the pH of the reaction to pH 4 using dilute HCl. Incubate for 10 min to ensure the complete destruction of excess borohydride. Hydrogen bubbles again will be evolved from the solution.
5. Readjust the pH to the optimal value for the bioconjugate application to be done. Use the reduced peptide immediately to prevent reoxidation of the thiols to disulfides.

Ellman's Assay for the Determination of Sulfhydryls

Ellman's reagent, 5,5′-dithio*bis*(2-nitrobenzoic acid), reacts with sulfhydryls under slightly alkaline conditions to release the highly chromogenic compound, 5-thio-2-nitrobenzoic acid (TNB) (Ellman, 1959; Riddles *et al.*, 1979) (Fig. 71). The reagent contains a disulfide bond between two TNB groups and reacts with free sulfhydryls to create a mixed disulfide product. The target of the reaction is the unprotonated, conjugate base form of the thiol R—S⁻. At pH 8, the release of one TNB group per available thiol provides a yellow-colored product with an extinction coefficient at 412 nm of 13,600 $M^{-1}cm^{-1}$. The increase in absorbance at this wavelength is directly proportional to the concentration of sulfhydryls in solution. Correlation to a standard curve of known sulfhydryl concentrations allows accurate measurement of the thiol content in unknown samples.

Ellman's reagent has been used not only for the determination of sulfhydryls in proteins and other molecules, but also as a precolumn derivatization reagent for the separation of thiol compounds by HPLC (Kuwata *et al.*, 1982), in the study of thiol-dependent enzymes (Masamune *et al.*, 1989; Tsukamoto and Wakil, 1988; Alvear *et al.*, 1989), and to create sulfhydryl-reactive chromatography supports for the coupling of affinity ligands (Jayabaskaran *et al.*, 1987). Another important use of the compound

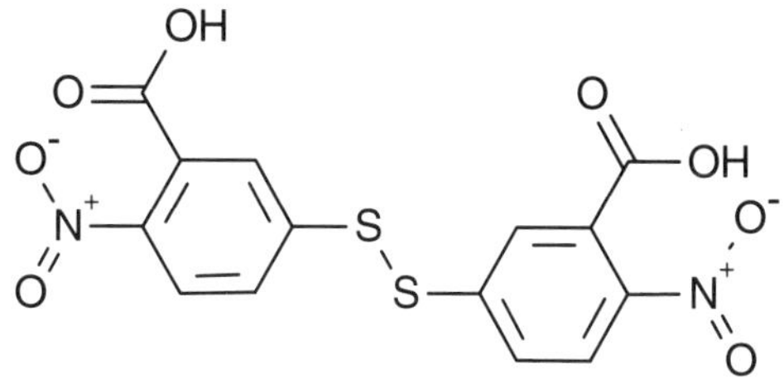

Ellman's Reagent
5,5'-Dithio-bis-(2-nitrobenzoic acid)
MW 396.4

is in the assessment of conjugation procedures using sulfhydryl-reactive cross-linking agents (Chapter 9, Section 5).

Depending on the conditions, an Ellman's assay can detect as little as 10 nM cysteine concentration. The linearity can extend into the mM range, making the test extremely flexible for different sample situations.

Protocol

1. Dissolve Ellman's reagent (Pierce) in 0.1 M sodium phosphate, pH 8, at a concentration of 4 mg/ml.
2. Prepare a set of standards by dissolving cysteine in 0.1 M sodium phosphate, pH 8, at an initial concentration of 2 mM (3.5 mg/ml) and serially diluting this solution (1:1) with reaction buffer down to at least 0.125 mM. This will produce five solutions of cysteine for generating a standard curve. If a more dilute concentration range is required, continue to dilute serially until a set of standards in the desired range is obtained.
3. Label a set of test tubes according to the standards and samples to be used. Add 250 μl of each standard and sample to the appropriate tubes. If the samples are in a buffer that may significantly change the pH of the reaction buffer, the samples should be buffer-exchanged or dialyzed into 0.1 M sodium phosphate, pH 8, before running the assay.
4. Add 50 μl of Ellman's reagent to each standard and sample tube. Mix well.
5. Incubate at room temperature for 15 min.
6. Measure the absorbance of each solution at 412 nm.
7. Plot the absorbance versus cysteine concentration for each of the standards.

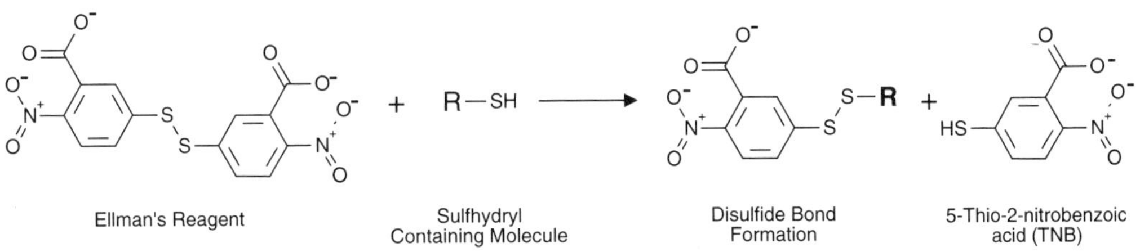

Ellman's Reagent   Sulfhydryl   Disulfide Bond   5-Thio-2-nitrobenzoic
                        Containing Molecule   Formation   acid (TNB)

Figure 71   The reaction of Ellman's reagent with a sulfhydryl group releases the chromogenic TNB anion, which can be quantified by its absorbance at 412 nm.

Determine the sulfhydryl concentration of the samples by comparison to the standard curve.

4.2. Introduction of Carboxylate Groups

Modification of various functional groups in macromolecules with the following types of reagents will introduce carboxylate functions for further derivatization purposes. Amines, sulfhydryls, and histidine and methionine side chains are readily modified to contain short molecules terminating in a carboxylic acid. The short chain can serve as a spacer to enhance steric accommodations and the terminal carboxylate group can facilitate subsequent couplings with amines or hydrazides. The introduction of carboxylates also affects the overall charge characteristics or pI of the molecule being derivatized. The modification of amine residues by acylation with anhydrides not only eliminates the positive charge contribution of the protonated amine, but also adds the negative charge contribution of the acid. The result may be a change of minus two in net charge per group modified. While the reactions involved in such derivatizations are conducted under relatively mild conditions, severe alterations in net charge may cause some macromolecules, like proteins, to denature or lose activity. In addition, if the group being modified happens to be critical for active center operation then the functional group may be compromised regardless of conditions. While the following reactions are facile and efficient, it should be kept in mind that in certain instances modification may lead to inactivity.

Modification of Amines with Anhydrides

Acid anhydrides, as their name implies, are formed from the dehydration reaction of two carboxylic acid groups (Fig. 72). Anhydrides are highly reactive toward nucleophiles and are able to acylate a number of the important functional groups of proteins and other macromolecules. Upon nucleophilic attack, the anhydride yields one carboxylic acid for every acylated product. If the anhydride was formed from monocarboxylic acids, such as acetic anhydride, then the acylation occurs with release of one carboxylate group. However for dicarboxylic acid anhydrides, such as succinic anhydride, upon reaction with a nucleophile the ring structure of the anhydride opens, forming the acylated product modified to contain a newly formed carboxylate group.

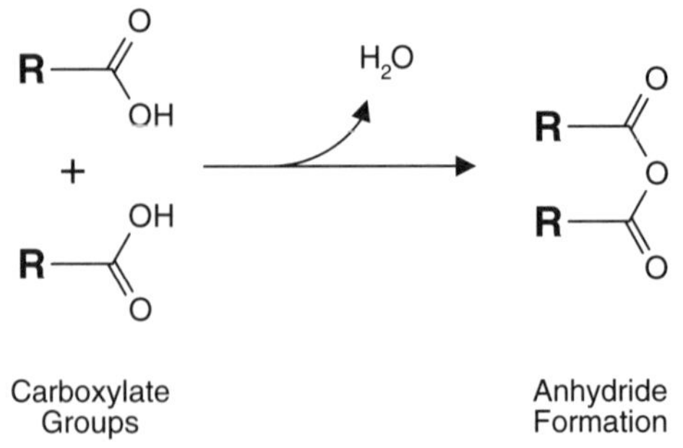

Carboxylate
Groups

Anhydride
Formation

Figure 72   Anhydrides are created from two carboxylate groups by the removal of one molecule of water.

Thus, anhydride reagents may be used both to block functional groups and to convert an existing functional group into a carboxylic acid.

Protein functional groups able to react with anhydrides include the α-amines at the N-terminals, the ε-amine of lysine side chains, cysteine sulfhydryl groups, the phenolate ion of tyrosine residues, and the imidazolyl ring of histidines. However, acylation of cysteine, tyrosine, and histidine side chains forms unstable complexes that are easily reversible to regenerate the original group. Only amine functional groups of proteins are stable to acylation with anhydride reagents (Fraenkel-Conrat, 1959; Smyth, 1967).

Another potential site of reactivity for anhydrides in protein molecules is modification of any attached carbohydrate chains. In addition to amino group modification in the polypeptide chain, glycoproteins may be modified at their polysaccharide hydroxyl groups to form esterified derivatives. Esterification of carbohydrates by acetic anhydride, especially cellulose, is a major industrial application for this compound. In aqueous solutions, however, esterification will be a minor product, since the oxygen of water is about as strong a nucleophile as the hydroxyls of sugar residues.

The major side reaction to the desired acylation product is hydrolysis of the anhydride. In aqueous solutions anhydrides may break down by the addition of one molecule of water to yield two unreactive carboxylate groups. The presence of an excess of the anhydride in the reaction medium usually is enough to minimize the effects of competing hydrolysis.

Since both hydrolysis and acylation yield the release of carboxylic acid functional groups, the medium becomes acidic during the course of the reaction. This requires either the presence of a strongly buffered environment to maintain the pH or periodic monitoring and adjustment of the pH with base as the reaction progresses.

Succinic Anhydride

Succinic acid is a four-carbon molecule with carboxylic acid groups on both ends. The anhydride has a five-atom cyclic structure that is highly reactive toward nucleophiles, especially amines. Attack of a nucleophile at one of the carbonyl groups opens the anhydride ring, forming a covalent bond with that carbonyl and releasing the other to create a free carboxylic acid (Klotz, 1967). Succinylation of positively charged amino groups of proteins and other molecules thus creates amide bond derivatives and converts the cationic site into a negatively charged carboxylate (Fig. 73). Succinylated proteins often experience dramatic changes in their three-dimensional structure. Subunits may dissociate (Klotz and Keresztes-Nagy, 1962), enzymatic activity may be compromised (Riordan and Valle, 1963, 1964), and the molecular radius and viscosity

Succinic Anhydride
MW 100

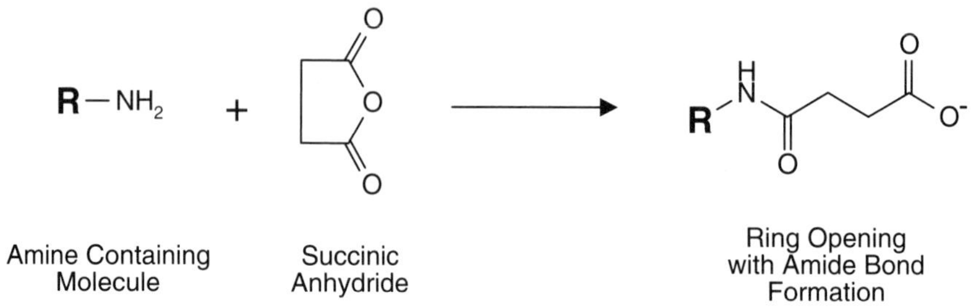

Figure 73 Succinic anhydride reacts with primary amine groups in a ring-opening process, creating an amide bond and forming a terminal carboxylate.

may be increased (Habeeb *et al.*, 1958). Other effects on protein conformation and function have been studied as well (Meighen *et al.*, 1971; Shetty and Rao, 1978; Shiao *et al.*, 1972).

Succinic anhydride also may react with protein phenolate side chains of tyrosine residues and the —OH group of aliphatic hydroxy amino acids (Fig. 74). The phenolate ester derivatives are unstable above pH 5, whereas the serine and threonine esters are relatively stable but may be specifically cleaved by treatment with hydroxylamine (Gounaris and Perlman, 1967).

A succinylated casein derivative that has nearly all its amines blocked can be used as a substrate in protease assays (Hatakeyama *et al.*, 1992). As the casein is degraded by a protease, free amines are created from α-chain cleavage and release of α-amino groups. The creation of amines can be monitored by an amine detection reagent such as trinitrobenzene sulfonic acid (TNBS; Section 4.3). The procedure forms the basis for a highly sensitive assay for protease activity.

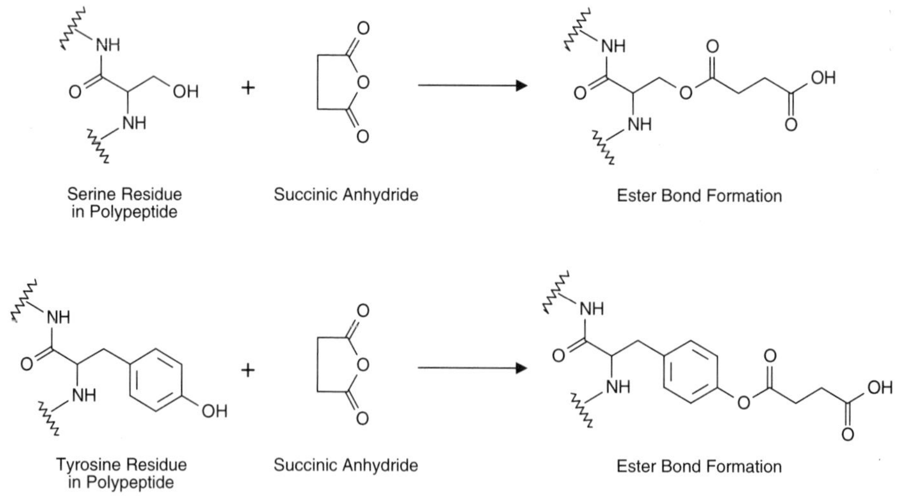

Figure 74 The hydroxyl group of serine residues and the phenolate ring of tyrosine groups may be modified with succinic anhydride to produce relatively unstable ester bonds. In aqueous conditions these reactions are minor due to competing hydrolysis by water.

Succinylated derivatives of nucleic acids may be prepared by reaction of the anhydride with available —OH groups. The reaction forms relatively stable ester derivatives that create carboxylates on the nucleotide for further conjugation or modification (Fig. 75). This method has been used in nucleic acid synthesis (Matteucci and Caruthers, 1980) and to derivatize nucleotide analogs such as AZT (Tadayoni *et al.*, 1993).

Succinic anhydride also is a convenient extender for creating spacer arms on chromatography supports. Supports derivatized with amine-terminal spacers may be succinylated to block totally the amine functional groups and form terminal carboxylic acid linkers for coupling amine-containing affinity ligands (Cuatrecasas, 1970).

Molecules modified with succinic anhydride to create terminal carboxylate functional groups may be further conjugated to amine-containing molecules by use of amide bond-forming reagents such as carbodiimides (Chapter 3, Section 1).

Protocol

1. Dissolve (or suspend in the case of insoluble polymers or support materials) the amine-containing molecule to be succinylated in a buffer having a pH between 6 and 9. Higher pH buffers will cause the reaction to occur faster and result in more amines in an unprotonated state. Suitable buffer salts include sodium acetate, sodium phosphate, and sodium carbonate in a 0.1–1.0 M concentration. Avoid buffers containing primary amine groups such as Tris. Alternatively, the substance may be dissolved in water and the pH maintained in the proper range by periodic addition of NaOH. This is conveniently done by means of a pH stat. Even in buffered reactions, the pH should be monitored to prevent severe acidification of the reaction solution, which could damage the molecule being modified.

2. Add a quantity of succinic anhydride to the reaction medium to provide at least a 5–10 molar excess of reagent over the amount of amines to be modified. Even greater molar excesses may be required for total blocking of all the amines of some proteins. When adding solid succinic anhydride, multiple additions may

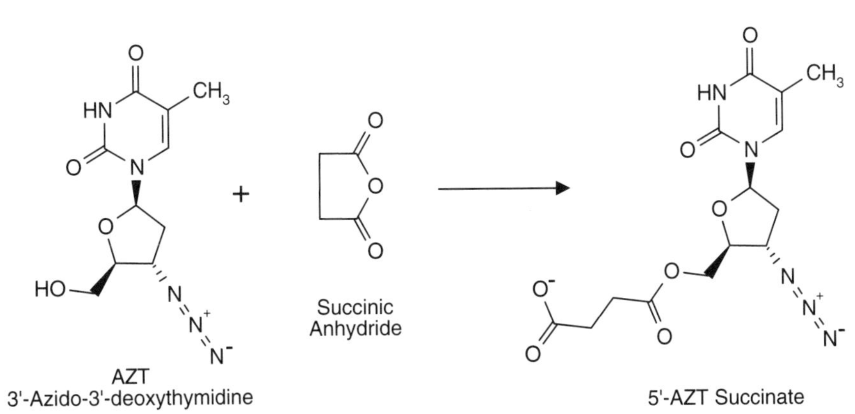

Figure 75 Succinic anhydride has been used in nonaqueous conditions to modify the 5'-hydroxyl group of nucleic acid derivatives such as AZT.

be done to maintain solubility of the reagent in the reaction solution. The anhydride also may be dissolved in dry dioxane before addition to aid in dissolution.

3. React for at least 1–2 h at room temperature. To ensure complete blocking of all amine groups, the reaction may be continued overnight.

4. Remove excess reactants from the succinylated molecule by dialysis, gel filtration, or some other suitable method. The efficiency of amine modification may be assessed by use of the TNBS test for amines (Section 4.3). A negative test for amines indicates complete succinylation.

Glutaric Anhydride

Glutaric acid is a linear, five-carbon molecule with carboxylic acid groups on both ends. It contains one additional carbon in length than the similar compound succinic acid. The anhydride of glutaric acid forms a cyclic structure containing six atoms. Attack of a nucleophile, such as an amino group, on one of the carbonyl groups of glutaric anhydride opens the ring, forming an amide linkage and liberating the other carboxylic acid (Fig. 76). Reaction with the phenolate of tyrosine or the sulfhydryl group of cysteine forms unstable linkages (an ester and a thioester, respectively) that can easily hydrolyze. As with succinic anhydride, however, aliphatic hydroxyl groups such as those of serine and threonine may be modified with glutaric anhydride to create more stable ester bonds (see above).

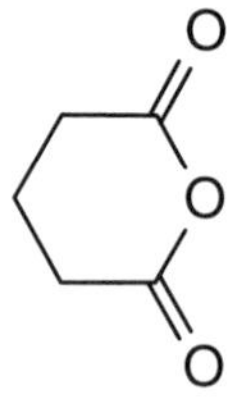

Glutaric Anhydride
MW 114

Protocol

The procedure for the modification of amine-containing compounds with glutaric anhydride is identical to that described for succinic anhydride, above.

Maleic Anhydride

Maleic acid is a linear four-carbon molecule with carboxylate groups on either end, similar to succinic acid, but with a double bond between the central carbon atoms. The anhydride of maleic acid is a cyclic molecule containing five atoms in its ring. Although the reactivity of maleic anhydride is similar other such reagents like succinic anhydride, the products of maleylation are much more unstable toward hydrolysis, and the site of unsaturation lends itself to additional side reactions. Acylation products of amino groups with maleic anhydride are stable at neutral pH and above, but they readily hydrolyze at acid pH values (around pH 3.5) (Butler *et al.*, 1967). Maleylation of sulfhydryls and the phenolate of tyrosine are even more sensitive to hydrolysis.

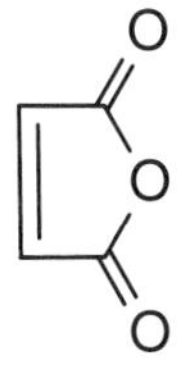

Maleic Anhydride
MW 98

As with other cyclic anhydrides, the acylation of an amine residue proceeds with elimination of the potential positive charge of the amine and addition of the negative charge created by the anhydride ring opening (Fig. 77). Thus, a molecule can undergo a change of minus two in net charge per site of maleylation. Proteins extensively modified with maleic anhydride may spontaneously dissociate into subunits or experience a general opening of their three-dimensional structures (Sia and Horecker, 1968; Uyeda, 1969).

The double bond of maleic anhydride may undergo free radical polymerization with the proper initiator. Polymers of maleic anhydride (or copolymers made with another monomer) are commercially available (Polysciences). They consist of a linear hydrocarbon backbone (formed from the polymerization of the vinyl groups) with cyclic anhydrides protruding from the chain. Such polymers are highly reactive toward amine-containing molecules.

Maleic acid imides (maleimides) are derivatives of the reaction of maleic anhydride and ammonia or primary amine compounds. The double bond of maleimides may undergo an alkylation reaction with sulfhydryl groups to form a stable thioether bond (Chapter 2, Section 2.2). Maleic anhydride may presumably undergo the same irreversible reaction with cysteine residues and other sulfhydryl compounds.

Proteins derivatized with maleic anhydride exhibit an increase in their absorptivity at wavelengths below 280 nm, likely due to the addition of the unsaturated carbon—carbon bond. The extent of maleylation may be estimated by measuring the absorbance increase before and after modification (Freedman et al., 1968).

Protocol
Modification of amines with maleic anhydride is done essentially the same as that described for succinic anhydride (this section), except the pH of the reaction should be kept alkaline (pH 8–9) at all times to prevent unwanted deacylation. Deblocking of

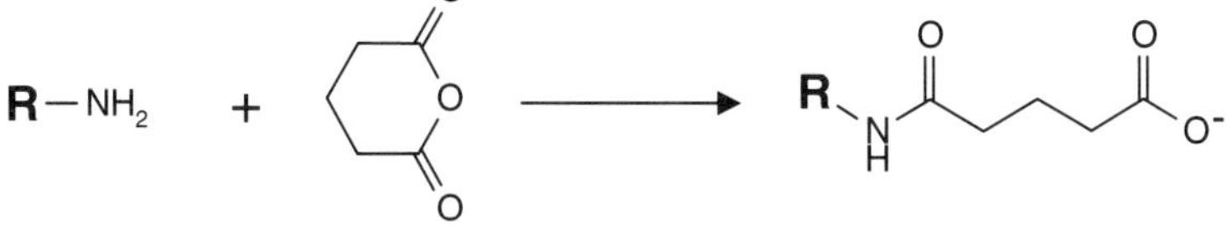

Amine Containing Molecule   Glutaric Anhydride   Ring Opening with Amide Bond Formation

Figure 76 Glutaric anhydride reacts with amines in a ring-opening process to create an amide bond linkage and a terminal carboxylate group.

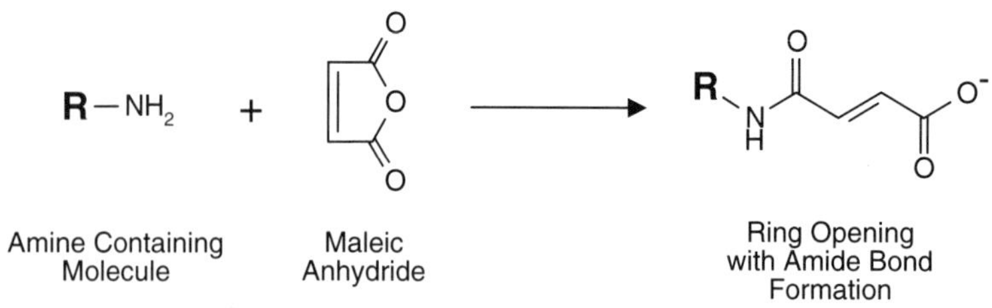

Figure 77 Maleic anhydride reacts with amine groups in a ring-opening process to create carboxylate derivatives.

maleylated amines can be accomplished according to the following procedure of Butler *et al.* (1967).

1. Adjust the pH of the maleylated protein or other molecule to pH 3.5 with formic acid and aqueous NH_3.
2. Incubate the solution at 37°C for 30 h.
3. Stop the deblocking reaction by the addition of NaOH to raise the pH back to neutrality.

Citraconic Anhydride

Citraconic anhydride (or 2-methylmaleic anhydride) is a derivative of maleic anhydride that is even more reversible after acylation than maleylated compounds. At alkaline pH values (pH 7–8) the reagent effectively reacts with amine groups to form amide linkages and a terminal carboxylate. However, at acid pH (3–4), these bonds rapidly hydrolyze to release citraconic acid and free the amine (Fig. 78) (Dixon and Perham, 1968; Klapper and Klotz, 1972; Habeeb and Atassi, 1970; Shetty and Kinsella, 1980). Thus, citraconic anhydride has been used to block temporarily amine groups while other parts of a molecule are undergoing derivatization. Once the modification is complete, the amines then can be unblocked to create the original structure.

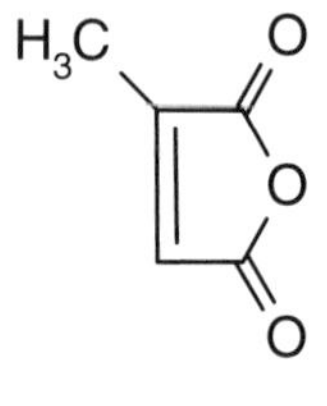

Citraconic
Anhydride
MW 113

Acid labile, heterobifunctional cross-linking reagents have been synthesized using 2-methylmaleic anhydride at one end (Blattler *et al.*, 1985a,b). Amines can be reacted with the anhydride end under alkaline conditions to form amide linkages. The other end, containing another functional group, in this case a maleimide group, is then made to react with a sulfhydryl-containing molecule. After the conjugation is complete, the citraconylamide end can be specifically released by lowering the pH.

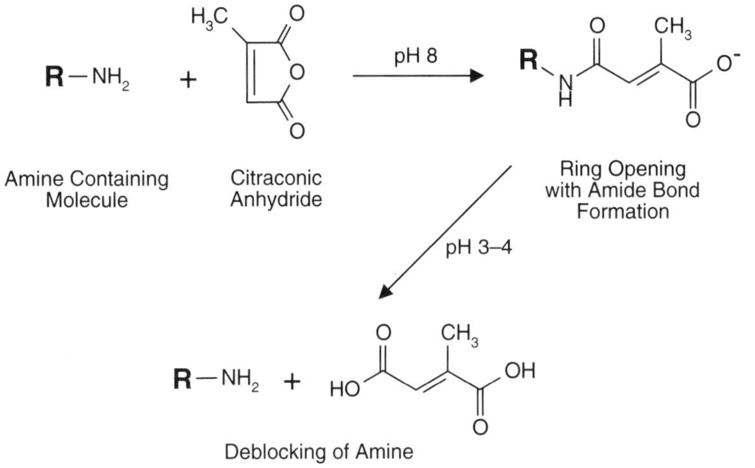

Figure 78 Citraconic anhydride can be used to block amine groups reversibly. The amide bond derivative is unstable to acidic conditions.

Protocol

1. Dissolve the amine-containing molecule to be modified in a buffer having a pH between 8 and 9. Maintenance of a high pH is necessary due to the high tendency of citraconylamides to hydrolyze at lower pH values. Suitable buffer salts include sodium phosphate and sodium carbonate in a $0.1–1.0\,M$ concentration. Avoid buffers containing primary amine groups such as Tris. Alternatively, the substance may be dissolved in water and the pH maintained in the proper range by periodic addition of NaOH. This is conveniently done by means of a pH stat.
2. Add a quantity of citraconic anhydride to the reaction medium to provide at least a $5–10$ molar excess of reagent over the amount of amines to be modified. Even greater molar excesses may be required for total blocking of all the amines of some proteins. When adding citraconic anhydride, multiple additions may be done to maintain solubility of the reagent in the reaction solution.
3. React for at least $1–2$ h at room temperature. To ensure complete blocking of all amine groups, the reaction may be continued overnight.
4. Remove excess reactants from the citraconylated molecule by dialysis, gel filtration, or some other suitable method. The efficiency of amine modification may be assessed by use of the TNBS test for amines (Section 4.3). A negative test for amines indicates complete modification.

To remove the citraconic modifications and free the amine groups, the protein may be treated in one of two ways:

1. Adjust the pH of the citraconylated molecule to 3.5–4.0 by addition of acid. Incubate at room temperature overnight or for at least 3 h at 30°C.

or

2. Treat the citraconylated molecule with 1 M hydroxylamine at pH 10 for 3 h at room temperature.

Modification of Sulfhydryls with Iodoacetate

Iodoacetate (and bromoacetate) can react with a number of functional groups within proteins: the sulfhydryl group of cysteine, both imidazolyl side chain nitrogens of histidine, the thioether of methionine, and the primary ε-amine group of lysine residues and N-terminal α-amines (Gurd, 1967). The relative rate of reaction with each of these residues is generally dependent on the degree of ionization and thus the pH at which the modification is done. The exception to this is methioninyl thioethers that react rapidly at nearly all pH values above about 1.7 (Vithayathil and Richards, 1960). The reaction products of these groups with iodoacetate are illustrated in Fig. 79. The only reaction resulting in one definitive product is that of the alkylation of cysteine sulfhydryls, giving the carboxymethylcysteinyl derivative (Cole *et al.*, 1958). Histidine groups may be modified at either nitrogen atom of its imidazolyl side chain. Both monosubstituted derivatives and disubstituted products of the imidazole ring are possible (Crestfield *et al.*, 1963). With primary amine groups such as in the side chain of lysine residues, the products of the reaction are either the secondary amine, mono-

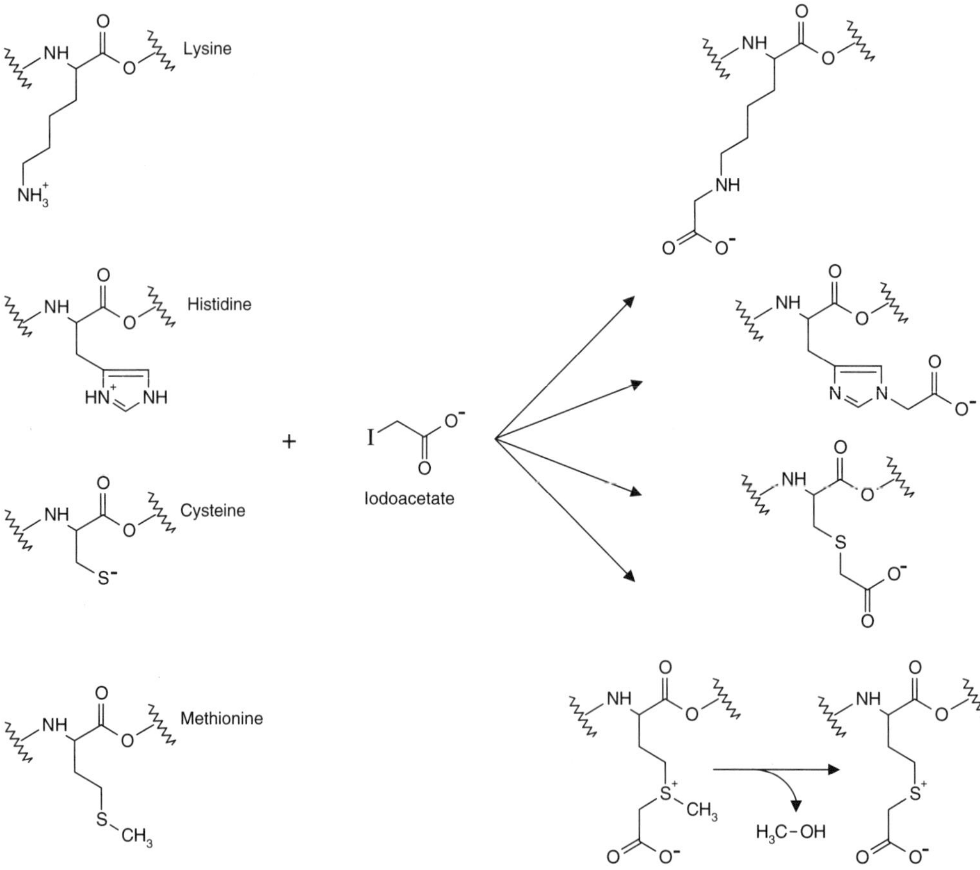

Figure 79   Iodoacetate can modify a number of amino acid side chains in proteins, forming alkylated derivatives containing a terminal carboxylate.

carboxymethyllysine, or the tertiary amine derivative, dicarboxymethyllysine. Methionine thioether groups give the most complicated products, some of which rearrange or decompose unpredictably. The only stable derivative of methionine is where the terminal methyl group is lost to form carboxymethylhomocysteine, the same product as that in the reaction of iodoacetate with homocysteine.

Iodoacetate
MW 185.9

The relative reactivity of α-haloacetates toward protein functionalities is sulfhydryl > imidazolyl > thioether > amine. Among halo derivatives the relative reactivity is I > Br > Cl > F, with fluorine being almost unreactive. The α-haloacetamides have the same trend of relative reactivities, but will obviously not create a carboxylate functional group. The acetamide derivatives typically are used only as blocking reagents.

Thus, iodoacetate has the highest reactivity toward sulfhydryl cysteine residues and may be directed specifically for —SH modification. If iodoacetate is present in limiting quantities (relative to the number of sulfhydryl groups present) and at slightly alkaline pH, cysteine modification will be the exclusive reaction. The specificity of this modification has been used in the design of heterobifunctional cross-linking reagents, where one end of the cross-linker contains an iodoacetamide derivative and the other end contains a different functional group directed at another chemical target (see SIAB, Chapter 5, Section 1.5).

Protocol

1. Dissolve the sulfhydryl-containing protein or macromolecule to be modified at a concentration of 1–10 mg/ml in 50 mM Tris, 0.15 M NaCl, 5 mM EDTA, pH 8.5. EDTA is present to prevent metal-catalyzed oxidation of sulfhydryl groups. The presence of Tris, an amine containing buffer, should not affect the efficiency of sulfhydryl modification. Not only do amines generally react slower than sulfhydryls, the amine in Tris buffer is of particularly low reactivity. If Tris does pose a problem, however, use 0.1 M sodium phosphate, 0.15 M NaCl, 5 mM EDTA, pH 8.

2. Add iodacetate to a concentration of 50 mM in the reaction solution. Alternatively, add a quantity of iodoacetate representing a 10-fold molar excess relative to the number of —SH groups present. An estimation of the sulfhydryl content in the protein to be modified can be accomplished by performing an Ellman's assay (Section 4.1). Readjust the pH if necessary. To aid in adding a small quantity of iodoacetic acid to the reaction, a concentrated stock solution may be made in the reaction buffer, the pH readjusted, and an aliquot added to the protein solution to give the desired concentration.

3. Mix and react for 2 h at room temperature.

4. Purify the modified protein from excess iodoacetate by dialysis or gel filtration.

5. An Ellman's assay comparing the unmodified protein to the iodoacetylated protein may be done to assess the degree of modification.

Modification of Hydroxyls with Chloroacetic Acid

Chloroacetic acid can be used to transform a rather unreactive hydroxyl into a carboxylate group that can be used in a variety of conjugation reactions. The reaction proceeds under basic conditions, yielding a stable ether bond terminating in a carboxymethyl group (Fig. 80) (Plotz and Rifai, 1982; Brunswick *et al.*, 1988). Side reactions will occur with other nucleophiles, such as amines, if they are present in the molecule to be modified. The reagent is used most often to modify pure polysaccharides or hydroxyl-containing polymers that contain no other functional groups.

The following protocol illustrates the modification of a dextran polymer with chloroacetic acid.

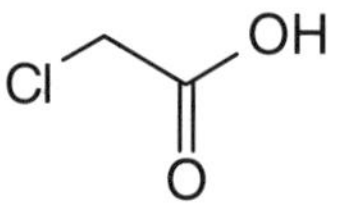

Chloroacetic Acid
MW 94.47

Protocol

1. In a fume hood, prepare a solution consisting of 1 M chloroacetic acid in 3 M NaOH.
2. Immediately add dextran polymer to a final concentration of 40 mg/ml. Mix well to dissolve.
3. React for 70 min at room temperature with stirring.
4. Stop the reaction by adding 4 mg/ml of solid NaH_2PO_4 and adjusting the pH to neutral with 6 N HCl.
5. Remove excess reactants by dialysis.

4.3. Introduction of Primary Amine Groups

Primary amine groups on proteins consisting of N-terminal α-amines and lysine side-chain ε-amines are typically present in abundant quantities for modification or conjugation reactions. Occasionally, however, a protein or peptide will not contain sufficient amounts of available amines to allow for an efficient degree of coupling to another molecule or protein. For instance, horseradish peroxidase (HRP), a popular enzyme to employ in the preparation of antibody conjugates, only possesses two free amines that

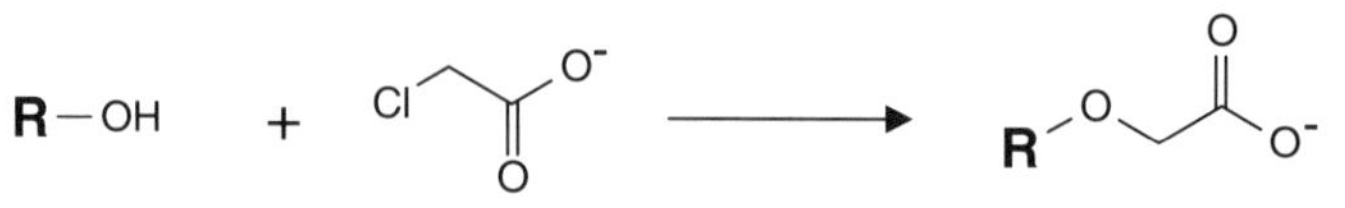

R—OH   +   Chloroacetic Acid   →   Ether Bond Formation with Conversion to Carboxylate

Hydroxyl Containing Compound

Figure 80   Chloroacetic acid can be used to create a carboxylate group from a hydroxyl.

can participate in conjugation protocols. Creating additional amines on HRP allows for higher amounts of modification and thus produces more active conjugates.

Other nonprotein molecules, such as nucleic acids and oligonucleotides, may not normally possess primary amines of sufficient nucleophilicity to react with common modification reagents. The ability to add amine functional groups to these molecules is sometimes the only route to successful conjugation. Creating amines at specific sites within these molecules allows for site-directed modification at known positions, thus better ensuring active conjugates once formed.

The following reagents and techniques can be used to transform directly carboxylates or sulfhydryl groups into reactive amine functional groups. In addition, sugars, polysaccharides, or carbohydrate-containing macromolecules may be modified to contain amines after mild periodate activation to form aldehyde groups.

Modification of Carboxylates with Diamines

Carboxylic acids may be covalently modified with short compounds containing primary amines at either end to form amide linkages. The result of such alterations is to block the carboxylates and form terminal amino groups. Reacting the diamine in excess ensures that only one end of the compound couples to each carboxylate and does not cross-link the molecule being modified. Amide bond formation may be accomplished by several methods including carbodiimide mediated coupling (Chapter 2, Section 1.11), active ester intermediates such as N-hydroxysuccinimide esters (Chapter 2, Section 1.4), and the use of carbonylating compounds like N,N'-carbonyldiimidazole (Chapter 2, Section 3.2). A combination of the water-soluble carbodiimide EDC and sulfo-NHS also is an efficient way of creating amide linkages (Chapter 2, Section 1.11).

Diamines that can be used for aminoalkylation include ethylene diamine, 1,3-diaminopropane, 3,3'-iminobispropylamine (also known as diaminodipropylamine), 1,6-diaminohexane, and the short-chain Jeffamine derivative EDR-148 containing a hydrophilic, polyether, 10-atom chain (Texaco Chemical Co.). Ethylene diamine is perhaps the most popular choice for protein carboxylate modification. Its short chain length ensures minimal steric effects and virtually no hydrophobic interactions. Diaminodipropylamine provides a longer spacer arm and has been used extensively as a bridging molecule for coupling carboxylate containing ligands to insoluble supports (Hermanson et al., 1992). The long hydrocarbon chain of 1,6-diaminohexane, however, may induce hydrophobic effects and probably should be avoided. The longest diamine of the group is the Jeffamine compound. Its chain is extremely hydrophilic and should function as an excellent modifier of carboxylates when a longer spacer is desired.

Diamine modification of proteins can have dramatic effects on the net charge of the molecule, usually significantly raising the pI from the native state. The amide linkage eliminates the negative potential of the carboxylate and the terminal amine adds a positive charge. Thus, diamine modification has a net effect of changing the overall charge by plus two for every carboxylate residue coupled. Heavily modified proteins may exhibit vital changes in activity due to the alteration of microenvironmental charge at each site of modification. In some cases, native conformation may be changed and activity completely lost.

Raising the pI of macromolecules also can significantly alter the immune response

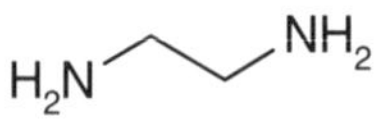

$$H_2N \diagdown\diagup NH_2$$

Ethylenediamine
MW 60

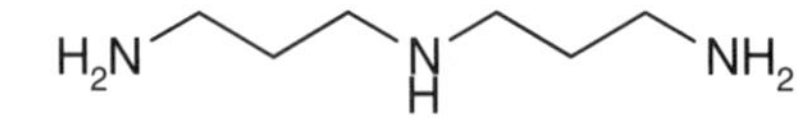

$$H_2N \diagup\diagdown N(H) \diagup\diagdown NH_2$$

3,3'-Iminobispropylamine
(diaminodipropylamine)
MW 131

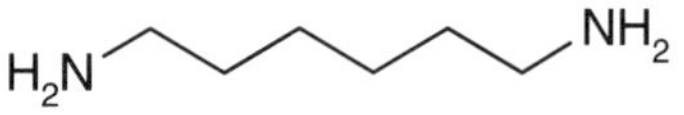

$$H_2N \diagdown\diagup\diagdown\diagup NH_2$$

1,6-Diaminohexane
MW 116

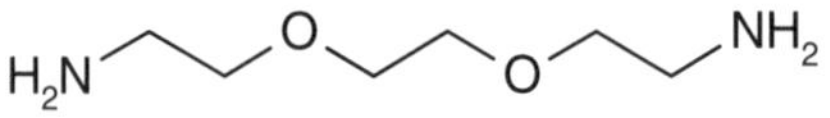

$$H_2N \diagdown O \diagup\diagdown O \diagup NH_2$$

Jeffamine EDR-148
MW 148

toward them on *in vivo* administration. Cationized proteins (those modified with diamines to increase their net charge or pI) are known to generate an increased immune response compared to their native forms (Muckerheide *et al.*, 1987a, Domen *et al.*, 1987; Apple *et al.*, 1988). The use of cationized BSA as a carrier protein for hapten conjugation can result in a dramatically higher antibody response toward a coupled hapten (Chapter 9, Section 2.1).

The following protocol using the carbodiimide EDC is an efficient way of modifying protein carboxylates with diamines either to increase the amount of amines present for further conjugation or to create a cationized protein having an increased net charge (Fig. 81). Note that glycoproteins containing sialic acid may be modified at this sugar's —COOH group in addition to coupling at C-terminal, aspartic acid, and glutamic acid functions on the polypeptide chain. Other carboxylate containing macromolecules may be modified using this procedure as well.

Protocol

1. Dissolve the protein to be modified at a concentration of 1–10 mg/ml in 0.1 M MES, pH 4.7 (coupling buffer). Other buffers may be used as long as they do not contain groups that can participate in the carbodiimide reaction. Avoid carboxylate- or amine-containing buffers such as citrate, acetate, glycine, or Tris. Higher pH conditions may be used up to about pH 7.5 (in sodium phosphate buffer) without severely affecting the yield of modification. The protein in solid form also may be added directly to the diamine solution prepared in (2).
2. Dissolve the diamine chosen for modification at a concentration of 1 M made up in the coupling buffer. If a free-base form of diamine is used, then the solution will become highly alkaline on dissolution. This operation also will generate

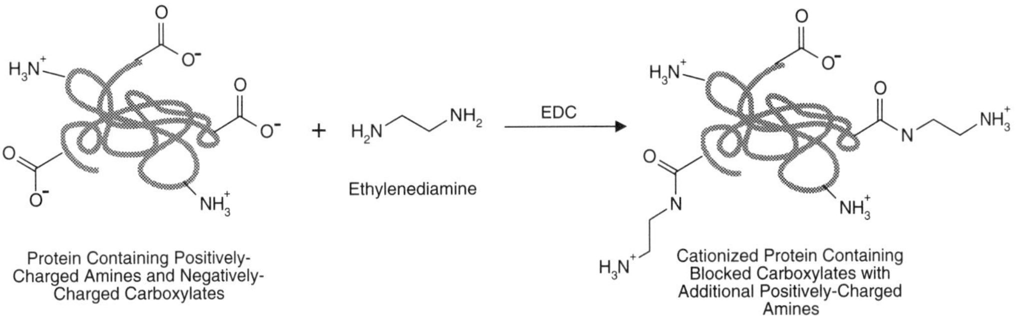

Figure 81 Cationization of protein molecules can be done using ethylene diamine to modify carboxylate groups using a carbodiimide reaction process.

heat—the solution process being highly exothermic. The easiest way to dissolve such a diamine is to add initially the correct amount to a beaker containing a quantity of crushed ice equal to the final solution volume desired. The ice should be made from deionized water or the equivalent to maintain purity. All operations should be done in a fume hood. Next, add an equivalent weight of concentrated HCl and mix. As the mixing becomes complete, the ice will almost totally melt and provide nearly the correct final solution volume. Finally, add an amount of MES buffer salt to bring its concentration to 0.1 M and adjust the solution pH to 4.7. In some cases, the dihydrochloride form of the diamine is commercially available and can be used to avoid such unpleasant pH adjustments. For instance, ethylenediamine dihydrochloride is available from Aldrich. It can be added to the 0.1 M MES buffer without a significant change in pH

3. Add the protein solution to an equal volume of diamine solution and mix. Alternatively, the solid protein can be dissolved directly in the diamine solution at the indicated concentration.
4. Add EDC (1-ethyl-3-(3-dimethylaminopropyl)carbodiimide hydrochloride; Pierce) to a final concentration of 2 mg/ml in the reaction solution. To aid in the addition of a small amount of EDC, a higher concentration stock solution may be prepared in water and an aliquot added to the reaction to give the proper concentration. Since EDC is labile in aqueous solutions, the stock solution must be made quickly and used immediately.
5. React for 1–2 h at room temperature.
6. Purify the modified protein by extensive dialysis against 0.02 M sodium phosphate, 0.15 M NaCl, pH 7.4 (PBS), or another suitable buffer.

The changes that occur in the pI of a protein modified with diamines may be assessed by isoelectric focusing or by general electrophoresis based on relative migration due to charge. A cationized protein will possess a higher pI value or migrate further toward the anode than its native form. Using the above protocol typically alters the net charge of bovine serum albumin from a native pI of 4.9 to the highly basic range of pI 9.5 to over pI 11.

Modification of carboxylate groups with diamines also may be done in organic solvent for those molecules insoluble in aqueous buffers. Some peptides are quite soluble in solvents such as DMF and DMSO, but relatively insoluble in water. Such molecules may be reacted in these solvents using the carbodiimide DCC using the same basic reactant ratios as given above for EDC in aqueous solutions (Chapter 3, Section 1.4).

Modification of Sulfhydryls with N-(β-Iodoethyl)trifluoroacetamide (Aminoethyl-8)

The conversion of sulfhydryl groups on cysteine residues or other molecules to amine-containing groups may be accomplished by aminoethylation with N-(β-iodoethyl) trifluoroacetamide (Schwartz *et al.*, 1980). The haloalkyl group specifically reacts with sulfhydryls to form the aminoalkyl derivative in one step. Under the conditions of the reaction, the trifluoroacetate amine-protecting group spontaneously hydrolyzes to expose the free primary amine without the need for a secondary deblocking step (Fig. 82). This reagent is commercially available from Pierce Chemical under the name Aminoethyl-8.

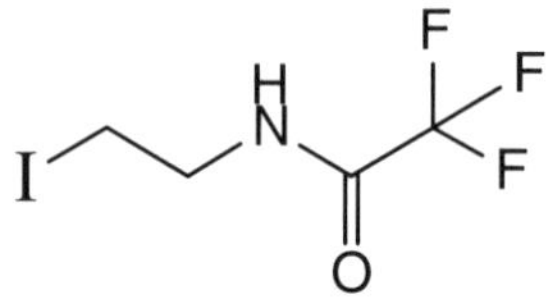

Aminoethyl–8™ Reagent
N-(iodoethyl)trifluoroacetamide
MW 267

Aminoethyl-8 has an advantage over ethylenimine modification (next section), due to the potential polymerization of ethylenimine in aqueous solutions. Such polymers are highly cationic and may nonspecifically block protein. The specificity of Amino-ethyl-8 for sulfhydryls makes it an optimum choice for modification.

For small molecules containing sulfhydryls or for low-molecular-weight peptides containing cysteine residues, modification may proceed without deforming agents. However, for intact proteins containing both disulfides and free sulfhydryls, a denaturant and a disulfide reducing agent may be required to open buried or structurally inaccessible groups if complete modification is desired.

Protocol

1. Dissolve the protein to be aminoalkylated at a concentration of 1–10 mg/ml in 6 M guanidine hydrochloride, 0.2 M N-ethylmorpholine acetate, pH 8.1. All water used in preparing buffers should be deoxygenated by boiling followed by cooling and bubbling with nitrogen. Small molecules that do not require denaturants to expose internal disulfides or sulfhydryls may be modified without using guanidine treatment.
2. Add dithiothreitol (DTT) to obtain a 20-fold molar excess over the amount of disulfides present.
3. React for 4 h at room temperature, maintaining a blanket of nitrogen over the solution.
4. Adjust the pH to 8.6 with NaOH, and heat the solution to 50°C.

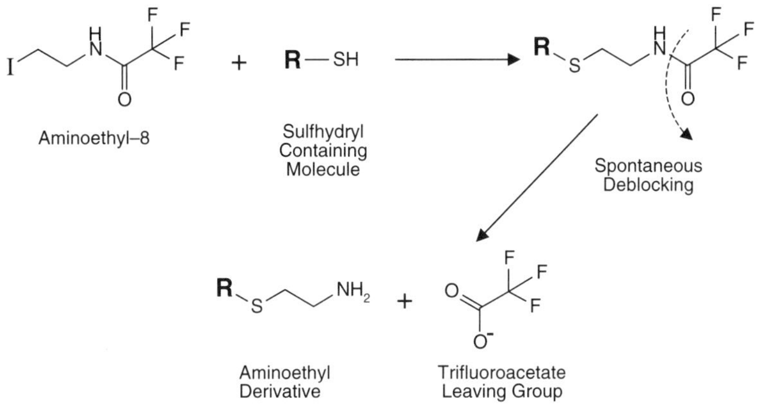

Figure 82 Aminoethyl-8 can be used to transform a sulfhydryl group into an amine. The intermediate spontaneously undergoes deblocking to release the primary amine group.

5. Add a quantity of Aminoethyl-8 in methanol to equal a 25-fold molar excess over the amount of sulfhydryl present (including the amount of DTT added). The solution in methanol should be made concentrated enough so only a small amount of methanol must be added to the reaction solution (i.e., no more than 10% of the final volume). A second addition of modifying agent may be made after 1 h to drive the reaction more completely toward total —SH aminoalkylation.

6. React for 3 h at 50°C.

7. Purify the modified protein or other macromolecule by gel filtration or dialysis. Occasionally, complete modification with Aminoethyl-8 will cause precipitation of the protein.

Modification of Sulfhydryls with Ethylenimine

The cyclic compound ethylenimine reacts with protein sulfhydryl groups causing ring opening and forming the aminoalkyl derivative, S-(2-aminoethyl)cysteine (Raftery and Cole, 1963, 1966) (Fig. 83). Under physiological conditions ethylenimine is virtually specific for sulfhydryls with no cross-reactivity toward other protein functional groups. At acid pH, a small degree of reactivity occurs with methionine residues, forming S-(2-aminoethyl)methionine sulfonium ion (Schroeder *et al.*, 1967). Since aminoethylated cysteine groups resemble the side-chain structure of lysine residues, except for the replacement of one methylene group with a thioether, these modifications make them susceptible to tryptic hydrolysis, although at an abbreviated rate (Plapp *et al.*, 1967; Wang and Carpenter, 1968).

Ethylenimine
MW 43

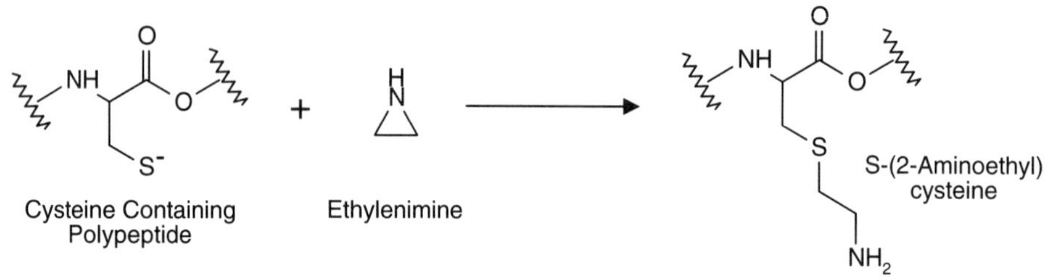

Figure 83   The small compound ethylenimine can react with sulfhydryls to form aminoethyl derivatives.

Ethylenimine may be used to introduce additional sites of tryptic cleavage for protein structural studies. In this case, complete sulfhydryl modification is usually desired. Proteins are treated with ethylenimine under denaturing conditions (6–8 *M* guanidine hydrochloride) in the presence of a disulfide reductant to reduce any disulfide bonds before modification. Ethylenimine may be added directly to the reducing solution in excess (similar to the procedure for Aminoethyl-8 described previously) to totally modify the —SH groups formed.

The disadvantage of using ethylenimine for protein modification stems from the fact that in the presence of water, slow formation of polyethylenimine occurs. The polymer is highly positively charged at physiological pH and can interact strongly with protein molecules, masking sites of potential sulfhydryl modification. Also, the polymer may have terminal aziridine residues (Chapter 2, Section 2.3), making it reactive and potentially forming a covalent attachment with the protein (Dermer and Ham, 1969).

Modification of Sulfhydryls with 2-Bromoethylamine

2-Bromoethylamine may undergo two reaction pathways in its modification of sulfhydryl groups in proteins (Fig. 84). In the first scheme, the thiolate anion of cysteine attacks the No. 2 carbon of 2-bromoethylamine to release the halogen and form a thioether bond (Lindley, 1956). This straightforward reaction mechanism is similar to the modification of sulfhydryls with iodoacetate (Section 4.2). In a two-step, secondary

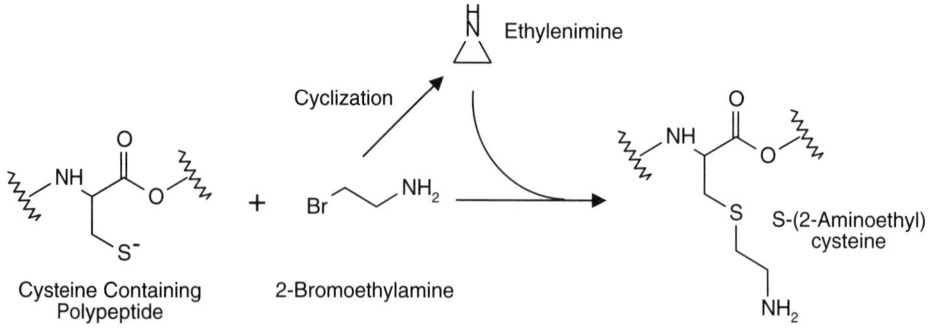

Figure 84   2-Bromoethylamine can be used to transform a thiol into an amine. The reaction may proceed through the intermediate formation of ethylenimine, yielding an aminoethyl derivative.

process, 2-bromoethylamine is converted under alkaline conditions to the cyclic ethylenimine derivative by the intramolecular attack of its primary amine on the number 2
carbon, causing release of the halogen and ring formation (Cole, 1967). Ethylenimine
then goes on to react with the sulfhydryl to form the aminoalkylated derivative (as
described in the previous section). The two-step reaction is slower than direct aminoalkylation by either 2-bromoethylamine or ethylenimine.

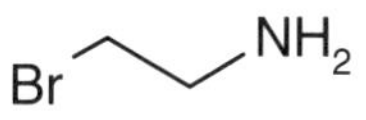

2-Bromoethylamine
MW 123.92

Protocol

1. Dissolve the protein or peptide to be aminoalkylated at cysteine sulfhydryls in
   0.5 M sodium carbonate. If cystine disulfides are present, add a 10- to 25-fold
   molar excess of DTT to reduce them fully to free sulfhydryls.
2. Add a quantity of 2-bromoethylamine to obtain a 10-fold molar excess over the
   number of sulfhydryls present in the sample, including any added DTT.
3. React overnight at room temperature.
4. Purify the modified protein by gel filtration or dialysis.

Modification of Carbohydrates with Diamines

Carbohydrates or oligosaccharides may be modified to contain primary amino groups
by selective reaction with a diamine compound. Several reaction pathways may be used
to accomplish this modification. In some cases, a particular carbohydrate may contain
sugar residues that possess potential amine coupling groups without prior derivatization to form such functional groups. For example, if carboxylate containing sugars are
present like sialic or uronic acid (Fig. 85), then direct modification with a diamine is
possible using the carbodiimide coupling protocol described previously in this section.
    If carboxylates are lacking in the carbohydrate molecule, then indigenous hy-

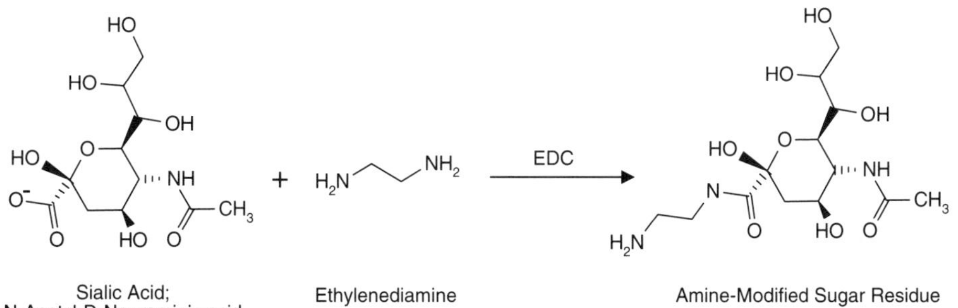

Sialic Acid;   Ethylenediamine   Amine-Modified Sugar Residue
N-Acetyl-D-Neuraminic acid

Figure 85   Carboxylate-containing sugars may be modified with diamines using a carbodiimide-mediated
reaction to create available amine groups for subsequent conjugation.

droxyls may be utilized to create aldehydes for coupling diamines by one of two routes. The simplest method of creating amine reactive groups in sugar molecules is by oxidation using sodium periodate (Section 4.4). Periodic acid cleaves adjacent hydroxyls to form highly reactive aldehyde groups (Rothfus and Smith, 1963). At a concentration of 1 mM sodium periodate specifically cleaves only at the adjacent hydroxyls between the No. 7, 8, and 9 carbon atoms of sialic acid residues (Van Lenten and Ashwell, 1971; Wilchek and Bayer, 1987). The product is the formation of one aldehyde group on the No. 7 carbon and liberation of two molecules of formaldehyde. The sialic acid aldehyde then can be coupled with diamines by Schiff base formation and reductive amination (Chapter 2, Section 5.3 and Chapter 3, Section 4).

Oxidation of polysaccharides using 10 mM or greater concentrations of sodium periodate results in the cleavage of adjacent diol containing carbon—carbon bonds on other sugars besides just sialic acid residues. Glycoproteins and polysaccharides may be modified using this procedure to form multiple formyl functional groups for coupling diamines or other amine containing molecules.

In some instances, reducing sugars are present that can be reductively aminated without prior periodate treatment. A reducing end of a monosaccharide, a disaccharide, or a polysaccharide chain may be coupled to a diamine by reductive amination to yield an aminoalkyl derivative bound by a secondary amine linkage (Fig. 86).

An alternative to the use of chemical means to create formyl groups is the specific modification afforded by sugar oxidasaes (Section 4.4). For instance, galactose oxidase may be reacted with a carbohydrate containing terminal D-galactose or N-acetyl-D-galactosamine residues to transform the C-6 hydroxyl group into an aldehyde (Avigad et al., 1962). Subsequent reaction with a diamine yields the desired amine modification.

The appropriate protocols for diamine modification of various carbohydrate or glycoprotein derivatives may be found in the indicated sections.

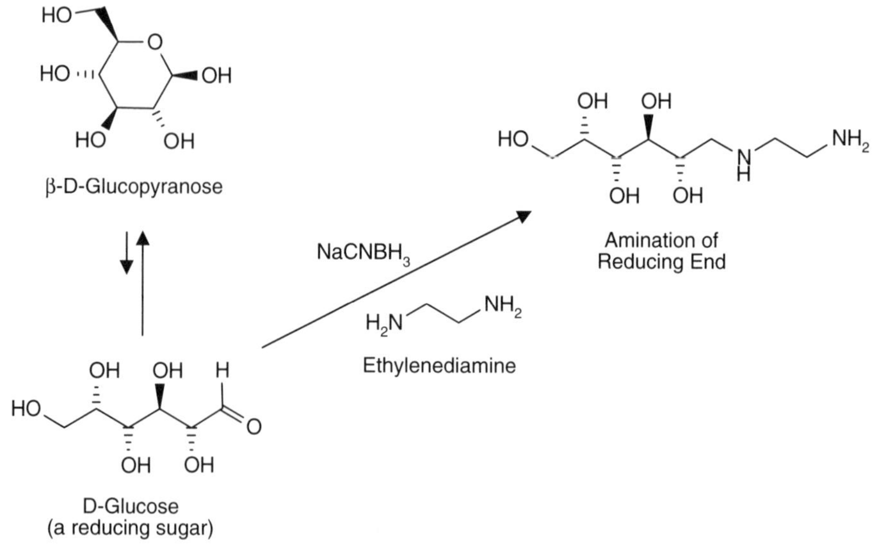

Figure 86   Reducing sugars may be aminated with diamines in the presence of sodium cyanoborohydride to produce amine modifications.

Modification of Alkylphosphates with Diamines

Alkylphosphate groups can be made to react with diamines to form aminoalkylphosphoramidate modifications. The primary amine thus formed then may be used to conjugate with other molecules containing amine reactive groups. In this sense, DNA or RNA may be modified with a diamine at the 5′ phosphate group mediated by a carbodiimide reaction. N-substituted carbodiimides can react with phosphate groups to form highly reactive phosphodiester derivatives that are extremely short-lived in aqueous solution (Chapter 3, Section 1) (Fig. 87). This active species then can react with a nucleophile such as a primary amine to form a phosphoramidate bond (Chu *et al.*, 1986). The process is analogous to the activation of a carboxylate by a carbodiimide with subsequent coupling to an amine-containing molecule to form an amide linkage (Williams and Ibrahim, 1981).

In most procedures, the water-soluble carbodiimide EDC (1-ethyl-3-(3-dimethylaminopropyl)carbodiimide hydrochloride) is the most effective mediator of this reaction. Both EDC and its reaction by-products are fully soluble in aqueous buffers and can be easily separated from the modified aminoalkylphosphate (Chapter 3, Section 1.1).

In some methods, the reaction is carried out in a two-step process by first forming an intermediate, reactive phosphorylimidazolide by EDC conjugation in an imidazole buffer. Next, the diamine, in this case cystamine, is reacted with the activated oligonucleotide, causing the imidazole to be replaced by the amine and creating a phosphoramidate linkage (Chu *et al.*, 1986). An easier protocol was described by Ghosh *et al.*, (1990) in which the oligo, cystamine, and EDC were all reacted together in an imidazole buffer. A modification of this method developed by Zanocco *et al.* (1993) is described in Chapter 17, Section 2.1.

Modification of Aldehydes with Ammonia or Diamines

Aldehyde groups can be converted into terminal amines by a reductive amination process with ammonia or a diamine compound. The reaction proceeds by initial formation of a Schiff base interaction—a dehydration step yielding an imine derivative. Reduction of the Schiff base with sodium cyanoborohydride or sodium borohydride produces the primary amine (in the case of ammonia) or a secondary amine derivative terminating in a primary amine (for a diamine compound) (Fig. 88).

This simple strategy can be used to add amine residues to polysaccharide molecules after formation of aldehydes by periodate or enzymatic oxidation (Section 4.4). Thus,

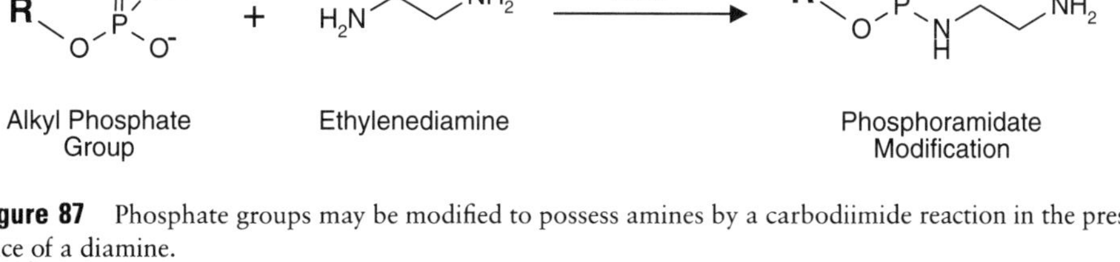

Alkyl Phosphate
Group

Ethylenediamine

Phosphoramidate
Modification

Figure 87 Phosphate groups may be modified to possess amines by a carbodiimide reaction in the presence of a diamine.

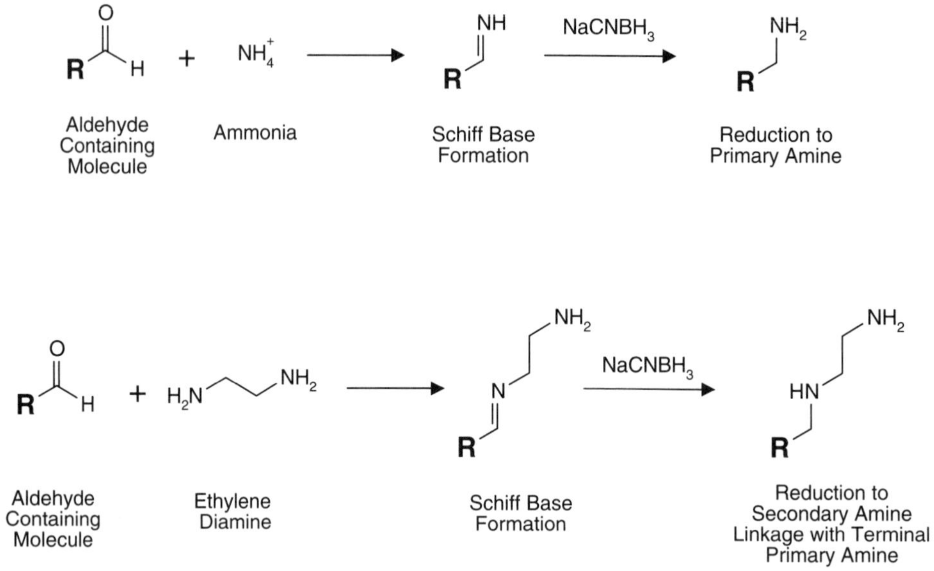

Figure 88 Aldehydes may be transformed into primary amines by reaction with ammonia or a diamine in the presence of a reducing agent.

glycoconjugates or carbohydrate polymers such as dextran may be derivatized to contain amines for further conjugation reactions.

The reaction occurs rapidly at alkaline pH (7–10), with higher pH values resulting in better yields due to faster Schiff base formation. To ensure complete conversion of available aldehydes to amines, add the ammonia or diamine compound to the reaction in at least a 10-fold molar excess over the expected number of formyl groups present, Diamines that are commonly used for this process include ethylene diamine, diamino-dipropylamine (3,3′-iminobispropylamine), 1,6-diaminohexane, and the Jeffamine derivative EDR-148 containing a hydrophilic, 10-atom chain (Texaco Chemical Co.).

Introduction of Arylamines on Phenolic Compounds

Compounds having phenol ring structures, such as tyrosine residues in proteins, often can be derivatized to contain aromatic amine groups through a two-stage reaction process. First, the phenolic ring is nitrated with tetranitromethane in aqueous solution to add a nitro group *ortho* (or *para*, if available) to the hydroxyl. This type of modification can be used to detect tyrosine residues by the strong absorptivity of the unprotonated (at pH 9), 3-nitrophenolate ring at 428 nm (extinction coefficient = 4200 $M^{-1}cm^{-1}$) (Sokolovsky *et al.*, 1967). The method has been used to quantify the tyrosine content in porcine trypsinogens and trypsins and to modify a variety of other proteins (Vincent *et al.*, 1970; Lundbald, 1991).

The nitrophenol group also may be reduced to an aminophenyl derivative in alkaline conditions with the use of sodium dithionite ($Na_2S_2O_4$). The amine then can be used to conjugate with an amine-reactive cross-linking reagent to label peptides or proteins at their tyrosine side chains. In addition, this strategy can be a route to creating modifiable amine groups on aromatic molecules other than just tyrosine. For instance, the Bolton–Hunter reagent (Chapter 8, Section 4.5) can be used to modify amine groups on proteins, leaving a phenolic end that is typically used as a site for radioiodination. However, such a derivative also could be used to create an arylamine for further transformation into a highly reactive diazonium group for coupling to tyrosines or phenolic functional groups in other molecules (Fig. 89) (Chapter 2, Section 6.1).

Protocol

1. Dissolve the protein-containing tyrosine residues (or another phenolic macro-molecule) in 0.02 M sodium phosphate, 0.15 M NaCl, pH 7.4, at a concentration of 2–4 mg/ml.
2. With stirring, add to each milliliter of the protein solution, 20 μl of 0.15 M tetranitromethane in 95% ethanol (Sigma). Make the addition in small aliquots if more than several milliliters of solution are to be derivatized.
3. React for 1 h at room temperature.
4. Quench the reaction by immediate gel filtration using a column of Sephadex G-25 (Pharmacia). Equilibrate the column and perform the chromatography using 0.2 M sodium borate, pH 9, so that the protein will be at the proper pH for the reduction step. After the separation, a determination of the modification level may be done by measuring its absorbance at 428 nm.

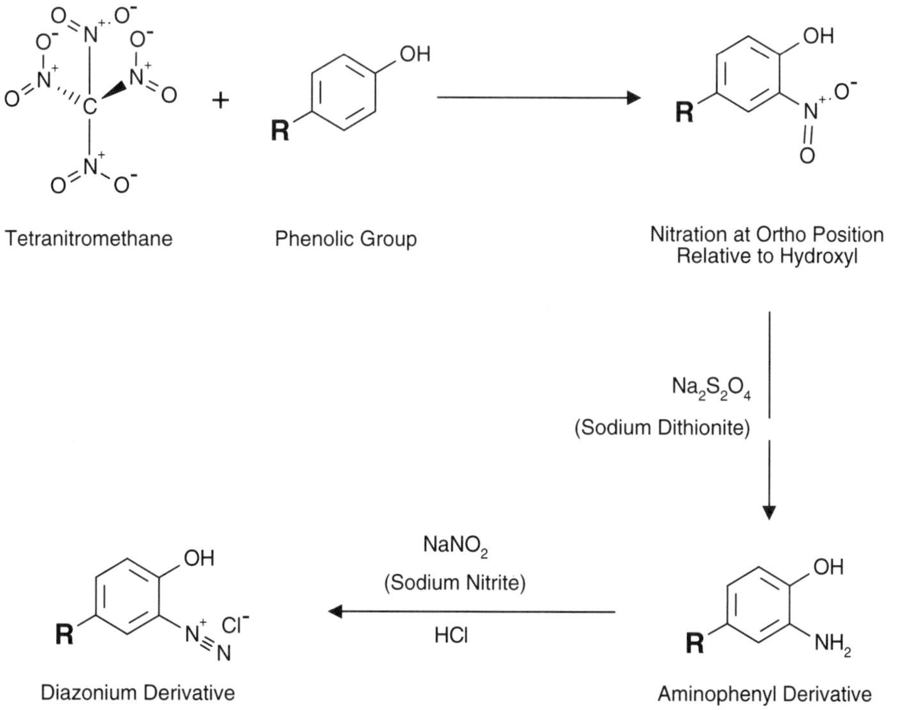

Figure 89 Phenolic compounds, such as the side chain of tyrosine residues, may be modified to contain an amine group by nitration followed by reduction to the aminophenyl derivative.

5. Add sufficient sodium dithionite to bring the final concentration in the reaction medium to 0.1 M.
6. React for 1 h at room temperature.
7. Purify the aminophenyl derivative by gel filtration or dialysis

The formation of a diazonium group from the arylamine derivative can be done by treatment with sodium nitrate in HCl (see protocol in Chapter 9, Section 6.1).

Amine Detection Reagents

There are several methods available for the detection or measurement of amine groups in proteins and other molecules. Accurate determination of target amine groups in molecules before or after modification may be important for assessing reaction yield or suitability for subsequent cross-linking procedures. The following methods use commercially available reagents and are easily employed to detect primary amines with simple spectrophotometric measurement.

TNBS

Molecules containing primary amines or hydrazide groups can react with 2,4,6-trinitrobenzenesulfonate (TNBS) to form a highly chromogenic derivative (Fig. 90). This reaction may be used to assay the amine content of compounds by measuring the absorbance of the orange-colored product at 335 nm.

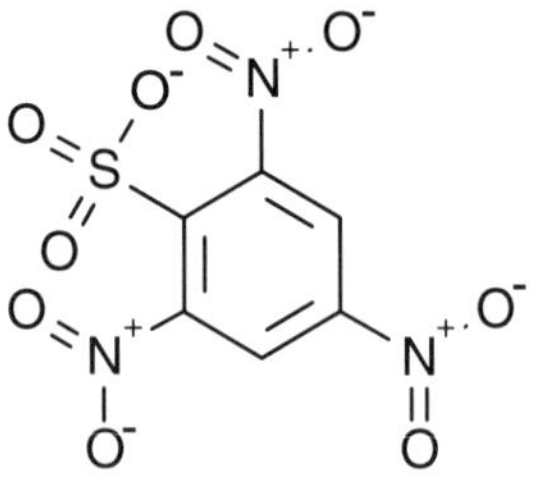

TNBS;
Trinitrobenzene sulfonic acid
MW 293

TNBS has been used to measure the free amino groups in proteins (Habeeb, 1966), as a qualitative check for the presence of amines, sulfhydryls, or hydrazides (Inman and Dintzis, 1969), and to determine specifically the number of ε-amino groups of L-lysine in carrier proteins (Sashidhar *et al.*, 1994).

The following protocol may be used for the measurement of amines in soluble molecules, such as proteins or other macromolecules.

Protocol

1. Dissolve or dialyze the molecule to be assayed into 0.1 M sodium bicarbonate, pH 8.5, at a concentration of 20–200 μg/ml (for large molecules like proteins) or 2–20 μg/ml (for small molecules like amino acids).
2. Dissolve TNBS in 0.1 M sodium bicarbonate, pH 8.5, at a concentration of 0.01% (w/v). Prepare fresh. Note: TNBS may be prepared as a stock solution in ethanol at a concentration of 1.5%. This solution is stable to long-term storage

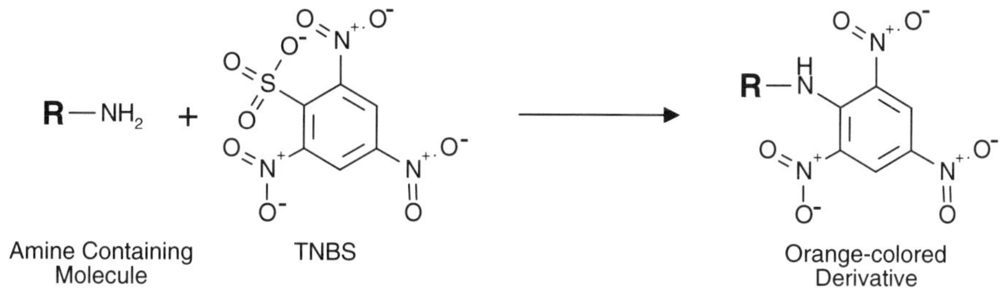

Figure 90 TNBS may be used to detect or quantify amine groups through the production of a chromogenic derivative.

and may be diluted as needed in the bicarbonate buffer to the required concentration.

3. Add 0.5 ml of TNBS solution to 1 ml of each sample solution. Mix well.
4. Incubate at 37°C for 2 h.
5. Add 0.5 ml of 10% SDS and 0.25 ml of 1 N HCl to each sample.
6. Measure the absorbance of the solutions at 335 nm. Determination of the number of amines present in a particular sample may be done by comparison to a standard curve generated by use of an amine containing compound (i.e., an amino acid) dissolved at a series of known concentrations in the bicarbonate sample buffer and assayed under identical conditions.

OPA

O-Phthaldialdehyde (OPA) is an amine detection reagent that reacts in the presence of 2-mercaptoethanol to generate a fluorescent product (for preparation, see Section 4.1, 2-mercaptoethanol) (Fig. 91). The resultant fluorophore has an excitation wavelength of 360 nm and an emission point at 455 nm. OPA can be used as a sensitive detection reagent for the HPLC separation of amino acids, peptides, and proteins (Fried *et al.,* 1985). It is also possible to measure the amine content in proteins and other molecules using a test tube or microplate format assay with OPA. Detection limits are typically in the microgram per milliliter range for proteins.

Protocol

1. Prepare a series of standards, preferably consisting of serial dilutions of the substance to be measured, dissolved in water or non-amine-containing buffer.

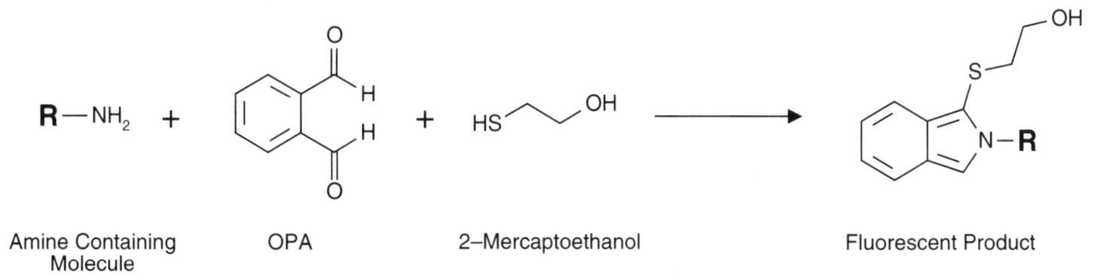

Figure 91 OPA reacts with amines to form a fluorescent product.

The concentration range of the standards can be anywhere between about 500 ng/ml and 1 mg/ml.

2. Prepare the samples dissolved in water or non-amine-containing buffer at an expected concentration level that falls within the standard curve range. The assay can tolerate the presence of most buffer salts, denaturants, and detergents. However, the standard curve should be run in the same buffer environment as the samples to obtain consistent response.

3. To a set of labeled tubes, add 2 ml of OPA reagent (Pierce) and 200 µl of the appropriate standard or sample. Mix well. If using a microplate format, scale back these quantities 10-fold to fit in the microwells.

4. Measure the fluorescence of each sample and standard using an excitation wavelength of 360 nm and an emission wavelength of 436 nm (or using a filter close to the 436 to 455-nm range).

5. Determine the concentration of the samples by comparison to the standard curve. Since the assay measures the presence of amine groups, the results may be correlated to the relative amount of amines available.

4.4. Introduction of Aldehyde Residues

The formation of an aldehyde group on a macromolecule can produce an extremely useful derivative for subsequent modification or conjugation reactions. In their native state, proteins, peptides, nucleic acids, and oligonucleotides contain no naturally occurring aldehyde residues. There are no aldehydes on amino acid side chains, none introduced by post-translational modifications, and no formyl groups on any of the bases or sugars of DNA and RNA. To create reactive aldehydes at specific locations within these molecules opens the possibility of directing modification reactions toward discrete sites within the macromolecule.

There are two basic ways of introducing aldehyde residues in biological macromolecules: (1) oxidation of carbohydrates or adjacent diol containing molecules and (2) modification of available amino groups with reagents that contain or produce aldehydes. In both cases, aldehydes can be created that will allow easy conjugation to amine containing molecules by Schiff base formation and reductive amination (Chapter 2, Section 5.3 and Chapter 3, Section 4). The following sections describe these methods.

Periodate Oxidation of Glycols and Carbohydrates

Carbohydrates and other biological molecules that contain polysaccharides, such as glycoproteins, can be specifically modified at their sugar residues to produce reactive formyl functionalities. With proteins, this method often allows modification to occur only at specific locales, usually away from critical active centers or binding sites.

Periodate oxidation is perhaps the simplest route to transforming the relatively unreactive hydroxyls of sugar residues into amine reactive aldehydes. Periodate cleaves carbon—carbon bonds that possess adjacent hydroxyls, oxidizing the —OH groups to form highly reactive aldehydes (Bobbit, 1956; Rothfus and Smith, 1963). Terminal *cis*-glycols result in the loss of one carbon atom as formaldehyde and the creation of an aldehyde group on the former No. 2 carbon atom. Varying the concentration of sodium periodate during the oxidation reaction gives some specificity with regard to what sugar residues are modified. Sodium periodate at a concentration of 1 mM at 0°C specifically cleaves only at the adjacent hydroxyls between carbon atoms 7, 8, and 9 of sialic acid residues (Van Lenten and Ashwell, 1971; Wilchek and Bayer, 1987). The product is the formation of one aldehyde group on the No. 7 carbon and liberation of two molecules of formaldehyde (Fig. 92).

Since sialic acid is a frequent terminal sugar constituent of the polysaccharide trees on glycoproteins, this method selectively forms reactive aldehydes on the most accessible parts for subsequent modifications. The carbohydrate polymer of a protein provides a long spacer arm that can be used to conjugate another large macromolecule, such as a second protein, with little steric problems.

Oxidation of polysaccharides using 10 mM or greater concentrations of sodium periodate results in the cleavage of adjacent hydroxyl-containing carbon—carbon bonds on other sugars besides just sialic acid residues (Lotan et al., 1975). High concentrations of periodate result in sugar ring opening and the creation of many aldehydes on each polysaccharide tree.

Using these methods, carbohydrate-containing proteins may be altered to contain aldehydes for conjugation with other proteins or for detection using hydrazide-containing probes (Chapter 13, Section 5). The aldehydes thus formed then can be coupled to other amine-containing molecules by Schiff base formation and reductive amination (chapter 2, Section 5.3 and Chapter 3, Section 4). For instance, the enzyme horseradish peroxidase (HRP) can be activated with periodate for conjugation with antibodies (Nakane and Kawaoi, 1974). Alternatively, such reactive formyl groups may be conjugated to hydrazide-containing molecules to form hydrazone bonds

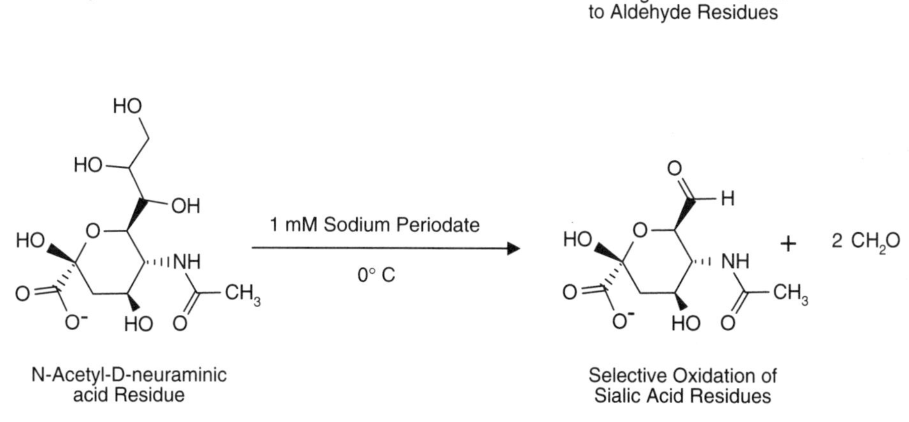

Figure 92   The reaction of sodium periodate with sugar residues can produce aldehydes for conjugation reactions.

(Chapter 4, Section 8, Chapter 10, Section 1.3, and Chapter 17, Section 2.1). Cell surface polysaccharides may be probed with hydrazide-containing reagents for sialic acid groups or total glycoconjugates. Glycoproteins or glycopeptides in solution also may be tagged in this manner. Gangliosides and other glycolipids may be modified with hydrazide reagents as well (Spiegal *et al.*, 1982).

Protocol

1. The glycoprotein or *cis*-diol-containing molecule is dissolved in deionized water or a buffer at physiological pH. Sodium phosphate buffer (0.01–0.1 *M*), pH 7, is an appropriate choice. When oxidizing cell surface glycoconjugates, use a buffer suitable for cellular stability requirements. Avoid amine-containing buffers such as Tris and glycine, because they may interact with the aldehyde groups as they are formed. For glycoproteins in solution, a concentration range of 1–10 mg/ml will produce acceptable results in this procedure. For sialic acid modification, place the sample in ice to cool to near 0°C.
2. Dissolve sodium periodate (MW 213.91) in water at a concentration of 10 mg/ml (0.046 *M*). Protect from light. To obtain approximately a 1 m*M* concentration of sodium periodate in the reaction solution (suitable for oxidizing only sialic acid residues), add 21.8 μl of this stock solution to each milliliter of the glycoprotein solution to be oxidized. Maintain the solution on ice. For general oxidation of carbohydrates other than just sialic acid, add 218 μl of the stock solution to obtain an approximate final concentration of 10 m*M* periodate in the reaction. Use room temperature conditions for general carbohydrate oxidation. Wrap the vial containing the reaction solution with aluminum foil to protect from light. The use of an amber vial is suitable for this purpose.
3. React for 15–30 min at room temperature.
4. Quench the reaction by the addition of 0.1 ml of glycerol per milliliter of reaction solution. Alternatively, the reaction may be stopped by immediate gel filtration on a Sephadex G-25 column. The dextran beads of the chromatography support will react with sodium periodate to quench excess reagent. To quench the reaction with cellular samples, wash the cells with buffer to remove remaining traces of periodate.

Oxidase Modification of Sugar Residues

Another method of forming aldehyde groups on carbohydrates and glycoproteins involves the use of specific sugar oxidases. These enzymes only affect the monosaccharide they are specific toward, leaving other sugar residues within polysaccharides alone. Probably the most often used oxidase for this purpose is galactose oxidase, which can form C-6 aldehydes on terminal D-galactose or *N*-acetyl-D-galactose residues (Avigad *et al.*, 1962) (Fig. 93). When galactose residues are penultimate to sialic acid residues, another enzyme, neuraminidase, must be used to remove the sialic acid sugars and expose galactose as the terminal residue (Wilchek and Bayer, 1987). The specificity of using glycosidases to create aldehyde residues on carbohydrates may be the method's greatest advantage. However, the use of a simple chemical reagent such as sodium periodate still may be the easiest way to create aldehydes on carbohydrates (Section 4.4).

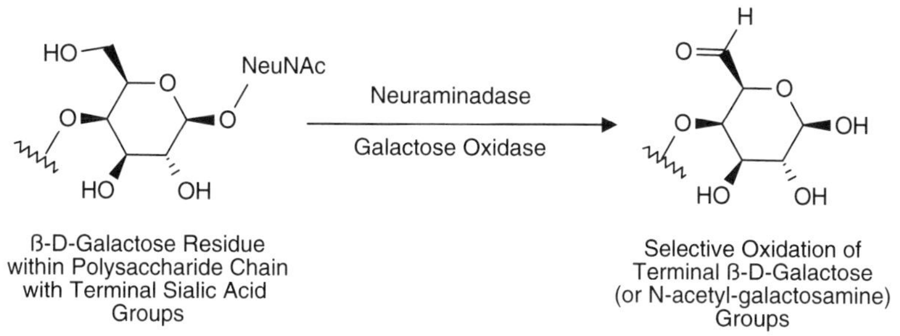

Figure 93 Galactose oxidase may be used to transform specifically the C-6 hydroxyl group of galactose residues into an aldehyde.

The following protocol was used by Wilchek and Bayer (1987) to label cell surface galactose residues.

Protocol

1. Prepare a 5% cell suspension in an appropriate buffer. Avoid amine-containing buffers as these will interact with aldehydes.
2. Add 0.05 units of *Vibrio cholerae* neuraminidase and 5 units of galactose oxidase per milliliter of cell suspension.
3. Incubate for 60 min at 37°C.

Modification of Amines with NHS-Aldehydes (SFB and SFPA)

Succinimidyl *p*-formylbenzoate (SFB) and succinimidyl *p*-formylphenoxyacetate (SFPA) are amine-reactive reagents that contain terminal aldehyde residues. Their NHS ester ends react with primary amines in proteins and other molecules at pH 7–9 to yield amide linkages (Chapter 2, Section 1.4) (Fig. 94.) The resulting formyl derivatives may be utilized to couple to other amine or hydrazide-containing molecules (Galardy *et al.*, 1978; Kraehenbuhl *et al.*, 1974). In particular, SFB can be used to produce aldehyde groups on alkaline phosphatase for conjugation with 5'-hydrazide-modified DNA for use in hybridization assays (Chapter 17, Section 2.4) (Ghosh *et al.*, 1989). SFB and SFPA are insoluble in water, but may be predissolved in DMF or acetonitrile before adding a small quantity to an aqueous reaction mixture. Both reagents contain aromatic phenyl rings and have absorptivity at wavelengths less than 300 nm. Their structures may contribute a significant degree of hydrophobicity to

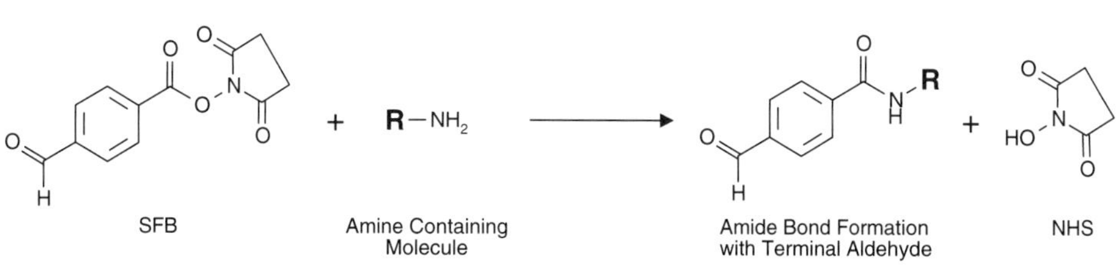

Figure 94 SFB reacts with primary amines to form amide bond derivatives containing aldehyde groups.

macromolecules being modified, especially if high-density couplings are achieved. For this reason, modified proteins and other soluble molecules may have a tendency to precipitate if modification is done too heavily. The optimal amount of modification may have to be adjusted to maintain solubility in each application.

Protocol

1. Dissolve a macromolecule containing amine groups at a concentration of 1–10 mg/ml in a buffer having a pH of 7–9 (i.e., 0.1 M sodium phosphate, pH 7.5). Avoid amine-containing or nucleophilic buffers such as Tris, glycine, or imidazole (see Chapter 2, Section 1.4).

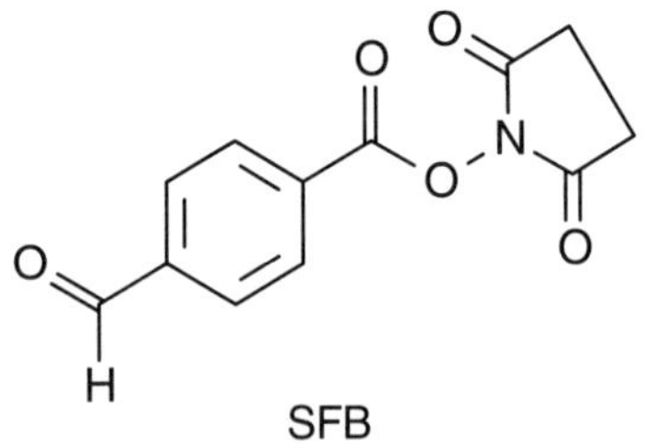

SFB
Succinimidyl-p-formyl benzoate
MW 247

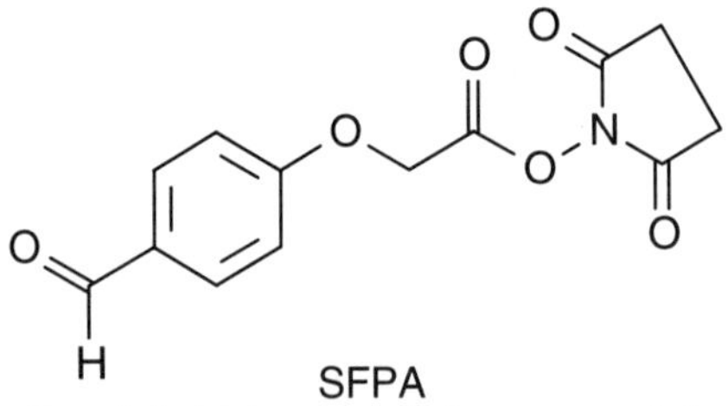

SFPA
Succinimidyl-p-formylphenoxyacetate
MW 277

2. Dissolve SFB or SFPA (Molecular Probes) in DMF. The concentration should be such that a small aliquot can be added to the reaction medium to obtain about a 10-fold excess of modifying reagent over the amount of amines to be modified. Add no more than 100 μl of the modifier/DMF solution to each milliliter of the macromolecule solution prepared in (1).
3. React for 2 h at room temperature.
4. Purify the modified macromolecule from excess reagent and reaction by-products by dialysis or gel filtration.

Modification of Amines with Glutaraldehyde

Amino groups on proteins may be reacted with the bis-aldehyde compound glutaraldehyde to form activated derivatives able to cross-link with other proteins. The reaction mechanism for this modification proceeds by one of several possible routes. In the first option, one of the aldehyde ends can form a Schiff base linkage with ε-amines or α-amines on proteins to leave the other aldehyde terminal free to conjugate with another molecule. Alternatively, a glutaraldehyde polymer may undergo vinyl addition to create stable secondary amine bonds, leaving the aldehydes exposed for subsequent reductive amination reactions. Finally, a cyclized form of glutaraldehyde also may react with the ε-amines of two neighboring lysine side chains to form a quaternary pyridinium cross-link (Fig. 95).

Schiff base interactions between aldehydes and amines typically are not stable enough to form irreversible linkages. These bonds may be reduced with sodium cyanoborohydride or a number of other suitable reductants (Chapter 3, Section 4) to form permanent secondary amine bonds. However, proteins cross-linked by glutaraldehyde without reduction nevertheless show stabilities unexplainable by simple Schiff base

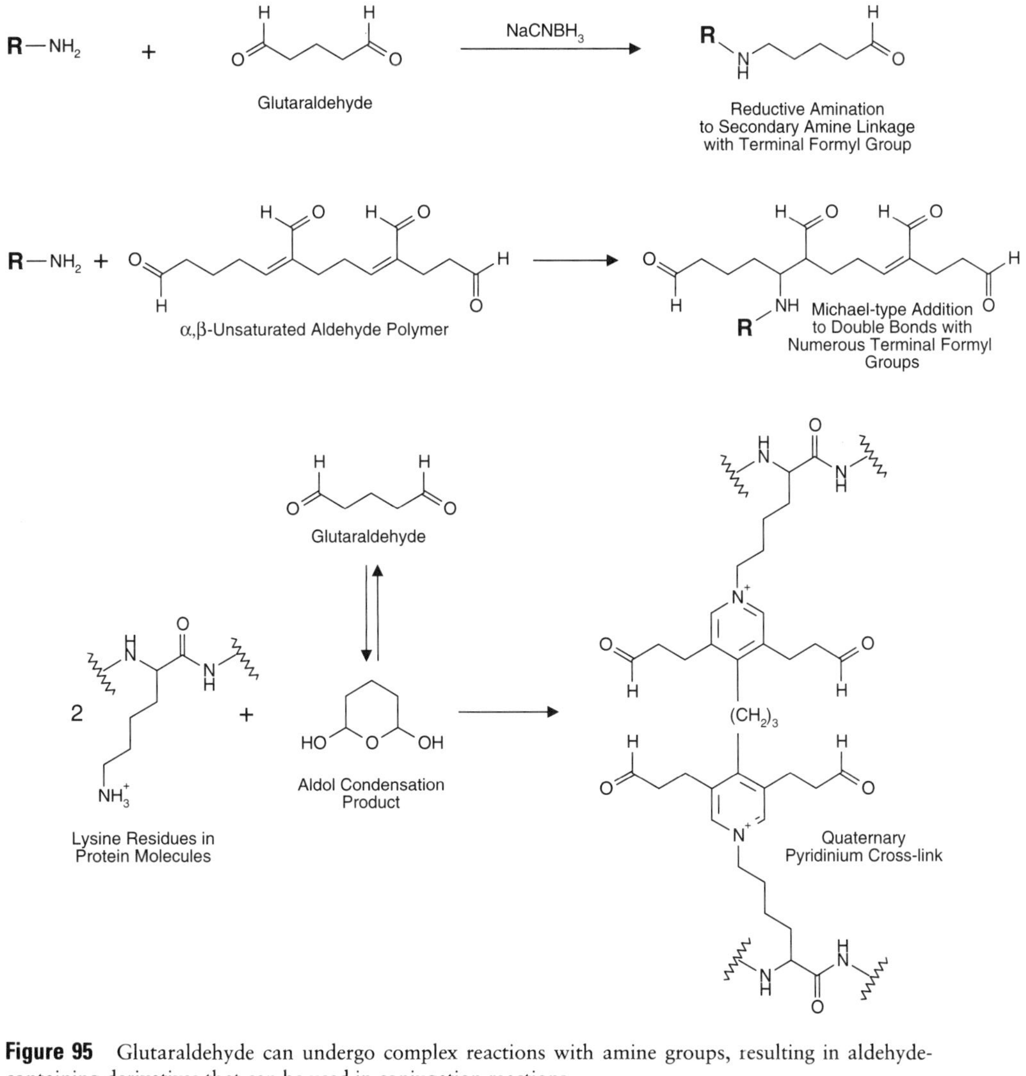

Figure 95 Glutaraldehyde can undergo complex reactions with amine groups, resulting in aldehyde-containing derivatives that can be used in conjugation reactions.

formation. The stability of such unreduced glutaraldehyde conjugates has been postulated to be due to the vinyl addition mechanism, which does not depend on the creation of Schiff bases.

Glutaraldehyde modification readily proceeds at alkaline pH. The higher the pH, the more efficient is Schiff base formation. Using a reductant like sodium cyanoborohydride that does not affect the aldehyde groups, while efficiently transforming the Schiff base into a secondary amine, provides the best possible yields. In many cases, the degree of glutaraldehyde-induced cross-links is so severe that conjugate precipitation occurs. This is especially well documented in antibody–enzyme conjugation schemes employing this reagent (Chapter 10, Section 1.2).

Glutaraldehyde also can be used to create aldehydes on amine-containing polymers. The use of this reagent in derivatizing chromatography supports and other soluble polymers is well known (Hermanson *et al.*, 1992).

The following protocol may be used as the first stage of a two-step glutaraldehyde conjugation reaction. In this initial reaction, glutaraldehyde modification converts available protein amines into reactive formyl groups. The subsequent addition of a second protein or another amine-containing molecule causes this activated protein to cross-link with the amines and form a conjugate. Glutaraldehyde also may be used in single-step conjugation procedures where the aldehyde-modified protein is not isolated before addition of a second protein. In single-step conjugations both proteins to be cross-linked are together in solution and glutaraldehyde is added to effect cross-linking (Chapter 10, Section 1.2).

Protocol

1. Dissolve the protein or other amine-containing macromolecule to be modified at a concentration of 1–10 mg/ml in a buffer having a pH from 7 to 10. The higher the pH, the more efficiently Schiff base formation will occur. Phosphate, borate, and carbonate buffers at 0.01–0.1 *M* are acceptable. Avoid amine-containing buffers like Tris and glycine, since they will react with glutaraldehyde.

2. Add a quantity of glutaraldehyde equal to a 10-fold molar excess over the amount of amines to be modified. A typical concentration of glutaraldehyde in the reaction mixture is 1.25%. In some cases, trial experiments will have to be done to check for solubility of the resultant modified protein. Scale back the quantity of glutaraldehyde added if precipitation occurs.

3. React for at least 2 h at 4°C.

4. Quickly isolate the modified protein by gel filtration using Sephadex G-25 or the equivalent.

In some cases, the modified protein may be stored for long periods before conjugation with another amine-containing molecule by immediate freezing and lyophilization. If stability is a problem, however, the modified protein should be conjugated immediately.

4.5. Introduction of Hydrazide Functional Groups

Hydrazide-containing reagents can be used for probing or conjugation of carbonyl-containing compounds, including macromolecules possessing aldehydes and ketones.

Fluorescent or enzymatic probes containing hydrazide functional groups can be used to assay or label carbohydrates, glycoproteins, the polysaccharide portion of cell surfaces, gangliosides, and glycoconjugates on blots (Wilchek and Bayer, 1987; Lotan *et al.*, 1975; Spiegal *et al.*, 1982; Hurwitz *et al.*, 1980; Gershoni *et al.*, 1985). Multivalent forms of hydrazide reagents created by modifying enzymes, ferritin, and polymers such as dextran and polypeptides with *bis*-hydrazides can be used to target formyl groups with high avidity and sensitivity (Roffman *et al.*, 1980; Kaplan *et al.*, 1983).

The creation of hydrazide probes most often is based on the derivatization of a detectable molecule with a *bis*-hydrazide compound. Although hydrazine itself (in the form of hydrazine hydrate) can be used in a methanolic solution to modify activated carboxylate molecules forming hydrazides, the availability of the bifunctional hydrazides provides a built-in spacer to accommodate greater steric accessibility.

The following protocols make use of the compounds adipic acid dihydrazide and carbohydrazide to derivatize molecules containing aldehydes, carboxylates, and alkylphosphates. The protocols are applicable for the modification of proteins, including enzymes, soluble polymers such as dextrans and poly-amino acids, and insoluble polymers used as microcarriers or chromatographic supports.

The addition of hydrazide groups into macromolecules containing aldehydes, carboxylates, or alkylphosphates has the effect of increasing the pI or net charge. In the case of carboxylates or alkylphosphates, blocking these groups with hydrazide compounds eliminates the negative charge contribution of the original functional group and adds a potential positive charge contribution due to the terminal hydrazide. The consequence of raising the pI of a macromolecule can have dramatic effects on the molecule's conformation and activity or on its relative nonspecificity in assay systems due to the presence of additional positive charge. For instance, the modification of avidin with adipic acid dihydrazide by coupling through the protein's carboxylate groups significantly increases the net charge of an already highly cationic molecule, and therefore increases its overall cross-reactivity in avidin–biotin assays (Chapter 13, Section 5).

Modification of Aldehydes with Bis-hydrazide Compounds

Aldehyde-containing macromolecules will react spontaneously with hydrazide compounds to form hydrazone linkages. The hydrazone bond is a form of Schiff base that is more stable than the Schiff base formed from the interaction of an aldehyde and an amine. The hydrazone, however, may be reduced and further stabilized by the same reductants utilized for reductive amination purposes (Chapter 3, Section 4). The addition of sodium cyanoborohydride to a hydrazide–aldehyde reaction drives the equilibrium toward formation of a stable covalent complex. Mallia (1992) has found that adipic acid dihydrazide derivatization of periodate-oxidized dextran (containing multiple formyl functionalities) proceeds with much greater yield when sodium cyanoborohydride is present.

The reaction of an excess of adipic acid dihydrazide with aldehyde groups present on proteins or other molecules will result in modified proteins containing alkylhydrazide groups (Fig. 96). Another bis-hydrazide compound, carbohydrazide, also may be employed with similar results, except that the spacer afforded through its use is considerably shorter. Target aldehydes may be created on macromolecules according

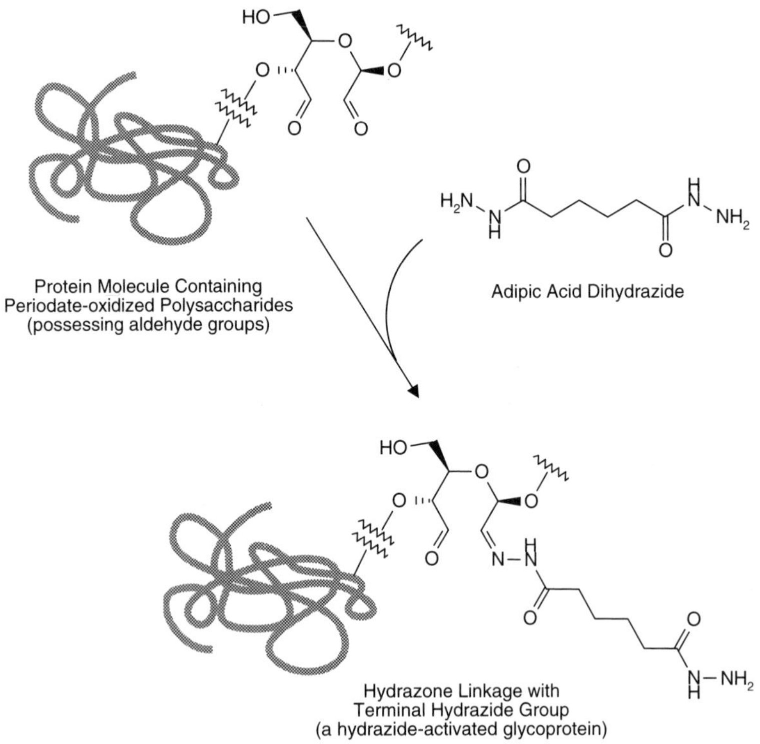

Figure 96   Glycoproteins that have been treated with sodium periodate to produce aldehyde groups can be further modified with adipic acid dihydrazide to result in a hydrazide derivative.

to the protocols described in Section 4.4. Thus, glycoproteins and other molecules containing polysaccharide may be periodate-oxidized to contain formyl groups and then modified with a *bis*-hydrazide compound to create the hydrazide-activated re-agent. Modification of proteins through carbohydrate residues obviates the blocking of negatively charged carboxylates and only adds limited numbers of hydrazides at discrete portions of a molecule. The enzyme horseradish peroxidase is conveniently modified with hydrazide functional groups using this approach (Chapter 16, Section 2.4).

Protocol

1. Dissolve a macromolecule (such as a protein) containing aldehyde functional groups in a buffered solution at a pH of about 7–8.5 and at a concentration of about 1–10 mg/ml. To modify a molecule to contain aldehyde groups, see Section 4.4. Phosphate, carbonate, borate, or similar buffers adjusted to this pH range work well. Avoid amine-containing buffers (i.e., glycine or Tris) or other components containing strong nucleophiles, since these may react with the aldehydes. Higher pH environments enhance the formation of hydrazone bonds and generally increase the yield of complex.

2. Add a quantity of adipic acid dihydrazide or carbohydrazide (Aldrich) to the

protein solution to obtain at least a 10-fold molar excess over the amount of aldehyde functional group present. If the concentration of aldehydes is unknown, the addition of 32 mg adipic acid dihydrazide per milliliter of the protein solution to be modified should work well.

3. React for 2 h at room temperature. Although hydrazone formation does not require the addition of a reductant to create a linkage, including sodium cyanoborohydride in the reaction considerably increases the yield and stability of bonds formed. If the presence of a reducing agent will not cause harm to the macromolecule being modified, the addition of 10 μl of 5 M sodium cyanoborohydride (Sigma) per milliliter of reaction solution may be done. Caution: cyanoborohydride is extremely toxic. All operations should be done with care in a fume hood. Also, avoid any contact with the reagent, as the 5 M solution is prepared in 1 N NaOH.

4. Purify the modified protein by dialysis or gel filtration.

Hydrazide-activated proteins are stable to long-term storage at 4°C in the presence of a preservative (0.05% sodium azide) or in a frozen or lyophilized state.

Modification of Carboxylates with Bis-hydrazide Compounds

Carboxylic acids may be covalently modified with adipic acid dihydrazide or carbohydrazide to yield stable imide bonds with extending terminal hydrazide groups. Hydrazide functionalities do not spontaneously react with carboxylate groups the way they do with formyl groups (Section 4.5). In this case, the carboxylic acid first must be activated with another compound that makes it reactive toward nucleophiles. In organic solutions, this may be accomplished by using a water-insoluble carbodiimide (Chapter 3, Section 1.4) or by creating an intermediate active ester, such as an NHS ester (Chapter 2, Section 1.4).

In aqueous solutions, the easiest method for forming this type of bond is by use of the water-soluble carbodiimide EDC (Chapter 3, Section 1.1). For proteins and other water-soluble macromolecules, EDC reacts with their available carboxylate groups to form an intermediate, highly reactive, O-acylisourea. This active ester species may further react with nucleophiles such as a hydrazide to yield a stable imide product (Fig. 97).

Most proteins contain an abundance of carboxylic acid groups from C-terminal functional groups and aspartic and glutamic acid side chains. these groups are readily modified with bis-hydrazide compounds to yield useful hydrazide-activated derivatives. Both carbohydrazide and adipic acid dihydrazide have been employed in forming these modifications using the carbodiimide reaction (Wilchek and Bayer, 1987).

Protocol

1. Dissolve 32 mg of adipic acid dihydrazide per milliliter of 0.1 M sodium phosphate, 0.15 M NaCl, pH 7.2.
2. Dissolve 5 mg of the protein or other macromolecule to be modified per milliliter of the above solution.
3. Add 16 mg EDC and react at room temperature for 4 h.
4. Purify the modified protein by dialysis or gel filtration.

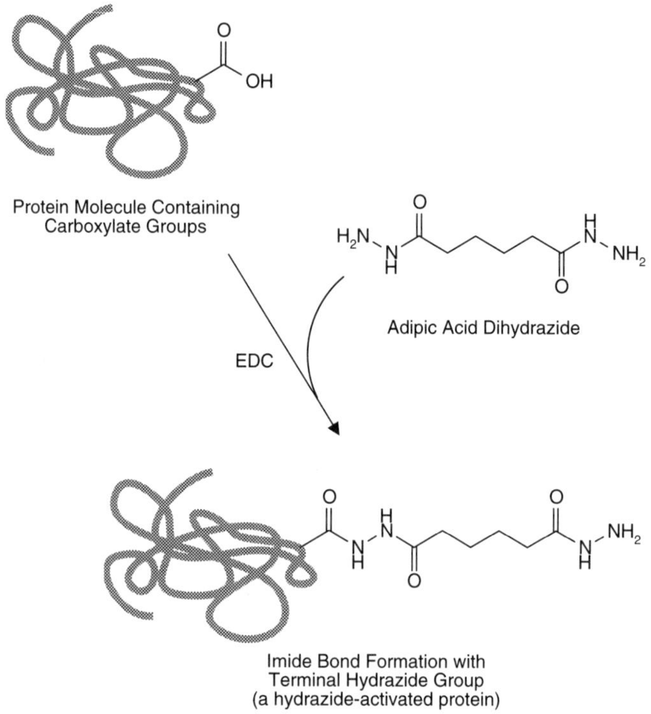

Figure 97 Carboxylate groups on proteins may be modified with adipic acid dihydrazide in the presence of a carbodiimide to produce hydrazide derivatives.

Modification of Alkylphosphates with Bis-hydrazide Compounds

Alkylphosphate groups such as those present at the 5' end of RNA and DNA molecules may be specifically modified with bis-hydrazide compounds. Mediated by the addition of the water-soluble carbodiimide EDC and imidazole, adipic acid dihydrazide or carbohydrazide will react with the phosphate group in a two-step process to form phosphoramidate bonds with short linker arms containing terminal hydrazides (Fig. 98) (Ghosh *et al.*, 1989). In the first stage, EDC activates the phosphate group forming a short-lived, but highly reactive, phosphodiester species, which in turn reacts with a molecule of imidazole to form a longer-lived, active phosphorimidazolide. The

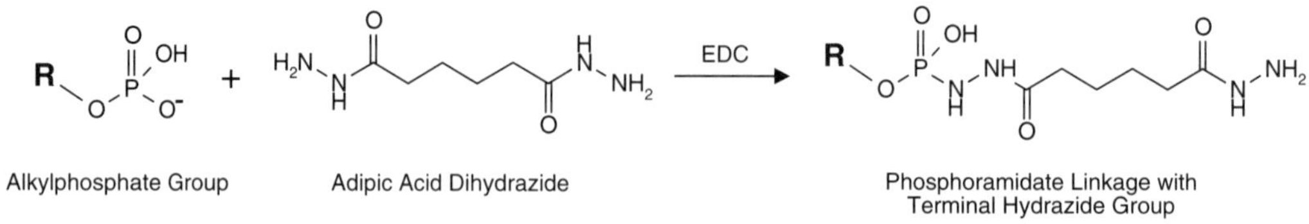

Figure 98 Phosphate groups may be modified with adipic acid dihydrazide in the presence of a carbodiimide to produce hydrazide derivatives. This is a common modification route for the 5'-phosphate group of oligonucleotides.

second stage involves addition and attack of the hydrazide nucelophile, releasing imidazole and forming the phosphoramidate bond. In a modification of the two-stage reaction, Zanocco *et al.* (1993) developed a single-pot reaction in which the alkylphosphate molecule is reacted in the presence of EDC, imidazole, and the bis-hydrazide compound. The modification reaction proceeds rapidly at room temperature.

Protocol

1. Weigh out 1.25 mg of the carbodiimide EDC (1-ethyl-3-(3-dimethylamino-propyl)carbodiimide hydrochloride; Pierce) into a microfuge tube.
2. Add to the tube 7.5 μl of RNA or DNA containing a 5′ phosphate group. The concentration of the oligonucleotide should be 7.5–15 nmol or total of about 57–115.5 μg. Also, immediately add 5 μl of 0.25 M adipic acid dihydrazide or carbohydrazide dissolved in 0.1 M imidazole, pH 6. Because EDC is labile in aqueous solutions, the addition of the oligo and bis-hydrazide/imidazole solutions should occur quickly.
3. Mix by vortexing, then place the tube in a microcentrifuge and spin for 5 min at maximal rpm.
4. Add an additional 20 ml of 0.1 M imidazole, pH 6. Mix and react for at least 2 h at room temperature. The additional buffer prevents pH drift during the carbodiimide reaction.
5. Purify the hydrazide-labeled oligo by gel filtration on Sephadex G-25 using 10 mM sodium phosphate, 0.15 M NaCl, 10 mM EDTA, pH 7.2. The hydrazide-containing probe now may be used to conjugate with a molecule containing an aldehyde reactive group.

5. Blocking Specific Functional Groups

It is often necessary to block specific groups on macromolecules to prevent them from participating in modification or conjugation reactions. In most blocking procedures, a chemical group is covalently coupled to an undesired functional group on the macromolecule to mask or eliminate its reactivity. In this sense, the modification is done with a compound that is relatively inert in whatever application for which the macromolecule is intended. The blocking agent is usually a small organic compound containing a functional group able to couple with the group to be masked. The blocking molecule may contain another functional group of its own, converting the blocked group into a chemical function of another type, but this conversion is all right, providing the newly created function does not interfere in subsequent reactions or applications.

In some cases, a blocking procedure is done to direct a conjugation reaction to discrete sites in a macromolecule. In other instances, blocking a group on one of two macromolecules can prevent self-polymerization and promote the desired intermolecular conjugation. For instance, HRP can be blocked with an amine-specific coupling reagent prior to periodate oxidation to prevent the reactivity of its two amino groups during subsequent conjugation with an antibody molecule (Chapter 10, Section 1.3).

In other uses of blocking reagents, proteins dissociated into subunits by the use of denaturants and disulfide reductants may be prevented from reassociation or oxidation of their sulfhydryls by blocking the —SH groups with the appropriate reagent. Alternatively, sulfhydryls may be blocked on a protein prior to activation with a heterobifunctional cross-linking agent that contains amine-reactive and sulfhydryl-reactive ends. The amine-reactive end will couple to the amines of the protein without reaction of the sulfhydryl-reactive end. This can prevent oligomer formation during the activation process and thus ensure that the sulfhydryl-reactive function is available for conjugation with the desired molecule.

Controlled functional studies of a protein's active center also may be done by blocking specific groups and observing its effect on activity. Often, this blocking procedure is performed through the use of a reversible blocking agent subsequently to regenerate activity, therefore demonstrating that the effect was directed at functional groups present in the active site (Perham and Jones, 1967).

Blocking also may be done to quench further modification or conjugation through a targeted group. In addition, after a conjugation reaction, excess functional groups may be masked from nonspecifically reacting with other molecules. For instance, periodate-oxidized glycoproteins may still contain aldehyde groups after conjugation with another protein by reductive amination. Blocking the aldehydes with a small amine-containing molecule prevents unwanted reactions from occurring when the conjugate is used in an assay or targeting operation. This is also true of excess sulfhydryl groups, which may undergo disulfide interchange with other sulfhydryl molecules subsequent to a conjugation reaction. Blocking these groups with the appropriate reagent prevents this type of side reaction from occurring.

Blocking of amine groups on proteins also has been used to create a sensitive reagent for measuring protease activity (Hatakeyama *et al.*, 1992). With nearly all the primary amines of casein blocked, an amine detection reagent such as trinitrobenzene sulfonic acid (TNBS) will only minimally react with the protein and form its typical orange derivative. As proteases cleave the protein, however, primary α-amines are created from cleavage of the α-chain peptide bonds, and TNBS can react with them. The more protease activity present, the more color is formed.

The choice and application of a specific blocking reagent can produce a modified macromolecule with unique and useful properties. Many of the common blocking reagents are discussed in this section. Beyond the scope of this book, however, is a discussion of the numerous blocking agents used in peptide or nucleic acid synthesis to block temporarily specific reactive groups during growth of the polymer chain.

5.1. Blocking Amine Groups

The amine functional groups most commonly found in macromolecules are primary amines such as those at the N-terminal of polypeptide chains (α-amines) and the side-chain ε-amino groups of lysine residues. Several acylation reagents can effectively block these primary amines, some of which are reversible under the right conditions. It should be noted that the cyclic anhydrides mentioned in this section react with amino groups to form amide bonds, opening the anhydride ring and effectively transforming the amine function into a carboxylate. There are additional cyclic anhydrides described in Section 4.2 that also create carboxylates from amines, but in this section the two discussed, maleic anhydride and citraconic anhydride, both are reversible and designed more for temporary masking than permanent blocking. For more stable blocking of amines, sulfo-NHS acetate and acetic anhydride are the best choices.

Sulfo-NHS Acetate

Sulfo-NHS acetate is the N-hydroxysulfosuccinimide ester of acetic acid. The NHS ester end provides high reactivity with the amino groups of proteins at a pH range of 7–9, acylating the amines and forming nonreversible acetamide modifications (Fig. 99). The sulfonate derivative of the NHS ester provides good water solubility to the reagent. Thus, the compound can be added directly to an aqueous solution of the protein to be blocked, or a stock solution may be prepared and a small aliquot added to the reaction medium. Stock solutions should be dissolved rapidly and use immediately. In aqueous solutions, the main competing reaction is hydrolysis of the active ester to release nonreactive sulfo-NHS and acetic acid. The use of a 10- to 50-fold molar excess of sulfo-NHS acetate over the molar amount of groups to be blocked should provide good yields of acylated amines. Reaction buffers should contain no extraneous amines that could cross-react with the sulfo-NHS acetate. Avoid Tris-, glycine-, and imidazole-containing buffers. Phosphate, borate, or bicarbonate buffers work well at a concentration of 0.05–0.1 M. React for at least 1 h at room temperature.

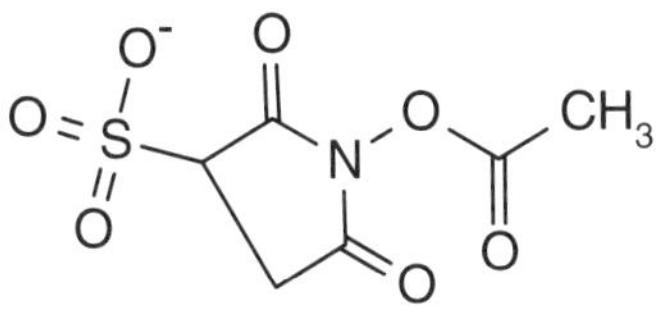

Sulfo–NHS-Acetate
MW 259.17

Protocol

1. Dissolve the protein or other amine-containing macromolecule at a concentration of 1–10 mg/ml in 0.1 M sodium phosphate, 0.15 M NaCl, pH 7.5.
2. Add a 25 molar excess of sulfo-NHS acetate over the amount of amines present in the sample. If the precise amount of amines is not known, adding an equal mass of reagent to the mass of protein will provide a large excess of reactivity to completely block all amines.
3. React at room temperature for at least 1 h.
4. Purify the modified protein by dialysis or gel filtration.

Acetic Anhydride

Acetic anhydride is the only monocarboxylic acid anhydride that is important in modification reactions. The acetylation of the amino groups of proteins can be made relatively specific if the reaction is done in saturated sodium acetate, since the O-acetyltyrosine derivative is unstable to an excess of acetate ions (Fraenkel-Conrat, 1959). The tyrosine derivative rapidly hydrolyzes in alkaline reaction conditions, even

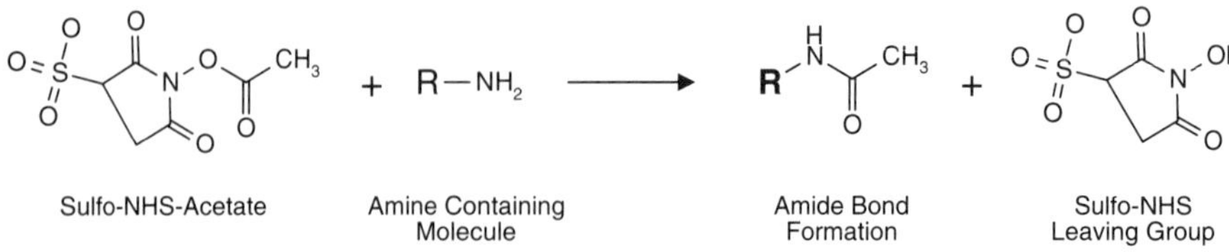

Figure 99 Sulfo-NHS acetate may be used to block amine groups, forming permanent amide bond derivatives.

in the absence of added acetate buffer (Uraki *et al.*, 1957; Smyth, 1967). Treatment with hydroxylamine also cleaves any *O*-acetyltyrosine modifications, forming acetylhydroxamate, which can be followed by its purple complex with Fe^{3+} at 540 nm (Balls and Wood, 1956).

At physiological pH values, acetylation of amine groups proceeds rapidly, requiring less than an hour to go to completion (Fig. 100).

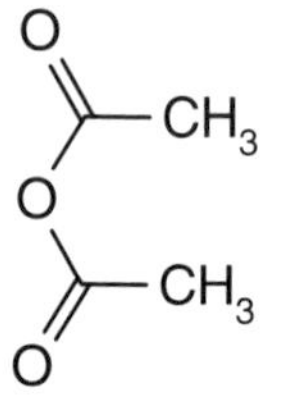

Acetic Anhydride
MW 102

Protocol

1. Dissolve the macromolecule to be modified at a concentration of 1–10 mg/ml in a buffered solution having a pH between 6.5 and 7.5. Avoid amine-containing buffers such as glycine and Tris. Sodium phosphate buffer at a concentration of 0.1 *M* works well. The addition of an equal volume of a saturated solution of sodium acetate may be done to prevent tyrosine derivatization.
2. Cool the solution on ice. With stirring, add an amount of acetic anhydride equal to the mass of macromolecule to be modified. Alternatively, add a 10-fold molar excess of acetic anhydride over the amount of amines present. The addition of the anhydride slowly or in several aliquots over the course of 1 h will ensure good yield of acetylation.
3. React with stirring for at least 1 h while cooling in an ice bath.
4. Purify the acetylated macromolecule by gel filtration or dialysis.

Citraconic Anhydride

Citraconic anhydride (or 2-methylmaleic anhydride) is a derivative of maleic anhydride that is reversible after acylation of amine groups. At alkaline pH values (pH 7–8) the reagent reacts with amines to form amide linkages with an extending terminal carboxylate. However, at acid pH (3–4), these bonds rapidly hydrolyze to release citraconic acid and free the amine (Dixon and Perham, 1968; Klapper and Klotz,

Figure 100 Acetic anhydride reacts with amines to form amide bond derivatives.

1972; Habeeb and Atassi, 1970; Shetty and Kinsella, 1980). Thus, citraconic anhydride is useful in temporarily blocking amine groups while other parts of a molecule are undergoing derivatization. Once the modification is complete, the amines can be then unblocked to create the original structure. See Section 4.2 for additional information and a protocol for modification of proteins with citraconic anhydride.

Maleic Anhydride

Maleic acid is a linear four-carbon molecule with carboxylate groups on both ends and a double bond between the central carbon atoms. The anhydride of maleic acid is a cyclic molecule containing five atoms. Although the reactivity of maleic anhydride is similar to that of other cyclic anhydrides, the products of maleylation are much more unstable toward hydrolysis, and the site of unsaturation lends itself to additional side reactions. Acylation products of amino groups with maleic anhydride are stable at neutral pH and above, but they readily hydrolyze at acid pH values around 3.5 (Butler *et al.*, 1967). Maleylation of sulfhydryls and the phenolate of tyrosine are even more sensitive to hydrolysis. Thus, maleic anhydride is an excellent reversible blocker of amino groups to mask them temporarily from reactivity while another reaction is being done. For additional information and a protocol for the modification of proteins with this reagent, see Section 4.2.

5.2. Blocking Sulfhydryl Groups

The sulfhydryl group is among the most highly reactive of nucleophiles found in biological macromolecules. Cysteine sulfhydryls in proteins undergo covalent reactions rapidly with most of the reactive groups utilized in modification and conjugation reagents. To prevent modification from occurring at these sites, it is often necessary to use a blocking agent that ties up the sulfhydryl and renders it inert toward further reactions.

There are two types of sulfhydryl blocking agents: permanent and reversible. The permanent ones form thioether linkages that do not readily break down. The reversible ones form disulfide bonds that are susceptible to cleavage by the addition of the appropriate reducing agent. Reversible sulfhydryl blockers can be used to mask an —SH group temporarily from modification while a reaction is done at another site. This is especially useful when the sulfhydryl forms a critical part of the active center of a protein. After the final modification is complete, the blocking agent can be removed to regenerate activity.

N-Ethylmaleimide

N-Ethylmaleimide (NEM) is an alkylating reagent that reacts with sulfhydryls to form stable thioether bonds (Smyth *et al.*, 1960). Maleimide reactions are specific for sulfhydryl groups in the pH range 6.5–7.5 (Heitz *et al.*, 1968; Smyth *et al.*, 1964; Gorin *et al.*, 1966; Partis *et al.*, 1983) (see Chapter 2, Section 2.2). At higher pH values some cross-reactivity with amino groups takes place (Brewer and Riehm, 1967). One of the carbons adjacent to the double bond undergoes nucleophilic attack by the thiolate anion to generate the addition product (Fig. 101). When sufficient quantities of —SH groups are being blocked, the reaction may be followed spectrophotometrically by the decrease in absorbance at 300 nm as the double bond reacts and disappears. The result is a stable, inert derivative that terminates in the ethyl group. NEM is useful for permanently blocking sulfhydryl residues in proteins and other macromolecules. It has been used for blocking sulfhydryl-containing reagents that interfere in a glucose oxidase assay system (Haugaard *et al.*, 1981).

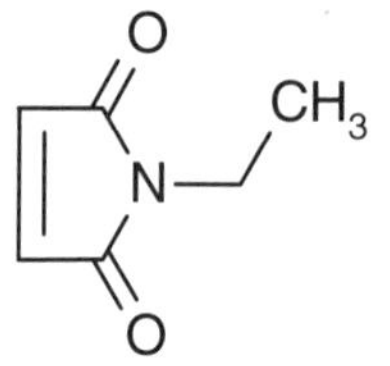

N-Ethylmaleimide
MW 125.12

Protocol

1. Dissolve the macromolecule containing sulfhydryl groups to be blocked in a buffer having a pH of 6.5–7.5. Sodium phosphate (0.01–0.1 *M*) at pH 7.2 works well. Avoid amine-containing buffers, since an excess of amines may cause some reactivity with the maleimide groups. Also, avoid the presence of sulfhydryl-containing disulfide reductants such as DTT or 2-mercaptoethanol, which will rapidly react with NEM.

2. Add at least a 10-fold molar excess of NEM over the amount of sulfhydryls present in the reaction. Alternatively, add an equal mass of NEM to the amount of macromolecule present. To facilitate the addition of a small quantity of reagent, a more concentrated stock solution may be prepared in buffer and an aliquot added to the reaction medium. Make the stock solution up fresh, and use it immediately to prevent loss of activity due to maleimide group breakdown.

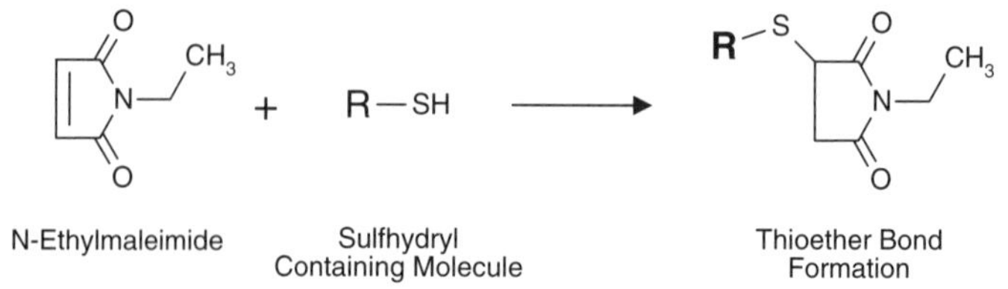

N-Ethylmaleimide   Sulfhydryl   Thioether Bond
                        Containing Molecule   Formation

Figure 101   The reaction of *N*-ethylmaleimide with sulfhydryl groups yields a thioether derivative, permanently blocking the thiol.

3. React for 2 h at room temperature.
4. Purify the modified protein by gel filtration or dialysis.

Iodoacetate Derivatives

Iodoacetate (and bromoacetate) can react with several nucleophilic functional groups within proteins. Their relative reactivity toward protein functional groups is sulfhydryl > imidazolyl > thioether > amine. Among α-haloacetate derivatives the relative reactivity is I > Br > Cl > F, with fluorine being almost unreactive. The α-haloacetamides have the same trend of relative reactivities, but will obviously not create a carboxylate functional group. The acetamide derivatives typically are used only as blocking reagents. The bond formed from the reaction of iodoacetamide and a sulfhydryl group is a stable thioether linkage that is not reversible under normal conditions.

Thus, iodoacetamide has the highest reactivity toward cysteine sulfhydryl residues and may be directed specifically for —SH blocking. If iodoacetamide is present in limiting quantities (relative to the number of sulfhydryl groups present) and at slightly alkaline pH, cysteine modification will be the exclusive reaction. For additional information on α-haloacetate reactivities and a protocol for blocking, see Section 4.2.

Sodium Tetrathionate

Sodium tetrathionate ($Na_2S_4O_6$) is a redox compound that under the right conditions can facilitate the formation of disulfide bonds from free sulfhydryls. The tetrathionate anion reacts with a sulfhydryl to create a somewhat stable active intermediate, a sulfenylthiosulfate (Fig. 102). Upon attack of the nucleophilic thiolate anion on this activated species, the thiosulfate (S_2O_3 =) leaving group is removed and a disulfide linkage forms (Pihl and Lange, 1962). The reduction of tetrathionate to thiosulfate *in vivo* was a subject of early study (Theis and Freeland, 1940; Chen *et al.*, 1934).

$$Na_2S_4O_6$$

Sodium Tetrathionate
MW 270.22

Depending on the proximity of cysteine sulfhydryl groups in proteins, intra- and interchain disulfide formation is possible on reaction with tetrathionate. When neighboring sulfhydryl groups are not close enough to create disulfide linkages, the sulfenylthiosulfate modification is sufficiently stable to block exposed —SH groups temporarily. For sulfhydryls present in the active centers of enzymes, tetrathionate may lead to reversible inactivation (Parker and Allison, 1969). Thus, the reagent may be used to protect certain sulfhydryl residues during modification reactions performed elsewhere on a protein. Using this approach, the enzyme ficin may be temporarily protected with tetrathionate during modification, conjugation, or immobilization reactions done through its amine groups (Liener and Friedenson, 1970). Subsequent treatment with thiol-containing disulfide reducing agents frees the sulfenylthiosulfate and regenerates the sulfhydryl with enzymatic activity. The following protocol is an adaptation of that of Englund *et al.* (1968), used in the purification of ficin.

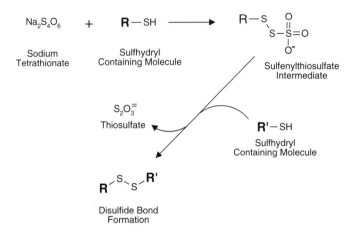

Figure 102   Sodium tetrathionate reacts with thiols to form reactive sulfenylthiosulfate intermediates. Another sulfhydryl-containing molecule may couple to this active group to create a disulfide linkage.

Protocol

1. The macromolecule containing sulfhydryl residues to be blocked or protected is dissolved in a buffer suitable for its individual stability requirements. The blocking process may be done on a purified protein or during the early stages of a purification process to protect sulfhydryl-active centers from oxidation. PBS buffers containing 1 mM EDTA work well.
2. Add sodium tetrathionate to obtain a final concentration of 10 mM.
3. React for 1 h at room temperature.
4. Excess tetrathionate may be removed by dialysis or gel filtration.
5. To remove the sulfenylthiosulfate blocking group, add a 300-fold excess of DTT over the amount of blocked sulfhydryls present. Alternatively, add DTT to obtain a 0.01–0.1 M final concentration. Cysteine also may be utilized to regenerate some enzymes to full activity.
6. Incubate for 2 h at room temperature.
7. For removal of excess DTT, a protein of molecular weight greater than 5000 may be isolated by gel filtration using Sephadex G-25. To maintain the stability of the exposed sulfhydryl groups, include 10 mM EDTA in the chromatography buffer. The presence of oxidized DTT can be monitored during elution by measuring the absorbance at 280 nm. The protein should elute in the first peak and the DTT reaction products in the second peak.

Ellman's Reagent

Ellman's reagent, 5,5′-dithio-bis-(2-nitrobenzoic acid), or DTNB, is a compound useful for the quantitative determination of sulfhydryls in solution (Ellman, 1958, 1959). The disulfide of Ellman's reagent readily undergoes disulfide exchange with a free sulfhydryl to form a mixed disulfide and release of one molecule of the chromogenic substance 5-sulfido-2-nitrobenzoate, also called 5-thio-2-nitrobenzoic acid (TNB). The intense yellow color produced by the TNB anion can be measured by its absorbance at 412 nm ($\varepsilon = 1.36 \times 10^4\ M^{-1}cm^{-1}$ at pH 8). Since each sulfhydryl present generates one molecule of TNB per molecule of Ellman's reagent, direct quantitation is easily done. This reagent has been used to measure the sulfhydryl content in peptides, proteins, and tissue samples (Anderson and Wetlaufer. 1975; Riddles *et al.*, 1979). See section 1.1.4.1 for the use of Ellman's reagent in the determination of sulfhydryl groups.

The same reaction between Ellman's reagent and the sulfhydryls of macromolecules can be used to block available —SH groups temporarily by the formation of a mixed disulfide bond. Treatment of a sulfhydryl-containing protein with an excess of Ellman's reagent blocks the accessible sulfhydryls with the TNB group, allowing chemical reactions to be done on other functional groups. Studies have shown that the rate of Ellman's reaction with the sulfhydryl groups in proteins is dependent on their accessibility (Damjanovich and Kleppe, 1966; Colman, 1969). The addition of a disulfide reducing agent then cleaves the TNB group and regenerates the free sulfhydryl. Enzymes containing sulfhydryls in their active sites may be reversibly blocked using this technique to preserve activity after modification or conjugation. Deblocking then restores catalytic activity in most instances.

Protocol

1. Dissolve the protein to be blocked at a concentration of 1–10 mg/ml in 0.1 *M* sodium phosphate, pH 8.
2. Dissolve the Ellman's reagent at a concentration of 4 mg/ml in 0.1 *M* sodium phosphate, pH 8.
3. Mix the protein solution with an equal volume of the Ellman's reagent solution and react for 15 min at room temperature.
4. Purify the modified protein from excess Ellman's reagent and reaction by-products by dialysis or gel filtration. A measurement of sulfhydryl content may be done by reading the absorbance of the modification reaction at 412 nm ($\varepsilon = 1.36 \times 10^4\ M^{-1}cm^{-1}$) versus a series of sulfhydryl standards treated in the same manner (e.g., cysteine).

To deblock the TNB-modified sulfhydryl residues, treat the protein with an excess of DTT according to the protocol described in section 1.1.4.1, DTT.

Dipyridyl Disulfide Reagents

The similar reagents 4,4'-dipyridyl disulfide (Grassetti and Murray, 1967) and 2,2'-dipyridyl disulfide (Brocklehurst *et al.*, 1974) react in a manner analogous to that of Ellman's reagent, both forming pyridyl disulfide bonds with free sulfhydryls and releasing a molecule of either pyridine-4-thione or pyridine-2-thione, respectively (Fig. 103). Both leaving groups are measurable spectrophotometrically at 324 nm (pyridine-4-thione) or 343 nm (pyridine-2-thione) to quantify the amount of sulfhydryl modification. The reagent 2,2'-dipyridyl disulfide is useful for creating sulfhydryl-reactive cross-linking agents, such as SPDP (Chapter 5, Section 1.1). Both reagents may be used to block sulfhydryl groups temporarily in macromolecules or to activate —SH groups for coupling to another sulfhydryl-containing molecule. The pyridine disulfide-modifying group can react with a sulfhydryl to form a disulfide

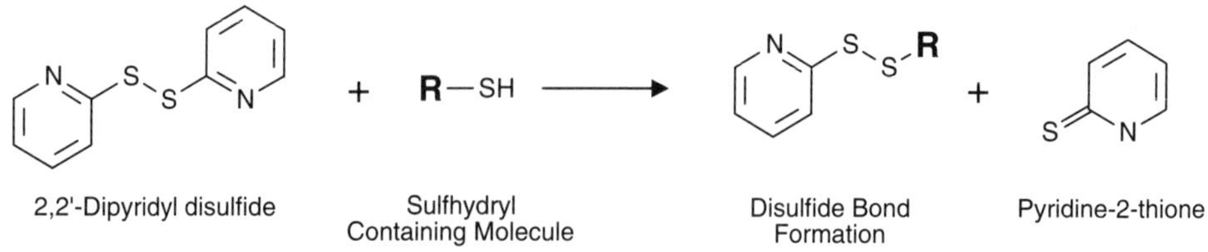

Figure 103   2,2'-Dipyridyl disulfide reacts with thiols to form an active pyridyl disulfide intermediate.

linkage. The pyridine disulfide also may be cleaved with an excess of disulfide reducing agents, such as DTT, making it a reversible blocking agent.

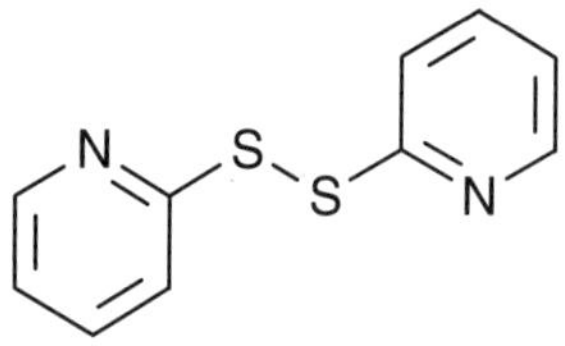

2,2'-Dipyridyl disulfide

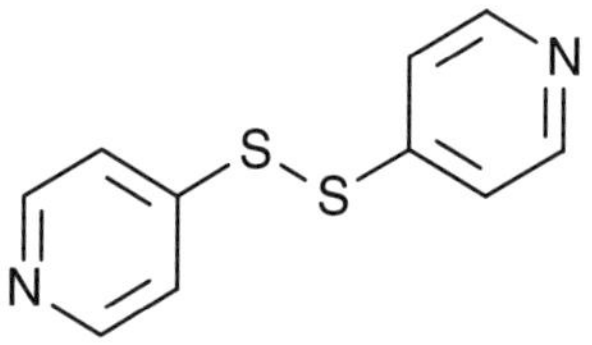

4,4'-Dipyridyl disulfide

Unfortunately, 2,2'-dipyridyl disulfide is relatively insoluble in aqueous buffers. The use of this compound to modify molecules usually involves prior dissolution in an organic solvent such as acetone and then performing the blocking reaction in an aqueous/organic mixture. Many proteins will not tolerate high concentrations of organic solvents without precipitation.

The 4,4'-dipyridyl disulfide can be used in aqueous solutions, but it has been found that modification of proteins with this reagent yields rapid disulfide bond formation. Only when 2-iminothiolane is used in tandem with 4,4'-dipyridyl disulfide can 4-dithiopyridyl groups be introduced into proteins (King *et al.*, 1978) (see Section 4.1). This is due to disulfide interchange reactions predominating without the addition of 2-iminothiolane.

For one-step methods, the use of Ellman's reagent (previous section) to yield a similar reversible sulfhydryl blocking group is probably a better choice with protein molecules.

5.3. Blocking Aldehyde Groups

Aldehyde groups are useful in facilitating modification or conjugation reactions, easily forming secondary amine linkages with amine-containing molecules in reductive amination procedures or hydrazone linkages with hydrazide-containing molecules. Macromolecules modified to contain aldehyde groups for use in these reactions (see Section 4.4) should be treated after conjugation to remove any excess formyl functionalities. The blocking step prevents subsequent nonspecific interactions when a conjugate is used in assay or targeting applications.

Reductive Amination with Tris or Ethanolamine

The simplest method for blocking aldehyde functionalities involves reductive amination with a small amine-containing molecule. The best such blockers do not have extra functional groups that may create additional sites of reactivity after blocking. Tris and ethanolamine are ideal in this regard. They both contain primary amines that readily react with aldehydes in the presence of a reductant, and they both possess relatively inert hydroxyl groups that maintain hydrophilicity after coupling. Reductive amination (Chapter 1, Section 5.3 and Chapter 3, Section 4) facilitated by the use of sodium cyanoborohydride can quickly block residual aldehyde groups and transform them into unreactive hydroxyls of low nonspecific binding potential (Fig. 104).

Protocol

1. Dissolve the macromolecuole containing aldehydes to be blocked (i.e., a glycoprotein that has been oxidized with sodium periodate to create formyl groups) at a concentration of 1–10 mg/ml in 0.1 M Tris buffer, pH 8. Alternatively, dissolve the macromolecule in 0.1 M sodium phosphate containing 0.1 M ethanolamine, pH 8. The use of other buffers having a pH between 7 and 10 will work as well, but the Tris or ethanolamine concentrations should be maintained in high excess to block efficiently all the aldehyde residues.

2. Add 10 μl of 5 M sodium cyanoborohydride in 1 N NaOH (Aldrich) per milliliter of the macromolecule solution volume prepared in (1). *Caution: Highly toxic compound. Use a fume hood and be careful to avoid skin contact with this reagent.*

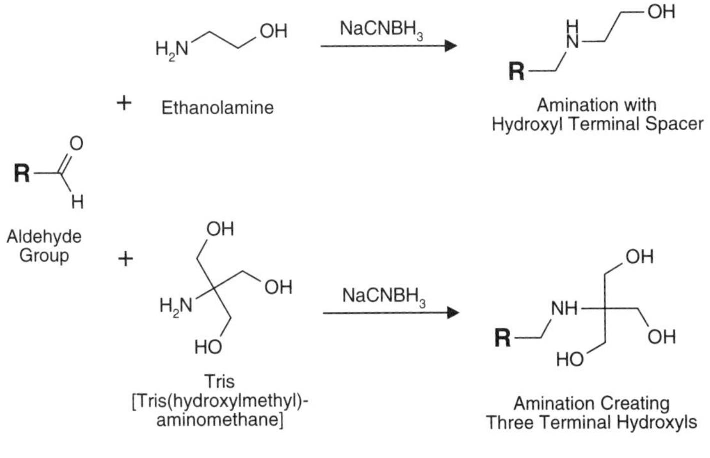

Figure 104 Aldehyde groups may be blocked with Tris or ethanolamine using a reductive amination process.

3. React for 15 min at room temperature.
4. Purify the derivatized macromolecule by dialysis or gel filtration using a buffer suitable for the nature of the substance being modified.

5.4. Blocking Carboxylate Groups

The presence of unwanted carboxylate groups in macromolecules may be easily blocked by the use of a small amine-containing molecule coupled via the carbodiimide procedure (Chapter 3, Section 1).

Tris or Ethanolamine plus EDC

Tris or ethanolamine are excellent choices for blocking procedures involving carboxylic acid groups, since they contain hydrophilic hydroxyls that mask the carboxylate and create an inert modification with low nonspecific binding potential. Using the water-soluble carbodiimide EDC to facilitate this reaction, the carboxylate is activated by forming an intermediate O-acylisourea. The amine-containing compound then reacts with this active species to create a stable amide linkage (Fig. 105).

Protocol

1. Dissolve the macromolecule containing carboxylate groups to be blocked at a concentration of 1–10 mg/ml in 0.1 M MES, pH 4.7, containing 0.1 M Tris or ethanolamine. Other conditions may be used to perform this reaction. See Chapter 3, Section 1 for further details.
2. Add 10 mg of EDC per milliliter of the solution prepared in (1).
3. React for 2–4 h at room temperature.
4. Purify by gel filtration or dialysis.

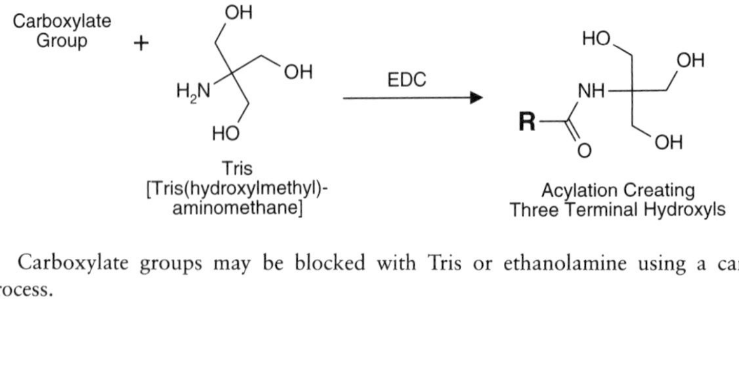

Figure 105 Carboxylate groups may be blocked with Tris or ethanolamine using a carbodiimide-mediated process.

2

The Chemistry of Reactive Groups

Every chemical modification or conjugation process involves the reaction of one functional group with another, resulting in the formation of a covalent bond. The creation of bioconjugate reagents with spontaneously reactive or selectively reactive functional groups forms the basis for simple and reproducible cross-linking or tagging of target molecules. Of the hundreds of reagent systems described in the literature or offered commercially, most utilize common organic chemical principles that can be reduced to a couple dozen or so primary reactions. An understanding of these basic reactions can provide insight into the properties and use of bioconjugate reagents even before they are applied to problems in the laboratory.

This section is designed to provide a general overview of activation and coupling chemistry. Some of the reagents discussed in this part are not themselves cross-linking or modification compounds, but may be used to form active intermediates with another functional group. These active intermediates subsequently can be coupled to a second molecule that possesses the correct chemical constituents that allow bond formation to occur.

Ultimately, this section is meant to function as a ready-reference database for learning or review of bioconjugate chemistry. In this regard a reaction can be quickly found, a short discussion of its properties and use read, and a visual representation of the chemistry of bond formation illustrated. What this section is not meant to be is an exhaustive discussion on the theory or mechanism behind each reaction, nor is it a review of every application in which each chemical reaction has been used. For particular applications where the chemistries are employed, cross-references are given to other sections in this book or to outside literature sources.

1. Amine-Reactive Chemical Reactions

Reactive groups able to couple with amine-containing molecules are by far the most common functional groups present on cross-linking or modification reagents. An amine-coupling process can be used to conjugate with nearly all protein or peptide molecules as well as a host of other macromolecules. The primary coupling chemical reactions for modification of amines proceed by one of two routes: acylation or alkylation (Chapter 1, Section 1.1). Most of these reactions are rapid and occur in high yield to give stable amide or secondary amine bonds.

1.1. Isothiocyanates

Isothiocyanates can be formed by the reaction of an aromatic amine with thiophosgene (Rifai and Wong, 1986). The group reacts with nucleophiles such as amines, sulfhydryls, and the phenolate ion of tyrosine side chains (Podhradsky *et al.*, 1979). The only stable product of these reactions, however, is with primary amine groups. Therefore, isothiocyanate compounds are almost entirely selective for modifying ε- and N-terminal amines in proteins or primary amines in other molecules (Jobbagy and Kiraly, 1966). The reaction involves attack of the nucleophile on the central, electrophilic carbon of the isothiocyanate group (Reaction 1). The resulting electron shift and proton loss creates a thiourea linkage between the isothiocyanate-containing compound and the amine with no leaving group involved.

$$
\mathbf{R}-NH_2 \;+\; \mathbf{R'}-N=C=S \;\longrightarrow\; \mathbf{R'}-HN-\overset{\displaystyle S}{\overset{\|}{C}}-NH-\mathbf{R} \tag{1}
$$

Amine   Isothiocyanate   Isothiourea Bond
Compound   Compound

Isothiocyanate compounds react best at alkaline pH values where the target amine groups are mainly unprotonated. Many reactions are done on 0.1 M sodium carbonate buffer at pH 9. Reaction times vary from 4 to 24 h at 4°C. Since the isothiocyanate group is relatively unstable in aqueous conditions, reagents containing this function should be stored desiccated at refrigerator or freezer temperatures.

1.2. Isocyanates

Isocyanates are similar to the isothiocyanates discussed above, except an oxygen atom replaces the sulfur. An isocyanate can be formed from the reaction of an aromatic amine with phosgene (Rifai and Wong, 1986). The group also can be created from acyl azides by treatment at 80°C in the presence of an alcohol (Chapter 8, Section 1.5). Under these conditions, the acyl azide rearranges to form an isocyanate. Isocyanates can react with amine-containing molecules to form stable isourea linkages (Reaction 2). The reactivity of isocyanates is greater than that of isothiocyanates, but for the same reason their stability can be a problem. Many commercial suppliers of bioconjugate reagents have found isocyanate compounds too unstable to offer them for sale, since moisture rapidly decomposes them, releasing CO_2 and leaving an aromatic amine.

$$
\mathbf{R}-NH_2 \;+\; \mathbf{R'}-N=C=O \;\longrightarrow\; \mathbf{R'}-HN-\overset{\displaystyle O}{\overset{\|}{C}}-NH-\mathbf{R} \tag{2}
$$

Amine   Isocyanate   Isourea Bond
Compound   Compound

Isocyanate-containing reagents also can be used to cross-link or label hydroxyl-containing molecules. A heterobifunctional compound containing a isocyanate group on one end and a maleimide group on the other end has been reported (Annunziato *et al.*, 1993). *p*-Maleimidophenyl isocyanate (PMPI) can be used to conjugate hydroxyl-containing compounds such as polysaccharides with sulfhydryl-containing molecules.

1.3. Acyl Azides

Acyl azides are activated carboxylate groups that can react with primary amines to form amide bonds. The azide function is a good leaving group similar to the *N*-hydroxysuccinimide group of NHS ester compounds. An acyl azide can be formed by treatment of a hydrazide with sodium nitrite at 0°C (Lowe and Dean, 1974). A coupling reaction with amine groups occurs by attack of the nucleophile at the electron-deficient carbonyl group (Reaction 3). Optimum conditions for the reaction are a pH range of 8.5–10 in buffers that contain no competing amines or other nucleophiles.

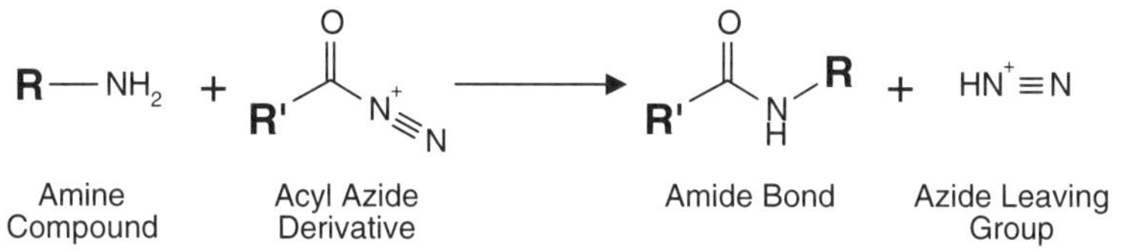

| Amine Compound | Acyl Azide Derivative | Amide Bond | Azide Leaving Group |

$$(3)$$

The major competing reaction in acyl azide coupling is hydrolysis. The higher the pH, the faster the reactivity, with regard to both amine conjugation and hydrolysis. Cross-linkers or modification reagents containing this compound must be kept dry to preserve activity. Reactions are complete in 2–4 h at room temperature.

1.4. NHS Esters

An *N*-hydroxysuccinimide (NHS) ester is perhaps the most common activation chemistry for creating reactive acylating agents. NHS esters were first introduced as reactive ends of homobifunctional cross-linkers (Bragg and Hou, 1975; Lomant and Fairbanks, 1976). Today, the great majority of amine-reactive cross-linking or modification reagents commercially available utilize NHS esters. An NHS ester may be formed by the reaction of a carboxylate with NHS in the presence of a carbodiimide. To prepare stable NHS ester derivatives, the activation reaction must be done in nonaqueous conditions using water-insoluble carbodiimides or condensing agents, such as DCC (Chapter 3, Section 1.4).

NHS or sulfo-NHS ester-containing reagents react with nucleophiles with release of the NHS or sulfo-NHS leaving group to form an acylated product (Reaction 4). The reaction of such esters with a sulfhydryl or hydroxyl group does not yield stable conjugates, forming thioesters or ester linkages, respectively. Both of these bonds hydrolyze in aqueous environments. Histidine side-chain nitrogens of the imidazolyl ring also may be acylated with an NHS ester reagent, but they too hydrolyze rapidly (Cuatrecasas and Parikh, 1972). Thus, the presence of imidazole in reaction buffers only serves to increase the hydrolysis rate of the active ester. Reaction with primary and

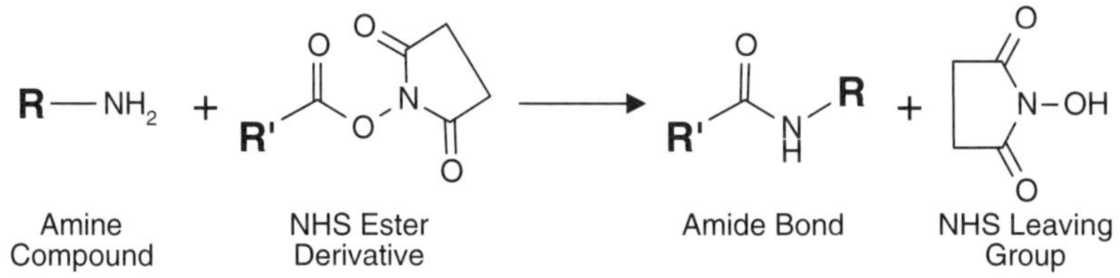

| Amine Compound | NHS Ester Derivative | Amide Bond | NHS Leaving Group |

$$(4)$$

secondary amines, however, creates stable amide and imide linkages, respectively, that do not readily break down. Thus, in protein molecules, NHS ester cross-linking reagents couple principally with the α-amines at the N-terminals and the ε-amines of lysine side chains.

NHS esters also may be formed *in situ* to react immediately with target molecules in aqueous reaction media. Using the water-soluble carbodiimide EDC (Chapter 3, Section 1.1) a carboxylate-containing molecule can be transformed into an active ester functional group by reaction in the presence of NHS or sulfo-NHS (*N*-hydroxy-sulfosuccinimide) (Chapter 3, Section 1.2). Sulfo-NHS esters are hydrophilic active groups that couple rapidly with amines on target molecules with the same specificity and reactivity as NHS esters (Staros, 1982). Unlike NHS esters that are relatively water-insoluble and must be first dissolved in organic solvent before being added to aqueous solutions, sulfo-NHS esters are relatively water-soluble and long-lived and hydrolyze more slowly in water. In the presence of amine nucleophiles that can attack at the electron-deficient carbonyl of the active ester, the sulfo-NHS group rapidly leaves, creating a stable amide linkage with the amine compound. Sulfhydryl and hydroxyl groups also will react with such active esters, but the products of such reactions, thioesters and esters, are unstable.

NHS esters have a half-life on the order of hours under physiological pH conditions. However, hydrolysis and amine reactivity both increase with increasing pH. At 0°C at pH 7, the half-life is typically 4–5 h (Lomant and Fairbanks, 1976). At pH 8 25°C it falls to 1 h (Staros, 1988), and at pH 8.6 and 4°C the half-life is only 10 min (Cuatrecasas and Parikh, 1972). The rate of hydrolysis may be monitored by measuring the increase in absorptivity at 260 nm as the NHS-leaving group is cleaved. The molar extinction coefficient of the NHS group in solution is $8.2 \times 10^3 \, M^{-1}cm^{-1}$ in Tris buffer at pH 9 (Carlsson *et al.*, 1978), but somewhat decreases to 7.5×10^3 $M^{-1}cm^{-1}$ in potassium phosphate buffer at pH 6.5 (Partis *et al.*, 1983). Unfortunately, the relatively low sensitivity of this absorptivity measurement does not allow for determining the rate of reaction in an actual cross-linking procedure.

To maximize the modification of amines and minimize the effects of hydrolysis, maintain a high concentration of protein or other target molecule in the reaction medium. By adjusting the molar ratio of cross-linker to the target molecule(s), the level of modification and conjugation may be controlled to create an optimal product. Water-insoluble cross-linkers containing NHS esters may be reacted in organic solvents, eliminating the hydrolysis problem, provided the target molecule is soluble and stable in such environments.

1.5. Sulfonyl Chlorides

Sulfonyl chlorides are reactive sulfonic acid derivatives similar in properties and reactivity to acid chlorides of carboxylates. The sulfonic acid group, however, is a highly hindered molecule, containing a tetrahedral configuration of substituents. The attack of a nucleophile on a sulfonyl chloride involves temporary formation of a pentavalent intermediate that is highly unstable. Unlike the capability of using other condensing agents such as carbodiimides when preparing amide linkages between carboxylate groups and amines, sulfonic acids are too hindered to allow such bulky active intermediates to be formed. Thus, the main activation chemistry employed with sulfonates is to create the sulfonyl chloride derivative. Reaction of a sulfonyl chloride compound with a primary amine-containing molecule proceeds with loss of the chlorine atom and formation of a sulfonamide linkage (Reaction 5).

$$R{-}NH_2 \;+\; R'{-}\overset{\displaystyle O}{\underset{\displaystyle O}{\overset{\|}{\underset{\|}{S}}}}{-}Cl \;\longrightarrow\; R'{-}\overset{\displaystyle O}{\underset{\displaystyle O}{\overset{\|}{\underset{\|}{S}}}}{-}NH{-}R \;+\; HCl \tag{5}$$

Amine Compound	Sulfonyl Chloride Derivative	Sulfonamide Bond

Sulfonic acids are frequent constituents of fluorescent probes (Chapter 8, Section 1). The sulfonyl chloride derivative allows simple conjugation of these molecules with proteins and other amine-containing compounds. The derivative is prepared by reaction of the sulfonate with thionyl chloride or phosphorus pentachloride in nonaqueous conditions. The reaction of a sulfonyl chloride with an amine proceeds under alkaline pH conditions (typically done at pH 9–10). It may also be done in organic solvent for the modification of water-insoluble compounds. Hydrolysis is the major competing reaction in aqueous environments, although the overall rate of sulfonyl chloride reactivity and hydrolysis is less than that of the corresponding acid chlorides of carboxylates. However, sulfonyl chloride-containing reagents should be stored under nitrogen or in a desiccator to prevent breakdown by moisture.

1.6. Aldehydes and Glyoxals

Carbonyl groups such as aldehydes, ketones, and glyoxals can react with amines to form Schiff base intermediates that are in equilibrium with their free forms. The interaction is pH-dependent, being more efficient at low pH and especially at high pH conditions. Certain compounds, particularly some reducing sugars, may undergo an Amadori rearrangement after Schiff base formation to a stable ketoamine structure (Chapter 1, Section 2.1). This occurs *in vivo* as glucose modifies amine-containing components in the blood to form glycated derivatives. Such modification is thought to be related to aging and is a signal for protein and cellular regeneration.

The rather labile Schiff base interaction can be chemically stabilized by reduction. The addition of sodium borohydride or sodium cyanoborohydride to a reaction medium containing an aldehyde compound and an amine-containing molecule will result in reduction of the Schiff base intermediate and covalent bond formation, creating a secondary amine linkage between the two molecules (Reaction 6).

$$R{-}NH_2 \;+\; R'{-}\overset{\displaystyle O}{\underset{\displaystyle H}{\big\langle}} \;\rightleftharpoons\; R'{\diagup}^{N{-}R} \xrightarrow{\;NaCNBH_3\;} R'{\diagup}^{\overset{H}{N}{-}R} \tag{6}$$

Amine Compound	Aldehyde	Schiff Base	Secondary Amine Bond

Although both borohydride and cyanoborohydride have been used for reductive amination purposes, borohydride will reduce the reactive aldehyde groups to hydroxyls at the same time it converts any Schiff bases present to secondary amines.

Cyanoborohydride, by contrast, is a milder reducing agent. It has been shown to be at least five times milder than borohydride in reductive amination processes with antibodies (Peng *et al.*, 1987). Although cyanoborohydride does not reduce aldehydes, it is very effective at Schiff base reduction. Thus, higher yields of conjugate formation can be realized using cyanoborohydride instead of borohydride. Other reducing agents also have been explored for reductive amination processes, including various amine boranes and ascorbic acid (Cabacungan *et al.*, 1982; Hornsey *et al.*, 1986). See Chapter 3, Section 4 for a protocol for reductive amination coupling.

1.7. Epoxides and Oxiranes

An epoxide or oxirane group will react with nucleophiles in a ring-opening process. The reaction can take place with primary amines, sulfhydryls, or hydroxyl groups to create secondary amine, thioether, or ether bonds, respectively. During the coupling process, ring opening forms a β-hydroxy group on the epoxy compound (Reaction 7). The reaction of the epoxide functional groups with hydroxyls requires high pH conditions, usually in the pH range of 11–12. Amine nucleophiles react at more moderate alkaline pH values, typically needing buffer environments of at least pH 9. Sulfhydryl groups are the most highly reactive nucleophiles with epoxides, requiring a buffered system closer to the physiological pH range of 7.5–8.5 for efficient coupling.

Amine Compound	Epoxide Derivative	Secondary Amine Bond

(7)

The principal side reaction to epoxide coupling is hydrolysis. At acid or alkaline pH values the epoxide ring can hydrolyze to form adjacent hydroxyls. The diol can be oxidized with periodate to create a terminal aldehyde residue with loss of one molecule of formaldehyde (Chapter 1, Section 4.4). The aldehyde then can be used in reductive amination reactions. The reaction of an epoxide group with an ammonium ion generates a terminal primary amine group that also can be used for further derivatization.

1.8. Carbonates

Carbonates are diester derivatives of carbonic acid formed from its condensation with hydroxyl compounds. These groups may be created from the reaction of a bifunctional carbonic acid compound like phosgene or carbonyl diimidazole (Chapter 3, Section 3) with two alcohols. Carbonates can rapidly react with nucleophiles to form carbamate linkages, which are extremely stable bonds (Reaction 8). A commonly used bifunctional carbonate compound, disuccinimidyl carbonate, can be used to activate hydroxyl-containing molecules to form amine-reactive succinimidyl carbonate intermediates (Section 4.3). This carbonate activation procedure can be used with great success in coupling polyethylene glycol to proteins and other amine-containing molecules (Chapter 15, Section 1.2).

Nucleophiles, such as the primary amino groups of proteins, can react with the

$$R-NH_2 \;+\; \underset{\substack{\text{Carbonate}\\\text{Compound}}}{R'\diagdown O \diagup \overset{O}{\underset{}{C}} \diagdown O \diagup R''} \longrightarrow \underset{\substack{\text{Carbamate}\\\text{Linkage}}}{R\diagdown \underset{H}{N} \diagup \overset{O}{\underset{}{C}} \diagdown O \diagup R''} \;+\; R'-OH \qquad (8)$$

succinimidyl carbonate functional groups to give stable carbamate (aliphatic urethane) bonds. The linkage is identical to that obtained through CDI activation of hydroxyl groups with subsequent coupling of amines (Chapter 3, Section 3, and Chapter 15, Section 1.4). However, the reactivity of the succinimidyl carbonate is much greater than that of the imidazole carbamate formed as the active species in CDI activation. A succinimidyl carbonate group may hydrolyze in aqueous solution to release NHS and CO_2, essentially regenerating the underivatized hydroxyl. Carbonates formed from esterification of two alcohol groups similarly hydrolyze to release CO_2 plus the original hydroxyl compounds.

The coupling reaction of a carbonate functional group with an amine is best done in slightly alkaline pH (7–9) and in the absence of any competing amine or sulfhydryl components.

1.9. Arylating Agents

Aryl halide compounds such as fluorobenzene derivatives can be used to form covalent bonds with amine-containing molecules like proteins. The reactivity of aryl halides, however, is not totally specific for amines. Other nucleophiles such as thiol, imidazolyl, and phenolate groups of amino acid side chains also can react (Zahn and Meinhoffer, 1958). Conjugates formed with sulfhydryl groups are reversible by cleaving with an excess of thiol (see Chapter 3, Section 2.5) (Shaltiel, 1967).

Fluorobenzene-type compounds have been used as functional groups in homobifunctional cross-linking agents (Chapter 4, Section 4). Their reaction with amines involves nucleophilic replacement of the fluorine atom with the amine derivative, creating a substituted aryl amine bond (Reaction 9). Detection reagents incorporating reactive aryl chemistry include 2,4-dinitrofluorobenzene and trinitrobenzenesulfonate (Eisen *et al.*, 1953). These compounds form colored complexes with target amine groups. The relative rate of reactivity for aryl compounds is: F > Cl ~ Br > Sulfonate.

$$R-NH_2 \;+\; \underset{\substack{\text{Fluorobenzene}\\\text{Derivative}}}{\text{[benzene ring]}-F} \longrightarrow \underset{\substack{\text{Arylamine}\\\text{Bond}}}{\text{[benzene ring]}-\underset{H}{N}-R} \;+\; HF \qquad (9)$$

1.10. Imidoesters

The imidoester (or imidate) functional group is one of the most specific acylating groups available for modifying primary amines. Unlike most other coupling chemistries, imidoesters posses minimal cross-reactivity toward other nucleophilic groups in proteins. The α-amines and ε-amines of proteins may be targeted and cross-linked by reacting with homobifunctional imidoesters at a pH of 7–10 (optimal pH 8–9). The product of this reaction, an imidoamide (or amidine) (Reaction 10), is protonated and thus carries a positive charge at physiological pH (Liu *et al.*, 1977; Kiehm and Ji, 1977; Ji, 1979; Wilbur, 1992).

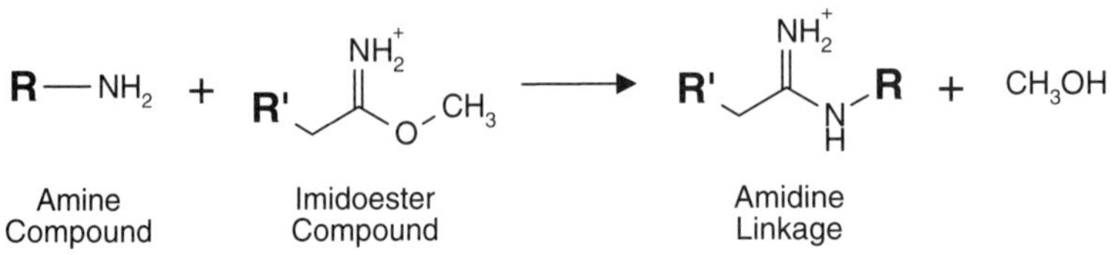

| Amine Compound | Imidoester Compound | | Amidine Linkage |

(10)

The amidine bond formed is quite stable at acid pH; however, it is susceptible to hydrolysis and cleavage at high pH. Typical reaction conditions for using imidate cross-linkers is a buffer system consisting of 0.2 *M* triethanolamine in 0.1 *M* sodium borate, pH 8.2. After conjugating two proteins with a bifunctional imidoester cross-linker, excess imidoester functional groups may be blocked with ethanolamine.

1.11. Carbodiimides

Carbodiimides are zero-length cross-linking agents used to mediate the formation of an amide or phosphoramidate linkage between a carboxylate and an amine or a phosphate and an amine, respectively (Hoare and Koshland, 1966; Chu *et al.*, 1986; Ghosh *et al.*, 1990). They are called zero-length reagents because in forming these bonds no additional chemical structure in introduced between the conjugating molecules.

N-substituted carbodiimides can react with carboxylic acids to form highly reactive, O-acylisourea derivatives that are extremely short-lived (Reaction 11). This active species then can react with a nucleophile such as a primary amine to form an amide bond (Reaction 12) (Williams and Ibrahim, 1981). Other nucleophiles are also reactive. Sulfhydryl groups may attack the active species and form thioester linkages, although these are not as stable as the bond formed with an amine.

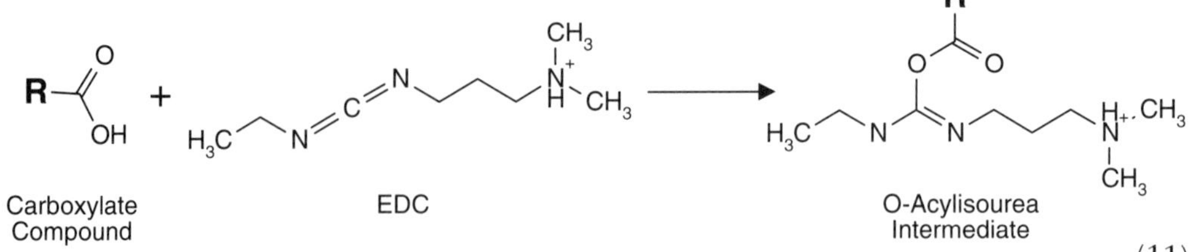

| Carboxylate Compound | EDC | O-Acylisourea Intermediate |

(11)

Hydrazide-containing compounds also can be coupled to carboxylate groups using a carbodiimide-mediated reaction. Using bifunctional hydrazide reagents, carboxylates can be modified to possess terminal hydrazide groups able to conjugate with other carbonyl compounds (Chapter 4, Section 8).

In addition, oxygen atoms may act as the attacking nucleophile, such as those in water molecules. In aqueous solutions, hydrolysis by water is the major competing

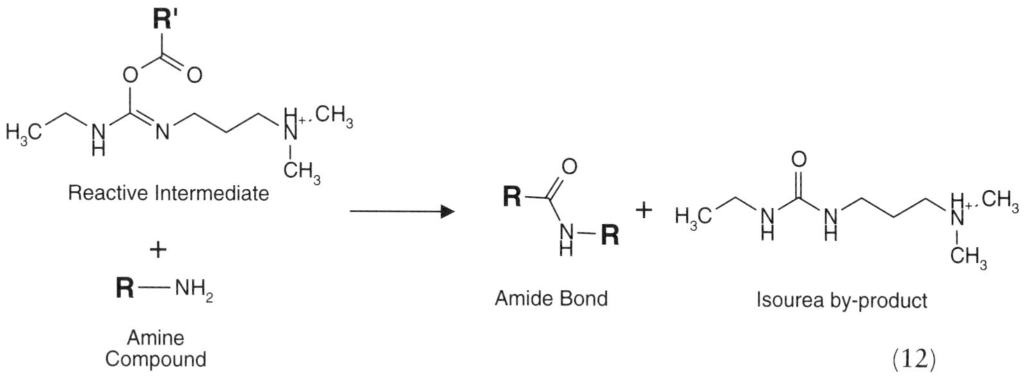

Reactive Intermediate

+

R—NH$_2$

Amine
Compound

Amide Bond   Isourea by-product

(12)

reaction, cleaving off the activated ester intermediate, forming an isourea, and regenerating the carboxylate group (Gilles *et al.*, 1990).

Carbodiimide-mediated amide bond formation effectively occurs between pH 4.5 and 7.5. Buffer systems using MES or phosphate may be used to stabilize the pH during the course of the reaction. For additional information on specific carbodiimides used in bioconjugate chemistry, see Chapter 3, Section 1.

Molecules containing phosphate groups, such as the 5′ phosphate of oligonucleotides, also may be conjugated to amine-containing molecules by using a carbodiimide-mediated reaction (Chapter 17, Section 2.1). The carbodiimide activates the phosphate to an intermediate phosphate ester, similar to its reaction with carboxylates (Chapter 3, Section 1). In the presence of an amine, the ester reacts to form a stable phosphoramidate bond (Reaction 13).

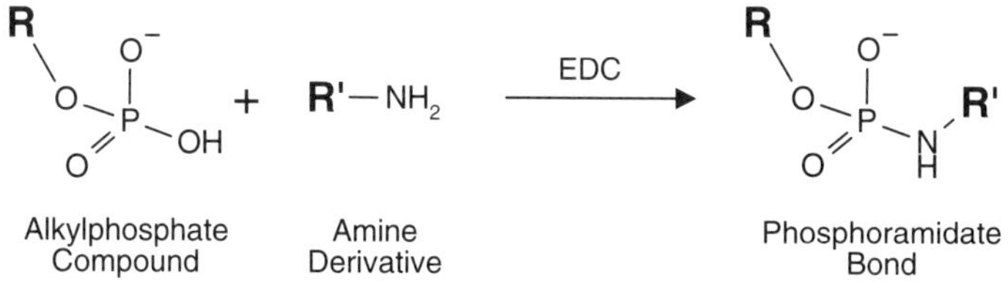

Alkylphosphate   Amine   Phosphoramidate
Compound   Derivative   Bond

(13)

1.12. Anhydrides

Acid anhydrides, as their name implies, are formed from the dehydration reaction of two carboxylic acid groups. Anhydrides are highly reactive toward nucleophiles and are able to acylate a number of the important functional groups of proteins and other macromolecules. On nucleophilic attack, the anhydride yields one carboxylic acid for every acylated product. If the anhydride was formed from monocarboxylic acids, such as acetic anhydride, then the acylation occurs with release of one carboxylate group. However, for dicarboxylic acid anhydrides, such as succinic anhydride, on reaction with a nucleophile the ring structure of the anhydride opens, forming the acylated product modified to contain a newly formed carboxylate group (Reaction 14). Thus, anhydride reagents may be used both to block functional groups and to convert an existing functional group into a carboxylic acid.

Protein functional groups able to react with anhydrides include the α-amines at the

$$R\!-\!NH_2 \;+\; \text{(succinic anhydride)} \longrightarrow R\!-\!\text{(amide bond structure)} \tag{14}$$

Amine	Succinic	Amide Bond
Compound	Anhydride	

N-terminals, the ε-amine of lysine side chains, cysteine sulfhydryl groups, the phenolate ion of tyrosine residues, and the imidazolyl ring of histidines. However, acylation of cysteine, tyrosine, and histidine side chains forms unstable complexes that are easily reversible to regenerate the original group. Only amine functional groups of proteins are stable to acylation with anhydride reagents, forming amide bonds (Fraenkel-Conrat, 1959; Smyth, 1967).

Another potential site of reactivity for anhydrides in protein molecules is modification of any attached carbohydrate chains. In addition to amino group modification in the polypeptide chain, glycoproteins may be modified at their polysaccharide hydroxyl groups to form esterified derivatives. Esterification of carbohydrates by acetic anhydride, especially cellulose, is a major industrial application for this compound. In aqueous solutions, however, esterification will be a minor product, since the oxygen of water is about as strong a nucleophile as the hydroxyls of sugar residues.

The major side reaction to the desired acylation product is hydrolysis of the anhydride. In aqueous solutions anhydrides may breakdown by the addition of one molecule of water to yield two unreactive carboxylate groups. The presence of an excess of the anhydride in the reaction medium usually is used to minimize the effects of competing hydrolysis.

Since both hydrolysis and acylation yield the release of carboxylic acid functional groups, the medium becomes acidic during the course of the reaction. This requires either the presence of a strongly buffered environment to maintain the pH or periodic monitoring and adjustment of the pH with base as the reaction progresses.

2. Thiol-Reactive Chemical Reactions

Reactive groups able to couple with sulfhydryl-containing molecules are perhaps the second most common functional groups present on cross-linking or modification reagents. Especially in the design of heterobifunctional cross-linkers, sulfhydryl-reactive groups frequently are present on one of the two ends. The other end of such cross-linkers is often an amine-reactive functional group that is coupled to a target molecule before the sulfhydryl-reactive end, due to the labile nature of amine acylation chemistries. The primary coupling chemical reactions for modification of sulfhydryls proceed by one of two routes: alkylation or disulfide interchange. Many of the reactive groups that undergo these reactions are stable enough in aqueous environments to allow a two-step conjugation strategy to be used (Chapter 5, Section 1). Once initiated, most of these reactions are rapid and occur in high yield to give stable thioether and disulfide bonds.

2.1. Haloacetyl and Alkyl Halide Derivatives

Three forms of activated halogen derivatives can be used to create sulfhydryl-reactive compounds; haloacetyl (see Chapter 1, Section 5.2), benzyl halides that react through a resonance activation process with the neighboring benzene ring, and alkyl halides that possess the halogen β to a nitrogen or sulfur atom, as in N- and S-mustards. In each of these compounds the halogen group is easily displaced by an attacking nucleophilic substance to form an alkylated derivative with loss of HX (where X is the halogen and the hydrogen comes from the nucleophile). Haloacetyl compounds and benzyl halides typically are iodo or bromo derivatives, whereas the halo-mustards mainly employ chloro and bromo forms (see Chapter 4, Section 10 for examples of homobifunctional reagents that employ reactive halogen groups).

Although the primary utility of active halogen compounds is to modify sulfhydryl groups in proteins or other molecules, the reaction is not totally specific. Iodoacetyl (and bromoacetyl) derivatives can react with a number of functional groups within proteins: the sulfhydryl group of cysteine, both imidazolyl side chain nitrogens of histidine, the thioether of methionine, and the primary ε-amine group of lysine residues and N-terminal α-amines (Gurd, 1967). The relative rate of reaction with each of these residues is generally dependent on the degree of ionization and thus the pH at which the modification is done. The exception to this rule is methioninyl thioethers, which react rapidly at nearly all pH values above 1.7 (Vithayathil and Richards, 1960). The only reaction resulting in one definitive product is that of the alkylation of cysteine sulfhydryls, giving the carboxymethylcysteinyl derivative (Cole *et al.*, 1958) (Reaction 15). Histidine groups may be modified at either nitrogen atom of its imidazolyl side chain, thus producing the possibility of either mono-substituted or di-substituted products (Crestfield *et al.*, 1963). With primary amine groups such as in the side chain of lysine residues, the products of the reaction are either the secondary amine monocarboxymethyllysine or the tertiary amine derivative dicarboxymethyllysine. Methionine thioether groups give the most complicated products, some of which rearrange or decompose unpredictably. The only stable carboxy derivative of methionine is where the terminal methyl group is lost to form carboxymethylhomocysteine, the same product as the reaction of iodoacetate with homocysteine. For a complete illustration of these reactions, see Chapter 4, Section 10.

$$\text{R}'\text{—SH} \quad + \quad \underset{\text{Iodoacetyl Derivative}}{\text{R}\overset{\text{O}}{\overset{\|}{\diagup}}\text{I}} \quad \longrightarrow \quad \underset{\text{Thioether Bond}}{\text{R}\overset{\text{O}}{\overset{\|}{\diagup}}\text{S}_{\text{R}'}} \quad + \quad \text{HI} \tag{15}$$

R'—SH   Sulfhydryl Compound

The relative reactivity of α-haloacetates toward protein functionalities is sulfhydryl > imidazolyl > thioether > amine. Among halo derivatives the relative reactivity is I > Br > Cl > F, with fluorine being almost unreactive. The α-haloacetamides have the same trend of relative reactivities, but will obviously not create a terminal carboxylate functional group.

Thus, iodoacetate has the highest reactivity toward sulfhydryl cysteine residues and may be directed specifically for —SH modification. If iodoacetate is present in limiting quantities (relative to the number of sulfhydryl groups present) and at slightly alkaline pH, cysteine modification will be the exclusive reaction. The specificity of this modification has been used in the design of heterobifunctional cross-linking reagents, where one end of the cross-linker contains an iodoacetamide derivative and the other end contains a different functional group directed at another chemical target (see SIAB, Chapter 5, Section 5).

2.2. Maleimides

Maleic acid imides (maleimides) are derivatives of the reaction of maleic anhydride and ammonia. This functional group is a popular constituent of many heterobifunctional cross-linking agents (Chapter 5). The double bond of maleimides may undergo an alkylation reaction with sulfhydryl groups to form stable thioether bonds. Maleimide reactions are specific for sulfhydryl groups in the pH range 6.5–7.5 (Heitz *et al.*, 1968; Smyth *et al.*, 1964; Gorin *et al.*, 1966; Partis *et al.*, 1983). At pH 7, the reaction of the maleimide with sulfhydryls proceeds at a rate 1000 times greater than its reaction with amines. At higher pH values some cross-reactivity with amino groups takes place (Brewer and Riehm, 1967). One of the carbons adjacent to the maleimide double bond undergoes nucleophilic attack by the thiolate anion to generate the addition product (Reaction 16). When sufficient quantities of —SH groups are being alkylated, the reaction may be followed spectrophotometrically by the decrease in absorbance at 300 nm as the double bond reacts and disappears.

$$R'—SH \quad + \quad \text{Maleimide Derivative} \quad \longrightarrow \quad \text{Thioether Bond} \tag{16}$$

Sulfhydryl Compound   Maleimide Derivative   Thioether Bond

The maleimide group also may undergo hydrolysis to an open maleamic acid form that is unreactive toward sulfhydryls (Chapter 9, Section 5). Hydrolysis may occur after sulfhydryl coupling to the maleimide, as well. This ring-opening reaction typically happens faster the higher the pH becomes. Hydrolysis is also dependent on the type of chemical group next to the maleimide function. For instance, the cyclohexane ring of SMCC (Chapter 5, Section 1.3) provides increased stability to maleimide hydrolysis probably due to its steric effects and its lack of aromatic character. However, the adjacent phenyl ring of MBS allows much greater rates of hydrolysis to occur at the maleimide ring (Chapter 5, Section 1.4).

2.3. Aziridines

An aziridine functional group is a small ring system composed of one nitrogen and two carbon atoms. The highly hindered nature of this heterocyclic ring gives it strong reactivity toward nucleophiles. Sulfhydryls will react with aziridine-containing reagents in a ring-opening process, forming thioether bonds (Reaction 17). The simplest aziridine compound, ethylenimine, can be used to transform available sulfhydryl groups into amines (Chapter 1, Section 4.3).

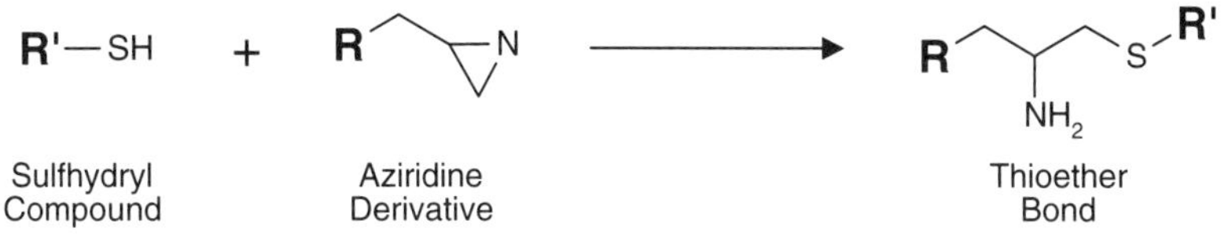

R'—SH + R⟨aziridine⟩N ⟶ R⟨...⟩NH₂ S—R'   (17)

Sulfhydryl   Aziridine   Thioether
Compound   Derivative   Bond

The reaction of an aziridine with a thiol is highly specific at slightly alkaline pH values. In aqueous solution, the major side reaction is hydrolysis.

Substituted aziridines have been used to form homobifunctional and trifunctional cross-linking agents, although their use has been limited (Ross, 1953; Alexander, 1954). The functional group has found use, however, in the design of the fluorescent probe dansyl aziridine (5-dimethylaminonaphthalene-2-sulfonyl aziridine) (Johnson *et al.*, 1978; Grossman *et al.*, 1981).

2.4. Acryloyl Derivatives

Reactive double bonds are capable of undergoing addition reactions with sulfhydryl groups. A popular example of this type of functional group is the maleimide group (Section 2.2). However, derivatives of acrylic acid are also able to participate in this reaction, although the rate of sulfhydryl addition is somewhat slower than that of maleimides. The reaction of an acryloyl compound with a sulfhydryl group occurs with the creation of a stable thioether bond (Reaction 18).

R'—SH + R⟨...⟩CH₂ ⟶ R⟨...⟩S—R'

                                                                        (18)

Sulfhydryl   Acryloyl   Thioether
Compound   Derivative   Bond

Although acryloyl cross-linking agents have not been common, the functional group has found use in the design of the sulfhydryl-reactive fluorescent probe 6-acryloyl-2-dimethylaminonaphthalene (acrylodan; Molecular Probes) (Yem *et al.*, 1992; Epps *et al.*, 1992).

2.5. Arylating Agents

Arylating agents are reactive aromatic compounds containing a constituent on the ring that can undergo nucleophilic substitution. The most common arylating agents are derivatives of benzene that possess either halogen or sulfonate groups on the ring. The presence of electron-withdrawing constituents, such as nitro groups, increases the reactivity of the replaceable group. Although aryl halides are commonly used to modify amine-containing molecules to form aryl amine derivatives, they also react quite readily with sulfhydryl groups.

Fluorobenzene-type compounds have been used as functional groups in homobi-

functional cross-linking agents (Chapter 4, Section 4). Their reaction with nucleophiles involves bimolecular nucleophilic substitution, causing the replacement of the
fluorine atom with the sulfhydryl derivative, creating a substituted aryl bond (Reaction
19). Conjugates formed with sulfhydryl groups are reversible by cleaving with an
excess of thiol (such as DTT) (Shaltiel, 1967). Detection reagents incorporating reactive aryl chemistry include 2,4-dinitrofluorobenzene and trinitrobenzenesulfonate
(Eisen *et al.*, 1953). The relative rate of reactivity for aryl compounds is F > Cl ~ Br >
Sulfonate.

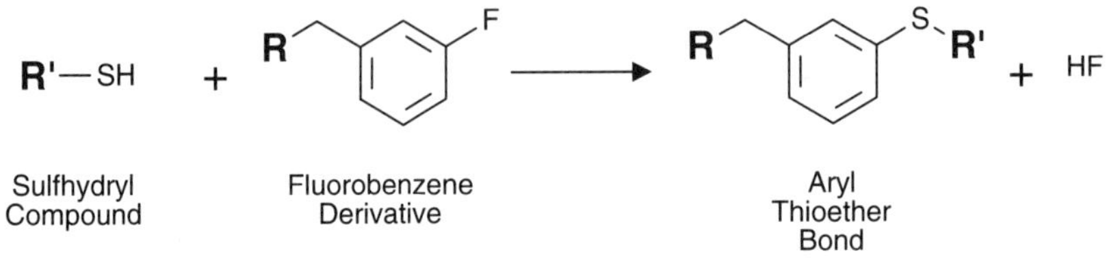

$$\tag{19}$$

| Sulfhydryl Compound | Fluorobenzene Derivative | Aryl Thioether Bond |

2.6. Thiol–Disulfide Exchange Reagents

Compounds containing a disulfide group are able to participate in disulfide exchange
reactions with another thiol. The disulfide exchange (also called interchange) process
involves attack of the thiol at the disulfide, breaking the -S—S- bond, with subsequent
formation of a new mixed disulfide constituting a portion of the original disulfide
compound (Reaction 20). The reduction of disulfide groups to sulfhydryls in proteins
using thiol-containing reductants proceeds through the intermediate formation of a
mixed disulfide (Chapter 1, Section 4.1). If the thiol is present in excess, the mixed
disulfide can go on to form a symmetrical disulfide consisting entirely of the thiol
reducing agent—thus completely reducing the original disulfide to free sulfhydryls. If
the thiol reductant is not present in large enough excess, the mixed disulfide product is
the end result.

$$\tag{20}$$

| Sulfhydryl Compound | Disulfide Derivative | Disulfide Interchange |

    Cross-linking or modification reactions using disulfide exchange processes form
disulfide linkages with sulfhydryl-containing molecules. These bonds are reversible
using disulfide reducing agents. Thus, conjugates may be created and later released for
analysis by incubation with DTT or other disulfide reductants. The disulfide bond
within these cross-links also permits important reactions to occur *in vivo*, such as the
release of the toxin component of immunotoxin conjugates, allowing the cytotoxic
portion to penetrate target cells and cause cell death (Chapter 11).
    Disulfide exchange reactions occur over a broad range of conditions—from acid to
basic pH—and in a wide variety of buffer constituents. Most cross-linking reactions
involving disulfide exchange are done under physiological conditions or those most
appropriate to maintain stability of the protein or other molecule being modified.

Pyridyl Disulfides

A pyridyl dithiol group is perhaps the most popular type of thiol-disulfide exchange functional group used in the construction of cross-linkers or modification reagents. Pyridyl disulfides can be created from available primary amines on molecules through the reaction of 2-iminothiolane in tandem with 4,4'-dipyridyl disulfide (King *et al.*, 1978). For instance, the simultaneous reaction between a protein, 2-iminothiolane, and 4,4'-dipyridyl disulfide yields a modification containing reactive pyridyl disulfide groups in a single step (Chapter 1, Section 4.1).

A pyridyl disulfide will readily undergo an interchange reaction with a free sulf-hydryl to yield a single mixed disulfide product. This is due to the fact that the pyridyl disulfide functional group contains a leaving group that is easily transformed into a nonreactive compound not capable of participating in further mixed disulfide forma-tion. Thus, the thiol-disulfide exchange reaction can be controlled to occur with only one-half of the original disulfide compound. For instance, a reagent system containing a pyridyl disulfide group, such as SPDP (Chapter 5, Section 1.1), is able to react with sulfhydryl groups by releasing the electron-stabilized compound pyridine-2-thione (Reaction 21). Since the leaving group does not possess a free thiol, it cannot disulfide exchange with another molecule of the attacking sulfhydryl compound. Thus, only one end of the reagent has potential for becoming attached to the sulfhydryl-containing molecule.

Sulfhydryl Compound	Pyridyl Disulfide Derivative	Disulfide Bond	Pyridine-2-thione

(21)

Pyridyl dithiol-containing cross-linking and modification reagents are highly effi-cient in forming disulfide bonds with sulfhydryl-containing molecules. In addition, the pyridine-2-thione leaving group has unique spectral properties that allow the mea-surement of sulfhydryl coupling by monitoring the increase in absorbance at 343 nm ($\varepsilon = 8.08 \times 10^3\,M^{-1}cm^{-1}$). Once a disulfide linkage is formed, it may be cleaved using standard disulfide reducing agents (Chapter 1, Section 4.1).

TNB-Thiol

Sulfhydryl groups activated with the leaving group 5-thio-2-nitrobenzoic acid can be used to couple free thiols by disulfide interchange similar to pyridyl disulfides, as discussed previously. A TNB-thiol-activated species may be created by reaction of a sulfhydryl group with Ellman's reagent, 5,5'-dithio-*bis*(2-nitrobenzoic acid) (DTNB), a compound useful for the quantitative determination of sulfhydryls in solution (Ell-man, 1958, 1959) (Chapter 1, Section 4.1). The disulfide of Ellman's reagent readily undergoes disulfide exchange with a free sulfhydryl to form a mixed disulfide with concomitant release of one molecule of the chromogenic substance 5-sulfido-2-nitrobenzoate, also called 5-thio-2-nitrobenzoic acid (TNB). The TNB-thiol group can again undergo interchange with a sulfhydryl-containing target molecule to yield a disulfide cross-link. Upon coupling with a sulfhydryl compound, the TNB group is released (Reaction 22). The intense yellow color produced by the TNB anion can be measured by its absorbance at 412 nm ($\varepsilon = 1.36 \times 10^4\,M^{-1}\text{cm}^{-1}$ at pH 8). Since each sulfhydryl that is coupled generates one molecule of TNB per molecule of Ellman's reagent, the possibility for quantifying the reaction exists.

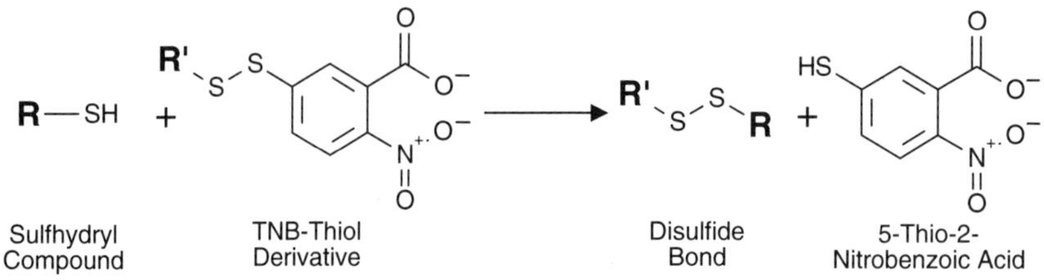

| Sulfhydryl Compound | TNB-Thiol Derivative | Disulfide Bond | 5-Thio-2-Nitrobenzoic Acid | (22) |

Disulfide exchange with a TNB-thiol group occurs efficiently at physiological to slightly alkaline pH conditions. Avoid the presence of disulfide reducing agents, as these will cleave the TNB group and prevent specific coupling.

Disulfide Reductants

Disulfide reduction by the use of disulfide interchange can be done using thiol-containing compounds such as DTT, 2-mercaptoethanol, or 2-mercaptoethylamine (Chapter 1, Section 4.1). The formation of free sulfhydryls from a disulfide group occurs in two stages. First, one molecule of the reducing agent undergoes disulfide exchange, cleaving the disulfide and forming a new, mixed disulfide. In the next stage, a second molecule of the thiol cleaves the mixed disulfide, releasing a free sulfhydryl and forming a molecule of oxidized reducing agent (Reaction 23).

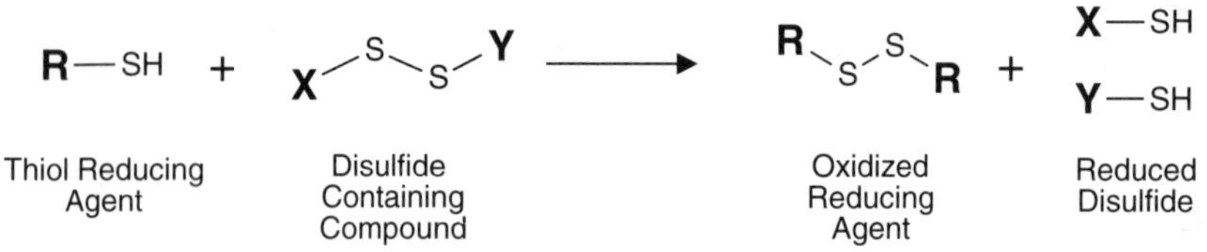

| Thiol Reducing Agent | Disulfide Containing Compound | Oxidized Reducing Agent | Reduced Disulfide | (23) |

Disulfide reduction occurs over a broad pH range and in a variety of buffer environments. The reaction can be done in denaturants, chaotropes, and detergents and under high salt conditions.

3. Carboxylate-Reactive Chemical Reactions

Chemical groups that specifically react with carboxylic acids are limited in variety. In aqueous solutions, the carboxylate functional group displays rather low nucleophilicity. For this reason, it is unreactive with the great majority of bioconjugate reagents that couple through a nucleophilic addition process.

Several important chemistries, however, have been developed that allow conjugation through a carboxylate group. The following sections briefly describe these reactions.

3.1. Diazoalkanes and Diazoacetyl Compounds

Diazomethane and other diazoalkyl derivatives have long been used to label carboxylate groups for analysis (Herriott, 1947; Riehm and Scheraga, 1965). A major application of such reagents has been in the HPLC analysis of low-molecular-weight compounds such as fatty acids (DeMar *et al.*, 1992). Several coumarin derivatives containing stable, carboxylate-reactive diazoalkane functionalities also are available for fluorescent-labeling of target molecules (Molecular Probes) (Ito and Sawanobori, 1982; Ito and Maruyama, 1983).

Diazoalkanes and diazoacetyl compounds (amides and esters) are spontaneously reactive with carboxylate groups without addition of other reactants or catalysts. The reaction mechanism involves attack of a negatively charged carboxylate oxygen atom on a protonated diazoalkyl group, liberating nitrogen gas and forming a covalent linkage (Reaction 24).

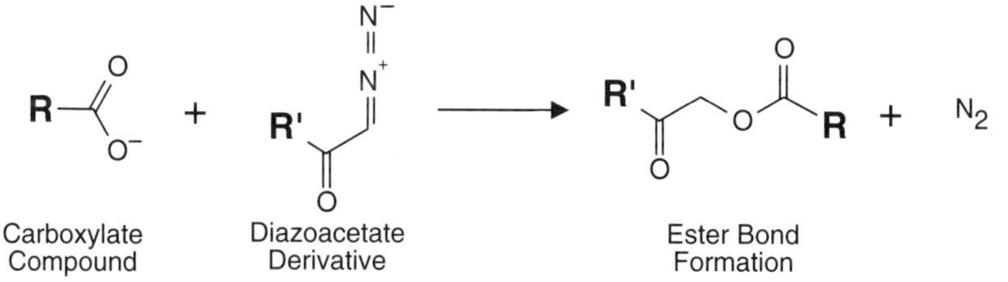

Carboxylate   Diazoacetate   Ester Bond
Compound   Derivative   Formation

$$\text{(24)}$$

The reaction with carboxylates occurs over a range of pH values, but is optimal at pH 5. Unfortunately the diazoalkyl compounds will cross-react with sulfhydryl groups at this pH. Under higher pH conditions, the reaction is even less specific due to reaction with other nucleophiles. In aqueous solution, the most likely side reaction is hydrolysis.

3.2. Carbonyldiimidazole

Carbonyldiimidazole (CDI), is a active carbonylating agent that contains two acylimidazole leaving groups (Chapter 3, Section 3). CDI reacts with carboxylic acids under nonaqueous conditions to form N-acylimidazoles of high reactivity (Reaction 25). The active intermediate forms in excellent yield due to the driving force created by the liberation of carbon dioxide and imidazole (Anderson, 1958). An active carboxylate then can react with amines to form amide bonds or with hydroxyl groups to form ester linkages (Reaction 26). The reaction has been used successfully in peptide synthesis (Paul and Anderson, 1960, 1962). In addition, activation of a styrene/4-vinylbenzoic acid copolymer with CDI was used to immobilize the enzyme lysozyme

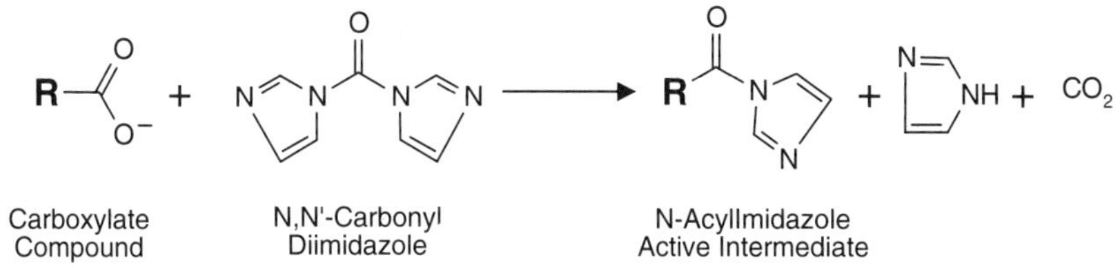

Carboxylate   N,N'-Carbonyl   N-Acylimidazole
Compound   Diimidazole   Active Intermediate

$$\text{(25)}$$

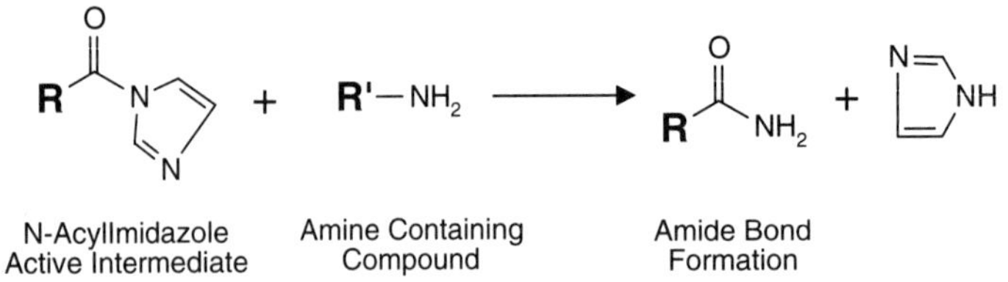

N-Acylimidazole   Amine Containing   Amide Bond
Active Intermediate   Compound   Formation

(26)

through its available amino groups to the carboxyl groups on the matrix (Bartling *et al.*, 1973).

CDI functions as a zero-length cross-linker if the activated species is a carboxylic acid, because the attack of another nucleophile liberates the imidazole leaving group. The conjugation reaction can be done in organic solvent or aqueous conditions, depending on the solubility of the nucleophile. For aqueous coupling of *N*-acylimidazoles to amine-containing compounds, optimal conditions include an alkaline pH environment from about pH 7 to 9 and in buffers containing no amines (avoid Tris or imidazole).

3.3. Carbodiimides

Carbodiimides function as zero-length cross-linking agents capable of activating a carboxylate group for coupling with an amine-containing compound. There are several major types of carbodiimide reagents commonly available that can be used for organic or aqueous reactions, depending on their individual solubility characteristics (Chapter 3, Section 1). The water-soluble reagents are used mainly for biological conjugations involving proteins and other macromolecules. The water-insoluble carbodiimides can be used in peptide synthesis or for the synthesis of other organic compounds.

Carbodiimides are used to mediate the formation of amide or phosphoramidate linkages between a carboxylate and an amine or a phosphate and an amine, respectively (Hoare and Koshland, 1966; Chu *et al.*, 1986; Ghosh *et al.*, 1990). Regardless of the type of carbodiimide, the reaction proceeds by the formation of an intermediate *o*-acylisourea that is highly reactive and short-lived in aqueous environments. The attack of an amine nucleophile on the carbonyl group of this ester results in the loss an isourea derivative and formation of an amide bond (see Reactions 11 and 12). The major competing reaction in water is hydrolysis.

4. Hydroxyl-Reactive Chemical Reactions

Hydroxyl-reactive chemical reactions include not only those modification agents able directly to form a stable linkage with an —OH group, but also a broad range of reagents that are designed to temporarily activate the group for coupling with a secondary functional group. Many of the chemical methods for modifying hydroxyls originally were developed for use with chromatography supports in the coupling of affinity ligands. Some of these same chemical reactions have found application in bioconjugate techniques for cross-linking a hydroxyl-containing molecule with another substance, usually containing a nucleophile. For instance, carbohydrate-containing molecules such as polysaccharides or glycoproteins can be coupled through their sugar residues using hydroxyl-specific reactions. In addition, polymers and other organic compounds containing hydroxyls (such as polyethylene glycol) may be conjugated with another molecule using these chemistries.

4.1. Epoxides and Oxiranes

An epoxide or oxirane group can react with nucleophiles in a ring-opening process. The reaction can take place with primary amines, sulfhydryls, or hydroxyl groups to create secondary amine, thioether, or ether bonds, respectively. See Section 1.7 for further information on this reaction.

4.2. Carbonyldiimidazole

Carbonyldiimidazole (CDI) is a active carbonylating agent that contains two acylimidazole leaving groups (Chapter 3, section 3). The compound can react with a carboxylate to form an active acylimidazole group capable of coupling with amine-containing molecules (Section 3.2). However, CDI also can react with hydroxyl groups to create a reactive intermediate. If CDI is used to activate a hydroxyl functional group, the reaction proceeds quite differently from its reaction with carboxylates. The active intermediate formed by the reaction of CDI with an —OH group is an imidazolyl carbamate (Reaction 27). Attack by an amine releases the imidazole, but not the carbonyl. Thus, hydroxyl-containing molecules may be coupled to amine-containing molecules with the result of a one-carbon spacer, forming stable urethane (N-alkyl carbamate) linkages (Reaction 28). This coupling procedure has been applied to the activation of hydroxyl-containing chromatography supports for the immobilization of amine-containing affinity ligands (Bethell et al., 1979; Hearn et al., 1979, 1983) and also to the activation of polyethylene glycol for the modification of amine-containing macromolecules (Beauchamp et al., 1983).

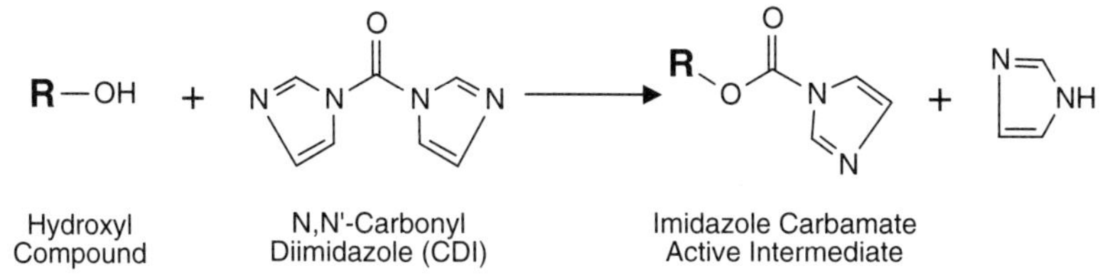

| Hydroxyl Compound | N,N'-Carbonyl Diimidazole (CDI) | Imidazole Carbamate Active Intermediate | |

(27)

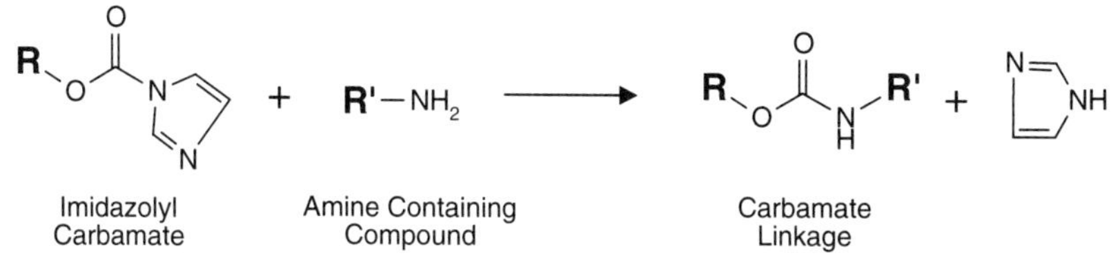

| Imidazolyl Carbamate | Amine Containing Compound | Carbamate Linkage | |

(28)

4.3. N,N'-Disuccinimidyl carbonate or N-Hydroxysuccinimidyl chloroformate

N,N'-Disuccinimidyl carbonate (DSC) consists of a carbonyl group containing, in essence, two NHS esters. The compound is highly reactive toward nucleophiles. In aqueous solutions, DSC will hydrolyze to form two molecules of N-hydroxysuccinimide (NHS) with release of one molecule of CO_2. In nonaqueous environments, the reagent can be used to activate a hydroxyl group to a succinimidyl carbonate derivative (Reaction 29). DSC-activated hydroxylic compounds can be used to conjugate with amine-containing molecules to form stable cross-linked products (Reaction 30). The linkage created from this reaction is a urethane derivative or a carbamate bond, displaying excellent stability.

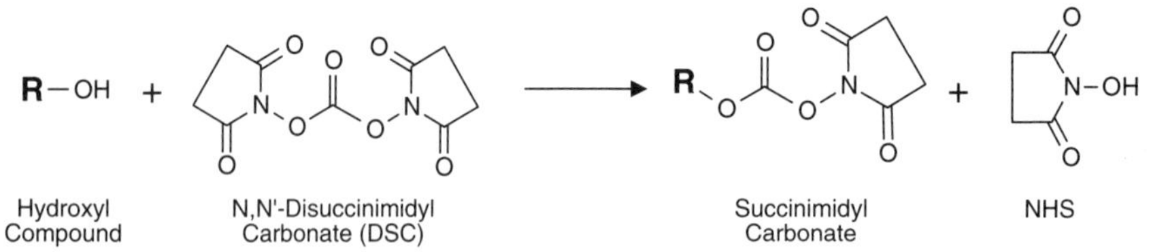

| Hydroxyl Compound | N,N'-Disuccinimidyl Carbonate (DSC) | Succinimidyl Carbonate | NHS | (29) |

| Amine Containing Compound | Succinimidyl Carbonate | Carbamate Linkage | NHS | (30) |

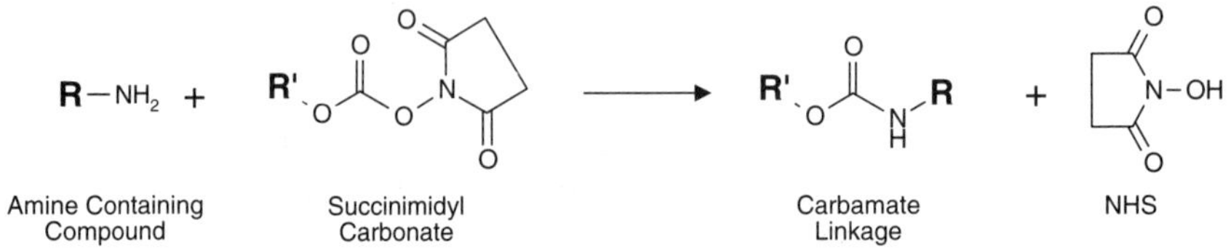

A related reagent, N-hydroxysuccinimidyl chloroformate, also is a bifunctional carbonyl derivative containing an NHS ester and an acid chloride. In aqueous solutions the compound is unstable to hydrolysis, rapidly breaking down to NHS, CO_2, and HCl. In nonaqueous environments, however, NHS-chloroformate may be used to activate a hydroxyl group in a manner similar to that of DSC. Reaction of the chloroformate with a hydroxylic residue forms the same succinimidyl carbonate derivative as the reaction of DSC with —OH groups (Reaction 31). Subsequent conjugation with an amine-containing compound yields a carbamate linkage. The bond is identical to that formed from the reaction of CDI-activated hydroxyls with amine-containing compounds (see previous section).

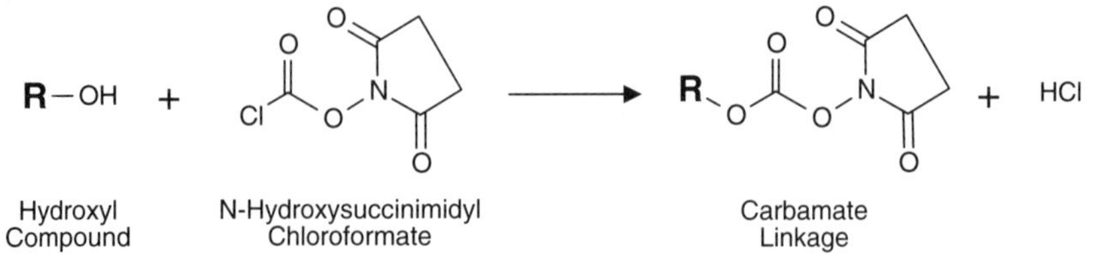

| Hydroxyl Compound | N-Hydroxysuccinimidyl Chloroformate | Carbamate Linkage | (31) |

4.4. Oxidation with Periodate

Sodium periodate can be used to oxidize hydroxyl groups on adjacent carbon atoms, forming reactive aldehyde residues suitable for coupling with amine- or hydrazide-containing molecules. The reaction occurs with two adjacent secondary hydroxyls to cleave the carbon—carbon bond between them and create two terminal aldehyde groups (Reaction 32). When one of the adjacent hydroxyls is a primary hydroxyl, reaction with periodate releases one molecule of formaldehyde and leaves a terminal aldehyde residue on the original diol compound (Reaction 33). These reactions can be used to generate cross-linking sites in carbohydrates or glycoproteins for subsequent conjugation of amine-containing molecules by reductive amination (Chapter 1, Section 4.4, and Chapter 3, Section 4). Sodium periodate also reacts with hydroxylamine derivatives—compounds containing a primary amine and a secondary hydroxyl group on adjacent carbon atoms. Oxidation cleaves the carbon—carbon bond, forming a terminal aldehyde group on the side that had the original hydroxyl residue (Reaction 34). This reaction can be used to create reactive aldehydes on N-terminal serine residues of peptides (Geoghegan and Stroh, 1992).

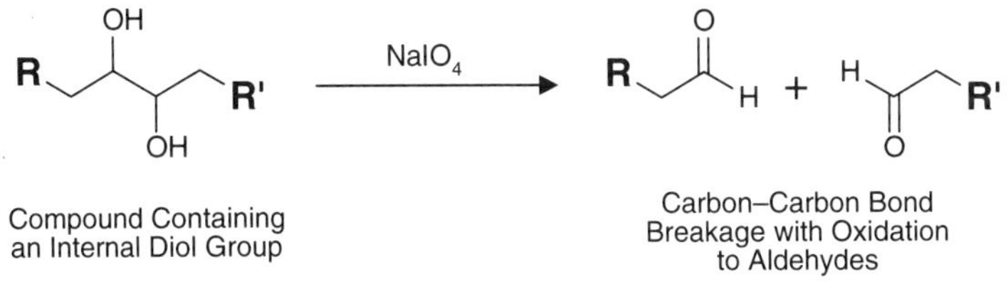

Compound Containing
an Internal Diol Group

Carbon–Carbon Bond
Breakage with Oxidation
to Aldehydes

(32)

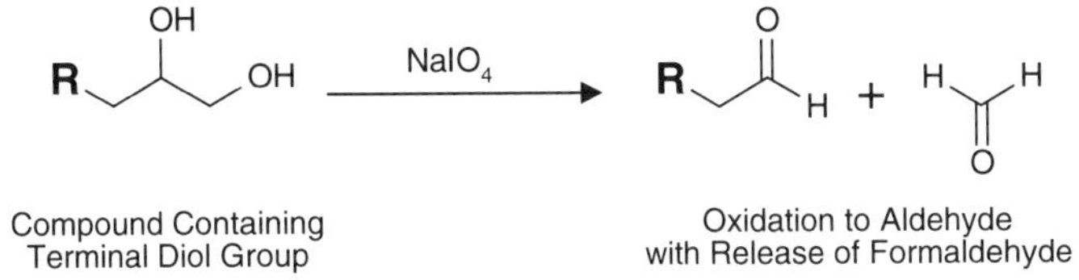

Compound Containing
Terminal Diol Group

Oxidation to Aldehyde
with Release of Formaldehyde

(33)

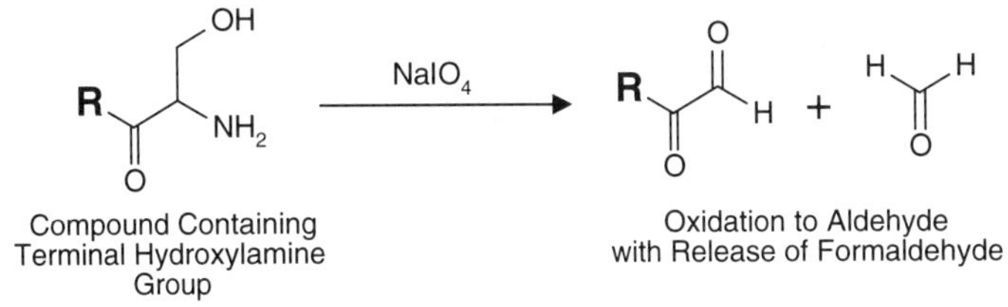

Compound Containing
Terminal Hydroxylamine
Group

Oxidation to Aldehyde
with Release of Formaldehyde

(34)

4.5. Enzymatic Oxidation

Certain enzymes may be used to oxidize hydroxyl-containing carbohydrates to create aldehyde groups (Chapter 1, Section 4.4). For example, the reaction of galactose

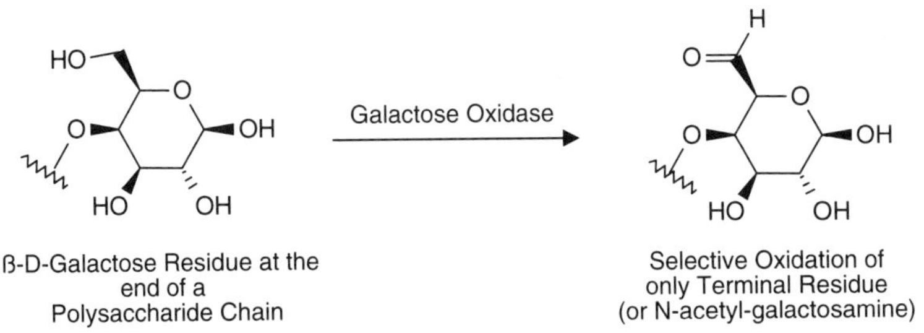

β-D-Galactose Residue at the
end of a
Polysaccharide Chain

Selective Oxidation of
only Terminal Residue
(or N-acetyl-galactosamine)

(35)

oxidase on terminal galactose or N-acetyl-D-galactose residues proceeds to form C-6 aldehyde groups on polysaccharide chains (Reaction 35). These groups then can be used for conjugation reactions with amine- or hydrazide-containing molecules.

4.6. Alkyl Halogens

Reactive alkyl halogen compounds can be used to modify specifically hydroxyl groups in carbohydrates, polymers, and other molecules. Chloro or bromo derivatives of short alkyl chains containing a second functional group on their other end (typically a carboxylate group) can be used to form spacer arms useful for conjugation with another substance. Brunswick *et al.* (1988) used chloroacetic acid to modify the hydroxyl groups of dextran, forming the carboxymethyl derivative (Reaction 36). The carboxylates then may be coupled with amine-containing molecules using a carbodiimide reaction scheme. In a somewhat similar approach, Noguchi *et al.* (1992) prepared a carboxylate spacer arm by reacting 6-bromohexanoic acid with a dextran polymer (Chapter 15, Section 2.2).

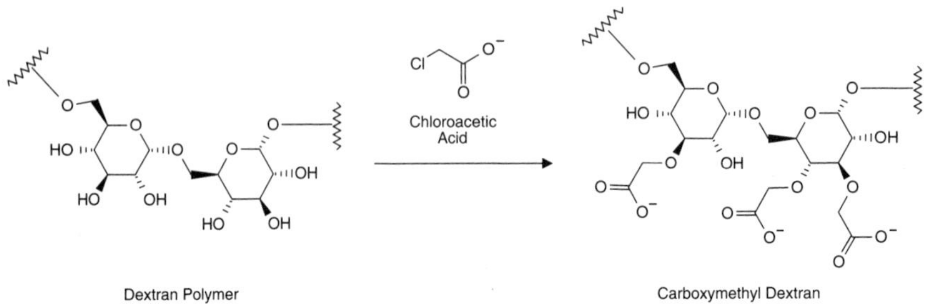

Dextran Polymer   Chloroacetic Acid   Carboxymethyl Dextran

(36)

Modification of hydroxyl groups with such compounds can be done in 3–10 M NaOH by reacting from 25 to 40°C for 1.5–4 h.

4.7. Isocyanates

Isocyanates can be formed from the reaction of an aromatic amine with phosgene (Rifai and Wong, 1986). They also can be created from acyl azides by treatment at 80°C in the presence of an alcohol (Chapter 8, Section 1.5). In the transformation, the acyl azide group rearranges to form an isocyanate that can react with hydroxyl-containing molecules to form a urethane (carbamate) linkage (Reaction 37). The reactivity of isocyanates is excellent, but for the same reason their stability can be a problem. Isocyanate compounds often are too unstable to allow commercial sale. In storage, moisture decomposes them, releasing CO_2 and leaving an aromatic amine in its place. In aqueous environments, the aromatic amine can react with another molecule of isocyanate to form a urea derivative (Annunziato *et al.*, 1993).

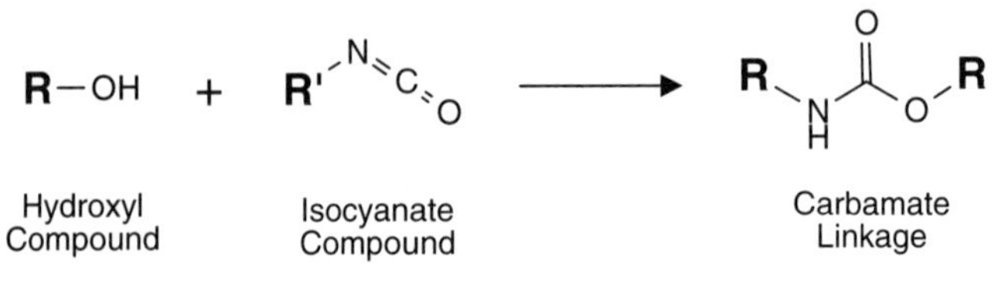

R—OH + R' N=C=O   R—N—C—O—R (37)

Hydroxyl Compound   Isocyanate Compound   Carbamate Linkage

Isocyanate-containing reagents can be used to cross-link or label hydroxyl-containing molecules, including polysaccharides. Carbohydrate modification can be done without the need for prior oxidation of sugar residues with periodate to form reactive aldehydes, as is common in many protocols (Chapter 1, Section 4.4). The reaction occurs best at alkaline pH values (e.g., pH 8.5). Many coupling protocols avoid the hydrolysis problem by performing the reaction in organic solvent (i.e., DMSO).

Annunziato *et al.* (1993) have reported on the synthesis and use of a novel heterobifunctional cross-linking reagent containing a hydroxyl-reactive isocyanate group on one end and a sulfhydryl-reactive maleimide group on the other end. The compound can be useful in labeling hydroxylic molecules for subsequent conjugation with thiol-containing molecules.

5. Aldehyde- and Ketone-Reactive Chemical Reactions

Aldehyde and ketone groups are important reactive sites in molecules for many bioconjugate strategies. Although some pharmacological agents contain ketones, these groups usually are not present in proteins and other biological molecules. Even when a molecule does not contain these functional groups, however, they may be created through a number of processes (Chapter 1, Section 4.4). The following sections discuss the major reactions that can be done with aldehydes and ketones to modify or cross-link molecules containing them.

5.1. Hydrazine Derivatives

Derivatives of hydrazine, especially the hydrazide compounds formed from carboxylate groups, can specifically react with aldehyde or ketone functional groups in target molecules. Reaction with either group creates a hydrazone linkage (Reaction 38)—a type of Schiff base. This bond is relatively stable if it is formed with a ketone, but somewhat labile if the reaction is with an aldehyde group. Even the linkages of hydrazones with aldehydes, however, are much more stable than the easily reversible Schiff base interaction of an amine with an aldehyde. To further stabilize the bond between a hydrazide and an aldehyde, the hydrazone may be reacted with sodium cyanoborohydride to reduce the double bond and form a secure covalent linkage.

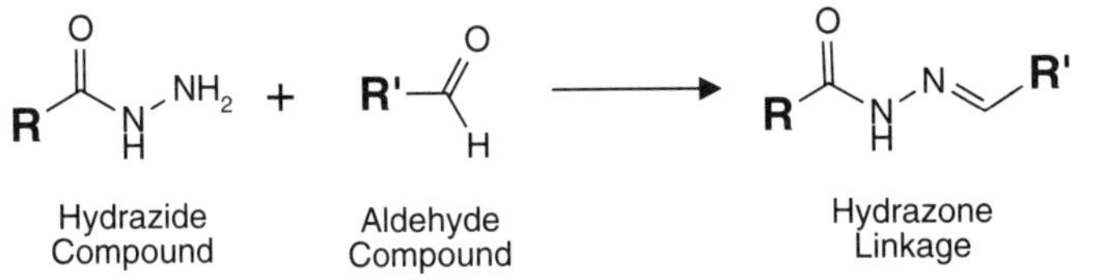

| Hydrazide Compound | Aldehyde Compound | Hydrazone Linkage |

(38)

5.2. Schiff Base Formation

Aldehydes and ketones can react with primary and secondary amines to form Schiff bases, a dehydration reaction yielding an imine (Reaction 39). However, Schiff base formation is a relatively labile, reversible interaction that is readily cleaved in aqueous solution by hydrolysis. The formation of Schiff bases is enhanced at alkaline pH values, but they are still not stable enough to use for cross-linking applications unless they are reduced by reductive amination (see below).

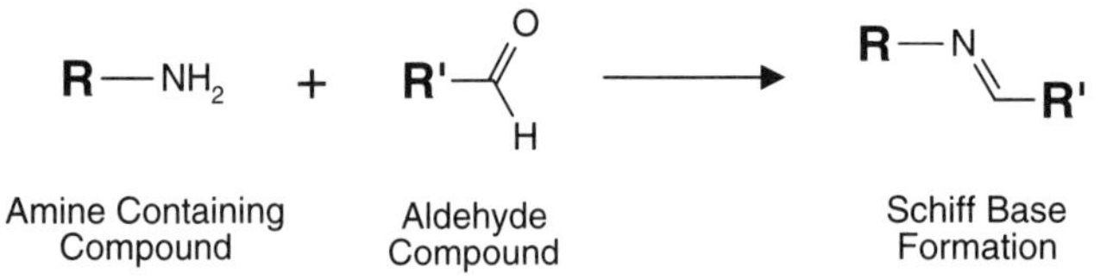

(39)

Amine Containing Compound   Aldehyde Compound   Schiff Base Formation

The reaction of dicarbonyl compounds, such as glyoxal or phenylglyoxal, with a guanidinyl group, such as that of an arginine residue, proceeds to yield a more stable linkage due to the formation of a cyclic derivative (Reaction 40).

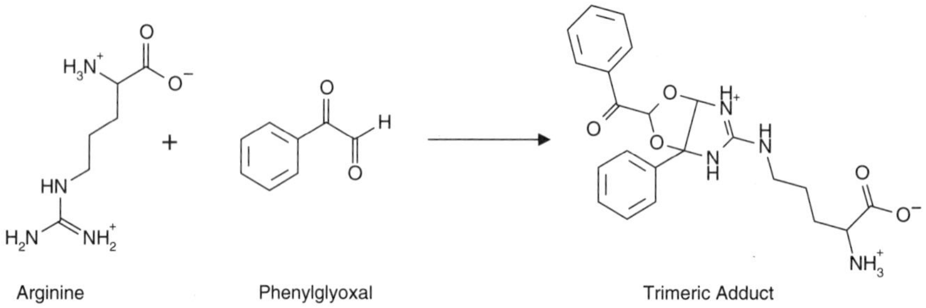

(40)

Arginine   Phenylglyoxal   Trimeric Adduct

5.3. Reductive Amination

Reductive amination (or alkylation) may be used to conjugate an aldehyde- or ketone-containing molecule with an amine-containing molecule. Schiff base formation between aldehydes and amines occurs readily in aqueous solutions, especially at elevated pH. This type of linkage, however, is not stable unless reduced to secondary or tertiary amine bonds. A number of reducing agents can be used to convert specifically the Schiff base interaction into an alkylamine linkage (Reaction 41). Once reduced, the bonds are highly stable and will not readily hydrolyze in aqueous environments. The use of reductive amination to conjugate an aldehyde-containing molecule to an amine-containing molecule results in a zero-length cross-linking procedure where no additional spacer atoms are introduced between the molecules (Chapter 3, Section 4). Reaction of ammonia or a diamine compound with an aldehyde by reductive amination is a method of creating a primary amine functional group (Chapter 1, Section 4.3).

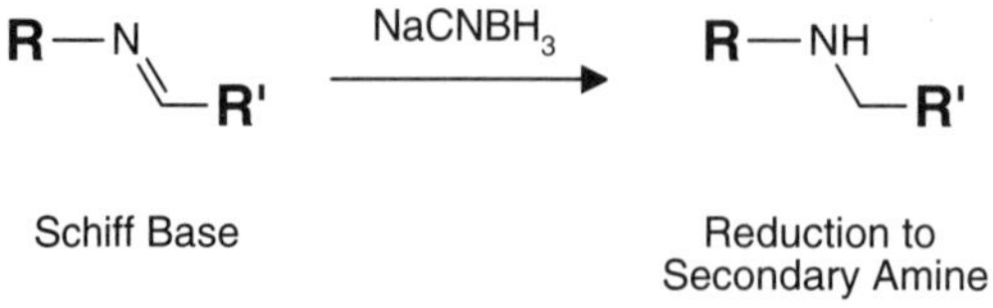

(41)

Schiff Base   Reduction to Secondary Amine

5.4. Mannich Condensation

Aldehydes may participate in a condensation reaction with an amine compound and a substance containing a sufficiently active hydrogen, yielding an alkylated derivative

that effectively cross-links the two molecules through the carbonyl group of the aldehyde. Strictly speaking, the Mannich reaction consists of the condensation of formaldehyde (or sometimes another aldehyde) with ammonia, in the form if its salt, and
another compound containing an active hydrogen. Instead of using ammonia, however, this reaction can be done with primary or secondary amines, or even with amides.
An example is illustrated in the condensation of phenol, formaldehyde, and a primary
amine salt (Reaction 42).

(42)

Phenol   Formaldehyde   Primary   Condensation
                                  Amine Salt   Products

The Mannich reaction provides an often-superior alternative to diazonium conjugation (Section 6.1), because of the disadvantages inherent in the instability of both the
diazonium group and the resultant diazo linkage. By contrast, conjugations done
through Mannich condensations result in stable covalent bonds.

The cross-linking scheme using this method can make use of the native ε- and
N-terminal amines on proteins as the source of primary amine for the condensation
reaction. Added to the conjugation reaction is formaldehyde and the desired molecule
to be coupled containing an appropriately active hydrogen. The Mannich reaction
should not be used for molecules containing both an amine and a reactive hydrogen,
since polymerization may occur. It is especially useful for preparing hapten–carrier
conjugates when the hapten contains no other available functional groups suitable for
cross-linking, but does contain an active hydrogen (Chapter 9, Section 6.2).

6. Active Hydrogen-Reactive Chemical Reactions

Many compounds contain reactive (or replaceable) hydrogens that are able to participate in conjugation reactions using certain chemical reactions. These hydrogens typically are associated with aromatic systems wherein an electron-donating group activates certain positions on the ring toward substitution reactions. At such carbons, the
hydrogen is easily displaced by an attacking electrophilic group able to form a new
covalent linkage. Several common modification reactions are used in bioconjugate
chemistry to label or cross-link molecules at active hydrogen sites. The following three
sections discuss these chemical reactions.

6.1. Diazonium Derivatives

Diazonium groups react with active hydrogen sites on aromatic rings to give covalent
diazo bonds. Generation of a diazonium functional group usually is done from an
aromatic amine by reaction with sodium nitrite under acidic conditions at 0°C (Chapter 1, Section 4.3, and Chapter 9, Section 6.1). The highly reactive and unstable
diazonium is reacted immediately with an active hydrogen-containing compound at
pH 8–10. In general, at pH 8 the diazonium group will react principally with histi-

dinyl residues, attacking the electron-rich nitrogens of the imidazole ring. At higher pH, the phenolic side chain of tyrosine groups can be modified (Reaction 43). The reaction proceeds by electrophilic attack of the diazonium group toward the electron-rich points on the target molecules. Phenolic compounds are modified at positions *ortho* and *para* to the aromatic hydroxyl group. For tyrosine side chains, only the *ortho* modification is available.

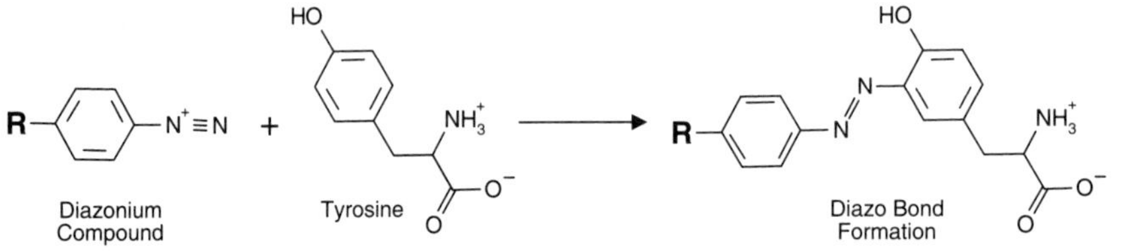

(43)

Cross-linking using diazonium compounds usually creates deeply colored products characteristic of the diazo bonds. Occasionally, the conjugated molecules may turn dark brown or even black. The diazo linkages are reversible by addition of 0.1 M sodium dithionite in 0.2 M sodium borate, pH 9. On cleavage, the color of the complex is lost.

6.2. Mannich Condensation

The Mannich reaction consists of the condensation of an active hydrogen-containing compound with an amine-containing compound in the presence of formaldehyde. See Section 5.4 for addition details.

6.3. Iodination Reactions

Radioiodination involves the substitution of radioactive iodine atoms for reactive hydrogen sites in target molecules. The process usually involves the action of a strong oxidizing agent to transform iodide ions into a highly reactive electrophilic iodine compound (typically I_2 or a mixed halogen species such as ICl). Formation of this electrophilic species leads to the potential for rapid iodination of aromatic compounds

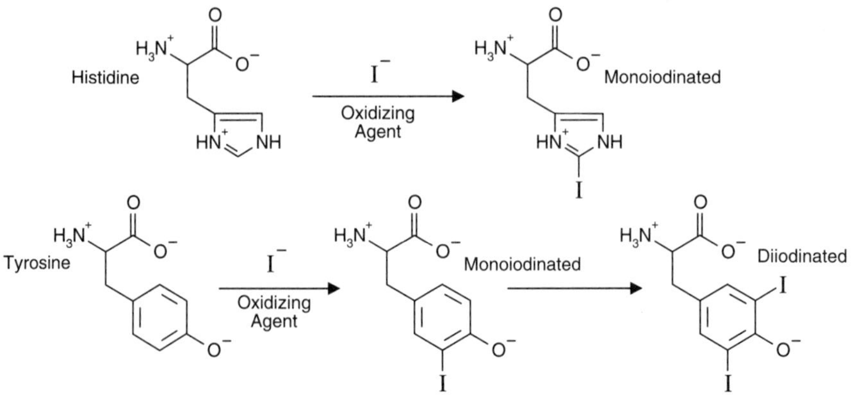

(44)

containing strong activating groups, such as aryl compounds. Particularly, aromatic constituents that have electron-donating groups can sufficiently activate the carbons on the ring to undergo electrophilic substitution reactions. Therefore, phenols, aniline derivatives, or alkyl anilines that contain OH, NH_2, or NHR constituents, respectively, are very susceptible to being iodinated. In proteins, this translates into tyrosine side-chain phenolic groups and histidine side-chain imidazole groups (Reaction 44). See Chapter 8, Section 4 for further details on iodination reactions.

7. Photoreactive Chemical Reactions

Photoreactive groups can be induced to couple with target molecules by exposure to UV light. Until they are photolyzed, photosensitive functional groups are relatively nonreactive in typical thermochemical processes. For this reason, reagents designed with a photoreactive group can be used in highly controlled reactions. The labeling reaction can be induced by a UV flash at predetermined points in an experimental protocol. For instance, covalent bond formation can be initiated after binding of photo-labeled ligands to receptors or after some other biochemical process has taken place. In this regard, photoreactive chemistry has become an important device for numerous bioconjugate applications. The following sections describe the major photosensitive groups that can be used in the design of modification or cross-linking reagents. Chapter 4, Section 5, Chapter 5, Sections 3–7, Chapter 6, and Chapter 8, Section 3.4 describe the reagents that utilize these functionalities.

7.1. Aryl Azides and Halogenated Aryl Azides

The most popular type of photosensitive functionality is the aryl azide derivative. On photolysis, phenyl azide groups form short-lived nitrenes that react rapidly with the surrounding chemical environment (Gilchrist and Rees, 1969). Nitrenes can insert nonspecifically into chemical bonds of target molecules, including undergoing addition reactions with double bonds and insertion reactions into active hydrogen bonds at C—H and N—H sites. Abundant evidence, however, indicates that the photolyzed intermediates of aryl azides principally undergo ring-expansion to create nucleophile-reactive dehydroazepines. Instead of inserting nonselectively at active carbon—hydrogen bonds, dehydroazepines have a tendency to react preferentially with nucleophiles, especially amines (Reaction 45).

Phenylazide   Nitrene Formation   Dehydroazepine   Nucleophilic
                                            Intermediate   Addition   (45)

However, some investigators have shown that aryl azides that possess a perfluorinated ring structure or are substituted completely with halogen atoms are quite efficient at forming the desired nitrene intermediate (Soundararajan *et al.*, 1993; Keana and Cai, 1990; Schnapp and Platz, 1993; Schnapp *et al.*, 1993; Cai *et al.*, 1993; Yan *et al.*, 1994). The ring substitution prevents ring expansion after nitrene formation, thus allowing the reactive intermediate to survive long enough to react with target molecules. Halogenated phenyl azides undergo the insertion reactions that were typically attributed to unsubstituted aryl azides in the past (Reaction 46).

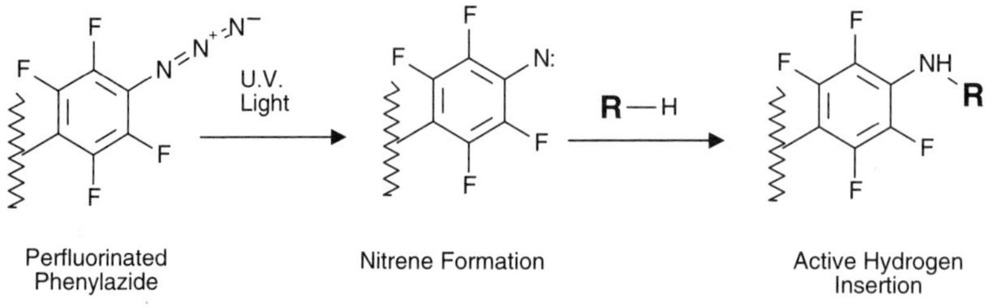

Perfluorinated   Nitrene Formation   Active Hydrogen
Phenylazide   Insertion

(46)

7.2. Benzophenones

A photoreactive group consisting of a benzophenone residue photolyzes upon exposure to UV light to give a highly reactive triplet-state ketone intermediate (Walling and Gibian, 1965). Similar to the reactive nitrene of photolyzed phenyl azides, the energized electron of an activated benzophenone can insert in hydrogen—carbon bonds and other active groups to give covalent linkages with target molecules (Reaction 47). Unlike phenyl azides, however, the decomposition or decay of the photoactivated species does not yield an inactive compound. Instead, benzophenones that have become deactivated without forming a covalent bond can be once again photolyzed to an active state. As a result of this multiple-activation characteristic, a benzophenone reagent has more than one chance to form a covalent bond with its intended target. Thus it typically gives much higher yields of photo-cross-linking than comparable phenyl azide cross-linkers.

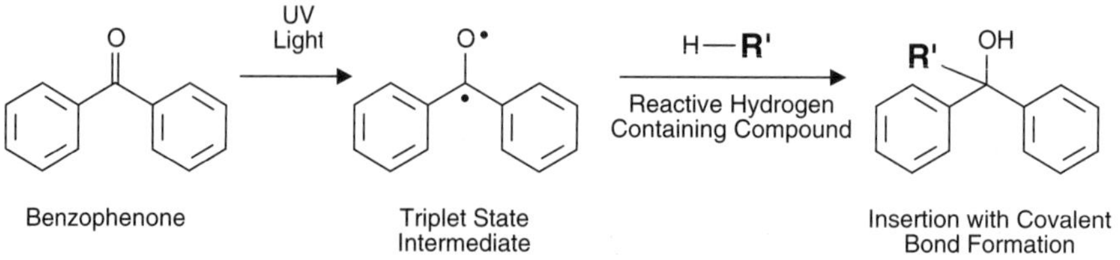

Benzophenone   Triplet State   Insertion with Covalent
                           Intermediate   Bond Formation

(47)

The use of a benzophenone photoactivatable group in the design of bioconjugate reagents is rare. Two sulfhydryl-reactive ones incorporating a maleimide group and a iodoacetyl group opposite the benzophenone are described in Chapter 5, Sections 4.3 and 4.4.

7.3. Certain Diazo Compounds

Certain diazo compounds can be photolyzed with UV light to generate highly reactive carbenes (Reaction 48). Similar to nitrenes, carbenes can insert into active C—H or

N—H bonds or add to double bonds, forming covalent linkages with target molecules (Gilchrist and Rees, 1969). Few diazo photoreactive reagents have been synthesized, probably due to their tendency to react with water molecules after photoactivation, thus severely decreasing coupling yields with intended molecules. One heterobifunctional cross-linker, PNP–DTP, containing an amine-reactive end and a photosensitive diazotrifluoropropionate group is available (Chapter 5, Section 3.12).

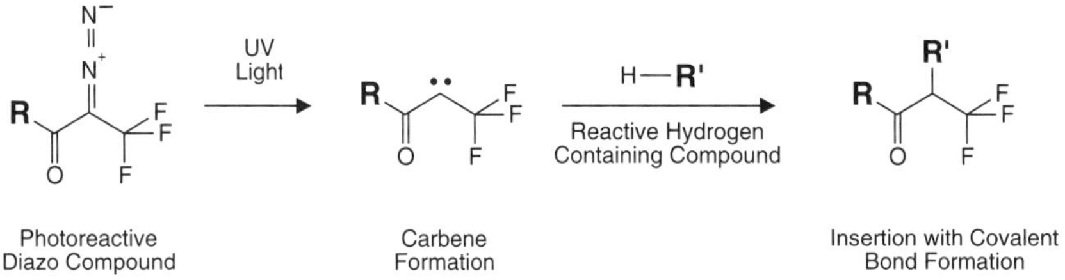

Photoreactive   Carbene   Insertion with Covalent
Diazo Compound   Formation   Bond Formation

(48)

Diazopyruvates are another class of photoreactive diazo compounds that have a unique coupling mechanism (Chapter 5, Section 3.11). The diazo functional group can be photolyzed by exposure to irradiation at 300 nm, forming a highly reactive carbene that can undergo a Wolff rearrangement to produce a ketene amide intermediate. In the presence a nucleophilic species on a target molecule, the ketene can undergo an acylation reaction to form a stable malonic acid derivative. The photolyzed product thus can couple to hydrazide- or amine-containing targets to form covalent linkages (Reaction 49).

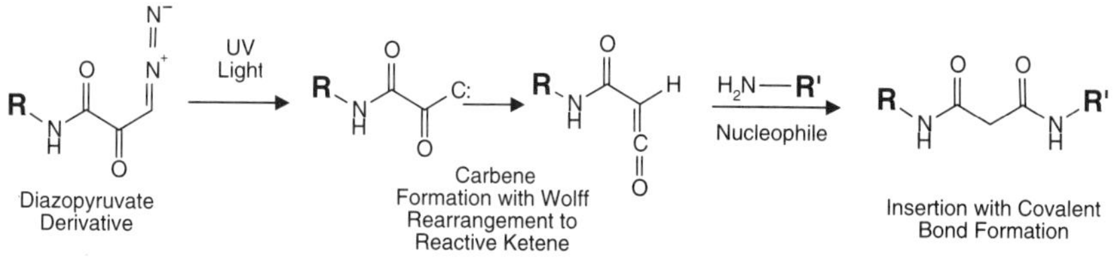

Diazopyruvate   Carbene   Insertion with Covalent
Derivative   Formation with Wolff   Bond Formation
                  Rearrangement to
                  Reactive Ketene

(49)

7.4. Diazirine Derivatives

Diazirine compounds are similar in their photoreactivity to diazo groups, forming highly reactive carbene intermediates on exposure to UV light of about 360 nm (Reaction 50). Diazirines consist of a three-member ring system containing two nitrogen atoms connected through a double bond. First developed by Smith and Knowles (1973), the photosensitive diazirine is perhaps second in popularity to phenyl azides in the design of photoreactive cross-linking agents.

Some diazirines, particularly the 3-trifluoromethyl-3-aryldiazirines, can rearrange upon photolysis to a linear diazo derivative, similar in structure to the photosensitive end of the cross-linker PNP–DTP (Chapter 5, Section 3.12). These isomerized products themselves can be photolyzed to the reactive carbene.

Carbene generation from photolysis of diazirine compounds leads to efficient insertion into C—H or N—H bonds and also causes addition reactions with points of unsaturation within target molecules. Diazirine-containing photoaffinity probes have

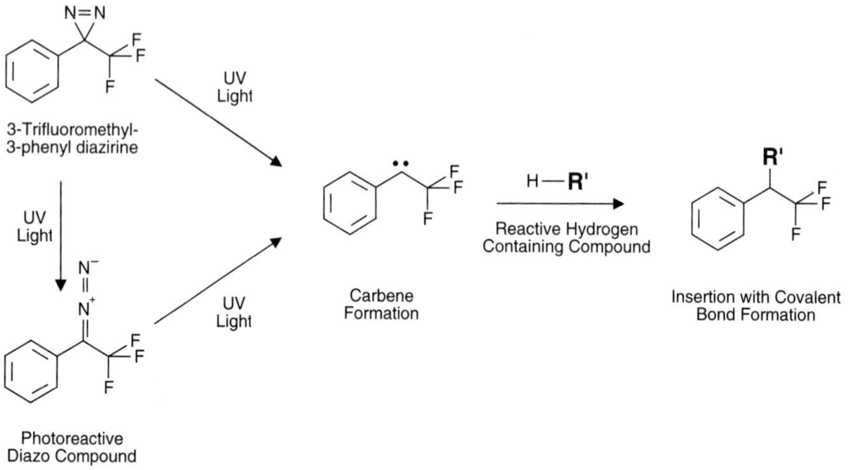
N=N
F
F
F
3-Trifluoromethyl-
3-phenyl diazirine
UV
Light
UV
Light
N⁻
N⁺
F
F
F
Photoreactive
Diazo Compound
UV
Light
F
F
F
Carbene
Formation
H—R'
Reactive Hydrogen
Containing Compound
R'
F
F
F
Insertion with Covalent
Bond Formation

(50)

been used to study numerous ligand–receptor interactions (Bergmann *et al.*, 1994). Heterobifunctional cross-linkers containing a diazirine photosensitive group also have been used to attach macromolecules to surfaces such as polystyrene and glass (Collioud *et al.*, 1993).

PART II

Bioconjugate Reagents

The reagent systems used in bioconjugate procedures are as varied as their intended applications. Whether it be for tagging proteins to make them chromogenic or fluorescent, labeling molecules with biospecific ligands for subsequent affinity interactions, or cross-linking two or more substances to create uniquely active conjugates, the choice of reagents available for use is limited only by the imagination.

Over the last 20 years, the selection of cross-linking and modifying agents has grown not only in shear number, but also in the availability of novel reactive groups and in the variety of their design. Today, regardless of the particular need, a workable reagent system that will yield a useful derivative can almost always be found. The best and most effective of the reported reagent systems now are available from commercial sources, and thus do not even have to be synthesized.

In Part II, the reagents of modification and conjugation have been categorized according to structural type, reactivity, and use. Where possible and appropriate, generalized protocols have been provided for each reagent's most likely application. The options described herein, combined with a thorough knowledge of the basic chemical reactions that their functional groups provide (as discussed in Part I), allow the creation of an intelligent design and plan of attack for any desired application. The labeling, tagging, cross-linking, or targeting of small ligands, peptides, proteins, carbohydrates, nucleic acids, oligonucleotides, lipids, and a host of other compounds may be accomplished by the judicious choice of the appropriate reagent system.

The following reagents have been used in everything from bench-scale experiments in research laboratories to process-optimized applications in the diagnostic and therapeutic industries. Conjugated or modified molecules have been applied in procedures designed to visualize target substances, as key components in clinical assay systems, and in the latest affinity-directed therapeutics, such as antitumor immunotoxins. Some of the reagent systems described in Part II have formed the basis for literally a multibillion dollar biotechnology industry.

3

Zero-Length Cross-linkers

The smallest available reagent systems for bioconjugation are the so-called zero-length cross-linkers. These compounds mediate the conjugation of two molecules by forming a bond containing no additional atoms. Thus, one atom of a molecule is covalently attached to an atom of a second molecule with no intervening linker or spacer. In many conjugation schemes, the final complex is bound together by virtue of chemical components that add foreign structures to the substances being cross-linked. In some applications, the presence of these intervening linkers may be detrimental to the intended use. For instance, in the preparation of hapten–carrier conjugates the complex is formed with the intention of generating an immune response to the attached hapten. Occasionally, a portion of the antibodies produced by this response will have specificity for the cross-linking agent used in the conjugation procedure. Zero-length cross-linking agents eliminate the potential for this type of cross-reactivity by mediating a direct linkage between two substances.

The reagents described in this chapter can initiate the formation of three types of bonds: an amide linkage made by the condensation of a primary amine with a carboxylic acid, a phosphoramidate linkage made by the reaction of a organic phosphate group with a primary amine, and a secondary or tertiary amine linkage made by the reductive amination of a primary or secondary amine with an aldehyde group. Therefore, using these reagent systems, substances containing amines can be conjugated with other molecules containing phosphates or carboxylates. Alternatively, substances containing amines can be cross-linked to molecules containing formyl groups. All of the reactions are quite efficient, and depending on the reagent chosen and the desired application, they may be performed in aqueous or nonaqueous environments.

1. Carbodiimides

Carbodiimides are used to mediate the formation of amide linkages between a carboxylate and an amine or phosphoramidate linkages between a phosphate and an amine (Hoare and Koshland, 1966; Chu *et al.*, 1986; Ghosh *et al.*, 1990). They are probably the most popular type of zero-length cross-linker in use, being efficient in forming conjugates between two protein molecules, between a peptide and a protein, between oligonucleotides and proteins, or any combination of these with small molecules. There are two basic types of carbodiimides: water-soluble and water-insoluble. The water-soluble ones are the most common choice for biochemical conjugations, because most macromolecules of biological origin are soluble in aqueous buffer solutions. Not only is the carbodiimide itself able to dissolve in the reaction medium, but the by-product of the reaction, an isourea, is also water-soluble, facilitating easy purification. Water-insoluble carbodiimides, by contrast, are used frequently in peptide synthesis and other conjugations involving molecules soluble only in organic solvents. Both the organic-soluble carbodiimides and their isourea by-products are insoluble in water.

1.1. EDC

EDC [or EDAC; 1-ethyl-3-(3-dimethylaminopropyl)carbodiimide hydrochloride] is perhaps the most popular carbodiimide for use in conjugating biological substances. Its water solubility allows for direct addition to a reaction without prior organic solvent dissolution. Excess reagent and the isourea formed as the by-product of the cross-linking reaction are both water-soluble and may be easily removed by dialysis or gel filtration (Sheehan *et al.*, 1961, 1965). The reagent is, however, labile in the presence of water. The bulk chemical should be stored desiccated at −20°C. Warm the bottle to room temperature before opening to prevent condensation that will cause decomposition of the reagent over time. A concentrated solution of EDC in water may be prepared to facilitate the addition of a small molar amount to a reaction, but the stock solution should be dissolved rapidly and used immediately to prevent extensive loss of activity.

EDC
1-Ethyl-3-(3-dimethylaminopropyl)
Carbodiimide Hydrochloride
MW 191.7

A variety of chemical conjugates may be formed using EDC (Yamada *et al.*, 1981; Chase *et al.*, 1983; Chu *et al.*, 1976, 1982; Chu and Ueno, 1977), provided one of the molecules contains a primary amine and the other a carboxylate group. N-substituted carbodiimides can react with carboxylic acids to form a highly reactive, O-acylisourea intermediates (Fig. 106). This active species can then react with a nucleophile such as a primary amine to form an amide bond (Williams and Ibrahim, 1981). Other nucleophiles are also reactive. Sulfhydryl groups may attack the active species and form thiol ester linkages, although these are not as stable as the bond formed with an amine. In addition, oxygen atoms may act as the attacking nucleophile, such as those in water molecules. In aqueous solutions, hydrolysis by water is the major competing reaction, cleaving off the activated ester intermediate, forming an isourea, and regenerating the carboxylate group (Gilles *et al.*, 1990). The potential for hydrolysis of the active ester is reflected in the rate constant at pH 4.7 for the water-soluble carbodiimide, EDC, of only 2–3 s^{-1}.

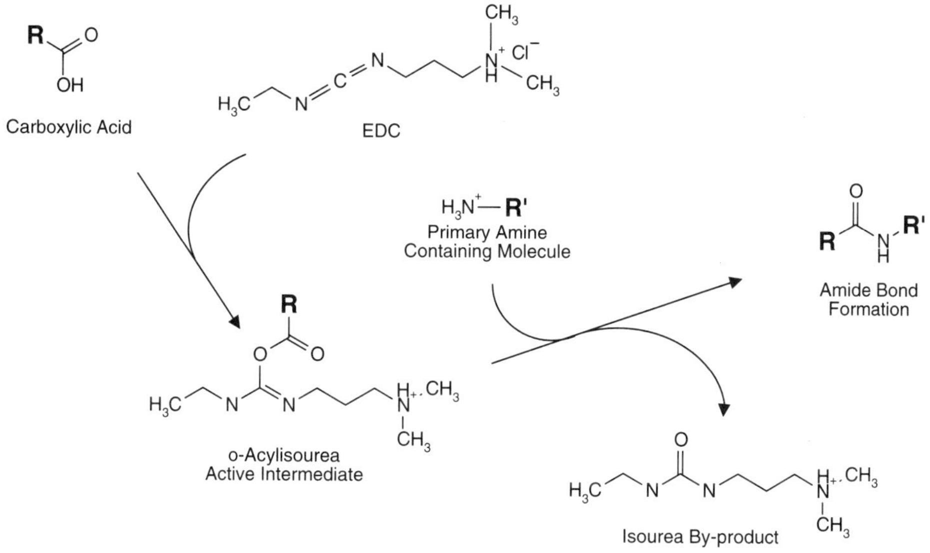

Figure 106 EDC reacts with carboxylic acids to create an active-ester intermediate. In the presence of an amine nucleophile, an amide bond is formed with release of an isourea by-product.

The presence of both carboxylates and amines on one of the molecules to be conjugated with EDC may result in self-polymerization, because the substance can then react with another molecule of its own kind instead of the desired target. For instance, when conjugating peptides to carrier proteins using EDC, the peptide usually contains both a carboxylate and an amine. The result typically is peptide polymerization in addition to coupling to the carrier (see Chapter 9, Section 3). For this type of immunogen conjugation, polymerization is not usually detrimental to its use, because polymerized peptide is also immunogenic. However, for other cross-linking applications where it may be more desirable to avoid oligomer formation, the use of a carbodiimide may not be the best choice of reagent, especially if one of the molecules being conjugated contains both a carboxylate and an amine.

Most references to the use of EDC describe the optimal reaction medium to be at a pH between 4.7 and 6. However, the carbodiimide reaction occurs effectively up to at least pH 7.5 without significant loss of yield. See Chapter 9, Section 3 for additional information on the properties of EDC conjugation using small peptides coupled to carrier proteins.

Some procedures utilize water as the solvent in an EDC reaction, while the pH is maintained constant by the addition of HCl. Buffered solutions are more convenient, because the pH does not have to be monitored during the course of the reaction. For low pH conjugations, MES [2-(N-morpholino)ethane sulfonic acid] buffer at 0.1 M works well. When doing neutral pH reactions, a phosphate buffer at 0.1 M is appropriate. Any buffers that do not interfere with the reaction may be used but avoid amine- or carboxylate-containing buffer salts or other components in the medium that may react with the carbodiimide.

There are some side reactions that may occur when using EDC with proteins. In addition to reacting with carboxylates, EDC itself can form a stable complex with exposed sulfhydryl groups (Carraway and Triplett, 1970). Tyrosine residues can react with EDC, most likely through the phenolate ionized form of its side chain (Carraway and Koshland, 1968). Finally, EDC may promote unwanted polymerization due to the usual abundance of both amines and carboxylates on protein molecules.

The following protocol is a generalized description of how to conjugate a small amine- or carboxylate-containing molecule to a protein. The protocol may be modified by changing the pH, buffer salts, and ratios of reactants to obtain the desired product. Specific protocols utilizing EDC in selected conjugation applications may be found in Part III. In some cases, the parameters of this generalized protocol may have to be modified to retain solubility or activity of the resulting conjugate. For instance, coupling hydrophobic molecules to the surface of proteins often causes partial or complete precipitation. This problem may be somewhat alleviated by decreasing either the amount of EDC or the amount the small molecule added to the reaction, thus resulting in a lower density of substitution.

Protocol

1. Dissolve the protein to be modified at a concentration of 10 mg/ml in one of the following reaction media: (*a*) water, (*b*) 0.1 *M* MES, pH 4.7–6.0, or *(c)* 0.1 *M* sodium phosphate, pH 7.3. NaCl may be added (i.e., 0.15 *M*) if desired. If lower or higher concentrations of the protein are used, adjust the amounts of the other reactants added as necessary to maintain the correct molar ratios. For the preparation of a peptide–protein immunogen conjugate, a 200-μl solution of the carrier protein at a concentration of 10 mg/ml in 0.1 *M* MES, pH 4.7, usually works well.

2. Dissolve the molecule to be coupled in the same buffer used in step 1. For small molecules, add them to the reaction in at least a 10-molar excess to the amount of protein present. If possible, the molecule may be added directly to the protein solution in the appropriate excess. Alternatively, dissolve the molecule in the buffer at a higher concentration, then add an aliquot of this stock solution to the protein solution. In the example of preparing a peptide–protein conjugate, dissolve the peptide in 0.1 *M* MES, pH 4.7, at a concentration of up to 2 mg/500 μl.

3. Add the solution prepared in step 2 to the protein solution to obtain at least a 10-fold molar excess of small molecule-to-protein. In the case of the peptide–protein immunogen conjugate, add the 500 μl of peptide solution to the 200 μl of protein solution.

4. Add EDC (Pierce) to the above solution to obtain at least a 10-fold molar excess of EDC to the protein. Alternatively, a 0.5–0.1 *M* EDC concentration in the reaction usually works well. To make it easier to add the correct quantity of EDC, a higher concentration stock solution may be prepared if it is dissolved and used rapidly. To prepare the peptide–protein conjugate, add the solution from step 3 to 10 mg of EDC in a test tube. Mix to dissolve. If this ratio of EDC to peptide or protein results in precipitation, scale back the amount of addition until a soluble conjugate is obtained. For some proteins, as little as 0.1 times this amount of EDC may have to be used to maintain solubility.

5. React for 2 h at room temperature.

6. Purify the conjugate by gel filtration or dialysis using the buffer of choice (for

many conjugates 0.01 *M* sodium phosphate, 0.15 *M* NaCl, pH 7.4, is appropriate). If some turbidity has formed during the conjugation procedure, it may be removed by centrifugation or filtration. When using EDC to prepare immunogen conjugates, the presence of some precipitated material is usually not of concern, because precipitated immunogens are often more immunogenic than soluble proteins.

1.2. EDC plus Sulfo-NHS

The water-soluble carbodiimide EDC may be used to form active ester functional groups with carboxylate groups using the water-soluble compound *N*-Hydroxysulfo-succinimide (sulfo-NHS) (Pierce). Sulfo-NHS esters are hydrophilic active groups that react rapidly with amines on target molecules (Staros, 1982; Anjaneyulu and Staros, 1987; Beth *et al.*, 1986; Kotite *et al.*, 1984; Donovan and Jennings, 1986; Denney and Blobel, 1984; Jennings and Nicknish, 1985; Ludwig and Jay, 1985). Unlike nonsulfo-nated NHS esters that are relatively water-insoluble and must be first dissolved in organic solvent before being added to aqueous solutions, sulfo-NHS esters typically are water-soluble and long-lived and hydrolyze relatively slowly in water. However, in the presence of amine nucleophiles that can attack at the carbonyl group of the ester, the *N*-hydroxysulfosuccinimide group rapidly leaves, creating a stable amide linkage with the amine. Sulfhydryl and hydroxyl groups also will react with such active esters, but the products of such reactions, thioesters and esters, are relatively unstable.

The advantage of adding sulfo-NHS to EDC reactions is to increase the stability of the active intermediate, which ultimately reacts with the attacking amine. EDC reacts with a carboxylate group to form an active ester (*O*-acylisourea) leaving group. Unfortunately, this reactive complex is subject to rapid hydrolysis in aqueous solutions, having a rate constant measured in seconds (Hoare and Koshland, 1967). If the target amine does not find the active carboxylate before it hydrolyzes, the desired coupling cannot occur. This is especially a problem when the target molecule is in low concentration compared to water, as in the case of protein molecules. Forming a sulfo-NHS ester intermediate from the reaction of the hydroxyl group on sulfo-NHS with the EDC active-ester complex extends the half-life of the activated carboxylate to hours. Since the concentration of added sulfo-NHS is usually much greater than the concentration of target molecule, the reaction preferentially proceeds through the longer-lived intermediate. However, the final product of this two-step reaction is identical to that obtained using EDC alone: the activated carboxylate reacts with an amine to give a stable amide linkage (Fig. 107).

EDC/sulfo-NHS-coupled reactions are highly efficient and usually increase the yield of conjugation dramatically over that obtainable solely with EDC. Staros *et al.* (1986) shows that the addition of just 5 m*M* sulfo-NHS to the EDC coupling of glycine to keyhole limpet hemocyanin increased the yield of derivatization about 20-fold compared to using EDC alone. This technique also can be used to create activated proteins containing sulfo-NHS esters (Grabarek and Gergely, 1990). A protein can be incubated in the presence of EDC/sulfo-NHS and the active ester form isolated and then mixed with a second protein or other amine-containing molecule for conjugation. This two-step process allows the active species to form only on one protein, thus gaining greater control over the conjugation (Fig. 108).

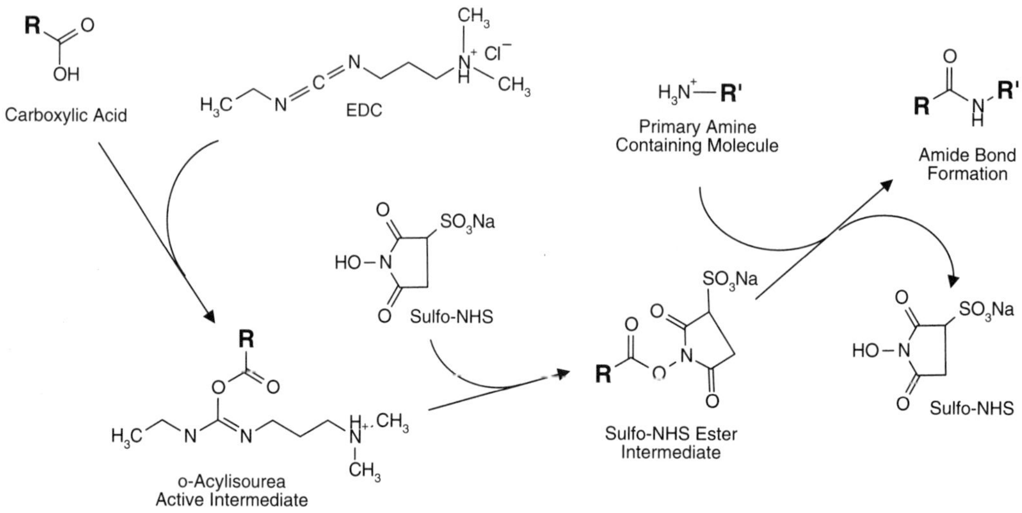

Figure 107 The efficiency of an EDC-mediated reaction may be increased through the formation of a sulfo-NHS ester intermediate. The sulfo-NHS ester survives in aqueous solution longer than the active ester formed from the reaction of EDC alone with a carboxylate. Thus, higher yields of amide bond formation may be realized using this two-stage process.

In addition to the potential side reactions of EDC as mentioned previously (Chapter 3, Section 1.1), the additional efficiency obtained by the use of a sulfo-NHS intermediate in the process may cause other problems. In some cases, the conjugation actually may be too efficient to result in a soluble or active complex. Particularly when coupling some peptides to carrier proteins, I have found that the use of EDC/sulfo-NHS often causes severe precipitation of the conjugate. Scaling back the amount of EDC/sulfo-NHS added to the reaction may solve this problem. However, eliminating the addition of sulfo-NHS altogether may have to be done in some instances to preserve the solubility of the product.

The following protocol is a generalized description of how to incorporate sulfo-NHS ester intermediates in EDC conjugation procedures. For specific applications of this technology, the amount of each reagent and unconjugated species may have to be adjusted to obtain an optimal conjugate.

Protocol

1. Dissolve the protein to be modified at a concentration of 1–10 mg/ml in 0.1 *M* sodium phosphate, pH 7.4. NaCl may be added to this buffer if desired. For the modification of keyhole limpet hemocyanin (KLH; Pierce), as described by Staros *et al.* (1986), include 0.9 *M* NaCl to maintain the solubility of this high-molecular-weight protein. If lower or higher concentrations of the protein are used, adjust the amounts of the other reactants as necessary to maintain the correct molar ratios.

2. Dissolve the molecule to be coupled in the same buffer used in step 1. For small

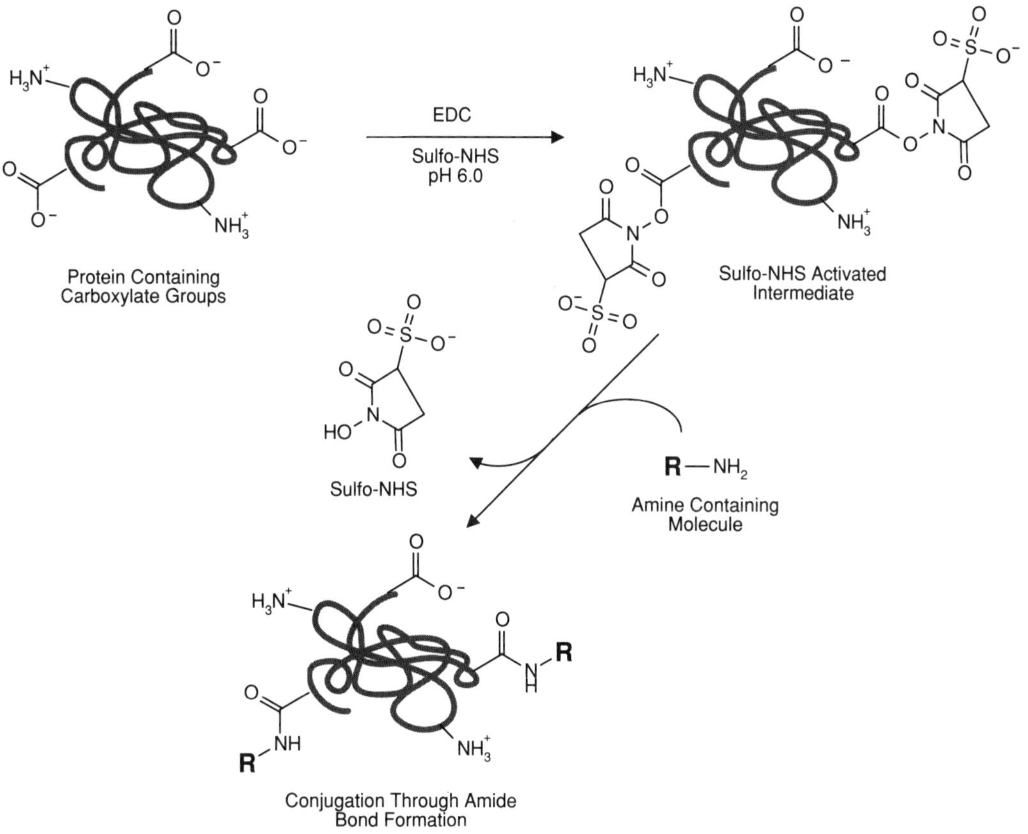

Figure 108 EDC may be used in tandem with sulfo-NHS to create an amine-reactive protein derivative containing active ester groups. The activated protein can couple with amine-containing compounds to form amide bond linkages.

molecules, add them to the reaction in at least a 10-fold molar excess over the amount of protein present. If possible, the molecule may be added directly to the protein solution in the appropriate excess. Alternatively, dissolve the molecule in the buffer at a higher concentration, then add an aliquot of this stock solution to the protein solution.

3. Add the solution prepared in step 2 to the protein solution to obtain at least a 10-fold molar excess of small molecule to protein.

4. Add EDC (Pierce) to the above solution to obtain at least a 10-fold molar excess of EDC to the protein. Alternatively, a 0.05–0.1 M EDC concentration in the reaction usually works well. Also, add sulfo-NHS (Pierce) to the reaction to bring its final concentration to 5 mM. To make it easier to add the correct quantity of EDC or sulfo-NHS, higher concentration stock solutions may be prepared if they are dissolved and used rapidly. Mix to dissolve. If this ratio of EDC/sulfo-NHS to peptide or protein results in precipitation, scale back the amount of addition until a soluble conjugate is obtained.

5. React for 2 h at room temperature.
6. Purify the conjugate by gel filtration or dialysis using the buffer of choice (for many conjugates 0.01 M sodium phosphate, 0.15 M NaCl, pH 7.4, is appropriate). If some turbidity has formed during the conjugation procedure, it may be removed by centrifugation or filtration.

A modification of a two-step protocol (Grabarek and Gergely, 1990) for the activation of proteins with EDC/sulfo-NHS and subsequent conjugation with amine-containing molecules is given below. The variation in the pH of activation from that described above provides greater stability for the active ester intermediate. At pH 6, the amines on the protein will be protonated and therefore be less reactive toward the sulfo-NHS esters that form. In addition, the hydrolysis rate of the esters is dramatically slower at acid pH. Thus, the active species may be isolated in a reasonable time frame without significant loss in conjugation potential. To quench the unreacted EDC, 2-mercaptoethanol is added to form a stable complex with the remaining carbodiimide, according to Carraway and Triplett (1970). In the following protocol, sulfo-NHS is used instead of NHS so that active ester hydrolysis is slowed (Thelen and Deuticke, 1988; Anjaneyulu and Staros, 1987).

Protocol

1. Dissolve the protein to be activated in 0.05 M MES, 0.5 M NaCl, pH 6 (reaction buffer), at a concentration of 1 mg/ml.
2. Add to the solution in step 1 a quantity of EDC and sulfo-NHS (Pierce) to obtain a concentration of 2 mM EDC and 5 mM sulfo-NHS. To aid in aliquoting the correct amount of these reagents, they may be quickly dissolved in the reaction buffer at a higher concentration, and then a volume immediately pipetted into the protein solution to obtain the proper molar quantities.
3. Mix and react for 15 min at room temperature.
4. Add 2-mercaptoethanol to the reaction solution to obtain a final concentration of 20 mM. Mix and incubate for 10 min at room temperature. Note: if the protein being activated is sensitive to this level of 2-mercaptoethanol, instead of quenching the reaction chemically, the activation may be terminated by desalting (step 5).
5. If the reaction was quenched by the addition of 2-mercaptoethanol, the activated protein may be added directly to a second protein or other amine-containing molecule for conjugation. Alternatively, or if no 2-mercaptoethanol was added, the activated protein may be purified from reaction by-products by gel filtration using Sephadex G-25 or equivalent. The desalting operation should be done rapidly to minimize hydrolysis and recover as much active ester function as possible. The use of centrifugal spin columns of some sort may afford the greatest speed in purification. After purification, add the activated protein to the second molecule for conjugation. The second protein or other amine-containing molecule should be dissolved in 0.1 M sodium phosphate, pH 7.5. This will bring the pH of the coupling medium above pH 7 to initiate the active ester reaction.
6. React for at least 2 h at room temperature.
7. Remove excess reactants by gel filtration or dialysis.

1.3. CMC

1-Cyclohexyl-3-(2-morpholinoethyl) carbodiimide (CMC) (usually synthesized as the metho *p*-toluene sulfonate salt) (Aldrich), is a water-soluble reagent used to form amide bonds between one molecule containing a carboxylate and a second molecule containing an amine. The presence of the positively charged morpholino group creates the water solubility. Along with EDC (Chapter 3, Section 1.1), CMC is the only other soluble carbodiimide commonly available for biological conjugations. It was first utilized in peptide synthesis (Sheehan and Hlavka, 1956) and found to be superior to other coupling agents used at the time (Ondetti and Thomas, 1965). It also has been used for the quantitative modification and estimation of total carboxyl groups in protein molecules (Hoare and Koshland, 1967) and for investigating the secondary structure of nucleic acids (Metz and Brown, 1969). Another early application area of CMC relates not to solution phase cross-linking of two molecules, but to coupling of ligands to insoluble support materials for use in affinity chromatography (Schmer, 1972; Lowe and Dean, 1971; Marcus and Balbinder, 1972).

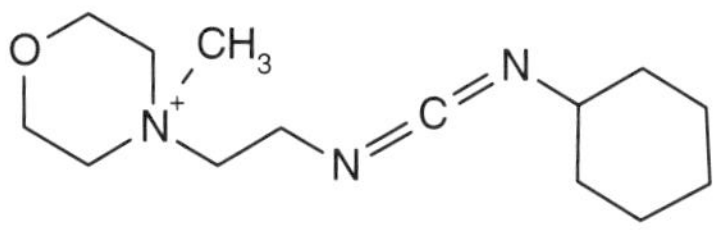

CMC
1-Cyclohexyl-3-(2-morpholinoethyl)carbodiimide
MW 423.58
(as the metho-p-toluene sulfonate salt)

CMC reacts with carboxylate groups by addition of the carboxyl across one of its diimide bonds, resulting in the characteristic active ester, O-acylisourea intermediate common to all carbodiimide mechanisms. Nucleophilic attack on this intermediate yields the acylated product—usually an amide bond, resulting from the reaction with a primary amine (Fig. 109). However, carbodiimide chemistry does create several potential side reactions. Sulfhydryl groups may react with CMC to form a stable covalent complex unreactive toward further conjugation. The reagent also may react with phenols, alcohols, and other nucleophiles to quench the cross-linking reaction. In aqueous solutions, hydrolysis of the active ester is by far the most frequent side reaction. Reaction of the group with water molecules regenerates the carboxylate and releases a soluble isourea by-product.

CMC should be able to participate in the two-step reaction using a sulfo-NHS ester intermediate similar to EDC; however, there are no reports in the literature to this effect. Protocols for the use of this reagent in biological cross-linking applications should be essentially the same as those given previously for EDC, except substituting a molar equivalent quantity of CMC. See Chapter 3, Sections 1.1 and 1.2 for additional information concerning carbodiimide reactions.

1.4. DCC

Dicyclohexyl carbodiimide (DCC) is one of the most frequently used coupling agents, especially in organic synthesis applications. It has been used for peptide synthesis since

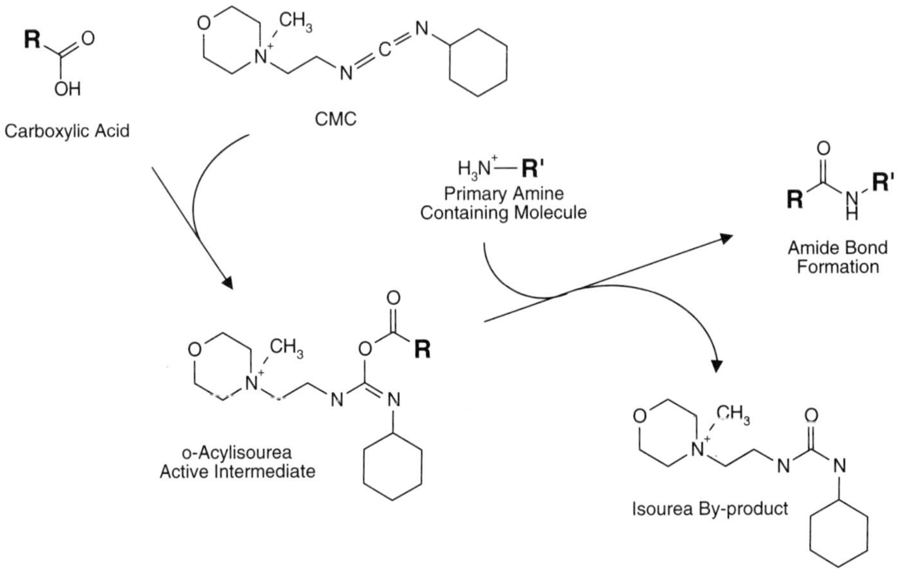

Figure 109   The water-soluble carbodiimide CMC reacts with carboxylates to form an active-ester intermediate. In the presence of amine-containing molecules, amide bond formation can take place with release of an isourea by-product.

1955 (Sheehan and Hess, 1955) and continues to be a popular choice for creating peptide bonds (Barany and Merrifield, 1980). DCC is water-soluble, but has been used in 80% DMF for the immobilization of small molecules onto carboxylate-containing chromatography supports for use in affinity separations (Larsson and Mosbach, 1971: Lowe *et al.*, 1973; Gutteridge and Robb, 1973). In addition to forming amide linkages, DCC has been used to prepare active esters of carboxylate-containing compounds using NHS or sulfo-NHS (Staros, 1982). Unlike the EDC/sulfo-NHS reaction described in Chapter 3, Section 1.2, active ester synthesis done with DCC is in organic solvent, and therefore does not have the hydrolysis problems of water-soluble EDC-formed esters. Thus, DCC is most often used to synthesize active ester-containing cross-linking and modifying reagents and not to perform biomolecular conjugations.

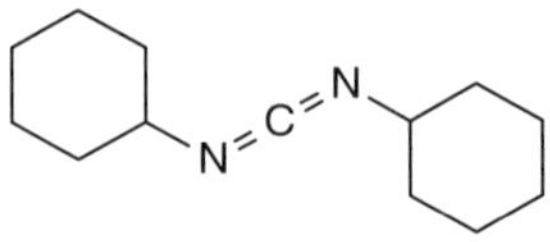

DCC
N,N'-Dicyclohexyl
carbodiimide
MW 206.32

DCC is a waxy solid that is often difficult to remove from a bottle. Its vapors are extremely hazardous to the eyes and to inhale. It should always be handled in a fume hood. The isourea by-product of a DCC-initiated reaction, dicyclohexyl urea (DCU) (Fig. 110), is also water-insoluble and must be removed by organic solvent washing.

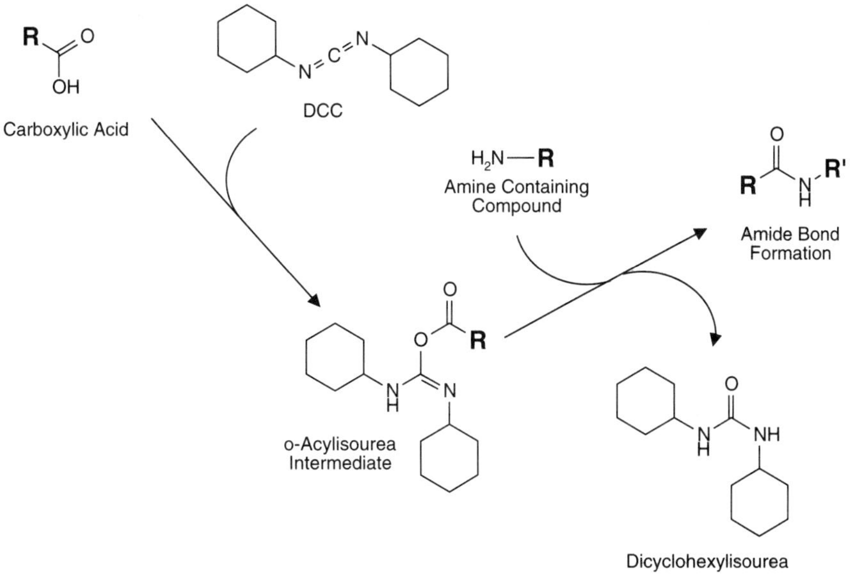

Figure 110 The organic-soluble carbodiimide DCC is often used to create amide bonds, especially between water-insoluble compounds.

For synthesis of peptides or affinity supports on insoluble matrices, this is not a problem because washing of the support material can be done without disturbing the conjugate coupled to the support. For solution phase chemistry, however, reaction products must be removed by solvent washings, precipitations, or recrystallizations.

A potential undesirable effect of DCC coupling reactions is the spontaneous rearrangement of the O-acylisourea to an inactive N-acylurea (Stewart and Young, 1984) (Fig. 111). The rate of this rearrangement is dramatically increased in aprotic organic solvents, such as DMF.

The activation efficiency of DCC is extraordinarily high, especially in anhydrous solutions that do not have competing hydrolysis problems. O-Acylisourea-activated

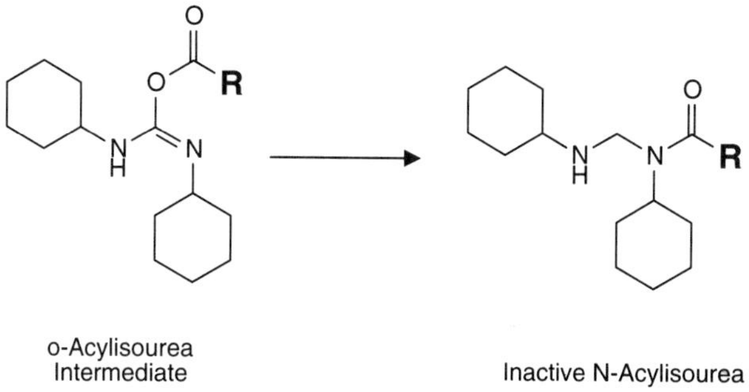

Figure 111 The active-ester intermediate formed from the reaction of DCC with a carboxylate group may undergo rearrangement to an inactive N-acylisourea product.

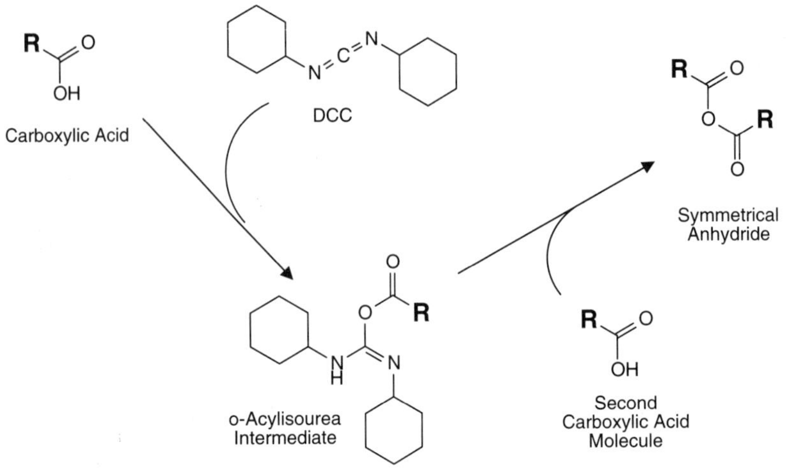

Figure 112   The reaction of DCC with a carboxylate compound in excess may create anhydride products in the absence of nucleophiles.

carboxylates may undergo two side reactions that form other active groups. If DCC is added to an excess of a carboxylate-containing molecule without the presence of an amine-containing target, then the activated carboxylate may react with another carboxylic acid to form a symmetrical anhydride (Fig. 112). The formation of an anhydride intermediate may be a frequent mechanism in route to the creation of an amide bond with an amine, especially under anhydrous conditions (Rebek and Feitler, 1974). In addition, a DCC-activated carboxylate may react with a amino acid to form an azlactone (Fig. 113) (Coleman *et al.*, 1990). Both the anhydride and the azlactone will react with amines to form covalent amide linkages. However, the ring-opening reaction of an azlactone will form a different product than the zero-length cross-linking result of coupling directly to an amine-containing molecule (Fig. 114).

1.5. DIC

Diisopropyl carbodiimide (DIC) is another water-insoluble amide bond-forming agent that has advantages over DCC (Chapter 3, Section 1.4). It is a liquid at room temperature and is therefore much easier to dispense than DCC. Its by-products, diisopropylurea and diisopropyl-*N*-acylurea, are more soluble in organic solvents than

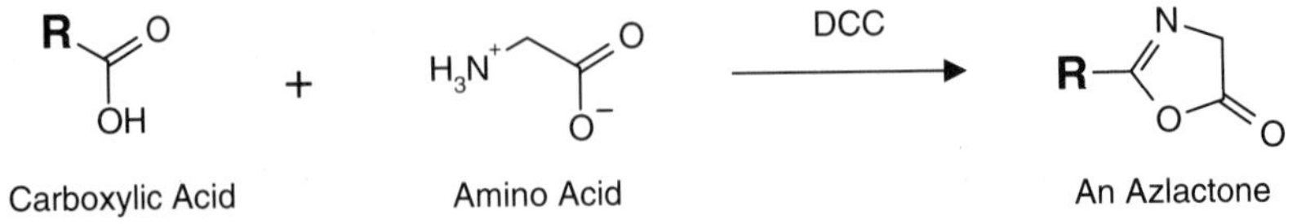

Figure 113   A DCC-mediated reaction with a carboxylate group in the presence of a small amino acid may form azlactone rings.

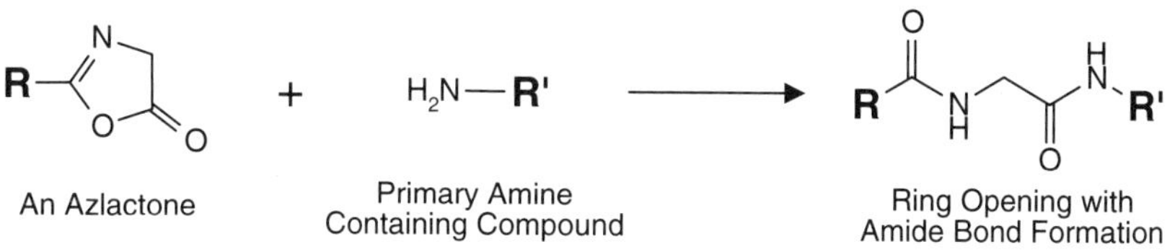

Figure 114 An azlactone reacts with amine groups through a ring-opening process, creating amide bond linkages with the attacking nucleophile.

the DCU by-product of a DCC reaction. DIC reacts in a manner similar to that of DCC, forming an active O-acylisourea intermediate with a carboxylic acid group (Fig. 115). This active species may then react with a nucleophile such as an amine to form an amide bond. Presumably, all the possible side reactions that DCC may undergo are also possible with DIC, although it is not well documented.

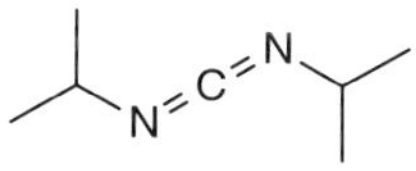

DIC
Diisopropyl carbodiimide
MW 126.2

2. Woodward's Reagent K

Woodward's reagent K is N-ethyl-3-phenylisoxazolium-3'-sulfonate, a zero-length cross-linking agent able to cause the condensation of carboxylates and amines to form amide bonds (Woodward *et al.*, 1961; Woodward and Olofson, 1961). The reaction

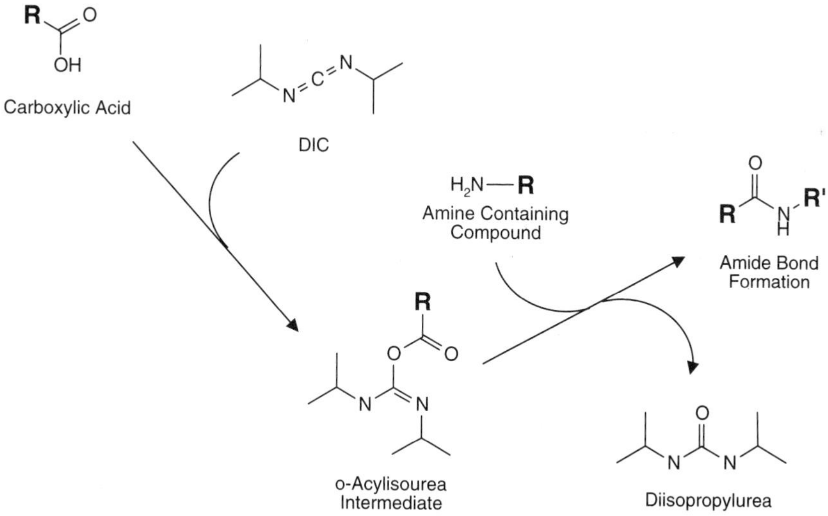

Figure 115 The symmetrical carbodiimide DIC reacts with carboxylates to form active-ester intermediates able to couple with amine-containing compounds to form amide bond linkages.

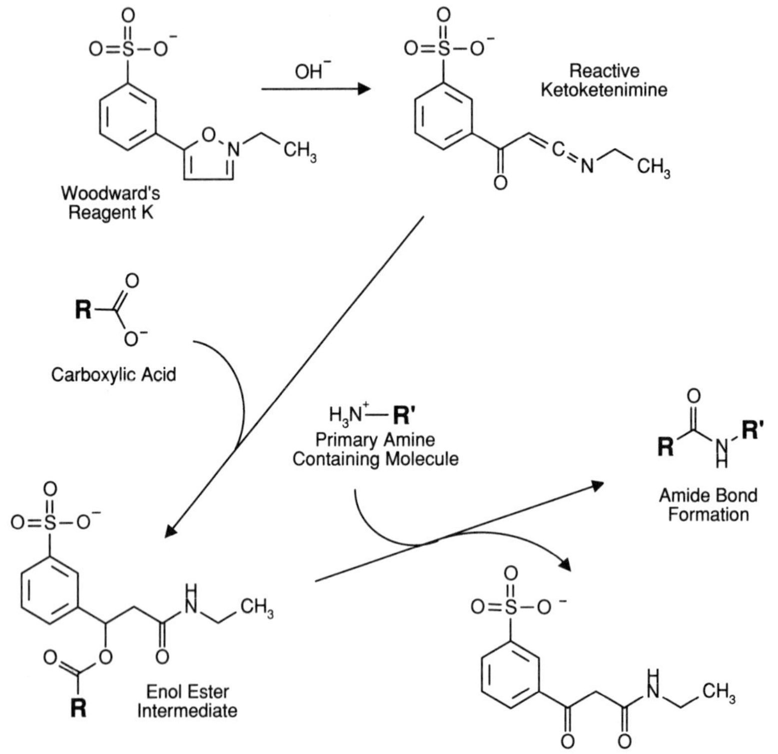

Figure 116 Woodward's reagent K undergoes a rearrangement in alkaline solution to form a reactive ketoketenimine. This active species can react with a carboxylate group to create another active group, a enol ester derivative. In the presence of amine nucleophiles, amide bond formation takes place.

mechanism involved in activating a carboxylate includes the conversion of the reagent under alkaline conditions to a reactive ketoketenimine. This intermediate then reacts with a carboxylate to create an enol ester. The enol ester is highly susceptible to nucleophilic attack. The reaction with an amine proceeds to amide bond formation with loss of the inactive diketo derivative (Fig. 116). In aqueous solution, the major side reaction is hydrolysis which occurs rapidly (Dunn and Affinsen, 1974). Although Woodward's reagent K has been used successfully for conjugation applications with proteins and other molecules (Pikuleva and Turko, 1989; Boyer, 1986), it is not available commercially, which severely limits its application.

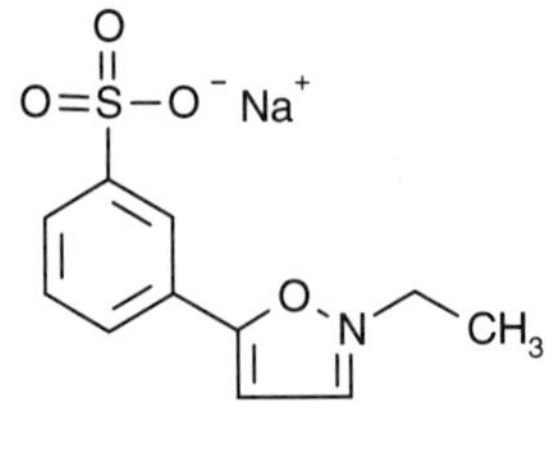

Woodward's Reagent K
N-Ethyl-5-phenylisoxazolium-3'-
sulfonate, sodium salt
MW 176

3. *N,N'*-Carbonyldiimidazole

N,N'-Carbonyldiimidazole (CDI) is a highly active carbonylating agent that contains two acylimidazole leaving groups (Aldrich). The result is that CDI can activate carboxylic acids or hydroxyl groups for conjugation with other nucleophiles, creating either zero-length amide bonds or one-carbon-length *N*-alkyl carbamate linkages between the cross-linked molecules. Carboxylic acid groups react with CDI to form *N*-acylimidazoles of high reactivity. The active intermediate forms in excellent yield due to the driving force created by the liberation of carbon dioxide and imidazole (Anderson, 1958). The active carboxylate then can react with amines to form amide bonds or with hydroxyl groups to form ester linkages (Fig. 117). Both reaction mechanisms have been used successfully in peptide synthesis (Paul and Anderson, 1960, 1962). In addition, activation of a styrene/4-vinylbenzoic acid copolymer with CDI was used to immobilize the enzyme lysozyme through its available amino groups to the carboxyl groups on the matrix (Bartling *et al.*, 1973).

CDI
N,N'-Carbonyldiimidazole
MW 162

CDI functions as a zero-length cross-linker if the activated species is a carboxylic acid, because the attack of another nucleophile liberates the imidazole leaving group. However, if CDI is used to activate a hydroxyl functional group, the reaction proceeds quite differently. The active intermediate formed by the reaction of CDI with an —OH group is an imidazolyl carbamate (Fig. 118). Attack by an amine releases the imidazole, but not the carbonyl. Thus, a hydroxyl-containing molecule may be coupled to an amine-containing molecule with the result of a one-carbon spacer, forming a

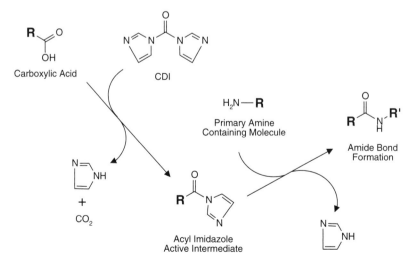

Figure 117 CDI reacts with carboxylate groups to form an active acylimidazole intermediate. In the presence of an amine nucleophile, amide bond formation can take place with release of imidazole.

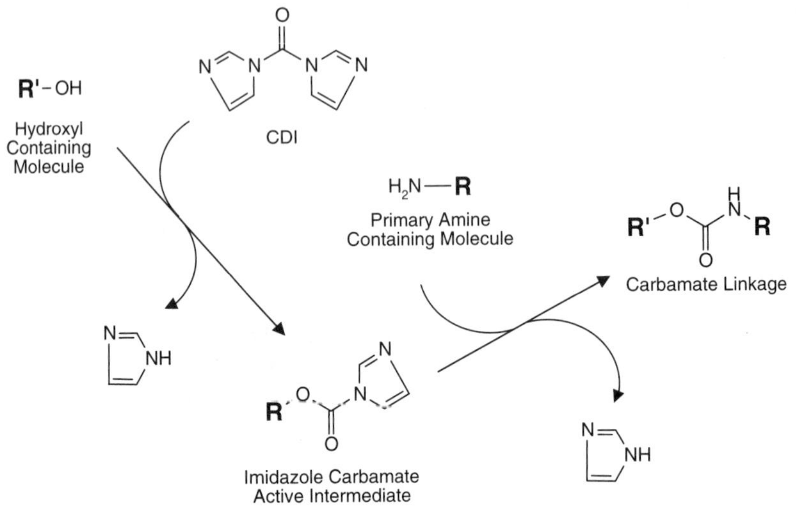

Figure 118 CDI reacts with hydroxyl groups to form an active imidazole carbamate intermediate. In the presence of amine-containing compounds, a carbamate linkage is created with loss of imidazole.

stable urethane (*N*-alkyl carbamate) linkage. This coupling procedure has been applied to the activation of hydroxyl-containing chromatography supports for the immobilization of amine-containing affinity ligands (Bethell *et al.*, 1979; Hearn *et al.*, 1979, 1983; M. T. W. Hearn, 1987) and also to the activation of polyethylene glycol for the modification of amine-containing macromolecules (Beauchamp *et al.*, 1983).

CDI-activated hydroxyls also may undergo a side reaction to form active carbonates. This occurs when an imidazolyl carbamate reacts with another hydroxyl group before the second hydroxyl has had a chance to get activated with CDI. Particularly with adjacent hydroxyls on the same molecule, this can be a problem if a defined reactive species is desired. Any carbonates formed, however, are still reactive toward amines to create carbamate linkages.

Formation of the activated species, whether with a carboxylate or a hydroxyl, must take place in nonaqueous environments due to the rapid breakdown of CDI by hydrolysis. Even in solvents containing small amounts of water, CDI quickly hydrolyzes to CO_2 and imidazole. It is best to use solvents with less than 0.1% water to prevent extensive CDI breakdown. Characteristic bubble formation is an indication of reagent hydrolysis, although CO_2 also is released upon reaction with a carboxylic acid. Activation of carboxylates or hydroxyls may be done in dry organic solvents such as acetone, dioxane, DMSO, THF, and DMF. If an excess of CDI is used during the activation step, it should be removed before adding the activated intermediate to an amine-containing molecule for conjugation. Alternatively, equal molar quantities of CDI and the molecule to be activated may be mixed to form the active species. After about an hour of activation, add an equivalent molar quantity of the amine-containing target molecule to be conjugated.

Aqueous reaction conditions that result in the best conjugation yields using CDI usually reflect the relative pK_a of the nucleophilic amine being coupled. Proteins are best coupled to CDI-activated supports or molecules in an environment at least one pH unit above their pI values. Frequently the greatest coupling yields occur in alkaline buffers within the pH range 8–10. In aqueous solutions, CDI-activated carboxylates or hydroxyls will hydrolyze and slowly lose activity. N-Acylimidazoles hydrolyze by loss of imidazole and regenerate the original carboxylate. The imidazole carbamate-active species hydrolyzes by loss of CO_2 and imidazole, regenerating, in this case, the original hydroxyl group. CDI-activated carboxylic acids hydrolyze faster in aqueous solutions than CDI-activated hydroxyls; however, both experience increasing hydrolysis with increasing pH.

Conjugation reactions using CDI also may be done in organic solutions. This is a distinct advantage if the reactants are not very soluble in aqueous environments. In addition, organic coupling will not experience the concomitant loss of activity due to hydrolysis as water-based reactions, thus nonaqueous reactions usually result in greater yields.

A protocol for the use of CDI in the activation of poly(ethylene glycol) is discussed in Chapter 15, Section 1.4.

4. Schiff Base Formation and Reductive Amination

Aldehydes and ketones can react with primary and secondary amines to form Schiff bases. A Schiff base is a relatively labile bond that is readily reversed by hydrolysis in aqueous solution. The formation of Schiff bases is enhanced at alkaline pH values, but they are still not completely stable unless reduced to secondary or tertiary amine linkages (Fig. 119). A number of reducing agents can be used to convert specifically the Schiff base bond into an alkylamine linkage. Once reduced, the bonds are highly stable. The use of reductive amination to conjugate an aldehyde-containing molecule to an amine-containing molecule results in a zero-length cross-link where no additional spacer atoms are introduced between the molecules.

Reductive amination (or alkylation) may be used to conjugate an aldehyde- or ketone-containing molecule with an amine-containing molecule. The reduction reaction is best facilitated by the use of a reducing agent such as sodium cyanoborohydride,

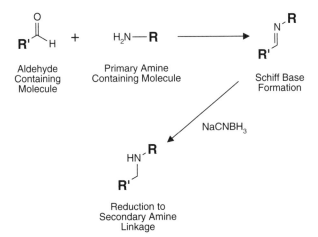

Figure 119 Carbonyl groups can react with amine nucleophiles to form reversible Schiff base intermediates. In the presence of a suitable reductant, such as sodium cyanoborohydride, the Schiff base is stabilized to a secondary amine bond.

because the specificity of this reagent is toward the Schiff base structure and will not affect the original aldehyde groups. By contrast, sodium borohydride also is used in this reaction, but its strong reducing power rapidly converts any aldehydes not yet reacted into nonreactive hydroxyls, effectively eliminating them from further participation in the conjugation process. Borohydride also may affect the activity of some sensitive proteins, whereas cyanoborohydride is gentler, effectively preserving the activity of even some labile monoclonal antibodies. Cyanoborohydride has been shown to be at least five times milder than borohydride in reductive amination processes with antibodies (Peng *et al.*, 1987). Other reducing agents that have been explored for reductive amination processes include various amine boranes and ascorbic acid (Cabacungan *et al.*, 1982; Hornsey *et al.*, 1986).

Immobilization by reductive amination of amine-containing biological molecules onto aldehyde-containing solid supports has been used for quite some time (Sanderson and Wilson, 1971). The reaction proceeds with excellent efficiency (Domen *et al.*, 1990). The optimum pH for the reaction is alkaline, although good yield can be realized from pH 7–10. At high pH (9–10) the formation of the Schiff bases is more efficient and the yield of conjugation or immobilization reactions can be dramatically increased (Hornsey *et al.*, 1986).

The introduction of aldehyde functional groups into protein and other molecules can be accomplished by a number of methods (Chapter 1, Section 4.4). Glyproteins may be oxidized at their carbohydrate residues using sodium periodate or a specific sugar oxidase. Amine groups may be modified to produce a formyl group by reacting with NHS-aldehydes or *p*-nitrophenyl diazopyruvate. The following generalized protocol assumes that the requisite groups are present on the two molecules to be conjugated.

Protocol

1. Dissolve the amine-containing protein to be conjugated at a concentration of 1–10 mg/ml in a buffer having a pH between 7 and 10. Higher pH reactions will result in greater yield of conjugate formation. Suitable buffers include 0.1 M sodium phosphate, 0.15 M NaCl, pH 7.2; 0.1 M sodium borate, pH 9.5; or 0.05 M sodium carbonate, 0.1 M sodium citrate, pH 9.5. Avoid amine-containing buffers like Tris.

2. Add a quantity of the aldehyde-containing molecule to the solution in step 1 to obtain the desired molar ratio for conjugation. For instance, if the amine-containing protein is an antibody and the aldehyde-containing protein is an enzyme such as horseradish peroxidase (HRP), a typical molar ratio for the reaction might be 2–4 mol of HRP per mole of antibody.

3. Add 10 μl of 5 M cyanoborohydride in 1 N NaOH (Aldrich) per milliliter of the conjugation solution volume. *Caution: Highly toxic compound. Use a fume hood and be careful to avoid skin contact with this reagent.*

4. React for 2 h at room temperature.

5. To block unreacted aldehyde sites, add 20 μl of 3 M ethanolamine (pH adjusted to desired value with HCl) per milliliter of the conjugation solution volume. React for 15 min at room temperature.

6. Purify the conjugate by dialysis or gel filtration using a buffer suitable for the nature of the proteins being cross-linked.

4

Homobifunctional Cross-linkers

The first cross-linking reagents used for modification and conjugation of macro-molecules consisted of bireactive compounds containing the same functional group at both ends (Hartman and Wold, 1966). Most of these homobifunctional reagents were symmetrical in design with a carbon chain spacer connecting the two identical reactive ends (Fig. 120). Like molecular rope, these reagents could tie one protein to another by covalently reacting with the same common groups on both molecules. Thus, the lysine ε-amines or N-terminal amines of one protein could be cross-linked to the same functional groups on a second protein simply by mixing the two together in the presence of the homobifunctional reagent.

The ability to link so easily two proteins or other molecules having different binding specificities or catalytic activities opened the potential for creating a new universe of unique and powerful reagent systems for use in assay and targeting applications. The variety and reactivity of homobifunctional reagents multiplied dramatically through-out the 1970s and 1980s. Today, there are dozens of commercially available cross-linkers possessing almost every length and reactivity desired.

The main disadvantage, however, of using simple homobifunctional reagents is the potential for creating a broad range of poorly defined conjugates (Avrameas, 1969). When cross-linking two proteins, for example, the reagent may react initially with either one of the proteins, forming an active intermediate. This activated protein may form cross-links with the second protein or react with another molecule of the same type. It also may react intramolecularly with other functional groups on part of its own polypeptide chain. In addition, other cross-linking molecules may continue to react with these intermediates to form various mixed oligomers, including severely polymerized products that may even precipitate (see Chapter 1, Section 1.2).

The problem of poorly defined conjugation products is exaggerated in single-step reaction procedures using homobifunctional reagents. Single-step procedures involve the addition of all reagents at the same time to the reaction mixture. This technique provides the least control over the cross-linking process and invariably leads to a multitude of products, only a small percentage of which represent the desired conju-gate. Excessive conjugation may cause the formation of insoluble complexes that consist of very high molecular weight polymers. For example, one-step glutaraldehyde conjugation of antibodies and enzymes (Chapter 10, Section 1.2) often results in significant oligomers and precipitated conjugates. To overcome this shortcoming, two-

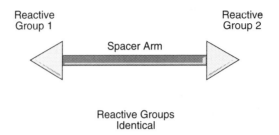

Figure 120   The general design of a homobifunctional cross-linking agent. The two reactive groups are identical and typically are located at the ends of an organic spacer group. The length of the spacer may be designed to accommodate the distance between two molecules to be conjugated.

step reaction procedures have been developed using homobifunctional reagents. Controlled, two-step conjugation protocols somewhat alleviate the polymerization problem with homobifunctional reagents, but can never totally avoid it.

In two-step protocols, one of the proteins to be conjugated is reacted with the homobifunctional reagent and excess cross-linker and by-products are removed. In the second stage, the activated protein is mixed with the other protein or molecule to be conjugated, and the final conjugation process occurs (Fig. 121).

One potential problem of such two-step procedures is hydrolysis of the activated intermediate before addition of the second molecule to be conjugated. For instance, NHS ester homobifunctionals hydrolyze rapidly and may degrade before the second stage of the cross-linking is initiated. In addition, the use of homobifunctional reagents in two-step protocols still produces many of the problems associated with single-step procedures, because the first protein can cross-link and polymerize with itself long before the second protein is added. Since the first protein to be activated has target functional groups on every molecule that can couple with the both reactive groups on the cross-linker, both ends of the reagent potentially can react. This inherent capacity to polymerize uncontrollably unfortunately is characteristic of all homobifunctional reagents, even in multistep protocols.

Although their shortcomings in this regard are clearly recognized, homobifunctional reagents continue to be popular choices for all kinds of conjugation applications. The fact is, in many cross-linking functions, they work well enough to form effective conjugates. Even glutaraldehyde-mediated antibody–enzyme conjugates still are commonly utilized in everything from research to diagnostics.

The particular cross-linkers discussed in this section are the types most often referred to in the literature or are commercially available. Many other forms of homobifunctional reagents containing almost every conceivable chain length and reactivity can be found mentioned in the scientific literature.

1. Homobifunctional NHS Esters

Carboxylate groups activated with *N*-hydroxysuccinimide (NHS) esters are highly reactive toward amine nucleophiles. In the mid-1970s, NHS esters were introduced as reactive ends of homobifunctional cross-linkers (Bragg and Hou, 1975; Lomant and

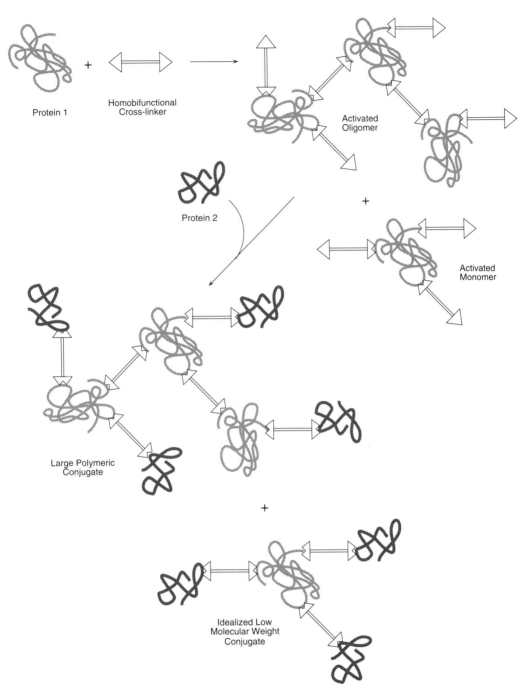

Figure 121   Homobifunctional cross-linkers may be used in a two-step process to conjugate two proteins or other molecules. In the first step, one of the two proteins is reacted with the cross-linker in excess to create an active intermediate. After removal of remaining cross-linker, a second protein is added to effect the final conjugate. Two-step reaction schemes somewhat limit the degree of polymerization obtained when using homobifunctional reagents.

Fairbanks, 1976). Their excellent reactivity at physiological pH quickly established NHS esters as viable alternatives to the imidoesters predominating at the time (Section 2).

Unfortunately, many NHS ester-containing cross-linkers are insoluble in aqueous buffers. Most protocols involve dissolving the compound at a relatively high concentration in an organic solvent and aliquoting the required quantity into the reaction medium. Prior dissolution helps to maintain at least some solubility in the buffered cross-linking environment. Most of the time, however, the addition of an organic-solubilized cross-linker into a buffered solution results in a microprecipitate that slowly goes into solution during the course of reaction.

In the early 1980s, Staros prepared a derivative of N-hydroxysuccinimide that aids in the water solubility of NHS ester cross-linkers (Staros, 1982). N-Hydroxysulfosuccinimide (sulfo-NHS) esters possess a negatively charged sulfonate group on carbon No. 2 or 3 of the succinimide ring (Fig. 122). Cross-linking reagents containing sulfo-NHS esters have half-lives of hydrolysis on the order of hours, sometimes even better than their NHS ester analogs (Anjaneyulu and Staros, 1987). The sulfo-NHS ester often lends enough polarity to a cross-linker to provide water solubility and thus eliminate the need for organic solvent dissolution. In addition, sulfo-NHS cross-linkers may be used for surface-only modification of membranes and cells, since they are more hydrophilic and will not penetrate the lipid environment of the membrane (Staros, 1982, 1988; Staros *et al.*, 1987). By contrast, many of the more hydrophobic NHS ester cross-linkers can be used to traverse the cell membrane and modify intracellular components.

NHS- or sulfo-NHS ester-containing homobifunctional cross-linkers react with

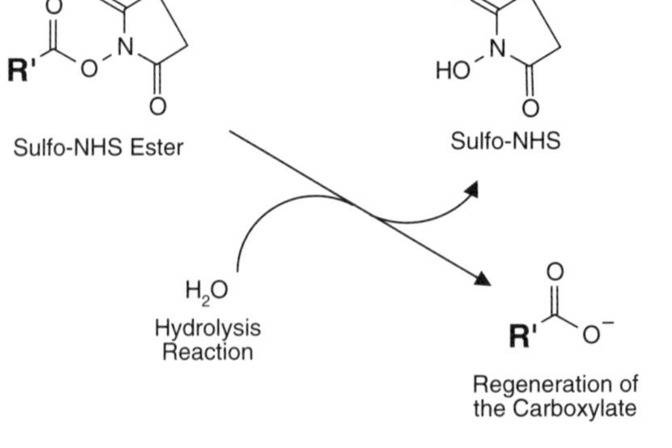

Figure 122   In aqueous solution, a sulfo-NHS ester can either couple to an amine group to form an amide bond or react with water to hydrolyze back to a carboxylate. Both processes release the sulfo-NHS leaving group.

nucleophiles to release the NHS or sulfo-NHS leaving group and form an acylated product. The reaction of such esters with a sulfhydryl or hydroxyl group is possible, but does not yield stable conjugates, forming thioesters and ester linkages that may hydrolyze in aqueous environments. Histidine side-chain nitrogens of the imidazolyl ring also may be acylated with an NHS ester reagent, but they too hydrolyze rapidly (Cuatrecasas and Parikh, 1972). Reaction with primary and secondary amines, however, creates stable amide and imide linkages, respectively, that do not readily break down. In protein molecules, NHS ester cross-linking reagents primarily react with the α-amines at the N-terminals and the abundant ε-amines of lysine side chains.

NHS ester cross-linking reactions in aqueous solutions consist of the potential for hydrolysis as well as the desired amide bond formation. Cross-linkers containing NHS esters have fairly good stability in aqueous solutions, despite their susceptibility to attack and break down by water. Studies done on the NHS ester-containing homobifunctional reagent dithio*bis*(succinimidylpropionate) (DSP) indicate that the activated carboxylates have half-lives on the order of hours at physiological pH. However, hydrolysis and amine reactivity both increase with increasing pH. At 0°C at pH 7, the half-life of the cross-linking reagent DSP is 4–5 h (Lomant and Fairbanks, 1976). At pH 8 at 25°C it falls to 1 h (Staros, 1988), and at pH 8.6 and 4°C the half-life is only 10 min (Cuatrecasas and Parikh, 1972).

The rate of hydrolysis may be monitored by measuring the increase in absorptivity at 260 nm as the NHS leaving group is cleaved. The molar extinction coefficient of the NHS group in solution is $8.2 \times 10^3\ M^{-1}cm^{-1}$ in Tris buffer at pH 9 (Carlsson *et al.*, 1978), but somewhat decreases to $7.5 \times 10^3\ M^{-1}cm^{-1}$ in potassium phosphate buffer at pH 6.5 (Partis *et al.*, 1983). Unfortunately, the sensitivity of this absorptivity usually does not allow for measuring the rate of reaction in an actual cross-linking procedure.

To maximize the modification of amines and minimize the effects of hydrolysis, maintain a high concentration of protein or other target molecule. By adjusting the molar ratio of cross-linker to target molecule(s), the level of modification and conjugation may be controlled to create an optimal product.

The reaction buffer chosen for the conjugation reaction should be free of extraneous amines. Avoid Tris or glycine buffers. Also, avoid imidazole buffers, since the nitrogens of the imidazole ring may react with the active ester and then quickly hydrolyze. The effect is that imidazole only acts to catalyze the hydrolysis process. The pH of the reaction should be in the range of 7 to 9 to promote the unprotonated state of primary amines, which is the nucleophilic species that most effectively attacks the activated carbonyl group. Dissolve NHS ester cross-linkers that are insoluble in water in an organic solvent such as DMF or DMSO prior to addition to the reaction medium. Sulfo-NHS cross-linkers may be added directly to the reaction mixture or predissolved in buffer at a higher concentration before an aliquot is added to the reaction. Aqueous stock solutions should be used immediately to prevent extensive hydrolysis of the active esters.

NHS ester cross-linking reagents also may be used in organic solvent-based reactions without the competing hydrolysis problem, provided the target molecules are soluble and stable in such environments. In this case, both molecules to be conjugated must be soluble in the solvent. DMF, DMSO, acetone, and dioxane are examples of solvents that can be utilized as the reaction medium. Refer to any solubility data on the cross-linking reagent of choice to see which solvents are most appropriate.

1.1. DSP and DTSSP

Lomant's reagent [dithiobis(succinimidylpropionate), DSP], is a homobifunctional NHS ester cross-linking agent containing an 8 atom spacer of 12 Å in length (Lomant and Fairbanks, 1976) (Pierce). It is symmetrically constructed around a central disulfide group that is cleavable after conjugation with typical disulfide reducing agents (Chapter 1, Section 4.1).

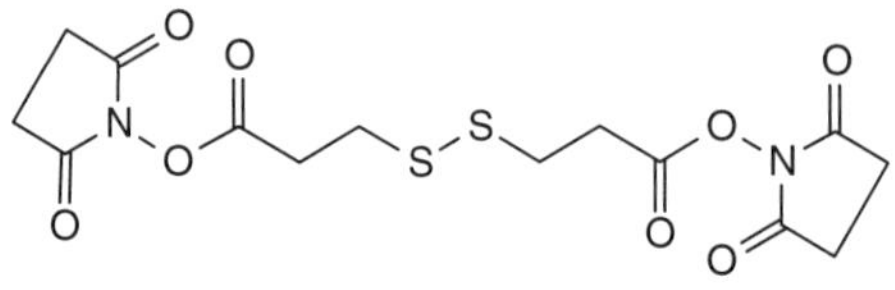

DSP
Dithiobis(succinimidylpropionate)
MW 404.42
12 Å

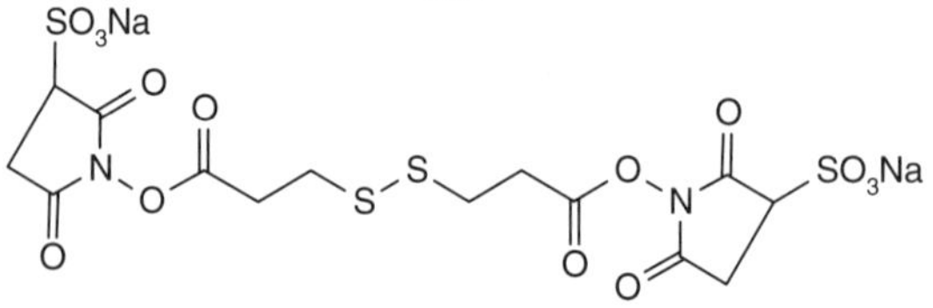

DTSSP
3,3'-Dithio*bis*(sulfosuccinimidylpropionate)
Water Soluble
MW 608.51
12 Å

DSP is water-insoluble and must be predissolved in an organic solvent before addition to a conjugation reaction. Concentrated stock solutions may be prepared in DMF or DMSO and an aliquot added to a buffered reaction medium. The NHS ester reaction occurs most efficiently at pH 7–9, with hydrolysis of the active species accelerating the greater the pH. The cross-linking buffers should be free of amine-containing components other than the target molecules to be conjugated. Avoid Tris or glycine buffers. A reaction buffer consisting of 0.1 M sodium phosphate, 0.15 M NaCl, pH 7.2–7.5, works well for most applications involving the cross-linking of two purified proteins. The relatively high concentration of sodium phosphate is used to prevent pH drift downward during the course of the reaction. For *in vitro* cross-linking of cellular components such as membrane proteins, a more dilute PBS buffer containing isotonic saline is more appropriate. Since DSP is a hydrophobic reagent, it is able to penetrate the cell membrane and conjugate membrane components. For this reason, it has become quite popular for use in investigating the interactions of membrane proteins.

DSP reacts with ε-amine groups on the side chains of lysine residues or the α-amine at the N-terminal of proteins to form amide linkages. Amine-containing macromolecules may be reversibly cross-linked with this reagent and later cleaved with DTT or 2-mercaptoethanol (Fig. 123). For reductive cleavage of conjugated molecules, add 10–50 mM DTT, and incubate at 37°C for 30 min. Alternatively, the conjugate may be reduced prior to electrophoresis using SDS sample buffer with 5% 2-mercaptoethanol at elevated temperatures.

Lomant's reagent is one of the most popular of all cross-linking agents, especially for the investigation of protein interactions. Hordern *et al.* (1979) used it to investigate

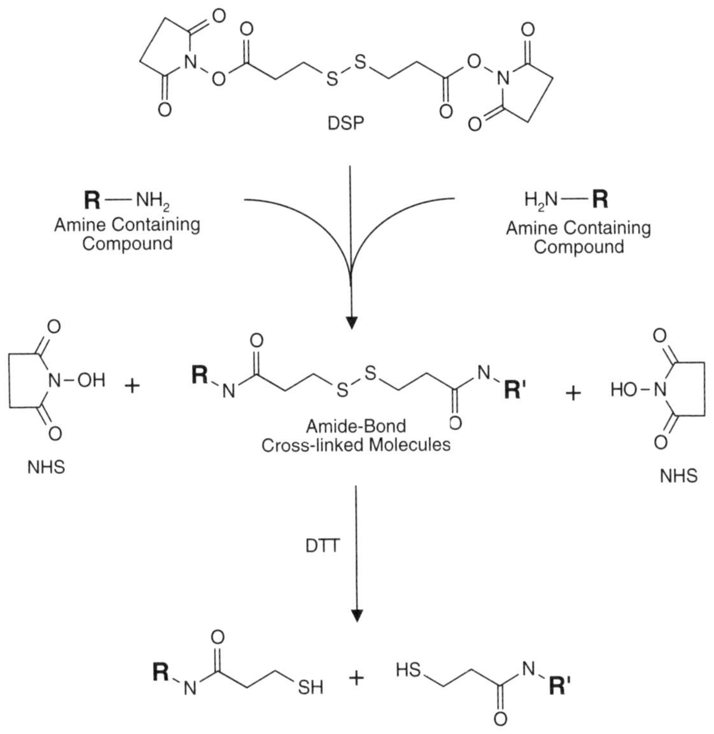

Figure 123 The reaction of DSP with amine-containing molecules yields amide bond cross-links. The conjugates may be cleaved by reduction of the disulfide bond in the cross-bridge with DTT.

the spatial relationships in the capsid polypeptides of the mengo virion. It has been used in studying the interactions of proteins involved with active transport (dePont *et al.*, 1980; Joshi and Burrows, 1990), in identifying cross-links involving cytochrome *P*-450 (Baskin and Yang, 1982), in characterization of cell surface receptors for colony-stimulating factor (Park *et al.*, 1986), in the determination of various membrane antigens by cross-linking to specific monoclonal antibodies (Hamada and Tsuro, 1987), in studying prothrombin self-association (Tarvers *et al.*, 1982), in investigating chemotaxis in *Escherichia coli* (Chelsky and Dahlquist, 1980), in molecular identification of receptors for vasoactive intestinal peptide in rat intestinal epithelium (Laburthe *et al.*, 1984), and in studying the cross-linking of the affinity-purified CCAAT transcription, α-CPl (Kim and Sheffrey, 1990).

The sulfo-NHS version of DSP, dithio*bis*(sulfosuccinimidyl propionate) (DTSSP), is a water-soluble analog of Lomant's reagent that can be added directly to aqueous reactions without prior organic solvent dissolution (Staros, 1982). DTSSP still contains the disulfide center portion that is cleavable with the proper reducing agents, and the sulfo-NHS ends have virtually the same reactivity as DSP (Fig. 124). Due to its hydrophilicity, however, DTSSP will not penetrate cellular membranes as does its more hydrophobic analog, DSP. It is therefore an excellent choice for the cross-linking of cell surface components without affecting intracellular substances.

DTSSP reportedly has been used to cross-link the extracytoplasmic domain of the

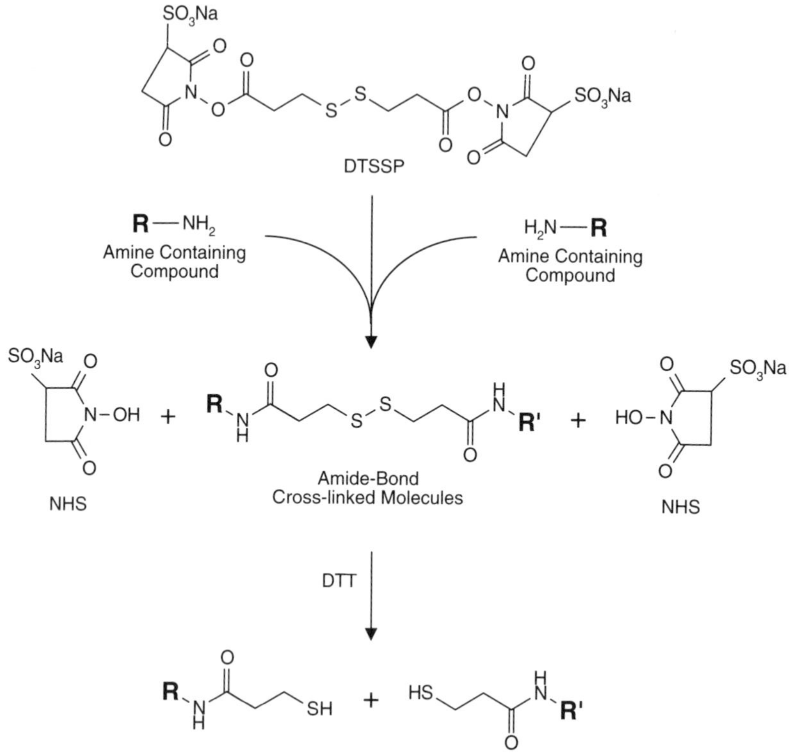

Figure 124 DTSSP can form cross-links between two amine-containing molecules through amide linkages. The conjugates may be cleaved by disulfide reduction using DTT.

anion exchange channel in human erythrocytes (Staros and Kakkad, 1983), for the characterization of a ribosomal complex in *Bacillus subtilis* (Caufield *et al.*, 1984), to investigate ascites hepatoma cytokeratin filaments (Knoller *et al.*, 1991), for the study of the B-lymphocyte Fc receptor for IgE (Waugh *et al.*, 1989), and to cross-link platelet glycoproteins (Jung and Moroi, 1983).

1.2. DSS and BS³

Disuccinimidyl suberate (DSS) is an amine-reactive, homobifunctional, NHS ester, cross-linking reagent that produces an 8-atom bridge (11.4 Å) between conjugated molecules (Fig. 125) (Pierce). Its hydrocarbon chain is noncleavable, so cross-links formed are irreversible. Many of the reported applications of DSS involve investigations of receptor–ligand binding on cell surfaces using radiolabeled molecules. The cross-linker is hydrophobic and must be solubilized in organic solvent prior to addition to a conjugation reaction. Predissolving in dry dioxane, DMF, or DMSO may be done at higher concentration and then an aliquot added to the aqueous reaction medium as needed. The final concentration of the organic solvent in the buffered reaction should not exceed 10%. Stock solutions should be prepared fresh. DSS is

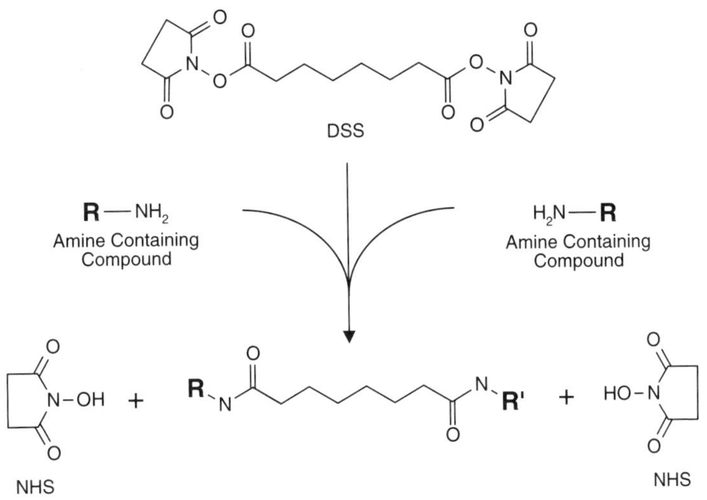
DSS
R—NH2
Amine Containing Compound
H2N—R
Amine Containing Compound
N-OH
NHS
HO-N
NHS
R
N
R'

Figure 125 DSS reacts with two amine-containing molecules to form amide bond cross-links. The cross-bridge is noncleavable.

membrane permeable and is therefore useful for intracellular and intramembrane conjugations. The optimum conditions for the cross-linking reaction are a pH range of 7 to 9 using buffers and other salts that contain no amines. Avoid the use of Tris or glycine. A phosphate buffer (PBS) at physiological pH works well. See Section 1 for additional information on NHS ester reactions.

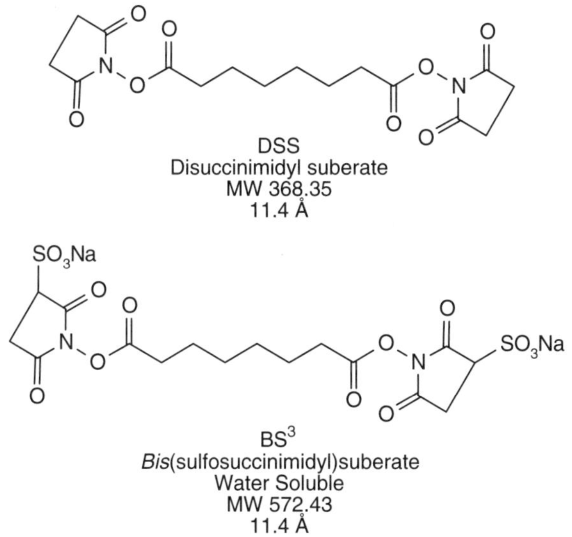
DSS
Disuccinimidyl suberate
MW 368.35
11.4 Å
SO3Na
SO3Na
BS3
Bis(sulfosuccinimidyl)suberate
Water Soluble
MW 572.43
11.4 Å

Reported applications of DSS include cross-linking the A and B subunits of ricin (Montesano *et al.*, 1982), studying human somatotropin and the components of the lactogenic binding sites of rat liver (Caamano *et al.*, 1983), cross-linking CSF-1 to its cell-surface receptor (Morgan and Stanley, 1984), studying angiotensin II interactions with its receptor (Petruzzelli *et al.*, 1985), cross-linking of vasoactive intestinal peptide to its receptor on human lymphoblasts (Wood and O'Dorisio, 1985), investigating insulin-dependent protein kinases (Petruzzelli *et al.*, 1985), identifying a cellular receptor for TNF (Kull *et al.*, 1985), affinity cross-linking of atrial natriuretic factor in aorta membranes (Vandelen *et al.*, 1985), studying the receptor for human interferon (Rashidbaigi *et al.*, 1986), cross-linking of endorphin to membranes rich in opioid receptors (Helmeste *et al.*, 1986), immunoprecipitation studies of the cross-linked complex of parathyroid hormone with its receptor (Wright *et al.*, 1987), binding of human interferon Y to its receptor (Novick *et al.*, 1987), identifying the erythropoietin receptor on Friend virus-infected erythroid cells (Sawyer *et al.*, 1987), and binding of the p75 peptide to an interleukin 2 receptor (Tsudo *et al.*, 1987).

Bis(sulfosuccinimidyl) suberate (BS[3]) is an analog of DSS that contains sulfo-NHS esters on both carboxylates. The effect of the negative charges provided by the sulfonate groups lends water solubility to the compound. Prior organic solvent dissolution (before addition to a reaction) is not necessary. The hydrophilicity of BS[3] also makes it membrane impermeable. Therefore, cell labeling with BS[3] results in hydrophilic region modification and cross-linking, targeting surface functional groups, whereas DSS is capable of targeting hydrophobic regions within the membrane structure itself. As with DSS, BS[3] is noncleavable, and thus all cross-links formed are irreversible. The reactivity of the sulfo-NHS esters is identical to that of NHS esters, being highly reactive toward amines in the pH range 7–9.

Reported applications of BS[3] include cross-linking of the β-endorphin–calmodulin interaction (Staros, 1982), cross-linking of the extracellular domain of intact human erythrocytes' anion exchange channel (Staros and Kakkad, 1983), cross-linking of hepatoma cytokeratin filaments (Ward *et al.*, 1985), investigating the β-lumphocyte Fc receptor for IgE (Lee and Conrad, 1985; Staros *et al.*, 1987), cross-linking of the tripeptide Arg–gly–Asp to an adhesion receptor on platelets (Souza *et al.*, 1988), cross-linking of the large and small subunits of cytochrome *b*559 (Knoller *et al.*, 1991), and for general receptor–ligand cross-linking (Waugh *et al.*, 1989).

1.3. DST and Sulfo-DST

Disuccinimidyl tartarate (DST) is a homobifunctional NHS ester cross-linking reagent that contains central *cis*-diols that are susceptible to cleavage with sodium periodate (Pierce) (Fig. 126). DST forms amide linkages with α-amines and ε-amines of proteins or other amine-containing molecules. The reagent is fairly insoluble in aqueous buffers, but may be predissolved in THF, DMF, or DMSO prior to addition of an aliquot to a reaction. Optimal conditions for reactivity include a pH range of 7–9 with no extraneous amines present that may cross-react with the NHS esters. Avoid Tris, glycine, or imidazole buffers. Subsequent to conjugating proteins, the cross-links may be broken for analysis by treatment with 0.015 *M* sodium periodate.

Reported applications of DST include the cross-linking of ubiquinone cytochrome *c* reductase (R. J. Smith *et al.*, 1978), characterization of the cell surface receptor for colony-stimulating factor (Park *et al.*, 1986), investigation of the Ca^{+2}, Mg^{+2}-activated ATPase of *E. coli* (Bragg and Hou, 1980), and characterization of human properdin polymers (Farries and Atkinson, 1989).

Disulfosuccinimidyl tartarate (sulfo-DST) is an analog of DST that contains sulfo-NHS esters. The negatively charged sulfonate groups contribute enough hydro-

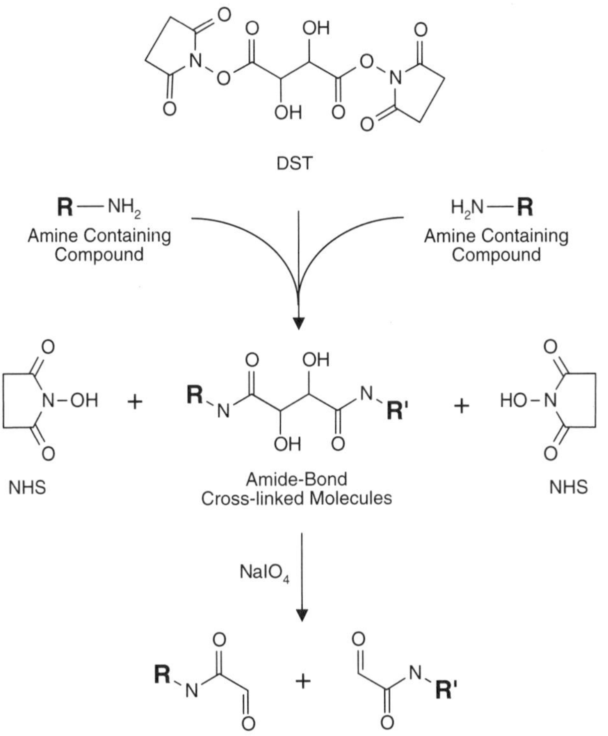

Figure 126 DST may be used to cross-link amine-containing molecules, forming amide bond linkages. The central diol of the cross-bridge is cleavable by treatment with sodium periodate.

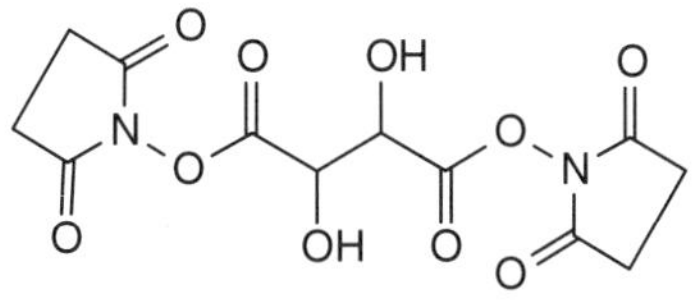

DST
Disuccinimidyl tartarate
MW 344.24
6.4 Å

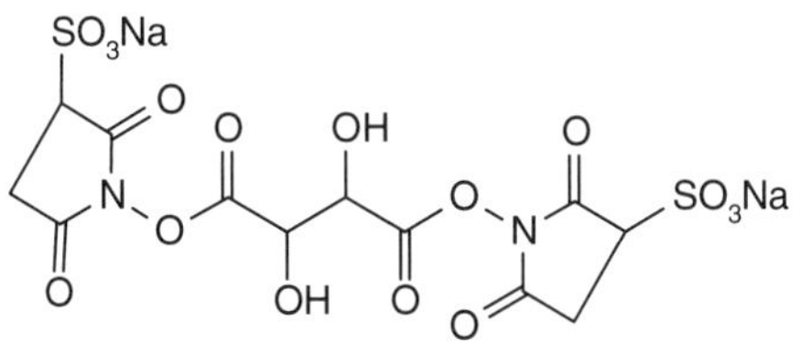

Sulfo-DST
Disulfosuccinimidyl tartarate
Water Soluble
MW 548.34
6.4 Å

philicity to provide water solubility for the reagent without the need for organic solvent dissolution before adding it to a cross-linking reaction. The conditions for conjugation are otherwise identical to DST.

1.4. BSOCOES and Sulfo-BSOCOES

Bis[2-(succinimidyloxycarbonyloxy)ethyl]sulfone (BSOCOES) is a water-insoluble, homobifunctional NHS ester cross-linking reagent that contains a central sulfone group that is cleavable under alkaline conditions (Fig. 127) (Pierce). The NHS ester ends are reactive with amine groups in proteins and other molecules to form stable amide linkages. Once proteins are cross-linked using this reagent, they may be dissociated for analysis by raising the pH to 11.6 and incubating for 2 h at 37°C. The sulfonate group is base labile under these conditions, and the conjugate cleaves at the center of the bridge.

BSOCOES is a hydrophobic cross-linker and therefore must be dissolved in organic solvent prior to its addition to an aqueous reaction medium. Preparing a stock solution in DMF or DMSO and then adding an aliquot to the cross-linking reaction is recommended. Do not exceed a concentration of more than 10% organic solvent in the buffered reaction.

Reported applications of BSOCOES include studying the polypeptide antigens on lymphocyte cell surfaces (Zarling *et al.*, 1980), cross-linking labeled β-endorphin to its opioid receptors (Howard *et al.*, 1985), and isolation and characterization of calcitonin receptors in rat kidney (Bouizar *et al.*, 1986).

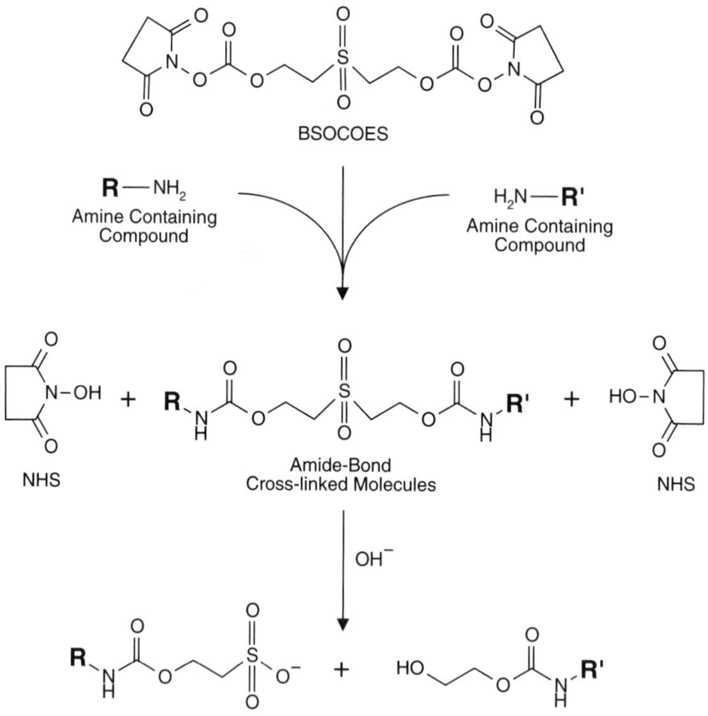

Figure 127 BSOCOES reacts with amine-containing molecules to create amide bond cross-links. The internal sulfone group is cleavable under alkaline conditions.

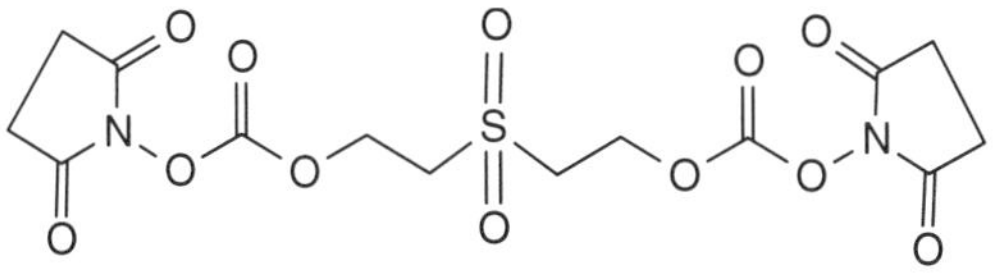

BSOCOES
Bis[2-(succinimidooxycarbonyloxy)ethyl]sulfone
MW 436.36
13 Å

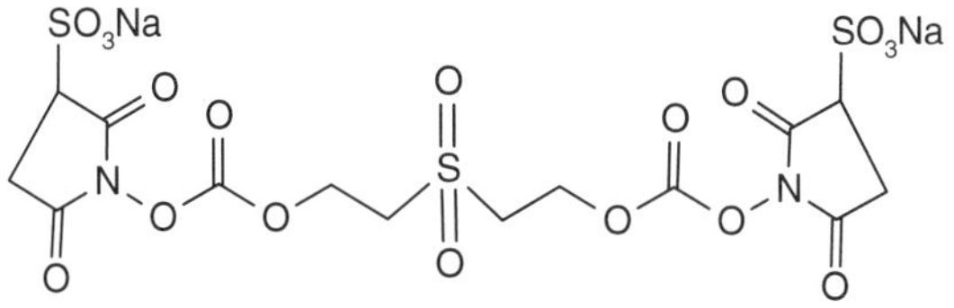

Sulfo-BSOCOES
Bis[2-(sulfosuccinimidooxycarbonyloxy)ethyl]sulfone
Water Soluble
MW 640.46
13 Å

There also is a water-soluble version of this reagent available. *Bis*[2-(sulfosuccinimidooxycarbonyloxy)ethyl]sulfone (sulfo-BSOCOES) is built on the same chemical structure as BSOCOES, but contains the negatively charged sulfonate groups on both of its succinimide rings. The presence of the sulfonates provides enough hydrophilicity to lend water solubility to the entire reagent. Thus, it may be added directly to aqueous reactions at concentrations of up to 10 mM. Prior dissolution in organic solvent, however, may provide solubility at greater concentrations in aqueous solutions.

1.5. EGS and Sulfo-EGS

Ethylene glycol*bis*(succinimidylsuccinate) (EGS) is a homobifunctional cross-linking agent that contains NHS ester groups on both ends (Pierce). Its central bridge is constructed from an ethylene glycol group esterified on either side with succinic acid, the terminal carboxylates of which are activated by forming N-hydroxysuccinimide esters. The two NHS esters are amine reactive, forming stable amide bonds between cross-linked molecules within a pH range of about 7–9. Avoid amine-containing buffers such as Tris or glycine, since they will cross-react with the NHS esters. Imidazole also should be avoided, because it has the effect of catalyzing the hydrolysis of the NHS ester groups. The internal structure of EGS provides two cleavable ester sites that may be broken at pH 8.5 by incubation with 1 M hydroxylamine for 3–6 h at 37°C (Abdella *et al.*, 1979) (Fig. 128). Thus, conjugates produced from the EGS cross-linking of the specific interaction of proteins or other molecules may be subsequently cleaved with hydroxylamine for analysis.

EGS is water insoluble and must be dissolved in an organic solvent prior to its addition to an aqueous reaction. Prepare a concentrated solution of EGS in DMF or DMSO and add an aliquot of the stock solution to the reaction. Do not exceed a concentration of more than about 10% organic solvent in the aqueous reaction buffer or precipitation of buffer salts or protein may occur.

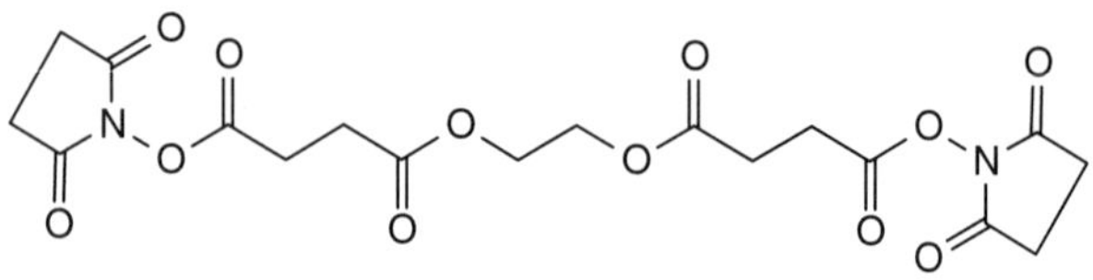

EGS
Ethylene glycolbis(succinimidylsuccinate)
MW 456.37
16.1 Å

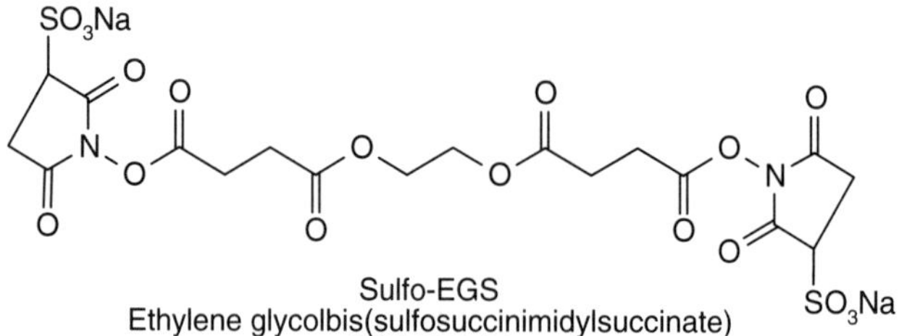

Sulfo-EGS
Ethylene glycolbis(sulfosuccinimidylsuccinate)
MW 660.47
16.1 Å

Reported applications of EGS include cross-linking studies of cytochrome *P*-450 (Baskin and Yang, 1980a), conjugation of tumor necrosis factor with lymphotoxin (Browning and Ribolini, 1989), the conversion of a gonadotropin-releasing hormone

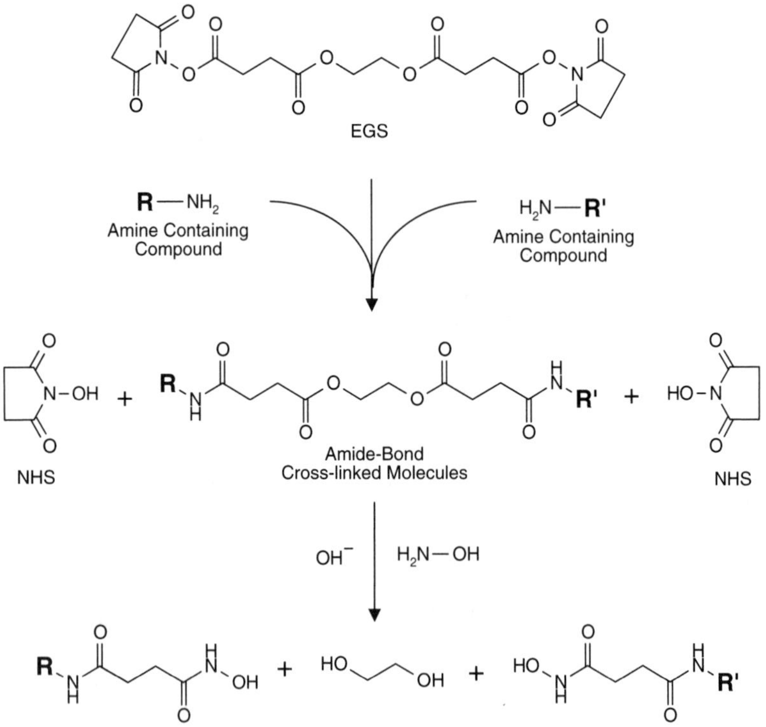

EGS

R—NH₂
Amine Containing
Compound

H₂N—R'
Amine Containing
Compound

NHS

**Amide-Bond
Cross-linked Molecules**

NHS

OH⁻   H₂N—OH

Figure 128 EGS reacts with amine-containing molecules to form amide linked conjugates. The ester groups within its cross-bridge are cleavable under alkaline conditions using hydroxylamine.

antagonist to an agonist (Conn *et al.*, 1982a), preparation of a conjugate of go-nadotropin releasing hormone with an agonist (Conn *et al.*, 1982b), covalent cross-linking of vasoactive peptide to its lymphoblast receptors (Wood and O'Dorisio, 1985), and study of bombesin receptors in Swiss 3T3 cells (Millar and Rozengur, 1990).

A water-soluble analog of EGS also is available commercially (Pierce). Sulfo-EGS contains negatively charged sulfonate groups on its NHS rings. The hydrophilicity of this modification provides water solubility to the compound so that prior dissolution in organic solvent is not necessary.

1.6. DSG

Disuccinimidyl glutarate (DSG) is a water-insoluble, homobifunctional cross-linker containing amine reactive NHS esters at both ends (Pierce). The active esters react with amino groups in protein molecules in the pH range of 7–9 to form amide linkages. DSG is a noncleavable reagent, forming stable 5-carbon bridges between amine-containing molecules (Fig. 129).

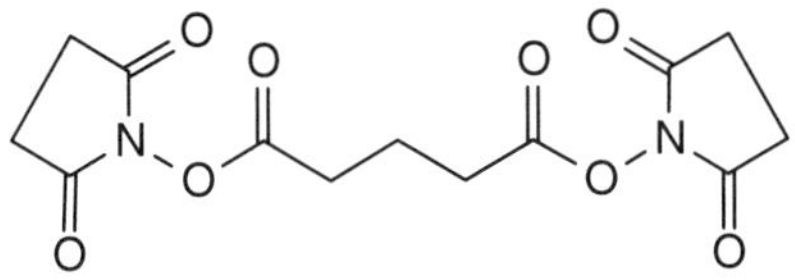

DSG
Disuccinimidyl glutarate
MW 326.26
7.7 Å

DSG should be dissolved in an organic solvent prior to addition to an aqueous reaction medium. Suitable solvents include DMF and DMSO. To initiate a reaction,

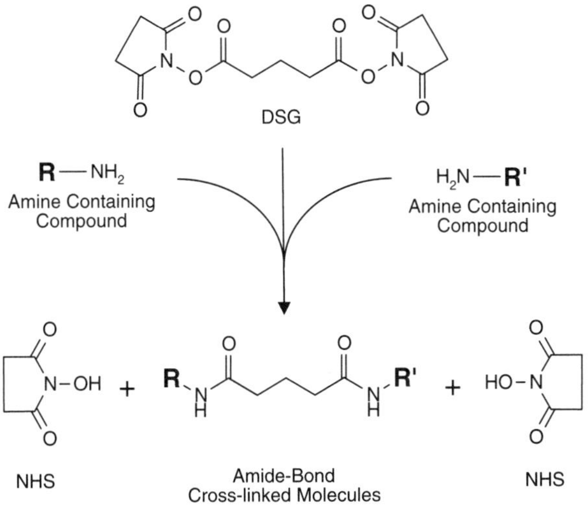

Figure 129 DSG is a noncleavable cross-linker that can react with two amine-containing molecules to form amide bonds.

add an aliquot of the organic solution to the buffered medium containing the molecules to be cross-linked. Reaction buffers should not contain any competing amine compounds such a Tris or glycine, as these will cross-react with the active esters. Avoid imidazole-containing buffers, also, since it catalyzes the hydrolysis of NHS esters.

The reported application of DSG relates to receptor–ligand studies by covalent cross-linking of their complexes (Waugh *et al.*, 1989).

1.7. DSC

N,N'-Disuccinimidyl carbonate (DSC) is the smallest homobifunctional NHS ester cross-linking reagent available (Aldrich). It is, in essence, merely a carbonyl group containing two NHS esters. The compound is highly reactive toward nucleophiles. In aqueous solutions, DSC will hydrolyze to form two molecules of N-hydroxysuccinimide (NHS) with release of CO_2. In nonaqueous environments it can react with two amine groups to form a substituted urea derivative with loss of two molecules of NHS. The reagent also can be used in anhydrous organic solvents to activate a hydroxyl group to an amine-reactive succinimidyl carbonate derivative. This activation procedure is commonly used to activate polyethylene glycol for conjugation with proteins and other molecules (Chapter 15, Section 1.2). In this case, DSC activated hydroxylic compounds can be used to conjugate with an amine-containing molecule to form a stable derivative (Fig. 130). The linkage created from this reaction is a urethane derivative or a carbamate bond, displaying excellent stability.

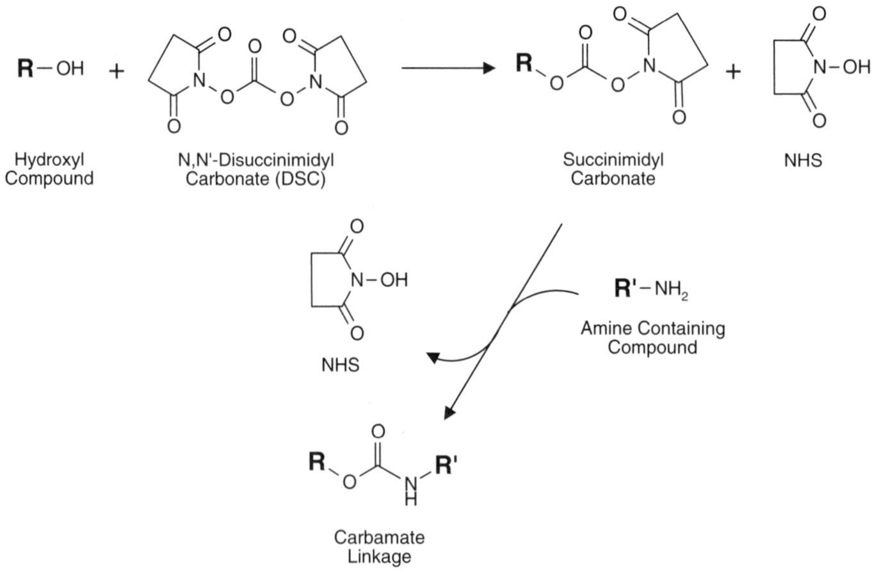

Figure 130   DSC can react with hydroxyl groups to create a succinimidyl carbonate intermediate that is highly reactive toward nucleophiles. In the presence of an amine-containing molecule, the active species can form stable carbamate linkages.

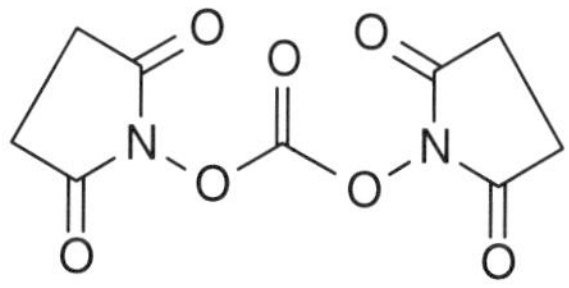

N,N'-Disuccinimidyl
Carbonate (DSC)
MW 256.17

Activation of hydroxyl groups with DSC can be done in acetone or dioxane by reacting for 4–6 h at room temperature. Subsequent conjugation with amine-containing molecules is done in organic or aqueous solutions. For buffered reactions, the optimal conditions include a pH range of 7–9 using common buffer salts (avoid amine-containing components, such as Tris). React for at least 4 h at room temperature or up to overnight at 4°C.

2. Homobifunctional Imidoesters

Cross-linking compounds containing imidoesters at both ends are among the oldest homobifunctional reagents used for protein conjugation (Hartman and Wold, 1966). The imidoester (or imidate) functional group is one of the most specific acylating groups available for the modification of primary amines, with minimal cross-reactivity toward other nucleophilic groups in proteins. The α-amines and ε-amines of proteins may be targeted and cross-linked by reacting with homobifunctional imidoesters at a pH of 7–10 (optimal pH 8–9). The product of this reaction, an imidoamide (or amidine) (Fig. 131), is protonated and thus carries a positive charge at physiological pH (Liu *et al.*, 1977; Kiehm and Ji, 1977; Ji, 1979; Wilbur, 1992). The result is no alteration of the charge characteristics of the cross-linked proteins, since the amines being modified were themselves protonated and originally contributed to the overall positive charge of the molecule. Imidoesters therefore can preserve the microenvironment within the vicinity of the cross-link bridge, possibly retaining native structure and activity better than reagents that modulate the net charge of a protein.

The amidine bond is quite stable at acid pH; however, it is susceptible to hydrolysis and cleavage at high pH. Derivatized proteins may be assayed by amino acid analysis after acid hydrolysis without loss of imidate modifications.

Imidoester cross-linkers are highly water soluble, but undergo continuous degradation due to hydrolysis. The half-life of the imidate functionality is typically less than 30

Figure 131 Imidoesters react with amine groups to form amidine bonds, which are positively charged at physiological pH.

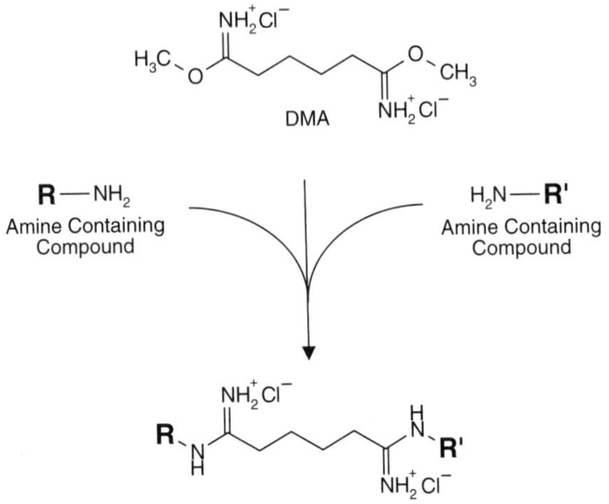

Figure 132 DMA can cross-link two amine-containing molecules to form charged amidine linkages.

min, especially in the alkaline conditions of the reaction medium (Hunter and Ludwig, 1962; Browne and Kent, 1975). Concentrated stock solutions may be prepared before addition of a small amount to a conjugation reaction, but they should be dissolved rapidly and used immediately.

The following list of homobifunctional imidoesters represent compounds that are commonly used for protein cross-linking and are currently available from commercial sources.

2.1. DMA

Dimethyl adipimidate (DMA) is a short-chain, homobifunctional cross-linking agent containing imidoesters at both ends (Pierce). After reaction with amine groups on target molecules, the compound creates a noncleavable, six-atom bridge with terminal amidine bonds (Fig. 132). DMA is water soluble and may be added directly to a cross-linking reaction or predissolved at higher concentration before addition of an aliquot to the reaction medium. Stock solutions should be used immediately to prevent break-down by hydrolysis. Reaction buffers having a pH of 8–9 are optimal. Avoid buffers containing primary amines (glycine and Tris), since these will cross-react with the imidoester groups. Borate or bicarbonate buffers adjusted to the optimum pH range work well. The addition of (or the exclusive use of) 0.1–0.2 M triethanolamine is often done to help catalyze the coupling reaction.

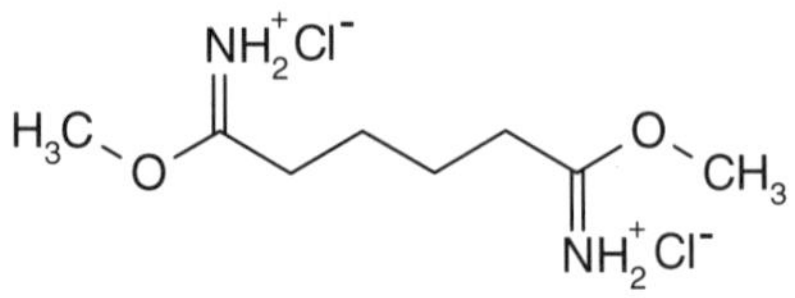

DMA
Dimethyl adipimidate dihydrochloride
MW 245.15
8.6 Å

Reported applications of DMA include the cross-linking of bovine pancreatic ribonuclease A (Hartman and Wold, 1967), treatment of erythrocyte membranes to reduce the effects of sickle cell anemia (Waterman *et al.*, 1975), conjugation and analysis of the outer membrane proteins of *Neisseria gonorrhoeae* (Newhall *et al.*, 1980), protein structural studies of bovine α-crystalline (Siezen *et al.*, 1980), cross-linking of hemoglobin S (Pennathur-Das *et al.*, 1982), and forming *S*-carbomethoxy-valeramidine during hydrolysis of DMA (Mentzer *et al.*, 1982).

2.2. DMP

Dimethyl pimelimidate (DMP) is a homobifunctional cross-linking agent that has imidoester groups on either end (Pierce). The imidoesters are amine reactive to give stable amidine linkages with target molecules. The seven-atom bridge created by DMP cross-links is noncleavable and positively charged at physiological pH due to the protonated amidine bonds (Fig. 133). The reagent is water soluble and may be reacted with proteins or other amine-containing macromolecules at a pH of 8–9 in aqueous media. Use buffers that contain no amine groups that may cross-react with the imidoesters. Avoid glycine or Tris buffers.

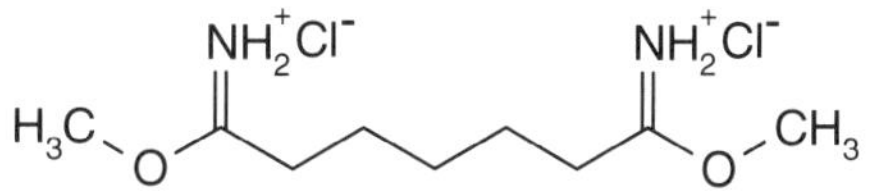

DMP
Dimethyl pimelimidate dihydrochloride
MW 259.18
9.2 Å

In protein cross-linking studies, DMP has been used to examine the subunit structure of muscle pyruvate kinase (Davies and Kaplan, 1972), for the cross-linking of lactose synthetase (Brew *et al.*, 1975), and to conjugate a fluorescent derivative of α-lactalbumin to glactosyltransferase (O'Keefe *et al.*, 1980). The reagent also has

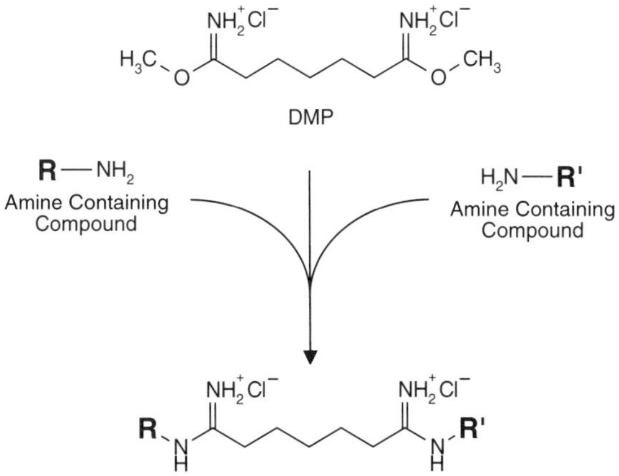

Figure 133 DMP reacts with amine-containing compounds to form amidine bonds.

found use in the immobilization of antibody molecules to insoluble supports containing bound protein A (Schneider *et al.*, 1982). The antibody molecules are first allowed to interact with the coupled protein A, orienting them with their antigen binding sites facing away from the matrix. DMP is then added to anchor covalently the antibodies to the protein A, forming a permanent immunoaffinity matrix.

2.3. DMS

Dimethyl suberimidate (DMS) is a homobifunctional cross-linking agent containing amine-reactive imidoester groups on both ends. The compound is reactive toward the ε-amine groups of lysine residues and N-terminal α-amines in the pH range of 7–10 (pH 8–9 is optimal). The resulting amidine linkages are positively charged at physiological pH, thus maintaining the positive charge contribution of the original amine. DMS creates eight atom bridges between conjugated molecules that are not cleavable (Fig. 134).

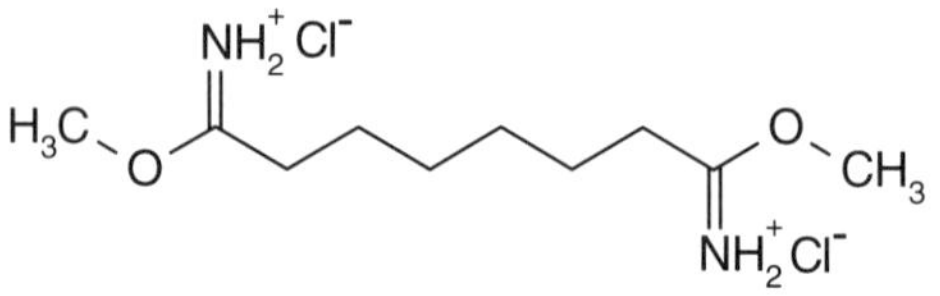

DMS
Dimethyl suberimidate dihydrochloride
MW 273.2
11 Å

DMS reportedly has been used as a tissue fixative for light and electron microscopy (Hassell and Hand, 1974), for the study of the subunit structure of oligomeric proteins (Davies and Stark, 1970), in the investigation of ATPase activity (Adolfson and Moudrianakis, 1976), in cross-linking ribonuclease A (Wang *et al.*, 1976), in binding

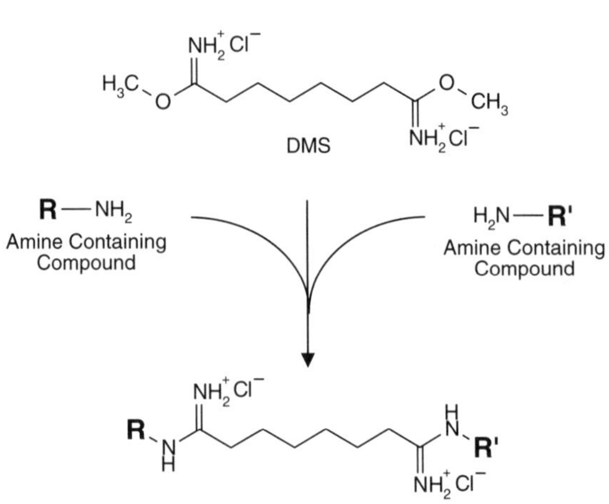

Figure 134   The reaction of DMS with amine-containing molecules yields amidine linkages.

studies of nerve growth factor to its receptor (Pulliam *et al.*, 1975), in the study of red cell shape (Mentzer and Lubin, 1979), in cross-linking of glycogen phosphorylase *b* (Hajdu *et al.*, 1979), in cross-linking of apo low-density lipoproteins (Ikai and Yanagita, 1980), in studying the mechanism of binding of multivalent immune complexes to Fc receptors (Dower *et al.*, 1981), in investigating the quaternary structure of the pyruvate dehydrogenase multienzyme complex of *Bacillus stearothermophilus* (Packman and Perham, 1982), in cross-linking studies of the protein topography of rat liver microsomes (Baskin and Yang, 1982), in affinity, cross-linking studies of the protein topography of rat liver microsomes (Pfeuffer *et al.*, 1985), and in the quantitative chemical cross-linking of CAD protein (Lee *et al.*, 1985).

2.4. DTBP

Dimethyl 3,3′-dithiobispropionimidate (DTBP) is a homobifunctional, reversible cross-linking agent containing imidoester groups on both ends (Pierce). The compound, commonly called Wang and Richards' reagent, is water soluble and reacts with amines in the pH range of 7–10 (optimum 8–9) to produce amidine linkages (Wang, 1974; Wang and Richards, 1974). Conjugated molecules subsequently may be cleaved by reduction of the internal disulfide group of the eight-atom bridge (Fig. 135). Cross-linked molecules may be analyzed by one- and two-dimensional electrophoresis, making use of the easy reversibility of the disulfide bond.

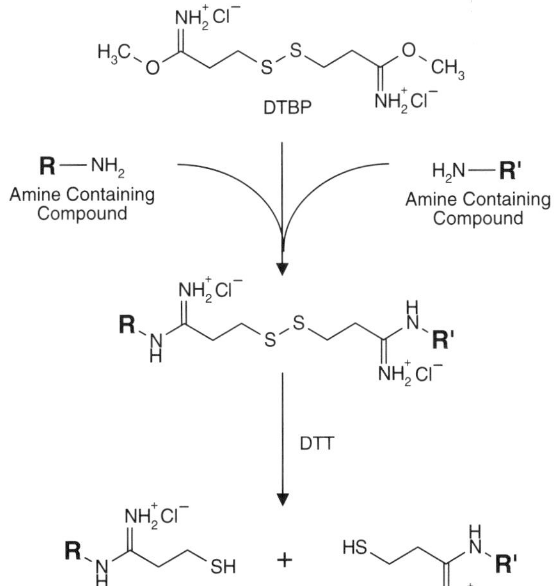

Figure 135 DTBP reacts with amine-containing molecules to form charged amidine bonds. The internal disulfide group can be cleaved with DTT to release the conjugate.

DTBP
Dimethyl-3,3'-dithiobispropionimidate dihydrochloride
MW 309.28
11.9 Å

Reported applications include studying protein–protein interactions with para-myxoviruses (Markwell and Fox, 1980), inhibiting adenylate cyclase activity (Young, 1979), studying red cell shape (Mentzer and Lubin, 1979), cross-linking phytochrome to its putative receptor (Yu and Schweinberger, 1979), and investigating the subunit structure of Na,K-ATPase, (DePont, 1979); studying Newcastle disease virus proteins (Nagai *et al.*, 1978), thylakoid membrane proteins (Novak-Hofer and Siegenthaler, 1978), glutamate dehydrogenase–amino transferase complexes (Fahien *et al.*, 1978), vesicular stomatitis virus proteins (Mudd and Swanson, 1978), hemagglutinin of influ-enza virus (Wiley *et al.*, 1977), pig heart lactate dehydrogenase crystals (Bayne and Ottesen, 1977), beef heart mitochondrial coupling factor 1 (Baird and Hammes, 1977), chloroplast coupling factor 1 (Baird and Hammes, 1976), rabbit muscle skele-tal sarcoplasmic reticulum protein (Louis *et al.*, 1977), human hemoglobin and eryth-rocyte membrane proteins (Wang and Richards, 1974, 1975; Miyakawa *et al.*, 1978), subunit interface of the *E. coli* ribosome (Cover *et al.*, 1981), rat liver 60S ribosomal subunits (Uchiumi *et al.*, 1980), proteins in avian sarcoma and leukemia viruses (Pepinsky *et al.*, 1980), outer membrane proteins of *Neisseria gonorrhoeae* (Newhall *et al.*, 1980), monooxygenase enzymes (Baskin and Yang, 1980b), and cytochrome *P*-450 and reduced NAD phosphate-cytochrome *P*-450 reductase (Baskin and Yang, 1980b); cross-linking initiation factor IF2 to proteins in 70S ribosomes of *E. coli* (Heinmark *et al.*, 1976); studying sheep red blood cell membranes (Brandon, 1980); conjugation of F-actin to skeletal muscle myosin subfragment 1 (Labbe *et al.*, 1982); studying decreased staining of proteins after electrophoresis (Leffak, 1983); and iden-tifying molecular association of IA antigens after T- and B-cell interaction (Shivdasani and Thomas, 1988).

3. Homobifunctional Sulfhydryl-Reactive Cross-linkers

Cross-linking agents that contain homobifunctional sulfhydryl-reactive groups at ei-ther end fall into two general categories: those that form permanent bonds with available sulfhydryls and those that create reversible linkages. Reactive groups yielding permanent links with the sulfhydryl usually form thioether bonds that are quite stable. Those that result in disulfide bonds can be cleaved with the use of a disulfide reducing agent like DTT (Chapter 1, Section 4.1). Mercurial-based coupling groups also can be reversed with reductant.

Many varieties of homobifunctional, sulfhydryl-reactive cross-linkers have been synthesized and described in the literature. Some have been based on bis-mercurial salts (Edsall *et al.*, 1954; Edelhoch *et al.*, 1953; Kay and Edsall, 1956; Singer *et al.*, 1960; Mandy *et al.*, 1961). Such mercurial reactive groups also have been used in reversible covalent chromatography applications to purify sulfhydryl-containing proteins (Cuatrecasas, 1970, 1972; Ruiz-Carrillo and Allfrey, 1973). Other homobifunctional sulfhydryl-reactive reagents have been based on forming a mixed disulfide active group with TNB (5-thio-2-nitrobenzoic acid). Reaction of the TNB active group with a sulfhydryl-containing macromolecule results in a reversible disulfide linkage (Gaffney *et al.,* 1983; Willingham and Gaffney, 1983). Active groups consisting of bis-thiosulfonates also have been used to create —SH-reactive cross-linkers (Bloxham and Sharma, 1979; Bloxham and Cooper, 1982). The thiosulfonate groups react with available sulfhydryls to form disulfide linkages, in this case with loss of the sulfonate groups. All of these disulfide cross-links are cleavable with disulfide reducing agents.

A number of bis-alkylhalide-reactive groups have been used to create homobifunctional sulfhydryl-reactive cross-linkers (Husain and Lowe, 1968; Ozawa, 1967; Wilchek and Givol, 1977). These react with sulfhydryls to create stable, nonreversible thioether bonds. Similar thioether bond formation has been realized using various bis-maleimide derivatives (Simon and Konigsberg, 1966; Moore and Ward, 1956; Fasold *et al.,* 1963; Zahn and Lumper, 1968; Kovacic and Hein, 1959; Heilmann and Holzner, 1981; Tawney *et al.,* 1961; Freedberg and Hardman, 1971; Chang and Flaks, 1972; Wells *et al.,* 1980; Moroney *et al.,* 1982; Chantler and Bower, 1988; Sato and Nakao, 1981; Srinivasachar and Neville, 1989; Partis *et al.,* 1983). Sulfhydryls add to the double bond of the maleimide to create the thioether linkage.

The differences within these families of reagents generally relate to the length of the spacer or bridging portion of the molecule. Occasionally, the bridging portion itself is designed to be cleavable by one of a number of methods (Chapter 7). The great majority of homobifunctional, sulfhydryl-reactive cross-linkers mentioned in the literature are not readily available from commercial sources and would have to be synthesized to make use of them. The ones listed in this section are obtainable from Pierce Chemical.

3.1. DPDPB

1,4-Di-[3′-(2′-pyridyldithio)propionamido]butane (DPDPB) is a homobifunctional cross-linking agent that contains sulfhydryl-reactive dithiopyridyl groups on both ends. These coupling groups are identical to the sulfhydryl-reactive end of the popular heterobifunctional cross-linking agent, SPDP (Chapter 5, Section 1.1). Available sulfhydryls of proteins and other molecules can react with the pyridyl disulfide groups to form disulfide linkages with release of pyridine-2-thione (Fig. 136). Conjugation of two macromolecules with DPDPB results in a 14-atom spacer of approximately 16 Å in length. Release of two molecules of pyridine-2-thione during the cross-linking reaction may be followed by their absorbance at 343 nm ($\varepsilon = 8.08 \pm 0.3 \times 10^3$ $M^{-1}cm^{-1}$) (Stuchbury *et al.,* 1975).

DPDPB
1,4-di-[3′-(2′-pyridyldithio)propionamido]butane
MW 482.7
19.9 Å

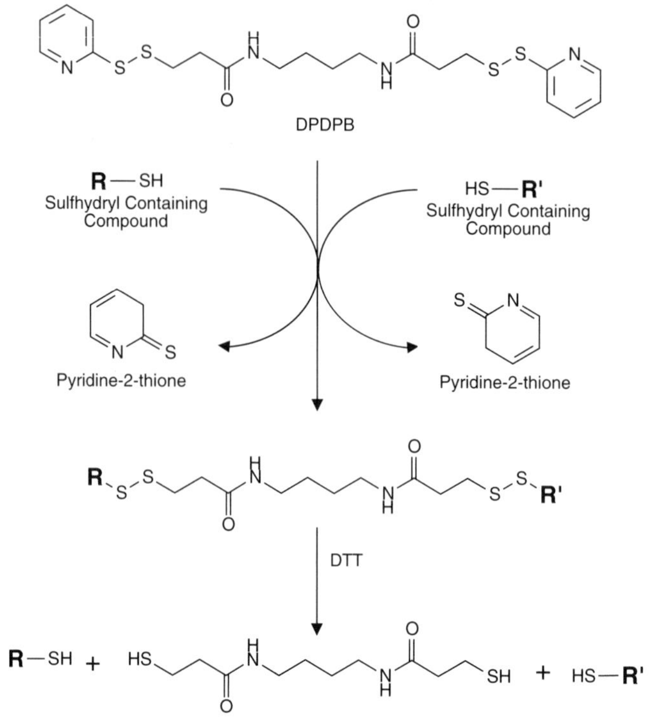

Figure 136 DPDPB is a sulfhydryl-reactive cross-linker that forms disulfide bonds with thiol-containing molecules. The conjugates may be disrupted using a disulfide reducing agent such as DTT.

DPDPB is insoluble in aqueous solutions and should be initially dissolved in an organic solvent prior to addition of a small aliquot to a buffered reaction medium. Preparation of a stock solution in DMSO at a concentration of 25 mM DPDPB works well. The addition of an aliquot of this stock solution to the conjugation reaction should not result in more than about 10% organic solvent by volume in the buffered mixture or protein precipitation may occur.

DPDPB has two absorbance maxima, one peak at 237 nm ($\varepsilon = 1.2 \times 10^4$ $M^{-1}cm^{-1}$) and another at 287 nm ($\varepsilon = 8.8 \times 10^3$ $M^{-1}cm^{-1}$) (Zecherle, 1990; Traut *et al.*, 1989). Reduction of the pyridyldithio groups causes a shift in the absorbance characteristics of the molecule, such that the peak at 237 nm is shifted to 272 nm and the peak at 287 nm is shifted to 343 nm. This absorbance shift correlates to the release of the pyridine-2-thione groups.

DPDPB can be used to conjugate reduced antibody molecules to β-D-galactosidase using essentially the same protocol as that described by O'Sullivan *et al.* (1979).

3.2. BMH

Homobifunctional cross-linking compounds containing maleimide groups on both ends have been used for quite some time (Simon and Konigsberg, 1966; Moore and

Ward, 1956; Fasold *et al.,* 1963; Zahn and Lumper, 1968; Kovacic and Hein, 1959; Heilmann and Holzner, 1981; Tawney *et al.,* 1961; Freedberg and Hardman, 1971; Chang and Flaks, 1972; Wells *et al.,* 1980; Moroney *et al.,* 1982; Chantler and Bower, 1988; Sato and Nakao, 1981; Srinivasachar and Neville, 1989; Partis *et al.,* 1983).

*Bis*maleimidohexane (BMH) is a homobifunctional reagent containing a noncleavable, six-atom spacer between terminal maleimides (Pierce). The maleimide groups can react with sulfhydryls to form stable thioether linkages (Fig. 137). Cross-links formed with this reagent create a relatively long 16.1-Å bridge between conjugated macromolecules. The reaction takes place optimally from pH 6.5 to 7.5. Within this pH range the reaction is very specific for sulfhydryls. At higher pH values cross-reactivity with amino groups may occur (see Chapter 2, Section 2.2).

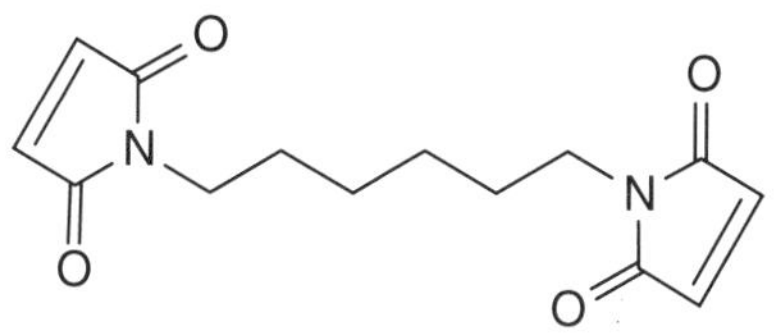

BMH
*Bis*maleimidohexane
MW 276.29
16.1 Å

4. Difluorobenzene Derivatives

Difluorobenzene derivatives are small homobifunctional cross-linkers that react with amine groups. Conjugation using these compounds results in bridges of only about 3 Å in length, potentially providing information concerning very close interactions between macromolecules.

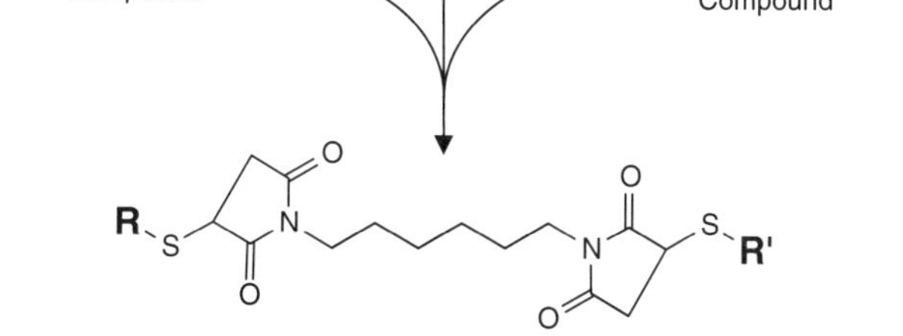

Figure 137 BMH contains two maleimide groups specific for cross-linking sulfhydryl-containing molecules. The thioether bonds that are formed are stable.

4.1. DFDNB

DFDNB is the acronym for an aryl halide-containing compound having the structural names 1,5-difluoro-2,4-dinitrobenzene or 1,3-difluoro-4,6-dinitrobenzene (Pierce). The reagent contains two reactive fluorine atoms that can couple to amine-containing molecules, yielding stable arylamine bonds (Fig. 138). However, the reactivity of aryl halides is not totally specific for amines. Thiol, imidazolyl, and phenolate groups of amino acid side chains also can react (Zahn and Meinhoffer, 1958). Conjugates formed with sulfhydryl groups, however, are reversible by cleaving with an excess of thiol (such as DTT) (Shaltiel, 1967). The compound is especially useful in cross-linking cellular membrane proteins, since it is able to penetrate the hydrophobic regions of the lipid bilayer.

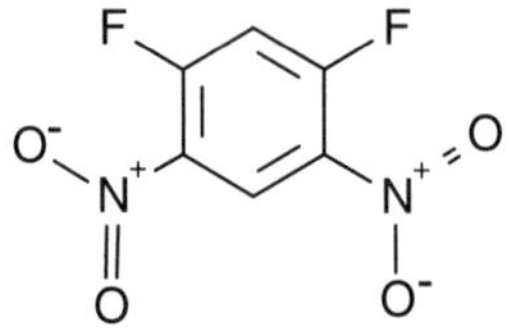

DFDNB
1,5-Difluoro-2,4-dinitrobenzene
MW 204.1
3 Å

Difluorobenzene reagents have been used for cross-linking phospholipids in human erythrocyte membranes (Berg *et al.*, 1965; Marfey and Tsai, 1975), conjugation of small peptides to the carrier protein albumin (Tager, 1976), studying the interaction of proteins in the myelin membrane (Golds and Braun, 1978), cross-linking of cytochrome oxidase subunits (Kornblatt and Lake, 1980), and studying the conformational effects of calcium on troponin C (Kareva *et al.*, 1986).

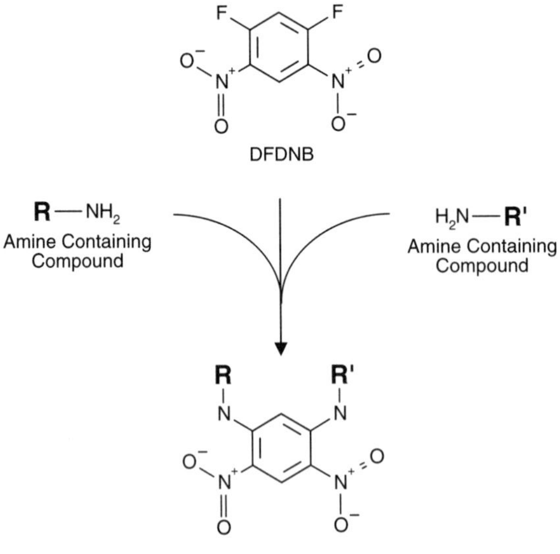

Figure 138   DFDNB is a small cross-linker able to form covalent bonds between amine-containing molecules. The aromatic fluorine atoms are readily displaced by nucleophiles.

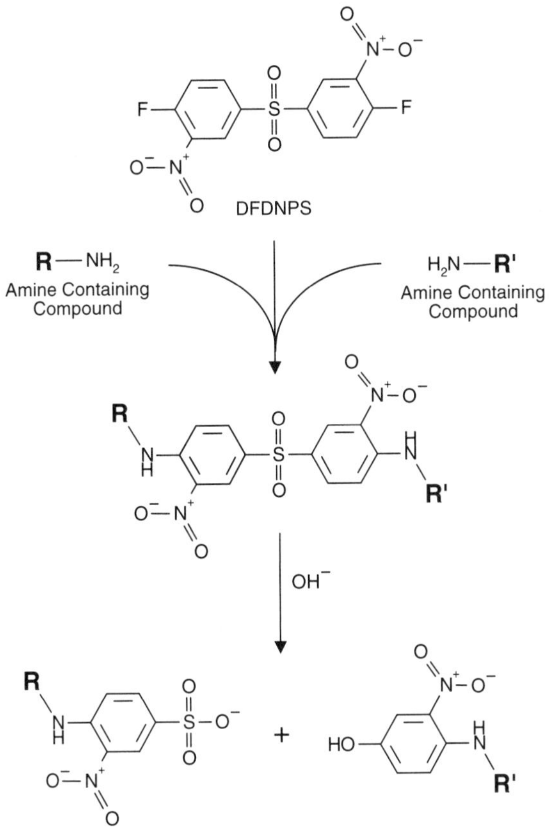

Figure 139 DFDNPS reacts with amine-containing molecules to form arylamine cross-links. The central sulfone group in the cross-bridge can be cleaved under alkaline conditions.

4.2. DFDNPS

4,4′-Difluoro-3,3′-dinitrophenylsulfone (DFDNPS) is a di-aryl halide reagent containing a central sulfone group (Fig. 139). The aromatic fluorines are reactive with amines, sulfhydryls, phenolates, and imidazolyl groups of proteins (see previous section). The reaction with amines forms stable arylamine linkages. The reaction with sulfhydryl groups, however, is reversible by treatment with an excess of thiol. The central sulfone group provides cleavability through hydrolysis with base at pH 11–12 at 37°C (Wold, 1961, 1972; Zarling *et al.*, 1980).

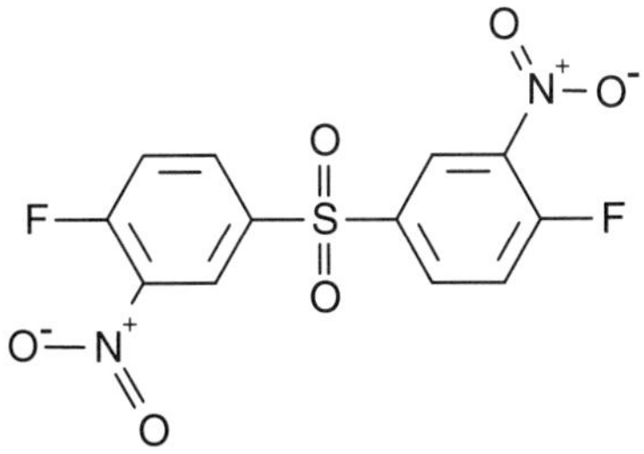

DFDNPS
4,4'-Difluoro-3,3'-dinitrodiphenylsulfone

5. Homobifunctional Photoreactive Cross-linkers

Although there are a number of photo-sensitive coupling chemistries that have been used in modification and conjugation reactions (Chapter 2, Section 7), only aryl azides have found application in homobifunctional cross-linkers. The photolysis reaction requires exposure of the phenyl azide to a bright light source at a wavelength of 265–275 nm (Ji, 1979). If the aromatic ring contains a nitro group *meta* to the azide functional group, then photolysis can occur at higher wavelengths (300–460 nm). The photolysis process initially forms a highly reactive aryl nitrene, but these quickly undergo ring expansion to create a dehydroazepine. This active species principally reacts with nucleophiles, rather than inserting in C—H or N—H bonds or adding to double bonds. Thus, instead of nonselective coupling into nearly any part of a molecular structure, aryl azides ultimately react with primary amines more than any other functionality (Schnapp *et al.*, 1993).

Reported structures for homobifunctional aryl azides include a biphenyl derivative and a naphthalene derivative (Mikkelsen and Wallach, 1976), a biphenyl derivative containing a central, cleavable disulfide group (Guire, 1976), and a compound containing a central 1,3-diamino-2-propanol bridge between phenyl azide rings that are nitrated (Guire, 1976). The only commercially available homobifunctional photoreactive cross-linker is BASED.

5.1. BASED

Bis-[β-(4-azidosalicylamido)ethyl]disulfide (BASED) is a homobifunctional photoreactive cross-linking agent containing phenyl azide groups at both ends (Pierce). Its central bridge contains a cleavable disulfide bond that may be broken after conjugation with the appropriate reducing agent (Chapter 1, Section 4.1). The aryl azides are salicylate derivatives that contain hydroxylic functions that activate the ring toward electrophilic reactions. Thus, the phenolic rings are modifiable with ^{125}I using traditional oxidative radioiodination reagents. Prior to the photoreactive conjugation step, the cross-linker may be iodinated with IODO-GEN or IODO-BEADS (Chapter 8, Section 4.2 and 4.3). After two proteins are cross-linked, cleavage of the conjugate with DTT releases the link but maintains a radiolabel on each of the molecules (Fig. 140).

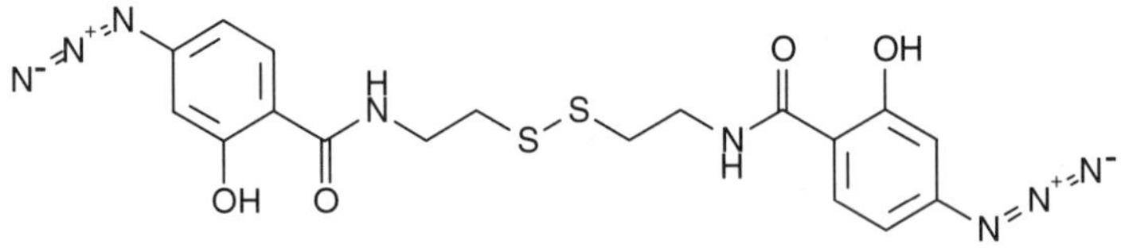

BASED
Bis-[β-(4-azidosalicylamido)ethyl]disulfide
MW 474.54

6. Homobifunctional Aldehydes

Numerous bis-aldehyde reagents have been used for the conjugation of biomolecules. Nearly every small organic compound containing two aldehyde groups has been at

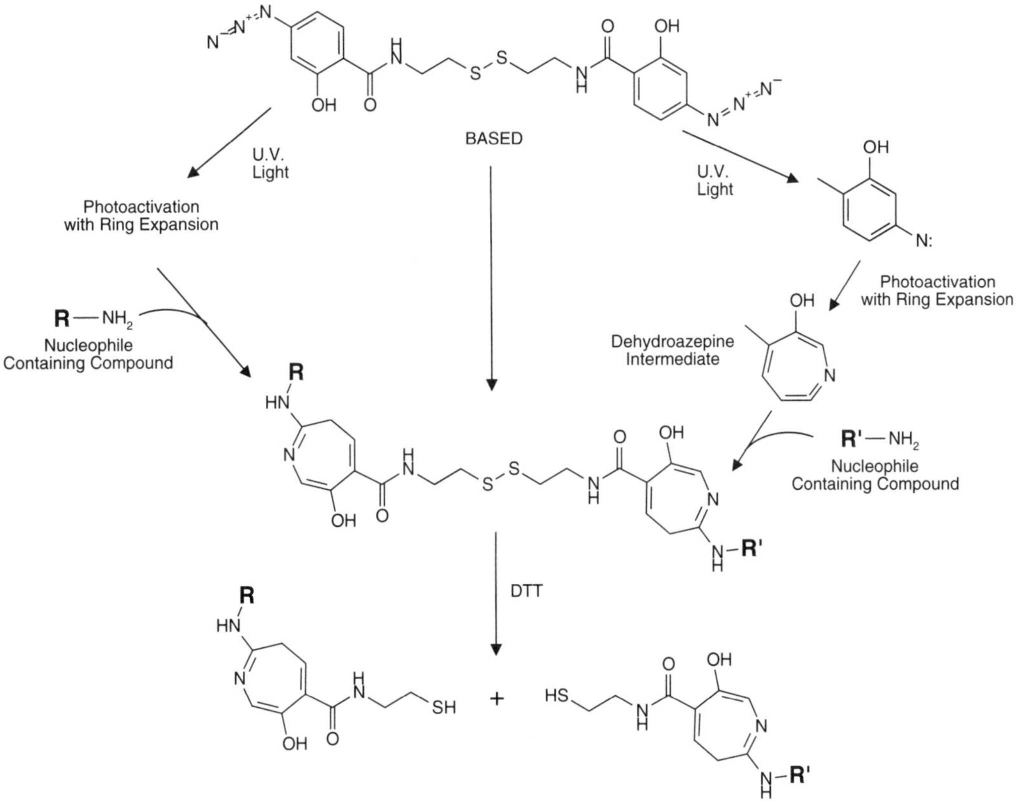

Figure 140 BASED can react with molecules after photoactivation to form cross-links with nucleophilic groups, primarily amines. Exposure of its phenyl azide groups to UV light causes nitrene formation and ring expansion to the dehydroazepine intermediate. This group is highly reactive with amines. The cross-bridge of BASED is cleavable using a disulfide reducing agent.

least tried in cross-linking reactions. The repertoire of available homobifunctional aldehydes ranges from the single-carbon formaldehyde (Section 6.1; yes, it behaves as if it were bifunctional) through the 2-carbon atom glyoxal (Brooks and Klamerth, 1968), the 3-carbon malondialdehyde (Cater, 1963), the 4-carbon succinaldehyde (Cater, 1963), the popular 5-carbon glutaraldehyde (Section 6.2), the 6-carbon adipaldehyde as well as its α-hydroxy derivative (Cater, 1963; Richard and Knowles, 1968; Fein and Filachione, 1957; Seligsberger and Sadlier, 1957; Hopwood, 1969), and several pyridoxal-polyphosphate derivatives that are internally cleavable with acid or base (Shimomura and Fukui, 1978; Benesch and Kwong, 1988).

By far, the two most popular *bis*-aldehyde reagents are formaldehyde and glutaraldehyde.

6.1. Formaldehyde

Formaldehyde is the smallest cross-linking reagent available that does not create a zero-length bridge between two molecules (Chapter 3). Although technically not a homobifunctional reagent, it undergoes cross-linking reactions as though it possessed two functional groups. In concentrated aqueous solutions, it can form the low-molecular-weight polymers typically observed in formalin preparations. In dilute solutions, it exists mainly in its monomeric state. Older solutions of formaldehyde may contain precipitated polymer that often can be resolubilized by heating. Commercial preparations of formaldehyde may be obtained as a 37% solution stabilized against polymerization by the addition of methanol.

Conjugation reactions using formaldehyde may proceed by one of two routes: the Mannich reaction or via an immonium cation intermediate. The Mannich reaction consists of the condensation of formaldehyde (or sometimes another aldehyde) with ammonia (in the form of its salt) and another compound containing an active hydrogen. For a review of this reaction mechanism, see Adams *et al.* (1942). Instead of ammonia, however, this reaction can be done with primary or secondary amines, or even with amides. An example of this is illustrated in Fig. 141 by the condensation of phenol, formaldehyde, and a primary amine salt (the active hydrogens are shown underlined). Figure 142 shows some active hydrogen-containing functional groups that can participate in the Mannich reaction.

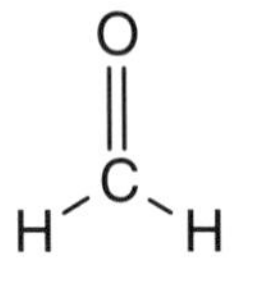

Formaldehyde
MW 30

The Mannich reaction can be used for the immobilization of certain drugs, steroidal compounds, dyes, or other organic molecules that do not possess the typical nucleophilic groups able to participate in traditional coupling reactions (Hermanson *et al.*, 1992). It can also be used to conjugate hapten molecules to carrier proteins when the hapten contains no convenient nucleophile for conjugation (Chapter 9, Section 6.2). In this case, the carrier protein contains the primary amines and the hapten contains at least one sufficiently active hydrogen to participate in the condensation reaction.

To obtain acceptable yields, the Mannich reaction must be done at elevated temperatures. Incubation at 37–57°C for at least 2–24 h usually is required to complete the reaction. Addition of formaldehyde is done by adding an aliquot of a 37% solution

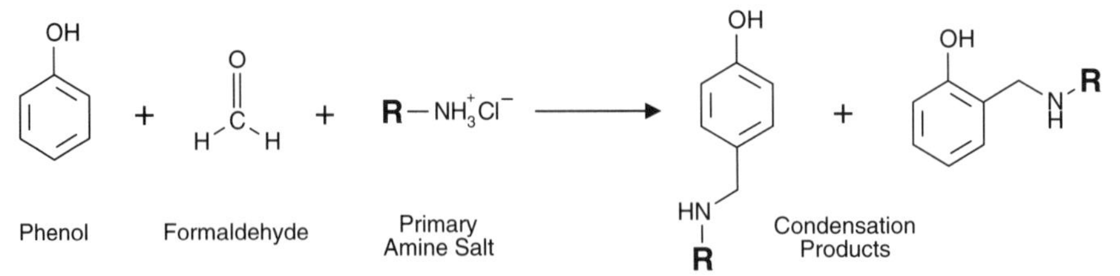

Figure 141   The Mannich reaction occurs between an active hydrogen-containing compound (phenol) and an amine-containing molecule in the presence of an aldehyde (formaldehyde). The condensation reaction forms stable cross-links.

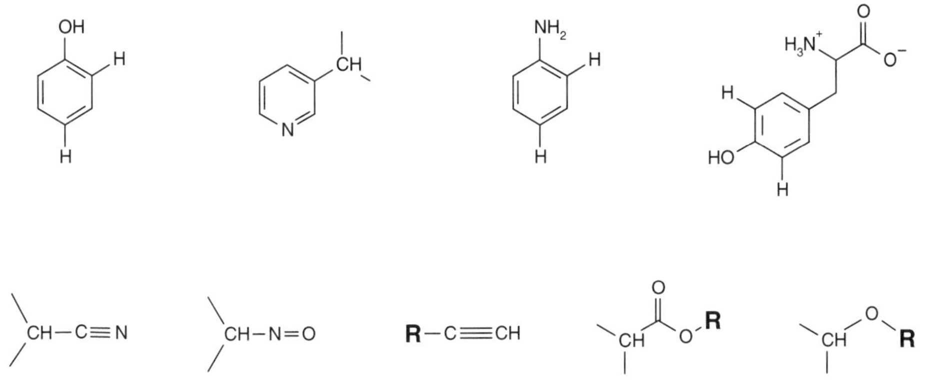

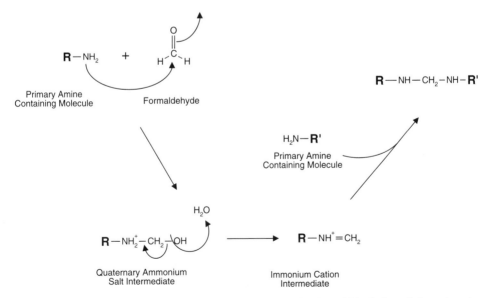

Figure 142 Examples of active hydrogen-containing compounds that can participate in the Mannich reaction. The points of reactivity are shown by the hydrogen atoms.

to the reaction to obtain about a 10- to 100-fold molar excess over the amount of active hydrogen-containing compound to be conjugated. Pierce Chemical has designed a kit for the conjugation of haptens to carrier proteins using the Mannich reaction mechanism.

A secondary reaction pathway also is possible in formaldehyde-facilitated conjugations. Formaldehyde may react with a primary amine to form a quaternary ammonium salt. This intermediate spontaneously reacts to create a highly active immonium cation with loss of one molecule of water (Ji, 1983; Blass *et al.*, 1965). The immonium cation

Figure 143 Two amine-containing molecules can be cross-linked by formaldehyde through formation of a quaternary ammonium salt with subsequent dehydration to an immonium cation intermediate. This active species then can react with a second amine compound to form stable secondary amine bonds.

is reactive toward nucleophiles in proteins and other molecules, including amines, sulfhydryls, phenolic groups, and imidazole nitrogens. The reaction yields methylene bridges between two nucleophiles, binding macromolecules with a one-carbon linker (Fig. 143).

It is obvious that the Mannich reaction pathway and the immonium ion mechanism may occur simultaneously, especially at conditions of room temperature or greater. Formaldehyde-facilitated cross-linking reactions between molecules that both contain nucleophiles probably occur primarily by the immonium ion pathway, since the Mannich reaction proceeds at a slower rate. In addition, the Mannich reaction will cause nondescript polymerization between molecules that possess both active hydrogens and amine groups. It is best to utilize the Mannich reaction only when one of the molecules contains no nucleophilic groups but at least one active hydrogen, and the other molecule contains a primary or secondary amine.

6.2. Glutaraldehyde

Glutaraldehyde is the most popular bis-aldehyde homobifunctional cross-linker in use today. However, a glance at glutaraldehyde's structure is not indicative of the complexity of its possible reaction mechanisms. Reactions with proteins and other amine-containing molecules would be expected to proceed through the formation of Schiff bases. Subsequent reduction with sodium cyanoborohydride or another suitable reductant would yield stable secondary amine linkages (Chapter 2, Section 5.3, and Chapter 3, Section 4). This reaction sequence certainly is possible, but other cross-linking reactions also are feasible.

Glutaraldehyde
MW 100.11

Glutaraldehyde in aqueous solutions can form polymers containing points of unsaturation (Hardy *et al.*, 1969, 1976; Monsan *et al.*, 1975) (Fig. 144). Such α,β-unsaturated glutaraldehyde polymers are highly reactive toward nucleophiles, especially primary amines. Reaction with a protein results in alkylation of available amines, forming stable secondary amine linkages. These glutaraldehyde-modified proteins still may react with other amine-containing molecules either through the Schiff base pathway or through addition at other points of unsaturation (Fig. 145). The proposed reaction mechanism of conjugation using these polymer conjugates explains the stability of proteins cross-linked by glutaraldehyde that have not been reduced. Schiff base formation alone would not yield stable cross-linked products without reduction.

Cross-linking using glutaraldehyde polymers is difficult to reproduce. Since the glutaraldehyde polymer size and structure is unknown, the exact nature of the conjugates formed by this method is indeterminable, as well. The age of a glutaraldehyde solution is another variable, because the older the solution the more polymer will be formed. Fresh glutaraldehyde often will not yield the same results as aged solutions.

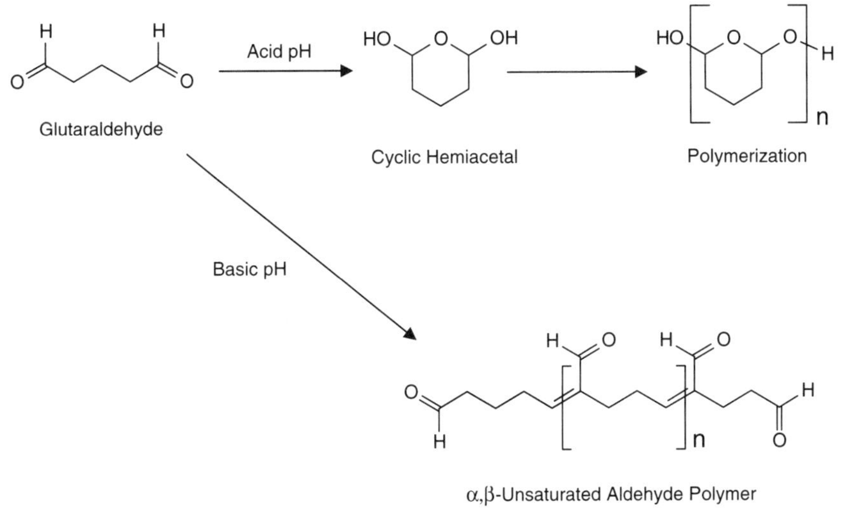

Figure 144   Glutaraldehyde in aqueous solution may polymerize at either acid or basic pH.

A third method of using glutaraldehyde in conjugation reactions is through its ability to rapidly react with hydrazide groups. A molecule containing hydrazide functional groups or modified to contain them (Chapter 1, Section 4.5) can be conjugated with another molecule containing either amines or hydrazides. Glutaraldehyde will react with the hydrazide groups to form hydrazone linkages. When cross-linking two macromolecules containing multiple sites of conjugation, the multivalent hydrazone

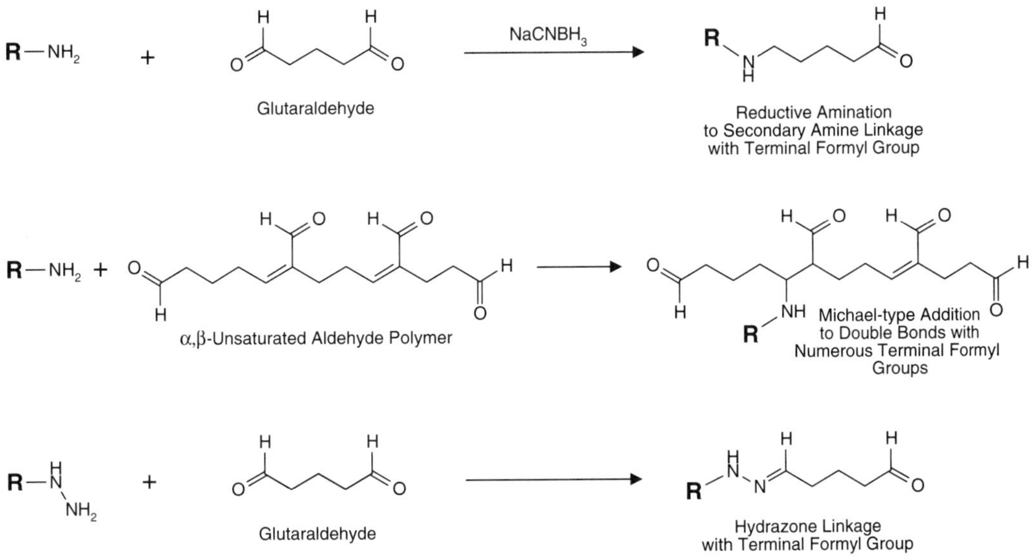

Figure 145   Glutaraldehyde may react by several routes to form covalent cross-links with amine-containing molecules.

bonds are stable enough to create a stable conjugate. If a small molecule is involved, however, reduction of the hydrazone with sodium cyanoborohydride is recommended to produce a leak-resistant bond.

Glutaraldehyde has been used extensively as a homobifunctional cross-linking reagent, especially for antibody–enzyme conjugations (Avrameas, 1969; Avrameas and Ternynck, 1971). To help overcome its tendency to form large-molecular-weight polymers upon cross-linking two proteins, a two-step protocol often is employed. In this regard, one protein is first reacted with glutaraldehyde and purified away from excess reagent. The second protein is then added to effect the conjugate formation. See the introduction to this chapter and Chapter 1, Section 1.2, and Chapter 10, Section 1.2, for additional information on the use of glutaraldehyde and two-step cross-linking procedures.

7. Bis-epoxides

Homobifunctional compounds containing epoxide groups on both ends can be used to cross-link molecules containing nucleophiles, including amines, sulfhydryls, and hydroxyls. The reaction proceeds with epoxide ring-opening to create secondary amine, thioether, or ether bonds with these functional groups (Fig. 146). During the ring opening process, a β-hydroxy group is created. Hydrolysis of the epoxy function without coupling to a nucleophile yields adjacent hydroxyls that can be oxidized with sodium periodate to create reactive aldehydes (Fig. 147). Expoxide groups, however, are quite stable in aqueous environments around neutral pH, being reactive only at alkaline pH values toward hydrolysis or other nucleophilic molecules.

Certain bis-epoxide reagents have been used to active hydroxylic matrices for coupling ligands containing amine, sulfhydryl, or hydroxyl groups for affinity chromatography purposes (Hermanson *et al.*, 1992). Conjugation reactions involving proteins also have been done using epoxide cross-linkers, but not to the extent of their use in immobilization.

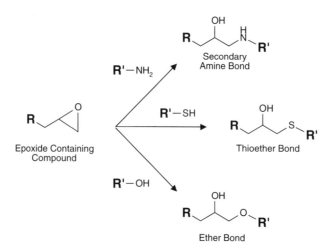

Figure 146   Epoxide groups are reactive toward sulfhydryls, amines, and hydroxyls.

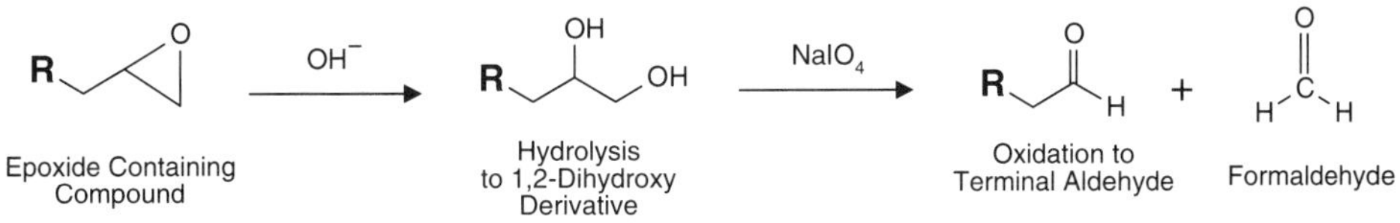

Figure 147 Hydrolysis of epoxy groups forms 1,2-dihydroxy derivatives that can be oxidized with periodate to create reactive aldehydes.

Bis-epoxy compounds that have been used for cross-linking purposes vary mainly in their chain length, ranging from the 4-carbon bridge of 1,2:3,4-diepoxybutane (Skold, 1983; Kohn *et al.*, 1966), the 6-carbon spacer of 1,2:5,6-diepoxyhexane (Fearnley and Speakman, 1950), the 7-atom bridge of bis(2,3-epoxypropyl)ether (Kohn *et al.*, 1966), to the 12-atom spacer of the popular 1,4-(butanedioil) diglycidyl ether (discussed below) (Sundberg and Porath, 1974; Porath, 1976). Longer chain polymeric bis-epoxide compounds also have been utilized in collagen cross-linking experiments (Murayama *et al.*, 1988).

7.1. 1,4-Butanediol Diglycidyl Ether

The most commonly used homobifunctional epoxide compound is 1,4-butanediol diglycidyl ether. The reagent can react with hydroxyls, amines, or sulfhydryl groups to produce ether, secondary amine, or thioether bonds, respectively. The reaction of the epoxide functional groups with hydroxyls requires high pH conditions, usually in the pH range of 11–12. Amine nucleophiles react at more moderate alkaline pH values, typically needing buffer environments of at least pH 9. Sulfhydryl groups are the most highly reactive nucleophiles with epoxides, requiring a buffered system in the pH range of 7.5–8.5 for efficient coupling.

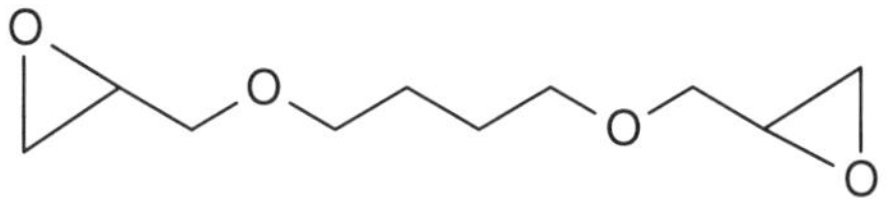

1,4-Butanediol Diglycidyl Ether
MW 202.25

1,4-Butanediol diglycidyl ether is a viscous liquid having a density of 1.4535 at 20°C. It is a hygroscopic, corrosive compound with a displeasing odor that should be handled with care in a fume hood. Aqueous solutions of the bis-epoxide usually possess a characteristic oily film on their surfaces, indicating the limited solubility of the reagent.

An example of the use of 1,4-butanediol diglylcidyl ether for the activation of soluble dextran polymers is given in Chapter 15, Section 2.3. One end of the bis-epoxide reacts with the hydroxylic sugar residues of dextran to form ether linkages terminating in epoxy functional groups. The epoxides of the activated derivative then can be used to couple additional molecules containing nucleophilic groups to the dextran backbone.

8. Homobifunctional Hydrazides

Homobifunctional cross-linking agents containing hydrazide groups at both ends can be used to conjugate molecules containing carbonyl or carboxyl groups. In one scheme, a bis-hydrazide compound can be reacted with the carboxylate groups of a protein in the presence of the water-soluble carbodiimide EDC to yield imide linkages containing terminal alkyhydrazides. The hydrazide-activated protein then can be used to conjugate with a glycoprotein that had been previously oxidized with sodium periodate to generate reactive aldehyde residues. The resulting hydrazone bonds can be further stabilized by reducing with sodium cyanoborohydride to give the secondary amine linkage.

These techniques have been used to target, detect, or assay glycoproteins in solution or on cell surfaces by using hydrazide-activated enzymes, avidin, or streptavidin (Chapter 13, Section 5) (Bayer and Wilchek, 1990; Bayer et al., 1987a,b; 1990) and to form conjugates with glycoproteins.

Bis-hydrazide-containing molecules also can be used to activate soluble polymeric substances containing aldehyde groups. For instance, dextran may be periodate-oxidized to create numerous formyl functional groups on each molecule. Subsequent reaction with a homobifunctional hydrazide in excess results in a hydrazide-activated polymer having multivalent binding capability toward aldehydes or ketones (Chapter 15, Section 2.2). Insoluble support matrices suitable for affinity chromatography have been activated in a similar fashion to create the hydrazide derivative (O'Shannessy and Wilchek, 1990).

8.1. Adipic Acid Dihydrazide

The dihydrazide derivative of adipic acid (Aldrich) is perhaps the most popular homobifunctional hydrazide compound in use. The reagent provides a 10-atom bridge between cross-linked molecules after conjugation. Adipic dihydrazide (ADH) is a solid that is soluble in aqueous solutions, but may need to be moderately heated to create concentrated solutions. Aldehyde-containing substances may be modified with this reagent to form hydrazone bonds with alkyl hydrazide spacers suitable for reactivity with other formyl-containing molecules (Fig. 148). In this sense, affinity chromatography matrices have been activated with ADH to produce a hydrazide derivative for coupling to aldehyde-containing ligands (O'Shannessy and Wilchek, 1990), enzymes have been modified at available carboxylate groups using an EDC-facilitated reaction to create hydrazide-activated derivatives suitable for targeting oxidized glycoproteins (Bayer et al., 1987a,b), and the biotin-binding proteins avidin and streptavidin have been activated to assay glycoconjugates using biotinylated enzymes (Bayer and Wilchek, 1990: Bayer et al., 1990). The cross-linker also has been utilized to study the carbohydrate portion of yeast acid phosphatase (Kozulic et al., 1984).

Adipic Acid Dihydrazide
MW 174.2

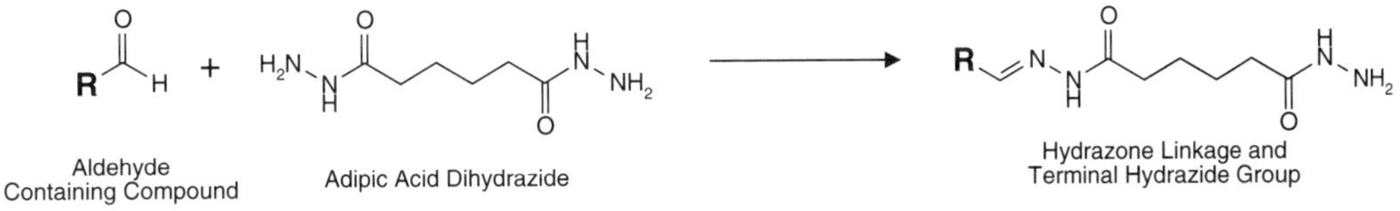

Figure 148 Adipic acid dihydrazide spontaneously reacts with aldehydes to form hydrazone linkages.

Protocols for the use of adipic acid dihydrazide in the modification of aldehyde or carboxylate functional groups can be found in Chapter 1, Section 4.5, and Chapter 13, Section 5.

8.2. Carbohydrazide

Carbohydrazide (carbonic dihydrazide or 1,3-diaminourea) is a small homobifunctional reagent containing reactive hyrazide groups on both ends. Its lack of an internal aliphatic bridge, as found in adipic dihydrazide, gives the compound excellent solubility characteristics in aqueous solutions. Carbohydrazide is freely soluble in water, but practically insoluble in ethanol and other organic solvents. The two hydrazide functional groups of the molecule can react with aldehyde or ketone groups to form hydrazone linkages. When reacted in excess with a molecule containing carbonyl groups, carbohydrazide modification results in short derivatives terminating in available hydrazides (Fig. 149). The compound has been used to modify microplate wells that have been graft polymerized with glycidyl methacrylate to form surfaces that would couple antibodies through their carbohydrate portions (Allmer *et al.*, 1990; Brillhart and Ngo, 1991). Although its use for protein modification has not been realized, carbohydrazide may be a superior alternative to adipic dihydrazide due to its hydrophilicity. Its only disadvantage may be in its shorter bridge (5-atom spacer as opposed to adipic dihydrazide's 10-atom bridge).

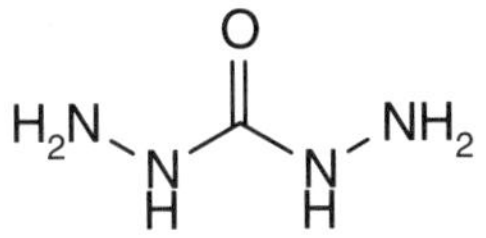

Carbohydrazide
MW 90.09

A protocol for the use of carbohydrazide in the modification of aldehyde or carboxylate functionalities can be found in Chapter 1, Section 4.5.

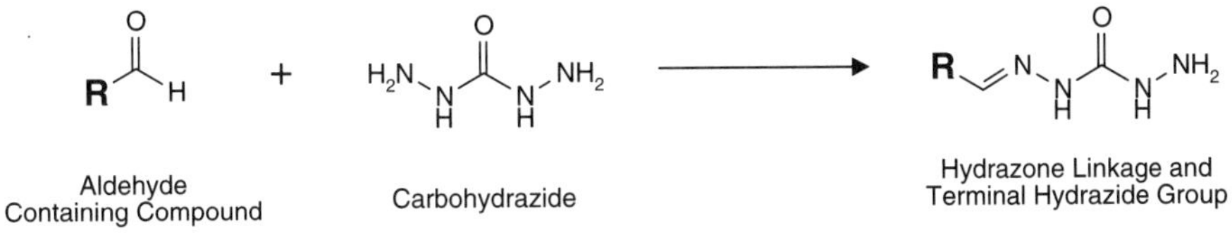

Figure 149 Carbohydrazide can be used to transform an aldehyde residue into a hydrazide group.

9. Bis-diazonium Derivatives

Diazonium groups react with active hydrogens on aromatic rings to give covalent diazo bonds. Generation of a diazonium functionality is usually done from an aromatic amine by reaction with sodium nitrite under acidic conditions at 0°C (see Chapter 1, Section 4.3, and Chapter 2, Section 6.1). The highly reactive and unstable diazonium is reacted immediately with an active hydrogen-containing compound at pH 8–10. In general, at pH 8 the diazonium group will principally react with histidinyl residues, attacking the electron-rich nitrogens of the imidazole ring. At higher pH, the phenolic side chain of tyrosine groups can be modified. The reaction proceeds by electrophilic attack of the diazonium group toward the electron-rich points on the target molecules. Phenolic compounds are modified at positions *ortho* and *para* to the aromatic hydroxyl group. For tyrosine side chains, only the *ortho* modification is available.

Bis-diazonium compounds are useful in cross-linking molecules containing no other convenient functional groups such as amines, carboxylates, or sulfhydryls. Conjugations done using these compounds usually create deeply colored products characteristic of the diazo bonds. Occasionally, the conjugated molecules may turn dark brown or even black. The diazo linkages are reversible by addition of 0.1 *M* sodium dithionite in 0.2 *M* sodium borate, pH 9. On cleavage, the color of the complex is lost.

9.1. *o*-Tolidine, Diazotized

o-Tolidine, or 3,3′-dimethylbenzidine, is a bis-aromatic-amine-containing compound that can be readily diazotized to a homobifunctional diazonium cross-linker by reaction with sodium nitrite (Fig. 150). The reagent is typically used in a one-step conjugation reaction wherein two active hydrogen-containing molecules are cross-linked through the addition of *o*-tolidine immediately after it has been diazotized under acidic conditions by reacting with sodium nitrite (Fig. 151). pH adjustment to aklaline conditions after diazonium formation rapidly causes cross-linking to occur. The diazotized form of *o*-tolidine must be used quickly due to its instability in aqueous solutions. The reagent has been used to couple active hydrogen-containing haptens to carrier proteins to form immunogens suitable for the production of antibodies (Chapter 9, Section 6.1).

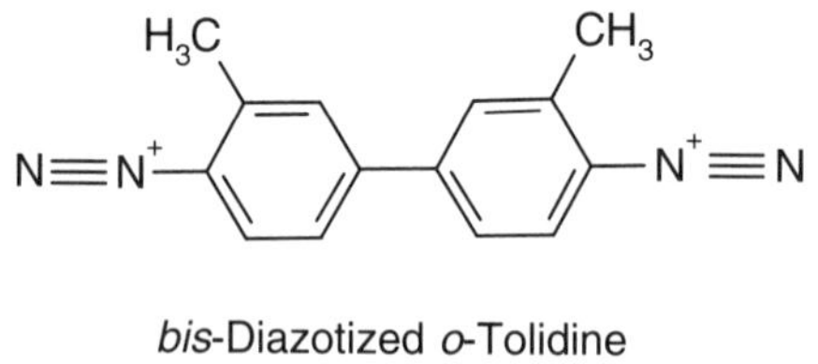

bis-Diazotized *o*-Tolidine

o-Tolidine is a benzidine derivative that should be considered a carcinogen. Handling should be done with proper safety precautions and with the use of a fume hood to avoid breathing in any dust particles. The reagent is sparingly soluble in water, but is more soluble under the dilute acidic condition necessary for activation to a diazonium derivative.

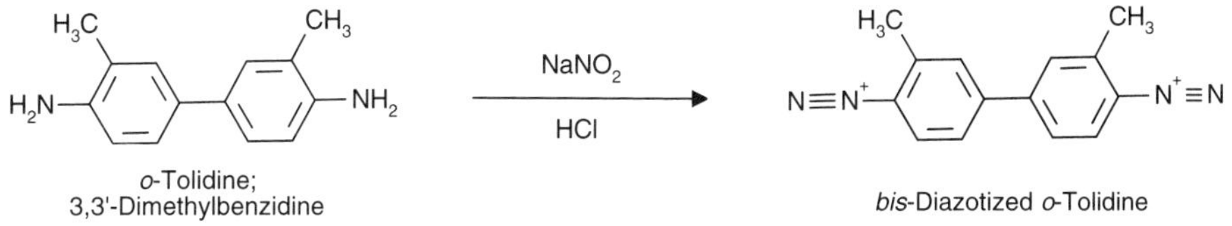

o-Tolidine;
3,3'-Dimethylbenzidine

bis-Diazotized *o*-Tolidine

Figure 150   Reaction of *o*-tolidine with sodium nitrite in the presence of HCl yields a highly reactive diazo derivative.

9.2. Bis-diazotized Benzidine

Benzidine, or *p*-diaminodiphenyl, may be diazotized with sodium nitrite to form a homobifunctional diazonium cross-linking agent useful in conjugating active hydrogen-containing molecules (Fig. 152). The coupling reaction proceeds via electrophilic attack on atoms containing extractable hydrogens. Particularly reactive are the phenolic side chains of tyrosine residues and the imidazole rings of histidine groups.

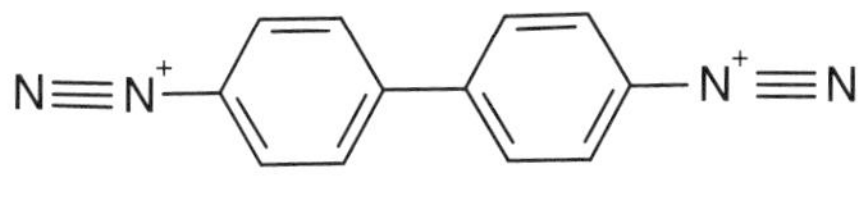

Bis-Diazotized Benzidine

Benzidine is a known carcinogen and should be handled with extreme caution (Fourth Annual Report on Carcinogens; NTP 85–002, 1985, p. 37). The solid and its

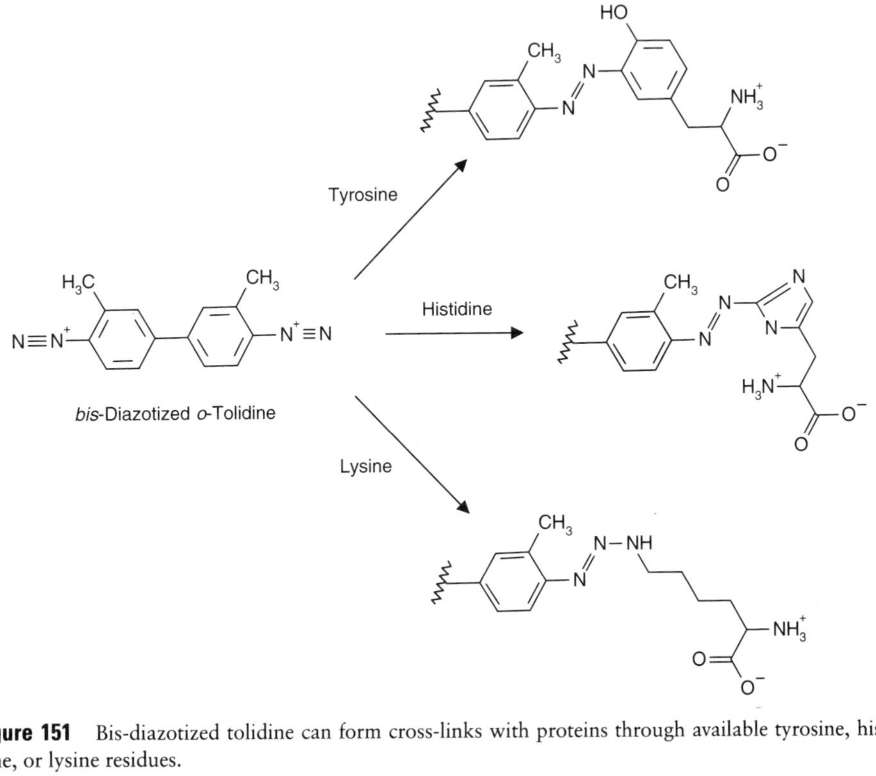

Figure 151   Bis-diazotized tolidine can form cross-links with proteins through available tyrosine, histidine, or lysine residues.

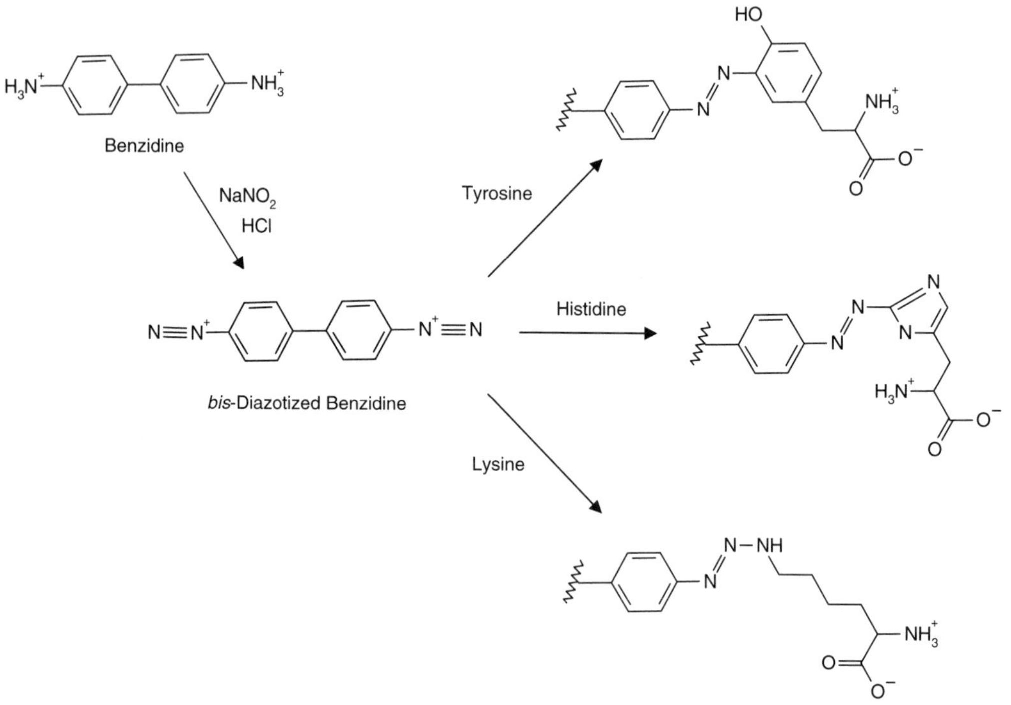

Figure 152 Benzidine can be diazotized with sodium nitrite and HCl for reaction with proteins through their tyrosine, histidine, or lysine side-chain groups.

vapors may be rapidly absorbed through skin. Protective clothing and the use of a fume hood are recommended. The compound is only sparingly soluble in water as the free base. The dihydrochloride form, however, is soluble in water and ethanol.

Bis-diazotized benzidine has been used to create active hydrogen-reactive spacer arms on chromatographic matrices (Lowe and Dean, 1971; Silman *et al.*, 1966). The compound may be used similarly to *o*-tolidine for the conjugation of active hydrogen-containing molecules (see Sections 9.1, this chapter, and Chapter 2, Section 6.1).

10. Bis-alkylhalides

Homobifunctional reagents containing reactive halogen groups on both ends are capable of cross-linking sulfhydryl-, amine-, or histidine-containing molecules by nucleophilic substitution. Three forms of activated halogen functionalities can be used to create these reagents: haloacetyl derivatives (see Chapter 1, Section 5.2), benzyl halides that react through a resonance activation process with the neighboring benzene ring, and alkyl halides that possess the halogen β to a nitrogen or sulfur atom, as in *N*- and *S*-mustards. Haloacetyl compounds typically are iodo or bromo derivatives, the simplest of which are 1,3-dibromoacetone (Husain and Lowe, 1968) and various iodoacetyl derivatives of short diamine alkyl spacers (Ozawa, 1967) (Fig. 153). Benzyl halides also are usually iodo or bromo derivatives, whereas the halo-mustards mainly employ chloro and bromo forms (Fig. 154).

Reactive halogen cross-linkers are mainly specific for sulfhydryl groups at physi-

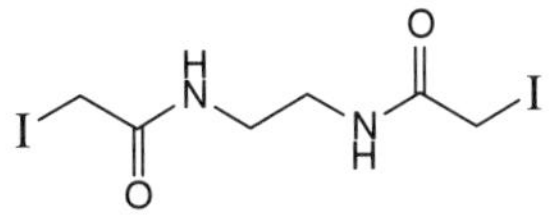

N,N'-Ethylene-bis(iodoacetamide)

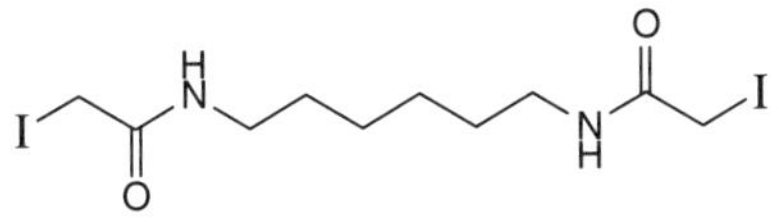

N,N'-Hexamethylene-bis(iodoacetamide)

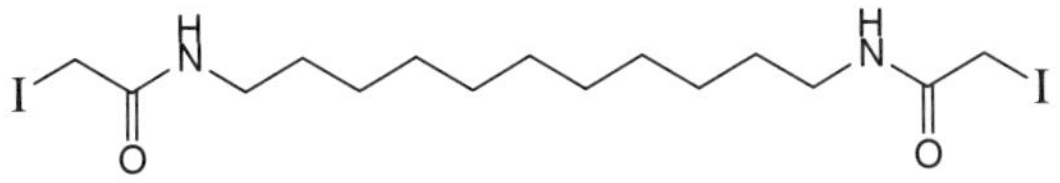

N,N'-Undecamethylene-bis(iodoacetamide)

Figure 153 Several varieties of iodoacetylated diamine compounds have been investigated for cross-linking proteins through sulfhydryl groups.

ological pH; however, at more alkaline pH values they can readily cross-react with amines and the imidazole nitrogens of histidine residues. Some reactivity with hydroxyl-containing compounds also may be realized, particularly with dichloro-S-triazine derivatives under alkaline conditions.

Most of the bis-alkyl halides referenced in the literature are unavailable commercially, and therefore must be synthesized. Some key references to the preparation and use of these compounds for the cross-linking of sulfhydryl-containing proteins and other molecules include Goodlad, 1957; Segal Hurwitz, 1976; Ewig and Kohn, 1977; Wilchek and Givol, 1977; Prestayko et al., 1981; Luduena et al., 1982; Hiratsuka, 1988; Aliosman et al., 1989.

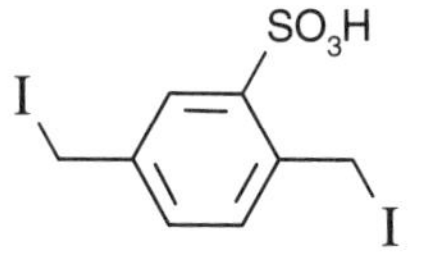

α,α'-Diiodo-p-xylene sulfonic acid

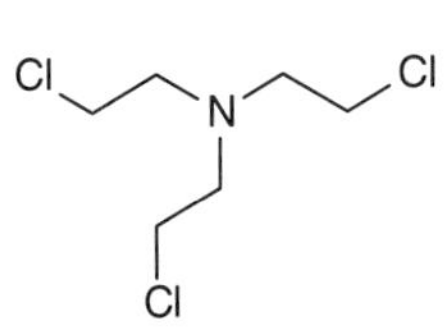

TCEA;
Tri(2-chloroethyl)amine

Figure 154 Benzylhalides and halomustards can be used as cross-linking agents reactive toward sulfhydryl groups.

5

Heterobifunctional Cross-linkers

Heterobifunctional conjugation reagents contain two different reactive groups that can couple to two different functional targets on proteins and other macromolecules (Fig. 155). For example, one part of a cross-linker may contain an amine-reactive group, while another portion may consist of a sulfhydryl-reactive group. The result is the ability to direct the cross-linking reaction to selected parts of target molecules, thus garnering better control over the conjugation process.

Heterobifunctional reagents can be used to cross-link proteins and other molecules in a two- or three-step process that limits the degree of polymerization often obtained using homobifunctional cross-linkers (Chapter 1, Section 1.2, and Chapter 4, Section 2.2). In a typical conjugation scheme, one protein is modified with a heterobifunctional using the cross-linker's most reactive or most labile end. The modified protein is then purified from excess reagent by gel filtration or rapid dialysis. Most heterobifunctionals contain at least one reactive group that displays extended stability in aqueous environments, therefore allowing purification of an activated intermediate before adding the second molecule to be conjugated. For instance, an NHS ester–maleimide heterobifunctional (for example, see this chapter, Section 1.3) can be used to react with the amine groups of one protein through its NHS ester end (the most labile functional group), while preserving the activity of its maleimide functional group. Since the maleimide group has greater stability in aqueous solution than the NHS ester group, a maleimide-activated intermediate may be created. After a quick purification step, the maleimide end of the cross-linker then can be used to conjugate to a sulfhydryl-containing molecule.

Such multistep protocols offer greater control over the resultant size of the conjugate and the molar ratio of components within the cross-linked product. The configuration or structure of the conjugate can be regulated by the degree of initial modification of the first protein and by adjusting the amount of second protein added to the final conjugation reaction. Thus, low- or high-molecular-weight conjugates may be obtained to better fashion the product toward its intended use.

Heterobifunctional cross-linking reagents also may be used to site-direct a conjugation reaction toward particular parts of target molecules. Amines may be coupled on one molecule while sulfhydryls or carbohydrates are targeted on another molecule. Directed coupling often is important in preserving critical epitopes or active sites within macromolecules. For instance, antibodies may be coupled to other proteins

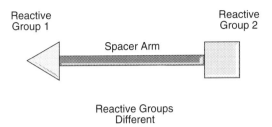

Figure 155   The general design of a heterobifunctional cross-linking agent includes two different reactive groups at either end and an organic cross-bridge of various length and composition. The cross-bridge may be constructed of chemically cleavable components for selective disruption of conjugates.

while directing the cross-linking reaction away from the antigen binding sites, thus maximizing antibody activity in the conjugate.

Heterobifunctional reagents containing one photoreactive end may be used to nonselectively insert into target molecules by UV irradiation. Ligands having specific affinity toward a receptor may be labeled with a photoreactive cross-linker, allowed to interact with its target, and then photolyzed to label permanently the receptor at its binding site. The photoreactive group is stable until exposed to high intensity light at UV wavelengths. Photoaffinity labeling techniques are an important investigative tool for determining binding site characteristics.

The third component of all heterobifunctional reagents is the cross-bridge or spacer that ties the two reactive ends together. Cross-linkers may be selected based not only on their reactivities, but also on the length and type of cross-bridge they possess. Some heterobifunctional families differ solely in the length of their spacer. The nature of cross-bridge also may govern the overall hydrophilicity of the reagent. A number of heterobifunctionals contain cleavable groups within their cross-bridge, lending greater flexibility to the experimental design. A few cross-linkers contain peculiar cross-bridge constituents that actually affect the reactivity of their functional groups. For instance, it is known that a maleimide group that has an aromatic ring immediately next to it is less stable to ring opening and loss of activity than a maleimide that has an aliphatic ring adjacent to it. In addition, conjugates destined for use *in vivo* may have different properties depending on the type of spacer on the associated cross-linker. Some spacers may be immunogenic and cause specific antibody production to occur against them. In other instances, the half-life of a conjugate *in vivo* may be altered by choice of cross-bridge, especially when using cleavable reagents.

The following heterobifunctional reagents are organized according to their reactivities. The majority are commercially available and their properties well documented in the literature.

1. Amine-Reactive and Sulfhydryl-Reactive Cross-linkers

Perhaps the most popular heterobifunctional reagents are those that contain amine-reactive and sulfhydryl-reactive ends. The amine-reactive group is usually an active ester, most often an NHS ester, while the sulfhydryl-reactive portion may be one of several different functional groups. The amine-reactive end of these cross-linkers is typically an acylating agent possessing a good leaving group that can undergo nucleophilic substitution to form an amide bond with primary amines. The sulfhydryl-reactive portion, by contrast, is usually an alkylating agent that is capable of creating either thioether or disulfide linkages with sulfhydryl-containing molecules. Depending on the chemistry chosen, linkages with a sulfhydryl-containing molecule may be either permanent covalent bonds or reversible disulfide bonds that can be cleaved by use of a suitable disulfide reductant.

The active ester chemistry of the amine-reactive end of these cross-linkers is characteristically the most labile functional group, being susceptible to rapid hydrolysis under the aqueous conditions of a conjugation reaction. The sulfhydryl-reactive group, however, is usually much more stable to breakdown in aqueous environments. Therefore, these reagents typically are used in multistep conjugation protocols wherein one protein or molecule is first modified through its amines to yield a sulfhydryl-reactive intermediate. After removal of excess cross-linker by gel filtration, a second protein or molecule containing a sulfhydryl group is added to effect the final conjugation. The stability of the sulfhydryl-reactive end of these cross-linkers allows greater control over the cross-linking process than is possible with single-step procedures.

1.1. SPDP, LC-SPDP, and Sulfo-LC-SPDP

N-Succinimidyl 3-(2-pyridyldithio)propionate (SPDP) may be the most popular heterobifunctional cross-linking agent available. The activated NHS ester end of SPDP

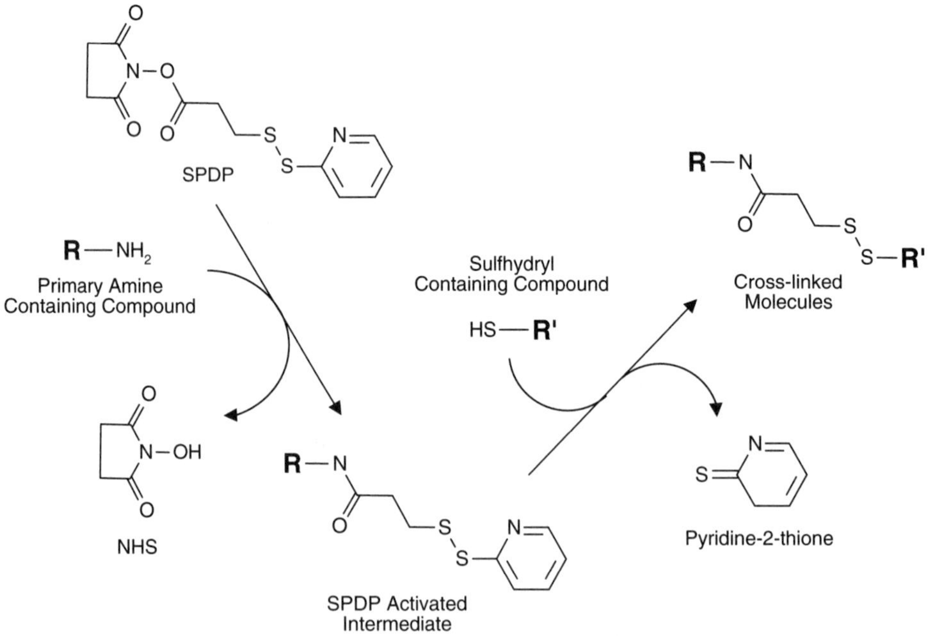

Figure 156 SPDP can react with amine-containing molecules through its NHS ester end to form amide bonds. The pyridyl disulfide group then can be coupled to a sulfhydryl-containing molecule to create a cleavable disulfide bond.

reacts with amine groups in proteins and other molecules to form an amide linkage. The 2-pyridyldithiol group at the other end reacts with sulfhydryl residues to form a disulfide linkage with sulfhydryl-containing molecules (Carlsson *et al.*, 1978) (Fig. 156). The cross-linker is used extensively to form enzyme conjugates for use in immunoassays or in labeled DNA probe techniques. It also is frequently used for the preparation of immunotoxin conjugates for *in vivo* administration (Chapter 17, Section 2.4, and Chapter 11, Section 2.1). In addition, the reagent is effective in creating sulfhydryls on proteins and other molecules (Chapter 1, Section 4.1). Once modified with SPDP, a protein can be treated with DTT (or another disulfide-reducing agent) to release the pyridine-2-thione leaving group and form the free sulfhydryl. The terminal —SH group then can be used to conjugate with any cross-linking agents containing sulfhydryl-reactive groups, such as maleimide or iodoacetyl (for covalent conjugation) or 2-pyridyldithiol groups (for reversible conjugation).

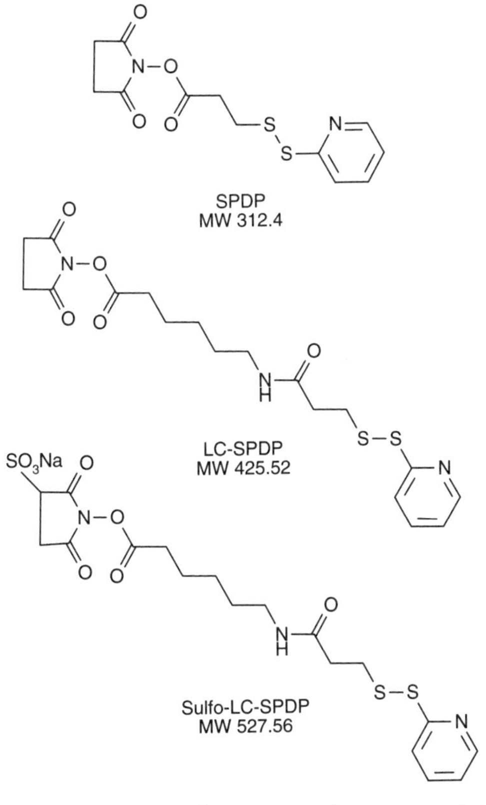

SPDP
MW 312.4

LC-SPDP
MW 425.52

Sulfo-LC-SPDP
MW 527.56

There are three forms of SPDP analogs currently commercially available (Pierce): the standard SPDP, a long-chain version designated LC-SPDP, and a water-soluble, sulfo-NHS form also containing an extended chain, called Sulfo-LC-SPDP. Both the standard SPDP and the LC-SPDP are insoluble in aqueous solutions and must be first solubilized in DMSO prior to addition to the reaction solution. The Sulfo-LC-SPDP may be solubilized directly in water or buffer. The long-chain versions extend the length of the cross-linker for those applications that require greater accessibility to react with sterically hindered functional groups. Since many sulfhydryl residues are found below the surface of a protein structure in more hydrophobic domains, the longer spacer arm of the LC versions may be more effective in conjugations with these groups.

The following procedure is a suggested multistep protocol involving the activation of one protein by modification of its amines through the NHS ester end of SPDP, purification of this active intermediate, and subsequent addition of a sulfhydryl-containing molecule for conjugation via the remaining pyridyl disulfide group.

Protocol

1. Dissolve a protein or macromolecule containing primary amines at a concentration of 10 mg/ml in 50 mM sodium phosphate, 0.15 M NaCl, pH 7.2. Other non-amine-containing buffers such as borate, Hepes, and bicarbonate also may be used in this reaction. Avoid sulfhydryl-containing components in the reaction mixture, as these will react with the pyridyl disulfide end of SPDP. The effective pH for the NHS ester modification reaction is in the range of 7 to 9.

2. Dissolve SPDP at a concentration of 6.2 mg/ml in DMSO (makes a 20 mM stock solution). Alternatively, LC-SPDP may be used and dissolved at a concentration of 8.5 mg/ml in DMSO (also makes a 20 mM solution). If the water-soluble Sulfo-LC-SPDP is used, a stock solution in water may be prepared just prior to addition of an aliquot to the thiolation reaction. In this case, prepare a 10 mM solution of Sulfo-LC-SPDP by dissolving 5.2 mg/ml in water. Since an aqueous solution of the cross-linker will degrade by hydrolysis of the sulfo-NHS ester, it should be used quickly to prevent significant loss of activity. If a sufficiently large amount of protein will be modified, the solid may be added directly to the reaction mixture without preparing a stock solution in water to allow accurate weighing of Sulfo-LC-SPDP.

3. Add 25 μl of the stock solution of either SPDP or LC-SPDP in DMSO to each milliliter of the protein to be modified. If Sulfo-LC-SPDP is used, add 50 μl of the stock solution in water to each milliliter of protein solution.

4. Mix and react for at least 30 min at room temperature. Longer reaction times, even overnight, will not adversely affect the modification.

5. Purify the modified protein from reaction by-products by dialysis or gel filtration using 50 mM sodium phosphate, 0.15 M NaCl, 10 mM EDTA, pH 7.2.

6. Add a sulfhydryl-containing protein or other molecule to the purified SPDP-modified protein to effect the conjugation reaction. Molecules lacking available sulfhydryl groups may be modified to contain them by a number of methods (Chapter 1, Section 4.1). The amount of this second protein added to the reaction should be governed by the desired molar ratio of the two proteins in the final conjugate. The conjugation reaction should be done in the presence of at least 10 mM EDTA to prevent metal-catalyzed sulfhydryl oxidation.

1.2. SMPT and Sulfo-LC-SMPT

Succinimidyloxycarbonyl-α-methyl-α-(2-pyridyldithio)toluene (SMPT) is a heterobifunctional cross-linking agent that contains an amine-reactive NHS ester on one end and a sulfhydryl-reactive pyridyl disulfide group on the other. SMPT is therefore an analog of SPDP that differs only in its cross-bridge which contains an aromatic ring and a hindered disulfide group (Thorpe *et al.*, 1987; Ghetie *et al.*, 1990a,b). The spacer arm of SMPT is slightly longer than SPDP (11.2 vs 6.8 Å), but the presence of the benzene ring and α-methyl group adjacent to the disulfide sterically hinders the structure sufficiently to provide increased half-life of conjugates *in vivo*.

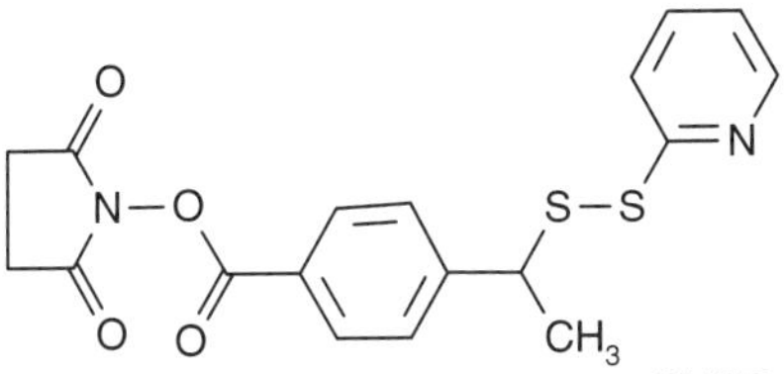

SMPT
4-Succinimidyloxycarbonyl-α-methyl-
α-(2-pyridyldithio)toluene
MW 388.5

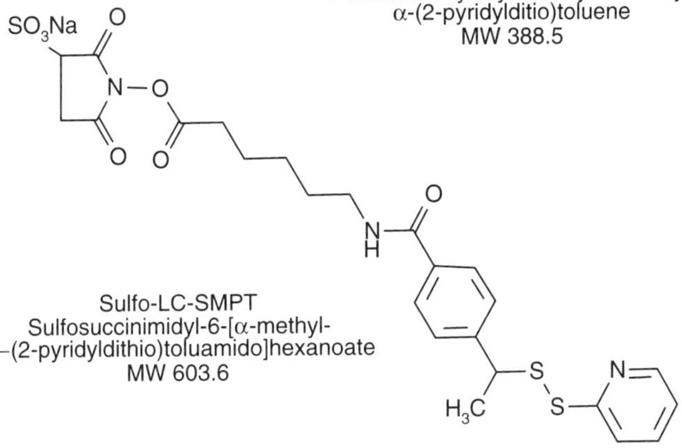

Sulfo-LC-SMPT
Sulfosuccinimidyl-6-[α-methyl-
α–(2-pyridyldithio)toluamido]hexanoate
MW 603.6

Conjugation reactions done using SMPT often proceed by a multistep protocol involving modification of one protein through its amine groups to create a pyridyl disulfide-activated intermediate. Since SMPT is not soluble in water, the reagent is first solubilized in DMF or DMSO and an aliquot of this stock solution added to the reaction. The NHS ester end of the reagent reacts with ε- and N-terminal amine groups to create stable amide linkages. After removal of excess cross-linker by gel filtration or dialysis, a second protein containing a sulfhydryl group is added to effect the conjugation (Fig. 157). The resultant protein–protein cross-link contains a disulfide bond that is susceptible to cleavage by reduction, although more slowly due to the hindered nature of the cross-bridge.

SMPT often is used for the preparation of immunotoxin conjugates that contain a monoclonal antibody directed against some cell-surface antigen (usually a tumor-associated antigen) cross-linked to a protein toxin molecule. In has been shown that a cleavable linkage between the antibody and toxin molecules helps to ensure a potent immunotoxin (Lambert *et al.*, 1985). Increased cytotoxicity is typically observed for immunotoxin conjugates containing cross-bridge disulfides as opposed to noncleavable linkages. Cleavability presumably facilitates the release of the toxin from the antibody after the conjugate has bound to the cell surface. However, the disulfide bonds formed from some cross-linkers, such as SPDP, are readily reduced and cleaved *in vivo*—often before they reach their target. The hindered disulfide of SMPT has distinct advantages in this regard. Thorpe *et al.* (1987) showed that SMPT conjugates had approximately twice the half-life *in vivo* as SPDP conjugates.

A water-soluble analog of SMPT, called sulfo-LC-SMPT or sulfosuccinimidyl-6-

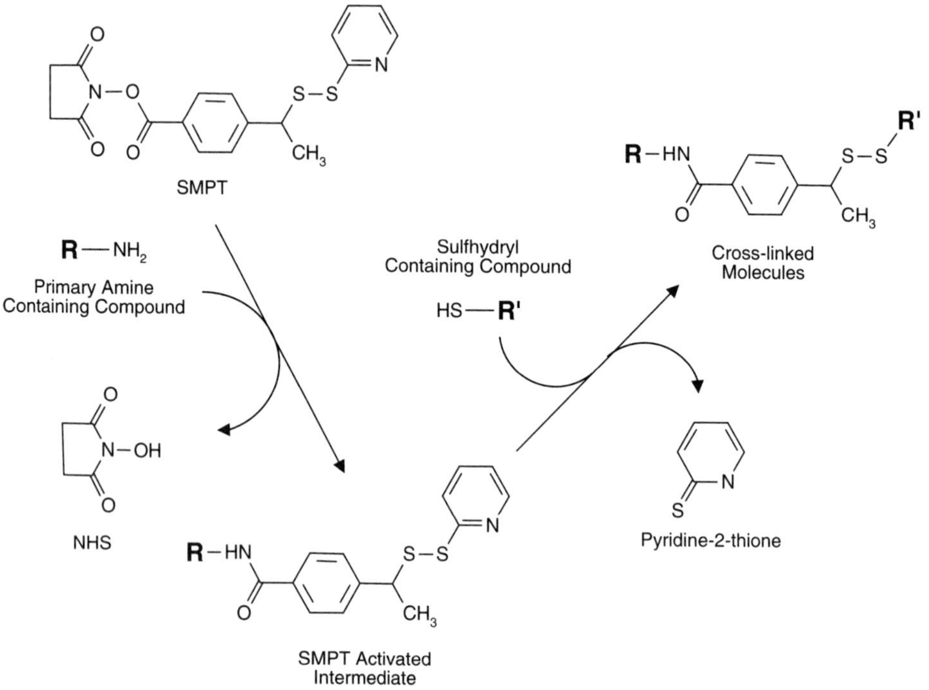

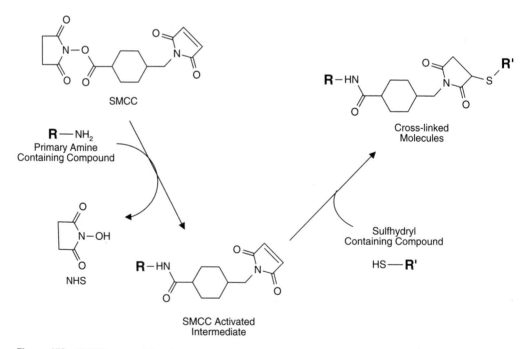

Figure 157 SMPT can form cross-links between an amine-containing molecule and a sulfhydryl-containing compound through amide and disulfide linkages, respectively. The hindered nature of the disulfide group provides better stability toward reduction and cleavage.

Figure 158 SMCC reacts with amine-containing molecules to form stable amide bonds. Its maleimide end then may be conjugated to a sulfhydryl-containing compound to create a thioether linkage.

[α-methyl-α-(2-pyridyldithio)toluamido]hexanoate, is available from Pierce Chemical. The sulfo-NHS ester end of the reagent provides the water solubility due to the negative charge of the sulfonate group. Although sulfo-LC-SMPT contains the same chemical reactivity as SMPT, its cross-bridge contains an additional 6-aminocaproic acid spacer providing a 20-Å cross-link as opposed to the 11.2-Å length of SMPT. The reactivity and use of sulfo-LC-SMPT is essentially the same as that of SMPT, except that the reagent may be added directly to aqueous reaction media or predissolved in water. A stock solution made in water should be used immediately to prevent extensive NHS ester hydrolysis.

1.3. SMCC and Sulfo-SMCC

Succinimidyl-4-(N-maleimidomethyl)cyclohexane-1-carboxylate (SMCC) is a heterobifunctional cross-linker with significant utility in cross-linking proteins, particularly in the preparation of antibody–enzyme and hapten–carrier conjugates (Hashida and Ishikawa, 1985; Dewey *et al.*, 1987). The NHS ester end of the reagent can react with primary amine groups on proteins to form stable amide bonds. The maleimide end of SMCC is specific for coupling to sulfhydryls when the reaction pH is in the range of 6.5–7.5 (Smyth *et al.*, 1964) (Fig. 158).

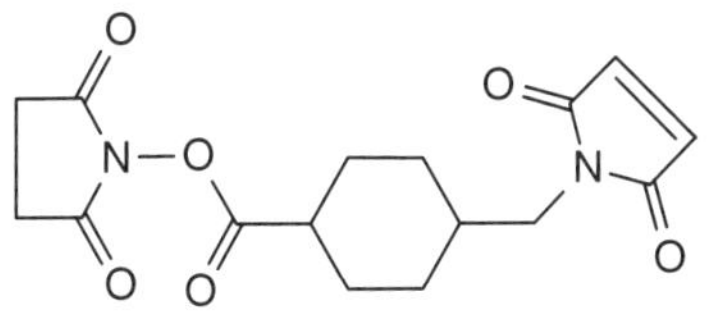

Sulfo-SMCC;
Succinimidyl 4-(N-maleimidomethyl)-
cyclohexane-1-carboxylate
MW 334.33
11.6 Å

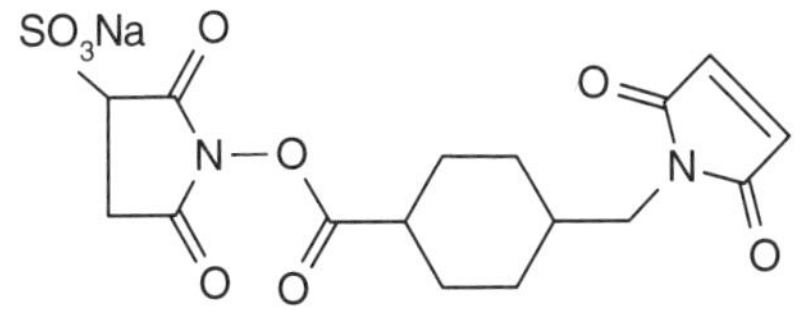

Sulfo-SMCC;
Sulfosuccinimidyl 4-(N-maleimidomethyl)-
cyclohexane-1-carboxylate
Water Soluble
MW 436.37
11.6 Å

At pH 7 the reaction of the maleimide group with sulfhydryls proceeds at a rate 1000 times greater than its reaction with amines. At more alkaline pH values, however, its reaction with amines becomes more evident. The maleimide end also may undergo hydrolysis to an open maleamic acid form that is unreactive toward sulfhydryls. Hydrolysis may occur after sulfhydryl coupling to the maleimide, as well. This ring-opening reaction typically happens faster the higher the pH becomes. However, the maleimide group of SMCC displays unusual stability up to pH 7.5 The increased stability of SMCC's maleimide group may be due to it not being attached directly to an aromatic ring structure. By contrast, some maleimide-containing reagents, such as *N,N'-o*-phenylenedimaleimide and *N,N'*-oxydimethylenedimaleimide are less stable under these conditions. Reportedly, only 4% of the maleimide groups of SMCC will decompose at neutral pH within 2 h at 30°C (Ishikawa *et al.*, 1983a,b). For this reason, proteins may be modified with SMCC to form relatively long-lived, maleimide-activated intermediates. The SMCC derivative then may be freeze-dried to provide a stock preparation of sulfhydryl-reactive protein.

SMCC frequently is used to effect hapten–carrier or antibody–enzyme conjugations. In both applications, one of the molecules is activated (usually the carrier or the enzyme) with the cross-linker, purified to remove excess reagents, and then mixed with the sulfhydryl-containing second molecule to effect the conjugation. Published applications using SMCC are numerous, but include conjugation of glucose oxidase to rabbit antibodies (Yoshitake *et al.*, 1979), cross-linking Fab' fragments to horseradish peroxidase (Ishikawa *et al.*, 1983a,b; Yoshitake *et al.*, 1982a,b; Imagawa *et al.*, 1982; Uto *et al.*, 1991), coupling anti-digoxin F(ab')$_2$ fragments to β-galactosidase (Freytag *et al.*, 1984a,b), preparing conjugates of alkaline phosphatase and human IgG F(ab')$_2$ fragments (Mahan *et al.*, 1987), and use in the preparation of immunogens (Peeters *et al.*, 1989).

Since SMCC is a water-insoluble cross-linker, it must be dissolved first in organic solvent (DMF) before it is added to a protein to be modified. In some cases, addition of even a small amount of organic solvent to a protein solution may be detrimental to activity. To be safe, no more than 10–20% solvent should be present in the aqueous reaction medium.

Sulfosuccinimidyl-4-(*N*-maleimidomethyl)cyclohexane-1-carboxylate   (sulfo-SMCC) is a water-soluble analog of SMCC that possesses a negatively charged sulfonate group on it *N*-hydroxysuccinimide ring. The charge gives just enough polarity to the molecule to provide water solubility at a level of at least 10 mg/ml at room temperature. This allows direct addition of the reagent to reaction mixtures without prior dissolution in organic solvent. The cross-linker is known to be soluble at a concentration of at least 10 m*M* in the following buffers: *(a)* 50 m*M* sodium acetate, pH 5; *(b)* 50 m*M* sodium borate, pH 7.6; *(c)* 0.1 *M* sodium phosphate, pH 6–7.5. Aqueous stock solutions also may be prepared using sulfo-SMCC, but these should be dissolved rapidly and used immediately to prevent extensive loss of sulfo-NHS coupling ability due to hydrolysis. Concentrated aqueous stock solutions (up to about 50 mg/ml) may be made by heating for a few minutes under hot running water. Quickly cool to room temperature before using.

The following is a generalized protocol for the activation of a protein with sulfo-SMCC with subsequent conjugation to a sulfhydryl-containing second molecule or protein. Specific examples of the use of this cross-linker to make antibody–enzyme or hapten–carrier conjugates may be found in Chapter 10, Section 1.1, and Chapter 9, Section 5, respectively.

Protocol

1. Dissolve 10 mg of a protein or other macromolecule to be activated with sulfo-SMCC in 1 ml of 0.1 *M* sodium phosphate, 0.15 *M* NaCl, pH 7.2.
2. Weigh out 2 mg of sulfo-SMCC and add it to the above solution. Mix gently to dissolve. To aid in measuring the exact quantity of cross-linker, a concentrated

stock solution may be made in water and an aliquot equal to 2 mg transferred to the reaction solution. If a stock solution is made, it should be dissolved rapidly and used immediately to prevent extensive hydrolysis of the active ester. As a general guideline of addition for a particular protein activation, the use of a 40- to 80-fold molar excess of cross-linker over the amount of protein present usually results in good activation levels.

3. React for 1 h at room temperature with periodic mixing.

4. Immediately purify the maleimide-activated protein by applying the reaction mixture to a desalting column packed with Sephadex G-25 or the equivalent. Do not dialyze the solution, since the maleimide activity will be lost over the time course required to complete the operation. To obtain good separation between the protein peak (eluting first) and the peak representing excess reagent and reaction by-products (eluting second), the applied sample size should be no more than 8% of the column bed volume. If complete separation of the activated protein from excess cross-linker is not obtained, then the maleimide content contributed from contaminating cross-linker may prevent subsequent conjugate formation. Perform the chromatography using 0.1 M sodium phosphate, 0.15 M NaCl, pH 7.2. Collect 1-ml fractions and pool the peak containing the protein. At this point, the maleimide-activated protein may be used immediately in a conjugation reaction with a sulfhydryl-containing protein or other molecule or freeze-dried to preserve the maleimide activity.

5. To effect the conjugation reaction, mix the maleimide-activated protein at the desired molar ratio with a sulfhydryl-containing molecule dissolved in 0.1 M sodium phosphate, 0.15 M NaCl, pH 7.2. The purified protein from step 4 may be concentrated if necessary using centrifugal concentrators, but this should be done quickly to avoid extensive loss of activity. The molar ratio of addition depends on the desired conjugate to be obtained. For instance, if coupling a sulfhydryl-containing small molecule to a protein, the molecule should be add- ed in excess to the amount of maleimide activity present on the protein. In such a case, a 10- to 100-fold molar excess may be appropriate (Chapter 9, Section 5). However, if preparing protein–protein conjugates, as in the case of antibody– enzyme conjugates, the ratio of maleimide-activated protein to the sulfhydryl- containing protein is a matter of choice. Often, when coupling enzymes to antibodies, the enzyme is in molar excess over the antibody (see Chapter 10, Section 1.1). Typical molar ratios of enzyme-to-antibody can range from 2:1 to 7:1.

6. React for 2–24 h at room temperature or 4–24 h at 4°C.

7. The conjugate may be isolated by gel filtration if the molecular weight of the complex is sufficiently different from that of the unconjugated molecules.

1.4. MBS and Sulfo-MBS

m-Maleimidobenzoyl-N-hydroxysuccinimide ester (MBS) is a heterobifunctional cross-linking agent containing an NHS ester on one end and a maleimide group on the other. The NHS ester can react with primary amines in proteins and other molecules to form stable amide bonds, while the maleimide end nearly exclusively reacts with sulfhydryl groups to create stable thioether linkages (Fig. 159). These characteristics

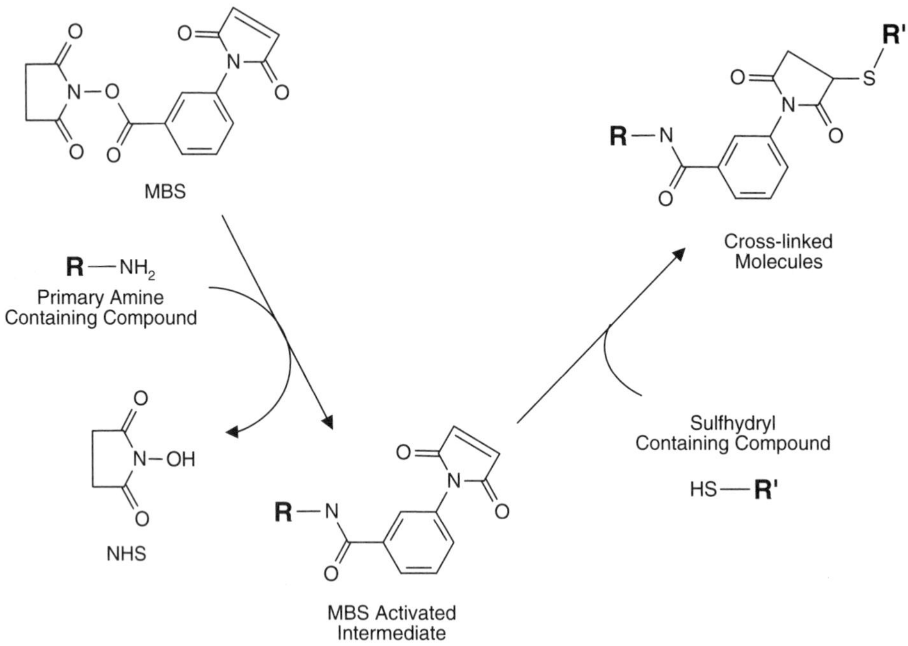

Figure 159 The two-step conjugation procedure for the MBS cross-linking of an amine-containing molecule with a sulfhydryl-containing molecule.

allow highly controlled conjugation reactions to be done with MBS using two- or three-step processes. In this sense, the NHS ester end of the reagent typically is reacted with the first protein to be cross-linked, forming a maleimide-activated intermediate. The maleimide group is more stable to breakdown by hydrolysis than the NHS ester, so the activated intermediate can be quickly purified from excess cross-linker and reaction by-products before it is added to the sulfhydryl-containing second molecule. However, due to the aromatic ring adjacent to its maleimide functional group, MBS displays less stability toward maleimide ring opening than SMCC (see this chapter Section 1.3). Unlike SMCC, MBS is therefore not recommended for preparing freeze-dried, maleimide-activated proteins, since during the processing necessary to purify and stabilize the derivative much activity can be lost by hydrolysis.

MBS contains a benzoic acid derivative as its cross-bridge, thus lending considerable hydrophobicity to the entire molecule. Because the reagent is water-insoluble, it must be first dissolved in organic solvent before it is added to an aqueous reaction medium. Making a concentrated stock solution of MBS in DMF of DMSO allows transfer of a small amount to a conjugation reaction (total concentration of the organic solvent should not exceed 10% in the reaction buffer). When these solvents are used, a microemulsion is formed in the aqueous solution, which provides cross-linker efficiently to the conjugating species. The reagent also is readily permeable to membrane structures due to its hydrophobic nature.

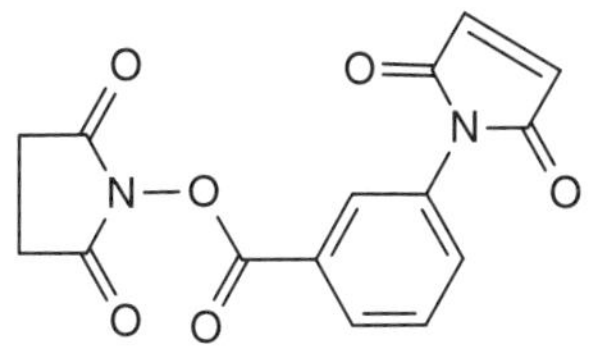

MBS;
m-Maleimidobenzoyl-N-hydroxy-
succinimide ester
MW 314.2
9.9 Å

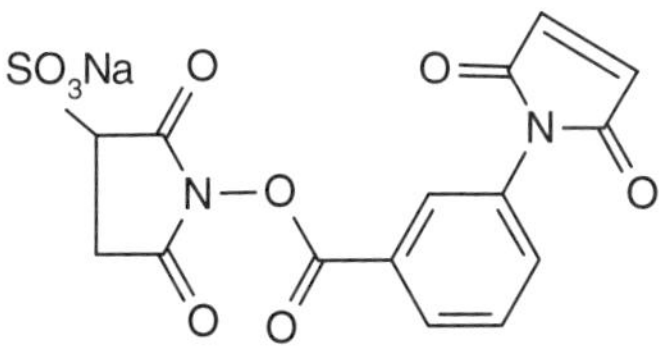

Sulfo-MBS;
m-Maleimidobenzoyl-N-hydroxy-
sulfosuccinimide ester
MW 416.24
9.9 Å

m-Maleimidobenzoyl-*N*-hydroxysulfosuccinimide ester (sulfo-MBS) is a water soluble analog of MBS that contains a negatively charged sulfonate group on its NHS ring (Aithal *et al.*, 1988; Martin and Papahadjopoulos, 1982). The negative charge lends enough hydrophilicity to the cross-linker to allow direct addition of the reagent to aqueous reaction media without prior dissolution in organic solvents. Sulfo-MBS has reactivity identical to of MBS.

MBS was one of the first and most popular of the family of NHS ester–maleimide heterobifunctionals (Kitagawa and Aikawa, 1976). It has been used extensively to produce antibody–enzyme and other enzyme conjugates (O'Sullivan *et al.*, 1979; Freytag *et al.*, 1984; Kitagawa *et al.*, 1978), in the preparation of hapten–carrier immunogens (Liu *et al.*, 1979; Lerner *et al.*, 1981; Chamberlain *et al.*, 1989; Edwards *et al.*, 1989; Miller *et al.*, 1989; Swanson *et al.*, 1991; Kitagawa *et al.*, 1982; Niman *et al.*, 1985), and for making immunotoxin conjugates (Youle and Nevelle, 1980; Myers *et al.*, 1989; Dell'Arciprete *et al.*, 1988).

The generalized protocol for performing a multistep conjugation reaction with MBS or sulfo-MBS is similar to that described for SMCC (this chapter, Section 1.3). Specific examples may be found in the cited references.

1.5. SIAB and Sulfo-SIAB

N-Succinimidyl(4-iodoacetyl)aminobenzoate (SIAB) is a heterobifunctional crosslinker containing amine-reactive and sulfhydryl-reactive ends (Weltman *et al.*, 1983). The NHS ester of SIAB can couple to primary amine-containing molecules, forming stable amide linkages (Chapter 2, Section 1.4). The other end contains an iodoacetyl group that is specific for coupling to sulfhydryl residues, creating stable thioether bonds (Chapter 2, Section 2.1). The aminobenzoate cross-bridge is a hydrophobic spacer that helps the reagent become fully permeable to membrane structures.

SIAB;
N-Succinimidyl(4-iodoacetyl)-
aminobenzoate
MW 402.15
10.6 Å

Sulfo-SIAB;
Sulfo-succinimidyl(4-iodoacetyl)-
aminobenzoate
MW 504.2
10.6 Å

As SIAB is water-insoluble, it must be first dissolved in organic solvent prior to addition to an aqueous reaction medium. The most commonly used solvents for this purpose include DMSO and DMF. Typically, a concentrated shock solution is prepared in one of these solvents and an aliquot added to the protein conjugation solution. Long-term storage of the reagent in these solvents is not recommended, however, due to slow uptake of water and breakdown of the NHS ester end.

Conjugations done with SIAB usually proceed by a multistep process. Because the cross-linker's NHS ester end is its most labile functional group, an amine-containing protein or molecule is reacted first to create an iodoacetyl-activated intermediate (Fig. 160). This iodoacetyl derivative is stable enough in aqueous solution to allow purification of the derivatized protein from excess reagent and other reaction by-products without significant loss of activity. The only consideration is to protect the iodoacetyl derivative from light, which may generate iodine and reduce the activity of the intermediate. Finally, the modified protein is mixed with a sulfhydryl-containing molecule to effect the conjugation through a thioether bond. The result of such two-step procedures is to direct the coupling toward only sulfhydryls on the second molecule while avoiding the polymerization that can occur with single-step protocols. Conjugations done with SIAB should avoid buffer components containing amines (i.e., Tris, glycine, or imidazole) or sulfhydryls (i.e., DTT, 2-mercaptoethanol, cysteine), since these will compete with the desired cross-linking reaction.

Sulfosuccinimidyl(4-iodoacetyl)aminobenzoate (sulfo-SIAB) is a water-soluble analog of SIAB that contains a negatively charged sulfonate on its NHS ring. The negative charge lends enough hydrophilicity to the entire molecule to provide good solubility in aqueous solutions (up to about 10 mM). Sulfo-SIAB may be added directly to reaction mediums without prior dissolution in organic solvent, or more concentrated solutions may be made in water before transfer of an aliquot to the reaction to facilitate easy

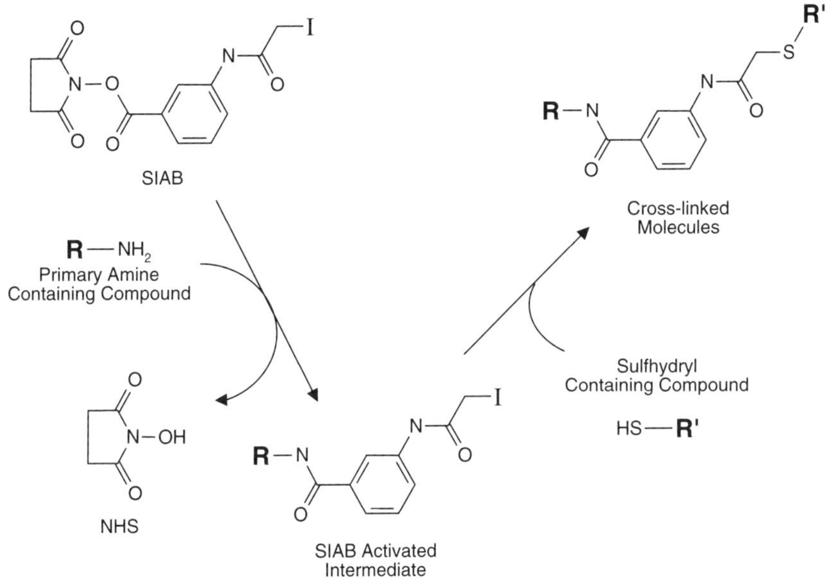

Figure 160 SIAB may be used to modify an amine-containing molecule for subsequent conjugation to a sulfhydryl-containing molecule.

addition of small quantities. Aqueous stock solutions should be dissolved rapidly and used immediately to avoid excessive hydrolysis of the NHS ester.

The following protocol illustrates the use of SIAB in preparing antibody–enzyme conjugates using β-galactosidase.

Protocol

1. Dissolve a specific antibody to be conjugated at a concentration of 10 mg/ml in 50 mM sodium borate, 5 mM EDTA, pH 8.3 (reaction buffer).
2. Dissolve SIAB (Pierce) in DMSO at a concentration of 1.4 mg/ml. Alternatively, dissolve sulfo-SIAB in deionized water at a concentration of 1.7 mg/ml. Prepare fresh and use immediately. Protect from light.
3. Add 100 μl of the SIAB stock solution to each milliliter of the antibody solution. Mix gently to dissolve.
4. React for 1 h at room temperature in the dark.
5. Purify the modified antibody by gel filtration on a Sephadex G-25 column. Perform the chromatography using the reaction buffer. To obtain good separation, apply sample at a ratio of no more than 8% of the total column gel volume. Monitor the eluting peak by using a small aliquot of each fraction and reacting it with a protein detection reagent such as Coomassie protein assay reagent (Pierce) in a microplate. This avoids exposure of the entire modified protein fractions to UV light from a spectrophotometer. Collect the first peak eluting from the column, which contains the protein.
6. Add β-galactosidase to the activated antibody solution at a ratio of 4 mg of enzyme per milligram of antibody.

7. React for 1 h at room temperature in the dark.
8. To block any remaining iodoacetyl sites, add cysteine to a final concentration of 5 mM and react for an additional 15 min at room temperature.
9. Purify the conjugate by gel filtration using a buffer of choice (i.e., PBS, pH 7.4).

1.6. SMPB and Sulfo-SMPB

Succinimidyl-4-(p-maleimidophenyl)butyrate (SMPB) is a heterobifunctional analog of MBS (this chapter, Section 1.4) containing an extended cross-bridge (Pierce). The reagent has an amine-reactive NHS ester on one end and a sulfhydryl-reactive maleimide group on the other end (Fig. 161). Conjugates formed using SMPB thus are linked by stable amide and thioether bonds. A comparison with SPDP produced conjugates concluded that SMPB formed more stable complexes that survive *in vivo* for longer periods (Martin and Papahadjopoulos, 1982).

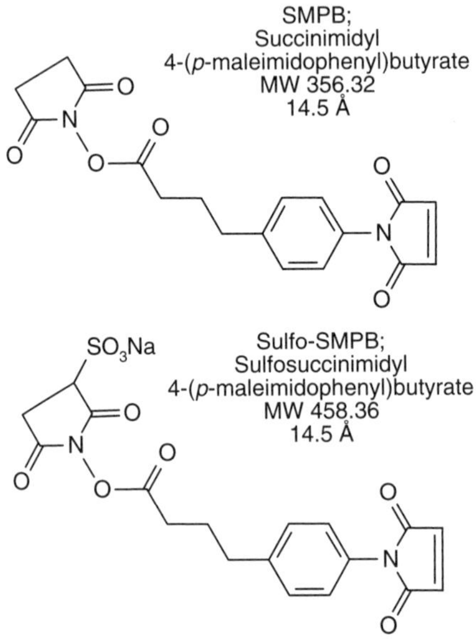

SMPB;
Succinimidyl
4-(p-maleimidophenyl)butyrate
MW 356.32
14.5 Å

Sulfo-SMPB;
Sulfosuccinimidyl
4-(p-maleimidophenyl)butyrate
MW 458.36
14.5 Å

Conjugation reactions done with SMPB typically are multistep procedures, wherein a protein is modified through its amine groups, purified to remove excess reagent, and then mixed with a sulfhydryl-containing molecule to effect the final conjugation. The maleimide group of SMPB is highly specific for coupling to sulfhydryl-containing proteins and other molecules, thus directing the conjugation to discrete points on the second molecule. This maleimide is, however, more labile to ring opening in aqueous solution than the maleimide group of SMCC due to its proximity to an aromatic ring. Therefore, the first protein modified with SMPB (to obtain a maleimide-activated intermediate) should be purified quickly to prevent extensive activity loss from hydrolysis and maleimide ring opening.

SMPB contains a hydrophobic cross-bridge and relatively nonpolar ends, which allows the reagent to permeate membrane structures. Due to its water-insolubility, it must be dissolved in an organic solvent prior to addition of an aliquot to a reaction mixture. The solvents DMF and DMSO work well for this purpose. A concentrated

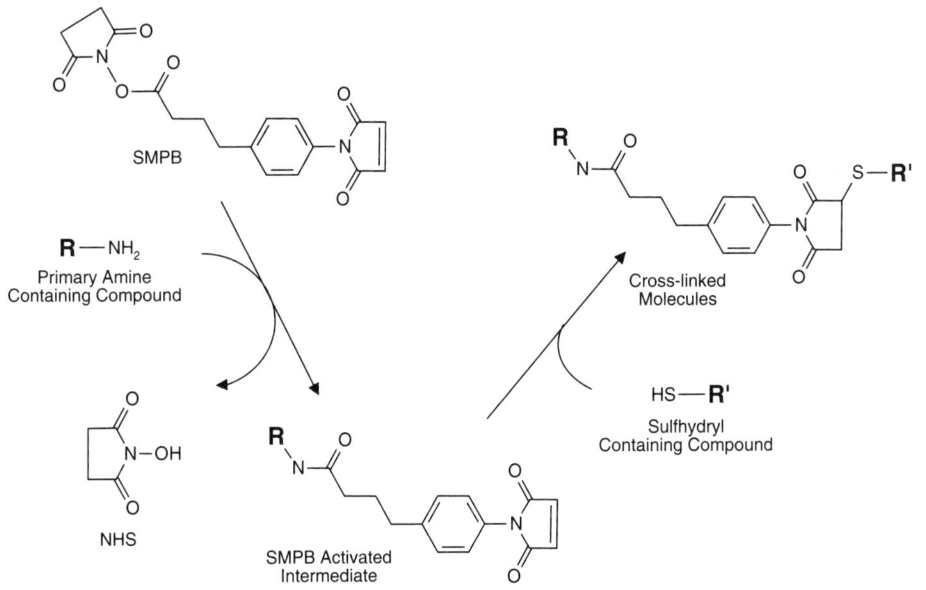

Figure 161 SMPB may be used in a two-step procedure to conjugate an amine-containing molecule to a sulfhydryl compound, forming amide and thioether bonds, respectively.

stock solution prepared in these solvents allows easy addition of a small amount to a conjugation reaction. Long-term storage in these solvents is not recommended due to slow water pickup and possible hydrolysis of the NHS ester end.

A water-soluble analog to SMPB, sulfosuccinimidyl-4-(*p*-maleimidophenyl)butyrate (sulfo-SMPB), contains a negatively charged sulfonate group that lends considerable hydrophilicity to the molecule (Pierce). Sulfo-SMPB may be added directly to aqueous reaction mixtures without prior dissolution in organic solvent. Concentrated stock solutions made in water should be dissolved quickly and used immediately to prevent hydrolysis of the NHS ester.

SMPB or sulfo-SMPB have been used to conjugate preformed vesicles and Fab′ fragments in a liposome carrier study (Martin and Papahodjopoulos, 1982), to attach insulin molecules to reconstituted Sendai virus envelopes (Gitman *et al.*, 1985a,b), for targeting of loaded virus envelopes by covalently attaching insulin molecules to receptor-depleted cells (Gitman *et al.*, 1985b), in forming alkaline phosphatase-Fab′ fragment conjugates for an ELISA (Teale and Kearney, 1986), in preparing peptide–protein immunogen conjugates (Iwai *et al.*, 1988), in studying the transport of the variant surface glycoprotein of Trypanosome brucia (Bangs *et al.*, 1986), and in preparing immunotoxin conjugates (Myers *et al.* 1989).

1.7. GMBS and Sulfo-GMBS

N-(γ-Maleimidobutyryloxy)succinimide ester (GMBS) is a heterobifunctional cross-linking agent that contains an NHS ester on one end and a maleimide group on the other (Fujiwara *et al.*, 1988) (Pierce). Its internal cross-bridge contains a linear 4-carbon spacer, resulting in a 10.2-Å cross-links between conjugated molecules (Fig. 162). GMBS is water-insoluble and therefore must be dissolved in organic solvent prior to use. Typically, a concentrated stock solution is prepared in DMF or DMSO just before use, and then an aliquot of the solution is transferred to the aqueous reaction medium. The result is the formation of a microemulsion that effectively supplies cross-linker to the aqueous phase.

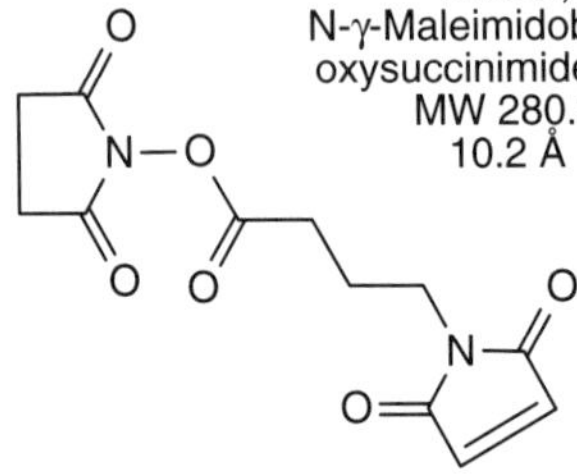

GMBS;
N-γ-Maleimidobutyryl-
oxysuccinimide ester
MW 280.2
10.2 Å

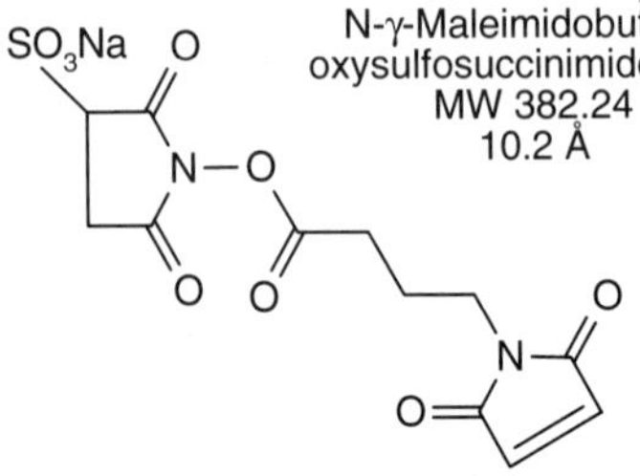

Sulfo-GMBS;
N-γ-Maleimidobutyryl-
oxysulfosuccinimide ester
MW 382.24
10.2 Å

GMBS can be used in multistep conjugation protocols wherein an amine-containing molecule or protein is first modified via the NHS ester end (its most labile functional group) to create a stable amide bond. The derivative at this point contains reactive

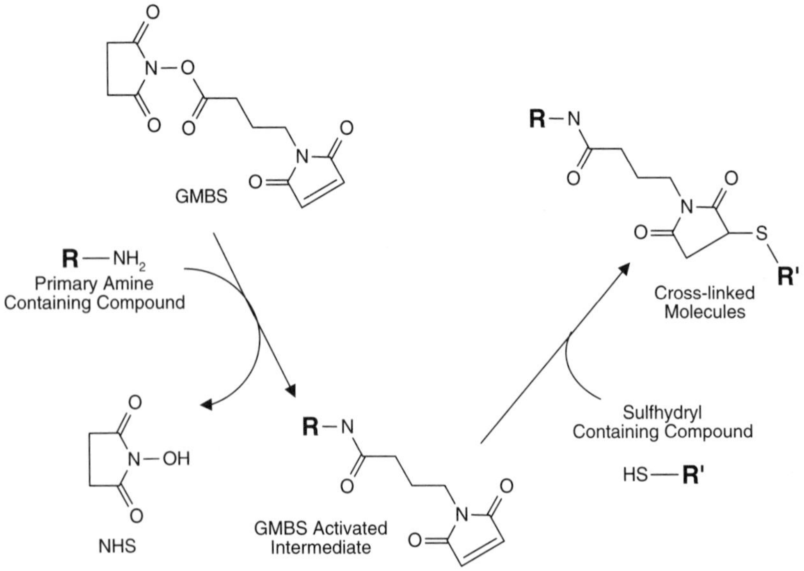

Figure 162   The reaction of GMBS with an amine-containing molecule yields a maleimide-activated intermediate that then can be used to cross-link with a sulfhydryl-containing compound.

maleimide groups able to couple with the available sulfhydryl groups of a second protein or molecule. This active intermediate then is purified to remove excess reagent and reaction by-products, and immediately added to the sulfhydryl-containing molecule to effect the final conjugation.

The maleimide group of GMBS is adjacent to an aliphatic spacer, so its stability toward ring opening should be better than cross-linkers like MBS, which contain adjacent aromatic groups. Hydrolysis of the maleimide group results in loss of sulfhydryl coupling capability. However, GMBS is not as stable as the hindered maleimide group of SMCC, since the cyclohexane ring of that reagent inhibits hydrolysis and ring opening.

N-(γ-Maleimidobutyryloxy)sulfosuccinimide ester (sulfo-GMBS) is a water-soluble analog of GMBS containing a negatively charged sulfonate group on its NHS ring (Pierce). The charge provides enough hydrophilicity to allow at least 10 mM concentrations of the cross-linker to be made in aqueous reaction mediums. Its reactivity is identical to that of GMBS.

The protocol for using GMBS or sulfo-GMBS in protein–protein cross-linking applications is similar to that of SMCC or sulfo-SMCC (see Section 1.3).

1.8. SIAX and SIAXX

Succinimidyl 6-((iodoacetyl)amino)hexanoate (SIAX) is a heterobifunctional reagent containing an NHS ester on one end and an iodoacetyl group on the other (Brinkley, 1992) (Molecular Probes). The NHS ester reacts with primary amines in proteins and other molecules to form stable amide bonds. The iodoacetyl group is highly specific for sulfhydryl groups, reacting to create stable thioether linkages (Fig. 163). The reactivity and use of this cross-linker is similar to that of SIAB, described previously. SIAX possesses a 6-aminohexanoic acid internal cross-bridge, providing a total of a 9-atom spacer between conjugated molecules.

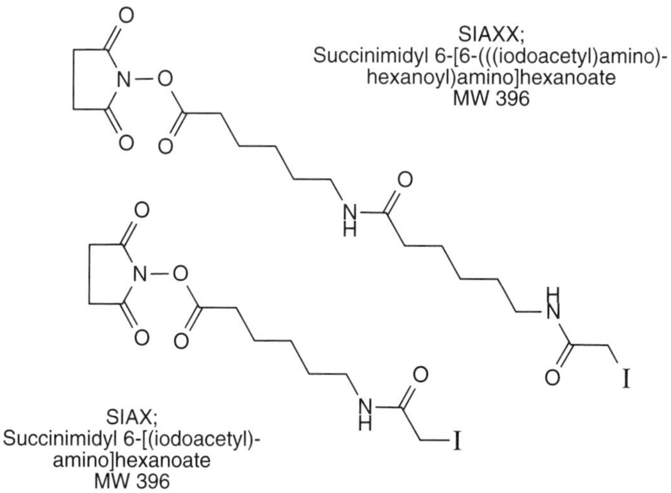

SIAXX;
Succinimidyl 6-[6-(((iodoacetyl)amino)-
hexanoyl)amino]hexanoate
MW 396

SIAX;
Succinimidyl 6-[(iodoacetyl)-
amino]hexanoate
MW 396

SIAX is a hydrophobic reagent that should penetrate membrane structures with good efficiency. The cross-linker must be solubilized in organic solvent (DMF or DMSO) prior to transfer of a small amount to an aqueous reaction medium.

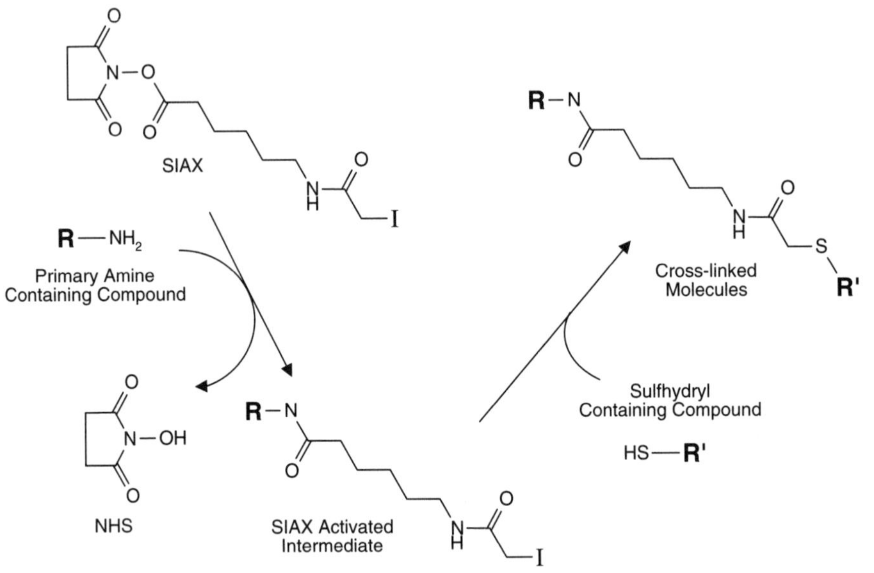

Figure 163 SIAX can be used to modify amine-containing molecules to produce sulfhydryl-reactive derivatives. Subsequent reaction with a thiol compound produces a thioether linkage.

Succinimidyl 6-(6-(((4-iodoacetyl)amino)hexanoyl)amino)hexanoate (SIAXX) is a long-chain analog of SIAX that contains two aminohexanoate spacer groups in its cross-bridge, instead of one (Molecular Probes). Conjugates prepared with this re-

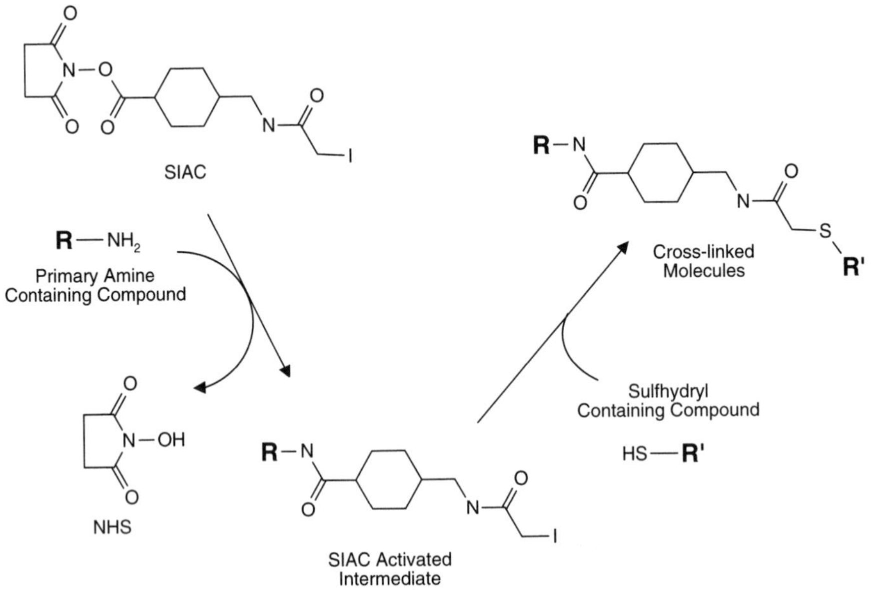

Figure 164 SIAC reacts with an amine-containing compound to yield an amide bond derivative that is reactive toward thiol-containing molecules. Secondary reaction with a sulfhydryl group gives a stable thioether bond.

agent are connected by a spacer arm containing 16 atoms. Like SIAX, SIAXX must be first solubilized in DMF or DMSO before it is added to a buffered reaction. The increased chain length SIAXX, however, does not affect its reactivity toward amines and sulfhydryls.

Conjugation reactions done with SIAX or SIAXX are usually multistep procedures similar to the protocol described previously for SIAB.

1.9. SIAC and SIACX

Succinimidyl 4-(((iodoacetyl)amino)methyl)cyclohexane-1-carboxylate (SIAC) is a heterobifunctional reagent containing an NHS ester on one end and a iodoacetyl group on the other (Molecular Probes). The cross-linker can react with amine groups via its NHS ester end to form stable amide bonds, while its iodoacetyl functional group can couple to sulfhydryl groups, creating stable thioether linkages (Fig. 164). SIAC contains a cross-bridge made from a cyclohexane derivative, which provides approximately an 8-atom spacer between conjugated species.

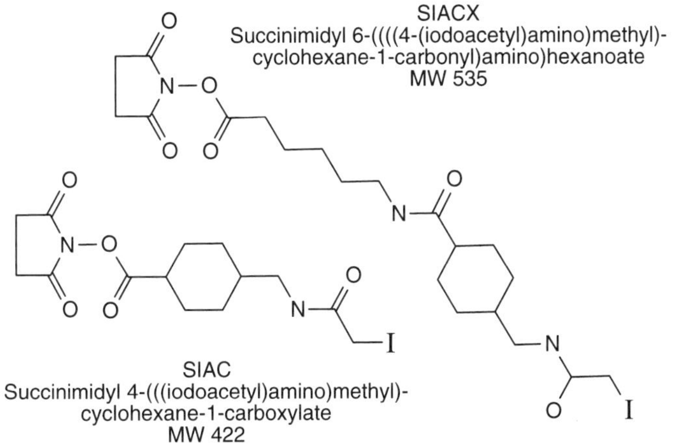

SIACX
Succinimidyl 6-((((4-(iodoacetyl)amino)methyl)-
cyclohexane-1-carbonyl)amino)hexanoate
MW 535

SIAC
Succinimidyl 4-(((iodoacetyl)amino)methyl)-
cyclohexane-1-carboxylate
MW 422

SIAC is a hydrophobic cross-linker that must be solubilized in organic solvent (DMF or DMSO) prior to addition of an aliquot to an aqueous reaction medium. It should exhibit good membrane permeability.

Succinimidyl 6-((((4-iodoacetyl)amino)methyl)cyclohexane-1-carbonyl)amino)-hexanoate (SIACX) is an analog of SIAC that contains an additional aminohexanoate spacer group next to its NHS ester end (Molecular Probes). The result is the creation of an approximately 16-atom spacer arm between conjugated molecules. All other properties of SIACX are similar to those of SIAC.

Conjugation reactions done with SIAC or SIACX are usually multistep procedures similar to the protocol described for SIAB, previously.

1.10. NPIA

p-Nitrophenyl iodoacetate (NPIA) is a heterobifunctional reagent based upon iodoacetate that has been activated at its carboxylic acid group with a p-nitrophenyl ester (Huang et al., 1975; Hudson and Weber, 1973) (Molecular Probes). This active

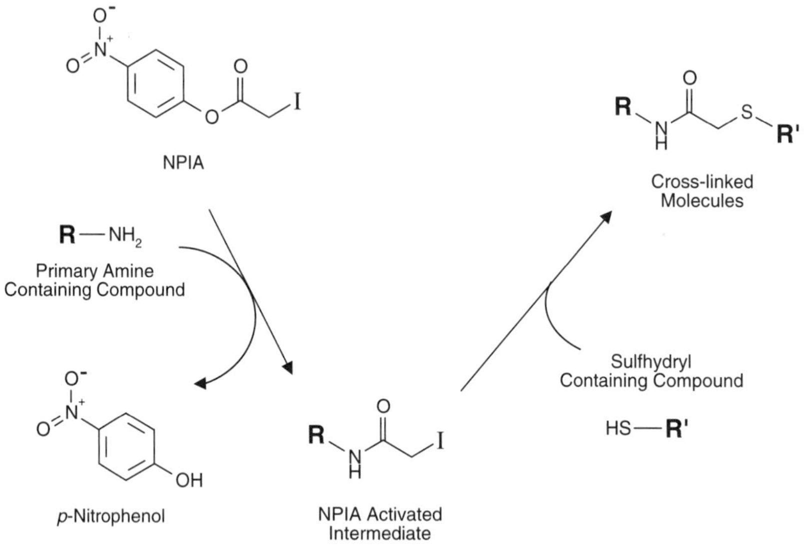

Figure 165 NPIA is one of the shortest heterobifunctional reagents. It reacts with amine-containing molecules through its *p*-nitrophenyl ester end to produce amide bonds. The iodoacetyl group then can be used to couple with thiol compounds to give stable thioether linkages.

ester species has much the same reactivity as an NHS ester, being highly reactive with amines at slightly basic pH values (pH 7–9). The *p*-nitrophenyl ester couples to amine-containing proteins and other molecules to form stable amide linkages. The other end of the short cross-linker can react with sulfhydryl groups to create thioether bonds. This is the smallest heterobifunctional iodoacetate-containing cross-linker available, forming only 2-atom cross-bridges between conjugated molecules (Fig. 165). NPIA has been used to investigate close interactions between biological molecules (Hiratsuka, 1987; Sutoh and Hiratsuka, 1988).

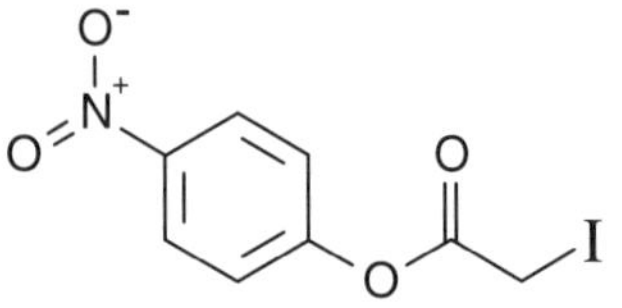

NPIA
p-Nitrophenyl iodoacetate
MW 307

NPIA is water-insoluble and should be dissolved in DMF or methylene chloride prior to addition of an aliquot to an aqueous reaction medium. Conjugation reactions done with NPIA are usually multistep procedures similar to the protocol previously described for SIAB.

2. Carbonyl-Reactive and Sulfhydryl-Reactive Cross-linkers

A relatively new set of heterobifunctional cross-linking agents now are available that contain a carbonyl-reactive group on one end and a sulfhydryl-reactive functional group on the other end. The main utility of these reagents is in conjugating carbohydrate-containing molecules, such as glycoproteins, to sulfhydryl-containing molecules. Both polysaccharide residues and sulfhydryl groups usually are present on proteins in limiting quantities and at discrete sites. In certain cases, conjugation through these groups can direct the coupling reaction away from critical active centers or binding sites, thus preserving activity of the proteins after cross-linking. A prime example of the advantages of this type of directed coupling can be seen when conjugating antibody molecules to other proteins, such as enzymes. The carbohydrate residues of immunoglobulin molecules usually occur on the Fc portion, away from the antigen binding sites. Coupling procedures that direct the cross-linking reaction to parts on the antibody far removed from the antigen binding sites have the best chance of retaining activity after conjugate formation.

The carbonyl-reactive functional group on these cross-linkers is a hydrazide group that can form hydrazone bonds with aldehyde residues. To utilize this functional group with carbohydrate-containing molecules, the sugars first must be mildly oxidized to contain aldehyde groups by treatment with sodium periodate. Oxidation with this compound will cleave adjacent carbon—carbon bonds that possess hydroxyl groups, as are abundant in polysaccharide molecules (Chapter 1, Sections 2 and 4.4).

Two types of sulfhydryl-reactive functions are available on these reagents: pyridyl disulfide groups and maleimide groups. The pyridyl disulfide group will react with a sulfhydryl residue to create a disulfide bond. This linkage is reversible by treatment with a disulfide reducing agent. Reaction of a maleimide group with a sulfhydryl, however, forms a permanent thioether bond of good stability. Thus, either reversible or permanent conjugates may be designed using these heterobifunctionals.

2.1. MPBH

4-(4-N-Maleimidophenyl)butyric acid hydrazide (MPBH) is a heterobifunctional cross-linking agent that contains a carbonyl-reactive hydrazide group on one end and a sulfhydryl-reactive maleimide on the other (Pierce). The cross-bridge between the two functional ends provides a long, 17.9-Å spacer. The hydrazide group is produced as the hydrochloride salt. The reagent as a whole has a good water solubility. It can be dissolved in 0.1 M sodium acetate, pH 5.5, up to a concentration of 327 mg/ml. It is also freely soluble in DMSO and may be stored as concentrated stock solution in this solvent without degradation.

MPBH
4-(4-N_Maleimidophenyl)butyric
acid hydrazide hydrochloride
MW 309.75
17.9 Å

The maleimide group of MPBH is adjacent to an aromatic ring and thus may exhibit instability to hydrolysis in aqueous solutions, especially at alkaline pH. Hydrolysis opens the maleimide ring and destroys its coupling ability with sulfhydryls. However, both reactive ends of the cross-linker are stable enough to survive a multistep coupling protocol without extensive loss of activity. Thus, a sulfhydryl-containing protein or molecule may be modified via the maleimide end of MPBH and the derivative purified by gel filtration to remove excess reactants, and then mixed with a glycoprotein (that had been previously oxidized to provide aldehyde residues) to effect the final conjugation (Fig. 166). The opposite approach also is possible: modification of the glycoprotein first, purification, and subsequent mixing with a sulfhydryl-containing molecule. With this second option, however, the purification step should be done quickly to prevent extensive hydrolysis of the maleimide group.

MPBH has been used to conjugate CD4 without loss of biological activity (Chamow *et al.*, 1992).

2.2. M$_2$C$_2$H

4-(*N*-Maleimidomethyl)cyclohexane-1-carboxyl-hydrazide (M$_2$C$_2$H) is a heterobifunctional cross-linking agent that contains a carbonyl-reactive hydrazide group on one end and a sulfhydryl-reactive maleimide group on the other (Pierce). The reagent is similar to MPBH (described previously), but the maleimide group on M$_2$C$_2$H is expected to be more stable in aqueous solutions, since it is adjacent to an aliphatic cyclohexane ring instead of an aromatic phenyl group. In this sense, the cross-bridge of M$_2$C$_2$H is nearly identical to that of SMCC, which contains one of the most stable maleimide groups known. The hindered environment of the cyclohexane ring should

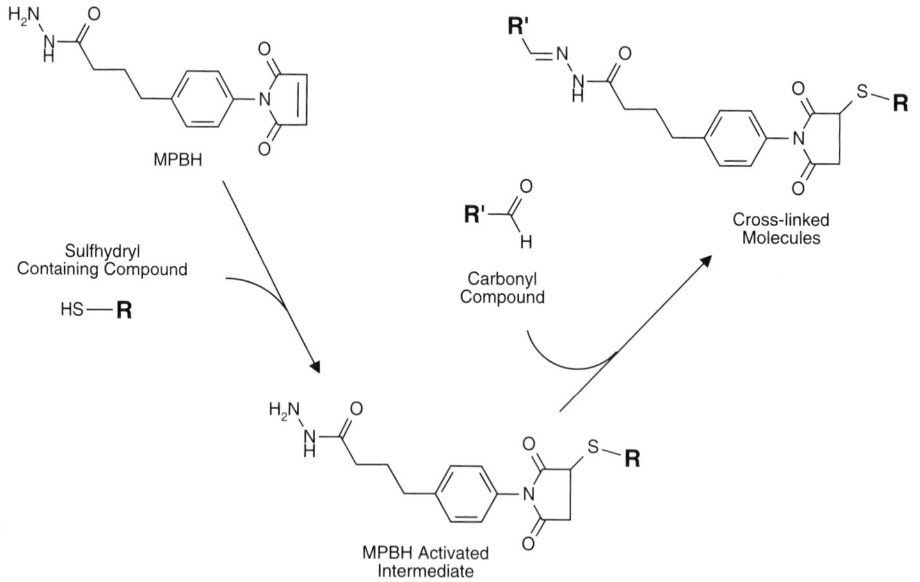

Figure 166 MPBH reacts with sulfhydryl-containing molecules through its maleimide end to produce thioether linkages. Its hydrazide group then can be used to conjugate with carbonyl-containing molecules (such as periodate oxidized carbohydrates which contain aldehydes) to give hydrazone bonds.

provide similar stability advantages to this reagent. Reaction of the maleimide group with a sulfhydryl residue results in the formation of a stable thioether bond.

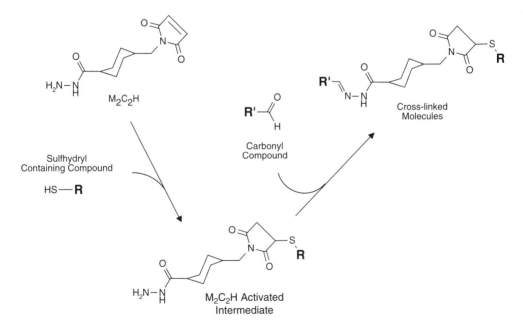

M$_2$C$_2$H
4-(N-Maleimidomethyl)cyclohexane-
1-carboxyl-hydrazide hydrochloride
MW 287.75
15.1 Å

On the other end of the cross-linker, the hydrazide functional group can react with periodate-oxidized carbohydrate molecules to form hydrazone linkages (Chapter 1, Sections 2 and 4.5). Thus, glycoproteins can be targeted specifically at their polysaccharide chains, avoiding cross-linking at active sites which can lead to activity losses (Fig. 167).

M$_2$C$_2$H is slightly soluble in aqueous solutions, reportedly having a maximal solubility of 3.2 mg/ml in 0.1 M sodium acetate at pH 5.5. It is also soluble in organic solvents, which allows for the preparation of concentrated stock solutions to be made

Figure 167 M$_2$C$_2$H can be used to cross-link a sulfhydryl-containing molecule with an aldehyde-containing compound. Glycoproteins may be conjugated using this reagent after treatment with sodium periodate to form reactive aldehyde groups.

prior to addition of a small aliquot to an aqueous reaction mixture. The cross-linker is particularly stable in acetonitrile.

2.3. PDPH

3-(2-Pyridyldithio)propionyl hydrazide (PDPH) is a heterobifunctional reagent that possesses a carbonyl-reactive hydrazide group on one end and a sulfhydryl-reactive pyridyl disulfide group on the other (Pierce). Thus, sulfhydryl-containing proteins or other molecules may be conjugated to carbohydrate-containing molecules (after treatment of the polysaccharide portion with sodium periodate) to create aldehyde residues (Fig. 168). Using this cross-linker, glycoproteins can be coupled specifically through their carbohydrate chains, in many cases better avoiding active centers or binding sites than when coupling through abundant polypeptide groups like amines. Since the pyridyl disulfide group reacts with sulfydryls to create disulfide bonds, the cross-linked proteins can be cleaved by reduction with DTT (Chapter 1, Section 4.1).

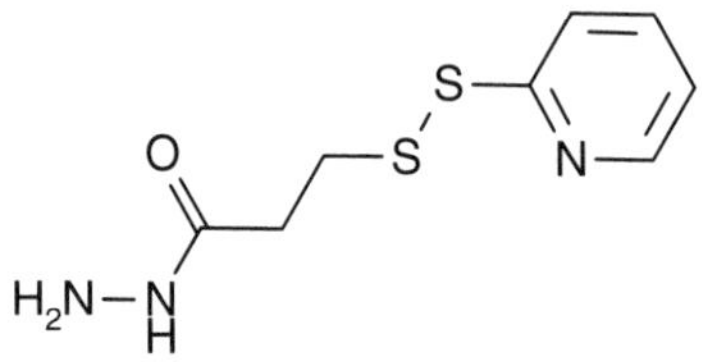

PDPH
3-(2-Pyridyldithio)
propionyl hydrazide
MW 229.31
9.2 Å

PDPH also may be used as a thiolation reagent to add sulfhydryl functional groups to carbohydrate molecules. The reagent can be used in this sense similar to the protocol described for AMBH (Chapter 1, Section 4.1). After modification of an oxidized polysaccharide with the hydrazide end of PDPH, the pyridyl group is removed by treatment with DTT, leaving the exposed sulfhydryl (Fig. 169).

PDPH is soluble in 0.1 M sodium acetate, pH 5.5, at a maximal concentration of 14.2 mg/ml. The reagent is particularly stable in acetonitrile for preparation of concentrated stock solutions.

PDPH has been used in the preparation of immunotoxin conjugates (Zara et al., 1991). It has also been used to create a unique conjugate of NGF with an antibody directed against the transferrin receptor OX-26, which could traverse the blood–brain barrier (Friden et al., 1993). Labeling of antibody molecules with PDPH at oxidized polysaccharide sites followed by reduction to free the sulfhydryl has been used to form a technetium-99m complex for radiopharmaceutical use (Ranadive et al., 1993) (Chapter 8, Section 2.5).

3. Amine-Reactive and Photoreactive Cross-linkers

An important class of heterobifunctional reagents is the photoreactive cross-linkers that have one end that can be photolyzed to initiate coupling. Photoreactive cross-

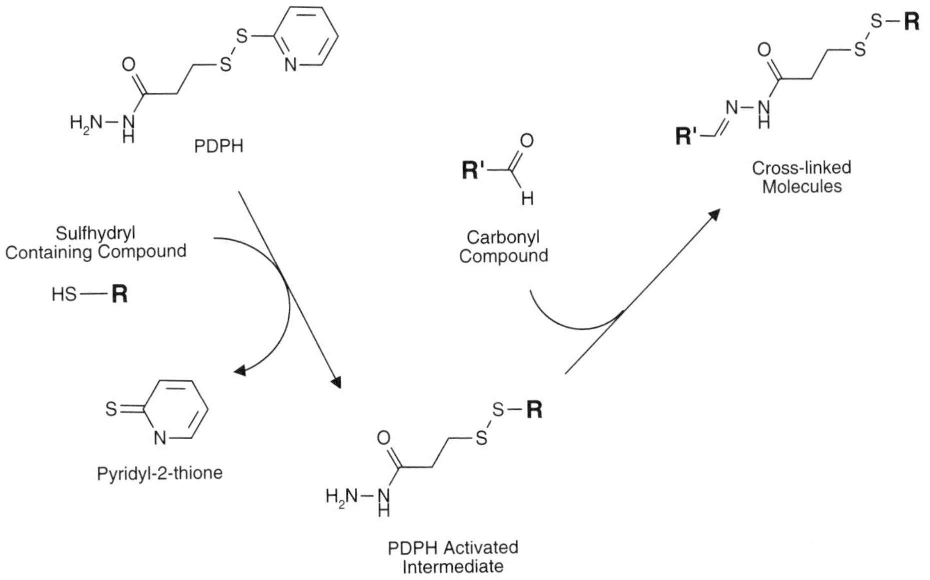

Figure 168 PDPH reacts with thiol-containing compounds through its pyridyl disulfide end to form reversible disulfide linkages. Its hydrazide end then may be subsequently conjugated with an aldehyde-containing molecule to form hydrazone bonds. Glycoproteins may be cross-linked using this approach after periodate activation to generate formyl groups.

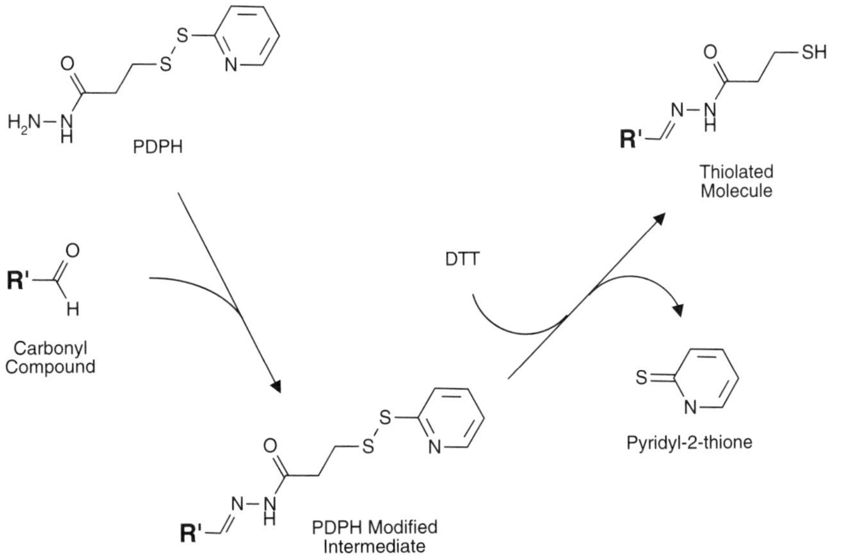

Figure 169 PDPH may be used to add a sulfhydryl group to an aldehyde-containing molecule. After reacting its hydrazide end with the aldehyde to form a hydrazone bond, the pyridyl disulfide may be reduced with DTT to create a free thiol.

linkers may be designed to utilize any one of a number of photosensitive groups, including aryl azides, fluorinated aryl azides, benzophenones, certain diazo compounds, and diazirine derivatives (Chapter 2, Section 7). The best photoreactive functional groups are stable in aqueous solution in the dark, and may be activated at the desired time by a pulse of light at the appropriate wavelength. The other end of these heterobifunctionals usually contains a spontaneously reactive functional group that will couple rapidly with certain groups present on target molecules. This secondary functionality is sometimes called *thermoreactive* to differentiate it from the photoreactive end and to emphasize its ready-reactivity or sometimes its labile nature in aqueous environments. The thermoreactive end is typically amine-reactive, sulfhydryl-reactive, carbonyl-reactive, carboxylate-reactive, or arginine-reactive. Still another class of photoreactive heterobifunctionals may use a biotin handle at one end to cross-link specifically, but noncovalently, with avidin or streptavidin molecules (Chapter 8, section 3.4).

Photoreactive groups can be categorized by the reactive species that is generated upon photolysis. The most popular type of photosensitive functional groups, aryl azide derivatives, form short-lived nitrenes that react extremely rapidly with the surrounding chemical environment (Gilchrist and Rees, 1969). Recent evidence, however, indicates that the photolyzed intermediates of aryl azides can undergo ring expansion to create nucleophile-reactive dehydroazepines. Instead of inserting nonselectively at active carbon—hydrogen bonds, dehydroazepines have a tendency to react preferentially with nucleophiles, especially amines (Fig. 170). However, some investigators have shown that aryl azides that possess a perfluorinated ring structure or are substituted completely with halogen atoms are quite efficient at forming the desired nitrene intermediate (Soundararajan *et al.*, 1993; Keana and Cai, 1990; Schnapp and Platz, 1993; Schnapp *et al.*, 1993; Cai *et al.*, 1993; Yan *et al.*, 1994). Unfortunately, at the present time, commercially available perfluorinated aryl azides are scarce.

One advantage of aryl azide photoreactive cross-linkers is that they have a relatively low energy of activation, which is optimal in the long UV region. In addition, many aryl azides possess nitro groups on their associated aromatic ring structures. These electron-withdrawing groups tend to increase the optimal wavelength for photolysis upward close to the 350-nm range. The benefit of this approach is that relatively low light exposure at higher energy UV wavelengths avoids potential bond breakage that may occur with some sensitive compounds upon photolysis.

Other phenyl azide-containing reagents possess hydroxyl groups on their aromatic rings. These electron-donating groups activate the ring system to allow electrophilic substitution reactions to occur on the cross-linker prior to its use. A major application of this ability is to radioiodinate the photoreactive reagent for detection purposes before the modification reaction with target molecules has taken place.

Suitable light sources for photolyzing include sunlamps manufactured by a number of companies, such as Philips Ultrapnil MLU 300 W, General Electric Sunlamp RSM 275 W, or National Self-Ballasted BHRF 240-250 V 250 W-P lamp. Irradiation for 15 min with such lamps while the sample is cooled in an ice bath will result in good photolysis of photoreactive cross-linkers and modification reagents.

Although photoreactive aryl azides are relatively inert to thermochemical reactions prior to photolysis, they are not stable in the presence of sulfhydryl compounds, which can reduce the azide functional group to an amine. Avoid, therefore, the use of reductants such as DTT or 2-mercaptoethanol before the photolyzing step, as these can react with the aryl azide within minutes (Staros *et al.*, 1978). Avoid also amine-containing

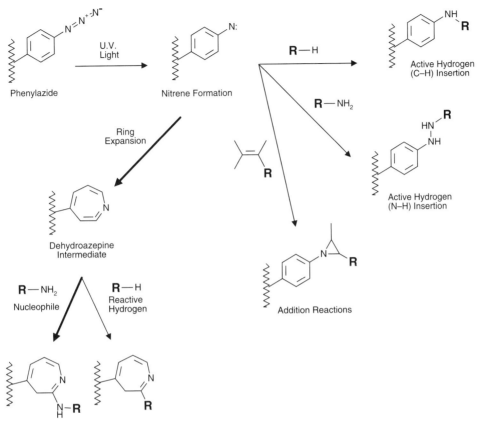

Figure 170 Photolyzing a phenyl azide group with UV light results in the formation of a short-lived nitrene. Nitrenes may undergo a number of reactions, including insertion into active carbon—hydrogen or nitrogen—hydrogen bonds and addition to points of unsaturation in carbon chains. The most likely route of reaction, however, is to ring-expand to a dehydroazepine intermediate. This group is highly reactive toward nucleophiles, especially amines.

buffer components such as Tris or glycine, because of the potential for nucleophilic reactivity of the photolyzed intermediate.

Of the following amine-reactive and photoreactive cross-linkers, the overwhelming majority use an aryl azide group as the photosensitive function. Only a few use alternative photoreactive chemical reactions, particularly perfluorinated aryl azide, benzophenone, or diazo compounds. For general background information on photoreactive cross-linkers see Das and Fox (1979), Kiehm and Ji (1977), Vanin and Ji (1981), Meijer *et al.*, 1988), and Brunner (1993).

3.1. NHS-ASA, Sulfo-NHS-ASA, and Sulfo-NHS-LC-ASA

N-Hydroxysuccinimidyl-4-azidosalicylic acid (NHS-ASA) is a heterobifunctional reagent containing an NHS ester on one end and a photoreactive aryl azide group on the other (Pierce). The amine-reactive NHS ester can be reacted with proteins or other primary amine-containing molecules to yield a photosensitive derivative suitable for probing biological interaction sites. Upon photolysis with a long UV light source, the aryl azide end is activated to complex covalently with closely associated target molecules (Fig. 171). The small cross-bridge of NHS-ASA is built from a salicylate derivative that contains a hydroxyl group on the aromatic ring. The ring-activating nature of this group provides an iodination site on the cross-linker to allow tracking of modified molecules (Ji and Ji, 1982) (Chapter 8, Section 4.5).

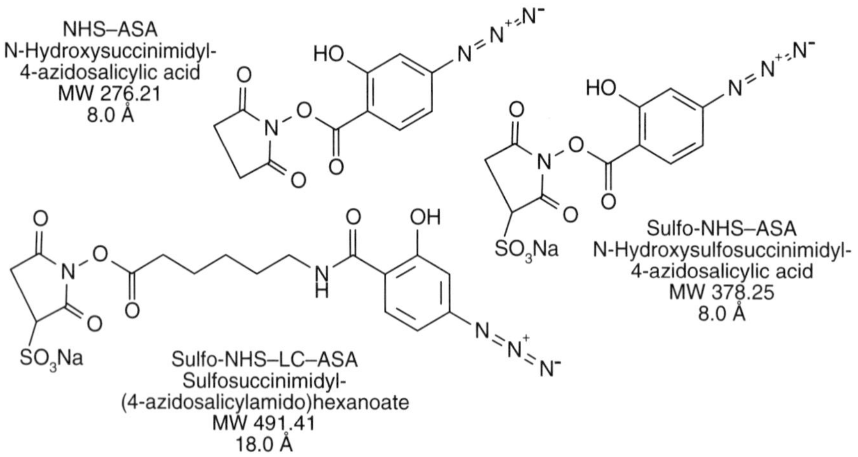

NHS–ASA
N-Hydroxysuccinimidyl-
4-azidosalicylic acid
MW 276.21
8.0 Å

Sulfo-NHS–ASA
N-Hydroxysulfosuccinimidyl-
4-azidosalicylic acid
MW 378.25
8.0 Å

Sulfo-NHS–LC–ASA
Sulfosuccinimidyl-
(4-azidosalicylamido)hexanoate
MW 491.41
18.0 Å

Reported applications of NHS-ASA include photoaffinity labeling of ^{125}I-ASA-Con A to erythrocyte ghosts (Ji and Ji, 1982), derivatization of human choriogonadotropin with ^{125}I-NHS-ASA with photo-initiated cross-linking of the α-β dimer (Ji *et al.*, 1985), radiolabeling of D-glucose and conjugation of the sugar to the human erythrocyte monosaccharide transporter protein (Shanahan *et al.*, 1985), and photoaffinity labeling of a bacterial sialidase (van der Horst *et al.*, 1990).

Two analogs of NHS-ASA that provide alternative physical characteristics are available. Sulfo-NHS-ASA is a water-soluble version of the cross-linker that contains a negatively charged sulfonate group on its NHS ring. Sulfo-NHS-LC-ASA also has the water-solubility advantage provided by a sulfonate, but possesses a longer cross-bridge made from a 6-aminocaproic acid chain in its internal structure. The longer spacer increases the potential distance between conjugated molecules, thus allowing more flexibility in the experimental design. Both analogs are still iodinatable to provide radiolabeling capability.

3.2. SASD

Sulfosuccinimidyl-2-(*p*-azidosalicylamido)ethyl-1,3′-dithiopropionate (SASD) is a heterobifunctional cross-linker containing a photoreactive group and an amine-reactive NHS ester (Pierce). The NHS ring possesses a negatively charged sulfonate group that lends water solubility to the reagent. The cross-bridge of SASD contains a central disulfide group that provides cleavability after conjugation. Reaction with a disulfide reducing agent such as DTT breaks the disulfide bond and releases the cross-

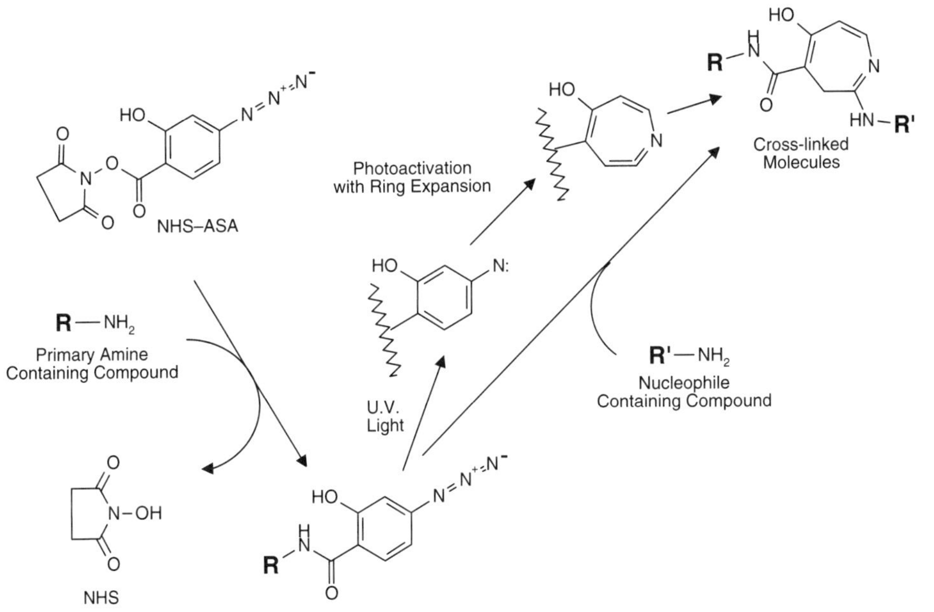

Figure 171 NHS-ASA reacts with amine-containing compounds to form stable amide linkages. Photolyzing with UV light results in ring expansion to a dehydroazepine intermediate, which can react with amines to form covalent bonds.

linked molecules. The photosensitive end of SASD is built from a salicylic acid derivative that contains a ring-activating hydroxyl group. Due to the presence of this group, the cross-linker can be radiolabeled with ^{125}I prior to a conjugation reaction. Iodination occurs *ortho* or *para* to the hydroxyl group on the phenyl ring, next to the aryl azide function (Fig. 172) (Chapter 8, Section 4.5).

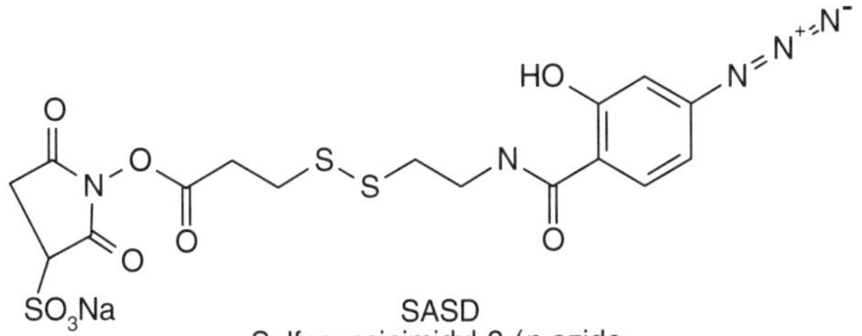

SASD
Sulfosuccinimidyl-2-(*p*-azido-
salicylamido)ethyl-1,3'-dithiopropionate
MW 541.51
18.9 Å

The combination of radiolabeling and cleavability provides the ability to detect the fate of the protein that retains the radiolabel after disulfide reduction. Thus, for investigations involving biomolecular interactions, a purified protein can be labeled with SASD through its amine groups via the NHS ester end of the cross-linker, allowed to interact *in vivo* with unknown target molecules, and photolyzed to effect a cross-link with these unknown substances. Subsequently the complex can be localized in the cell or effectively isolated by following the radiolabel. Alternatively, the conjugate can

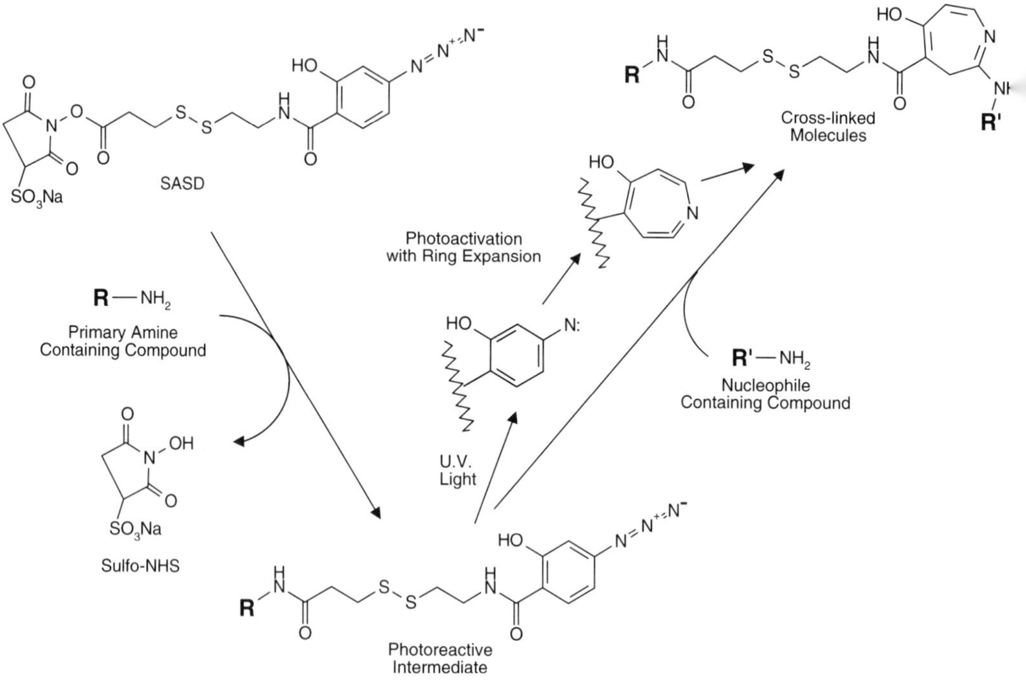

Figure 172   SASD is a photoreactive cross-linker that can be used to modify amine-containing compounds through its NHS ester end and subsequently photolyzed to initiate coupling with nucleophiles (after ring expansion to an intermediate dehydroazepine derivative). The cross-links may be selectively cleaved at the internal disulfide group using DTT.

be cleaved by reduction and the unknown molecule's fate identified through the radio-label (Fig. 173).

Reported applications of SASD involve modification of lipopolysaccharide (LPS) molecules and studying their interaction with albumin and an antibody directed against LPS (Wollenweber and Morrison, 1985), identification of the murine inter-leukin-3 receptor and an *N*-formyl peptide receptor (Sorenson *et al.*, 1986), cross-linking of factor V and Va to iodinated peptides (Chattopadhyay *et al.*, 1992), and a comparison of radiolabeling techniques for the cross-linker (Shephard *et al.*, 1988).

The best radiolabeling technique for SASD is to use the IODO-GEN method (Shep-hard *et al.*, 1988) described in Chapter 8, Section 4.3. The following suggested proto-col for using SASD was based on the method described in the Pierce Catalog.

Protocol

The following operations should be done using standard safety procedures for work-ing with radioactive compounds. All steps involving SASD prior to initiation of the photoreaction should be done protected from light to avoid loss of phenyl azide activity. The radiolabeling procedure should be done quickly to prevent excessive loss of NHS ester activity due to hydrolysis.

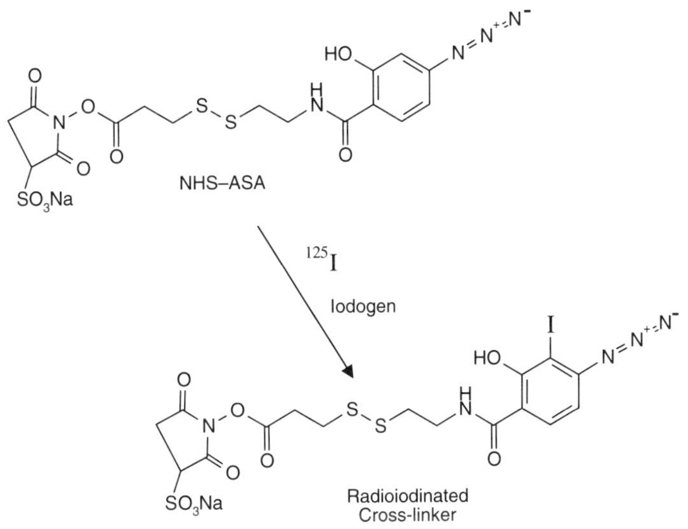

Figure 173 The hydroxyl group on the phenyl azide ring of SASD may be iodinated with ^{125}I to allow radiolabeling studies to be done on photolyzed conjugates.

1. Radiolabel 55 nmol of SASD using IODO-GEN (Pierce Chemical) and 40 μCi Na[125]I for 30 s. Do not use Chloramine-T, since termination of the iodination reaction with this reagent involves addition of a reducing agent that may cleave the disulfide bonds of the cross-linker.

2. Terminate the iodination by removing the SASD solution from the IODO-GEN reagent using a transfer pipette. Be careful not to carry any solid IODO-GEN reagent with the transfer. Since free radioactive iodine still may be present in the solution, it may be necessary to add an iodine scavenger to prevent the possibility of radiolabels being incorporated into the proteins being cross-linked. Suitable scavengers include tyrosine or *p*-hydroxyphenylacetic acid. Adding these compounds in molar excess to the amount of iodine present will prevent any secondary modifications from occurring. Immediately add the radiolabeled SASD solution to the equivalent of 16 nmol of a protein to be modified. The protein should be dissolved previously in a minimum quantity of 0.1 *M* sodium borate, pH 8.4 (conjugation buffer). The more concentrated the protein, the more efficient will be the modification reaction.

3. React for 30 min to create the SASD derivative, coupled through the NHS ester functional groups of the cross-linker onto available amine groups of the protein (forming amide bonds).

4. Purify the modified protein by desalting using a Sephadex G-25 column or the equivalent and performing the chromatography using a buffer of choice. Pool fractions containing protein. The protein should be radiolabeled at this point and also contain photoreactive phenyl azide groups from the SASD modification.

5. Add the SASD-modified protein to a second protein or other molecule to be

conjugated. After mixing, photolyze the solution with long-wave UV light for 10 min at room temperature to effect the conjugation.

3.3. HSAB and Sulfo-HSAB

N-Hydroxysuccinimidyl-4-azidobenzoate (HSAB) is a heterobifunctional reagent containing an amine-reactive NHS ester on one end and a photoreactive phenyl azide group on the other (Pierce). The small cross-bridge, built from a benzoic acid group, provides cross-linking ability at short intermolecular distances. Reaction of one protein via the NHS ester end of the cross-linker provides a stable derivative that can be incubated with a target molecule and then photolyzed to effect the final conjugation (Fig. 174).

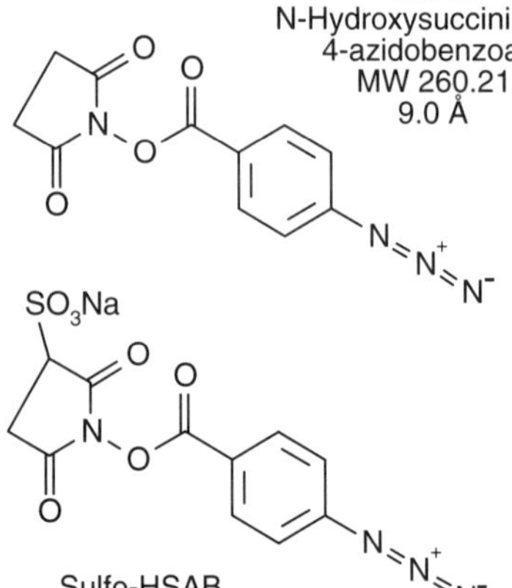

HSAB
N-Hydroxysuccinimidyl-
4-azidobenzoate
MW 260.21
9.0 Å

Sulfo-HSAB
N-Hydroxysulfosuccinimidyl-
4-azidobenzoate
MW 362.25
9.0 Å

Reactions done with HSAB should involve predissolution of the cross-linker in organic solvent prior to addition to a molecule to be modified. DMSO or DMF are suitable solvents to prepare concentrated stock solutions. Protect all solutions from light to avoid loss of photoreactive phenyl azide groups prior to the desired point of photolysis.

Reported applications of HSAB include photoaffinity labeling of peptide hormone binding sites (Galardy et al., 1974), photoaffinity labeling of the insulin receptor with derivatized insulin analog (Yeung et al., 1980), identifying nerve growth factor receptor proteins in sympathetic ganglia membranes (Massague et al., 1981), labeling of the hormone receptor of both α and β subunits of human choriogonadotropin (Ji and Ji, 1981), isolation of in situ cross-linked ligand–receptor complexes (Ballmer-Hofer et al., 1982), and cross-linking vasoactive intestinal polypeptide to its receptors on intact human lymphocytes (Wood and O'Dorisio, 1985).

N-Hydroxysulfosuccinimidyl-4-azidobenzoate (sulfo-HSAB) is a water-soluble analog of HSAB possessing a negatively charged sulfonate group on its NHS ring. This cross-linker may be added directly to aqueous reaction media without prior dissolution in organic solvent. To aid in the addition of small quantities of the reagent, a concentrated solution of sulfo-HSAB may be made in water and then an aliquot added

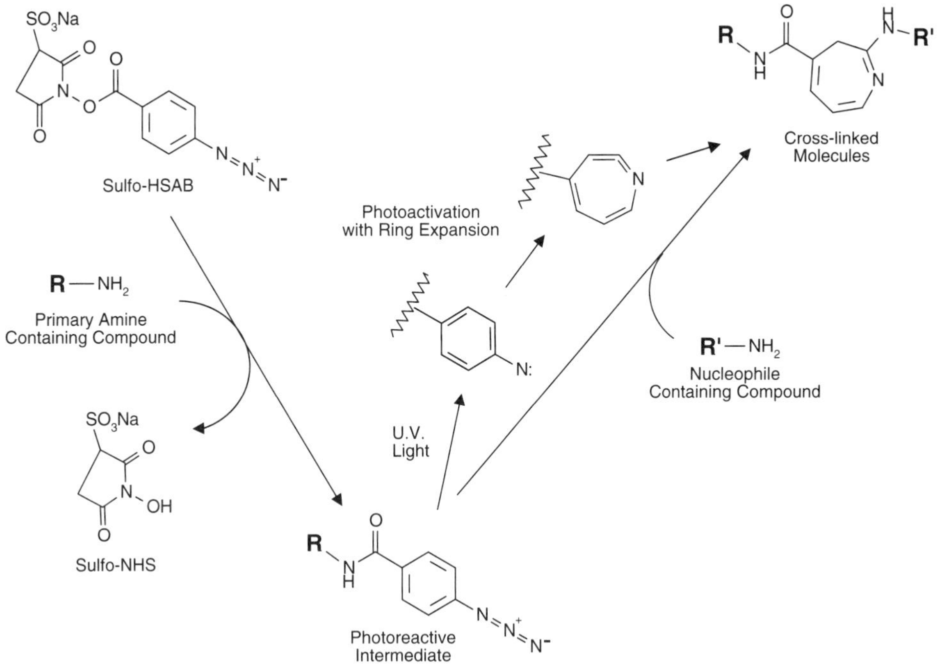

Figure 174   Sulfo-HSAB is a short photoreactive cross-linker that can be used to modify amine-containing molecules through its NHS ester end to form amide linkages. After photolysis, the phenyl azide group can react with amines to create a covalent bond.

to the reaction. Aqueous stock solutions should be dissolved quickly and used immediately to prevent extensive hydrolysis of the NHS ester.

3.4. SANPAH and Sulfo-SANPAH

N-Succinimidyl-6-(4′-azido-2′-nitrophenylamino)hexanoate (SANPAH) is a hetero-bifunctional cross-linking agent containing an NHS ester and a photoreactive phenyl azide group (Pierce). The NHS ester end can react with amine groups in proteins and other molecules, forming stable amide linkages. The photoreactive end is sensitive to long UV light, being selectively activated to a highly reactive nitrene or dehydroazepine intermediate. Either of these photolyzed species can couple to molecules within van der Waals contact, rapidly forming covalent bonds (Fig. 175). The cross-bridge of SANPAH is a noncleavable 6-aminohexanoic acid derivative that provides a long spacer between conjugated molecules. The phenyl azide group also contains a nitro group on the ring that has the effect of increasing the wavelength of optimal photolysis. Exposure to light at a wavelength in the range of 320–350 nm promotes the photo-reaction process. SANPAH is a water-insoluble cross-linker that will permeate membrane structures efficiently. The reagent should be dissolved in DMSO or DMF prior to addition of an aliquot to an aqueous reaction medium.

SANPAH
N-Succinimidyl-6-(4'-azido-
2'-nitrophenylamino)hexanoate
MW 390.95
18.2 Å

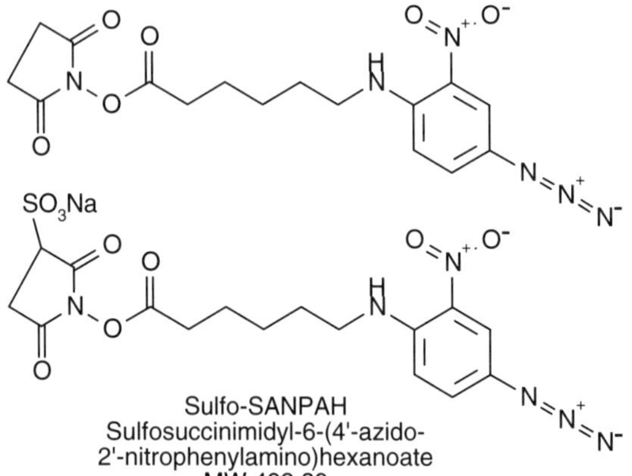

Sulfo-SANPAH
Sulfosuccinimidyl-6-(4'-azido-
2'-nitrophenylamino)hexanoate
MW 492.39
18.2 Å

Reported applications of SANPAH include the cross-linking of ligand–receptor complexes *in situ* (Ballmer-Hofer *et al.*, 1982), preparing photoactivatable glycopeptide derivatives for site-specific labeling of lectins (Baenziger and Fiete, 1982), photo-

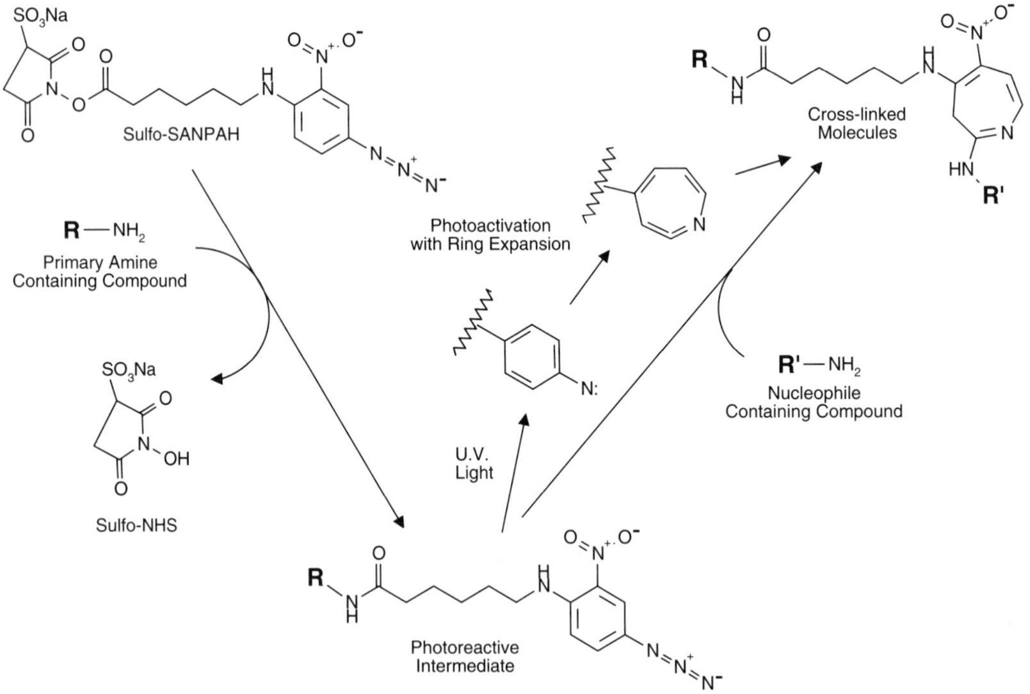

Figure 175 The reaction sequence of cross-linking with sulfo-SANPAH involves first derivatizing an amine-containing molecule using its NHS ester end to create an amide bond. Exposure to UV light then causes ring expansion to the dehydroazepine derivative, which can couple with amines to form the final conjugate.

affinity labeling of the *N*-formyl peptide receptor binding site of intact human poly-morphonuclear leukocytes (Schmitt *et al.*, 1983), and the cross-linking of vasoactive intestinal peptide to receptors on intact human lymphoblasts (Wood and O'Dorisio, 1985).

A water-soluble version of this cross-linker also exists. Sulfosuccinimidyl-6-(4'-azido-2'-nitrophenylamino)hexanoate (sulfo-SANPAH) contains the negatively charged sulfonate group on its NHS ring, lending greater hydrophilicity to the compound.

3.5. ANB-NOS

N-5-Azido-2-nitrobenzoyloxysuccinimide (ANB-NOS) is a photoreactive, heterobi-functional cross-linker containing an amine-reactive NHS ester group (Pierce). Its cross-bridge is formed form a benzoic acid derivative, allowing molecules to be conju-gated at relatively short 7.7-Å distances apart. The phenyl ring of ANB-NOS contains a nitro group that has the effect of shifting the optimal wavelength of activation to longer UV regions. The photoreaction is initiated by exposure to light in the range of 320–350 nm. Without the presence of the nitro group, activation would occur at much lower wavelengths, around 265–275 nm—wavelengths that potentially can damage biological molecules when exposed under high-photon irradiation. ANB-NOS typ-ically is used to label an amine-containing protein or molecule by its NHS ester end. The resultant derivative is allowed to interact with other molecules that potentially can bind specifically to it and photolyzed to effect the final conjugation (Fig. 176).

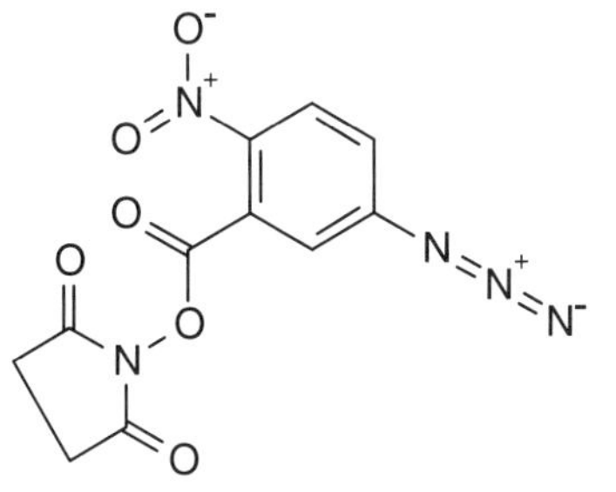

ANB–NOS
N-5-Azido-2-nitrobenzoyloxy-
succinimide
MW 305.21
7.7 Å

Reported applications using this reagent include cross-linking of the aggregation state of cobra venom phospholipase A2 (Lewis *et al.*, 1977) and conjugation of the signal sequence of nascent preprolactin to a polypeptide of the signal recognition particle (Krieg *et al.*, 1986).

3.6. SAND

Sulfosuccinimidyl-2-(*m*-azido-*o*-nitrobenzamido)-ethyl-1,3'-dithiopropionate (SAND) is a heterobifunctional reagent containing an amine-reactive sulfo-NHS ester at one end and a photoreactive phenylazide group on the other (Pierce). The presence of the sulfonate group on the NHS ring lends water solubility to the reagent due to its

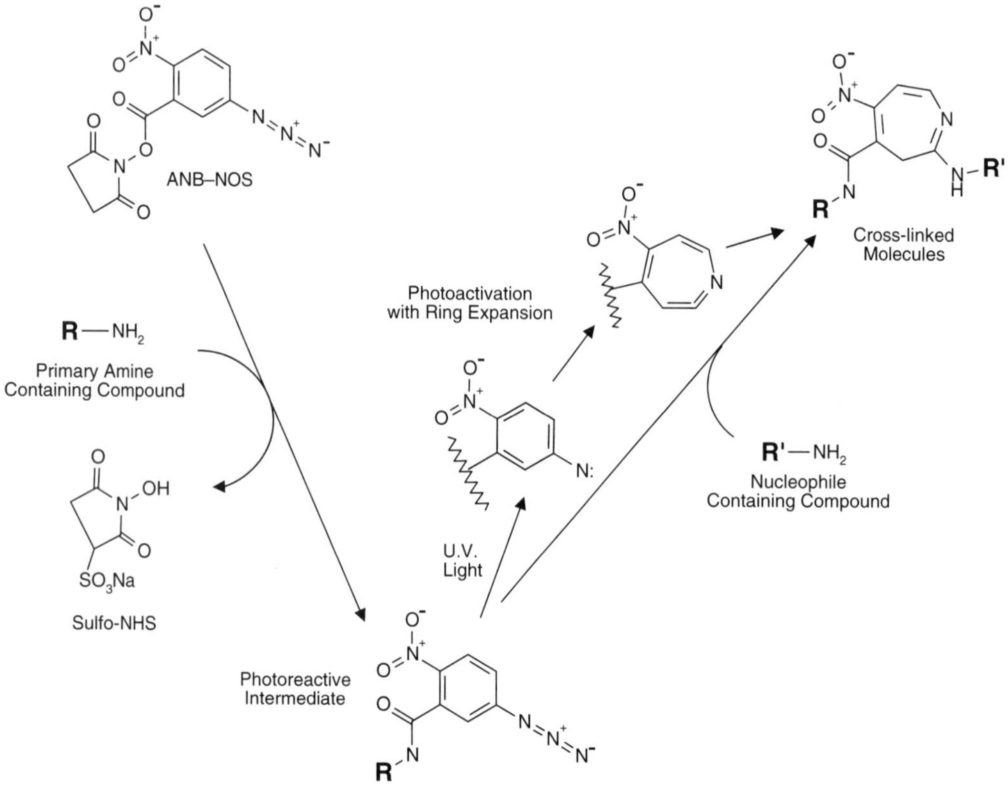

Figure 176 The NHS ester of ANB-NOS reacts with amines to form amide bonds. Subsequent photolyzing of the complex with UV light causes phenyl azide ring expansion and reaction with neighboring amines.

negative charge in aqueous solutions. In addition, the phenylazide group contains a nitro constituent that shifts the optimal range of photoactivation toward higher wavelengths—into the 320 to 350-nm region, thus decreasing the potential of photolytic damage to other sensitive groups that may be present during cross-linking. The extended cross-bridge of SAND (18.5 Å) provides a long spacer to accommodate even relatively distant sites between interacting molecules. The presence of a disulfide bond within the cross-bridge means that the reagent also is cleavable by the use of a disulfide reductant, allowing the potential for disruption of the cross-links after purification of the conjugate.

In use, SAND is first reacted with an amine-containing protein or other molecule—being careful to protect the photoreactive functional group from inadvertent degradation by exposure to excessive room light or sun. The modified intermediate then is allowed to interact with a target molecule. Finally, the photolyzing process is done to effect a nonselective cross-link between the modified molecule and any target molecules within van der Waals distance to the cross-linker (Fig. 177). Its use may be similar to that reported for sulfo-SANPAH, and its cleavability similar to that reported for SADP.

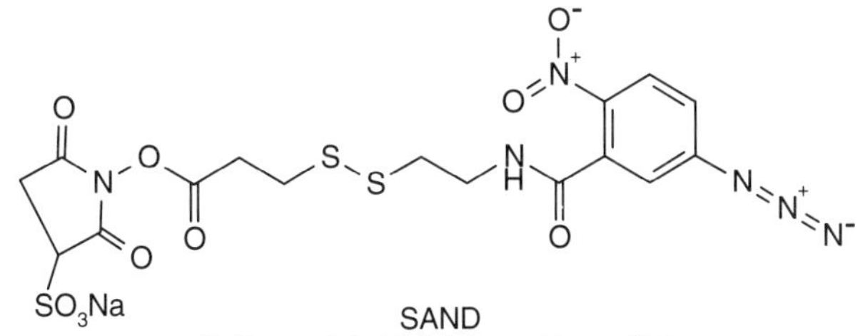

SAND
Sulfosuccinimidyl-2-(*m*-azido-*o*-nitro-
benzamido)-ethyl-1,3'-dithiopropionate
MW 570.52
18.5 Å

3.7. SADP and Sulfo-SADP

N-Succinimidyl-4(4-azidophenyl)1,3'-dithiopropionate (SADP) is a photoreactive heterobifunctional cross-linker that is cleavable by treatment with a disulfide reducing agent (Pierce). The cross-linker contains an amine-reactive NHS ester and a photoactivatable phenylazide group, providing specific, directed coupling at one end and nonselective insertion at the other end.

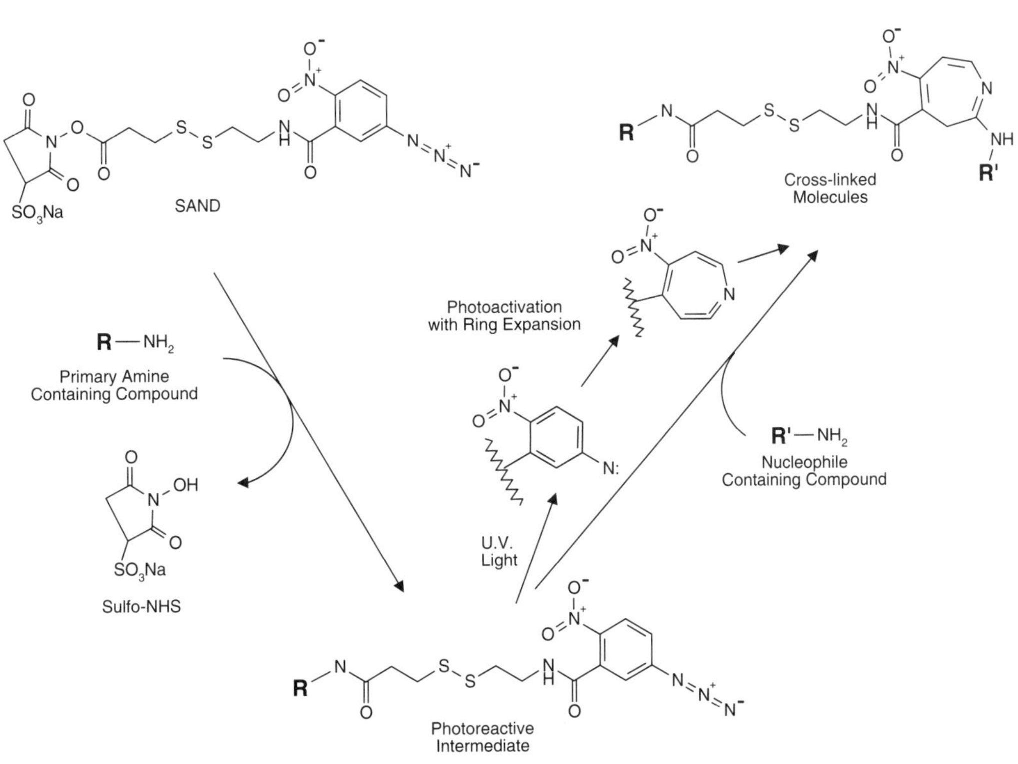

Figure 177   SAND can be used to modify amine-containing molecules, and then photoinitiate cross-linking to another amine-containing molecule via a ring-expansion process. The conjugates may be disrupted by reduction of the cross-bridge disulfide with DTT.

SADP is first used to modify a protein via its amine groups through the reactive NHS ester end of the cross-linker. After allowing for interaction of the modified protein with target molecules, the photoreactive group is used to couple with any molecules within van der Waals distance. The photolysis reaction requires UV exposure in the range of 265–275 nm to effect the final linkage. The presence of the disulfide group in SADP's cross-bridge allows disruption of cross-links with 50 mM DTT after the conjugation reaction is complete (Fig. 178).

SADP
N-Succinimidyl(4-azidophenyl)-
1,3'-dithiopropionate
MW 352.38
13.9 Å

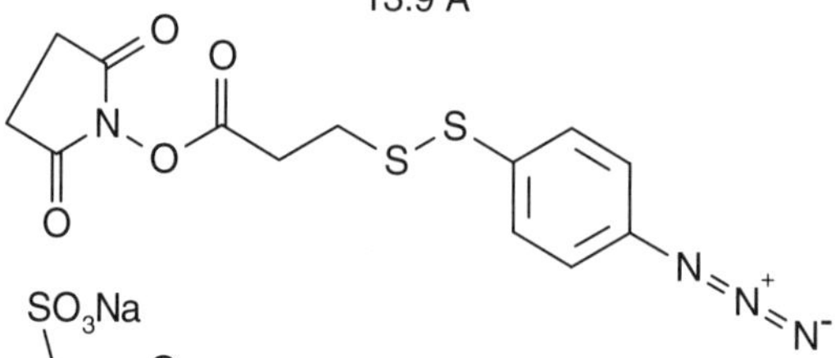

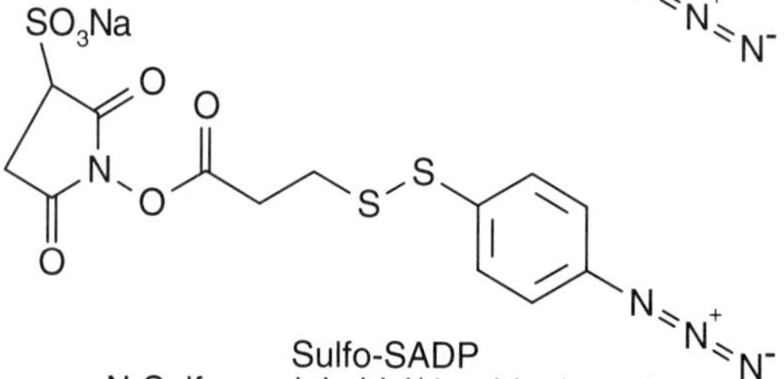

Sulfo-SADP
N-Sulfosuccinimidyl(4-azidophenyl)-
1,3'-dithiopropionate
MW 454.45
13.9 Å

SADP is hydrophobic and should be dissolved in organic solvent prior to addition of a small aliquot to an aqueous reaction. Concentrated stock solutions can be prepared in DMSO or DMF. Final concentration of the organic solvent in a cross-linking reaction should not exceed about 10% to prevent protein precipitation or denaturation.

Reported applications of SADP include the cross-linking of Con A to receptors on human erythrocyte membranes (Vanin and Ji, 1981), site-specific labeling of lectins using modified glycopeptides (Baenziger and Fiete, 1982), conjugation of a mouse cell-surface polypeptide with a Sendai virion envelope on newly infected cells (Zarling *et al.*, 1982), and cross-linking of platelet glycoprotein Ib (Jung and Moroi, 1983).

Sulfo-SADP is a water-soluble analog of SADP that contains a negatively charged sulfonate group on its NHS ring. The reagent may be added directly to aqueous reaction mixtures without prior dissolution in an organic solvent. Concentrated stock solutions prepared in water should be used immediately to prevent extensive hydrolysis of the sulfo-NHS ester group.

3.8. Sulfo-SAPB

Sulfosuccinimidyl 4-(*p*-azidophenyl)butyrate (sulfo-SAPB) is a photoreactive heterobifunctional reagent containing an amine-reactive sulfo-NHS ester at one end (Pierce).

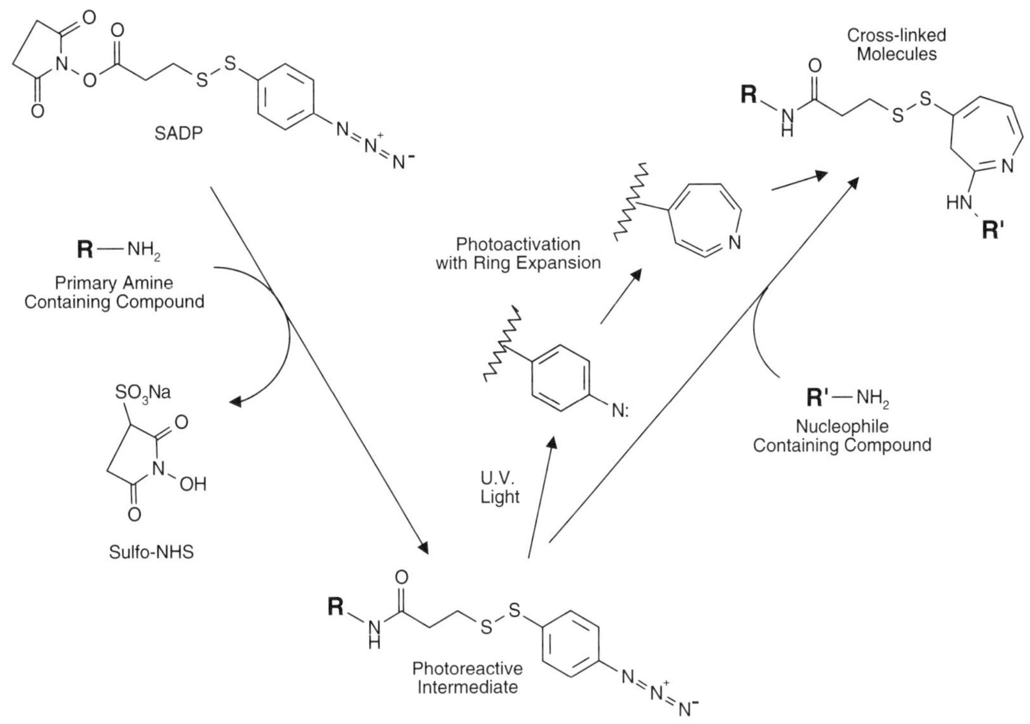

Figure 178 SADP reacts with amines via its NHS ester end to produce amide bonds. The modified molecule then may be photolyzed to create a nucleophile-reactive dehydroazepine intermediate able to covalently couple with amine-containing compounds.

The cross-linker is similar in design to sulfo-HSAB (Section 3.3), but containing a 3-carbon-longer cross-bridge. The sulfo-NHS ester provides water solubility to the reagent due to the negative charge of the sulfonate group. The phenylazide end can be photolyzed by exposure to UV light in the wavelength range of 265–275 nm (Fig. 179). Although there are no reported applications for the cross-linker, its reactivity and use is similar to that of sulfo-HSAB. The commercial availability of the reagent provides additional options for spacer length to study the interactions between two proteins or other molecules.

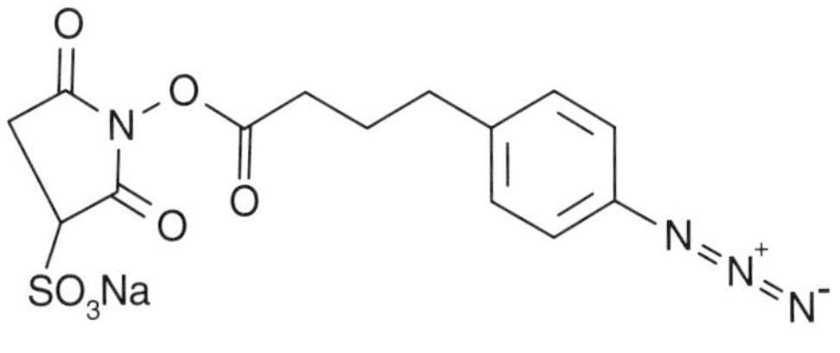

Sulfo-SAPB
Sulfosuccinimidyl-
4-(*p*-azidophenyl)butyrate
MW 404.32
12.8 Å

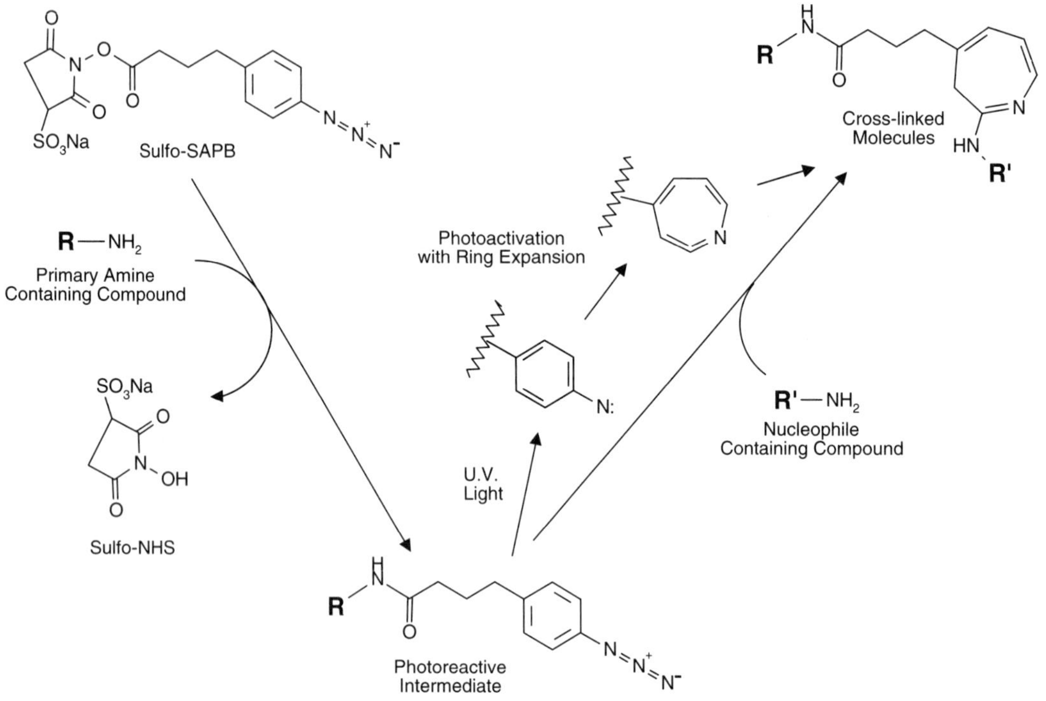

Figure 179   The reaction of sulfo-SAPB with an amine group is done first to form amide bond derivatives through its NHS ester end. Subsequent exposure to UV light causes the phenyl azide group to ring expand to a highly reactive dehydroazepine, which can couple to nucleophiles, such as amines.

3.9. SAED

Sulfosuccinimidyl   2-(7-azido-4-methylcoumarin-3-acetamide)ethyl-1,3′-dithiopro-pionate (SAED) is a photoreactive heterobifunctional cross-linking agent that also contains a fluorescent group (Pierce). The sulfo-NHS ester end of the reagent reacts with primary amines in proteins and other molecules to form stable amide linkages. The photoreactive end is an AMCA derivative (Chapter 8, Section 1.3) containing a light-sensitive azide group on the aromatic ring. Photolyzing with light in the range of long UV to within the visible spectrum will result in nonselective bond formation with nucleophiles and active carbon—hydrogen bonds within van der Waals distance (Fig. 180).

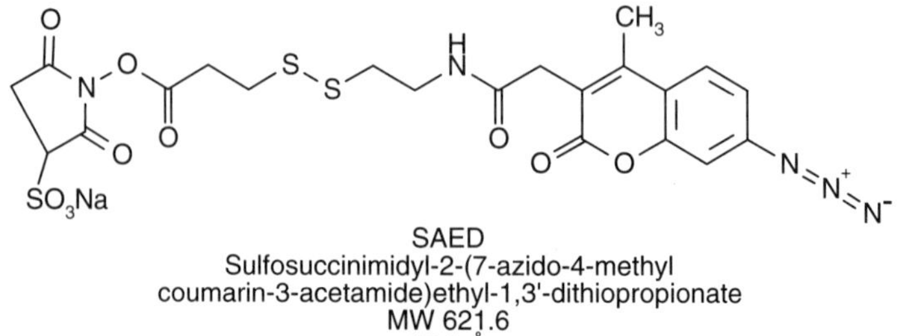

SAED
Sulfosuccinimidyl-2-(7-azido-4-methyl
coumarin-3-acetamide)ethyl-1,3′-dithiopropionate
MW 621.6
23.6 Å

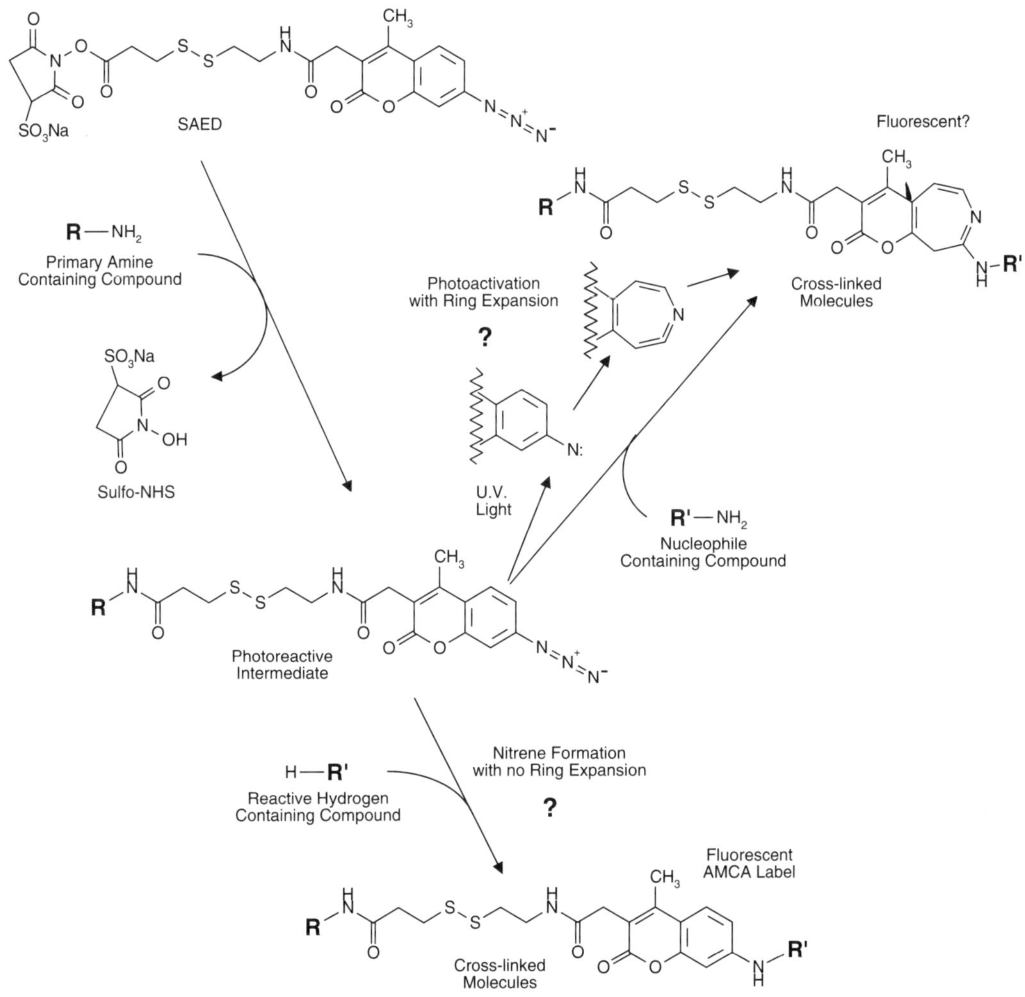

Figure 180 SAED can be used to modify amine-containing molecules through its NHS ester end. Subsequent exposure to UV light causes bond formation with nearby nucleophilic groups, such as amines. The photosensitive phenyl azide group is created on the aromatic ring of an AMCA fluorophore. Before photolysis, the azide group makes the cross-linker nonfluorescent. After photolyzing, however, the azide group is either lost by N_2 generation or couples to a target molecule. Either way, the AMCA portion becomes fluorescent to allow tracking of the conjugate. The photoreaction may occur through ring expansion to an intermediate dehydroazepine or might happen through nitrene formation.

SAED is a relatively large cross-linker containing a long (22.5 Å) cross-bridge. The central portion of its cross-bridge contains a disulfide bond, making the reagent susceptible to cleavage with disulfide reducing agents. The aromatic character of the coumarin derivative creates a maximal UV absorptivity at 327 nm with an extinction coefficient of 18,200 $M^{-1}cm^{-1}$ for a 1 mg/ml solution in acetonitrile:water (15:2 v/v). The extinction coefficient at 298 nm for the same concentration of SAED in the identical solvent is 13,625 $M^{-1}cm^{-1}$.

SAED contains a sulfo-NHS ester with a negatively charged sulfonate group on its ring. The presence of this negative charge does lend some expected water solubility to the reagent (3 mg/ml at room temperature), but because of the reagent's large size it does not provide the same water solubility benefits as with other smaller cross-linkers. It is also sparingly soluble in acetonitrile (2.5 mg/ml), but only if a small amount of water is present (15:2 acetonitrile:water, v/v). However, SAED is very soluble in DMSO and DMF (about 50 mg/ml). Stock solutions may be prepared in dry DMSO or DMF while maintaining fairly good stability of the reagent's functional groups. The addition of a small quantity of these stock solutions to an aqueous reaction medium facilitates the amine-modification process via the sulfo-NHS ester end of the cross-linker. The final concentration of organic solvent in the aqueous reaction should not exceed 10%.

The coumarin derivative of SAED is not fluorescent until the photolysis reaction is initiated. A protein modified with SAED will fluoresce after activation with UV light whether or not the photoreactive end actually couples to the intended target, since breakdown of the azide group on the ring is all that is required to initiate fluorescence. Thus, the level of SAED incorporation into a macromolecule may be assessed by the resultant coumarin fluorescence after separation of the derivative from excess reagent. Native AMCA has an excitation optimum at 345–350 nm and an emission wavelength range of 440–460 nm. The excitation and emission properties of SAED may change somewhat upon its attachment to macromolecules due to fluorescent quenching; however, the coumarin tag will still remain fluorescently active even after cross-linking.

Since the cross-linker is cleavable, SAED provides a means of fluorescent transfer of the coumarin tag to a second molecule, which interacts with the initially modified protein (Fig. 181). For example, soybean trypsin inhibitor (STI) was labeled with SAED and then allowed to interact with trypsin. After photoreactive cross-linking of the two interacting molecules, the complex was reduced with DTT, breaking the conjugate and transferring the fluorescent tag to trypsin near the STI binding site (Thevenin et al., 1991).

In another study, SAED was used to investigate the role of the foot protein moiety of the triad and its relationship to Ca^{2+} release from sarcoplasmic reticulum (Kang et al., 1991). Modification of poly-L-lysine (a Ca^{2+} release inducer) and neomycin with the cross-linker was done followed by subsequent incubation with the foot protein and photoreactive conjugation. Cleavage of the cross-links with a disulfide reductant allowed transfer of the fluorescent tag to the foot protein in areas near the binding sites. Fluorescent monitoring of conformational changes within the protein upon varying the Ca^{2+} concentration was then possible.

Since the photoreactive cross-linking step with SAED occurs rapidly on exposure to even bright light within the visible spectrum, UV lamps are not required. However, special care should be taken to protect the reagent from exposure to light before the photolysis reaction is initiated. The solid should be stored in amber bottles and any stock solutions prepared in organic solvent should be wrapped to exclude light totally. In addition, the initial derivatization of an amine-containing molecule should be done in the dark in wrapped containers.

3.10. Sulfo-SAMCA

Sulfosuccinimidyl 7-azido-4-methylcoumain-3-acetate (sulfo-SAMCA) is a heterobifunctional reagent similar in design to SAED (this chapter, Section 3.9) (Pierce). One

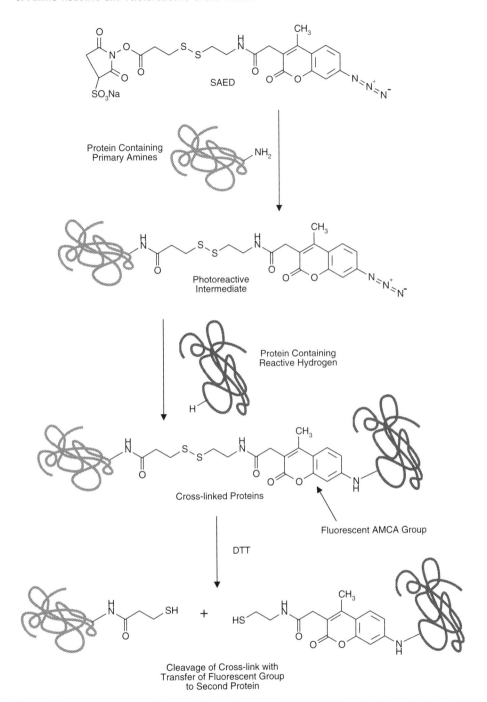

Figure 181 SAED may be used to transfer the fluorescent AMCA label from the first molecule modified with the cross-linker to the second molecule cross-linked with it by reduction of its internal disulfide bond with DTT. Thus, unknown target molecules may be fluorescently tagged to follow them *in vivo*.

end of the cross-linker contains an amine-reactive sulfo-NHS ester, while the other end is an AMCA derivative (Chapter 2, Section 1.3) that contains a photosensitive phenylazide group. Unlike SAED, however, sulfo-SAMCA contains a short noncleavable cross-bridge (12.8 Å) where the active ester functional group is constructed directly off the carboxylate of AMCA acid without any other intervening spacer groups. Conjugated molecules will retain the fluorescent label, thus providing detectability to the complexes formed (Fig. 182). However, since cross-links formed with this reagent are not cleavable, sulfo-SAMCA cannot function as a fluorescent transfer agent in the fashion of SAED.

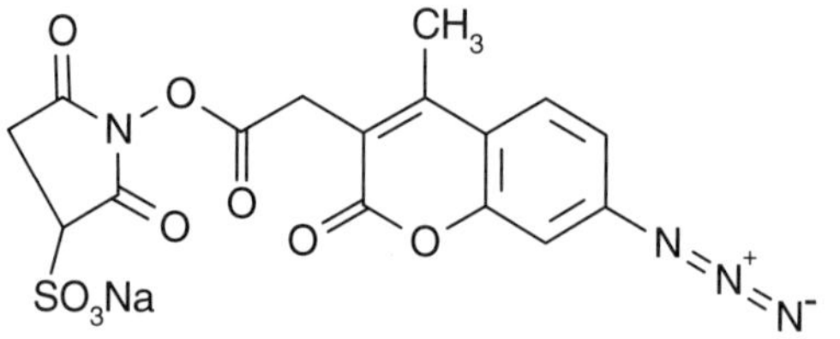

Sulfo-SAMCA
Sulfosuccinimidyl-7-azido-
4-methylcoumarin-3-acetate
MW 458.34
12.8 Å

3.11. p-Nitrophenyl Diazopyruvate

Diazopyruvates are a relatively new class of photoreactive reagents that can be used in heterobifunctional cross-linking experiments. The p-nitrophenyl ester derivative of diazopyruvate provides amine-reactive, acylating potential, while the photosensitive group can be activated with UV light to generate reactive aldehydes. More specifically, the diazo functional group can be photolyzed by exposure to irradiation at 300 nm, forming a highly reactive carbene that can undergo a Wolff rearrangement that produces a ketene amide intermediate. In the presence of a nucleophilic species on a target molecule, the ketene can undergo an acylation reaction to form a stable malonic acid derivative. The photolyzed product thus can couple to hydrazide- or amine-containing targets to form covalent linkages (Fig. 183).

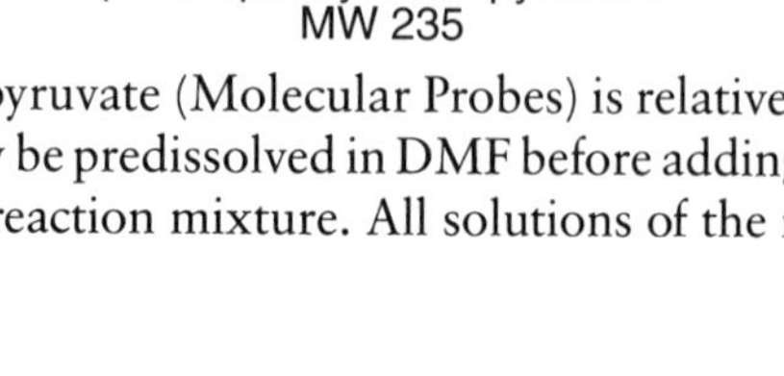

pNPDP
p-Nitrophenyl diazopyruvate
MW 235

p-Nitrophenyl diazopyruvate (Molecular Probes) is relatively insoluble in water or aqueous buffers, but may be predissolved in DMF before adding an aliquot of the stock solution to an aqueous reaction mixture. All solutions of the reagent should be care-

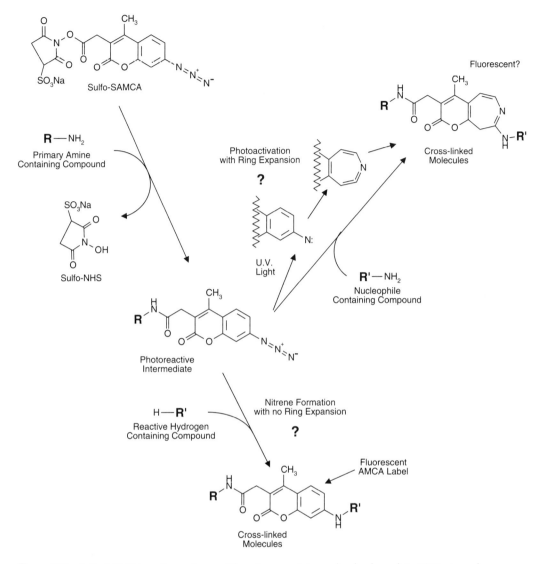

Figure 182 Sulfo-SAMCA can be used to modify amine-containing molecules through its NHS ester end. Subsequent exposure to UV light causes bond formation with nearby nucleophilic groups, such as amines. The photosensitive phenyl azide group is created on the aromatic ring of an AMCA fluorophore. Before photolysis, the azide group makes the cross-linker nonfluorescent. After photolyzing, however, the azide group is either lost by N_2 generation or couples to a target molecule. Either way, the AMCA portion becomes fluorescent to allow tracking of the conjugate. The photoreaction may occur through ring expansion to an intermediate dehydroazepine or might happen through nitrene formation.

fully protected from light to prevent premature photolysis. *p*-Nitrophenyl diazopyruvate has an absorbance maximum at 390 nm with a molar extinction coefficient of about 19,000 $M^{-1}cm^{-1}$ in methanol.

p-Nitrophenyl diazopyruvate has been used in the photoreactive cross-linking of

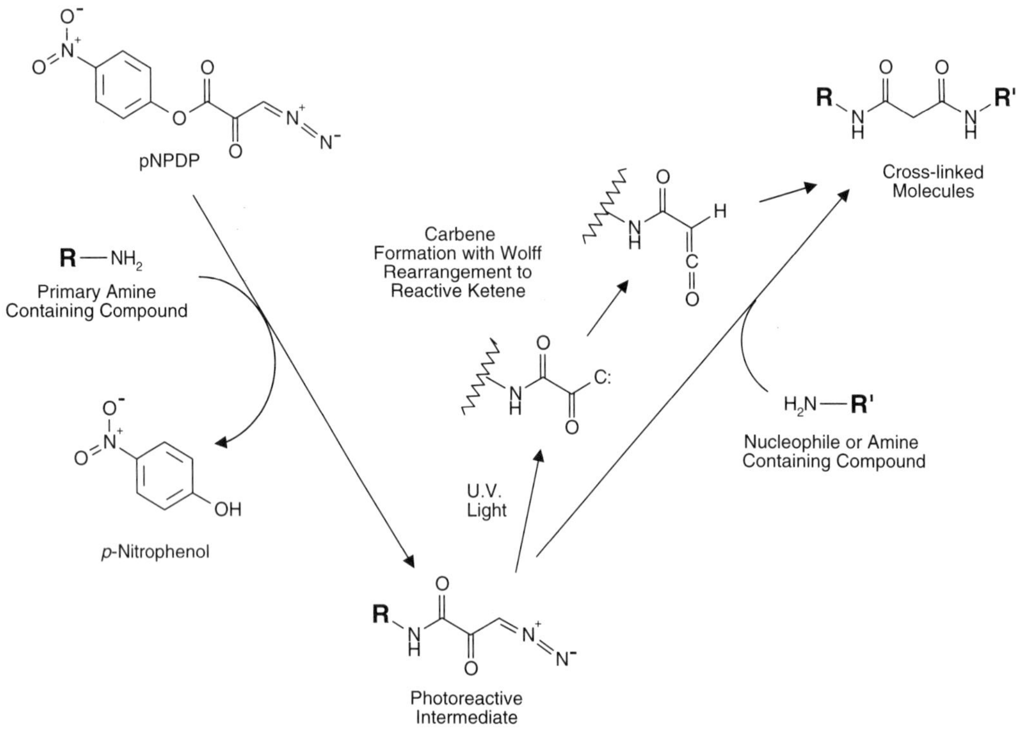

Figure 183 pNPDP reacts with amine-containing compounds by its *p*-nitrophenyl ester group to form amide bonds. After photolyzing the diazo derivative with UV light, a Wolff rearrangement occurs to a highly reactive ketene intermediate. This group can couple to nucleophiles such as amines.

calmodulin with adenylate cyclase from bovine brain (Harrison *et al.*, 1989) and to cross-link aldolase (Goodfellow *et al.*, 1989).

3.12. PNP-DTP

p-Nitrophenyl-2-diazo-3,3,3-trifluoropropionate (PNP-DTP) is a unique photoreactive heterobifunctional cross-linker that contains an amine-reactive group on one end and a photosensitive diazo group on the other (Chowdhry *et al.*, 1976) (Pierce) (Fig. 184). *p*-Nitrophenyl esters react in a manner similar to that of NHS esters (Chapter 4, Section 1, and Chapter 2, Section 1.4) with *p*-nitrophenol as the leaving group upon reaction with a nucleophile. Amine-containing target molecules such as proteins can be modified with this reagent to form amide bond derivatives possessing photoactivatable functional groups. The reagent is small enough to probe deep within the active centers of receptor molecules and other sites of biomolecular interactions.

PNP-DTP has been used to photoaffinity-label the thyroid hormone nuclear receptors in intact cells by preparing a derivative of 3,5,3′-triiodo-L-thyronine with the cross linker (Pascual *et al.*, 1982; Casanova *et al.*, 1984). Effective photoreactive conjugation was found to occur after irradiation with UV light at 254 or 310 nm.

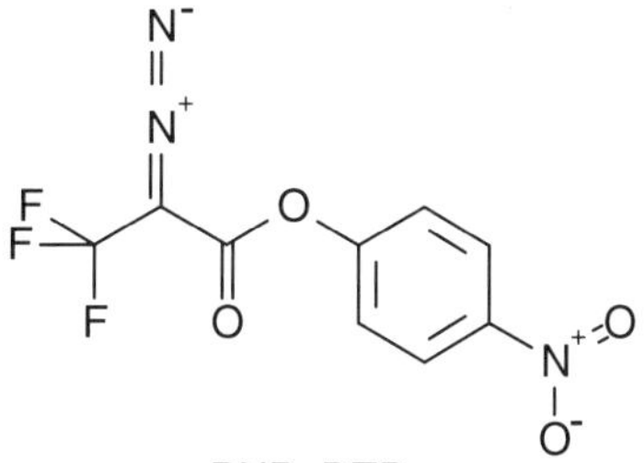

PNP–DTP
p-Nitrophenyl-2-diazo-
3,3,3-trifluoropropionate
MW 276.15

4. Sulfhydryl-Reactive and Photoreactive Cross-linkers

The benefits of nonselective photoreactive cross-linking can be merged with the directed coupling ability of sulfhydryl-reactive functional groups to create heterobifunctional reagents possessing greater utility than the standard amine- and photoreactive agents discussed previously. Having a sulfhydryl-reactive group on one end of the

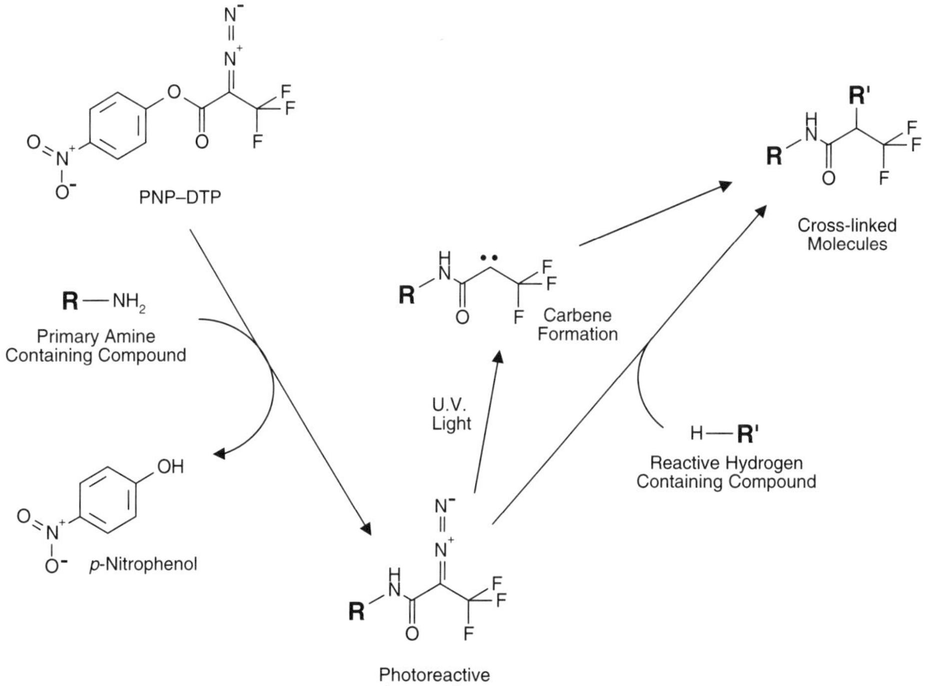

PNP–DTP

R—NH₂
Primary Amine
Containing Compound

OH
p-Nitrophenol

Carbene
Formation

U.V.
Light

H—R'
Reactive Hydrogen
Containing Compound

Cross-linked
Molecules

Photoreactive
Intermediate

Figure 184 PNP-DTP can modify amine-containing molecules through its p-nitrophenyl ester group to form amide bonds. Exposure of its photosensitive diazo group with UV light generates a highly reactive carbene that can insert into active C—H or N—H bonds.

cross-linker allows the initial conjugation to take place at more limited sites on proteins and other molecules before irradiation to effect the final photosensitive reaction.

The following reagents contain a variety of sulfhydryl-reactive groups, including iodoacetyl derivatives, maleimide compounds, and pyridyl disulfide chemistries. The iodoacetyl and maleimide functions form permanent thioether bonds with target molecules containing free sulfhydryls. The pyridyl disulfide derivative reacts with —SH groups to form reversible disulfide linkages, which can be cleaved with disulfide reducing agents like DTT.

The photoreactive end of the following cross-linkers also varies from the traditional aryl azide group to the newer benzophenone and fluorinated aryl azide derivatives. The fluorinated phenyl azide functional groups photolyze to true nitrenes without the ring expansion side-reaction characteristic of aryl azides. The result is that fluorinated aryl azides more effectively insert into active carbon—hydrogen bonds, rather than potentially undergoing nucleophilic reactions like phenyl azides. Benzophenone groups generally have higher degrees of bond formation with the intended target molecule compared to the yields obtained using traditional phenyl azides, due to their ability to be repeatedly photolyzed without breakdown of the precursor to an inactive form.

The number of commercially available cross-linkers for sulfhydryl- and photoreactive conjugations provide enough variety to design successful experiments in photolabeling active centers and studying macromolecular interactions.

4.1. ASIB

1-(*p*-Azidosalicylamido)-4-(iodoacetamido)butane (ASIB) is a heterobifunctional cross-linker containing a sulfhydryl-reactive iodoacetyl group on one end and a photosensitive phenyl azide group on the other (Pierce). The phenyl azide ring is substituted with a ring-activating hydroxyl group that provides the ability to radioiodinate the compound before the conjugation reaction is performed. Since both the iodoacetyl and the phenyl azide functional groups are relatively stable in aqueous solutions, the steps involved in iodination and cross-linking do not detrimentally affect the subsequent reactivity of the reagent. All operations should be done protected from light, however, to prevent premature photolysis before the desired cross-linking reaction is initiated. The cross-bridge of the reagent provides a long 18.8-Å spacer between cross-linked molecules.

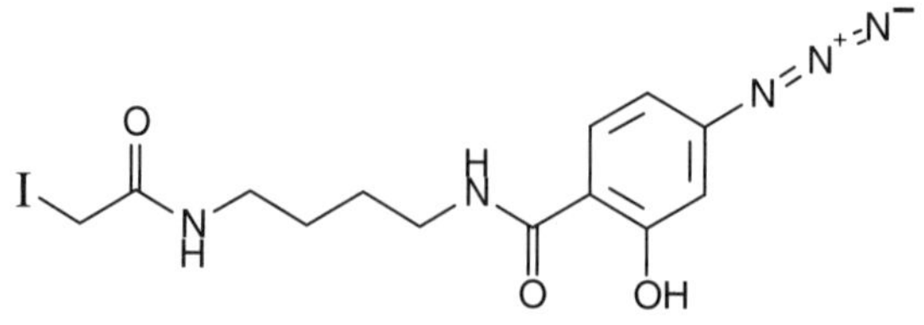

ASIB
1-(*p*-Azidosalicylamido)-
4-(iodoacetamido)butane
MW 417.21
18.8 Å

The reaction of ASIB with sulfhydryl-containing molecules can be done at mildly alkaline pH with excellent specificity. Higher pH conditions may cause cross-reactivity with amines. Photolyzing with UV light may result in immediate reaction of the nitrene intermediate with a target molecule within Van der Waals distance, or may result in ring expansion to the nucleophile-reactive dehydroazepine. The ring-expanded product is primarily reactive toward amine groups (Fig. 185).

4.2. APDP

N-[4-(p-Azidosalicylamido)butyl]-3'-(2'-pyridyldithio) propionamide (APDP) is a radioiodinatable, heterobifunctional cross-linking agent that contains a sulfhydryl-reactive pyridyl disulfide group on one end and a photosensitive phenyl azide on the other end (Pierce). Radioiodinatable cross-linkers eliminate the need to radiolabel one of the reacting proteins, thus avoiding potential activity losses due to modification of important residues (Chapter 2, Section 4.5). They also allow radiolabeling of unknown target molecules that interact with the initially modified protein. APDP reacts with sulfhydryl-containing proteins and other molecules to form a reversible disulfide bond. If the cross-linker is radiolabeled prior to conjugation, cleavage of the disulfide group with DTT after cross-linking effectively transfers the iodinated portion to the

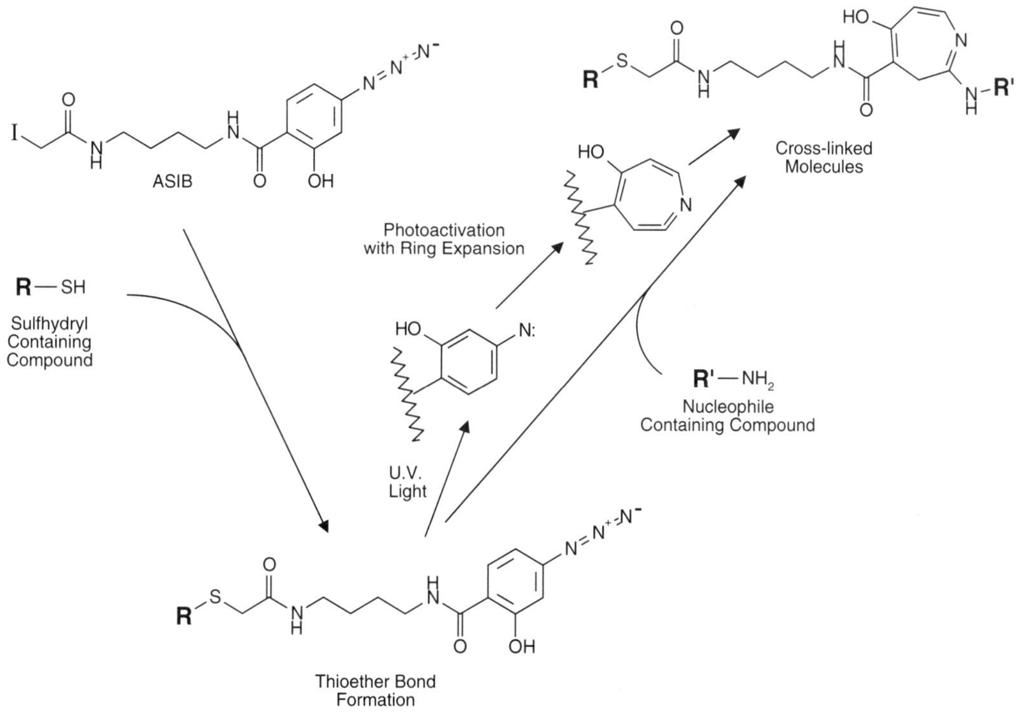

Figure 185 ASIB can react with sulfhydryl-containing molecules through its iodoacetate group to form thioether linkages. Subsequent exposure to UV light causes a ring-expansion process to occur, creating a highly reactive dehydroazepine intermediate that can couple to amine-containing molecules.

secondary, photocoupled protein. This radiolabel transfer process allows tracking of a specific receptor or other interacting species after conjugation with its complementary ligand (Fig. 186).

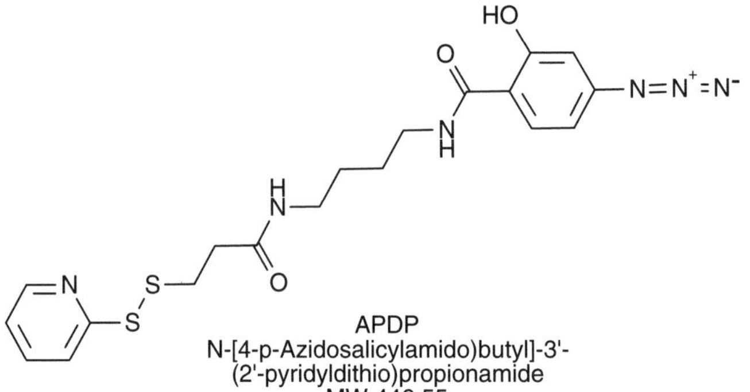

APDP
N-[4-p-Azidosalicylamido)butyl]-3'-
(2'-pyridyldithio)propionamide
MW 446.55

The reactions of APDP are similar to that of the reported compound N-(4-azidophenyl)thiophthalimide, a nonradioiodinatable cross-linker (Moreland *et al.*, 1982). Both the phenyl azide group and the pyridyl disulfide portion are stable in

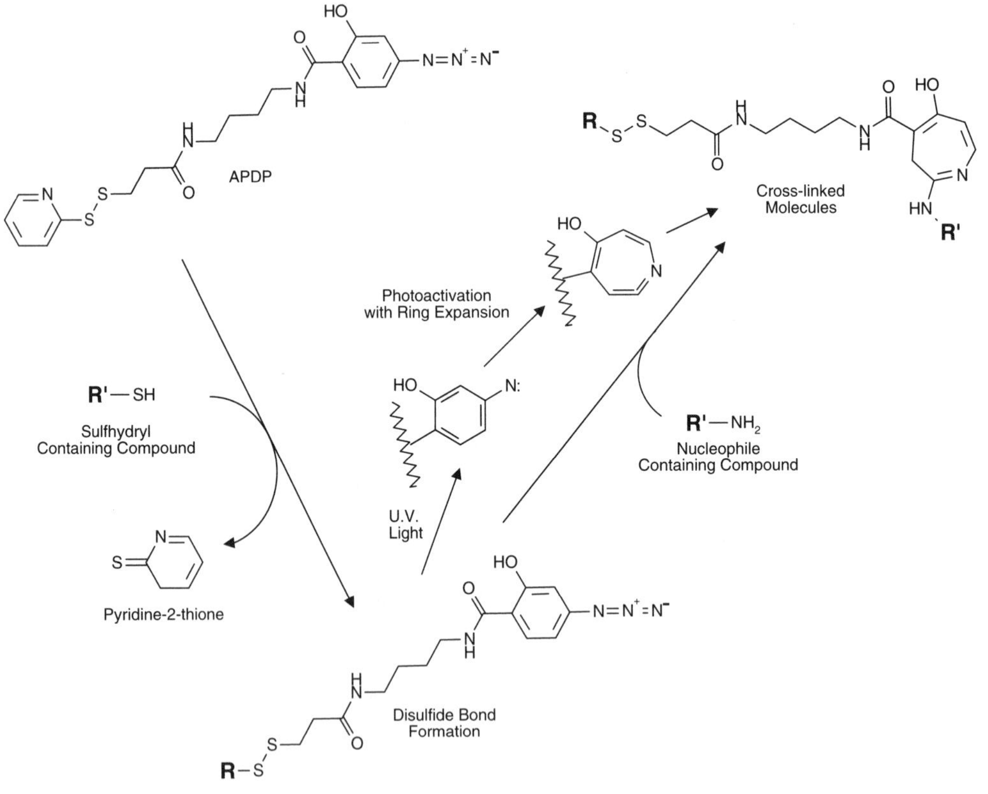

Figure 186 APDP can modify sulfhydryl-containing compounds through its pyridyl disulfide group to form disulfide bonds. Its phenyl azide end then can be photolyzed with UV light to couple with nucleophiles via a ring expansion process. The disulfide group of the cross-link can be selectively cleaved using DTT.

aqueous environments prior to the cross-linking reaction. The initial modification with a sulfhydryl-containing protein should be done protected from light to preserve the activity of the photosensitive functional group. Avoid, also, in the reaction medium disulfide reducing agents that can react with the pyridyl disulfide group as well as inactivate the phenyl azide portion.

The cross-bridge of APDP provides an extremely long, 21.02-Å spacer that is able to reach distant points between two interacting molecules. Cleavage of the cross-link with a disulfide reducing agent regenerates the original sulfhydryl-modified protein without leaving any other chemical groups behind. The remainder of the cross-linker stays attached to the second, photolabeled protein.

APDP is soluble in DMSO and DMF, but almost insoluble in acetone or water. Stock solutions may be prepared in DMSO or DMF and a small aliquot added to an aqueous reaction mixture. Do not exceed 10% organic solvent in the buffered reaction. Both functional groups of APDP will react in a variety of salt conditions and pH values. For reaction with a sulfhydryl-containing protein, a buffer at physiological pH containing a metal chelating agent to protect the free sulfhydryl groups from oxidation is recommended (i.e., 0.01–0.1 M sodium phosphate, 0.15 M NaCl, pH 7.2, containing 10 mM EDTA).

Iodination of the cross-linker may be done according to the procedures discussed in Chapter 8, Section 4, or similar to that described for SASD (this chapter, Section 3.2).

4.3. Benzophenone-4-iodoacetamide

A photoreactive group consisting of a benzophenone residue photolyzes upon exposure to UV light to give a highly reactive triplet-state ketone intermediate (Walling and Gibian, 1965). Similar to the reactive nitrene of photolyzed phenyl azides, the energized electron of an activated benzophenone can insert in active hydrogen—carbon bonds and other reactive groups to give covalent linkages with target molecules. Unlike phenyl azides, however, the decomposition or decay of the photoactivated species does not yield an inactive compound. Instead, benzophenones that have become deactivated without forming a covalent bond can be once again photolyzed to an active state. The results of this multiple-activation characteristic are more than one chance to form a cross-link with the intended target and much higher yields of photo-cross-linking.

The heterobifunctional cross-linker benzophenone-4-iodoacetamide is a photoreactive reagent containing a sulfhydryl-reactive iodoacetyl derivative at one end and a benzophenone group on the other end (Hall and Yalpani, 1980; Tao et al., 1984; Lu and Wong, 1989) (Molecular Probes). The iodoacetyl group has reactivity similar to that of the same group on the heterobifunctional reagent SIAB (this chapter, Section 1.5). Under alkaline pH conditions (pH 8–9) the iodoacetyl reaction is highly specific for sulfhydryl residues in proteins and other molecules, forming stable thioether linkages. The initial modification reaction of sulfhydryl-containing compounds should be done protected from light to avoid premature photolysis of the benzophenone functionality. After purification of the modified protein from excess reagent (by dialysis or gel filtration) mix the modified protein with a second target molecule, allow the interaction to take place, and photolyze with UV light to effect the final cross-link (Fig. 187). Since repeated photolysis of the benzophenone species is possible, the yield of

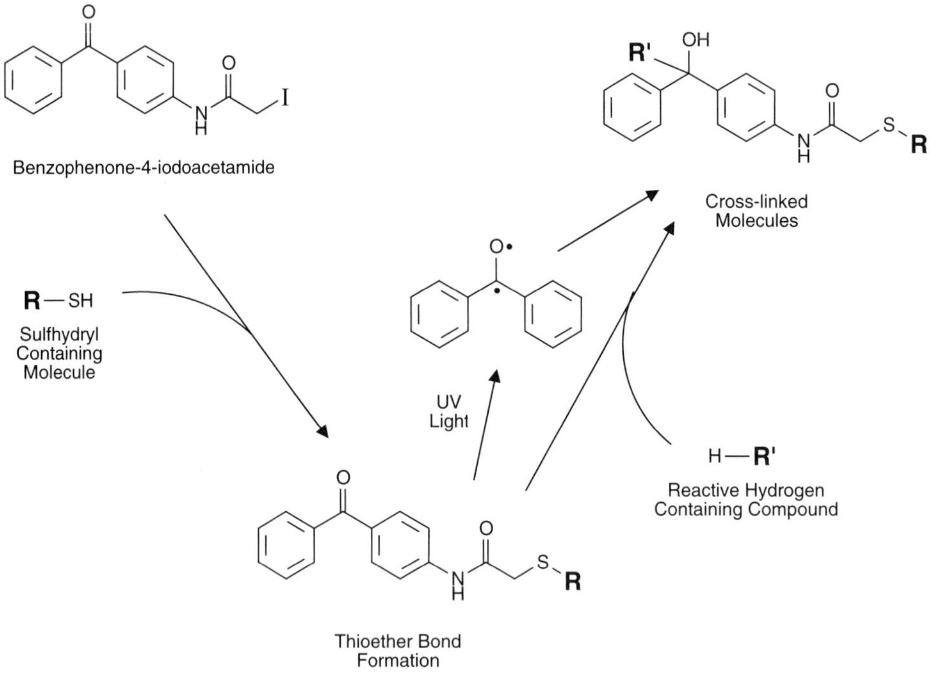

Benzophenone-4-iodoacetamide

R—SH
Sulfhydryl
Containing
Molecule

UV
Light

Cross-linked
Molecules

H—R'
Reactive Hydrogen
Containing Compound

Thioether Bond
Formation

Figure 187 Benzophenone-4-iodoacetamide reacts with sulfhydryl-containing compounds to give thioether linkages. Subsequent photolyzing of the benzophenone residue gives a highly reactive triplet-state ketone intermediate. The energized electron can insert in active C—H or N—H bonds to give covalent cross-links.

such conjugation reactions can be significantly higher than that when using other photoreactive groups. One report indicated that cross-linking with chymotrypsin approached 100% efficiency (Campbell and Gioannini, 1979).

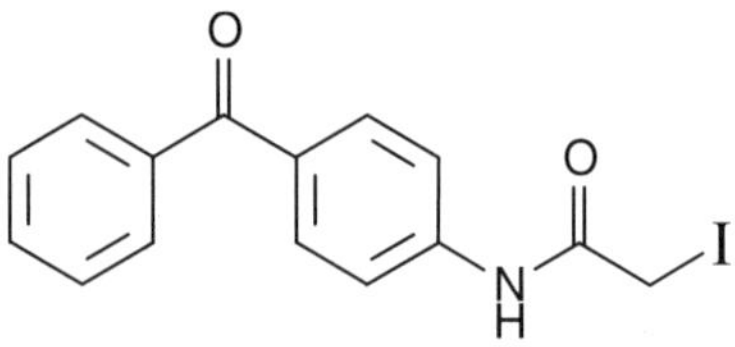

Benzophenone-4-iodoacetamide
MW 365

Benzophenone-4-iodoacetamide is water-insoluble and should be predissolved in DMF or another organic solvent prior to adding an aliquot to an aqueous reaction mixture. Stock solutions may be prepared and stored successfully if protected from light.

4.4. Benzophenone-4-maleimide

Benzophenone-4-maleimide is a heterobifunctional photoreactive cross-linker that has sulfhydryl reactivity similar to benzophenone-4-iodoacetamide discussed in the

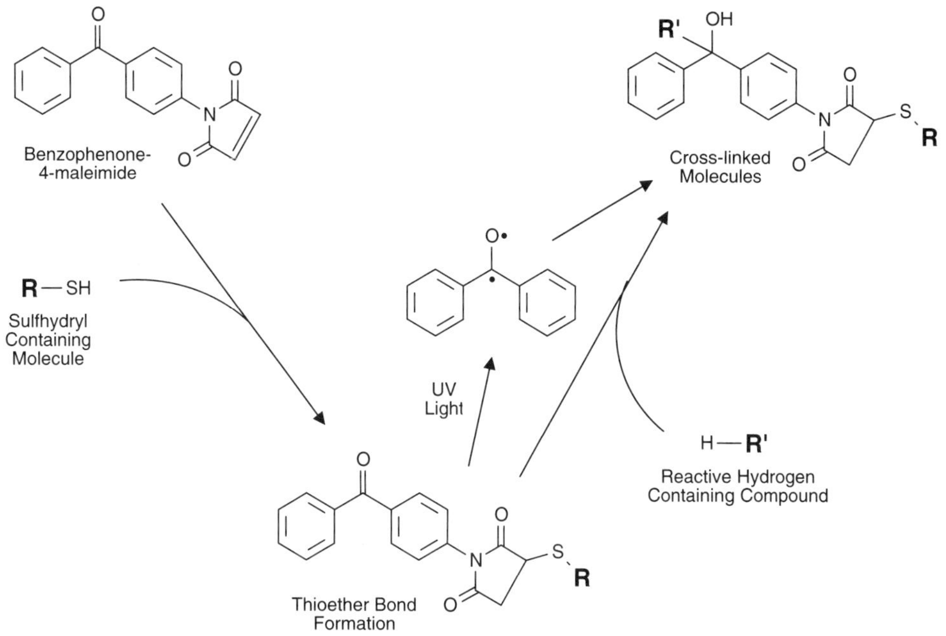

Figure 188 Benzophenone-4-maleimide can couple to thiol-containing molecules to form stable thioether bonds. Exposure of the benzophenone group to UV light causes transition to a triplet-state ketone of high reactivity for insertion into C—H or N—H bonds.

previous section (Molecular Probes). In this case, the sulfhydryl-reactive portion is provided by the presence of a maleimide functional group, which couples to —SH groups by addition to the double bond (Chapter 2, Section 2.2). The maleimide group is reasonably specific for sulfhydryls, and the reaction results in a thioether linkage that is quite stable. Sulfhydryl-containing proteins and other molecules modified with this reagent may be used in photoaffinity-labeling studies to investigate the specific interactions between two molecules. After mixing, the solution may be photolyzed to create a covalent cross-link between the two interacting substances. Ultraviolet photolysis of the benzophenone group results in a highly reactive triplet-state intermediate that can rapidly insert or add to organic functional groups within van der Waals distance (Fig. 188). Decay of the active-state intermediate returns the photosensitive group to its orignial chemical form, thus allowing repeated photoactivations without losing the potential for coupling to its intended target.

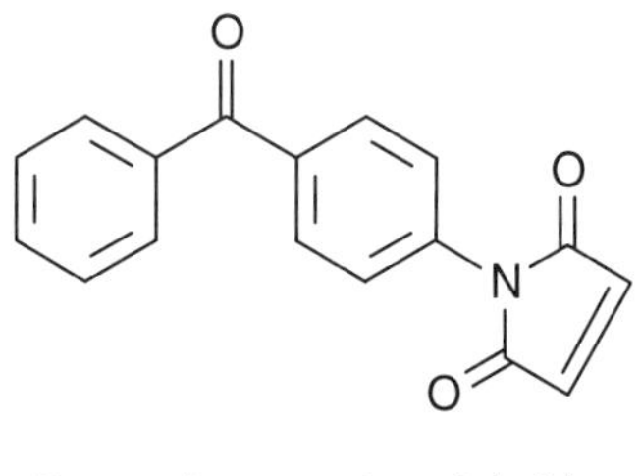

Benzophenone-4-maleimide
MW 277

Benzophenone-4-maleimide is water-insoluble and should be predissolved in DMF or another organic solvent prior to adding an aliquot to an aqueous reaction mixture. Stock solutions may be prepared and stored successfully if protected from light. The hydrophobicity and bulkiness of the benzophenone group may cause insolubility problems in the initial protein that is modified if the derivatization is done at too high a level. Fortunately, the use of a sulfhydryl-reactive reagent can limit the degree of derivatization, since —SH groups usually are present in lower quantities and in more discrete locations than groups like amines.

5. Carbonyl-Reactive and Photoreactive Cross-linkers

Cross-linking reagents containing a photoreactive function on one end and a carbonyl-reactive group on the other end are rare. The use of an amine group on one end of a photosensitive heterobifunctional reagent has been described (Gorman and Folk, 1980; Drafler and Marinetti, 1977; Das and Fox, 1979), but the presence of a hydrazide is required for spontaneous reactivity toward carbonyls. The following compound is the only commercially available reagent containing a phenyl azide photoreactive group and a hydrazide functional group.

5.1. ABH

p-Azidobenzoyl hydrazide (ABH) is a small, heterobifunctional cross-linker containing a photoreactive phenyl azide group on one end and a hydrazide functional group on the other end (Pierce). The hydrazide can react with carbohydrate-containing molecules after oxidation with sodium periodate (Chapter 1, Section 4.4) to create aldehyde residues. The reaction forms a hydrazone linkage. Thus, glycoproteins may be specifically labeled on their polysaccharide chains for subsequent investigation of their interaction with receptor molecules (Fig. 189). In this sense, lectin–carbohydrate interactions may be studied through direct modification of the sugar groups at or adjacent to the binding site. Other amine- or sulfhydryl-reactive probes may not be suitable for such studies due to the lack of amine or sulfhydryl groups near enough to a polysaccharide structure.

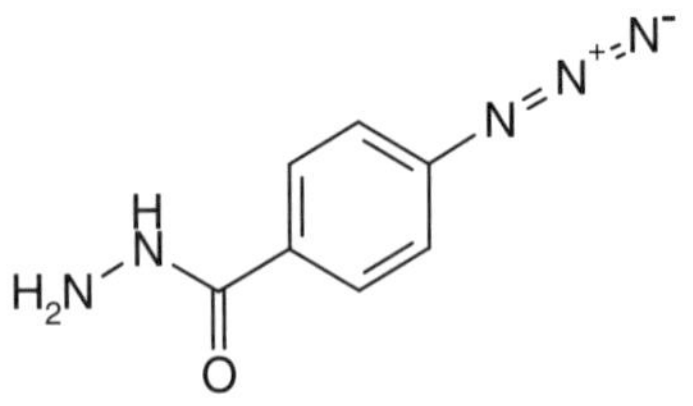

ABH
p-Azidobenzoyl hydrazide
MW 177.17
11.9 Å

The cross-bridge of ABH consists of a benzoic acid derivative and thus provides a short spacer between conjugated molecules. After ABH modification of a glycoprotein and incubation with a potential target molecule, the solution may be photolyzed with

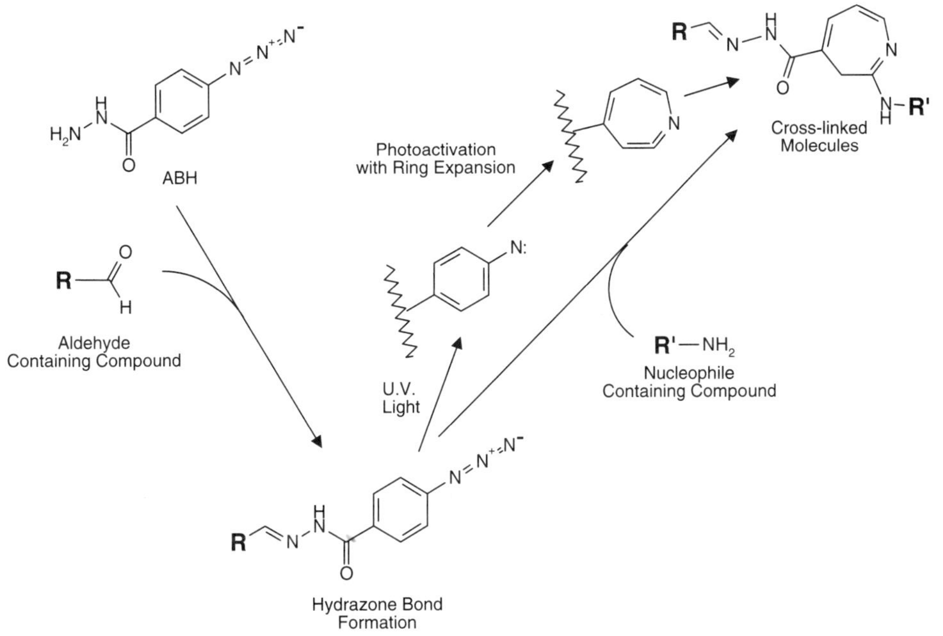

Figure 189 ABH reacts with aldehyde-containing compounds through its hydrazide end to form hydrazone linkages. Glycoconjugates may be labeled by this reaction after oxidation with sodium periodate to form aldehyde groups. Subsequent photolysis with UV light causes photoactivation of the phenyl azide to a nitrene. The nitrene undergoes rapid ring expansion to a dehydroazepine, which can couple to nucleophiles, such as amines.

UV light to initiate the final cross-link. Prior to photolysis, the reagent and all modified species should be protected from light to prevent degradation of the phenyl azide group.

ABH is relatively insoluble when directly added to water or buffer, and therefore it should be predissolved in DMSO prior to addition of an aliquot to an aqueous reaction medium. Stock solutions at a concentration of 50 mM ABH in DMSO work well. Since both functional groups of ABH are stable in aqueous environments as long as the solution is protected from light, a secondary stock solution may be made from the initial organic preparation by adding an aliquot to the hydrazide reaction buffer (0.1 M sodium acetate, pH 5.5; O'Shannessy and Quarles, 1985; O'Shannessy *et al.*, 1984). Make a 1:10 dilution of the ABH/DMSO solution in the reaction buffer. This solution may be stored in the dark at 4°C without decomposition.

6. Carboxylate-Reactive and Photoreactive Cross-linkers

A carboxylate-reactive cross-linking compound typically contains a primary amine functional group that can be coupled to a carboxylic acid group in a protein or other molecule through the use of a suitable activating agent, such as a carbodiimide. The carbodiimide forms an active ester intermediate that then reacts with the amine to create an amide bond (Chapter 3, Section 1). Recent reported use of diazoalkyl derivatives that spontaneously react with carboxylates have been tried with fluorescent probes, but not yet applied to heterobifunctional cross-linking agents (DeMar *et al.*, 1992; Schneede and Ueland, 1992) (Chapter 2, Section 3.1). The following heterobifunctional reagent is the only carboxylate-reactive photosensitive cross-linker currently available commercially.

6.1. ASBA

4-(*p*-Azidosalicylamido)butylamine (ASBA) is a carboxylate-reactive cross-linking agent containing a primary amine on one end and a photosensitive phenyl azide group on the other (Pierce). The cross-linker is not spontaneously reactive with carboxylates, but must be used with another agent that facilitates bond formation. For instance, it can be used in conjunction with a carbodiimide or other such reagent system that can initiate covalent bond formation with a recipient carboxylic acid. A water-soluble carbodiimide like EDC (Chapter 3, Section 1.1) is able to activate the carboxylates on a target protein, forming active ester intermediates (Fig. 190). In the presence of ASBA, derivatization will occur resulting in amide bond formation, thus leading to modification with a photoreactive group.

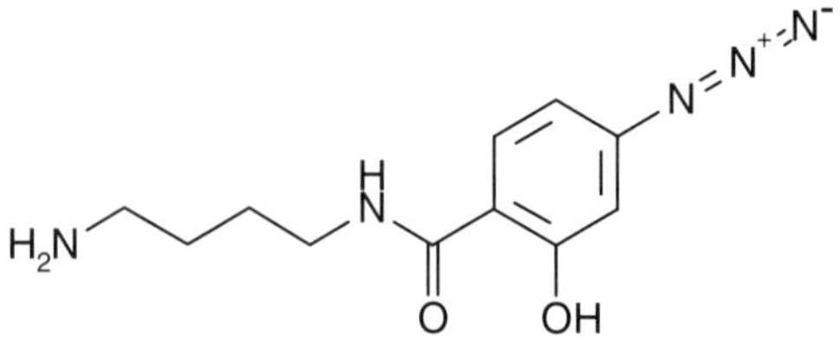

ASBA
4-(*p*-Azidosalicylamido)
butylamine
MW 249.27
16.3 Å

The cross-bridge of ASBA provides a reasonably long spacer (16.3 Å). The phenyl azide portion is constructed from a salicylic acid derivative and thus possesses a ring-activating hydroxyl group. The presence of this group allows radioiodination of the ring prior to cross-linking (Chapter 8, Section 4.5).

Before the photolyzing step is initiated, the reagent should be handled in the dark or protected from light to avoid decomposition of the phenyl azide group.

7. Arginine-Reactive and Photoreactive Cross-linkers

The guanidinyl group on arginine's side chain can be specifically targeted by the use of 1,2-dicarbonyl reagents, such as the diketone group of glyoxal (Chapter 2, Section 5.2). Under alkaline conditions, this type of group can condense with the guanidinyl residue to form a Schiff base-like complex. The presence of other chemical compounds in the reaction can cause further structural rearrangements, such as stabilization by boronate (Pathy and Smith, 1975). Derivatives such as phenylglyoxal and *p*-nitro-

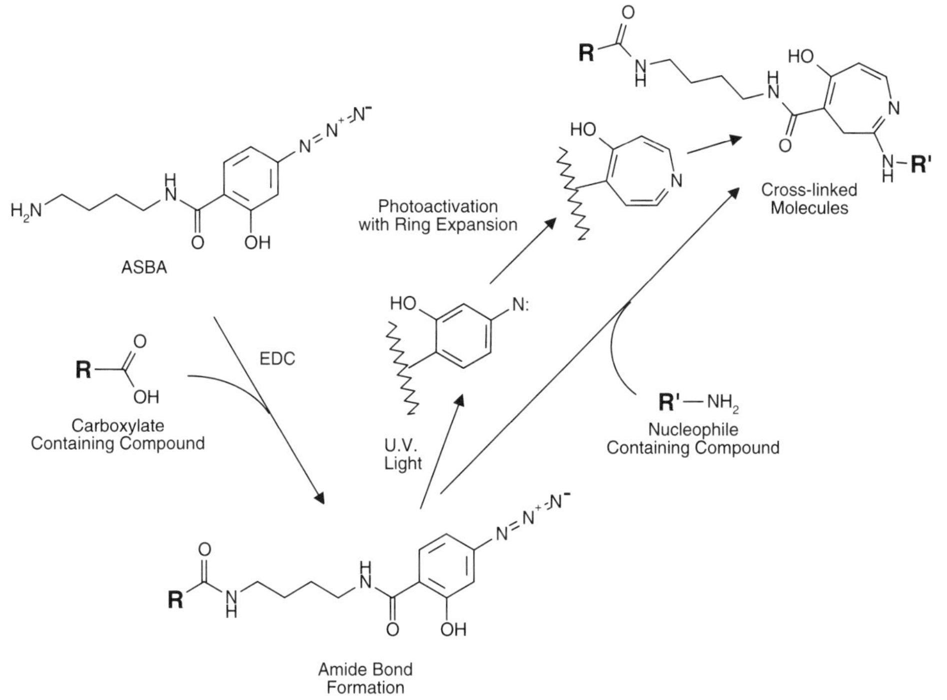

Figure 190 ASBA contains a primary amine group that may be conjugated to carboxylate compounds using the carbodiimide EDC. Subsequent exposure to UV light initiates the photoreaction leading to covalent cross-links.

phenylglyoxal can be used to block or quantitatively determine the amount of arginine in a protein (Yamasaki *et al.*, 1981). Studies have shown that if the reaction is done with a 2:1 ratio of glyoxal compound to arginine residues then the modification that results is reversible (Takahashi, 1968). If the modification is done at a 1:1 stoichiometry, then it is irreversible (Konishi and Fujioka, 1987).

The ability to direct conjugation or modification specifically through arginine residues using this chemistry has been exploited in the availability of the only photoreactive glyoxal derivative, APG.

7.1. APG

p-Azidophenyl glyoxal (APG) is a heterobifunctional cross-linker containing an arginine-specific diketone group on one end and a photosensitive phenyl azide group on the other end (Pierce). The reagent is a derivative of phenylglyoxal, a compound long used as an arginine guanidinyl modifier. Reaction of APG with proteins at pH 7–8 results in selective modification of arginine, leaving photoreactive groups available for subsequent cross-linking with interacting molecules (Fig. 191). Exposure to UV light effects the final cross-link. The cross-bridge of an APG cross-link is only 9.3 Å in

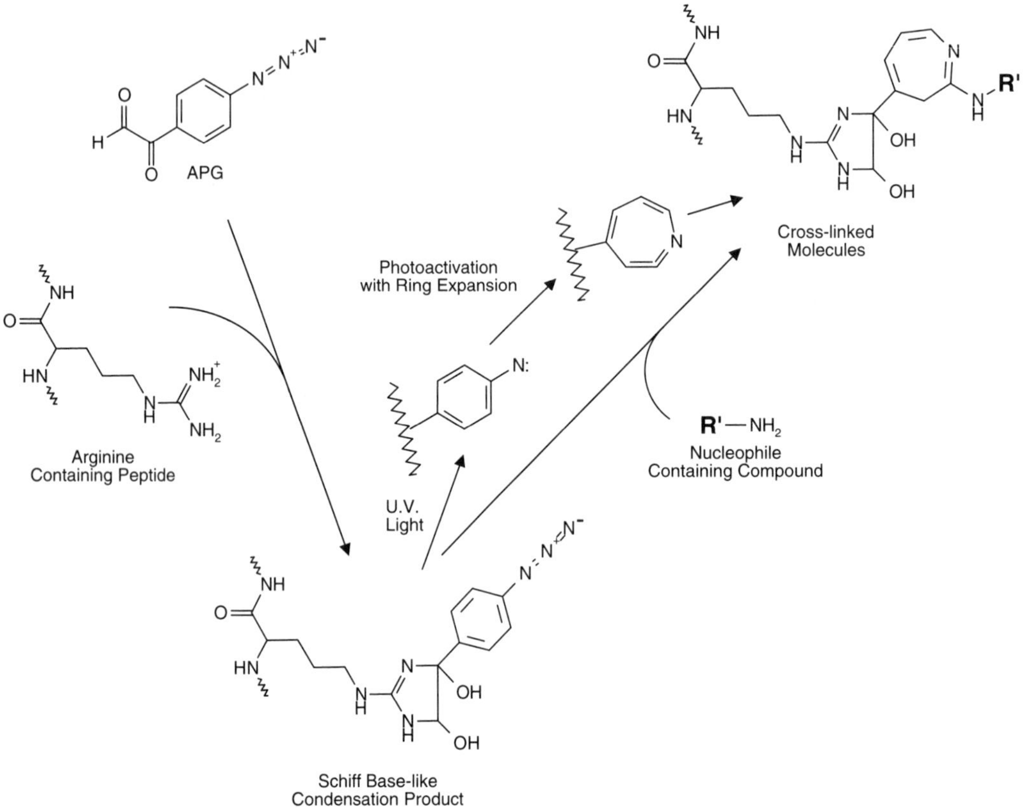

Figure 191 APG can be used to label specifically arginine residues in proteins, producing stable Schiff base interactions with the side-chain guanidino groups. Photolyzing with UV light then causes ring expansion of the phenyl azide group, initiating covalent bond formation with amines.

length, allowing proximity interactions to be studied or the irreversible labeling of arginine areas in proteins.

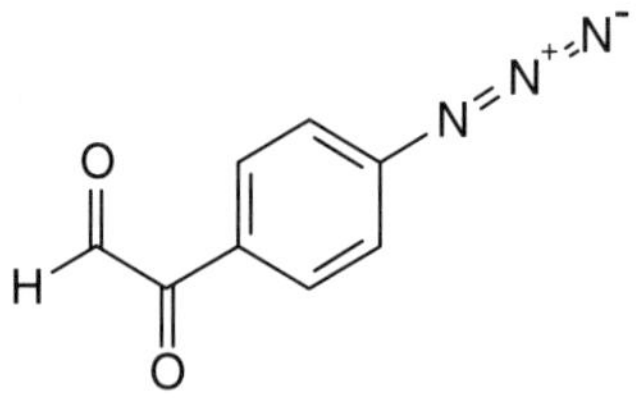

APG
p-Azidophenyl glyoxal
MW 193.16
(as the monohydrate)
9.3 Å

APG has been used to inhibit bovine heart lactic dehydrogenase, egg white lysozyme, horse liver alcohol dehydrogenase, and yeast alcohol dehydrogenase (Ngo *et al.*, 1981), in cross-linking RNA–protein interactions in *E. coli* ribosomes (Politz *et al.*, 1981), and in identifying regions of brome mosaic virus coat protein chemically cross-linked *in situ* to viral RNA (Sgro *et al.*, 1986).

6

Trifunctional Cross-linkers

Trifunctional cross-linkers are a relatively new form of conjugation reagent, possessing three different reactive or complexing groups per molecule. The design of this type of reagent is more elaborate than multifunctional cross-linkers such as polyaldehyde dextran (Chapter 15, Section 2.1) or small organic molecules like trichloro-*S*-triazine (Chapter 15, Section 1.1), which merely contain more than two groups of the same functional group per molecule. The trifunctional approach incorporates elements of the heterobifunctional concept wherein two ends of the linker contain reactive groups able to couple with two different functional groups on target molecules. A trifunctional reagent, however, has a third arm terminating in still another group able to link specifically to a third chemical or biological target.

A convenient molecule from which to build trifunctionals is the amino acid L-lysine. Its three functional groups, α-carboxy, α-amino, and ε-amino, can be derivatized independently to contain three arms. Each arm can be designed to terminate in a complexing group able to participate in a particular type of conjugation reaction.

The initial attempts at producing trifunctional reagents have used biocytin as the starting point. Biocytin is the lysine derivative of biotin having its valeric acid side chain amide-bonded to the ε-amino group of the amino acid (Chapter 8, Section 3.1). Thus, cross-linkers built on this compound have one of their trifunctional arms ending in a biotin label that is able to complex specifically with avidin or streptavidin probes. Creating two additional reactive arms from the α-carboxy and α-amino groups of biocytin results in the completed trifunctional. The following two cross-linkers are examples of this approach.

1. 4-Azido-2-nitrophenylbiocytin-4-nitrophenyl ester

Wedekind *et al.* (1989) designed a trifunctional reagent for studying the hormone binding site of the insulin receptor. The cross-linker 4-azido-2-nitrophenylbiocytin-4-nitrophenyl ester (ABNP) contains a nitrophenyl ester group that can react with amine functions in proteins and peptides, similar to the reaction of NHS esters with amines (Chapter 2, Section 1.4). This group can be used to modify a ligand (such as insulin) prior to its binding to a specific receptor molecule. The second chemically reactive

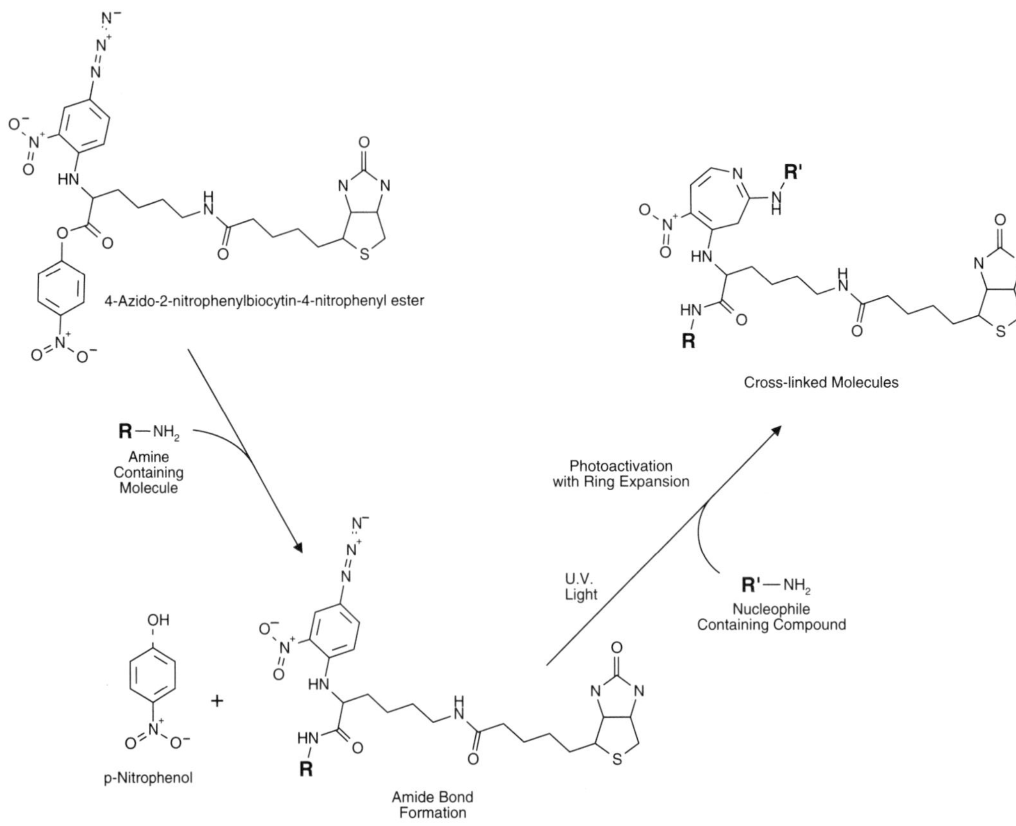

Figure 192   The Wedekind trifunctional cross-linker can react with amine groups via its *p*-nitrophenyl ester to form amide bond linkages. The phenyl azide group then can be photolyzed to generate covalent bond formation with a second molecule. The biotin side chain provides binding capability with avidin or streptavidin probes.

functional group on ABNP is a photosensitive phenylazide group capable of being activated by exposure to UV light. After the labeled ligand is allowed to interact with its receptor, forming a complex, the mixture is photolyzed to effect a covalent attachment point (Fig. 192). The third arm of the trifunctional reagent is the biotin handle (from biocytin). This component allows the complex to be purified by affinity chromatography on immobilized avidin or immobilized streptavidin. Alternatively, the biotin group could be used to visualize the binding of the ligand to its receptor using labeled avidin or streptavidin reagents (Chapter 13).

ABNP is soluble in DMF. Insulin labeling was done in DMF:water at a ratio of 9:1. For molecules not soluble in organic solvent, such as proteins, the trifunctional may be dissolved in DMF and a small aliquot added to an aqueous reaction medium. The nitrophenyl ester functional group can be coupled to amine groups at alkaline pH (7–9) and in buffers containing no extraneous amines (avoid Tris). Unfortunately, ABNP is not commercially available at the time of this writing.

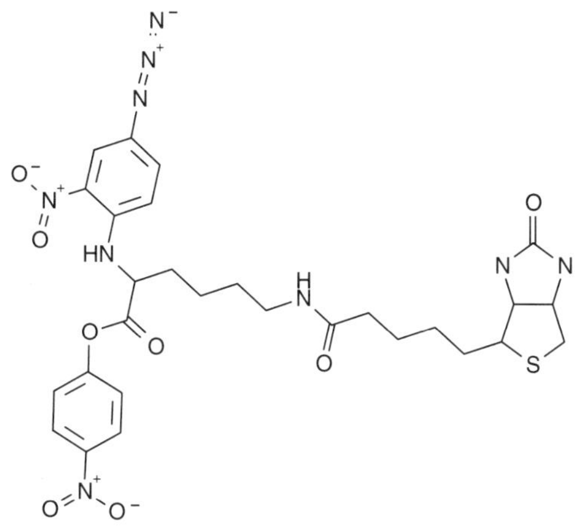

4-Azido-2-nitrophenylbiocytin-4-nitrophenyl ester

2. Sulfo-SBED

Another trifunctional cross-linking agent is sulfosuccinimidyl-2-[6-(biotinamido)--
2-(p-azidobenzamido)hexanoamido]ethyl-1,3'-dithiopropionate (sulfo-SBED) developed by Ed Fujimoto at Pierce Chemical. Like ABNP, discussed previously, sulfo-
SBED is built on a biocytin backbone. Thus, one arm of the trifunctional compound
consists of a biotin handle that can be used for purification or detection purposes using
avidin or streptavidin probes. The chemically reactive groups of sulfo-SBED include a
sulfo-NHS ester and a phenyl azide group. The sulfo-NHS ester provides amine-
coupling capability, forming amide bond linkages with target molecules (Chapter 2,
Section 1.4). The phenyl azide is photosensitive and may be activated by exposure to
UV light at wavelengths > 300 nm. Most phenyl azides react by ring expansion to
dehydroazepines with subsequent reactivity toward nucleophiles (Chapter 2, Section
7.1) (Fig. 193).

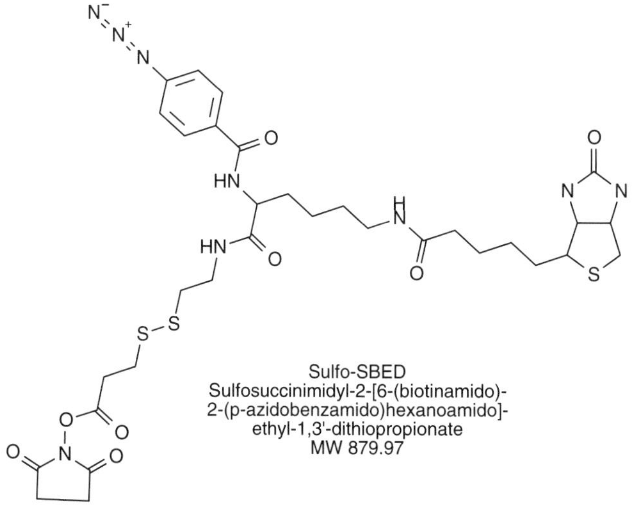

Sulfo-SBED
Sulfosuccinimidyl-2-[6-(biotinamido)-
2-(p-azidobenzamido)hexanoamido]-
ethyl-1,3'-dithiopropionate
MW 879.97

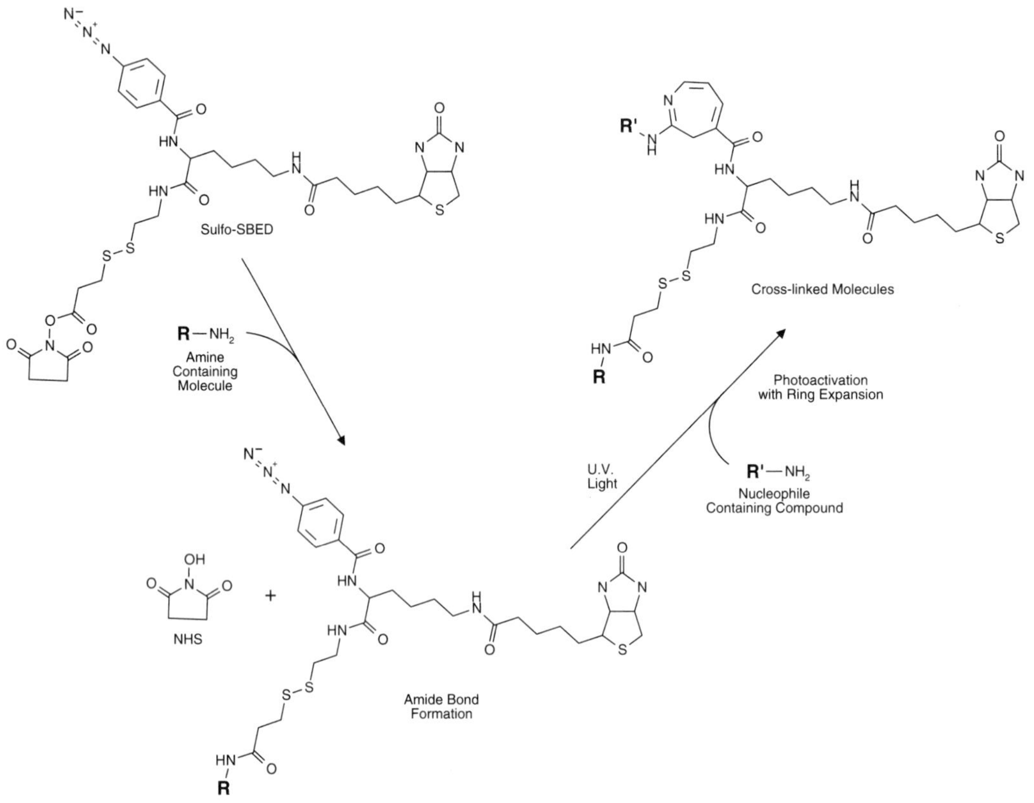

Figure 193 The trifunctional reagent sulfo-SBED reacts with amine-containing molecules via its NHS ester side chain. Subsequent exposure to UV light can cause cross-link formation with a second interacting molecule. The biotin portion provides purification or labeling capability using avidin or streptavidin reagents. The disulfide bond on the NHS ester arm provides cleavability using disulfide reductants.

The sulfo-NHS ester of sulfo-SBED is negatively charged and provides a degree of water solubility (about 5 mM maximum concentration) for the entire molecule. Limited water solubility is all that can be expected due to the large size of the trifunctional, most of it consisting of relatively hydrophobic structures. However, the reagent is much more soluble in organic solvents such as DMF (170 mM) and DMSO (125 mM). Concentrated stock solutions may be prepared in these solvents prior to addition of a small aliquot to an aqueous reaction mixture.

Since the active ester end of the molecule is subject to hydrolysis (half-life of about 20 min in phosphate buffer at room temperature conditions), it should be coupled to an amine-containing protein or other molecule before the photolysis reaction is done. During the initial coupling procedure, the solutions should be protected from light to avoid decomposition of the phenyl azide functional group. The degree of derivatization should be limited to no more than a 5- to 20-fold molar excess of sulfo-SBED over the quantity of protein present to prevent possible precipitation of the modified molecules. For a particular protein, studies may have to be done to determine the optimal level of modification.

An additional feature of sulfo-SBED is the presence of a cleavable disulfide group in the cross-bridge of the NHS ester arm of the molecule. After a conjugation reaction has taken place, the complexes may be first purified using immobilized avidin or immobilized streptavidin and then the conjugates released by treatment with a disulfide reducing agent. This allows analysis of the complexed molecules, for example, after the binding of a ligand to its receptor.

Since sulfo-SBED has three functional arms, the length of each portion should be considered when doing conjugation studies. The biotin handle has an effective length of 19.1 Å, including the side-chain length for the lysine component. The sulfo-NHS ester arm is approximately 13.7 Å long, measuring from the same point in the lysine group. The phenyl azide arm is the shortest, only 9.1 Å long.

The following suggested protocol was developed by Barb Olson at Pierce Chemical for the labeling of soybean trypsin inhibitor with its subsequent complexation with trypsin. Modifications to this procedure may have to be done for other proteins.

Protocol

1. Dissolve 5 mg of soybean trypsin inhibitor (STI) in 0.5 ml 0.1 M sodium phosphate, 0.15 M NaCl, pH 7.2.
2. Dissolve 1.12 mg of sulfo-SBED in 25 μl of DMSO. Prepare fresh.
3. Add 11 μl of the sulfo-SBED solution to the STI solution. Mix well.
4. React for 30 min at room temperature or for 2 h at 4°C.
5. If some precipitation occurs, clarify the solution by centrifugation using a microfuge. Remove excess reactant by gel filtration using a column of Sephadex G-25 (Pharmacia).
6. Mix the purified sulfo-SBED modified STI with 5 mg of trypsin dissolved in 0.1 M sodium phosphate, 0.15 M NaCl, pH 7.2.
7. Incubate at room temperature 3.5 min to allow the specific binding of the two molecules to occur.
8. Photolyze the solution with long UV light (about 365 nm) at a distance of about 5 cm for 15 min. This process may be done with the solution on ice to prevent heating of the sample.

Isolation of complexed molecules may be done by affinity chromatography using a column of immobilized avidin or immobilized streptavidin. Cleavage of the disulfide bond of the cross-linker may be done by treatment with 50 mM DTT.

7

Cleavable Reagent Systems

Cross-linking and modification reagents may possess functions other than their reactivity toward certain chemical groups. In particular, the cross-bridge of the molecule can be designed to contain constituents that allow cleavage of the reagent after use. Occasionally, the coupling reaction itself provides linkages susceptible to subsequent cleavage. Why would you want to break a conjugate apart after having formed it? The ability can be very important in studies involving the biospecific interactions between two molecules, especially if only one of the molecules is known or characterized. Cleavability allows the conjugation reaction to be verified through identification of the cross-linked molecules after conjugation and purification of the complex. Precise points of modification can be determined by amino acid analysis after cleavage. In some cases, purification of unknown target molecules is facilitated by the ability to cleave the cross-linking bonds after isolation of the complex. For instance, a protein modified near its binding site with a photoreactive heterobifunctional cross-linker can be incubated with its receptor or specific binding molecule, a covalent linkage then can be formed by photolyzing, and the complex can be analyzed by subsequent cleavage of the cross-link.

In another example, ligands can be biotinylated with a cleavable biotinylation reagent and then incubated with receptor molecules. The resulting complex can be isolated by affinity chromatography on immobilized avidin. Final purification of the ligand–receptor can be accomplished by cleaving the biotin modification sites while the complex is still bound to the support. The receptor complex thus can be eluted from the column without the usual harsh conditions required to break the avidin–biotin interaction.

The ability to cleave a cross-link also can provide a means of transferring a label from one protein to another. For instance, a photoreactive heterobifunctional cross-linker that is iodinatable and cleavable can be used to tag an unknown receptor molecule after conjugation. For example, the cross-linker SASD (Chapter 5, Section 3.2) can be iodinated before it is employed in a cross-linking reaction. It is then used to modify a protein through its amine-reactive NHS ester end and purified from excess cross-linker. After incubation of the modified protein with specific binding molecules (receptors) and photoreactive cross-linking, the conjugate can be broken by reduction of the disulfide group within the cross-bridge of the reagent. Since the radiolabeled part of the cross-linker now is attached to the receptor molecule, the tracer is effectively transferred from the initially modified molecule. Thus, unknown interacting molecules can be tracked after their binding to an SASD-labeled substance.

Another cross-linker, SAED (Chapter 5, Section 3.9), can be used in a similar fashion, but instead of transferring a radioactive label, it contains a fluorescent portion that is transferred to a binding molecule after cleavage.

The ability to break a cross-link can be an important feature of a modification or conjugation reagent. This chemistry typically is built into the cross-bridge or reactive ends of a reagent using disulfides, glycol groups, diazo bonds, esters, sulfone groups, or acetal linkages. The following sections describe these chemical characteristics and their respective cleavage conditions.

1. Cleavage of Disulfides by Reduction

The formation of a disulfide linkage between cross-linked molecules is an important option for many conjugation chemistries. Examples of reagents that have this capability include the pyridyl disulfide-containing heterobifunctionals like SPDP (Chapter 5, Section 1.1) and SMPT (Chapter 5, Section 1.2). Other non-sulfhydryl-reactive cross-linkers may still possess a disulfide group within their cross-bridge construction. The presence of such disulfide groups, whether designed in the cross-linker or created as a product of their reactions, allows for specific cleavage of the complex or modified molecule after conjugation. Disulfide bonds can be broken by a number of methods (Chapter 1, Section 4.1), utilizing either direct hydrogenolysis by a strong reductant such as sodium borohydride or through a disulfide interchange process with a compound containing one or more free sulfhydryls (Fig. 194).

Cleavage of disulfide bonds is easily done by incubation with a reducing agent at a level of 10–100 mM concentration. If the disulfides in the cross-links are the only ones present in the complexed molecules, then reduction will yield unconjugated molecules—one or both of which will contain a portion of the cross-linker, and on both molecules a free sulfhydryl will be created. Caution should be used with this

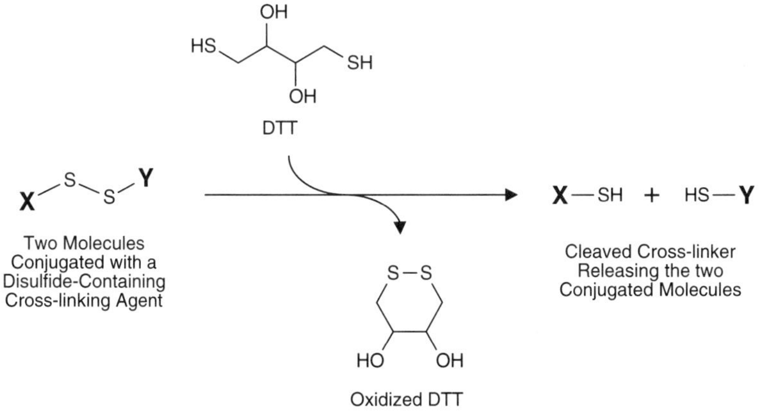

Figure 194 Cleavage of disulfide-containing cross-linking agents can be done using a reducing agent such as DTT. Reduction causes the conjugates to break apart in to their original components.

method of cleavage, however, if other disulfides are present in the conjugated molecules. Some protein disulfides, for instance, also may be affected by the reduction step. Complete cleavage of all disulfides in cross-linked proteins by inclusion of unfolding agents (i.e., guanidine) may yield additional protein fragments of lower molecular weight due to subunit disassociation.

2. Periodate-Cleavable Glycols

Cross-linking agents can be designed to contain adjacent carbon atoms possessing hydroxyl groups. Cross-bridges containing such diols or glycol residues can be constructed from the inclusion of an internal tartaric acid spacer or similar compound in their synthesis (e.g., DST, Chapter 4, Section 1.3). These groups can be easily cleaved by oxidation with sodium periodate (Chapter 1, Section 4.4). Treatment with 15 mM periodate at physiological pH for 4 h will break the carbon—carbon bond between the glycol portion, oxidizing each hydroxyl to an aldehyde, and cleaving the associated cross-linked molecules (Fig. 195). Under these conditions, glycosylated portions of glycoproteins or other carbohydrate-containing molecules also will be affected, forming additional aldehyde groups. In some cases, the production of aldehyde residues may cause secondary reactions to occur, especially Schiff base formation with available amine groups. To avoid unexpected cross-links that form through such intermolecular Schiff base formation, Tris or ethanolamine may be included to tie up the aldehydes as they form.

Sodium periodate also may affect tryptophan residues in some proteins. The oxidation of tryptophan can result in activity losses if the amino acid is an essential component of the active site. For instance, avidin and streptavidin may be severely inactivated by treatment with a large excess of periodate, since tryptophan is important in forming the biotin-binding pocket.

The use of periodate as a cleavage agent does have advantages, however. Unlike the use of cleavable cross-linkers that contain disulfide bonds which require a reductant to break the conjugate, cleavage of diol-containing cross-links with periodate typically preserves the indigenous disulfide bonds and tertiary structure of proteins and other molecules. As a result, with most proteins bioactivity usually remains unaffected after periodate treatment.

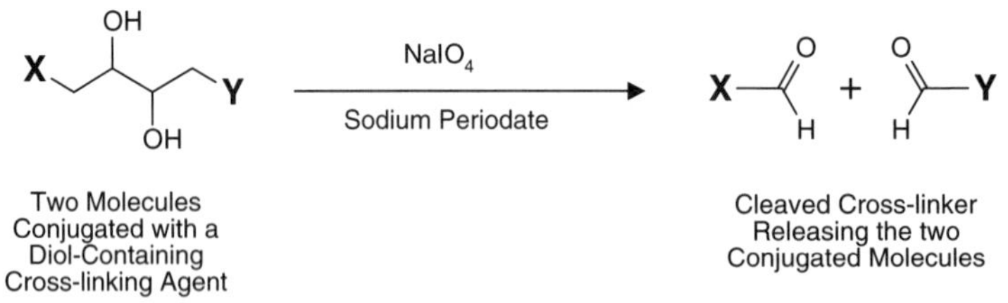

Figure 195   Cross-linkers containing a diol group in their cross-bridge design may be cleaved by oxidation with sodium periodate.

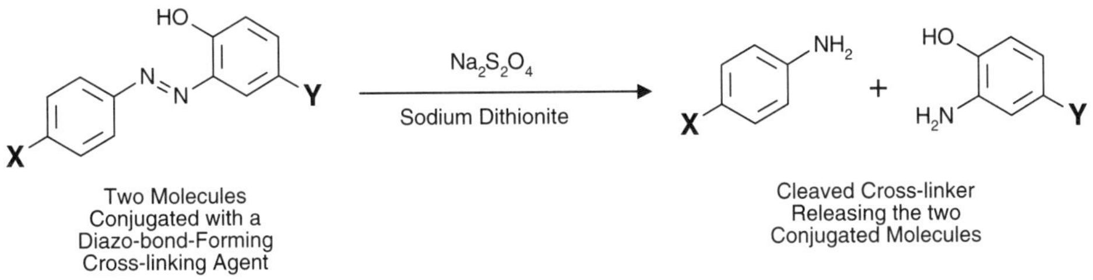

Figure 196 Cross-linking agents that form diazo bonds may be cleaved using sodium dithionite.

3. Dithionite-Cleavable Diazo Bonds

Cross-linking compounds containing diazo bonds within their structures can be specifically cleaved with dithionite (Jaffe *et al.,* 1980). In addition, cross-links formed by the reaction of a diazonium compound (Chapter 2, Section 6.1) with a tyrosine residue also can be broken using this reagent. Sodium dithionite (also called sodium hydrosulfite) reduces the diazo linkage, breaking the bond between the nitrogens, and leaving a primary amine on both fragments of the cross-linker (Fig. 196). The reaction usually is carried out in alkaline conditions; a 25-min incubation with 0.1 M dithionite in 0.2 M sodium borate, pH 9, works well. As the diazo bonds are broken, any color associated with the reagent will disappear.

4. Hydroxylamine-Cleavable Esters

Hydroxylamine is a powerful nucleophile which, under alkaline conditions, is effective in breaking ester bonds. Cross-linking agents containing esterified spacer components can be cleaved after undergoing a conjugation reaction by incubation with 0.1 N hydroxylamine, pH 8.5, for 3–6h at 37°C (Abdella *et al.,* 1979). The reaction results in the formation of an amide derivative on one fragment of the cleaved cross-linker and a hydroxyl group on the other (Fig. 197). Thioester bonds also are susceptible to cleavage under these conditions. Thioesters may be broken with the production of an amide and a sulfhydryl group on either side of the cross-linker fragments.

An example of an hydroxylamine-cleavable reagent is EGS (Chapter 4, Section 1.5),

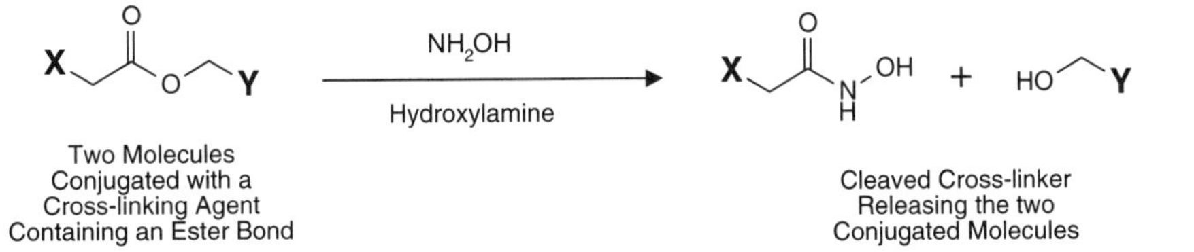

Figure 197 Cross-linkers containing an ester bond in their cross-bridge are susceptible to cleavage under alkaline conditions using hydroxylamine.

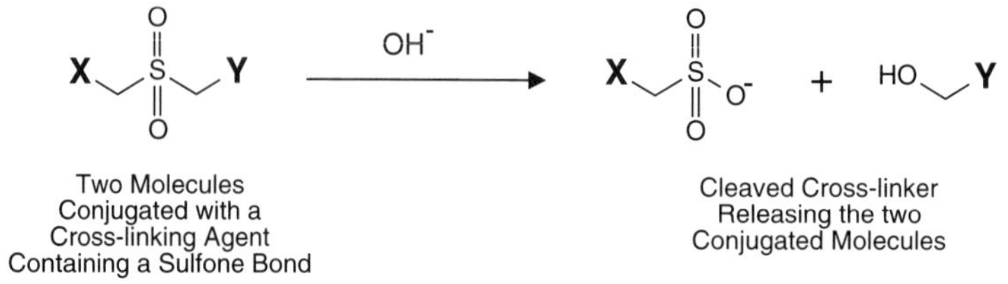

Figure 198 Cross-linkers that have an internal sulfone group in their cross-bridge may be cleaved using base.

which contains two ester bonds made by the esterification of ethylene glycol with succinic acid. Cleavage with hydroxylamine yields two fragments terminating with an amide bond and concomitant release of ethylene glycol.

5. Base Labile Sulfones

The presence of a sulfone group in a cross-linking reagent can allow for cleavage of a conjugate through hydrolysis of the linkage under basic conditions. In 0.1 M sodium phosphate, adjusted to pH 11.6 by addition of Tris base, containing 6 M urea, 0.1% SDS, and 2 mM DTT, sulfone groups were successfully cleaved after incubation at 37°C for 2 h (Zarling *et al.,* 1980). In that study peptide antigens on the surface of lymphocyte receptors were cross-linked with the homobifunctional, amine-reactive reagent BSOCOES (Chapter 4, Section 1.4), purified, and cleaved for analysis. The presence of urea, SDS, and DTT were not absolutely necessary for breaking the sulfone bond, rather they served to disrupt completely protein/peptide structure for complete dissociation of the complex.

In addition to BSOCOES, the amine-reactive, *bis*fluorobenzene reagent DFDNPS (Chapter 4, Section 4.2) also contains an internal sulfone group that is easily cleaved under basic conditions (Wold, 1961, 1972). Hydrolysis of the sulfone yields two cross-linker fragments, one terminating in a sulfonic acid group and the other containing a hydroxyl group (Fig. 198).

8

Tags and Probes

Tags and probes are relatively small modifying agents that can be used to label proteins, nucleic acids, and other molecules. These compounds often contain groups that provide sensitive detectability by virtue of some intrinsic chemical or atomic property such as fluorescence, visible chromogenic character, radioactivity, or bioaffinity toward another protein. Most probes can be designed to contain a reactive portion capable of coupling to the functional groups of biomolecules. After modification of a protein via this reactive part, the probe becomes covalently attached, thus permanently tagging it with a unique detectable property. Subsequent interactions that the protein is allowed to undergo can be followed through the tag's visibility.

Detection of a probe usually takes one of three general forms: spectrophotometric or radiosensitive devices, or indirectly through another labeled substance. Spectral probes can be of two types, chromogenic or fluorescent. Chromogenic labels typically are reserved for noncovalent staining of gross structural features within cells and tissues. The sensitivity of visible wavelength dyes often is not good enough to provide detectability for low-concentration antigens. Even if a protein is covalently modified with a chromogen, the number of associated dye molecules needed to detect subsequent interactions with another target molecule could be prohibitively large to make it viable.

Fluorescent labels, by contrast, can provide tremendous sensitivity due to their large quantum emission yield upon excitation. Even the absorptivity of some fluorophores is great enough to provide excellent detectability for sensitive measurements. Proteins, nucleic acids, and other molecules can be labeled with fluorescent probes to provide highly receptive reagents for numerous *in vitro* assay procedures. For instance, fluorescently tagged antibodies can be used to probe cells and tissues for the presence of particular antigens, and then detected through the use of fluorescence microscopy techniques. Since each probe has its own fluorescent character, more than one labeled molecule—each tagged with a different fluorophore—can be used at the same time to detect two or more target molecules.

Tagging molecules by adding a radioactive component was one of the first means of creating highly sensitive detection capabilities. Covalent modification of activated aromatic rings with ^{125}I or ^{131}I easily can be done through tyrosine side chains in proteins or by the use of a phenolic-ring-containing modification agent such as the Bolton-Hunter reagent (Section 4.5). Other methods of introducing a radioactive isotope involve the intermediary use of metal-chelating modification reagents such as DTPA (Section 2.1). Heavy metal isotopes may be held in coordination complexes on a protein or other molecule and provide extreme sensitivity for *in vivo* diagnostic procedures involving the detection of malignancies. Such complexes coupled to monoclonal antibodies also are being used in the treatment of cancer by their ability to cause cell death in proximity to the bound radiolabel.

Finally a biomolecule may be detected through the use of a modification reagent having strong affinity for another labeled molecule. For instance, a protein or oligonucleotide probe may be modified with a biotinylation reagent to produce an avidin- or streptavidin-binding derivative (Section 3). If the avidin is labeled with a detectable tag or enzyme, then the resultant complex can be tracked. This type of two-stage detection system can result in higher sensitivities than using direct labels, since more than one biotin label per protein or oligo can create multisite interactions with the tagged avidin molecules (Chapter 10, Section 2.3).

The following sections describe the principal reagents available for producing tagged molecules using fluorescent probes, radiolabels, and biotinylation techniques. In many cases, suggested protocols are included (or other sections referenced) for the use of these probes.

1. Fluorescent Labels

A fluorescent molecule has the ability to absorb photons of energy at one wavelength and subsequently emit the energy at another wavelength. The absorption process is also called excitation, since the quantum energy levels of some of the compound's electrons increase with photon uptake. The absorption band is not isolated at a discrete photon energy level, but spread out over a range of wavelengths with at least one peak of maximal absorbance within this spectral region. The extinction coefficient (ε, expressed as $M^{-1}cm^{-1}$) at the absorbance peak maximum is a unique characteristic of each fluorophore under a given environmental condition.

The excess energy of a photolyzed fluorophore can be lost as heat or through collisions with adjacent molecules or released as photons of light as the electrons return to the lower, ground-state energy level (Fig. 199). This process of emission occurs in less than 10^{-4}s of excitation and is known as fluorescence. Fluorescence takes place from the lowest excited singlet state. According to Stoke's Law, the emission wavelength is always longer and thus of lower energy than the wavelength of excitation (Kawamura, 1977). The ratio of total photon emission over the entire range of fluorescence to the total photon absorption is called the quantum yield (Q). Quantum yield values range from 0 to 1. The larger the Q value the stronger the photon emission or luminescence. For a particular fluorophore under fixed environmental conditions, both its extinction coefficient and its quantum yield are constants characteristic of its photochemical behavior.

Not only should a fluorescent compound suitable for analytical studies have a high Q, but its fluorescence emission spectrum should be separated sufficiently from its excitation spectrum to ensure good signal isolation. A fluorophore's Stoke's shift is a measure of the separation of its maximal absorbance wavelength from its emission wavelength maxima (Fig. 200). The greater the Stoke's shift, the better the signal isolation and therefore less interference from Rayleigh-scattered excitation light.

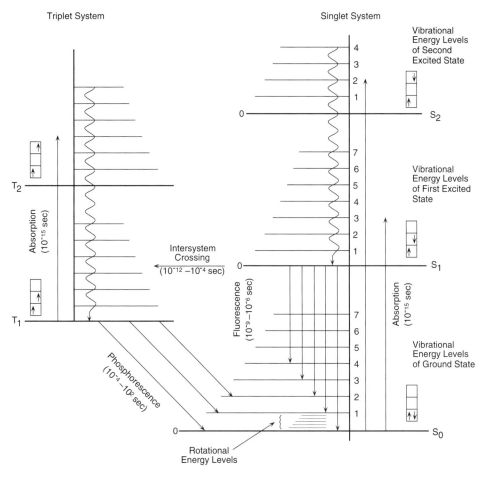

Figure 199   Energy diagram showing the transition states involved in the absorption and decay of electromagnetic energy. Energy may be released through heat or internal collisions, or it may be emitted as photons of light. Fluorescence occurs from within the singlet system as light energy is released, returning the electrons to the ground state. Phosphorescence occurs from the triplet system and involves a longer emission process at lower energies than that of fluorescence.

The majority of reported fluorophores of practical use in labeling biomolecules contain an aromatic ring system as the generator of luminescence. In general, as the ring system gets larger, the emission wavelength is shifted to the red. Also the Quantum yield of larger ring systems typically is greater than that of small aromatic compounds.

Aromatic ring constituents can have a pronounced effect on fluorescence. The presence of ring activators or electron-donating groups (e.g., *ortho* and *para* directors) generally increases Q, while the presence of electron-withdrawing groups (e.g., *meta* directors) reduces Q. An example of this effect can be found in the photoreactive heterobifunctional cross-linker SAED (Chapter 5, Section 3.9). Before photolysis SAED possesses a photosensitive phenyl azide group on its AMCA-derived end. This electron-withdrawing group quenches the fluorescence of the compound so that the

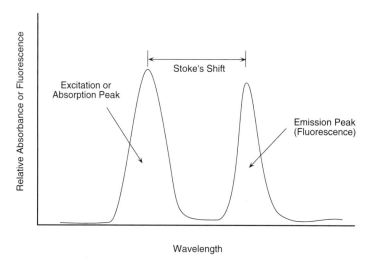

Figure 200 Typical spectral scan of a fluorescent compound showing its absorbance peak or wavelengths of most efficient excitation and its emission peak or wavelengths where light emission occurs. The Stokes shift is the distance in nanometers between the absorbance peak and the emission peak. The larger the Stokes shift, the less interference will occur from excitation light when measuring fluorescence.

AMCA group does not behave as it characteristically does in the underivatized state. Upon photolysis, however, the phenyl azide either gets coupled to a target molecule or breaks down to an amine. In either case, the presence of an amine (or its derivative) provides enough ring-activating properties to restore AMCAs fluorescent character.

Another potential ring constituent having dramatic effect on fluorescence is the presence of heavy atoms. Aromatic rings possessing heavy atoms diminish Q by enhancing the probability of the excited singlet state going on to triplet transition. Energy decay from a triplet excited state causes phosphorescence instead of fluorescence. The phosphorescent band is located at longer wavelengths (and thus lower energies) relative to the fluorescent spectrum. The energy transition to the triplet state therefore is in direct contention with fluorescence, and has the effect of decreasing potential luminescence. In this regard, the relative fluorescent-quenching effect of halogen substitution on aromatic-ring-containing fluorophores is I > Br > Cl. In practice, the presence of a chloride atom on a fluorescent aromatic ring may still allow some luminescence to occur; however, the presence of bromide or iodine atoms nearly guarantees complete elimination of fluorescence. Thus, radioiodination of a fluorescent ring structure is not possible without dramatically decreasing (or eliminating) its quantum yield.

The aromatic ring systems of most of the following fluorophores consist of polycyclic structures. To maintain fluorescence in such compounds, it is important that the entire system be coplanar or the rings be in the same dimensional plane. In fact, the differences between some chromogenic dyes and corresponding, structurally similar fluorophores are minor, but the planar nature of the fluorophores gives them their luminescent properties. For instance, Fig. 201 illustrates the similarity of the dyes phenolphthalein and malachite green to the nearly identical fluorescent molecules fluorescein and rhodamine, respectively. The only difference among these dyes and

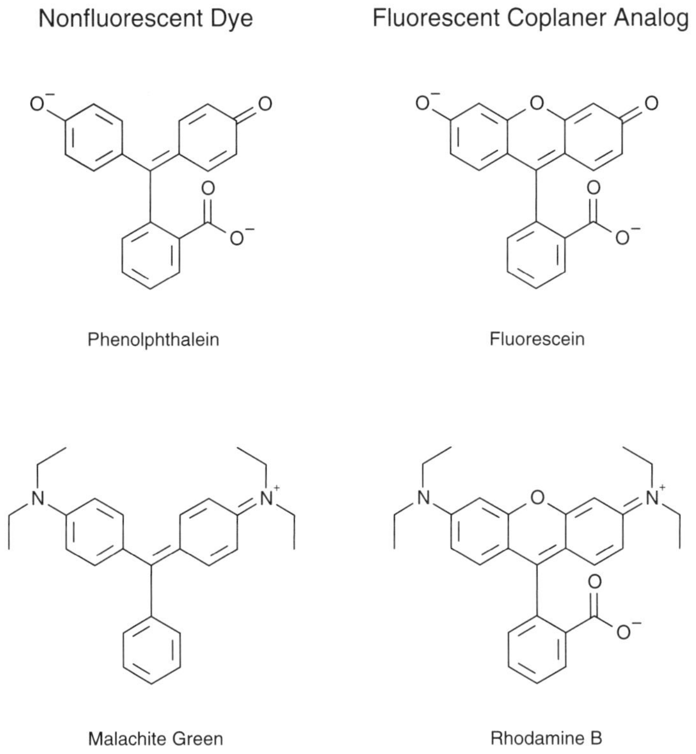

Nonfluorescent Dye   Fluorescent Coplaner Analog

Phenolphthalein   Fluorescein

Malachite Green   Rhodamine B

Figure 201 Fluorescent character in organic compounds is often determined by the presence of an almost planar aromatic ring system. The fluorescent compounds differ only in the closure of their central ring system, which produces the required constraints to create a planar triple-ring configuration.

fluorophores is the presence of the oxygen bridge between the upper phenyl rings, which constricts the molecule to a planar shape, thus conferring luminescent qualities.

Fluorescent compounds are sensitive to changes in their chemical environment. Alterations in media pH, buffer components, solvent polarity, or dissolved oxygen can affect and quench the quantum yield of a fluorescent probe (Bright, 1988). The presence of absorbing components in solution that absorb light at or near the excitation wavelength of the fluorophore will have the effect of decreasing luminescence. In addition, noncovalent interactions of the probe with other components in solution can inhibit rotational freedom and quench fluorescence.

Overlabeling with a fluorescent probe also causes decreases in quantum yield. Quenching caused by interactions between fluorescent molecules often occurs as the level of probe substitution reaches about 8–10 fluorophores per protein. Not only can the emission intensity decrease severely at high substitution levels, but the degree of nonspecific binding caused by the number of aromatic groups attached to the biomolecule can increase severely. Fluorophore self-quenching at high substitution or concentration levels can be due to energy transfer from excited-state molecules to ground-state dimers (Chen and Knutson, 1988).

The following sections describe some of the most popular fluorescent probes for use in labeling biomolecules. These fluorophores are available from a number of manufacturers in several analog forms, each with a different reactive group able to couple to specific functional groups or target molecules. By far, the most often used fluorescent tags are derivatives of fluorescein, rhodamine, coumarin, and Texas red—probably in that order. However, new fluorescent molecules are being developed and reported constantly. Careful study of their individual properties can lead to the successful labeling of a biomolecule for a particular application.

Suggested references on fluorescent spectroscopy and the use of fluorescent probes include Lakowicz (1991); Bright (1988); Dewey (1991); McGown and Warner (1990); Ploem and Tanke (1987); Darzynkiewicz and Crissman (1990); Haugland (1991); and Waggoner (1990).

1.1. Fluorescein Derivatives

Fluorescein is perhaps the most popular of all fluorescent labeling agents. Its fluorescent character is created by the presence of a multiring aromatic structure due to the planar nature of its upper, fused three-ring system (Fig. 202). In its most elementary form, the molecule has a relative molecular mass of about 332 Da, but most modification reagents based on fluorescein are derivatives of this basic structure prepared through substitutions off the No. 5 or 6 carbons of its lower ring. The derivatives provide reactivity toward particular functional groups in biomolecules, allowing rapid labeling of proteins and nucleic acids.

Fluorescein has an effective excitation wavelength range of about 488–495 nm, closely matching the photon emission for an argon laser (488 nm). Its emission spectrum occurs between 518 and 525 nm, depending on the derivative chosen. Under ideal conditions, its quantum yield can be as high as 0.75; however, its fluorescent intensity fades when it is dissolved in buffers, exposed to light, or stored for extended periods (Kawamura, 1977). In environments below pH 7, fluorescein's quantum yield is significantly quenched. In addition, when derivatives of fluorescein are conjugated to proteins, the degree of fluorescent quenching can be as high as 50%. Even so, the fluorophore usually maintains excellent detectability in assay systems.

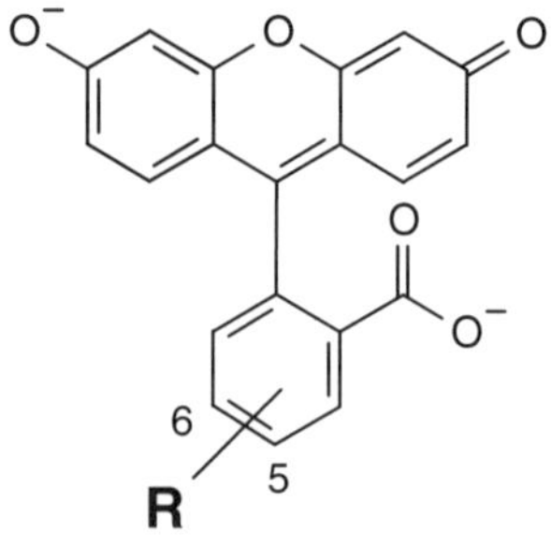

Reactive Fluorescein Derivatives

Figure 202   Fluorescein derivatives are produced through modification at the C-5 or C-6 positions on the lower ring.

The following sections describe the most important fluorescein derivatives commonly used to label biomolecules.

Amine-Reactive Fluorescein Derivatives

Two forms of amine-reactive fluorescein probes are available. Both of them react under alkaline conditions with primary amines in proteins and other molecules to form stable, highly fluorescent derivatives.

Fluorescein Isothiocyanate (FITC)

FITC is probably the most popular fluorescent probe ever created. An isothiocyanate derivative of fluorescein is synthesized by modification of its lower ring at the 5- or 6-carbon positions. The two resulting isomers are nearly identical in their reactivity and spectral properties, including excitation and emission wavelengths and intensities. Their chemical differences, however, may affect the separation of modified proteins from excess reagent or the analysis of tagged molecules by electrophoresis. For this reason, most manufacturers purify the carbon-5 derivative as the FITC reagent of choice.

Isothiocyanates react with nucleophiles such as amines, sulfhydryls, and the phenolate ion of tyrosine side chains (Podhradsky *et al.*, 1979). The only stable product, however, is with primary amine groups, and so FITC is almost entirely selective for modifying ε- and N-terminal amines in proteins (Jobbagy and Kiraly, 1966). The reaction involves attack of the nucleophile on the central, electrophilic carbon of the isothiocyanate group (Fig. 203). The resulting electron shift creates a thiourea linkage between FITC and the protein with no leaving group.

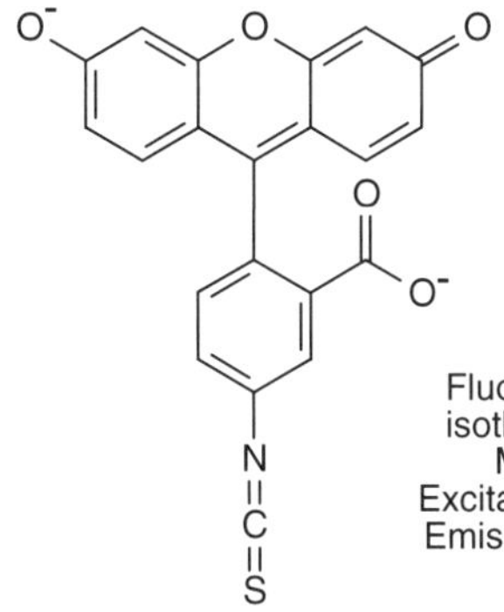

FITC;
Fluorescein-5-
isothiocyanate
MW 389
Excitation 494 nm
Emission 520 nm

FITC can be dissolved in DMF as a concentrated stock solution prior to its addition to an aqueous reaction mixture. This may make aliquoting small quantities of the compound easier. The reagent is water-soluble above pH 6. The isothiocyanate group is reasonably stable in aqueous solution for short periods, but will degrade. FITC also can break down and lose activity upon storage. It is best, therefore, to use only fresh reagent for modification purposes. Storage should be done under desiccated conditions, protected from light, and at −20°C.

The fluorescent properties of FITC include an absorbance maximum at about 495 nm and an emission wavelength of 520 nm. Fluorescent quenching of the molecule is possible. Under concentrated conditions, fluorescein-to-fluorescein interactions result

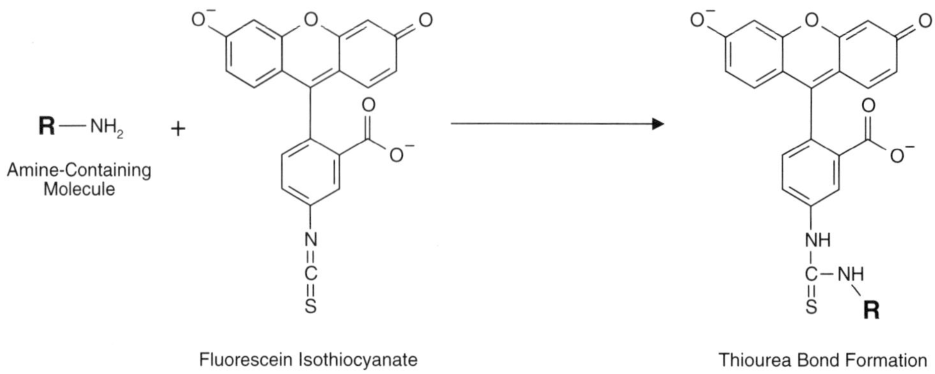

R—NH₂ +

Amine-Containing
Molecule

Fluorescein Isothiocyanate

Thiourea Bond Formation

Figure 203 FITC reacts with amine-containing compounds to produce an isothiourea linkage.

in self-quenching, which reduces the luminescence yield. This phenomenon can occur with fluorescein-tagged molecules, as well. If derivatization of a protein is done at too high a level, the resultant quantum yield of the conjugate will be depressed. Typically, modifications of proteins involve adding no more than 8–10 fluorescein molecules per protein molecule, with a 4–5 substitution level considered optimal.

FITC has been used in numerous applications involving fluorescence detection. Antibodies or their fragments can be labeled to detect antigens in cells, tissues sections, or blots or on surfaces (Clausen, 1988). Tagging molecules with FITC also is useful in detecting proteins after electrophoretic separations (Strottmann *et al.*, 1983), for microsequencing analysis of proteins and peptides (Muramoto *et al.*, 1984), in analysis of molecules using capillary zone electrophoresis (Cheng and Dovichi, 1988), and in tracking and detecting molecules involved in various biointeractions (Friedman and Ball, 1989; Burtnick, 1984).

The level of fluorescein modification in a macromolecule can be determined by measuring its absorbance at or near its characteristic excitation maximum (~498 nm). The number of fluorochrome molecules per molecule of protein is known as the F/P ratio. This value should be measured for all derivatives prepared with fluorescent tags. The ratio is important in predicting the behavior of antibodies labeled with FITC (Hebert *et al.*, 1967; Beutner, 1971). Using the known extinction coefficient of FITC in solution at pH 13 ($\epsilon_{498 \text{ nm}} = 8.1$–$8.5 \times 10^4$; Jobbagy and Jobbagy, 1972; McKinney *et al.*, 1964), a determination of derivatization level can be made after excess FITC is removed. At pH 7.8, the absorbance of protein-coupled FITC decreases by 8%.

A general protocol for the modification of proteins, particularly immunoglobulins, with FITC is given below. Slight modifications to the amount of reagent added to the reaction may be done to optimize the F/P ratio.

Protocol

1. Prepare a protein solution in 0.1 *M* sodium carbonate, pH 9, at a concentration of at least 2 mg/ml.
2. In a darkened lab, dissolve FITC (Pierce) in dry DMSO at a concentration of 1 mg/ml. Do not use old FITC, as breakdown of the isothiocyanate group over

time may decrease coupling efficiency. Protect from light by wrapping in aluminum foil or using amber vials.

3. In a darkened lab, slowly add 50–100 µl of FITC solution to each milliliter of protein solution (at 2 mg/ml concentration). Gently mix the protein solution as the FITC is added.

4. React for at least 8 h at 4°C in the dark.

5. The reaction may be quenched by the addition of ammonium chloride to a final concentration of 50 mM. Some protocols also include at this point the addition of 0.1% xylene cylanol and 5% glycerol as a photon absorber and protein stabilizer, respectively. React for a further 2 h to stop the reaction by blocking remaining isothiocyanate groups.

6. Purify the derivative by gel filtration using a PBS buffer or another suitable buffer for the particular protein being modified. The use of Sephadex G-25 or similar matrices with low exclusion limits work well. To obtain complete separation, the column size should be 15–20 times the size of the applied sample. Fluorescent molecules often nonspecifically stick to the gel filtration support, so reuse of the column is not recommended.

NHS-Fluorescein and NHS-LC-Fluorescein

NHS-fluorescein is another amine-reactive fluorescent probe that contains a carboxysuccinimidyl ester group off the No. 5 or 6 carbons on fluorescein's lower ring structure (Brinkley, 1992; Vigers *et al.*, 1988; Khanna and Ullman, 1980). The 5- and 6-isomers are virtually identical in their reactivity and fluorescent characteristics. Similar to FITC (above), NHS-fluorescein can be used to label proteins and other macromolecules that contain primary amine groups. This reagent is more stable than FITC, especially in storage. The NHS ester reaction proceeds rapidly at slightly alkaline pH values, resulting in a stable, amide-linked derivative (Chapter 2, Section 1.4, and Chapter 4, Section 1; Fig. 204).

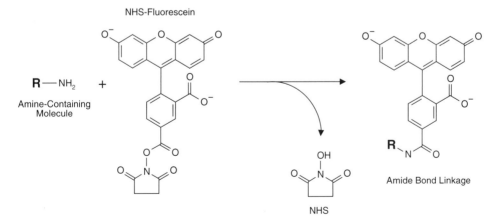

Figure 204 NHS-fluorescein reacts with amine-containing compounds via its NHS ester to form amide bonds.

NHS-Fluorescein;
5-Carboxyfluorescein,
succinimidyl ester
MW 457
Excitation 491 nm
Emission 518 nm

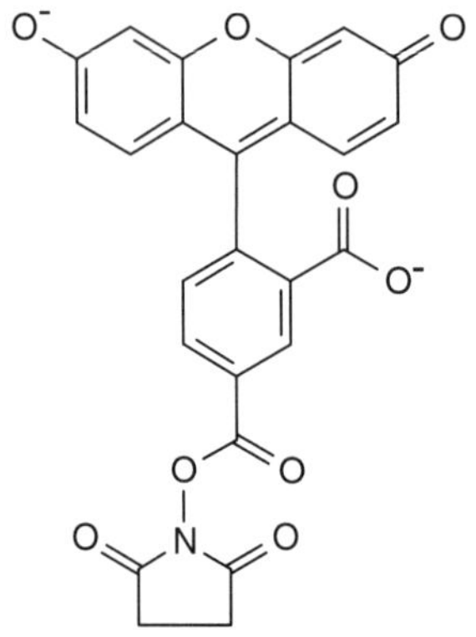

NHS-LC-Fluorescein;
6-(Fluorescein-5-carboxamido)
hexanoic acid, succinimidyl ester
MW 587
Excitation 494 nm
Emission 519 nm

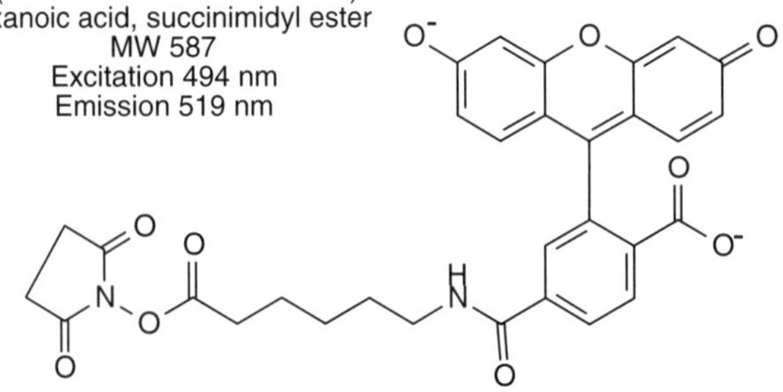

The fluorescent properties of NHS-fluorescein are similar to those of FITC. The wavelength of maximal absorbance or excitation for the reagent is 491 nm and its emission maximum is 518 nm, exhibiting a visual color of green (Sheehan and Hrapchak, 1980). Its molar extinction coefficient at 491 nm in a pH 8 buffer environment is $66,000\ M^{-1}cm^{-1}$. Other components in solution as well as the pH can change this value.

NHS-fluorescein is insoluble directly in aqueous solution and should be dissolved in organic solvent prior to addition of a small aliquot to a buffered reaction medium. Concentrated stock solutions may be prepared in DMSO or DMF. Such solutions are relatively stable if protected from light. Reaction conditions should be maintained at the optimal reactivity for NHS esters—pH 7–9.

NHS-LC-fluorescein (Molecular Probes) is an analog of NHS-fluorescein that contains a 6-aminocaproic acid spacer group, extending the NHS ester group away from the fluorescein portion. The longer length of the coupling arm may decrease steric hindrance around the fluorescent head of the molecule, thus reducing any fluorescence quenching due to its attachment to a macromolecule. All other properties of the long-chain version are virtually identical to that described above for NHS-fluorescein.

The following procedures is a suggested method for using NHS-fluorescein to label immunoglobulins.

Protocol

1. Dissolve the IgG to be labeled in 50 mM sodium bicarbonate buffer, pH 8.5, at a concentration of 10 mg/ml.
2. Dissolve 0.5 mg of NHS-fluorescein (Pierce) in 0.5 ml DMSO. Protect from light.

3. In a darkened lab, slowly add 50–100 μl of the NHS-fluorescein solution to the antibody solution, while mixing. Protect from light by wrapping the reaction vessel in aluminum foil.
4. React for 2 h on ice.
5. Remove unreacted NHS-fluorescein and reaction by-products by gel filtration or dialysis.

A spectrophotometric assessment of the F/P ratio should be done after purification of the tagged antibody. The measurement of absorbance at 495 nm (for fluorescein) divided by the absorbance at 280 nm should be between 0.3 and 1 to obtain a good fluorescent derivative of acceptable activity and low background. This usually translates into a ratio of about four to seven fluorescein molecules per protein molecule.

Sulfhydryl-Reactive Fluorescein Derivatives

Fluorescein derivatives containing a sulfhydryl-reactive group off the lower ring structure are available to direct the modification reaction to more limited sites on target molecules. Coupling through sulfhydryls instead of amines can help to avoid active centers in proteins, thus preserving activity in the fluorescent probe complex. Sulfhydryl reaction sites can be naturally available through free cysteine side chains, generated by reduction of disulfides, or created by the use of thiolation reagents (Chapter 1, Section 4.1).

The first two compounds discussed in this section are truly sulfhydryl-reactive, using the common iodoacetyl and maleimide functionalities, respectively. The third derivative, however, is not reactive directly with sulfhydryl groups, but contains a protected sulfhydryl which, after deprotection, can be used to react with other sulfhydryl-reactive cross-linkers.

5- (and 6)-Iodoacetamidofluorescein

The iodoacetamido derivatives of fluorescein possess a sulfhydryl-reactive iodoacetyl group (Chapter 1, Section 4.2, and Chapter 2, Section 2.1) at either the 5- or 6-carbon

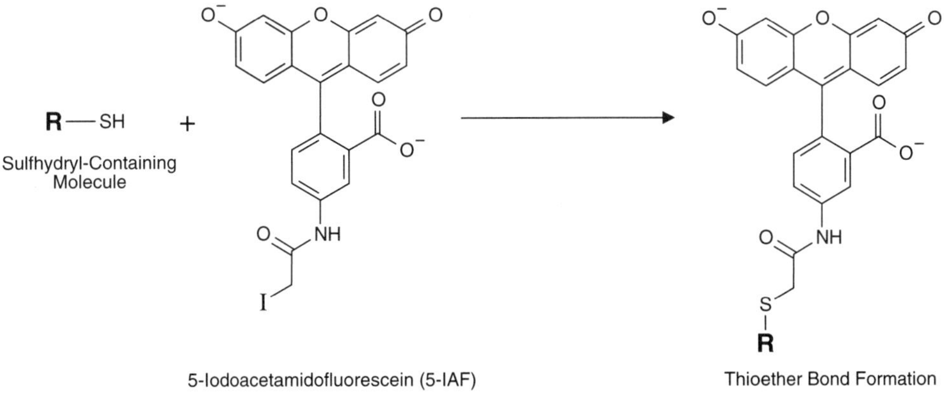

R—SH   +
Sulfhydryl-Containing
Molecule

5-Iodoacetamidofluorescein (5-IAF)   Thioether Bond Formation

Figure 205   5-IAF can be used to modify sulfhydryl-containing molecules, creating stable thioether linkages.

position on their lower ring. The isomers are commercially available in purified form, since some reactivity and specificity differences between the 5- and 6-derivatives toward various sulfhydryl sites in proteins may be observed. Both iodoacetamido derivatives are among the most intense fluorophores available for labeling biomolecules.

The iodoacetyl group of both isomers reacts with sulfhydryls under slightly alkaline conditions to yield stable thioether linkages (Fig. 205). They do not react with unreduced disulfides in cystine residues or with oxidized glutathione (Gorman *et al.*, 1987). The thioether bonds will be hydrolyzed under conditions necessary for complete protein hydrolysis prior to amino acid analysis.

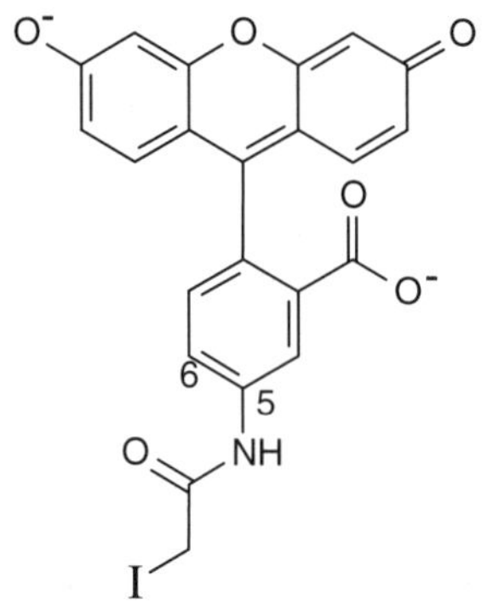

5-(and 6-)-Iodoacetamidofluorescein
MW 515.26
Excitation: 490–495 nm
Emission: 515–520 nm
ε at 492 nm = 80,000–85,000 $M^{-1}cm^{-1}$

Care must be taken to protect these reagents from light, not only to maintain the fluorescent yield of the fluorescein probe, but also to protect the iodoacetyl group from light-catalyzed breakdown. Iodoacetamidofluorescein is soluble in DMF and also in aqueous solutions maintained above pH 6. Concentrated stock solutions may be prepared in DMF prior to addition of a small aliquot to a reaction mixture. Protect solutions from light by wrapping in aluminum foil and working in subdued light.

The spectral properties of these derivatives are similar to those of native fluorescein. The excitation maximum occurs at about 490–495 nm and its emission peak at 515–520 nm, producing light in the green region of the spectrum. The extinction coefficient of 5-iodoacetamidofluorescein at its wavelength of maximum absorbance, 491 nm, is 82,000 $M^{-1}cm^{-1}$ (pH 9), whereas the extinction coefficient of 6-iodoacetamidofluorescein is 77,000 $M^{-1}cm^{-1}$ at 493 nm (pH 9).

The 5-iodoacetamido derivative of fluorescein (5-IAF) has been used to label numerous proteins and other biomolecules, including actin (Plank and Ware, 1987), myosin (Aguirre *et al.*, 1986), troponin (Greene, 1986), hemoglobin (Hirsch *et al.*, 1986), and sulfhydryl-containing proteins separated by SDS electrophoresis (Gorman, 1984).

The 6-iodoacetamido derivative (6-IAF) has been used to label myosin (Ando, 1984), actin (Konno and Morales, 1985), microtubule-associated proteins (Scherson *et al.*, 1984), and histones (Cocco *et al.*, 1986).

The following protocol for labeling proteins with 5-IAF is adapted from Gorman (1984). It is a bit unusual in that it involves reduction of disulfides with DTT and immediate reaction with 5-IAF in excess without removal of excess reductant. The procedure can be changed to include a gel filtration step after disulfide reduction to remove excess DTT, but in any case, it should be optimized for each protein to be modified. An alternative to the use of DTT to produce sulfhydryls is thiolation with a compound that can generate free thiols upon reaction with a protein (Chapter 1, Section 4.1).

Protocol

1. Dissolve a protein containing disulfide residues at a concentration of 5–10 mg/ml in 0.1 M ammonium carbonate containing 1% SDS and 20 mM DTT.
2. React for 16 h at 0°C and 2 h at room temperature.
3. Add a five-fold molar excess of 5-IAF (Pierce) over the amount of DTT present. The fluorescent probe may be solubilized in DMF prior to addition of a small aliquot to the reaction mixture. Do not exceed 10% DMF in the final aqueous solution.
4. React for 2 h at room temperature in the dark. Maintain the pH at 7.5–8 by the addition of small quantities of 6 M NaOH as necessary.
5. Precipitate the protein by the addition of 9 Vol of methanol at −20°C.
6. Collect the protein pellet by centrifugation at 8000 × g for 15 min (4°C).

An alternative protocol for labeling sulfhydryl-containing proteins that does not require DTT reduction can be found in a method adapted from Ando (1984).

Protocol

1. Prepare a 20 mM 6-IAF (Pierce) solution by dissolving 10.3 mg per ml of DMF. Prepare fresh and protect from light.
2. Dissolve the protein to be modified at a concentration of 5–10 mg/ml in 20 mM 2-[[tris(hydroxymethyl)methyl]amino]ethanesulfonic acid (TES), pH 7.
3. Slowly add 25–50 μl of the 6-IAF solution to each milliliter of the protein solution while mixing.
4. React for 2 h at 4°C in the dark.
5. Remove excess reactant and reaction by-products by gel filtration using Sephadex G-25 or the equivalent.

Fluorescein-5-maleimide

Fluorescein-5-maleimide is a fluorescent probe containing a sulfhydryl-reactive maleimide group on its lower ring structure. Modification of sulfhydryl-containing molecules under physiological pH conditions results in stable thioether bonds (Chapter 2, Section 2.2) (Fig. 206). The derivative thus possesses fluorescent properties closely characteristic of fluorescein molecules: excitation wavelength = 490 nm; emission wavelength = 515 nm, in the green spectral region. Conjugates prepared by fluorescein-5-maleimide are among the most intensely fluorescent probes available. The reactivity of the maleimide group is similar to that of the iodoacetyl derivative, discussed previously.

Thus, this reagent can be used to label fluorescently proteins and other biomolecules containing free sulfhydryl residues. If there are no —SH groups available,

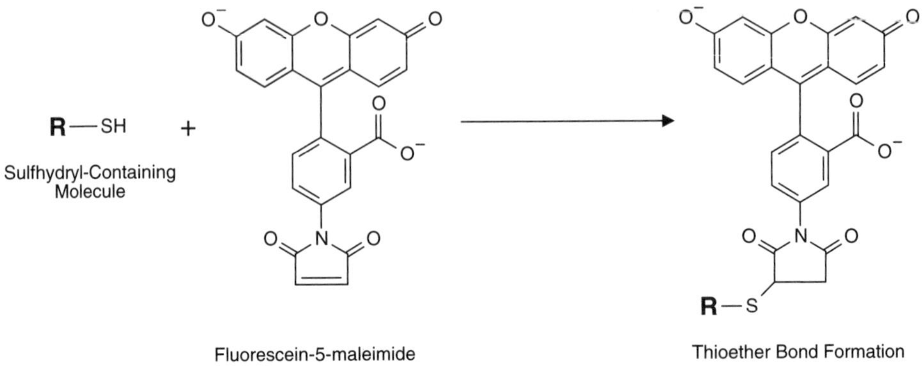

R—SH   +

Sulfhydryl-Containing
Molecule

Fluorescein-5-maleimide

Thioether Bond Formation

Figure 206 Fluorescein-5-maleimide can be used to modify sulfhydryl groups, forming thioether bonds.

their creation can be accomplished by reduction of indigenous disulfides or through the use of various thiolation reagents (Chapter 1, Section 4.1).

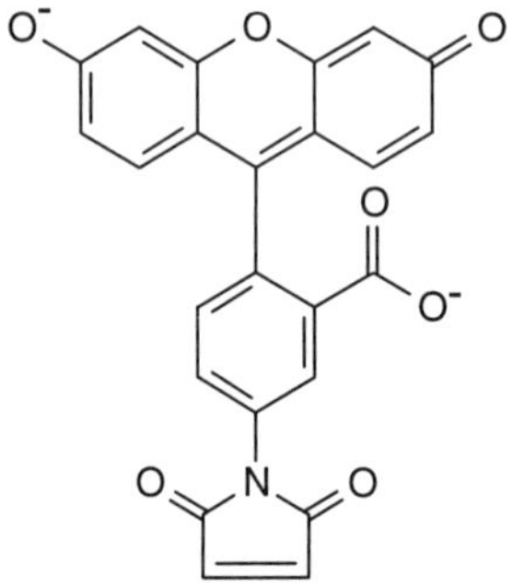

Fluorescein-5-maleimide
MW 427
Excitation: 490–495 nm
Emission: 515–520 nm
ε at 490 nm = 83,000 $M^{-1}cm^{-1}$

Fluorescein-5-maleimide is slightly soluble in aqueous solutions above pH 6 ($\sim$1 mM concentration). It may be dissolved in DMF at higher concentrations and a small addition of this solution made to an aqueous reaction mixture to initiate labeling. Do not exceed 10% DMF in the reaction buffer to avoid protein precipitation. At pH 8 the reagent has an extinction coefficient at 490 nm of about 78,000 $M^{-1}cm^{-1}$.

Fluorescein-5-maleimide has been used in numerous applications, including labeling the transmembrane glycoprotein H-2K^k on both the N- and the C-terminal regions to investigate the structure of the molecule (Cardoza et al., 1984), for the determination of two different conformations of the protein actin (Konno and Morales, 1985), in the study of a bacterial sensory receptors (Falke et al., 1988), in the structural mapping of chloroplast coupling factor (Snyder and Hammes, 1984, 1985) for localization of the stilbenedisulfonate receptor on human erythrocytes (Rao et al., 1979), in investigating the calcium-dependent ATPase protein structure of sarcoplasmic reticulum (Bigelow and Inesi, 1991), and to study the movement of tRNA during peptide bond formation on ribosomes (Odom et al., 1990).

Protocol

1. Dissolve a sulfhydryl-containing protein or other macromolecule at a concentration of 1–10 mg/ml in 20 mM sodium phosphate, 0.15 M NaCl, pH 7.2. Other buffers within the pH range 6.5–7.5 may be used as long as they do not contain extraneous sulfhydryls.
2. Dissolve fluorescein-5-maleimide (Pierce) in DMF at a concentration of 10 mM (4.25 mg/ml). Protect from light.
3. In subdued lighting conditions, add 25–50 μl of the fluorescein solution to each milliliter of protein solution while mixing. Alternatively, determine the exact molar quantity of protein present and add a 25-fold molar excess of fluorescein-5-maleimide solution.
4. React for 2–4 h at room temperature in the dark. The reaction also may be done at 0–4°C, but allow at least 8 h for completion.
5. Immediately purify the derivative using gel filtration on a Sephadex G-25 column. Protect the solutions from light during the chromatography.

SAMSA-Fluorescein

5-{[2(and 3)-5-(Acetylmercapto)-succinoyl]amino}fluorescein (SAMSA-fluorescein) is a fluorescent probe containing a protected sulfhydryl group. In its protected state, the compound is unreactive. The acetyl protecting group can be removed by treatment with dilute NaOH at pH 10 (Fig. 207). The resulting free sulfhydryl derivative can be used to label —SH reactive cross-linkers or to couple with sulfhydryl residues on proteins and other molecules. After activating proteins with cross-linkers containing terminal maleimide, pyridyl disulfide, or iodoacetyl groups, SAMSA-fluorescein can be used to assess the level of modification. For instance, a maleimide-activated protein that has been derivatized with SMCC could be reacted with this reagent to yield a fluorescein derivative that can be assayed spectrofluorometrically for its level of fluorescence. Using the molar extinction coefficient for SAMSA-fluorescein (ε at 495 nm = 80,000 $M^{-1}cm^{-1}$), the molar level of incorporation of the label can be calculated. This

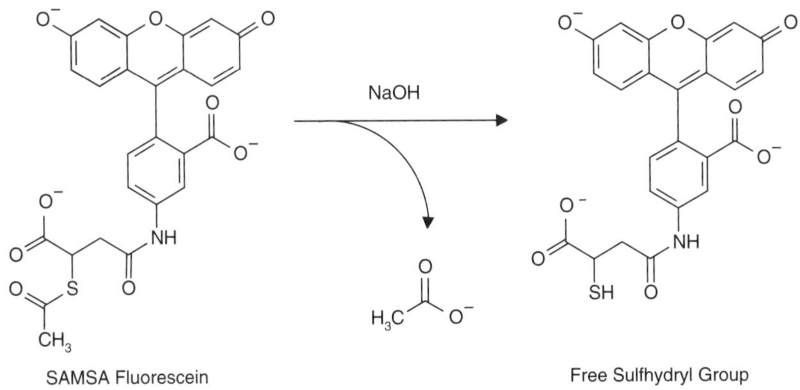

SAMSA Fluorescein   Free Sulfhydryl Group

Figure 207   SAMSA fluorescein contains a protect thiol that can be deblocked by treatment with hydroxylamine. The reagent then can be used to modify molecules containing sulfhydryl-reactive groups.

determination directly correlates to the original level of maleimide groups present on the protein.

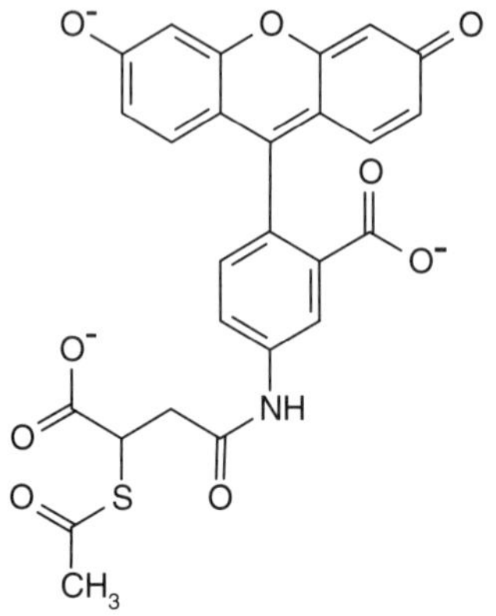

SAMSA Fluorescein
MW 522
Excitation = 495 nm
Emission = 518 nm

SAMSA-fluorescein is an orange solid compound. Dissolved in buffer at pH 9, its maximal wavelength of absorption or excitation is at 495 nm, while its emission wavelength maximum is 520 nm. The reagent and all solutions and derivatives made from it are light sensitive and should be stored in the dark. SAMSA-fluorescein is soluble in aqueous solutions above pH 6, but it can be dissolved in DMF to prepare a concentrated stock solution prior to addition of a small amount to a buffered reaction mixture.

Protocol

1. To deprotect the acetylated sulfhydryl, dissolve the desired amount of SAMSA-fluorescein (Pierce) in 100 mM NaOH, pH 10, at a concentration of 10 mg/ml.
2. React for 15 min at room temperature.
3. Lower the pH to 7–8 by the addition of solid sodium phosphate.
4. Add the required amount of deprotected SH-fluorescein to a protein or other macromolecule that had been modified to contain a sulfhydryl-reactive group. Use a 5- to 10-fold molar excess of SH-fluorescein to the expected amount of sylfhydryl-reactivity present.
5. React for 2 h at room temperature, protected from light.
6. Remove excess fluorescent probe by gel filtration using Sephadex G-25 or the equivalent.
7. Measure the absorbance of the derivative at 495 nm. Determine the level of fluorophore incorporation by using its molar extinction coefficient.

Aldehyde/Ketone- and Cytosine-Reactive Fluorescein Derivatives

Hydrazide groups directly react with aldehyde and ketone functional groups to form relatively stable hydrazone linkages (Chapter 2, Section 5.1). Two fluorescein derivatives are commonly available that contain hydrazide groups off their No. 5 carbons on the lower ring structure. Both may be used to label fluorescently aldehyde- or ketone-containing molecules. Although most biomolecules do not contain aldehyde or ketone groups in their native state, carbohydrates, glycoproteins, RNA, and other molecules containing sugar residues can be oxidized with sodium periodate to produce reactive formyl functional groups. The use of modification reagents that generate aldehydes upon coupling to a molecule also can be used to produce a hydrazide-reactive site (Chapter 1, Section 4.4).

DNA and RNA may be modified with hydrazide-reactive probes by reacting their cytosine residues with bisulfite to form reactive sulfone intermediates. These derivatives undergo transamination to couple hydrazide- (or amine)-containing probes (Draper and Gold, 1980) (Chapter 17, Section 2.1).

Fluorescein-5-thiosemicarbazide

Fluorescein-5-thiosemicarbazide is a hydrazide derivative of fluorescein that can spontaneously react with aldehyde- or ketone-containing molecules to form a covalent, hydrazone linkage (Fig. 208) (Pierce). It also can be used to label cytosine residues in DNA or RNA by use of the bisulfite activation procedure (Chapter 17, Section 2.1). The resulting fluorescent derivative exhibits an excitation maximum at a wavelength of 492 nm and a maximal emission wavelength of 519 nm when dissolved in buffer at pH 8.6. In the same buffered environment, the compound has an extinction coefficient of approximately 78,000 $M^{-1}cm^{-1}$ at 492 nm.

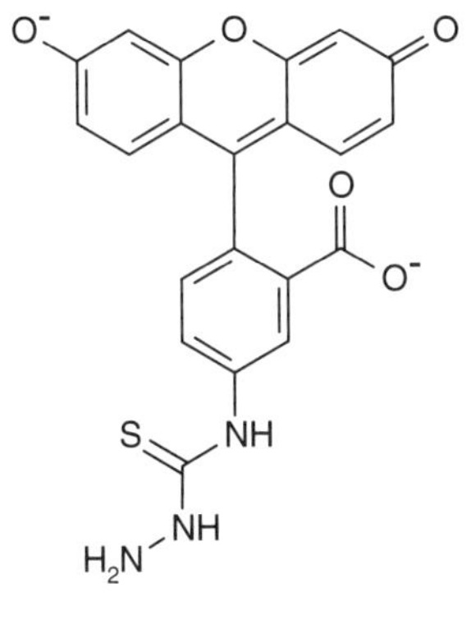

Fluorescein-5-thiosemicarbazide
MW 421
Excitation: 492 nm
Emission: 516 nm
ε at 492 nm = 85,000 $M^{-1}cm^{-1}$

Fluorescein-5-thiosemicarbazide is soluble in DMF or in buffered aqueous solutions at pH values above 7. The reagent may be dissolved in DMF as a concentrated stock solution before a small aliquot is added to an aqueous reaction medium. The compound itself and all solutions made with it should be protected from light to avoid decomposition of its fluorescent properties.

This hydrazide derivative of fluorescein has been used in a number of applications, including site-directed labeling of antibodies through their carbohydrate chains (Duijndam et al., 1988); labeling thrombin and anti-thrombin (Atha et al., 1964), the Na⁺/K⁺–ATPase glycoprotein (Lee and Fortes, 1985), and periodate-oxidized RNA (Odom et al., 1980, 1984; Ferguson and Yang, 1986; Friedrich et al., 1988); for the

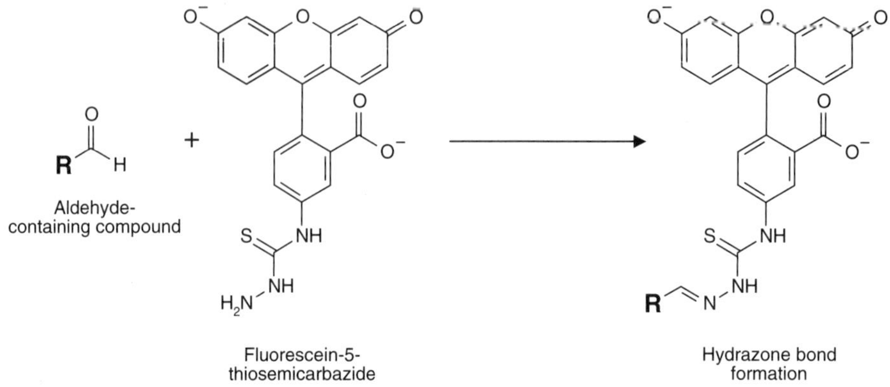

Figure 208 Fluorescein-5-thiosemicarbazide reacts with aldehyde groups to produce hydrazone linkages.

determination of carbonyl groups in proteins; and to detect oxidized glycoproteins in gels (Ahn *et al.*, 1987).

The following protocols are generalized for the labeling of cell surface glycoproteins or glycoproteins in solution. Some optimization may be necessary to achieve the best level of fluorescent modification for each particular application.

Protocol for Labeling Cell Surfaces

1. Add 10^6–10^8 cells/ml in a PBS solution (10 mM sodium phosphate, 0.15 M NaCl, pH 7.4) containing 1 mM sodium periodate and incubate on ice for 30 min in the dark. This level of periodate addition will target the oxidation only to sialic acid residues (Chapter 1, Section 4.4). If additional sites of glycosylation also are to be labeled, increase the periodate concentration to 10 mM and do the reaction at room temperature in the dark.

2. Centrifuge and wash cells several times with PBS. Some protocols include a quench step wherein excess sodium periodate is eliminated by addition of glycerol. This can be accomplished by adding 3 Vol of 0.1 M glycerol in PBS prior to centrifugation.

3. Resuspend cells in PBS containing 0.5 mg/ml fluorescein-5-thiosemicarbazide.

4. Incubate 30 min in the dark at room temperature.

5. To reduce the hydrazone bonds to more stable linkages, cool the cell suspension to 0°C and add an equal volume of 30 mM sodium cyanoborohydride in PBS. Incubate for 40 min. Note: if the presence of a reducing agent is detrimental to protein activity, eliminate the reduction step. In most cases, the hydrazone linkage is stable enough for fluorescent labeling experiments.

6. Centrifuge and wash cells extensively with PBS.

Protocol for Labeling Glycoproteins in Solution

1. Dissolve the glycoprotein(s) to be labeled in ice-cold 1 mM sodium periodate, 10 mM sodium phosphate, 0.15 M NaCl, pH 7.4, for the exclusive oxidation of sialic acid residues. For general carbohydrate oxidation, increase the periodate concentration to 10 mM in PBS at room temperature.

2. React for 30 min on ice (for sialic acids) or at room temperature (for other polysaccharide residues).

3. Remove excess reactant by gel filtration using a column of Sephadex G-25 using PBS, pH 7.4. Some protocols use a quenching agent to remove excess periodate prior to gel filtration. This can be done by adding glycerol to a final concentration of 0.1 M.

4. To the purified, oxidized glycoprotein(s), add fluorescein-5-thiosemicarbazide to a final concentration of 0.5 mg/ml.

5. React for 30 min at room temperature in the dark.

6. To reduce the hydrazone bonds to more stable linkages, cool the solution to 0°C and add an equal volume of 30 mM sodium cyanoborohydride in PBS. Incubate for 40 min. Note: if the presence of a reducing agent is detrimental to protein activity, eliminate the reducing step. In most cases, the hydrazone linkage is stable enough for fluorescent labeling experiments.

7. Purify the fluorescently labeled glycoprotein(s) by gel filtration using a Sephadex G-25 column.

5-(((2-(Carbohydrazino)methyl)thio)acetyl)-aminofluorescein

Another hydrazine derivative of fluorescein, 5-(((2-(Carbohydrazino)methyl)thio)-acetyl)-aminofluorescein, contains a longer spacer arm off its No. 5 carbon atom of its lower ring than fluorescein-5-thiosemicarbazide, described previously (Molecular Probes). The reagent can be used to react spontaneously with aldehyde- or ketone-containing molecules forming a hydrazone linkage (Fig. 209). It also can be used to label cytosine residues in DNA or RNA by use of the bisulfite activation procedure (Chapter 17, Section 2.1). The resulting fluorescent derivative exhibits a maximal excitation at 490 nm and a maximal luminescence emission peak at 516 nm when dissolved in buffer at pH 8. In the same buffered environment, the compound has an extinction coefficient of approximately 75,000 $M^{-1}cm^{-1}$ at 490 nm.

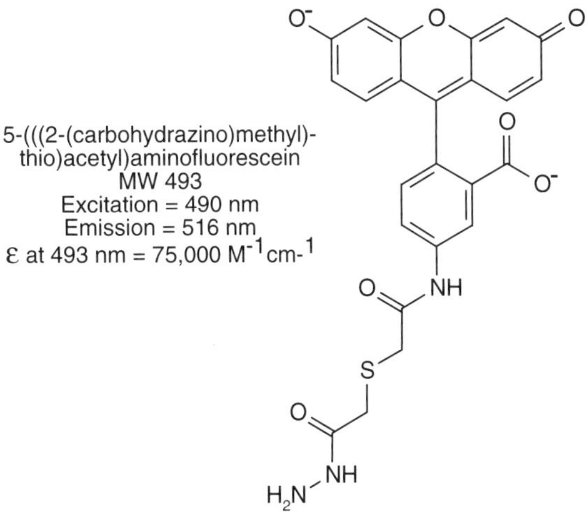

5-(((2-(carbohydrazino)methyl)-
thio)acetyl)aminofluorescein
MW 493
Excitation = 490 nm
Emission = 516 nm
ε at 493 nm = 75,000 $M^{-1}cm^{-1}$

The fluorescent probe, 5-(((2-(carbohydrazino)methyl)thio)acetyl)-aminofluorescein, is soluble in DMF or in buffered aqueous solutions at pH values above 7. The reagent may be dissolved in DMF as a concentrated stock solution before adding a

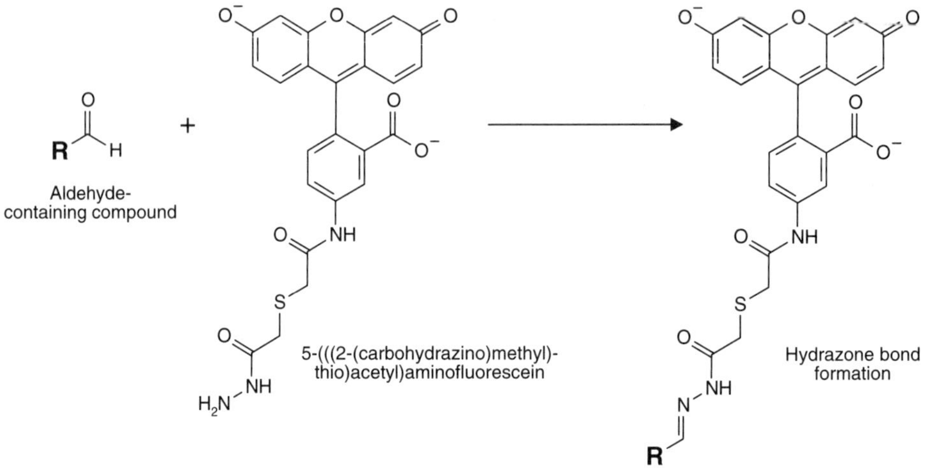

Figure 209 This carbohydrazide-containing fluorescein derivative can be used to modify aldehyde-containing molecules. Glycoconjugates may be labeled with this reagent after treatment with sodium periodate to produce aldehydes.

small aliquot to an aqueous reaction medium. The compound itself and all solutions made with it should be protected from light to avoid decomposition of its fluorescent properties.

The methods for using this reagent in labeling glycoproteins on cell surfaces or in solution are similar to those described for fluorescein-5-thiosemicarbazide, above.

1.2. Rhodamine Derivatives

Rhodamine and its derivatives are popular fluorescent probes for labeling all types of biomolecules. Their fluorescent character is created by the presence of a planar, multi-ring aromatic structure similar to fluorescein, but with nitrogen atoms replacing the oxygens on the outer rings (Fig. 210). Fluorescent modification reagents based on rhodamine are derivatives of this basic structure. Activated rhodamine probes have reactive groups prepared through substitutions off the No. 5 or 6 carbons of its lower

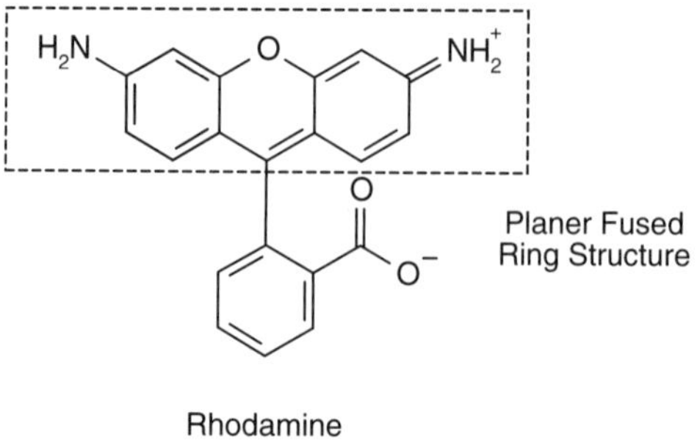

Figure 210 The basic structure of rhodamine derivatives.

ring. These derivatives provide reactivity toward particular functional groups in bio-molecules, allowing rapid labeling of proteins and nucleic acids. Other alterations to the basic rhodamine structure modulate its fluorescent character, creating more intense or stable fluorophores, or changing its wavelength of excitation and emission. Many such derivatives are now commercially available.

The tetramethylrhodamine derivative, for instance, has two methyl groups attached to each nitrogen on its outer rings. Activated forms of tetramethylrhodamine may be the most common derivatives of rhodamine currently employed for fluorescent labeling. Another useful derivative is rhodamine B, which contains two ethyl groups on each nitrogen as well as a carboxylate group at the No. 3 position on its lower ring. Rhodamine 6G adds two methyl groups on the outer rings as well as an ethyl ester group off rhodamine B's carboxylate. Rhodamine 110 contains no substituents on the upper nitrogens and only the carboxylate on the lower ring. Sulforhodamine B possesses rhodamine B's two ethyl groups on each nitrogen of the upper rings, but has two sulfonates at the No. 3 and 5 positions of its lower ring. This derivative is often called Lissamine rhodamine B—Lissamine being a trademark of Imperial Chemical Industries. Another popular derivative of rhodamine, sulforhodamine 101, goes by the name of Texas Red (a trademark of Molecular Probes). This derivative has intense luminescent properties that take it the farthest into the red region of the spectrum. The basic structures of these rhodamine derivatives are shown in Fig. 211. The corresponding commercially available rhodamine fluorophores usually contain additional reactive groups, on the No. 5 or 6 carbons of the lower ring, to permit coupling to target molecules (Kodak, Molecular Probes, Pierce).

Rhodamine derivatives have effective excitation wavelengths within the visible light spectrum in the approximate range from the low- to high-500 nm, depending on the particular derivative. Their associated emission wavelengths occur from the mid- to high-500 nm—with Texas Red derivatives typically over 600 nm—within the orange-to-red visible spectrum. The quantum yield of rhodamine derivatives is generally less than that of fluorescein—only about 25%. However, its fluorescent intensity fades more slowly than fluorescein when it is dissolved in buffers, exposed to light, or stored for extended periods. In addition, its orange-to-red luminescence is in stark contrast to the green of fluorescein, thus these two types of probes form an ideal pair for use in double staining techniques, especially in fluorescent microscopy.

The following sections describe the most important rhodamine derivatives commonly used to label biomolecules.

Amine-Reactive Rhodamine Derivatives

Four forms of amine-reactive rhodamine probes are commonly available. Two of them are based on the tetramethyl derivatives of the fundamental rhodamine structure, one is based on the sulforhodamine B or Lissamine derivative, and the last is the sulforhodamine 101 or Texas Red derivative. All of them react under alkaline conditions with primary amines in proteins and other molecules to form stable, highly fluorescent complexes.

Tetramethylrhodamine-5- (and 6)-isothiocyanate (TRITC)

TRITC is one of the most popular fluorescent probes available. The isothiocyanate derivative of tetramethylrhodamine is synthesized by modification of its lower ring at

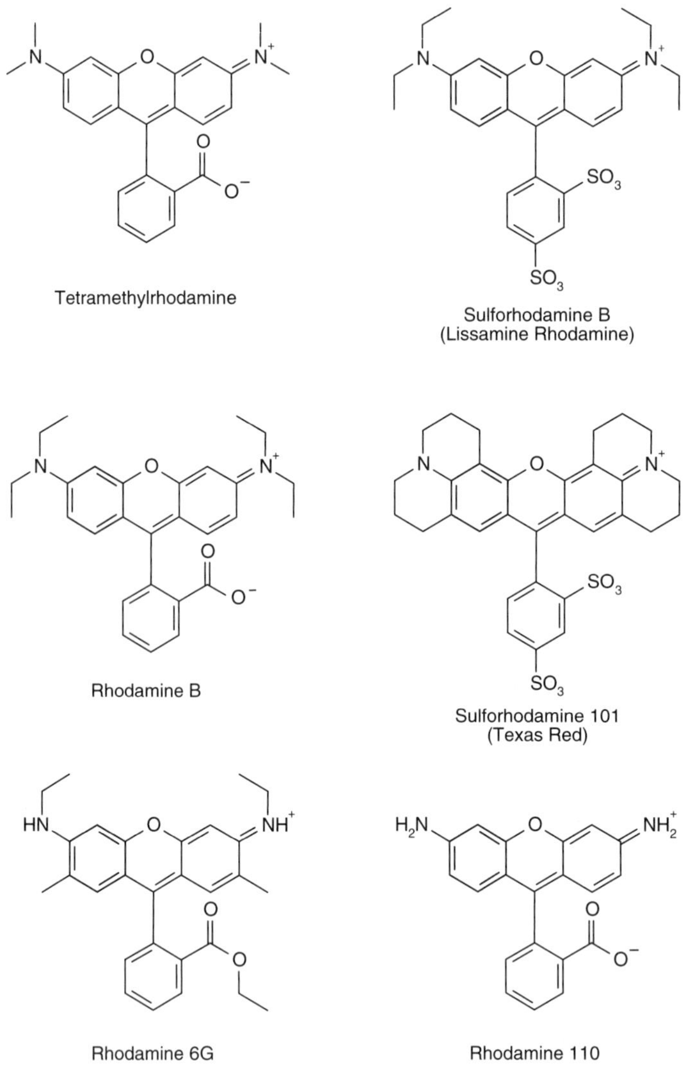

Figure 211   The primary rhodamine derivatives useful for fluorescent labeling.

the 5- or 6-carbon positions. The two resulting isomers are almost identical in their reactivity but slightly different in their spectral properties, including excitation and emission wavelengths and intensities. The chemical differences in the isomers, however, may affect the separation of modified proteins from excess probe or the analysis of tagged molecules by electrophoresis. For this reason, most manufacturers offer the mixed isomers as well as the purified 5- or 6-isothiocyanate derivatives individually.

Isothiocyanates react with nucleophiles such as amines, sulfhydryls, and the phenolate ion of tyrosine side chains (Podhradsky *et al.*, 1979). The only stable product, however, is with primary amine groups, and so TRITC is almost entirely selective for modifying ε- and N-terminal amines in proteins. The reaction involves attack of the nucleophile on the central, electrophilic carbon of the isothiocyanate group (Fig. 212).

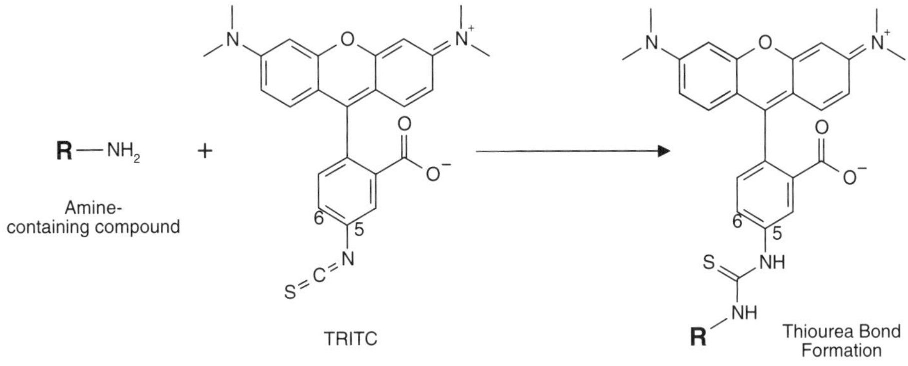

Figure 212 TRITC reacts with amine-containing molecules to create an isothiourea linkage.

The resulting electron shift creates a thiourea linkage between TRITC and the protein with no leaving group.

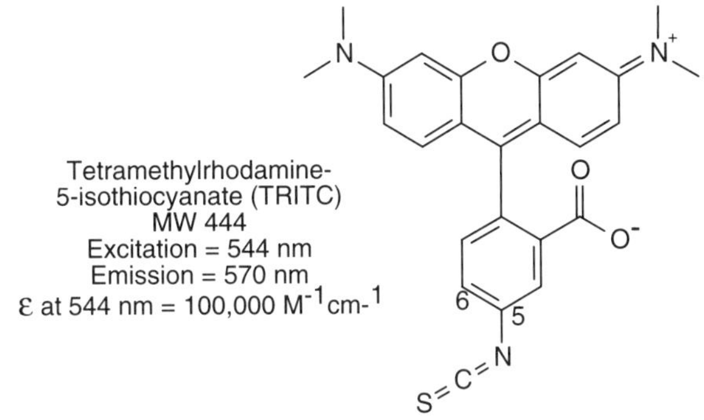

Tetramethylrhodamine-
5-isothiocyanate (TRITC)
MW 444
Excitation = 544 nm
Emission = 570 nm
ε at 544 nm = 100,000 M^{-1}cm^{-1}

TRITC is relatively insoluble in water, but it can be dissolved in DMF or DMSO as a concentrated stock solution prior to its addition to an aqueous reaction mixture. The isothiocyanate group is reasonably stable in aqueous solution for short periods, but will degrade by hydrolysis. TRITC also is more stable to photobleaching than FITC (This chapter, Section 1.1), and its absorption and emission spectra are less sensitive to environmental conditions, such as pH. It is best, however, to use only fresh reagent for modification purposes. Storage should be done under desiccated conditions and protected from light, at −20°C.

The fluorescent properties of TRITC (mixed isomers) include an absorbance maximum at about 544 nm and an emission wavelength of 570 nm. Fluorescent quenching of the molecule is possible. Under concentrated conditions, rhodamine-to-rhodamine interactions result in self-quenching, which reduces its luminescence yield. This phenomenon can occur with TRITC-tagged molecules, as well. If derivatization of a protein is done at too high a level, the resultant quantum yield of the conjugate will be depressed from expected values. Typically, modifications of proteins involve adding no more than 8–10 rhodamine molecules per molecule of protein, with a 4–5 substitution level considered optimal.

TRITC has been used in numerous applications involving fluorescence detection, including double-staining techniques with fluorescein-labeled probes (Mossberg and

Ericsson, 1990), the synthesis of fluorescently labeled DNA probes (L. M. Smith *et al.*, 1985), as a label in homogeneous immunoassay systems (Nithipatikom and McGown, 1987), to investigate specific interactions of proteins with cells surfaces (Hochman *et al.*, 1988), and as an important fluorescent tag of antibodies in immunohistochemical staining techniques (Davidson and Hilchenbach, 1990).

The level of TRITC modification in a macromolecule can be determined by measuring its absorbance at or near its characteristic absorption maximum (~575 nm) (Van Dalen and Haaijman, 1974). The number of fluorochrome molecules per molecule of protein is known as the *F/P* ratio. This value should be measured for all derivatives prepared with fluorescent tags. The ratio is especially important in predicting the behavior of antibodies labeled with TRITC. For a TRITC-labeled protein, the ratio of absorbency at 575 to 280 nm should be between 0.3 and 0.7.

A general protocol for the modification of proteins, particularly immunoglobulins, with TRITC is given below. Modifications to the amount of reagent added to the reaction may be done to optimize the *F/P* ratio.

Protocol

1. Prepare a protein solution in 0.1 *M* sodium carbonate, pH 9, at a concentration of at least 2 mg/ml.
2. In a darkened lab, dissolve TRITC (Pierce) in dry DMSO at a concentration of 1 mg/ml. Do not use old TRITC, as breakdown of the isothiocyanate group over time may decrease coupling efficiency. Protect from light by wrapping in aluminum foil or using amber vials.
3. In a darkened lab, slowly add 50 μl of TRITC solution to each milliliter of protein solution. Gently mix the protein solution as the TRITC is added.
4. React for at least 8 h at 4°C in the dark.
5. The reaction may be quenched by the addition of ammonium chloride to a final concentration of 50 m*M*. Some protocols also include at this point the addition of 0.1% xylene cylanol and 5% glycerol as a photon absorber and protein stabilizer, respectively. React for a further 2 h to stop the reaction by blocking remaining isothiocyanate groups.
6. Purify the derivative by gel filtration using a PBS buffer or another suitable buffer for the particular protein being modified. The use of Sephadex G-25 (Pharmacia) or similar matrices with low exclusion limits work well. To obtain complete separation, the column size should be 15–20 times the size of the applied sample. Fluorescent molecules often nonspecifically stick to the gel filtration support, so reuse of the column is not recommended.

NHS-Rhodamine

NHS-rhodamine is an amine-reactive fluorescent probe that contains a carboxysuccinimidyl ester group off the No. 5 or 6 carbons on rhodamines's lower ring structure (Kellogg *et al.*, 1988). The 5- and 6-isomers are virtually identical in their reactivity and fluorescent characteristics. Similar to TRITC (described previously), NHS-rhodamine can be used to label proteins and other macromolecules that contain primary amine groups. The isomeric forms of the fluorescent probe are available in mixed and purified forms (Molecular Probes, Pierce). The pure forms are particularly important for labeling nucleic acid probes that will be separated by electrophoresis (Chehab and Kan, 1989). The NHS ester labeling reaction proceeds rapidly at slightly alkaline pH values, resulting in a stable, amide-linked derivative (Chapter 2, Section 1.4, and Chapter 4, Section 1; Fig. 213).

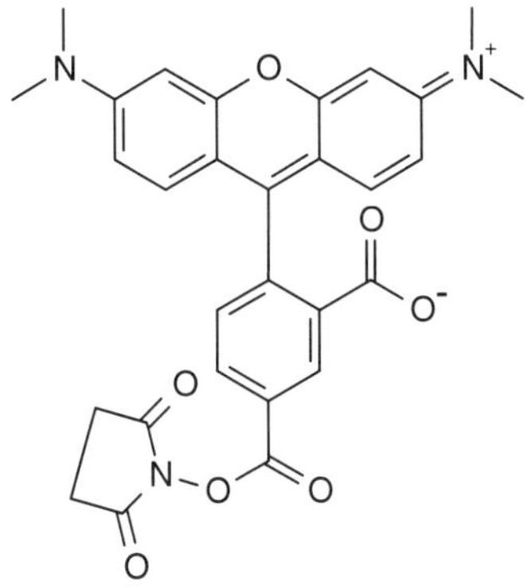

NHS-Rhodamine;
5-Carboxytetramethyl-
rhodamine, succinimidyl ester
MW 528
Excitation = 546 nm
Emission = 579 nm
ε at 546 nm = 100,000 $M^{-1}cm^{-1}$

The fluorescent properties of NHS-rhodamine are similar to those of TRITC. The wavelength of maximal absorbance or excitation for the reagent is 544 nm and its emission maximum is 576 nm, exhibiting a visual color of orange-red. Its molar extinction coefficient at 546 nm in a methanol environment is 63,000 $M^{-1}cm^{-1}$. Other components in solution as well as the pH (in aqueous buffers) can change this value.

NHS-rhodamine is insoluble directly in aqueous solution and should be dissolved in organic solvent prior to addition of a small aliquot to a buffered reaction medium. Concentrated stock solutions may be prepared in DMSO or DMF. Such solutions are relatively stable for short periods if protected from light, but should be prepared fresh.

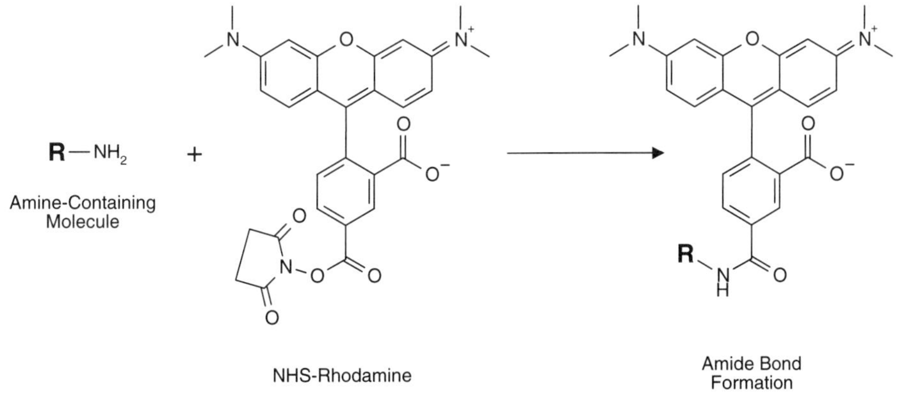

R—NH₂   +

Amine-Containing
Molecule

NHS-Rhodamine

Amide Bond
Formation

Figure 213   NHS-rhodamine can be used to label amine-containing molecules via its NHS ester group.

Reaction conditions should be maintained at the optimal reactivity for NHS esters—pH 7–9.

NHS-rhodamine has been used in numerous applications, including the detection of specific DNA sequences (Chehab and Kan, 1989), studying the behavior of microtubules and actin filaments in living *Drosophila* embryos (Kellogg *et al.*, 1988), investigation into the light-initiated breakup of microtubules (Vigers *et al.*, 1988), studying ω-conotoxin-sensitive channels in neurons (Jones *et al.*, 1989), investigating growth cones during axon elongation (Tanaka and Kirschner, 1991), and studying the pathways of mitotic spindle assembly *in vitro* (Sawin and Mitchison, 1991).

The following generalized protocol relates to the labeling of IgG with NHS-rhodamine. Optimization of the level of rhodamine incorporation may have to be done with other proteins or macromolecules.

Protocol

1. Dissolve the immunoglobulin to be labeled in ice-cold, 50 mM sodium bicarbonate, pH 8.5, at a concentration of 10 mg/ml.
2. Dissolve NHS-rhodamine at a concentration of 1 mg/ml in DMSO. Protect from light.
3. In a darkened lab, slowly add 50–100 μl of the NHS-rhodamine solution to each milliliter of the antibody solution with mixing. Wrap the vessel with aluminum foil to protect from light.
4. Place the sample on ice and react for 2 h.
5. Remove unreacted NHS-rhodamine and reaction by-products by gel filtration or dialysis.

Lissamine Rhodamine B Sulfonyl Chloride

The Lissamine form of rhodamine B consists of diethyl modifications on the two nitrogens of the upper rings of the basic rhodamine molecule as well as two sulfonate groups added at the 3- and 5-carbon positions of the lower ring. Lissamine rhodamine

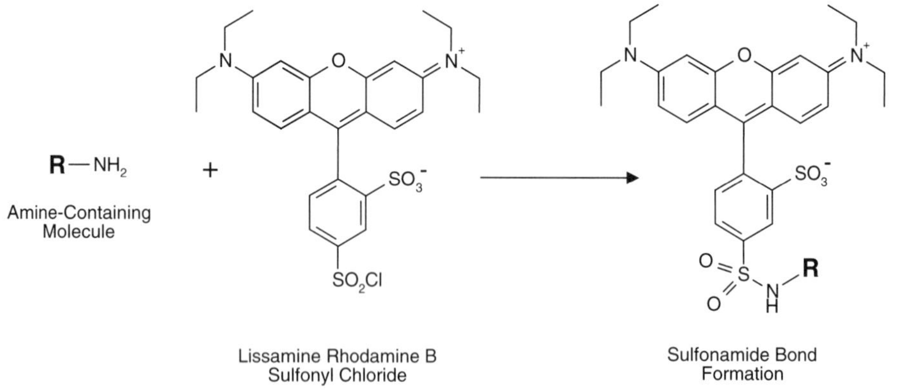

Figure 214 Lissamine rhodamine B sulfonyl chloride reacts with amine-containing molecules to produce stable sulfonamide bonds.

B sulfonyl chloride is an amine-reactive reagent made by converting the 5-sulfonate group over to the reactive sulfonyl halide. Reaction with proteins and other amine-containing molecules results in the formation of sulfonamide bonds (Fig. 214). Lissamine is a trademark of Imperial Chemical Industries.

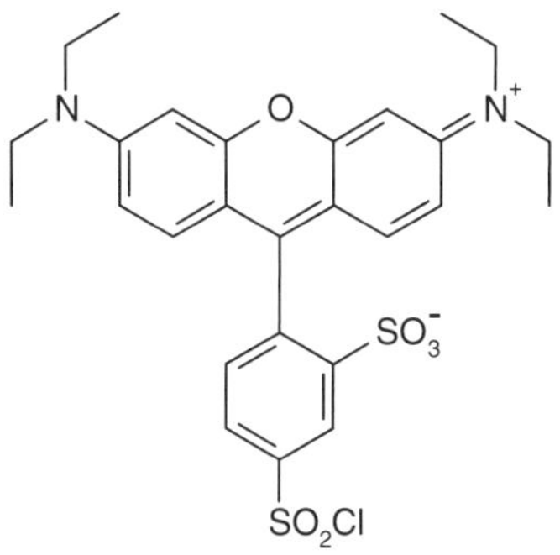

Lissamine Rhodamine B
Sulfonyl Chloride
MW 577
Excitation = 556 nm
Emission = 576 nm
ε at 556 nm = 100,000 $M^{-1}cm^{-1}$

The spectral characteristics of protein conjugates made with Lissamine rhodamine B derivatives are of longer wavelength than those of tetramethylrhodamine—more toward the red region of the spectrum. In addition, modified proteins have better chemical stability and are somewhat easier to purify than those made from TRITC (discussed previously). Lissamine derivatives also make more photostable probes than the fluorescein derivatives (Chapter 8, Section 1.1).

Lissamine rhodamine B sulfonyl chloride is relatively insoluble in water, but may be dissolved in DMF prior to the addition of a small aliquot to an aqueous reaction. Do not dissolve in DMSO, as sulfonyl chlorides will readily react with this solvent (Boyle, 1966). The compound has a maximal absorptivity at 556 nm with an extremely high extinction coefficient of up to 93,000 $M^{-1}cm^{-1}$ (in methanol) in highly purified form. Its emission maximum occurs at 576 nm, emitting red luminescence.

A sulfonyl chloride rapidly reacts with amines on proteins and other molecules to form stable sulfonamide bonds. It also may react with tyrosine —OH groups, aliphatic alcohols, thiols, and imidazole groups (such as histidine side chains). Conjugates of sulfonyl chlorides with sulfhydryls and imidazole rings are unstable, while esters formed with alcohols are subject to nucleophilic displacement (Nillson and Mosbach, 1984; Scouten and Van der Tweel, 1984). The only stable derivative with proteins therefore is the sulfonamide, formed by reaction with ε-lysine and N-terminal amines. Optimal conditions for coupling are non-amine-containing buffers in the pH range 9–10. Phosphate, bicarbonate, or borate buffers are recommended for the modification reaction. Avoid the presence of other nucleophiles that can cross-react with the sulfonyl chloride (e.g., amine-containing components or sulfhydryl-containing reducing agents). In aqueous solutions, hydrolysis is a competing reaction, but occurs more slowly with sulfonyl halides than with the acid chlorides of carboxylate groups. The unreacted, hydrolyzed probe is the water-soluble sulforhodamine B fluorophore, which is easily removed by gel filtration or dialysis.

Lissamine rhodamine B sulfonyl chloride has been used in numerous applications, including multiple-labeling techniques in microscopy (Wessendorf, 1990), for confocal microscopy techniques (Isien and Waggoner, 1990),in the study of fibronectin receptors (Duband *et al.*, 1988), for investigations into microtubule and intermediate filament association (Geiger and Singer, 1980), for the labeling of glycoconjugates (Wilchek *et al.*, 1980), for studying regulation of the $Na^+,K^+-ATPase$ system (Sipe *et al.*, 1991), and for investigations into the redox potential within mitochondria (Chazotte and Hackenbrock, 1991).

The following protocol is a general guide for labeling biological macromolecules with Lissamine rhodamine B sulfonyl chloride. Optimization of the fluorophore incorporation level (F/P ratio) may have to be done for specific labeling experiments.

Protocol

1. Dissolve the amine-containing macromolecule to be labeled (i.e., a protein) in 0.1 M sodium carbonate/bicarbonate buffer, pH 9, at a concentration of 1–5 mg/ml.
2. Dissolve Lissamine rhodamine B sulfonyl chloride (Kodak, Molecular Probes) in DMF at a concentration of 1–2 mg/ml. Protect from light and use immediately.
3. In a darkened lab and with gentle mixing, slowly add 50–100 μl of the fluorophore solution to the protein solution.
4. React for 1 h at room temperature in the dark.
5. Remove excess fluorophore by gel filtration using a column of Sephadex G-25 or by dialysis.

Texas Red Sulfonyl Chloride

Texas Red sulfonyl chloride is the active halogen derivative of sulforhodamine 101. This important derivative of the basic rhodamine molecule possesses dual aliphatic rings off the upper-ring nitrogens and sulfonate groups on the No. 3 and 5 carbon

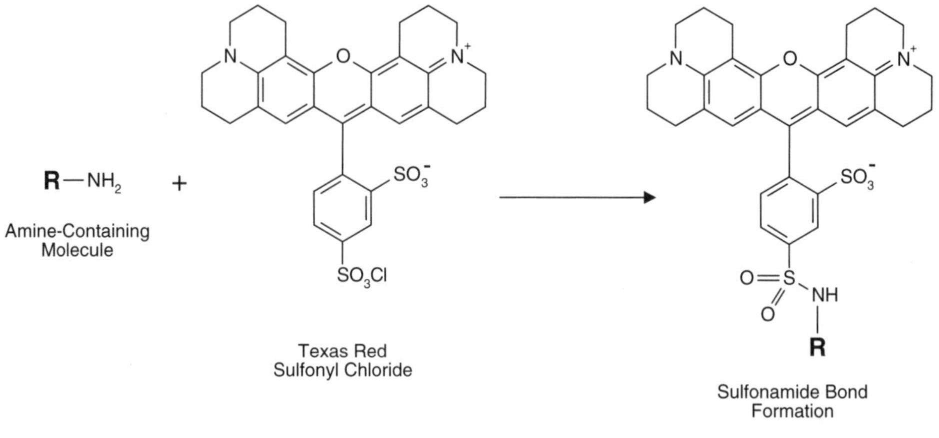

Figure 215 Texas Red sulfonyl chloride can be used to label amine-containing molecules through sulfonamide bond formation.

atoms of its lower ring component. The sulfonyl chloride group can react with primary amines in proteins and other molecules to form stable sulfonamide bonds (Fig. 215). The group, however, can hydrolyze in the presence of moisture. For this reason, only fresh Texas Red sulfonyl chloride should be used for modification experiments.

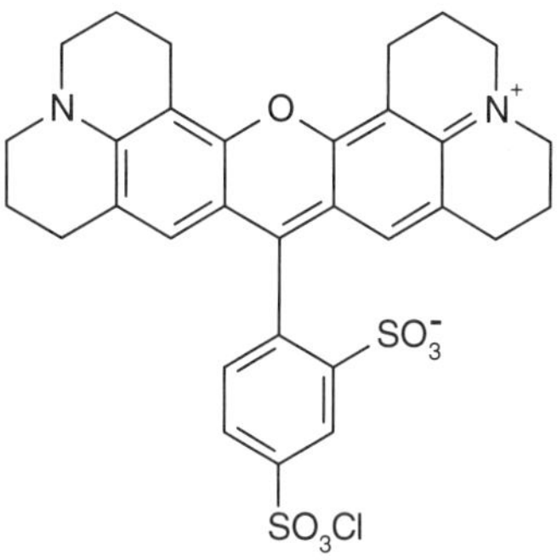

Texas Red Sulfonyl Chloride
MW 577
Excitation = 556 nm
Emission = 576 nm
ε at 556 nm = 93,000 $M^{-1}cm^{-1}$

The intense Texas Red fluorophore has a quantum yield that is inherently higher than the tetramethylrhodamine or Lissamine rhodamine B derivatives. Texas Red's luminescence is shifted maximally into the red region of the spectrum, and its emission peak only minimally overlaps with that of fluorescein. This makes Texas Red derivatives among the best choices of labels for use in double staining techniques.

Texas Red sulfonyl chloride has a maximal excitation at 589 nm and a maximum emission at 615 nm when dissolved in methanol. The extinction coefficient of the compound dissolved in acetonitrile is 85,000 $M^{-1}cm^{-1}$ at 596 nm. The only disadvantage of this fluorophore is its poor excitation by the standard argon laser at 488 nm. However, since both Texas Red and fluorescein are weakly excited by an argon laser at 514 nm, it makes them fairly good pairs for use in laser confocal microscopy or flow cytometry (Mossberg and Ericsson, 1990). The fluorophore is particularly appropriate for excitation by the 568-nm line produced by an argon–krypton mixed laser used on some confocal devices. Compared to other rhodamine derivatives, Texas Red fluorophores display low background in staining techniques and are among the most photostable probes available.

Texas Red sulfonyl chloride is soluble in DMF or acetonitrile and may be dissolved as a concentrated stock solution in either solvent prior to the addition of a small aliquot to an aqueous reaction medium. Avoid the use of DMSO, as sulfonyl chlorides react with this solvent (Boyle, 1966). The solid and all solutions made from it must be protected from light to avoid photodecomposition. Prepare the stock solution fresh immediately before use.

A sulfonyl chloride group rapidly reacts with amines in the pH range 9–10 to form stable sulfonamide bonds. Under these conditions, it also may react with tyrosine —OH groups, aliphatic alcohols, thiols, and histidine side chains. Conjugates of sulfonyl chlorides with sulfhydryls and imidazole rings are unstable, while esters formed with alcohols are subject to nucleophilic displacement (Nillson and Mosbach, 1984; Scouten and Van der Tweel, 1984). The only stable derivative with proteins therefore is the sulfonamide, formed by reaction with ε-lysine and N-terminal amines. For coupling reaction media use non-amine-containing buffers, including phosphate, borate, or bicarbonate (avoid Tris, imidazole, or glycine).

A suggested protocol on the use of this fluorescent probe is described below. Optimization may be necessary to achieve the best level of fluorescent modification (F/P ratio) for a particular application.

Protocol

1. Dissolve the protein or macromolecule to be labeled in 0.1 M sodium carbonate, pH 9, at a concentration of 1–5 mg/ml.
2. Dissolve Texas Red sulfonyl chloride (Pierce) in acetonitrile at a concentration of 20 mg/ml. Prepare fresh and protect from light.
3. In subdued lighting conditions, add 50 μl of the Texas Red sulfonyl chloride solution to each milliliter of the protein solution. Mix well.
4. React for 1 h at room temperature.
5. Remove excess fluorophore and reaction by-products by gel filtration using Sephadex G-25 or by dialysis.

Determine the level of fluorophore incorporation (the F/P ratio) by measuring the absorbance of the labeled protein at 520 and 280 nm. Labeled proteins having a 520/280 nm ratio of absorbancy of 0.3–0.8 should perform well in most applications (Titus *et al.*, 1982).

Sulfhydryl-Reactive Rhodamine Derivatives

Rhodamine derivatives containing a sulfhydryl-reactive group off the lower ring structure are available to direct the modification reaction to more limited sites on target molecules. Coupling through sulfhydryls instead of amines can help to avoid active centers in proteins, thus preserving activity in the fluorescent probe complex. Sulfhydryl reaction sites can be naturally available through free cysteine side chains, generated by reduction of disulfides, or created by the use of thiolation reagents (Chapter 1, Section 4.1).

Tetramethylrhodamine-5- (and 6)-iodoacetamide

The iodoacetamido derivatives of tetramethylrhodamine possess a sulfhydryl-reactive iodoacetyl group (Chapter 1, Section 4.2, and Chapter 2, Section 2.1) at either the 5- or 6-carbon position on their lower ring. The isomers are commercially available only in mixed form, but some reactivity and specificity differences between the purified 5 and 6 derivatives toward various sulfhydryl sites in proteins may be observed (Ajtai, 1992) (Molecular Probes, Pierce).

The iodoacetyl group of both isomers reacts with sulfhydryls under slightly alkaline conditions to yield stable thioether linkages (Fig. 216). They do not react with unreduced disulfides in cystine residues or with oxidized glutathione (Gorman *et al.*, 1987). The thioether bonds are hydrolyzed under conditions necessary for complete protein hydrolysis prior to amino acid analysis.

Care must be taken to protect these reagents from light, not only to maintain the fluorescent yield of the rhodamine derivative, but also to protect the iodoacetyl group

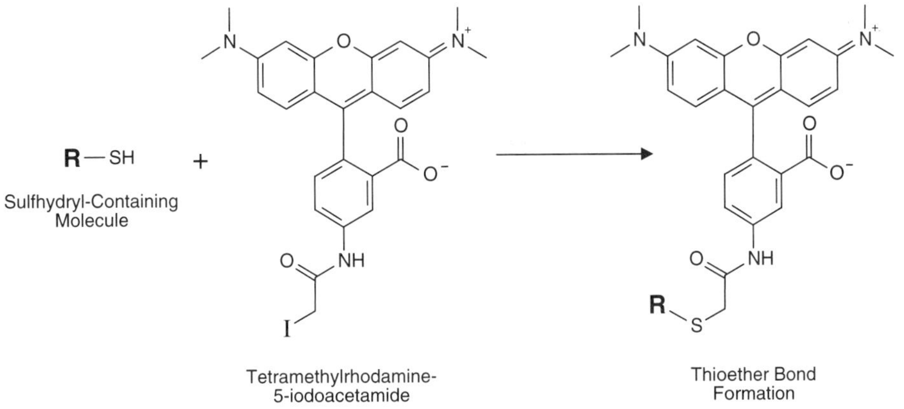

Figure 216 This iodoacetamide derivative of tetramethylrhodamine can be used to label sulfhydryl groups via thioether bond formation.

from light-catalyzed breakdown. Tetramethylrhodamine-5- (and 6)-iodoacetamide is soluble in DMF and DMSO. Concentrated stock solutions may be prepared in these solvents prior to addition of a small aliquot to an aqueous reaction mixture. Protect solutions from light by wrapping in aluminum foil and working in subdued light. Quenching reactions with cysteine, glutathione, or mercaptosuccinic acid will sometimes facilitate removal of unconjugated fluorophore by dialysis or gel filtration.

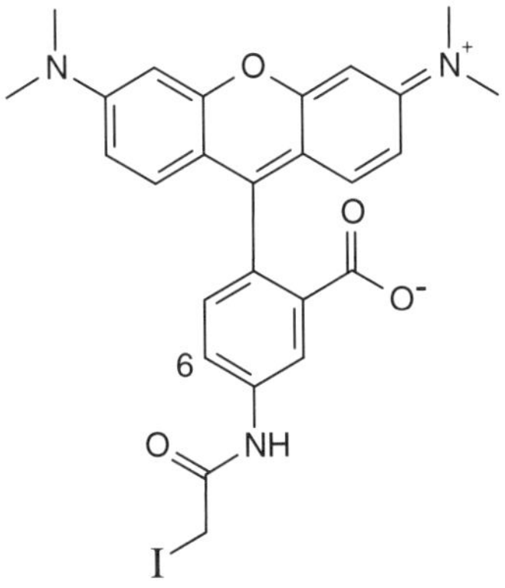

Tetramethylrhodamine-
5-iodoacetamide
MW 569
Excitation = 540 nm
Emission = 567 nm
ε at 540 nm = 76,000 $M^{-1}cm^{-1}$

The spectral properties of these derivatives are similar to native rhodamine. The excitation maximum occurs at about 543 nm and its emission peak at 567 nm, producing light in the orange-red region of the spectrum. The extinction coefficient of tetramethylrhodamine-5- (and 6)-iodoacetamide in methanol at its wavelength of maximum absorptivity, 542 nm, is 81,000 $M^{-1}cm^{-1}$.

The fluorescent probe has been used extensively to label numerous proteins and other biomolecules, including actin (Meige and Wang, 1986; Wang, 1985; Glacy, 1983), myosin light chains (Mittal *et al.*, 1987), α-actin (Simon and Taylor, 1988; Stickel and Wang, 1988), blood coagulation factor Va (Isaacs *et al.*, 1986), and histones (Murphy *et al.*, 1982).

The following protocol for labeling proteins with tetramethylrhodamine-5- (and 6)-iodoacetamide represents a general guideline. The procedure should be optimized for each macromolecule being labeled to obtain the best *F/P* ratio to produce intense fluorescence and high activity in the final complex.

Protocol

1. Prepare a 20 m*M* tetramethylrhodamine-5- (and 6)-iodoacetamide (Pierce) solution by dissolving 11.3 mg per milliliter of DMF. Prepare fresh and protect from light.
2. Dissolve the protein to be modified at a concentration of 5–10 mg/ml in 20 m*M* 2-[[tris(hydroxymethyl)methyl]amino]ethanesulfonic acid (TES), pH 7.
3. Slowly add 25–50 μl of the tetramethylrhodamine-5 (and 6)-iodoacetamide solution to each milliliter of the protein solution while mixing.
4. React for 2 h at 4°C in the dark.
5. Remove excess reactant and reaction by-products by gel filtration using Sephadex G-25 or the equivalent.

Aldehyde/Ketone- and Cytosine-Reactive Rhodamine Derivatives

Hydrazide groups can be coupled directly to aldehyde and ketone functional groups to form relatively stable hydrazone linkages (Chapter 2, Section 5.1). Two rhodamine derivatives are commonly available that contain a sulfonylhydrazine group off their No. 5 carbon on the lower ring structure (Molecular Probes, Pierce). They are based on the Lissamine and Texas Red structures and may be used to label aldehyde- or ketone-containing molecules with an intensely fluorescent probe. Although most biomolecules do not contain aldehyde or ketone groups in their native state, carbohydrates, glycoproteins, RNA, and other molecules containing sugar residues can be oxidized with sodium periodate to produce reactive formyl functional groups. The use of modification reagents that generate aldehydes on coupling to a molecule also can be used to produce a hydrazide-reactive site (Chapter 1, Section 4.4).

DNA and RNA also may be modified with hydrazide-reactive probes by reacting their cytosine residues with bisulfite to form reactive sulfone intermediates. These derivatives undergo transamination to couple with hydrazide- (or amine)-containing probes (Draper and Gold, 1980) (Chapter 17, Section 2.1).

Lissamine Rhodamine B Sulfonyl Hydrazine

Lissamine rhodamine B sulfonyl hydrazine is a hydrazide derivative of sulforhodamine B that can spontaneously react with aldehyde- or ketone-containing molecules to form a covalent, hydrazone linkage (Fig. 217). It also can be used to label cytosine residues in DNA or RNA by use of the bisulfite activation procedure (Chapter 17, Section 2.1). The resulting fluorescent derivative exhibits an excitation maximum at a wavelength of 556 nm and a maximal emission wavelength of 580 nm when dissolved in methanol. In

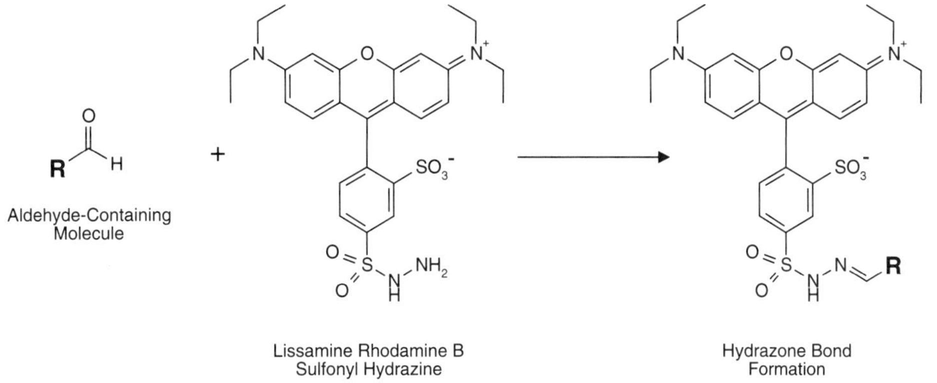

Figure 217 Lissamine rhodamine B sulfonyl hydrazine can react with aldehyde groups to form hydrazone linkages.

the same solvent, the compound has an extinction coefficient of approximately 75,000 $M^{-1}cm^{-1}$.

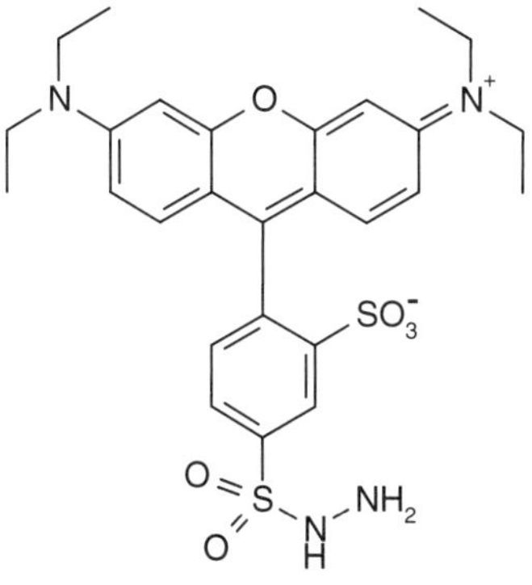

Lissamine Rhodamine B
Sulfonyl Hydrazine
MW 573
Excitation = 560 nm
Emission = 585 nm
ε at 560 nm = 95,000 $M^{-1}cm^{-1}$

Lissamine rhodamine B sulfonyl hydrazine is soluble in DMF. The reagent may be dissolved in this solvent as a concentrated stock solution before a small aliquot is added to an aqueous reaction medium. The compound itself and all solutions made with it should be protected from light to avoid decomposition of its fluorescent properties.

Generalized protocols for the use of hydrazine-probes reactive toward aldehyde residues can be found in Section 1.1. These procedures are directed at the labeling of cell surface glycoproteins or glycoproteins in solution. Substitution of Lissamine rhodamine B sulfonyl hydrazine for the fluorescein-5-thiosemicarbazide reagent described in that section can be done without difficulty. Some optimization may be necessary to achieve the best level of fluorescent modification for each particular application.

Texas Red Hydrazide

Texas Red hydrazide is a derivative of Texas Red sulfonyl chloride made by reaction with hydrazine (Molecular Probes, Pierce). The result is a sulfonyl hydrazine group on the No. 5 carbon position of the lower ring structure of sulforhodamine 101. The intense Texas Red fluorophore has a quantum yield that is inherently higher than either the tetramethylrhodamine or the Lissamine rhodamine B derivatives of the basic rhodamine molecule. Texas Red's luminescence is shifted maximally into the red region of the spectrum, and its emission peak only minimally overlaps with that of fluorescein. This makes derivatives of this fluorescent probe among the best choices of labels for use in double staining techniques.

The hydrazide derivative can be used to modify aldehyde- or ketone-containing molecules, including cytosine residues using the bisulfite activation procedure described in Chapter 17, Section 2.1. The sulfonyl hydrazine group of Texas Red hydrazide reacts with aldehydes or ketones in target functional groups to form hydrazone bonds (Fig. 218). Carbohydrates and glycoconjugates can be specifically labeled at the polysaccharide portion if the required aldehydes are first formed by periodate oxidation or another such method (Chapter 1, Section 4.4).

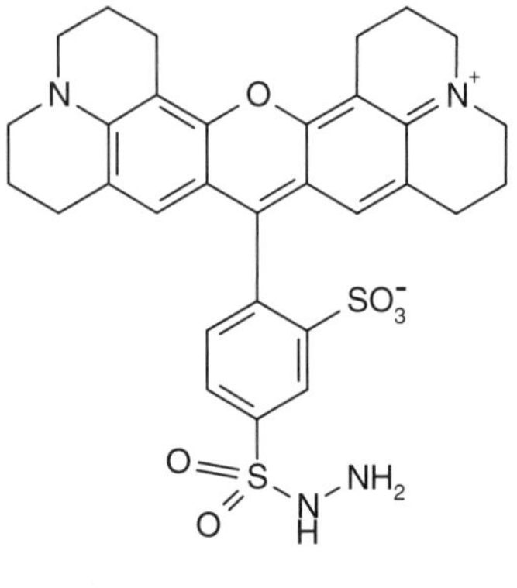

Texas Red Hydrazide
MW 621
Excitation = 580 nm
Emission = 604 nm
ε at 580 nm = 80,000 $M^{-1}cm^{-1}$

Texas Red hydrazide has a maximal excitation wavelength of 580 nm and a maximum emission at 604 nm when dissolved in methanol. Its extinction coefficient in the same solvent is 80,000 $M^{-1}cm^{-1}$ at 580 nm. The only disadvantage of this fluorophore is its poor excitation by an argon laser at 488 nm. However, since both Texas Red and fluorescein are weakly excited by an argon laser at 514 nm, it makes them fairly good pairs for use in laser confocal microscopy or flow cytometry (Mossberg and Ericsson, 1990). The fluorophore is particularly appropriate for excitation by the 568 nm line produced by an argon–krypton mixed laser used on some confocal devices. Compared to other rhodamine derivatives, Texas Red fluorophores display low background in staining techniques and are among the most photostable probes available.

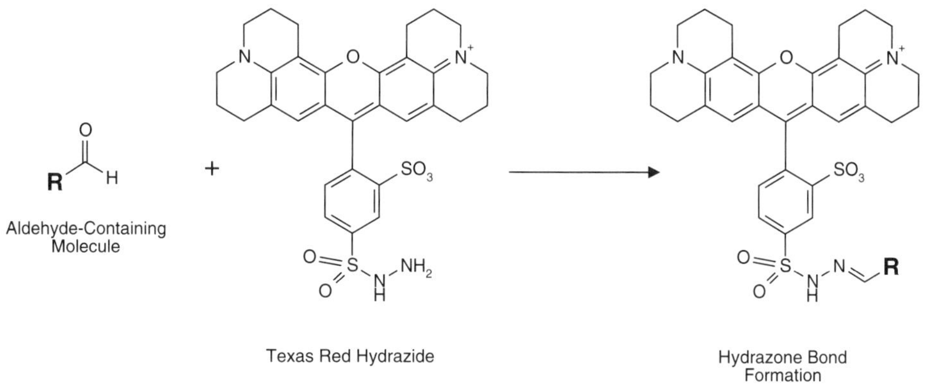

Texas Red Hydrazide

Hydrazone Bond
Formation

Aldehyde-Containing
Molecule

Figure 218   Texas Red hydrazide reacts with aldehydes to create hydrazone bonds.

Texas Red hydrazide is soluble in DMF and may be dissolved as a concentrated stock solution in this solvent prior to the addition of a small aliquot to an aqueous reaction medium. The solid and all solutions made from it must be protected from light to avoid photodecomposition. Prepare the stock solution fresh immediately before use. A suggested protocol on the use of this fluorescent probe may be obtained by following the method outline for fluorescein-5-thiosemicarbazide in Section 1.1. Optimization may be necessary to achieve the best level of fluorescent modification (F/P ratio) for a particular application.

1.3. Coumarin Derivatives

Coumarin (2H-1-benzopyran-2-one) is a naturally occurring substance found in tonka beans, lavender oil, and sweet clover (Merck Index **11**:2563) (Fig. 219). Many of its derivatives are highly fluorescent compounds that are widely studied (Baranowska-Kortylewicz and Kassis, 1993a, b; Eschrich and Morgan, 1985; Jones *et al.*, 1984, 1985; Ernsting *et al.*, 1982; Schimitschek *et al.*, 1974). The 7-amino-4-methyl-coumarin derivatives have excellent fluorescent properties useful for labeling biological molecules with a detectable tag (Uchino *et al.*, 1979; Bos, 1981). Particularly, the 3-acetic acid derivative of this molecule, known as AMCA, provides a carboxylate group from which to create easily other reactive functional groups suitable for coupling to proteins and other molecules.

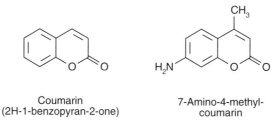

Coumarin
(2H-1-benzopyran-2-one)

7-Amino-4-methyl-
coumarin

Figure 219   The basic structural characteristics of coumarin fluorophores.

Aminomethylcoumarin derivatives possess intense fluorescent properties within the blue region of the visible spectrum. Their emission range is sufficiently removed from other common fluorophores that they are excellent choices for double-labeling techniques. In fact, coumarin fluorescent probes are very good donors for excited-state energy transfer to fluoresceins.

The following sections describe the most popular derivatives of aminomethyl-coumarin used to label proteins and other biological macromolecules.

Amine-Reactive Coumarin Derivatives

Three main forms of amine-reactive AMCA probes are commonly available. One of them is simply the free acid form of AMCA, which can be used to couple to amine-containing molecules using the carbodiimide reaction (Chapter 3, Section 1). The other two are active-ester derivatives of AMCA—the water-insoluble NHS ester and the water-soluble sulfo-NHS ester forms—both of which spontaneously react with amines to create stable amide linkages. All of them react under mild conditions with primary amines in proteins and other molecules to form highly fluorescent derivatives.

AMCA

7-Amino-4-methylcoumarin-3-acetic acid (AMCA) is a fluorescent probe that exhibits a spectacular blue fluorescence (Khalfan *et al.*, 1986) (Pierce). AMCA absorbs light at a wavelength of 345 nm and luminesces in the range 440–460 nm. Its emission wavelength is in a region that does not overlap with the emission spectra of other major fluorescent probes. This makes double staining techniques particularly effective with this fluorophore. AMCA also has pronounced stability toward photoquenching effects, retaining its full fluorescence more than three times longer than fluorescein-based probes. AMCA derivatives and labeled molecules fluoresce with a bright blue color upon excitation. This color is easily visualized and photographed. The blue emission color avoids problems of autofluorescence associated with high background. In addition, the large stokes shift of the molecule minimizes interference from Raleigh scatter effects during excitation. The fluorescent intensity of AMCA is not affected by changes in pH in the range 3 to 10. This is in marked contrast to other fluorescent probes, such as fluorescein, which display considerable variability in their emission spectra with pH.

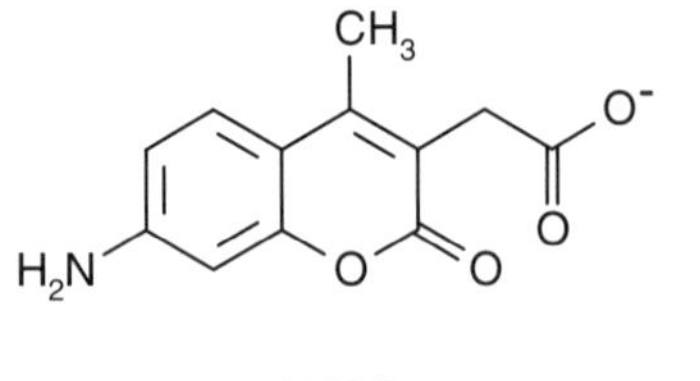

AMCA
7-Amino-4-methyl-
coumarin-3-acetic acid
MW 233
Excitation = 345-350 nm
Emission = 440-460 nm

AMCA may be coupled to amine-containing molecules through the use of the carbodiimide reaction using EDC (Chapter 3, Section 1.1). EDC will activate the

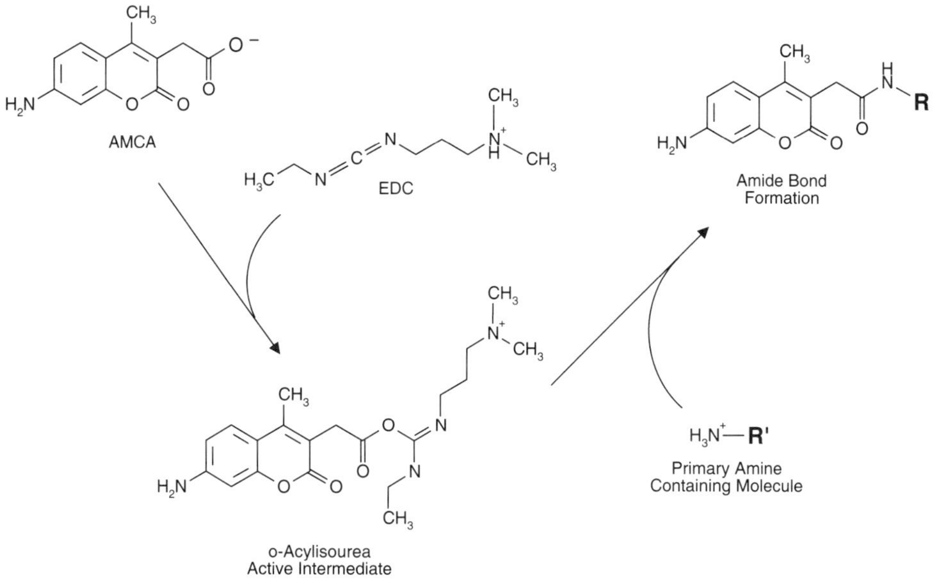

Figure 220 AMCA may be linked to amine-containing molecules through its carboxylate group using a carbodiimide reaction with EDC.

carboxylate on AMCA to a highly reactive O-acylisourea intermediate. Attack by a nucleophilic primary amine group results in the formation of an amide bond (Fig. 220). Derivatization of AMCA off its carboxylate group causes no major effects on its fluorescent properties. Thus, proteins and other macromolecules may be labeled with this intensely blue probe and easily detected by fluorescence microscopy and other techniques.

AMCA-NHS and AMCA-Sulfo-NHS

Succinimidyl-7-amino-4-methylcoumarin-3-acetic acid (AMCA-NHS) is an amine-reactive derivative of AMCA containing an NHS ester on its carboxylate group (Pierce). The result is directed reactivity toward amine-containing molecules, forming amide linkages with the AMCA fluorophore (Fig. 221). Proteins labeled with AMCA show little to no effect on the isoelectric point of the molecule.

Reaction of AMCA-NHS with proteins proceeds efficiently in the pH range 7–9. Avoid buffers containing amines that can compete in the coupling reaction, such as Tris or glycine, and avoid imidazole buffers since they promote hydrolysis of the NHS ester. AMCA-NHS is relatively insoluble in aqueous buffers. The compound must be first dissolved in organic solvent prior to addition of a small aliquot to the reaction mixture. A concentrated stock solution may be prepared in DMSO and stored up to 2 weeks refrigerated or frozen without loss of activity. The solid and all solutions of AMCA-NHS should be protected from light to avoid photodecomposition of the fluorophore.

AMCA-sulfo-NHS is an analog of AMCA-NHS that contains a sulfonate group on its NHS ring (Pierce). The negative charge of this group provides enough polarity to

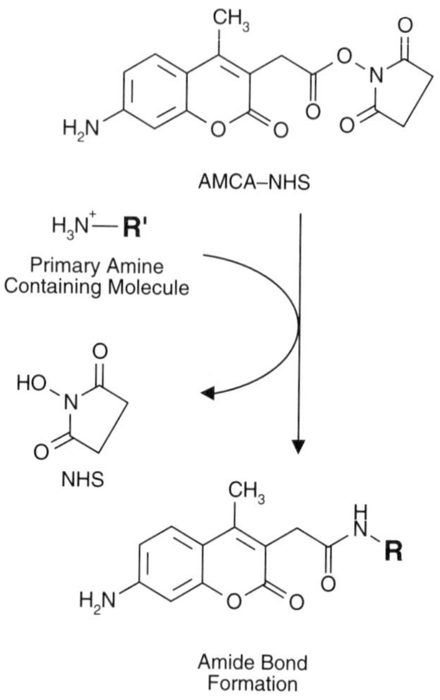

Figure 221 AMCA-NHS reacts with amines to form amide bonds.

promote water solubility for the entire reagent. The reactivity and properties of AMCA-sulfo-NHS are identical to those of AMCA-NHS.

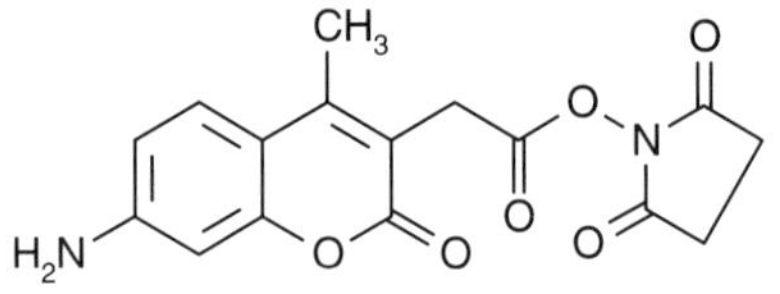

AMCA–NHS
Succinimidyl-7-amino-
4-methylcoumarin-3-acetic acid
MW 330

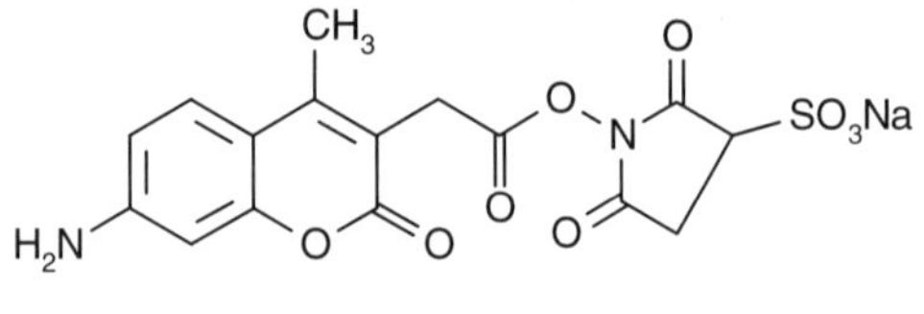

AMCA–Sulfo-NHS
Sulfosuccinimidyl-7-amino-
4-methylcoumarin-3-acetic acid
MW 431

Preparing an optimal fluorescent conjugate is largely dependent upon the degree of modification with the label. The following protocol is generalized for the labeling of a protein with AMCA-NHS. For particular labeling experiments, it is often necessary to vary the amount of fluorophore added to the reaction mixture to obtain the best combination of protein activity and fluorescent intensity in the conjugate. If there is too much label, nonspecific binding or fluorescent quenching may result; if there is too little label, the complex will not possess enough fluorescent intensity to be sufficiently detectable.

Protocol

1. Dissolve the protein to be modified in 50 mM sodium borate, pH 8.5, at a concentration of 10 mg/ml. Other buffers may be used for an NHS ester reaction, including 0.1 M sodium phosphate, pH 7.5 (Chapter 2, Section 1.4).
2. Dissolve AMCA-NHS (Pierce) in DMSO at a concentration of 2.6 mg/ml. Protect from light.
3. In subdued lighting conditions, slowly add 50–100 μl of the AMCA-NHS stock solution to each ml of the protein solution, with gentle mixing.
4. React for 1 h at room temperature in the dark.
5. Remove excess reagent and reaction by-products by gel filtration using a column of Sephadex G-25. The sample volume should be no more than about 5–8% of the column volume.

The F/P ratio of the purified, labeled protein may be determined by measuring the absorbance at 345 and 280 nm. Ratios between 0.3 and 0.8 usually produce labeled molecules having acceptable levels of fluorescent intensity and good retention of protein activity. AMCA-labeled proteins may be lyophilized without significant loss of fluorescence. The addition of bovine serum albumin (15 mg/ml) or another such stabilizer is often necessary to retain solubility of the freeze-dried, labeled protein after reconstitution.

Sulfhydryl-Reactive Coumarin Derivatives

Two aminomethylcoumarin derivatives are available for labeling sulfhydryl-containing molecules. The ability to label—SH groups in proteins provides a means of directing the modification reaction to a limited number of sites, possibly avoiding active centers or binding regions better than when using amine-reactive probes. The first sulfhydryl-reactive probe discussed in this section makes use of a pyridyl disulfide group on the AMCA derivative. The second probe is an iodoacetyl compound made from a diethyl- and aminophenyl derivative of the basic aminomethylcoumarin structure.

AMCA-HPDP

AMCA-HPDP   is   N-[6-(7-amino-4-methylcoumarin-3-acetamido)hexyl]-3'-(2'-pyridyldithio)propionamide. It is formed from AMCA plus a 1,6-diaminohexyl spacer off the carboxylate that has been additionally modified at its other end with SPDP (Chapter 5, Section 1.1). The result is a long spacer arm terminating in a pyridyl disulfide group reactive toward free sulfhydryl residues. The reaction of this group with a thiol creates a disulfide bond between the AMCA fluorophore and the molecule being modified. Thus, the fluorescent tag can be specifically cleaved by reduction with DTT or other disulfide reducing agents (Fig. 222).

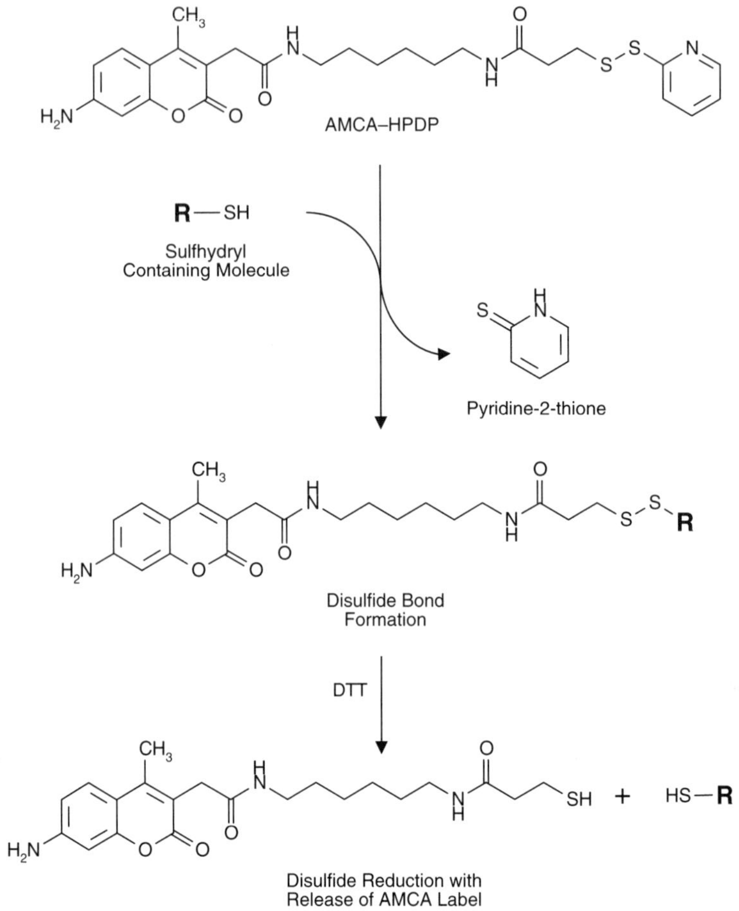

Figure 222 AMCA-HPDP reacts with sulfhydryl groups through its pyridyl disulfide end to form reversible disulfide bonds.

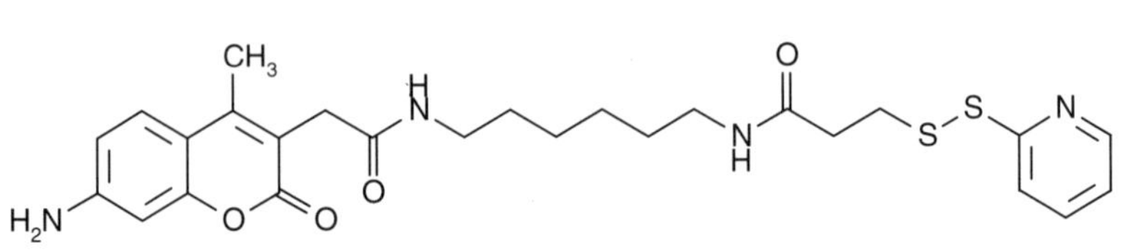

AMCA–HPDP
N-[6-(7-amino-4-methylcoumarin-
3-acetamido)hexyl]-3'-(2'-pyridyldithio)propionamide
Excitation = 345 nm
Emission = 440-460 nm

The required sulfhydryl residue targets can be naturally occurring on a protein or created by reduction of cystine cross-links or by thiolation (Chapter 1, Section 4.1). For the labeling of antibody molecules, mild reduction with 2-mercaptoethylamine results in two half-antibody fragments containing free sulfhydryl groups in the hinge region. Labeling in this area is advantageous to direct the modification away from antigen binding regions. Sulfhydryl residues also may be created on oligonucleotides without difficulty (Chapter 17, Section 2.2).

Although the reaction of a sulfhydryl-containing molecule with AMCA-HPDP results in the release of the chromogenic leaving group, pyridine-2-thione, using it to quantify the extent of modification may be difficult, because it absorbs at 343 nm—in the same region as AMCA itself (345 nm). The emission range of the AMCA probe is about 440–460 nm, in the blue region of the spectrum.

The following protocol is a suggested method for labeling a protein with AMCA-HPDP. It is assumed that the presence of a sulfhydryl on the protein has been documented or created. The reaction conditions can be carried out in a variety of buffers between pH 6 and 9. Avoid the presence of extraneous sulfhydryl-containing compounds (such as disulfide reductants) that will compete in the reaction. The inclusion of EDTA in the modification buffer prevents metal-catalyzed sulfhydryl oxidation. Optimization for a particular labeling experiment should be done to obtain the best level of fluorophore incorporation.

Protocol

1. Dissolve the sulfhydryl-containing protein to be labeled in 0.1 M sodium phosphate, 0.15 M NaCl, 10 mM EDTA, pH 7.2, at a concentration of 10 mg/ml.
2. Dissolve AMCA-HPDP in DMSO at a concentration of 0.5 mg/ml. Protect from light.
3. In subdued lighting conditions, add 50–100 μl of the AMCA-HPDP stock solution to each milliliter of sulfhydryl-containing protein solution. Mix.
4. React for 1 h at room temperature in the dark with occasional mixing.
5. Remove excess fluorophore and reaction by-products by gel filtration using a column of Sephadex G-25.

To determine the F/P ratio of the labeled protein, measure the absorbance of the purified preparation at 345 and 280 nm. Ratios of 345/280 nm within the range 0.3 to 0.8 usually result in fluorescent conjugates with a good balance of high-intensity luminescence, low nonspecific binding, and excellent retention of biological activity within the protein component.

DCIA

7-Diethylamino-3-[(4′-(iodoacetyl)amino)phenyl]-4-methylcoumarin (DCIA) is a derivative of the basic aminomethylcoumarin structure that contains a sulfhydryl-reactive iodoacetyl group and diethyl substitutions on its amine. This particular coumarin derivative is among the most fluorescent UV-excitable iodoacetamide probes available (Sippel, 1981) (Molecular Probes).

The iodoacetyl group of DCIA reacts with sulfhydryls under slightly alkaline conditions to yield stable thioether linkages (Fig. 223). They do not react with unreduced disulfides in cystine residues or with oxidized glutathione (Gorman *et al.*, 1987). The thioether bonds will be hydrolyzed under conditions necessary for complete protein hydrolysis prior to amino acid analysis.

Care must be taken to protect iodoacetyl reagents from light, not only to maintain the fluorescent yield of the coumarin component, but also to protect the iodoacetyl group from light-catalyzed breakdown. DCIA is soluble in DMF and DMSO. Concen-

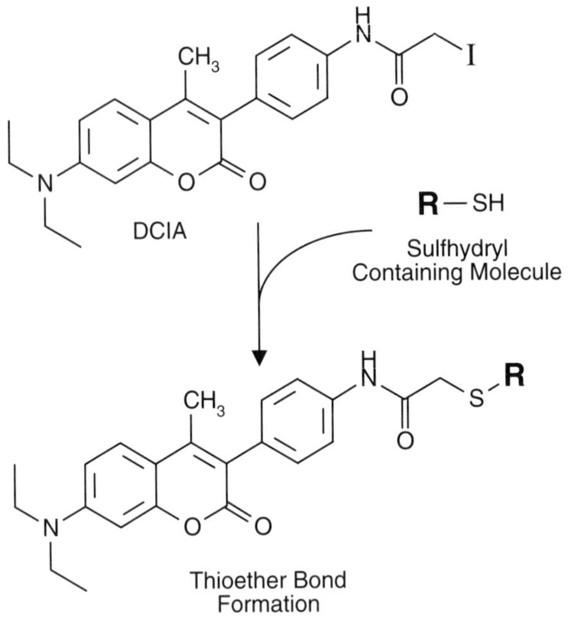

Figure 223   DCIA can modify sulfhydryl groups through its iodoacetamide group to form thioether linkages.

trated stock solutions may be prepared in either solvent prior to addition of a small aliquot to a reaction mixture. Protect solutions from light by wrapping vessels in aluminum foil and working in subdued light.

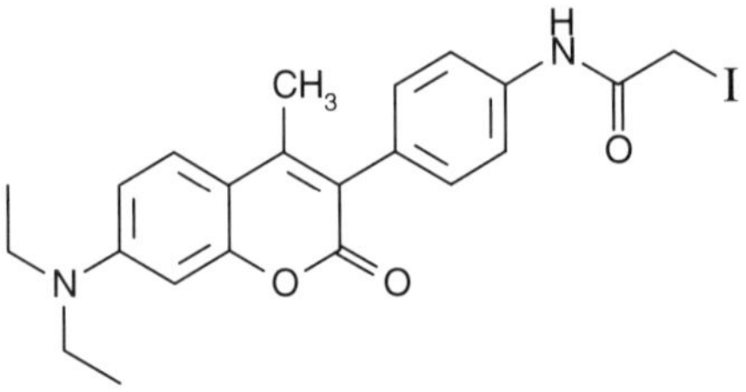

DCIA
7-Diethylamino-3-((4'-(iodoacetyl)-
amino)phenyl)-4-methylcoumarin
MW490
Excitation = 382 nm
Emission = 472 nm
ε at 382 nm = 33,000 $M^{-1}cm^{-1}$

The spectral properties of this fluorophore are similar to those of other coumarin derivatives. The excitation maximum occurs at about 382 nm, and its emission peak at 472 nm, producing light in the blue region of the spectrum. The extinction coefficient of DCIA at its wavelength of maximum absorbance, 382 nm, is 33,000 $M^{-1}cm^{-1}$ (in methanol).

DCIA has been used to label numerous proteins and other biomolecules, including phospholipids (Silvius *et al.*, 1987), to study the interaction of mRNA with the 30S ribosomal subunit (Czworkowski *et al.*, 1991), in the investigation of cellular thiol components by flow cytometry (Durand and Olive, 1983), in the detection of carboxylate compounds using peroxyoxalate chemiluminescence (Grayeski and DeVasto, 1987), and for general sulfhydryl labeling (Sippel, 1981).

A general protocol for the use of DCIA for fluorescently labeling proteins that contain sulfhydryl residues may be obtained by following the method discussed for AMCA-HPDP (previous section). After purification of the labeled protein, the *F/P* ratio of fluorophore incorporation may be determined by measuring its 382/280 nm absorbance ratio.

Aldehyde- and Ketone-Reactive Coumarin Derivatives

Hydrazide groups can be coupled directly to aldehyde and ketone functional groups to form relatively stable hydrazone linkages (Chapter 2, Section 5.1). One AMCA derivative is commonly available that contains a hydrazine group modification of its carboxylate. Although most biomolecules do not contain aldehyde or ketone groups in their native state, carbohydrates, glycoproteins, RNA, and other molecules that contain sugar residues can be oxidized with sodium periodate to produce reactive formyl functional groups. The use of modification reagents that generate aldehydes on coupling to a molecule also can be used to produce a hydrazide-reactive site (Chapter 1, Section 4.4).

DNA and RNA may be modified with hydrazide-reactive probes by reacting their cytosine residues with bisulfite to form reactive sulfone intermediates. These derivatives can undergo transamination reactions with hydrazide- (or amine)-containing probes to yield covalent bonds (Draper and Gold, 1980) (Chapter 17, Section 2.1).

AMCA-Hydrazide

7-Amino-4-methylcoumarin-3-acetyl hydrazide (AMCA-hydrazide) is a hydrazine derivative off the carboxyl group of the basic AMCA molecule (Pierce). AMCA-based fluorophores are highly stable toward photobleaching. Molecules labeled with this probe retain their fluorescent intensity over three times longer than a fluorescein label when exposed to light. In addition, AMCA derivatives exhibit a large Stokes shift of over 100 nm; thus they are minimally affected by Rayleigh scattering effects during excitation. The blue light emitted by these labels is in a region of the spectrum well removed from the emission characteristics of other major fluorescent probes. This means that double staining techniques easily can be used with an AMCA label. AMCA also exhibits little luminescence dependency on pH over the range 3–10.

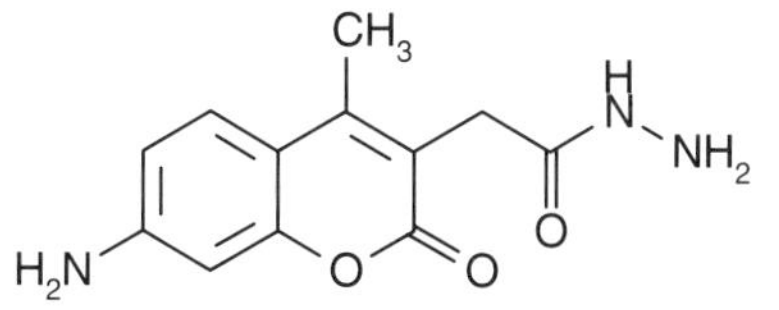

AMCA–Hydrazide
7-Amino-4-methylcoumarin-
3-acetyl hydrazide
MW 247. 1
Excitation = 345 nm
Emission = 440–460 nm

The hydrazide derivative can be used to modify aldehyde- or ketone-containing molecules, including cytosine residues using the bisulfite activation procedure described in Chapter 17, Section 2.1. AMCA-hydrazide reacts with these target functional groups to form hydrazone bonds (Fig. 224). Carbohydrates and glycoconjugates can be specifically labeled at the polysaccharide portion if the required aldehydes are first formed by periodate oxidation or another such method (Chapter 1, Section 4.4).

AMCA-hydrazide has a maximal excitation wavelength of 345 nm and a maximum emission wavelength in the range 440–460 nm. A solution of AMCA in PBS at a concentration of 16.7 ng/ml (71.61 nmol/ml) gives an absorbance at 345 nm of about 1.28. This translates into a molar extinction coefficient at this wavelength of about 13,900 $M^{-1}cm^{-1}$. Different solvents and conditions may alter this value somewhat.

AMCA-hydrazide is soluble in DMSO or DMF and may be dissolved as a concentrated stock solution in either of these solvents prior to the addition of a small aliquot to an aqueous reaction medium. The solid and all solutions made from the fluorophore must be protected from light to avoid photodecomposition. Prepare the stock solution fresh immediately before use. A suggested protocol on the use of this fluorescent probe may be obtained from the following method on the labeling of periodate-oxidized IgG. Optimization may be necessary to achieve the best level of fluorescent modification (*F/P* ratio) for a particular application.

Protocol
Oxidation of IgG Carbohydrate Residues with Sodium Periodate

1. Dissolve the antibody to be labeled in 0.1 *M* sodium phosphate, 0.15 *M* NaCl, pH 7.5, at a concentration of at least 10 mg/ml. The immunoglobulin must be glycosylated to work in this procedure.
2. Dissolve sodium periodate in water to a final concentration of 100 m*M*. Protect from light. Add 0.1 ml of this stock periodate solution to each milliliter of the antibody solution.

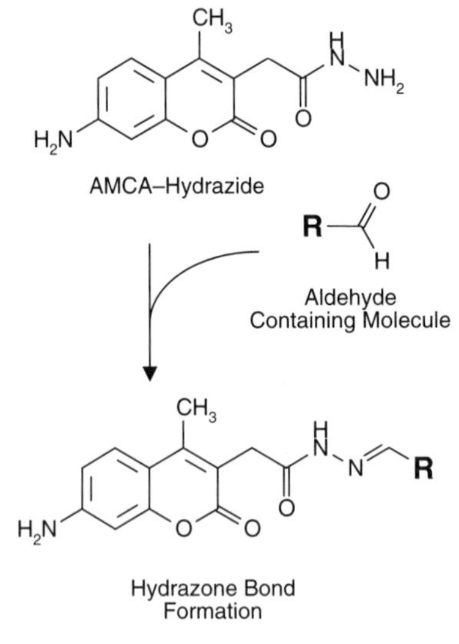

Figure 224 AMCA-hydrazide can be used to label aldehyde-containing molecules, such as periodate-oxidized carbohydrates.

3. React for 30 min at room temperature, protected from light.
4. Quench the reaction by addition of 0.1 ml of glycerol per milliliter of reaction volume, mix, and then react for an additional 15 min. Remove excess reagents by gel filtration using a column of Sephadex G-25, performing the chromatography using the phosphate buffer.
5. Adjust the concentration of IgG in the purified preparation to 1 mg/ml by the addition of 0.1 M sodium phosphate, 0.15 M NaCl, pH 7.5.

Modification of oxidized IgG with AMCA-hydrazide.
1. Dissolve AMCA-hydrazide in DMF at a concentration of 0.4 mg/ml. Protect from light.
2. Add 50–100 µl of the AMCA-hydrazide stock solution to each milliliter of the oxidized antibody solution. Note: at a level of 50 µl probe addition, polyclonal human IgG will be modified at a level that gives an F/P ratio of about 0.113. Since the labeling occurs only at the oxidized carbohydrate sites, the fluorophore incorporation typically is less than that observed when using amine-reactive probes.
3. React for 30 min at room temperature in the dark.
4. Remove excess fluorophore by dialysis or gel filtration using Sephadex G-25. Protect the labeled immunoglobulin from light.

Determine the F/P ratio by measuring the absorbance at 345 and 280 nm.

1.4. BODIPY Derivatives

BODIPY fluorophores are a relatively new class of probes based on the fused, multiring structure 4,4-difluoro-4-bora-3a,4a-diaza-s-indacene (Fig. 225) (note: BODIPY is a registered trademark of Molecular Probes; U.S. Patent 4,774,339). This fundamental molecule can be modified, particularly at its 1,3,5,7 and 8 carbon positions, to produce new fluorophores with different characteristics. The modifications cause spectral shifts in its excitation and emission wavelengths and can provide sites for chemical coupling to label other molecules.

The BODIPY derivatives typically have high extinction coefficients and excellent quantum yields, often greater than 0.8. Their spectral characteristics are relatively insensitive to changes in pH. Luminescent changes with shifts in pH usually are due to

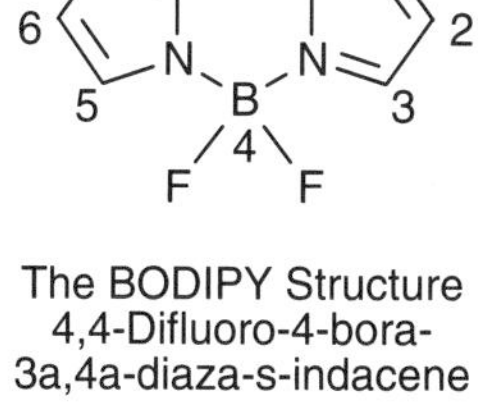

The BODIPY Structure
4,4-Difluoro-4-bora-
3a,4a-diaza-s-indacene

Figure 225 The basic structure of BODIPY fluorophores.

reconfiguration of a fluorophore's π-electron cloud if an atom on the ring system becomes protonated or unprotonated. Since the BODIPY structure lacks an ionizable group, alterations in pH have no effect on its spectral attributes.

The emission spectra of BODIPY derivatives normally display narrow bandwidths, providing intensely fluorescent labels for biomolecules. Unfortunately, they also have very small Stokes shifts, typically on the order of only 10–20 nm. Excitation at the optimal wavelength may cause some interference in measurements at the emission wavelength due to light scattering or cross-over from the wide bandwidth of the excitation source. The dyes usually require excitation at suboptimal wavelengths to prevent this problem.

The following sections discuss the major BODIPY derivatives that are reactive toward particular functional groups in proteins and other molecules.

Amine-Reactive BODIPY Derivatives

A number of BODIPY derivatives that contain functional groups able to couple with amine-containing molecules are commonly available. The derivatives contain either a carboxylate group, which can be reacted with an amine in the presence of a carbodiimide to create an amide bond, or an NHS ester derivative of the carboxylate, which can react directly with amines to form amide linkages. The three mentioned in this section are representative of this amine-reactive BODIPY family. The two NHS ester derivatives react under alkaline conditions with primary amines in molecular targets to form stable, highly fluorescent derivatives. The carboxylate derivative can be coupled to an amine using the EDC/sulfo-NHS reaction discussed in Chapter 3, Section 1.2.

The only disadvantage of using BODIPY fluorophores to label amines in macromolecules is the severe fluorescence quenching the probe exhibits if multiple sites on one molecule are modified. Especially with proteins, using an amine-reactive probe usually results in a number of sites being modified on each molecule. All fluorophores experience some quenching effect if the degree of substitution is high, because probe–probe interactions are possible that can transfer energy from an excited-state fluorophore to a ground-state fluorophore before luminescence occurs. BODIPY probes, however, are especially notorious for probe–probe quenching effects. For this reason, the manufacturer (Molecular Probes) recommends that the amine-reactive BODIPY probes only be used to modify substances that have the potential for just one substitution per molecule. In this sense, BODIPY fluorophores are particularly well suited for tagging DNA probes at the 5′ end. Oligonucleotides modified to contain an amine on their 5′ phosphate group (Chapter 17, Section 2.1) are particularly good candidates for labeling with this fluorophore. Other BODIPY probes that contain reactivity toward non-amine functional groups such as sulfhydryls or polysaccharides may be more effective at labeling macromolecules like proteins, since these groups occur at limited sites within the molecules and the modification level can be better controlled.

BODIPY FL C$_3$-SE

BODIPY FL C$_3$-SE is 4,4-difluoro-5,7-dimethyl-4-bora-3a,4a-diaza-S-indacene-3-propionic acid, succinimidyl ester (Molecular Probes). The derivatizations to the base BODIPY molecule result in fluorescent properties that mimic fluorescein in emission

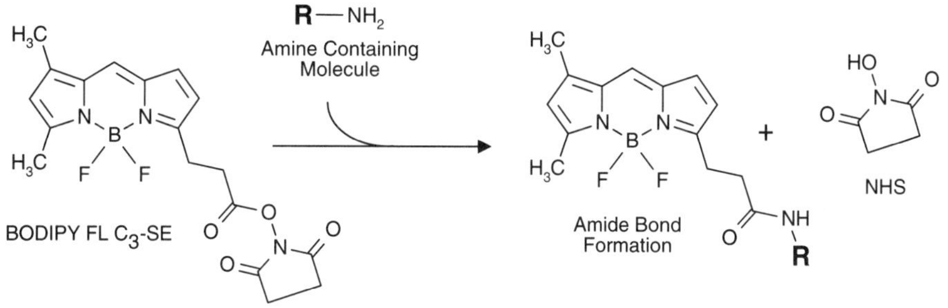

Figure 226 The side-chain NHS ester of this BODIPY derivative can be used to modify amine-containing molecules, forming amide bond linkages.

wavelength. The molecule emits light in the green region of the spectrum. The NHS ester on its propionic side chain provides amine-reactivity, resulting in amide bond linkages with modified molecules (Fig. 226).

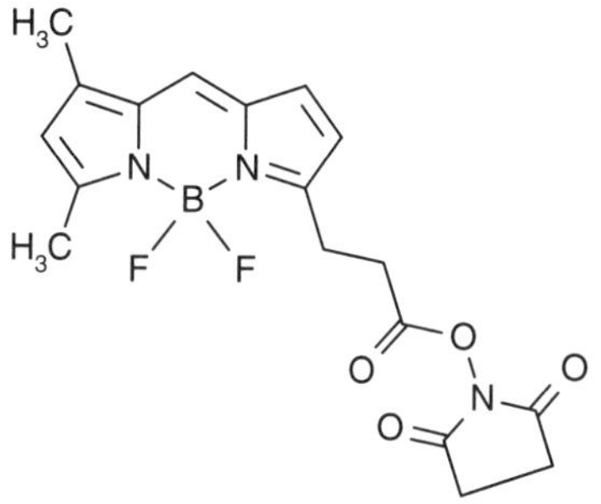

BODIPY FL C$_3$-SE
4,4-Difluoro-5,7-dimethyl-4-bora-3a,4a-
diaza-sindacene-3-propionic acid,
succinimidyl ester
MW 389
Excitation = 502 nm
Emission = 510 nm
ε at 502 nm = 77,000 M^{-1}cm^{-1}

This fluorophore has an excitation maximum at 502 nm and an emission maximum at 510 nm. The small Stokes shift of only 8 nm creates some difficulty in discrete excitation without contaminating the emission measurement with scattered or overlapping light. The extinction coefficient of the molecule in methanol is about 77,000 M^{-1}cm^{-1} at 502 nm.

BODIPY FL C$_3$-SE is insoluble in aqueous solution, but may be dissolved in DMF or DMSO as a concentrated stock solution prior to addition of a small aliquot to a reaction. For aqueous reactions, a pH range of 7–9 is optimal. Avoid amine-containing buffers. The reaction also may be done in organic solvent.

Since BODIPY fluorophores are easily quenched if substitutions on a molecule exceed a 1:1 stoichiometry, modification of proteins with this fluorophore probably will not yield satisfactory results. However, for labeling molecules that contain only one amine group, BODIPY FL C$_3$-SE will give intensely fluorescent derivatives.

BODIPY 530/550 C$_3$

BODIPY 530/550 C$_3$ is 4,4-difluoro-5,7-diphenyl-4-bora-3a,4a-diaza-s-indacene-3-propionic acid (Molecular Probes). This derivative of the basic BODIPY structure contains two phenyl rings off the No. 5 and 7 carbon atoms and a propionic acid group on the No. 3 carbon atom. The carboxylate group may be used to attach the fluorophore to amine-containing molecules via a carbodiimide reaction to create an amide bond. The substituents on this BODIPY fluorophore result in alterations to its spectral properties, pushing its excitation and emission maximums up to higher wavelengths.

The excitation maximum for the molecule occurs at 535 nm and its emission at 552 nm. Its Stokes shift is slightly greater than some of the other BODIPY fluorophores, but a 17-nm shift still may not be enough to avoid completely problems of excitation-light interference in emission measurements. BODIPY 530/550 C$_3$ has an extinction coefficient in methanol of about 62,000 $M^{-1}cm^{-1}$ at 535 nm.

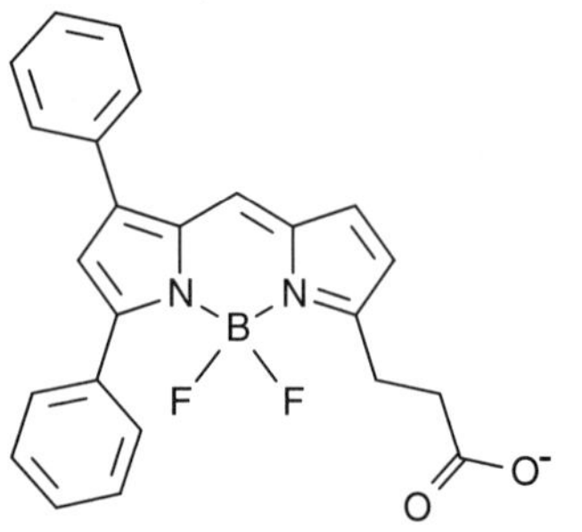

BODIPY 530/550 C3
4,4-Difluoro-5,7-diphenyl-4-bora-3a,4a-
diaza-s-indacene-3-propionic acid
MW 416
Excitation = 535 nm
Emission = 552 nm
ε at 535 nm = 62,000 $M^{-1}cm^{-1}$

BODIPY 530/550 C$_3$ is insoluble in aqueous solution, but it may be dissolved in DMF or DMSO as a concentrated stock solution prior to addition of a small aliquot to a reaction. Coupling to amine-containing molecules may be done using the EDC/sulfo-NHS reaction discussed in Chapter 3, Section 1.2 (Fig. 227). However, modification of proteins with this fluorophore probably will not yield satisfactory results, since BODIPY fluorophores are easily quenched if substitutions on a molecule exceed a 1:1 stoichiometry. For labeling molecules that contain only one amine group, such as DNA probes modified at the 5′ end to contain an amine (Chapter 17, Section 2.1), BODIPY 530/550 C$_3$ will give intensely fluorescent derivatives.

BODIPY 530/550 C$_3$-SE

BODIPY 530/550 C$_3$-SE is 4,4-difluoro-5,7-diphenyl-4-bora-3a,4a-diaza-S-indacene-3-propionic acid, succinimidyl ester (Molecular Probes). The compound is an analog of BODIPY 530/550 C$_3$, which contains an active NHS ester on its propionic acid side chain (Chapter 2, Section 1.4). The ester reacts with primary amines and other nucleophiles to form stable amide bonds.

The excitation maximum for BODIPY 530/550 C$_3$-SE occurs at 533 nm and its emission at 550 nm. Its Stokes shift is relatively small and may not be enough to avoid

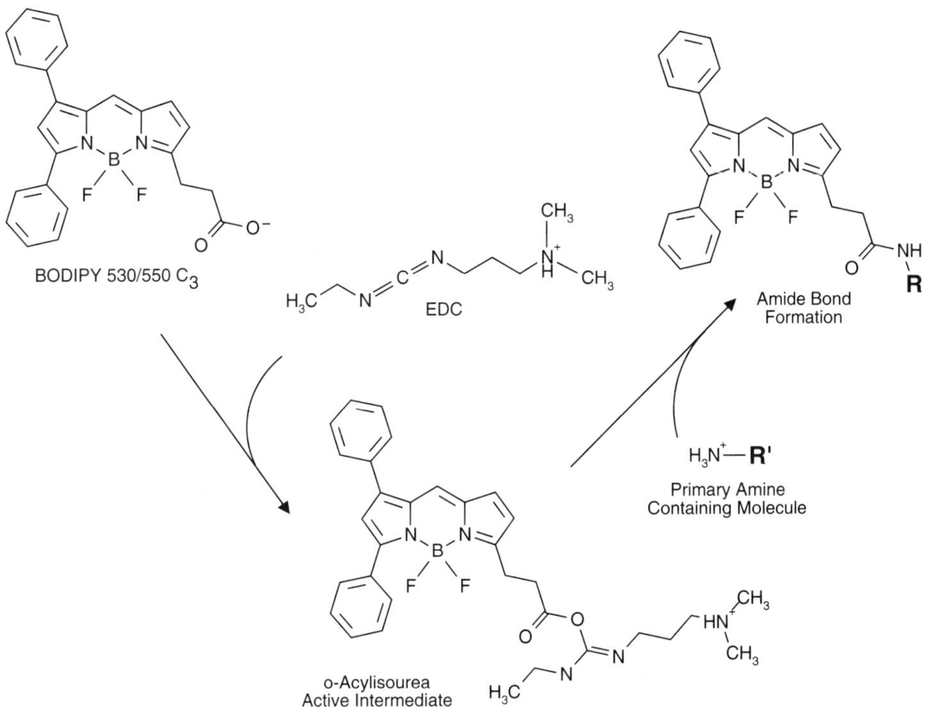

Figure 227 This BODIPY fluorophore contains a carboxylate group that can be attached to amine-containing molecules using a carbodiimide reaction.

completely problems of excitation-light interference in emission measurements. The molecule has an extinction coefficient in methanol of about 70,000 $M^{-1}cm^{-1}$ at 533 nm.

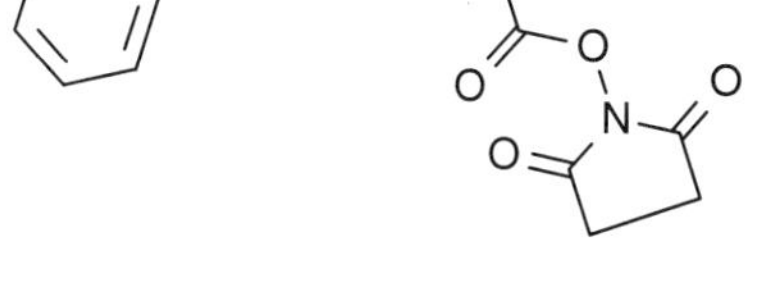

BODIPY 530/550 C$_3$-SE
4,4-Difluoro-5,7-diphenyl-4-bora-3a,4a-
diaza-s-indacene-3-propionic acid, succinimidyl
ester
MW 513
Excitation = 533 nm
Emission = 550 nm
ε at 533 nm = 70,000 $M^{-1}cm^{-1}$

BODIPY 530/550 C_3-SE is insoluble in aqueous solution, but may be dissolved in DMF or DMSO as a concentrated stock solution prior to addition of a small aliquot to a reaction. Coupling to amine-containing molecules proceeds by nucleophilic attack at the carbonyl group, release of the NHS leaving group, and formation an amide linkage (Fig. 228). The reaction may be done in buffered environments having a pH range of 7–9. However, modification of proteins with this fluorophore may not yield satisfactory results, since BODIPY fluorophores are easily quenched if substitutions on a molecule result in a high molar ratio of incorporation. For labeling molecules that contain only one amine group, such as DNA probes modified at the 5′ end to contain an amine (Chapter 17, Section 2.1), BODIPY 530/550 C_3-SE will give intensely fluorescent derivatives.

Aldehyde/Ketone-Reactive BODIPY Derivatives

Hydrazide groups react with aldehyde and ketone functional groups to form hydrazone linkages (Chapter 2, Section 5.1). Three BODIPY derivatives that contain a hydrazine group modification of carboxylate side chains are available. Biomolecules such as proteins that do not possess aldehyde residues can be modified to contain them by a number of chemical means (Chapter 1, Section 4.4).

In addition, DNA and RNA may be modified with hydrazide-containing fluorophores by a transamination reaction of their cytosine residues using bisulfite as a catalyst (Chapter 17, Section 2.1) (Draper and Gold, 1980).

BODIPY 530/550 C_3 Hydrazide

BODIPY 530/550 C_3 hydrazide is 4,4-difluoro-5,7-diphenyl-4-bora-3a,4a-diaza-s-indacene-3-propionyl hydrazide, a derivative of the basic BODIPY structure, which contains two phenyl rings off the No. 5 and 7 carbon atoms and a propionic acid hydrazide group on the No. 3 carbon atom (Molecular Probes). The hydrazide functional group reacts with aldehyde- or ketone-containing molecules to form hydrazone linkages (Fig. 229). The compound may be used to label glycoproteins or other carbohydrate-containing molecules after oxidation of their polysaccharide portions with sodium periodate to yield aldehyde residues.

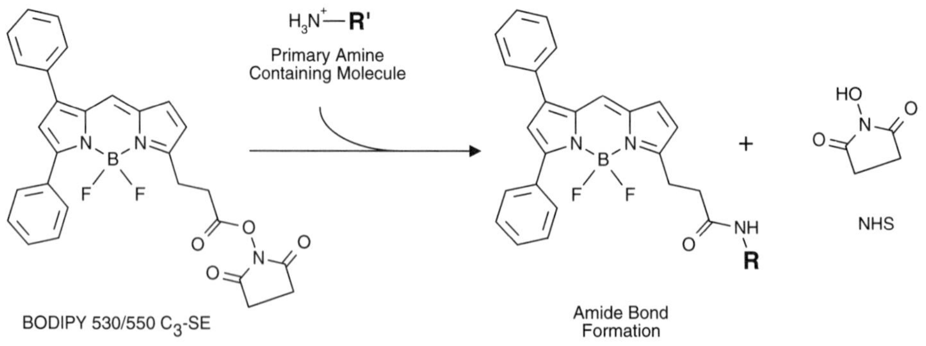

Figure 228 The NHS ester group of this BODIPY compound provides amine reactivity.

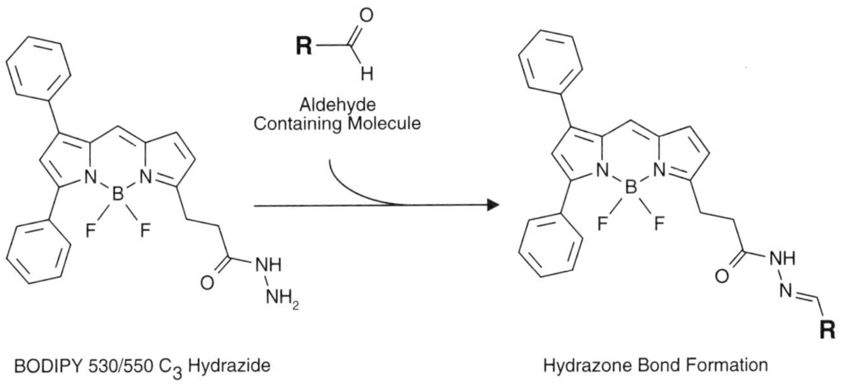

Figure 229 The side-chain hydrazide group of this BODIPY derivative can be used to label aldehyde-containing molecules. Glycoconjugates may be labeled after oxidation of carbohydrate with sodium periodate to produce the required aldehydes.

The excitation maximum for BODIPY 530/550 C$_3$ hydrazide occurs at 534 nm and its emission at 551 nm. Since this is a relatively small Stokes shift, it may be difficult to avoid completely problems of excitation-light interference in critical emission measurements. The molecule has an extinction coefficient in methanol of about 79,000 M^{-1}cm^{-1} at 534 nm.

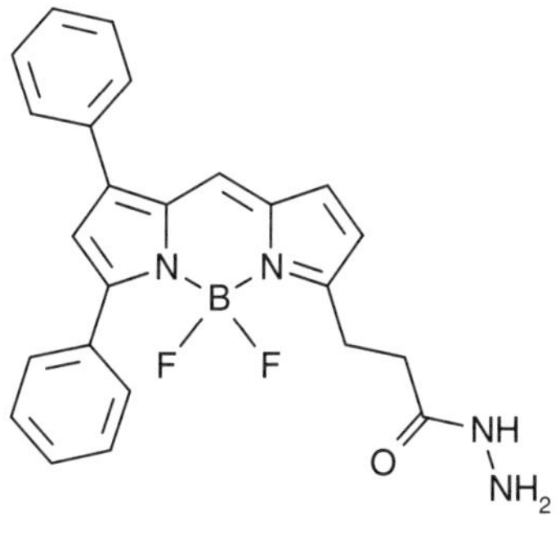

BODIPY 530/550 C3 Hydrazide
4,4-difluoro-5,7-diphenyl-4-bora-3a,4a-
diaza-s-indacene-3-propionyl hydrazide
MW 430
Excitation = 534 nm
Emission = 551 nm
ε at 534 nm = 79,000 M^{-1}cm^{-1}

BODIPY 530/550 C$_3$ hydrazide is insoluble in aqueous solution, but may be dissolved in DMF or methanol as a concentrated stock solution prior to addition of a small aliquot to a reaction. Coupling to aldehyde-containing molecules proceeds rapidly with the formation of a hydrazone linkage. The reaction may be done in buffered environments having a pH range of 5–10. However, modification of glycoproteins with this fluorophore may not yield satisfactory results, since BODIPY fluorophores are easily quenched if substitutions on a molecule result in a high molar ratio of incorporation.

BODIPY 493/503 C₃ Hydrazide

BODIPY 593/503 C₃ hydrazide is 4,4-difluoro-1,3,5,7-tetramethyl-4-bora-3a,4a-diaza-s-indacene-8-propionyl hydrazide (Molecular Probes). Unlike BODIPY 530/550 C₃ hydrazide, this BODIPY derivative contains substituents that shift to lower wavelengths the spectral characteristics of its fluorescent properties. The molecule is highly reactive toward aldehyde-containing compounds, including glycoproteins that have been oxidized with sodium periodate to create the requisite groups (Fig. 230).

BODIPY 493/503 C3 Hydrazide
4,4-difluoro-1,3,5,7-tetramethyl-4-bora-
3a,4a-diaza-s-indacene-8-propionyl hydrazide
MW 334
Excitation = 498 nm
Emission = 506 nm
ε at 498 nm = 92,000 M⁻¹cm⁻¹

The excitation maximum for BODIPY 493/503 C₃ hydrazide occurs at 498 nm, and its emission at 506 nm. As this is an extremely small Stokes shift, it may be difficult to avoid completely problems of excitation-light interference in critical emission measurements unless suboptimal excitation wavelengths are used. The molecule has an extinction coefficient in methanol of about 92,000 $M^{-1}cm^{-1}$ at 493 nm.

BODIPY 493/503 C₃ hydrazide is insoluble in aqueous solution, but may be dissolved in DMF or DMSO as a concentrated stock solution prior to addition of a small

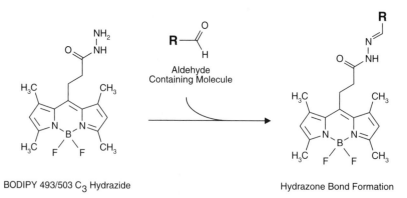

BODIPY 493/503 C₃ Hydrazide

Aldehyde Containing Molecule

Hydrazone Bond Formation

Figure 230 Reaction of this BODIPY fluorophore with aldehyde groups creates hydrazone linkages.

aliquot to a reaction mixture. Coupling to aldehyde-containing molecules proceeds rapidly with the formation of a hydrazone linkage. The reaction may be done in buffered environments having a pH range of 5–10. However, modification of glycoproteins with this fluorophore may not yield satisfactory results, since BODIPY fluorophores are easily quenched if substitutions on a molecule result in a high molar ratio of incorporation. Limiting the modification level by reacting no more than a two- to four-fold molar excess of probe to the amount of glycoconjugate present may overcome this quenching problem.

BODIPY FL C₃ Hydrazide

BODIPY FL C_3 hydrazide is 4,4-difluoro-5,7-dimethyl-4-bora-3a,4a-diaza-s-indacene-3-propionyl hydrazide (Molecular Probes). Unlike the two BODIPY hydrazide derivatives discussed above, this derivative contains substituents that produce luminescent characteristics similar to that of fluorescein, particularly with regard to fluorescing in the green region of the spectrum. The molecule is highly reactive toward aldehyde-containing compounds, including glycoproteins that have been oxidized with sodium periodate to create the requisite groups (Fig. 231).

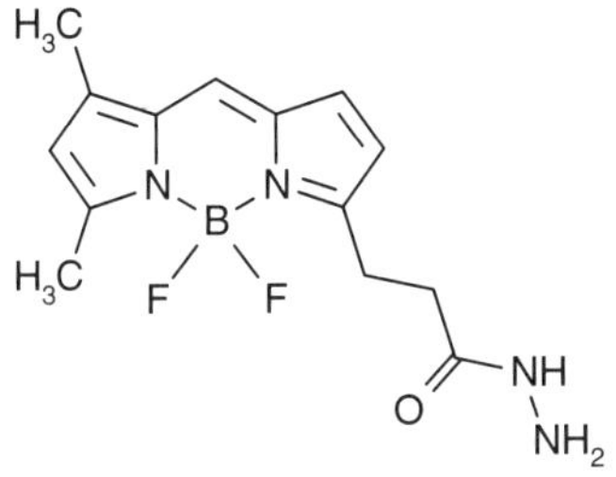

BODIPY FL C3 Hydrazide
4,4-difluoro-5,7-dimethyl-4-bora-3a,4a-
diaza-s-indacene-3-propionyl hydrazide
MW 306
Excitation = 503 nm
Emission = 510 nm
ε at 503 nm = 71,000 $M^{-1}cm^{-1}$

The excitation maximum of BODIPY FL C_3 hydrazide occurs at 503 nm, and its emission at 510 nm. The extremely small Stokes shift makes it difficult to avoid problems of excitation-light interference in critical emission measurements unless suboptimal excitation wavelengths are used. The molecule has an extinction coefficient in methanol of about 71,000 $M^{-1}cm^{-1}$ at 503 nm.

BODIPY FL C_3 hydrazide is insoluble in aqueous solution, but may be dissolved in DMF or methanol as a concentrated stock solution prior to addition of a small aliquot to a reaction mixture. Coupling to aldehyde-containing molecules proceeds rapidly with the formation a hydrazone linkage. The reaction may be done in buffered environments having a pH range of 5–10. However, modification of glycoproteins with this fluorophore may result in fluorescent quenching effects if substitutions on a molecule are at a high molar ratio of incorporation. Limiting the modification level by reacting no more than a two- to four-fold molar excess of probe to the amount of glycoconjugate present may help to overcome the quenching problem.

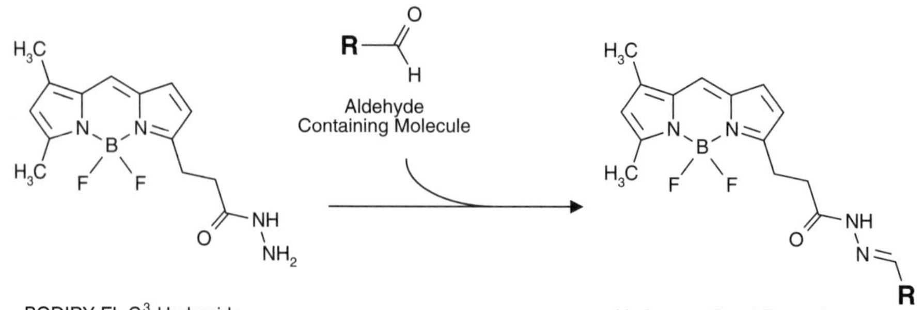

BODIPY FL C³ Hydrazide

Hydrazone Bond Formation

Figure 231 Modification of aldehyde-containing molecules can be done through this BODIPY derivative's hydrazide group.

Sulfhydryl-Reactive BODIPY Derivatives

Three BODIPY derivatives are available for labeling sulfhydryl-containing molecules. The ability to label —SH groups in proteins with sulfhydryl-reactive probes provides a means of directing the modification reaction to a more limited number of sites than occurs when using amine-reactive chemistries. Directed coupling potentially can avoid active centers or binding regions. The first two sulfhydryl-reactive probes discussed in this section make use of iodoacetyl derivatives off the basic BODIPY molecule. The third probe is a bromomethyl derivative that also has good reactivity toward sulfhydryls.

BODIPY FL IA

BODIPY FL IA is *N*-(4,4-difluoro-5,7-dimethyl-4-bora-3a,4a-s-indacene-3-propionyl)-*N'*-iodoacetylethylenediamine, an intensely fluorescent derivative of the basic BODIPY structure, which is useful in modifying sulfhydryl groups (Molecular Probes). The iodoacetyl group reacts with —SH groups in proteins and other molecules to form a stable thioether linkage (Fig. 232). The reactive functional group is at the end of a reasonably long spacer arm, providing enough length to avoid steric problems in modifying sulfhydryls not easily accessible at the surface of macromolecules.

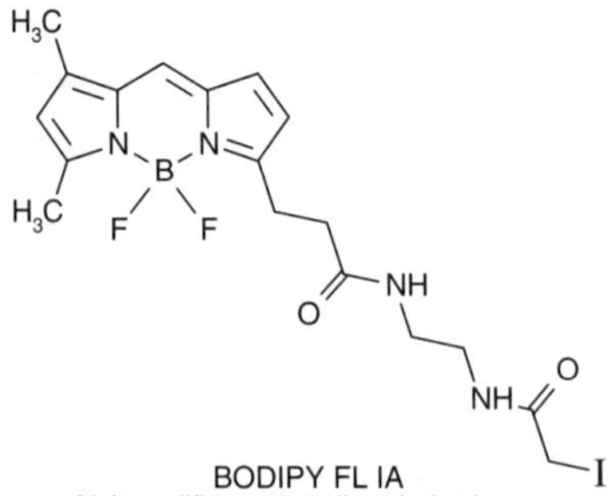

BODIPY FL IA
N-(4,4-difluoro-5,7-dimethyl-4-bora-
3a,4a-s-indacene-3-propionyl)-
N'-iodoacetylethylenediamine
MW 502
Excitation = 504 nm
Emission = 510 nm
ε at 504 nm = 79,000 $M^{-1}cm^{-1}$

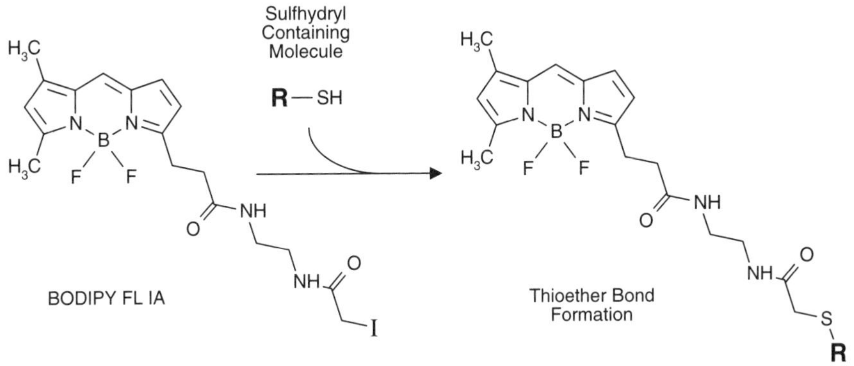

Figure 232 The long side chain of this BODIPY derivative contains a sulfhydryl-reactive iodoacetamide group that can couple to a thiol group to form a thioether bond.

The spectral characteristics of BODIPY FL IA somewhat mimic the green lumines-cence of fluorescein, thus the FL designation in its name. The excitation maximum for the probe occurs at 504 nm, and its emission at 510 nm. The extremely small Stokes shift makes it difficult to avoid problems of excitation-light interference in critical emission measurements unless suboptimal excitation wavelengths below its excitation maximum are used. The molecule has an extinction coefficient in methanol (con-taining 1% sodium acetate and 1% 2-mercaptoethanol) of about 79,000 $M^{-1}cm^{-1}$ at 504 nm.

BODIPY FL IA is insoluble in aqueous solution, but may be dissolved in DMF or DMSO as a concentrated stock solution prior to addition of a small aliquot to a reaction mixture. Coupling to sulfhydryl-containing molecules proceeds rapidly with the formation a thioether linkage. The reaction may be done in 50 mM sodium borate, 5 mM EDTA, pH 8.3. The main consideration is to protect the iodoacetyl derivative from light, which may generate iodine and reduce the reactivity of the probe. In addition, to avoid the fluorescence quenching effects that are often a problem with BODIPY probes and to better limit the modification to sulfhydryls (avoiding any amine modification), react no more than a two-fold molar excess of probe to the amount of sulfhydryl groups present. Oligonucleotides containing a sulfhydryl mod-ification at their 5' ends (Chapter 17, Section 2.2) may be coupled with BODIPY FL IA, yielding highly fluorescent probes.

BODIPY 530/550 IA

BODIPY 530/550 IA is N-(4,4-difluoro-5,7-diphenyl-4-bora-3a,4a-diaza-s-indacene-3-propionyl)-N'-iodoacetylethenediamine, a derivative similar to that of BODIPY FL IA, but containing two phenyl groups rather than two methyl substituents in the 5 and 7 positions. The change in groups results in modulation of the spectral characteristics such that its excitation and emission wavelengths and its Stokes shift are all increased. The spacer arm off the No. 3 carbon atom of its basic BODIPY structure contains a terminal iodoacetyl functional group that reacts with sulfhydryl groups in proteins and other macromolecules to create stable thioether linkages (Fig. 233).

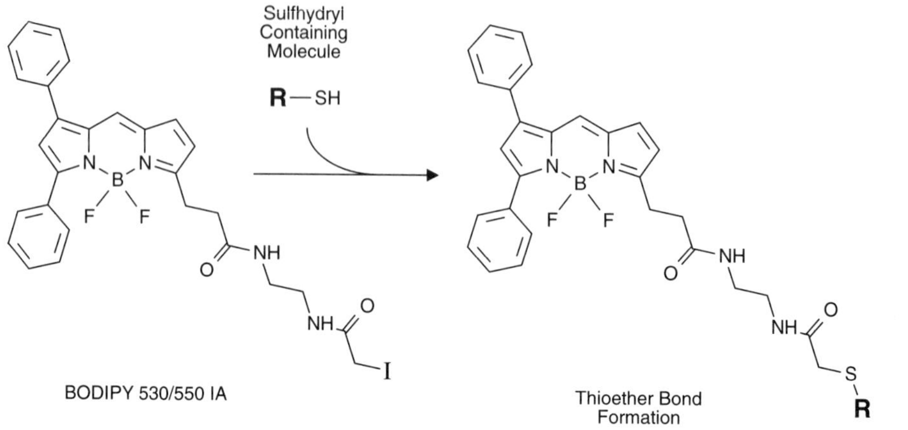

Figure 233 The iodoacetamide group of this BODIPY fluorophore can react with sulfhydryl-containing molecules to form thioether linkages.

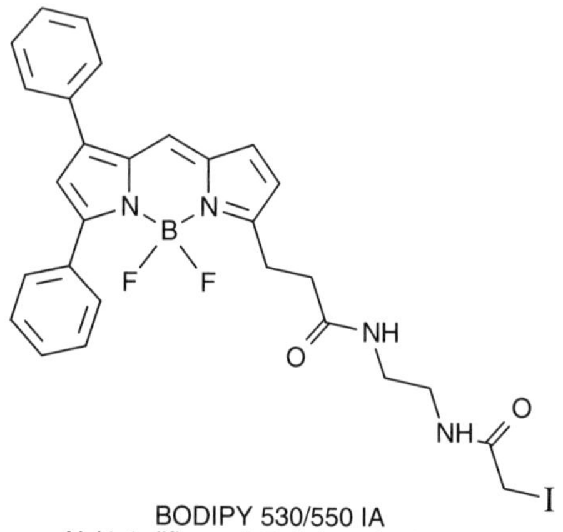

BODIPY 530/550 IA
N-(4,4-difluoro-5,7-diphenyl-4-bora-
3a,4a-diaza-s-indacene-3-propionyl)-
N'-iodoacetylethylenediamine
MW 626
Excitation = 534 nm
Emission = 552 nm
ε at 534 nm = 69,000 $M^{-1}cm^{-1}$

The excitation maximum for BODIPY 530/550 IA occurs at 534 nm and its emission at 552 nm. The Stokes shift is greater than the "FL" BODIPY derivatives, but the relatively small 18-nm differential makes it difficult to avoid completely problems of excitation-light interference in critical emission measurements unless suboptimal excitation wavelengths below the excitation maximum are used. The molecule has an extinction coefficient in methanol (containing 1% sodium acetate and 1% 2-mercaptoethanol) of about 69,000 $M^{-1}cm^{-1}$ at 534 nm.

BODIPY 530/550 IA is insoluble in aqueous solution, but may be dissolved in DMF or DMSO as a concentrated stock solution prior to addition of a small aliquot to a reaction mixture. Coupling to sulfhydryl-containing molecules proceeds rapidly with the formation of a thioether linkage. The reaction may be done in 50 mM sodium borate, 5 mM EDTA, pH 8.3. The main consideration is to protect the iodoacetyl derivative from light, which may generate iodine and reduce the reactivity of the probe. To limit the degree of fluorescent quenching in the resultant conjugate, react the probe in no more than a two- to four-fold molar excess over the amount of target molecule present.

Br-BODIPY 493/503

Br-BODIPY 493/503 is 8-bromomethyl-4,4-difluoro-1,3,5,7-tetramethyl-4-bora--3a,4a-diaza-3-indacene, a small BODIPY derivative containing a short, sulfhydryl-reactive bromomethyl group. The substituent groups on the molecule result in modulation of its spectral characteristics such that, after conjugation, its excitation and emission wavelengths are reduced somewhat from other BODIPY probes. The reagent can be coupled to —SH-containing molecules to produce a thioether linkage (Fig. 234).

Br-BODIPY 493/503
8-Bromomethyl-4,4-difluoro-1,3,5,7-
tetramethyl-4-bora-3a,4a-diaza-3-indacene
MW 341
Excitation = 515 nm
Emission = 525 nm
ε at 515 nm = 55,000 M^{-1}cm^{-1}

The excitation maximum for Br-BODIPY 493/503 is 515 nm and its emission occurs at 525 nm when dissolved in methanol. On coupling to a sulfhydryl compound, however, the excitation wavelength of the adduct decreases to 493 nm and its emission drops to 503 nm. The very small 10-nm Stokes shift may be a problem, particularly in avoiding interference due to excitation-light scattering in critical emission measure-

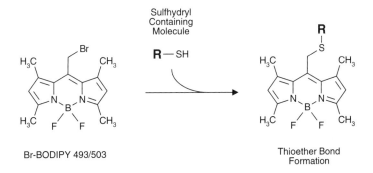

Figure 234   Br-BODIPY can be used to modify sulfhydryl-containing molecules to form thioether linkages.

ments. Suboptimal excitation wavelengths below the excitation maximum may be used to reduce extraneous light contamination. The molecule has an extinction coefficient in methanol of about 55,000 $M^{-1}cm^{-1}$ at 515 nm.

Br-BODIPY 493/503 is insoluble in aqueous reaction mixtures, but may be dissolved in DMF or DMSO as a concentrated stock solution prior to addition of a small amount to a buffered solution. Coupling to sulfhydryl-containing molecules proceeds rapidly with the formation of a thioether linkage. The reaction may be done in 50 mM sodium borate, 5 mM EDTA, pH 8.3.

1.5. Cascade Blue Derivatives

Cascade Blue (registered trademark of Molecular Probes) derivatives are fluorescent probes having strong luminescence in the blue region of the spectrum. The basic Cascade Blue molecule is derived from a trisulfonated pyrene backbone (Fig. 235) (Whitaker et al., 1991). It is a fixable analog of 8-methoxypyrene-1,3,6-trisulfonic acid, which is a blue fluorescent neural tracer. The fluorophore emits light in a region removed form the luminescent signal of fluorescein or Lucifer Yellow, making it a good choice for multilabeling applications. The dye can be used along with Lucifer Yellow CH and sulforhodamine 101 for three-color mapping of neuronal components and processes. Cascade Blue derivatives have relatively high absorptivity, good quantum yields (typically about 0.54), excellent water solubility due to the presence of the negatively charged sulfonate groups, and good photostability. Labeling proteins and other macromolecules with Cascade Blue derivatives can be done with little fluorescent quenching due to dye–dye interactions.

The following sections discuss the most important Cascade Blue derivatives that are available for covalent modification purposes.

Amine-Reactive: Cascade Blue Acetyl Azide

One Cascade Blue derivative is available for creating linkages with amine-containing molecules. The acetyl azide functional group of this reagent reacts with primary

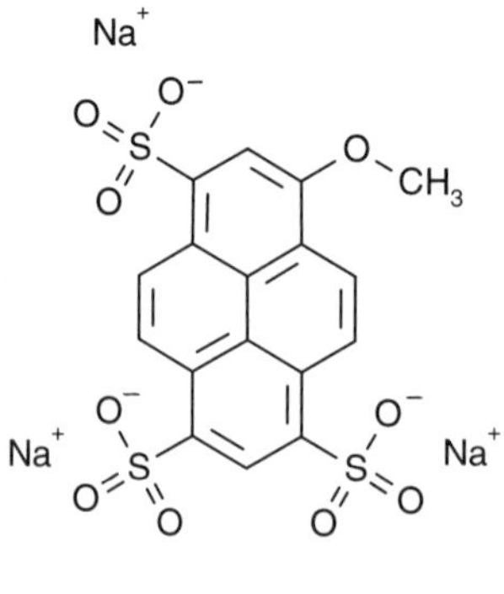

8-Methoxypyrene-1,3,6-trisulfonic acid,
trisodium salt

Figure 235   The basic structure of Cascade Blue fluorophores.

amines at ambient temperatures or below to create amide-bond derivatives (Oparka *et al.*, 1991; Lanier and Recktenwald, 1991). At elevated temperatures (80°C in DMF), the acetyl azide group rearranges to form an isocyanate that can react with hydroxyl-containing molecules to form a urethane linkage (Fig. 236). The Cascade Blue urethane derivatives of macromolecules are extremely fluorescent and can be detected down to femtogram quantities (Takadate *et al.*, 1985).

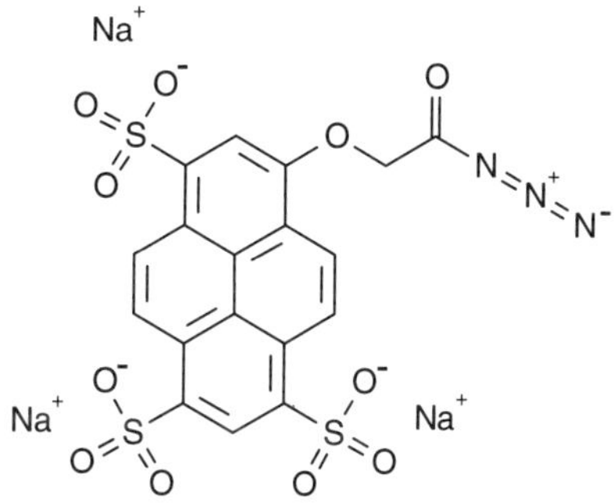

Cascade Blue Acetyl Azide
MW 607
Excitation = 375, 400 nm
Emission = 410 nm
ε at 375 nm = 27,000 M^{-1}cm^{-1}

This fluorophore has excitation maxima at 375 and 400 nm and an emission maximum at 410 nm. The small Stokes shift may create some difficulty in discrete excitation without contaminating the emission measurement with scattered or over-lapping light. The extinction coefficient of the molecule in water is about 27,000 M^{-1}cm^{-1}. Cascade Blue and Lucifer Yellow derivatives can be simultaneously excited by light of less than 400 nm, resulting in two-color detection at 410 and 530 nm.

Cascade Blue acetyl azide is soluble in aqueous solution, but the reactive azide group will hydrolyze and should be used directly. A concentrated stock solution may be prepared in water and dissolved quickly, and an aliquot immediately added to a buffered reaction medium. For aqueous reactions, a pH range of 7–9 is optimal. Avoid amine-containing buffers.

Carboxylate-Reactive: Cascade Blue Cadaverine and Cascade Blue Ethylenediamine

Cascade Blue cadaverine and Cascade Blue ethylenediamine both contain a carboxamide-linked diamine spacer off the 8-methoxy group of the pyrene trisulfonic acid backbone. The cadaverine version contains a 5-carbon spacer, while the eth-ylenediamine has only a 2-carbon arm. Both can be coupled to carboxylic acid-containing molecules using a carbodiimide reaction (Chapter 3, Section 1). Because Cascade Blue derivatives are water-soluble, the carbodiimide EDC can be used to couple these fluorophores to proteins and other carboxylate-containing molecules in aqueous solutions having a pH range of 4.5–7.5. The reaction forms amide-bond linkages (Fig. 237).

These fluorophores have excitation maxima at 377–378 nm and at 398–399 nm, and emission maxima at 422–423 nm. The extinction coefficients of the molecules in

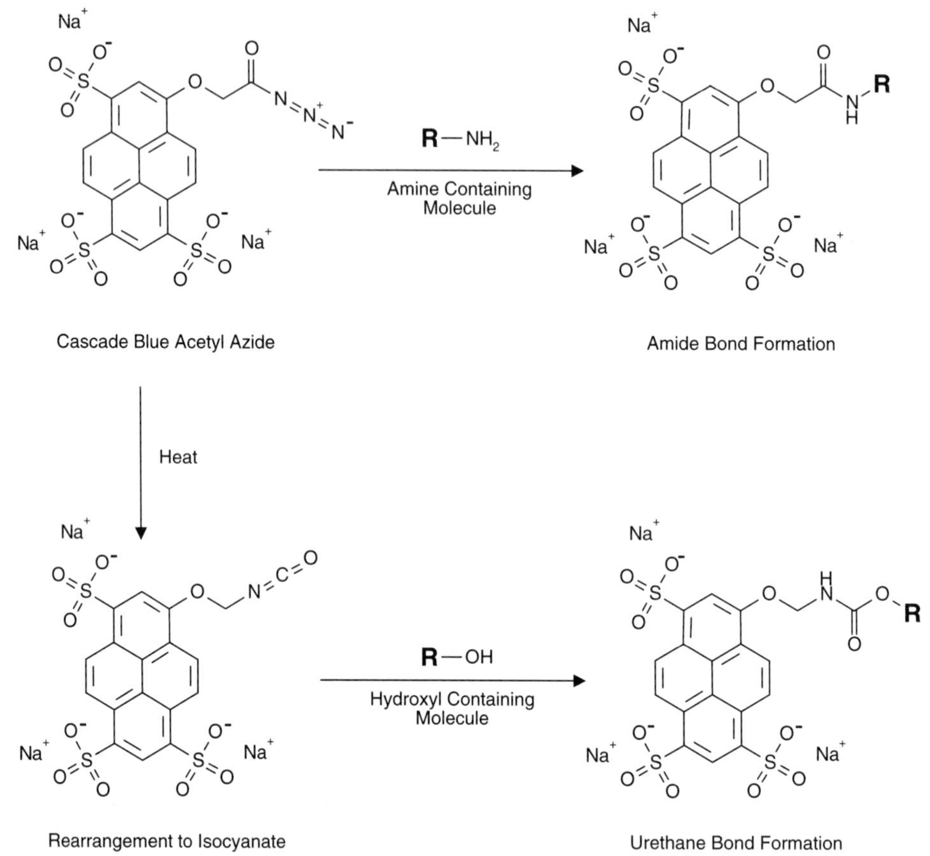

Cascade Blue Acetyl Azide

Amide Bond Formation

Rearrangement to Isocyanate

Urethane Bond Formation

Figure 236 The acetyl azide group of this Cascade Blue derivative has dual functions. It can react with amine groups to form amide bonds, or it can be converted to an isocyanate at high temperatures to couple with hydroxyl functional groups, creating a carbamate linkage.

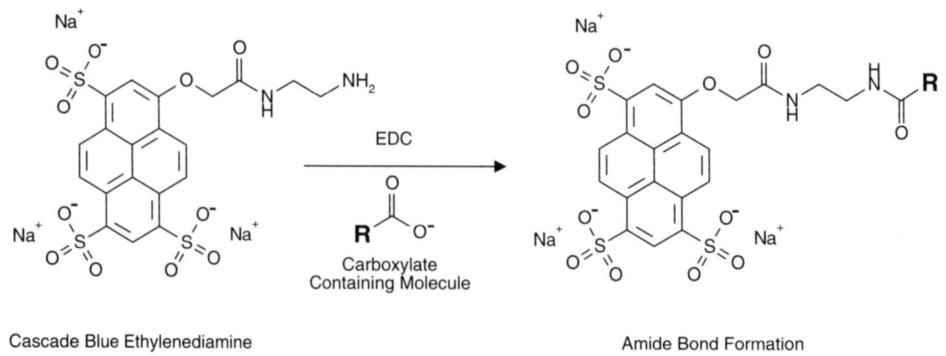

Cascade Blue Ethylenediamine

Amide Bond Formation

Figure 237 The side-chain primary amine group of this Cascade Blue derivative can be coupled to carboxylate-containing molecules using a carbodiimide reaction.

water are about 27,000 $M^{-1}cm^{-1}$. These Cascade blue derivatives can be used along with Lucifer Yellow derivatives and simultaneously excited by light of less than 400 nm, resulting in two-color detection at 422 and 530 nm.

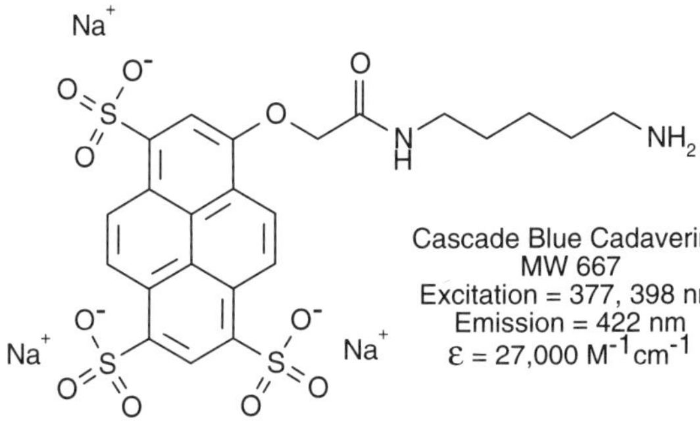

Cascade Blue Cadaverine
MW 667
Excitation = 377, 398 nm
Emission = 422 nm
$\varepsilon = 27,000\ M^{-1}cm^{-1}$

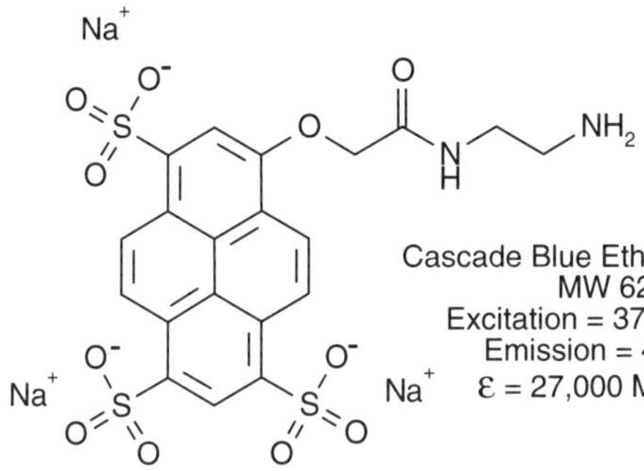

Cascade Blue Ethylenediamine
MW 624
Excitation = 378, 399 nm
Emission = 423 nm
$\varepsilon = 27,000\ M^{-1}cm^{-1}$

Cascade Blue diamine derivatives are soluble in aqueous solution. A concentrated stock solution may be prepared in water and dissolved quickly, and an aliquot immediately added to a buffered reaction medium. For aqueous reactions, 0.1 M MES, pH 4.7–6, may be used to stabilize the pH during the coupling process. Avoid amine- or carboxylate-containing buffers such as Tris or glycine, since these can compete with the coupling reaction.

Aldehyde/Ketone-Reactive: Cascade Blue Hydrazide

Cascade Blue hydrazide is a carboxy-hydrazine derivative of the 8-methoxy group on the pyrene trisulfonic acid fluorophore. Hydrazide groups react with aldehyde and ketone functional groups to form relatively stable hydrazone linkages (Chapter 2, Section 5.1) (Fig. 238). Although most biomolecules do not contain aldehyde or ketone groups in their native state, carbohydrates, glycoproteins, RNA, and other molecules that contain sugar residues can be oxidized with sodium periodate to produce reactive formyl functional groups. The use of modification reagents that generate aldehydes upon coupling to a molecule also can be used to produce hydrazide-reactive sites (Chapter 1, Section 4.4).

In addition, DNA and RNA may be modified with hydrazide-reactive probes by reacting their cytosine residues with bisulfite to form reactive sulfone intermediates (Chapter 17, Section 2.1). These derivatives can undergo transamination reactions with hydrazide- (or amine)-containing probes to yield covalent bonds (Draper and Gold, 1980).

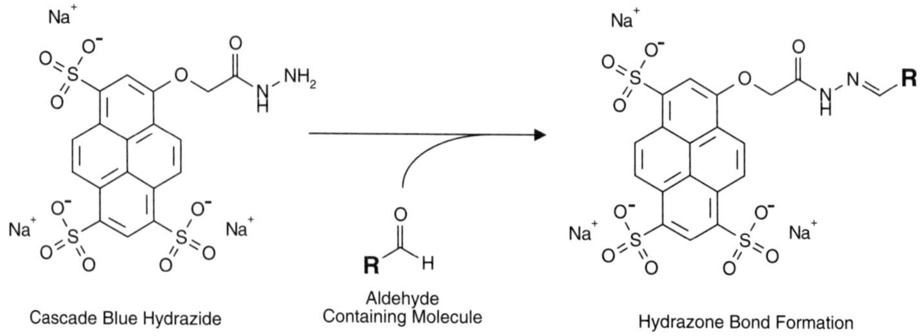

Figure 238 Cascade Blue hydrazide can be used to modify aldehyde-containing molecules to form hydrazone bonds.

This fluorophore has an excitation maximum at 400 nm, and an emission maximum at 420 nm. The extinction coefficient of the molecule in aqueous solution at pH 7 is about 31,000 $M^{-1}cm^{-1}$. Cascade Blue hydrazide and Lucifer Yellow derivatives can be simultaneously excited by light of less than 400 nm, resulting in two-color detection at 420 and 530 nm.

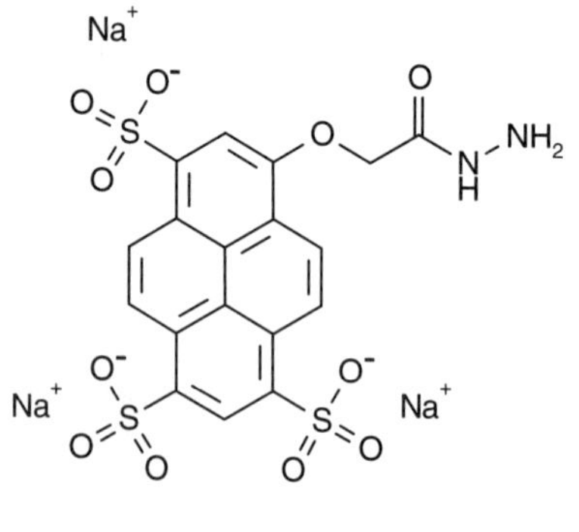

Cascade Blue Hydrazide
MW 645
Excitation = 400 nm
Emission = 420 nm
ε at 400 nm = 31,000 $M^{-1}cm^{-1}$

Cascade Blue hydrazide is soluble in aqueous solution, and it should be stable for a while if protected from light. A concentrated stock solution of the reagent may be prepared in water, and an aliquot added to a buffered reaction medium to facilitate the transfer of small quantities. For aqueous reactions, a pH range of 5–9 will result in efficient hydrazone formation.

1.6. Lucifer Yellow Derivatives

Lucifer Yellow derivatives are used extensively for cytochemical staining applications, especially in neurophysiology (Stewart, 1981a, b). The fluorophores are 3.6-disul-

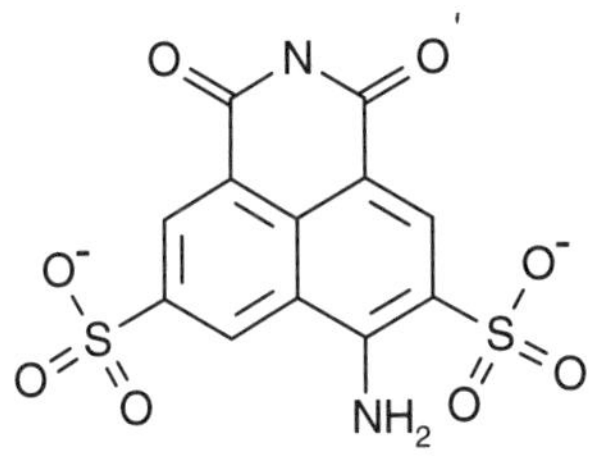

3,6-Disulfonate-4-amino-
naphthalimide

Figure 239   The basic structure of Lucifer Yellow fluorophores.

fonate 4-aminonaphthalimide derivatives that can be further modified at their imide nitrogen to contain reactive groups suitable for conjugation with biomolecules (Fig. 239). Cell staining with membrane-impermeant Lucifer Yellow dyes is usually done by osmotic shock, microinjection, or pinocytosis (Swanson *et al.*, 1987). Derivatives containing amines or hydrazide functional groups are fixable with formaldehyde or glutaraldehyde, coupling to nearby proteins or other amine-containing molecules intracellularly. Glycoconjugates on cell surfaces or in solution may be labeled after periodate oxidation with the hydrazide derivative (Stewart, 1978). Sulfhydryl-containing molecules may be tagged with the iodoacetamide derivative.

Lucifer Yellow probes are water-soluble to at least 1.5%. The absorbance maximum of the derivatives occurs at about 426–428 nm with an emission peak at about 530–535 nm, in the yellow region of the spectrum. The quantum yield of Lucifer dyes is about 0.25. The good intensity of luminosity from these dyes makes possible detection of small quantities of labeled molecules intracellularly. The fluorescent conjugates are readily visible in living cells at concentrations that are nontoxic to cell viability. The low molecular weight and water solubility of these dyes allow passage of labeled compounds from one cell to another, potentially revealing molecular relationships between cells.

Sulfhydryl-Reactive: Lucifer Yellow Iodoacetamide

One Lucifer Yellow derivative is available for labeling sulfhydryl-containing molecules. Lucifer Yellow iodoacetamide is a 4-ethyliodoacetamide derivative of the basic disulfonate aminonaphthalimide fluorophore structure (Molecular Probes). The iodoacetyl groups react with —SH groups in proteins and other molecules to form stable thioether linkages (Fig. 240).

The spectral characteristics of Lucifer Yellow iodoacetamide produce luminescence at somewhat higher wavelengths than the green luminescence of fluorescein, thus the yellow designation in its name. The excitation maximum for the probe occurs at 426 nm, and its emission at 530 nm. The rather large Stokes shift makes sensitive measurements of emission intensity possible without interference by scattered excitation-light. The 2-mercaptoethanol derivative of the fluorophore has an extinction coefficient at pH 7 of about 13,000 $M^{-1}cm^{-1}$ at 426 nm.

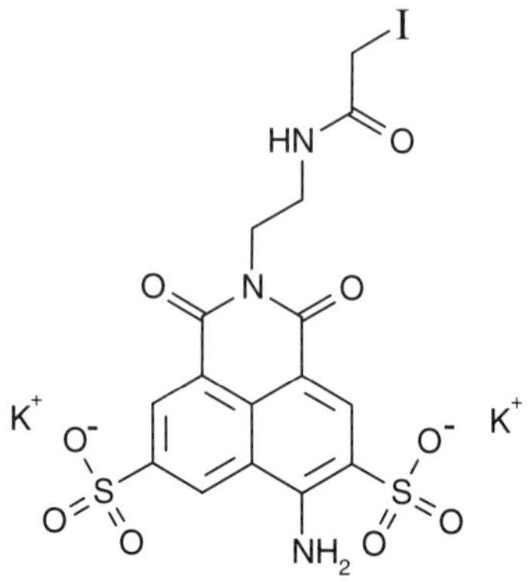

Lucifer Yellow Iodoacetamide,
dipotassium salt
MW 649
Excitation = 426 nm
Emission = 530 nm
ε at 426 nm = 13,000 $M^{-1}cm^{-1}$

Lucifer Yellow iodoacetamide is soluble in aqueous solution due to its negatively charged sulfonate groups. A concentrated stock solution may be prepared in water prior to addition of a small aliquot to a reaction mixture. Coupling to sulfhydryl-containing molecules proceeds rapidly with the formation a thioether linkage. The reaction may be done in 50 mM sodium borate, 5 mM EDTA, pH 8.3. The main consideration is to protect the iodoacetyl derivative from light, which may generate iodine and reduce the reactivity of the probe. The reaction may be limited to sulf-hydryls (avoiding any amine derivatization) by reacting no more than a two-fold molar excess of probe to the amount of sulfhydryl groups present. In addition, oli-gonucleotides containing a sulfhydryl modification at their 5′ ends (Chapter 17, Section 2.2) may be coupled with Lucifer Yellow iodoacetamide, yielding highly fluo-rescent, yellow probes.

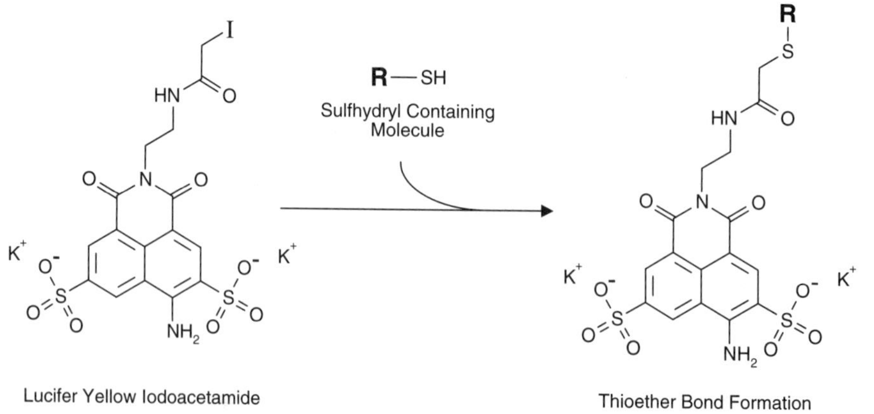

Lucifer Yellow Iodoacetamide

Thioether Bond Formation

Figure 240 Lucifer Yellow iodoacetamide can be used to label sulfhydryl-containing molecules, forming thioether bonds.

Aldehyde/Ketone-Reactive: Lucifer Yellow CH

Lucifer Yellow CH is a carbohydrazide derivative of the basic disulfonate aminonaphthalimide fluorophore structure (Molecular Probes). Hydrazide groups react with aldehyde and ketone functional groups to form relatively stable hydrazone linkages (Chapter 2, Section 5.1) (Fig. 241). Although most biomolecules do not contain aldehyde or ketone groups in their native state, carbohydrates, glycoproteins, RNA, and other molecules that contain sugar residues can be oxidized with sodium periodate to produce reactive formyl functional groups. The use of modification reagents that generate aldehydes on coupling to a molecule also can be used to produce hydrazide-reactive sites (Chapter 1, Section 4.4). In addition, DNA and RNA may be modified with hydrazide-reactive probes by reacting their cytosine residues with bisulfite to form reactive sulfone intermediates (Chapter 17, Section 2.1). These derivatives can undergo transamination reactions with hydrazide- (or amine)-containing probes to yield covalent bonds (Draper and Gold, 1980).

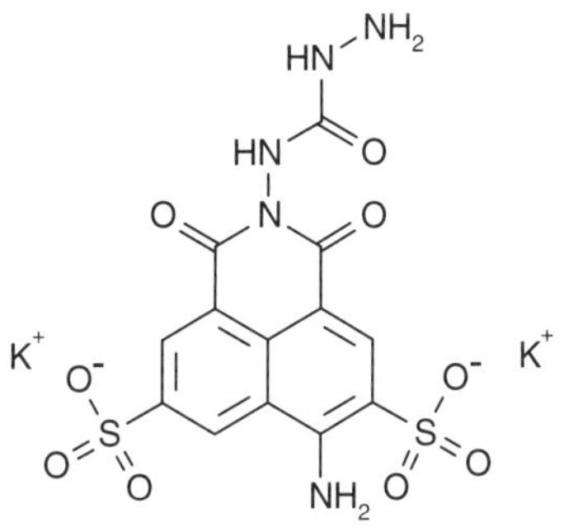

Lucifer Yellow CH
MW 522 (potassium salt)
MW 457 (lithium salt)
MW 479 (ammonium salt)
Excitation = 428 nm
Emission = 533-535 nm
ε at 428 nm = 12,000 $M^{-1}cm^{-1}$

Lucifer Yellow CH is commonly used as a neuronal tracer by staining cells and then fixing them with formaldehyde or glutaraldehyde. It also can be used to label glycoproteins or glycolipids on cell surfaces after periodate oxidation (Lee and Fortes, 1985; Spiegel et al., 1983). The labeling of oxidized ribonucleotides and gangliosides can be done similarly (Sun et al., 1988; Spiegel et al., 1985).

This fluorophore has an excitation maximum at 428 nm, and an emission maximum at 534 nm. The extinction coefficient of the molecule in aqueous solution is about 12,000 $M^{-1}cm^{-1}$. Cascade Blue hydrazide and Lucifer Yellow CH derivatives can be simultaneously excited by light of less than 400 nm, resulting in the possibility for two-color detection at 420 and 534 nm.

Lucifer Yellow CH is soluble in aqueous solution, and it should be stable for a while if protected from light. The reagent is available as three different salts of the sulfonate groups. The ammonium salt of the fluorophore is soluble to a level of 9% in water, while the lithium and potassium salts have solubilities of 5 and 1%, respectively. A concentrated stock solution of the fluorophore may be prepared in water, and an aliquot added to a buffered reaction medium to facilitate the transfer of small quan-

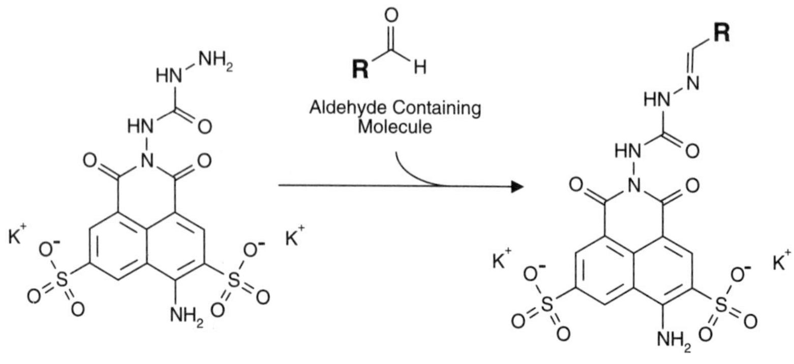

Lucifer Yellow CH   Hydrazone Linkage

Figure 241 The hydrazide group of this Lucifer Yellow derivative can react with aldehyde-containing molecules to form hydrazone bonds.

tities. For aqueous reactions, a pH range of 5–9 will result in efficient hydrazone formation with aldehyde or ketone residues.

1.7. Phycobiliprotein Derivatives

Phycobiliproteins are intensely fluorescent proteins that function as components in the photosynthetic apparatus of eukaryotic, blue-green algae and cyanobacteria (Glazer, 1981). The proteins are found as aggregates in phycobilisome particles near the chlorophyll regions (Kronick, 1986). In the native state, phycobiliproteins do not fluoresce; rather excitation energy is designed to be efficiently transferred to chlorophyll molecules for utilization in synthetic processes within the cell. Once purified, however, excitation energy is released from phycobiliproteins as strong luminosity. The fluorescent quantum efficiencies of these proteins can be as high as 0.98, far better than most synthetic probes (Grabowski and Gantt, 1978). In addition, each biliprotein contains multiple chromophoric bilin prosthetic groups, inferring extremely high absorbance coefficients to each protein molecule. B-phycoerythrin, for example, typically contains 34 chromophoric groups giving an effective, combined extinction coefficient at 545 nm of $2.4 \times 10^6 \ M^{-1}\mathrm{cm}^{-1}$ (Glazer and Hixson, 1977). The strong absorption bands are in the visible region of the spectrum, extending from the green to the far-red wavelengths. These absorption spectra extend over a broad range of potential excitation wavelengths, allowing for versatility in the excitation source employed and creating large Stokes shifts, thus minimizing interference from Rayleigh-scattered light (Loken *et al.*, 1987).

Due to the presence of multiple fluorescent groups in each phycobiliprotein, conjugates of these molecules form extraordinarily luminescent probes. Labeling of macromolecules with phycobiliprotein derivatives can provide absorption coefficients 30-fold higher than labeling with small, synthetic fluorophores. Their ability to be monitored by fluorescing in the red region of the spectrum decrease potential interferences from indigenous biological fluorescence. The protected bilin (tetra-pyrrole) prosthetic groups are not easily affected by their external environment. They are not readily quenched by conjugation to another molecule or affected by other components in solution. The prosthetic group orientation within the protein molecules enables fluorescence to take place independent of pH or ionic strength. The excellent solubility of phycobiliproteins in aqueous solution allows easy chemical manipulation for modification or conjugation reactions, and their hydrophilic nature provides low non-specific binding character in fluorescent detection applications.

There are three main classes of phycobiliproteins, differing in their protein structure, bilin content, and fluorescent properties. These are phycoerythrin, phycocyanin, and allophycocyanin. There are two main forms of phycoerythrin proteins commonly in use: B-phycoerythrin isolated from *Porphyridium cruentum* and R-phycoerythrin from *Gastroclonium coulteri*. There also are three main forms of pigments found in these proteins: phycoerythrobilin, phycourobilin, and phycocyanobilin (Glazer, 1985). The relative content of these pigments in the phycobiliproteins determines their spectral properties. All of them, however, have extremely high absorption coefficients ranging from a magnitude of 10^5 to 10^6, and excellent quantum yields ranging from 0.51 up to 0.98.

The spectral properties of four major phycobiliproteins used as fluorescent labels can be found in Table 3. The bilin content of these proteins ranges from a low of 4 prosthetic groups in C-phycocyanin to the 34 groups of B- and R-phycoerythrin. Phycoerythrin derivatives, therefore, can be used to create the most intensely fluorescent probes possible using these proteins. The fluorescent yield of the most luminescent phycobiliprotein molecule is equivalent to about 30 fluoresceins or 100 rhodamine molecules. Streptavidin–phycoerythrin conjugates, for example, have been used to detect as little as 100 biotinylated antibodies bound to receptor proteins per cell (Zola *et al.*, 1990).

Conjugation of phycobiliproteins to targeting components such as antibodies, avidin, biotin, or other molecules preserves the binding or activity of the attached constituent and does not alter the spectral characteristics of the bilin prosthetic groups. Common heterobifunctional cross-linking agents can be used to create phycobiliprotein conjugates, including SPDP (Chapter 5, Section 1.1), SMCC (Chapter 5, Section 1.3), and SMPB (Chapter 5, Section 1.6) (Oi *et al.*, 1982). These cross-linkers react with

Table 3 Properties of the Phycobiliproteins

Property	B-Phycoerythrin	R-Phycoerythrin	C-Phycocyanin	Allophycocyanin
Source	*Porphyridium cruentum*	*Gastroclonium coulteri*	*Anabaena variabilis*	*Anabaena variabilis*
Subunit structure	$(\alpha\beta)_6\gamma$	$(\alpha\beta)_6\gamma$	$(\alpha\beta)_2$	$(\alpha\beta)_3$
Molecular weight	240,000	240,000	72,000	110,000
Pigment content (bilin groups)	34	34	4	6
Absorbance maximum	546 nm	566 nm	614 nm	650 nm
Molar extinction coefficient	2.4×10^6	2.0×10^6	5.8×10^5	7.0×10^5
Emission maximum	575 nm	574 nm	643 nm	660 nm

amine groups on the phycobiliproteins, producing activated intermediates able to couple with sulfhydryl-containing molecules. Thiolating reagents such as 2-im-inothiolane, SATA, and SAMSA (Chapter 1, Section 4.1) can be used to create thiols on the secondary molecule to effect the final coupling reaction.

Glazer and Stryer (1983) report on the preparation and use of tandem phy-cobiliprotein conjugates, wherein B-phycoerythrin is cross-linked to allophycocyanin to create an energy donor–acceptor pair. The B-phycoerythrin component can be excited at 545 nm, emitting energy at 575 nm that can be accepted by allophycocyanin, which in turn emits light at 660 nm. The result is a large shift in the spectral charac-teristics from that of the individual proteins, increasing the effective Stokes shift to over 100 nm. Conjugation of such tandem pairs to other proteins can create superior fluorescent reagents.

The following protocol is a generalized method for creating sulfhydryl-reactive phycobiliprotein reagents for coupling to —SH-containing molecules. The procedure uses the heterobifunctional cross-linker SPDP (Chapter 5, Section 1.1). Other amine and sulfhydryl-reactive cross-linking agents may be used in a similar manner.

Protocol

1. Dialyze the phycobiliprotein into 50 mM sodium borate, 0.3 M NaCl, pH 8.5 (note: commercial preparations of these proteins come as an ammonium sulfate suspension). After dialysis, adjust the protein solution to a concentration of 1 mg/ml. Higher protein concentrations may be used, but the amount of cross-linking reagent added to each milliliter of the reaction should be proportionally scaled up, as well. Protect the protein solution from undue exposure to light.

2. Dissolve SPDP at a concentration of 6.2 mg/ml in DMSO (makes a 20 mM stock solution). Alternatively, LC-SPDP may be used and dissolved at a concentration of 8.5 mg/ml in DMSO (also makes a 20 mM solution). If the water-soluble Sulfo-LC-SPDP is used, a stock solution in water may be prepared just prior to adding an aliquot to the reaction. In this case, prepare a 10 mM solution of Sulfo-LC-SPDP by dissolving 5.2 mg/ml in water. Since an aqueous solution of the crosslinker will degrade by hydrolysis of the sulfo-NHS ester, it should be used quickly to prevent significant loss of activity. If a sufficiently large amount of phycobiliprotein will be modified, the solid may be added directly to the reaction mixture without preparing a stock solution in water to allow accurate weighing of Sulfo-LC-SPDP.

3. Add 25 μl of the stock solution of either SPDP or LC-SPDP in DMSO to each milliliter of the protein solution. If Sulfo-LC-SPDP is used,, add 50 μl of the stock solution in water to each milliliter of protein solution.

4. Mix and react for at least 30 min at room temperature. Longer reaction times, even overnight, will not adversely affect the modification.

5. Purify the modified protein from reaction by-products by dialysis or gel filtra-tion using Sephadex G-25 and a buffer consisting of 50 mM sodium phosphate, 0.15 M NaCl, 10 mM EDTA, pH 7.2.

The SPDP-activated phycobiliprotein may be reacted with a sulfhydryl-containing protein to create a fluorescent conjugate linked through disulfide bonds.

2. Bifunctional Chelating Agents and Radioimmunoconjugates

Monoclonal antibodies provide extremely high antigen specificity that can be useful as cancer therapeutic reagents *in vivo* (Waldmann, 1991). Radiolabeled monoclonals are currently undergoing developmental clinical trials for their use in the diagnosis or treatment of cancer. The antigen binding specificity of the radioimmunoconjugate provides the targeting capability to localize in tumor sites, while the associated radio-label provides cytotoxic properties or visibility for imaging applications (Schlom, 1986).

Iodine-131 was among the first radioactive isotopes used for radioimmunoconju-gate preparation (Order, 1982; Regoeczi, 1984). Since the earliest studies on the efficacy of radiotherapy, additional isotopes have been employed, such as iodine-125, bismuth-212, yttrium-90, yttrium-88, technetium-99m, copper-67, rhenium-188, rhenium-186, galium-66, galium-67, indium-111, indium-114m, indium-115, and boron-10.

There are several methods commonly used to label monoclonal antibodies with radionuclides. In a direct labeling process, a radioactive atom is attached to functional groups on the antibody without the use of an intervening chemical group. For instance, radioiodination can be done through modification of tyrosine side chains using estab-lished techniques and reagents (Section 4). Another direct method uses indigenous sulfhydryl groups or those formed through disulfide reduction to covalently couple certain metal nuclides (Holmberg and Meurling, 1993; Ranadive *et al.*, 1993). Thiola-tion reagents also can be used in this regard to create the requisite —SH groups by modification of other protein functional groups (Joiris *et al.*, 1991) (Chapter 1, Section 4.1).

Indirect methods of protein labeling with radiolabels utilize organic compounds able to chelate metal ions in a coordination complex. Bifunctional chelating agents (BCAs) contain a chemical-reactive group for coupling to proteins or other molecules and a strong metal-chelating group for complexing certain radioactive metals. Their extensive use with monoclonal antibodies that are able to target specific cellular antigens has resulted in important radiopharmaceutical applications for the diagnosis and treatment of cancer (Meares, 1986; Otsuka and Welch, 1987; Liu and Wu, 1991; Hnatowich, 1990; Subramanian and Meares, 1991; Wessels and Rogus, 1984). The BCAs may be loaded with the radioactive metal before or after their conjugation with a monoclonal antibody (Frytak *et al.*, 1993). If they are loaded with radionuclides prior to modifying the antibody, the BCA–metal pair is called a preformed complex (Kasina *et al.*, 1991).

The following sections describe the major methods and BCAs used to create radio-immunoconjugates.

2.1. DTPA

Diethylenetriaminepentaacetic anhydride (DTPA) is a bifunctional chelating agent containing two amine-reactive anhydride groups. The compound reacts with N-termi-nal and ε-amine groups of proteins to form amide linkages. The anhydride rings open to create multivalent, metal-chelating arms able to bind tightly metals in a coordina-

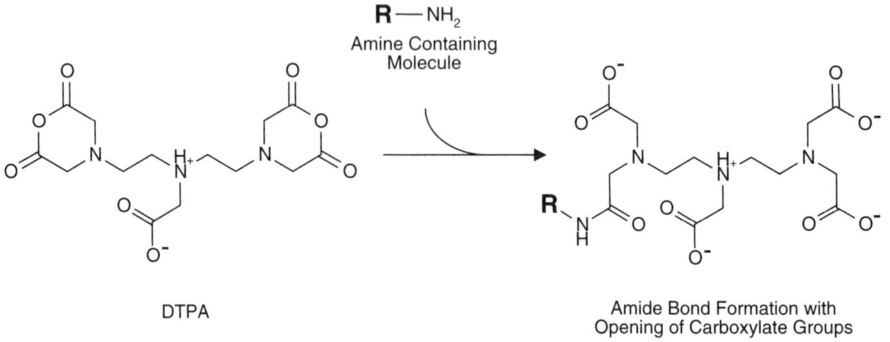

Figure 242 DTPA reacts with amine-containing molecules via ring-opening of its anhydride groups to create amide bond linkages.

tion complex (Fig. 242) (Hnatowich *et al.*, 1982). Optimal reaction conditions for antibody modification are neutral to slightly alkaline pH environments containing no extraneous amines. A pH of 7–8 may be used with buffering provided by phosphate or bicarbonate buffers at 0.1 *M*. As two anhydride groups are present on each DTPA molecule there is potential for creating cross-links between two amine-containing molecules. Conjugation of antibody through DTPA cross-links may be a major reason some immunoglobulins lose antigen binding activity after modification (Lanteigne and Hnatowich, 1984). Optimization of the amount of protein present and the quantity of DTPA added to the reaction may have to be done to avoid this type of cross-linking and polymerization.

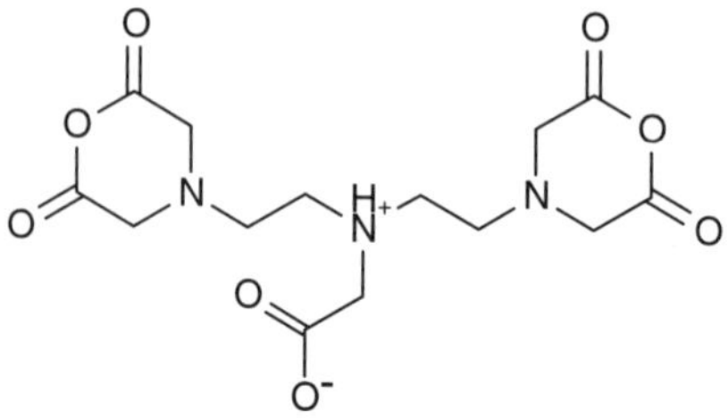

DTPA
Diethylenetriamine-
pentaacetic anhydride
MW 357. 33

DTPA also can be used to modify amine-containing polymers, such as poly-L-lysine, to create a chelating polymer possessing multiple metal binding sites (Trubetskoy *et al.*, 1993). Subsequent polymer modification of antibodies provides much higher radioactivity per molecule than when DTPA is directly coupled to the protein. Such chelating polymers can introduce into proteins as much as 100 DTPA residues able to complex radiolabels for each 55,000-Dal poly-L-lysine chain attached (Torchilin *et al.*, 1993). Directed coupling to the antibody through only the N-terminal of the poly-L-lysine chain limits the modification to a single point along the polymer, thus avoiding cross-linking or multisite attachment that can affect antibody activity (Slinkin *et al.*, 1991). This approach can dramatically increase the radioactivity level at tumor sites, thus increasing cytotoxicity or enhancing imaging capability.

Although DTPA has been used extensively as a BCA to prepare radiopharmaceutical reagents, newer metal chelators such as those discussed below may show greater promise for use in *in vivo* applications.

2.2. DOTA, NOTA, and TETA

1,4,7,10-Tetraazacyclododecane-*N*,*N'*,*N''*,*N'''*-tetraacetic acid (DOTA) is a chelating ring structure containing four acetic acid carboxylate groups off the four nitrogens of its 12-atom cyclic structure. C- or N-functionalized derivatives of this basic structure produce a BCA capable of modifying proteins and binding radioactive metal ions in strong coordination complexes (Cox *et al.*, 1990; Renn and Meares, 1992). Perhaps the simplest method of DOTA functionalization is through modification of one of its carboxylates to contain a short spacer terminating in a reactive group capable of being coupled to proteins (Li and Meares, 1993). However, modification of carbons on its cyclic backbone also is possible (Brechbiel *et al.*, 1993). Complexes of metal ions with DOTA have been studied in detail (Sherry *et al.*, 1989; Aime *et al.*, 1992).

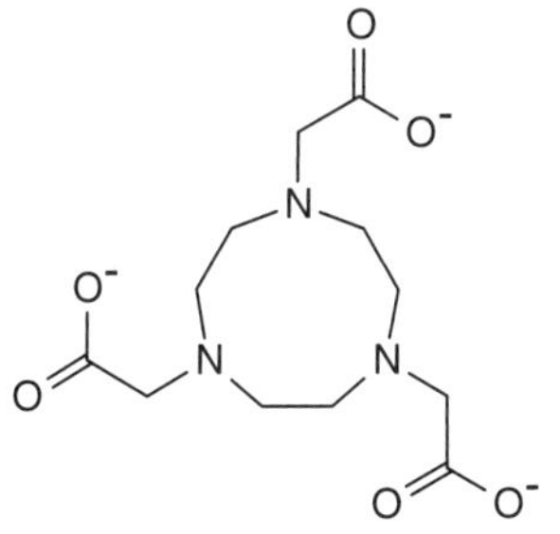

NOTA
1,4,7-Triazacyclononane-
N, N', N''-triacetic acid

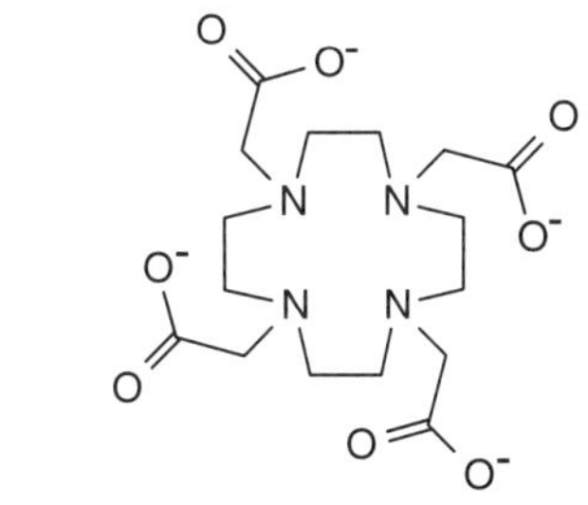

DOTA
1,4,7,10-Tetraazacyclodo-
decane-N, N', N'', N'''-tetraacetic acid

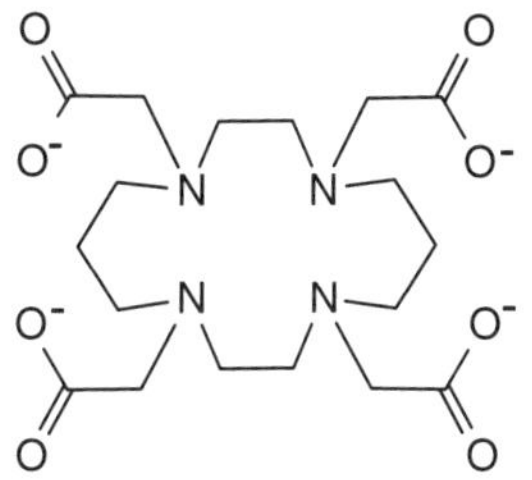

TETA
1,4,8,11-Tetraazacyclotetra-
decane-N, N', N'', N'''-tetraacetic acid

A BCA similar to DOTA is 1,4,7-triazacyclononane-*N*,*N'*,*N''*-triacetic acid (NOTA). NOTA contains a smaller ring structure in comparison to DOTA and only has three chelating nitrogen groups. Synthesis of functional derivatives can be done through C- or N-modifications, creating reactive groups able to couple to proteins and other molecules (Cox *et al.*, 1990). Cox and co-workers (1990) have prepared an (*S*)-lysine derivative of a benzamide-protected, C-substituted NOTA.

A third compound in the same category as DOTA and NOTA is 1,4,8,11-tetraazacyclotetradecane-*N,N',N'',N'''*-tetraacetic acid (TETA). TETA contains a larger ring structure than the other two BCAs and has four chelating nitrogen groups. Similar to the other two chelators, C- and N-functionalized derivatives can be prepared with TETA (Brechbiel *et al.*, 1993). A *p*-bromoacetamidobenzyl-TETA derivative could be used to label antibodies through sulfhydryl groups and securely bind radioactive copper for probing biological systems *in vivo* (Moi *et al.*, 1985).

These BCAs chelate metal atoms by holding them in the center of their nitrogenous ring structure, in a manner similar to the chelation of iron at the core of a porphyrin prosthetic group. The excellent stability of this complex allows the use of DOTA, NOTA, and TETA conjugates *in vivo* with minimal potential for leaching and deleterious side effects.

2.3. DTTA

DTTA is N^1-(*p*-isothiocyanatobenzyl)-diethylenetriamine-N^1,N^2,N^3,N^3-tetraacetic acid. This BCA contains four carboxylate groups that can hold metals tightly in a coordination complex. The compound is especially good at chelating lanthanide-series elements, such as europium, samarium, terbium, and dysprosium. Unlike the previous bifunctional chelating agents, which are used to prepare radiopharmaceutical reagents, this one is used primarily for complexing metals to form fluorescent probes for time-resolved fluoroimmunoassays (Hemmila, 1988). The most commonly used lanthanides for this purpose are europium (Eu^{3+}) and samarium (Sm^{3+}). Proteins modified with DTTA and complexed with Eu^{3+} or Sm^{3+} form the basis for unique fluorescent probes possessing long-lived signals on excitation.

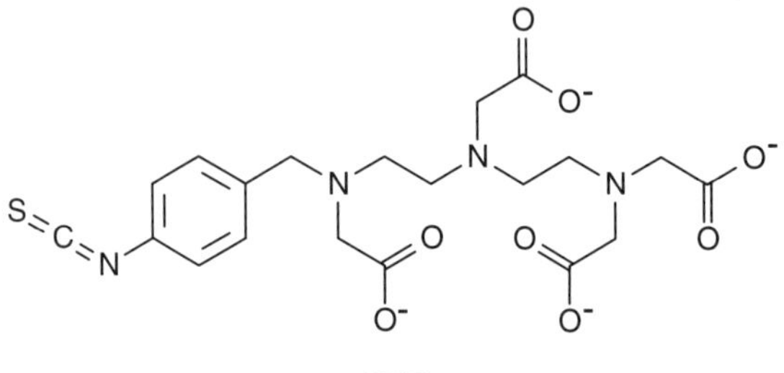

DTTA
N^1-(p-isothiocyanatobenzyl)-diethylene-
triamine-N^1, N^2, N^3, N^3-tetraacetic acid

Time-resolved fluorescence is based on the property that these metal-chelated labels can be excited and their luminescence measured after a short delay period. The delay time allows the nonspecific background fluorescence to decay before the label's signal is measured. This eliminates all interference from naturally fluorescing compounds in biological samples. Further enhancing the sensitivity is the use of what is called a dissociation-enhancement procedure. After a labeled targeting molecule has bound its intended target, the metal complex is disrupted by a special low-pH solution containing amphipathic chelating molecules. The Eu^{3+} (or other lanthanide) then rapidly forms another coordination complex within chelating components in the dissociation solution and becomes surrounded and protected by a micelle. The resultant fluorescent intensity of this micelle-chelate can be up to 10^6 times more than that of the original labeled molecule.

The isothiocyanate group of DTTA reacts with primary amines in proteins and other molecules to form stable thiourea bonds (Fig. 243) (Mukkala *et al.*, 1989). The reagent is water-soluble and can be reacted under relatively mild conditions (in 0.1 *M* sodium carbonate, pH 9). The isothiocyanate group is reasonably stable in aqueous solution for short periods, but will degrade. Best results will be obtained if fresh DTTA is used. The reaction involves attack of the nucleophile on the central, electrophilic carbon of the isothiocyanate group. The resulting electron shift creates a thiourea linkage between the chelating compound and the protein with no leaving group. Modification of antibodies can be done without loss of significant antigen binding activity, even when up to 10–15 DTTA chelates are substituted per immunoglobulin molecule (Stahlberg *et al.*, 1993). Diagnostic assays using DTTA-Eu^{3+} chelates are commercially available employing the DELFIA (Wallac Oy, Turku, Finland) time-resolved fluoroimmunoassay system.

2.4. DFA

DFA or deferoxamine is *N'*-[5-[[4-[[5-(acetylhydroxyamino)pentyl]amino]-1,4-dioxobutyl]hydroxyamino]pentyl]-*N*-(5-aminopentyl)-*N*-hydroxybutanediamide, a

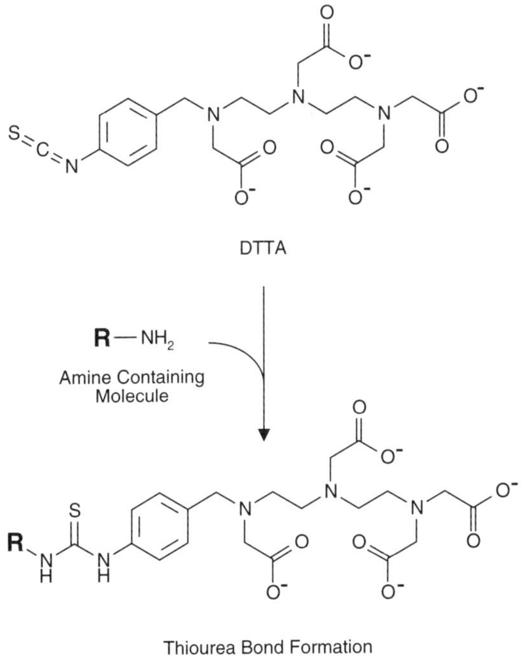

DTTA

Amine Containing
Molecule

Thiourea Bond Formation

Figure 243 The isothiocyanate group of DTTA can react with amine-containing molecules to form isothiourea bonds.

naturally occurring product is isolated from *Streptomyces pilosus*. Its native activity is forming iron complexes, but it is also very proficient at forming coordination chelates with other metals, particularly galium-68 (Motta-Hennessy *et al.*, 1985). Radiopharmaceutical agents for imaging can be produced by modification of targeting molecules with DFA complexes containing radioactive metals. The amine group of DFA can be utilized for direct labeling of antibodies and other proteins through coupling with available carboxylates using carbodiimide-mediated conjugation with EDC (Chapter 3, Section 1.1). The use of amine-reactive, homobifunctional cross-linkers (Chapter 4) also can be employed to modify proteins with DFA at their amino groups.

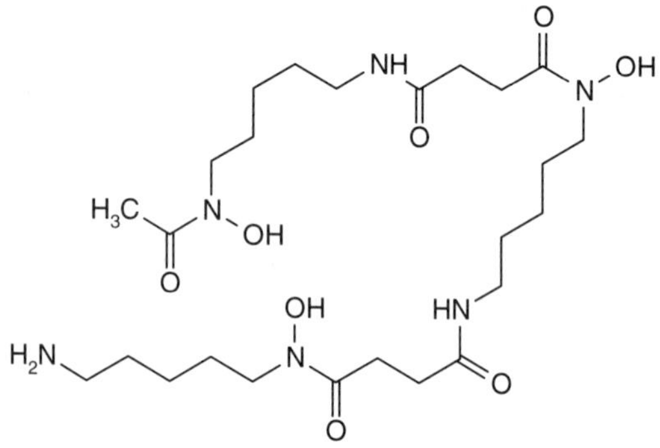

DFA
Deferoxamine
MW 560.71

DFA–polymer conjugates can be made containing multiple chelating groups along the length of the polymer. The polymer backbone utilized for this synthesis can be either activated dextran (Torchilin *et al.*, 1989) or succinylated poly-L-lysine (Slinkin *et al.*, 1990; Torchilin *et al.*, 1993). These DFA–polymer constructs can be attached to antibody molecules through additional functional groups on the polymer. Chelating polymers can provide much higher signals than direct attachment of DFA to antibodies, since each coupled polymer derivative can possess dozens of chelated radioactive metals. In addition, high substitution levels of DFA directly coupled to antibodies can significantly affect activity by denaturation or blocking of the antigen binding sites.

2.5. Use of Thiolation Reagents for Direct Labeling to Sulfhydryl Groups

Proteins containing sulfhydryl residues can be labeled with a radioactive element by direct complexation to the —SH group, avoiding entirely the use of a bifunctional chelating agent. Particularly, reduced sulfhydryls in antibody molecules can be coupled with ^{99m}Tc to yield thiol–metal derivatives (Thakur and DeFulvio, 1991; Rhodes, 1991). However, cleavage of disulfide linkages within the antibody can lead to activity losses and fragmentation (Pimm *et al.*, 1991). The required sulfhydryl groups can be introduced into antibodies without disulfide reduction through the use of a thiolating reagent that modifies amine residues within the antibody (Joiris *et al.*, 1991). Thiolating agents such as 2-iminothiolane or SATA provide efficient ways of

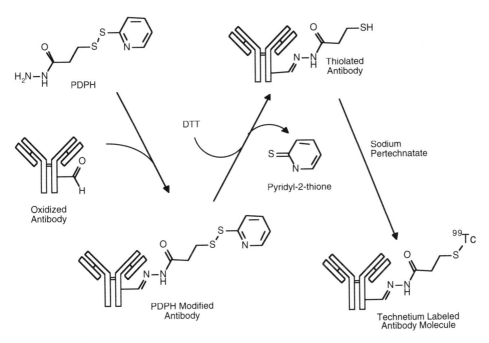

Figure 244 Antibody molecules oxidized with sodium periodate to create aldehyde groups on their polysaccharide chains can be modified with PDPH to produce thiols after reduction of the pyridyl disulfide. Direct labeling of the sulfhydryls with ^{99}Tc produces a radioactive complex.

introducing multiple sulfhydryl groups for this type of radiopharmaceutical preparation (Chapter 1, Section 4.1).

Site-directed thiolation at carbohydrate residues within the Fc region of antibody molecules may prove to be the best choice for —SH group introduction while maintaining antigen binding activity. Ranadive and co-workers (1993) have used the heterobifunctional cross-linking agent PDPH (Chapter 5, Section 2.3) to react specifically with oxidized polysaccharide components of monoclonals. The polysaccharide chains are first treated with sodium periodate (Chapter 1, Section 4.4) to generate reactive aldehyde residues. PDPH is then coupled to these aldehydes via its hydrazide end to create stable hydrazone linkages. The other end of the cross-linker, containing a pyridyl disulfide group, is then reduced with DTT under mild conditions (25 mM DTT, pH 4.5, 30 min) to produce the free sulfhydryl groups. Since the thiolation occurs only at carbohydrate locations within the antibody, the modification is away from the antigen binding sites, thus preserving immunoglobulin activity. Subsequent treatment with sodium pertechnatate yields the ^{99m}Tc derivative on the sulfhydryl groups (Fig. 244).

3. Biotinylation Reagents

The highly specific interaction of avidin with the small vitamin biotin can be a useful tool in designing assay, detection, and targeting systems for biological analytes (see

Chapter 13). The extraordinary affinity of avidin's interaction with biotin allows biotin-containing molecules in complex mixtures to be discretely bound with avidin conjugates. If the avidin–biotin complex contains detection components, then the targeted analytes can be located or quantified. This assay concept is made possible through the ability of biotin to be covalently attached to other targeting molecules, such as antibodies. In this sense, biotin derivatives may be prepared that contain reactive portions able to couple with particular functional groups in proteins and other molecules. Biotin modification of secondary molecules, called "*biotinylation*," results in covalent derivatives containing one or more bicyclic biotin rings extending from the parent structure. These biotinylation sites are still capable of binding avidin or streptavidin with the specificity and avidity of free biotin in solution. Since the biotin components are relatively small, macromolecules can be modified with these reagents without significantly affecting their physical or chemical properties (Della-Penna *et al.*, 1986).

The basic design of a biotin-labeling compound is illustrated in Fig. 245. Common to all such modification reagents is the presence of the bicyclic biotin ring at one end of the structure and a reactive functional group at the other end that can be used to couple with other molecules. Biotinylation reagents also possess various cross-bridges or spacer groups built off the valeric acid side chain of the molecule. As the binding sites for biotin on avidin and streptavidin are pockets buried about 9 Å beneath the surface of the proteins, spacers can affect the accessibility of biotinylated compounds for efficiently binding avidin or streptavidin conjugates (Green *et al.*, 1971). In some applications, the use of a long spacer arm in the biotinylation reagent will result in the greatest potential assay sensitivity. The rate of binding of an avidin or streptavidin probe to a biotinylated molecule also is affected by the length of spacer in the biotinylation reagent used. When longer spacers are utilized to make biotinylated macromolecules, it potentially can result in a five-fold greater rate of streptavidin interaction (Bonnard *et al.*, 1984).

Another variable to consider in choosing biotinylation reagents is the use of a biotin analog such as iminobiotin that has a moderated affinity constant in its binding of avidin or streptavidin (Section 3.1). Analogs may be useful if release of the avidin–

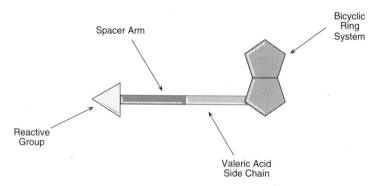

Figure 245 The basic design of a biotinylation reagent includes the bicyclic rings and valeric acid side chain of D-biotin at one end and a reactive functional group to couple with target groups at the other. Spacer groups may be included in the design to extend the biotin group away from modified molecules, thus ensuring better interaction capability with avidin or streptavidin probes.

biotin bond is important for isolating a targeted analyte. Using native biotin, the interaction with avidin is so strong that up to 6–8 M guanidine at pH 1.5 is required to break the bond, possibly causing extensive denaturation of any other complexed molecules. By contrast, iminobiotinylated molecules can be released simply by adjusting the pH down to 4.

The following sections discuss some of the more common biotinylation reagents available for modification of proteins and other biomolecules. Each biotin derivative contains a reactive portion (or can be made to contain a reactive group) that is specific for coupling to a particular functional group on another molecule. Careful choice of the correct biotinylation reagent can result in directed modification away from active centers or binding sites, and preserve the activity of the modified molecule.

3.1. Amine-Reactive Biotinylation Agents

Amine-reactive biotinylation reagents contain functional groups off biotin's valeric acid side chain that are able to form covalent bonds with primary amines in proteins and other molecules. Two basic types are commonly available: NHS esters and carboxylates. NHS esters spontaneously react with amines to form amide linkages (Chapter 2, Section 1.4). Carboxylate-containing biotin compounds can be coupled to amines via a carbodiimide-mediated reaction using EDC (Chapter 3, Section 1.1).

D-*Biotin and Biocytin*

D-Biotin (hexahydro-2-oxo-lH-thieno[3,4-d]imidazole-4-pentanoic acid) is a naturally occurring growth factor present in small amounts within every cell. It is a key component in numerous processes involving carboxylation reactions, wherein it functions as a cofactor and transporter of CO_2 (coenzyme R). Biotin is mainly found covalently attached to lysine ε-amine groups of proteins via its valeric acid side chain. The compound was originally discovered through symptoms of deficiency caused by eating too many raw egg whites. Biotin (or vitamin H) was found to be complexed and inactivated by the egg-white protein avidin (Boas, 1927; du Vigneaud, 1940). Treatment with additional vitamin H alleviated the symptoms.

Biotin's interaction with the proteins avidin and streptavidin is among the strongest noncovalent affinities known (K_a = 10^{15} M^{-1}). The binding occurs between the bicyclic ring of biotin and a pocket within each of the four subunits of the proteins. The valeric acid portion is not involved or required for the interaction (Green, 1975; Wilchek and Bayer, 1988). This characteristic allows modification of the valeric acid side chain without affecting the binding potential toward avidin or streptavidin.

D-Biotin is thus the basic building block for constructing biotinylation reagents. The molecule may be attached directly to a protein via its valeric acid side chain or derivatized at this carboxylate with other organic components to create spacer arms and various reactive groups. Reaction of biotin with primary amine groups on proteins can be done using the water-soluble carbodiimide EDC (Chapter 3, Section 1.1). EDC activates the carboxylate to create a highly reactive, intermediate ester. The ester then can couple to amines to form stable amide bond derivatives (Fig. 246). Biotinylated molecules thus formed retain the ability to bind avidin or streptavidin with high affinity.

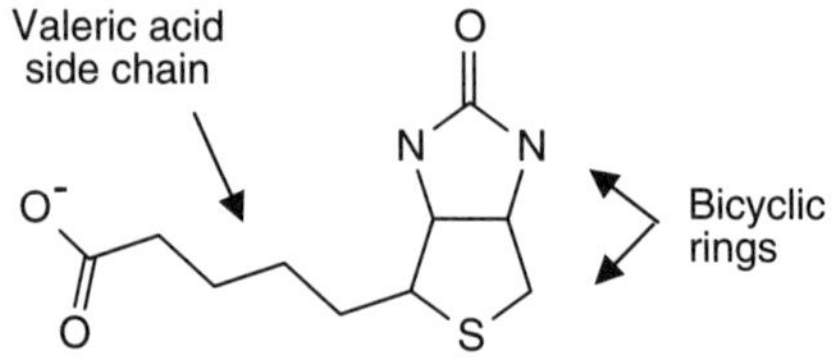

D-Biotin
MW 244.31

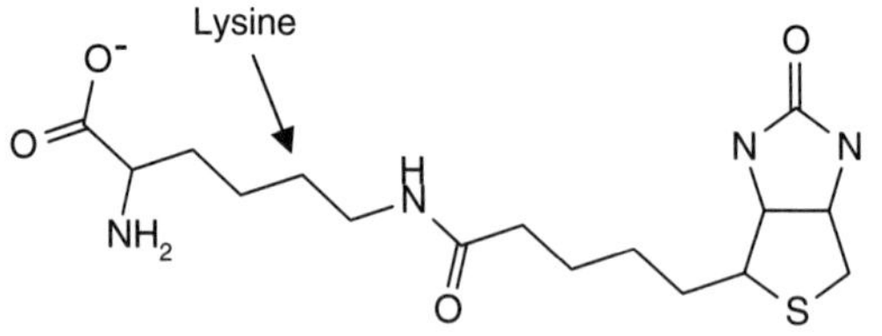

Biocytin
MW 372.48

The only potential deficiency in using D-biotin to modify directly a protein is the relatively short spacer arm afforded by the indigenous valeric acid group. Some applications may require longer spacers to maintain good binding potential toward avidin or streptavidin.

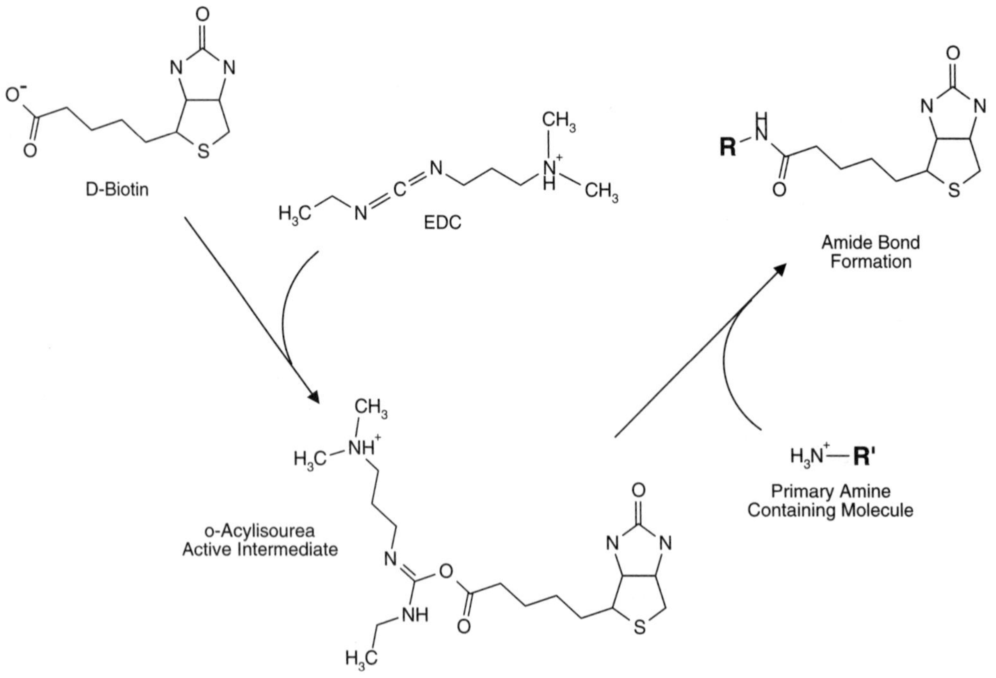

Figure 246 D-Biotin can be directly coupled to amine-containing molecules using the water-soluble carbodiimide EDC to form an amide bond linkage.

Biocytin is ε-N-biotinyl-L-lysine, a derivative of D-biotin containing a lysine group coupled at its ε-amino side chain to the valeric acid carboxylate. It is a naturally occurring complex of biotin that is typically found in serum and urine, and probably represents breakdown products of recycling biotinylated proteins. The enzyme biotinidase specifically cleaves the lysine residue and releases the biotin component from biocytin (Ebrahim and Dakshinamurti, 1986, 1987).

Biocytin has been used extensively as a labeling reagent for intracellular components within neurons (Horikawa and Armstrong, 1988; King *et al.*, 1989; Izzo, 1991; Granata and Kitai, 1992). It is particularly good for anterograde tracing studies in the central nervous system, since it can be easily injected into neurons using micropipettes. Subsequent visualization of biocytin locations may be done using an avidin–enzyme conjugate (Chapter 13).

Biocytin should not be used in a carbodiimide reaction to modify proteins or other molecules, since it contains both a carboxylate and an amine group. A carbodiimide-mediated reaction, as suggested for D-biotin, would cause self-conjugation and polymerization of this reagent.

Biocytin, however, can form the basis for constructing trifunctional cross-linking reagents (Chapter 6). The lysine component of the molecule contains a free carboxylate and an α-amine group that can be used to build spacers and reactive groups for cross-linking purposes. The biotin component is the third arm of the trifunctional system, retaining its ability to bind avidin probes after conjugation has occurred at its other two ends. Such a trifunctional derivative has been used to study the hormone binding site of the insulin receptor (Wedekind *et al.*, 1989). This compound, 4-azido-2-nitrophenyl-biocytin-4-nitrophenyl ester, contains an amine-reactive group and a photoreactive phenyl azide functional group (Chapter 6, Section 1). The nitrophenyl ester reacts with amines on proteins and other molecules to from stable amide linkages. Once a molecule is modified in this manner, it contains both a photosensitive group and a biotin handle for conjugation and detection, respectively. Interaction of the modified protein with another biospecific receptor and subsequent photolyzing with UV light results in covalently cross-linking. Localization and detection of the cross-linked molecules then can be done using an avidin or streptavidin conjugate. Another trifunctional compound, sulfo-SBED, also is based on a biocytin core (Chapter 6, Section 2).

NHS-Biotin and Sulfo-NHS-Biotin

The valeric acid carboxylate of D-biotin may be activated to an NHS ester for direct modification of amine groups in proteins and other molecules. NHS esters react by nucleophilic attack of an amine on the carbonyl group, releasing the NHS group, and forming a stable amide linkage (Chapter 2, Section 1.4) (Fig. 247). NHS-biotin is the simplest biotinylation reagent available. Modification reactions are carried out under mildly alkaline conditions and usually result in a high efficiency of biotin incorporation.

NHS-biotin is insoluble in aqueous environments. It must be first dissolved in organic solvent as a concentrated stock solution, and an aliquot added to an aqueous reaction medium to facilitate dissolution. Organic solvents such as DMF or DMSO are suitable for this purpose. Addition of an NHS-biotin solution to a reaction should not exceed a level of about 10% organic solvent in the buffer to avoid precipitation problems. Once added to the reaction medium, the NHS-biotin may appear as a

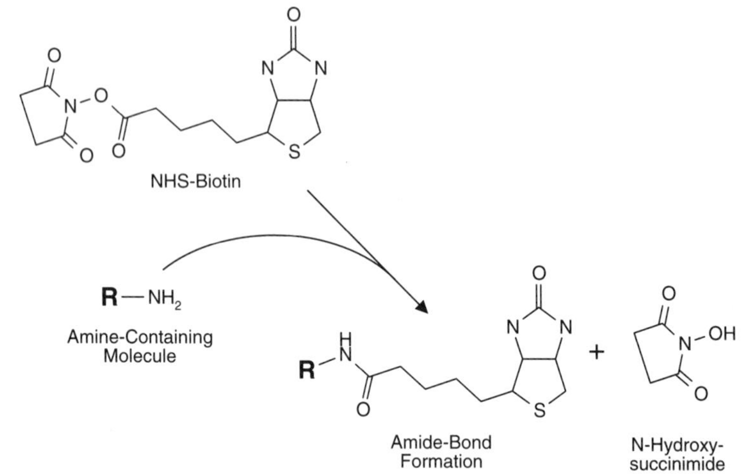

Figure 247 The active ester group of NHS-biotin reacts with amine-containing compounds to form amide bond linkages.

cloudy or hazy suspension, indicating incomplete solubility. However, such micro-dispersions still are effective at modification, often driving the bulk of the reagent into solution ast the NHS ester reacts. Biotinylation of peptides or other molecules that are water-insoluble may be done completely in organic solvent. For example, insulin can be biotinylated with NHS-biotin in an organic medium (Hofmann *et al.,* 1977).

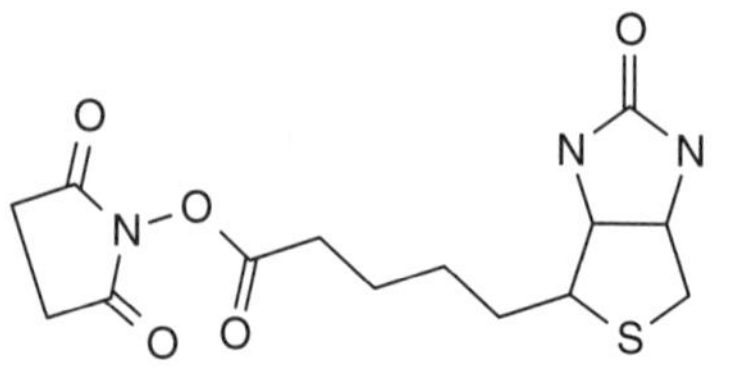

NHS-Biotin
MW 341.38
13.5 Å

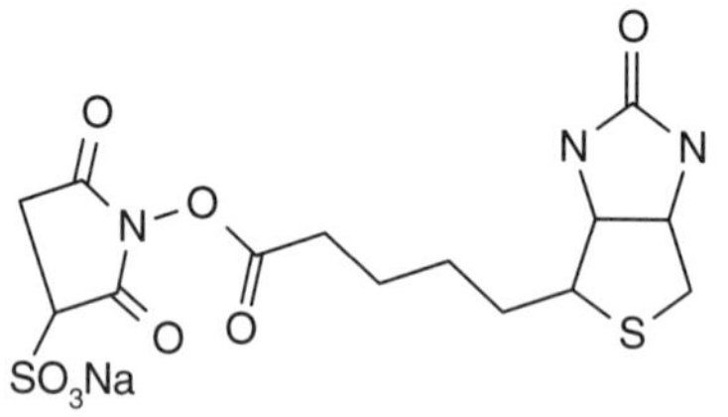

Sulfo-NHS-Biotin
MW 443.42
13.5 Å

A water-soluble analog of NHS-biotin containing a negatively charged sulfonate group on its NHS ring structure also is available. Sulfo-NHS-biotin may be added directly to aqueous reactions without the need for organic solvent dissolution. A concentrated stock solution may be prepared in water to facilitate the addition of a small quantity to a reaction, but hydrolysis of the NHS ester will occur at a rapid rate, so the solution must be used immediately.

The only disadvantage to the use of NHS-biotin or sulfo-NHS-biotin is the lack of a long spacer group off the valeric acid side chain. Because the binding site for biotin on avidin and streptavidin is somewhat below the surface of the proteins, some biotinylated molecules may not interact as efficiently with avidin (or streptavidin) as when longer cross-bridges are used (Green *et al.*, 1971; Bonnard *et al.*, 1984).

NHS esters of D-biotin have been used in many applications, including the biotinylation of rat IgE to study receptors on murine lymphocytes (Lee and Conrad, 1984), in the development of an immunochemical assay for a postsynaptic protein and its receptor (LaRochelle and Froehner, 1986a), in the study of plasma membrane domains by biotinylation of cell surface proteins in *Dictyostelium disoideum* amoebas (Ingalls *et al.*, 1986), and for the detection of blotted proteins on nitrocellulose membranes after transfer from polyacrylamide electrophoresis gels (LaRochelle and Froehner, 1986b).

The following protocol is a generalized method for the biotinylation of a protein using sulfo-NHS-biotin.

Protocol

1. Dissolve the protein to be biotinylated in 0.1 M sodium phosphate, 0.15 M NaCl, pH 7.2, at a concentration of 1–10 mg/ml.
2. Immediately before use, dissolve sulfo-NHS-biotin (Pierce) in water at a concentration of 20 mg/ml. Adjust the quantity of this stock solution to be prepared according to the amount of reagent needed to biotinylate the required amount of protein. The sulfo-NHS-biotin solution must be used immediately, since the NHS ester is subject to hydrolysis in aqueous environments.
3. With mixing, add a quantity of the sulfo-NHS-biotin solution to the protein solution to obtain a 12- to 20-fold molar excess of biotinylation reagent over the quantity of protein present. For instance, for an immunoglobulin (MW 150,000) at a concentration of 10 mg/ml, 20 µl of the sulfo-NHS-biotin solution (8×10^{-4} mmol) should be added per milliliter of antibody solution to obtain a 12-fold molar excess. For more dilute protein solutions (i.e., 1–2 mg/ml), increased amounts of biotinylation reagent may be required (i.e., 20-fold molar excess) to obtain similar incorporation yields as when using more concentrated protein solutions.
4. React for 30–60 min at room temperature.
5. Purify the biotinylated protein from excess reagent and reaction by-products by gel filtration using a column of Sephadex G-25 or by dialysis against PBS.

Determination of the degree of biotinylation can be done using the HABA assay (Chapter 13, Section 7).

NHS-LC-Biotin and Sulfo-NHS-LC-Biotin

NHS-LC-biotin is a derivative of D-biotin containing a spacer arm off the valeric acid side chain, terminating in an NHS ester. The compound is also known as succinimidyl-6-(biotinamido)hexanoate or NHS-X-biotin. The 6-aminocaproic acid spacer provides greater length between a convalently modified molecule and the bicyclic biotin rings. The total distance from an attached molecule to the biotin component is about 22.4 Å, significantly greater than the 13.5 Å length of NHS-biotin without a spacer arm. This increased distance can result in better binding potential for avidin or streptavidin probes, because the binding sites on these proteins are buried relatively deep inside the surface plane.

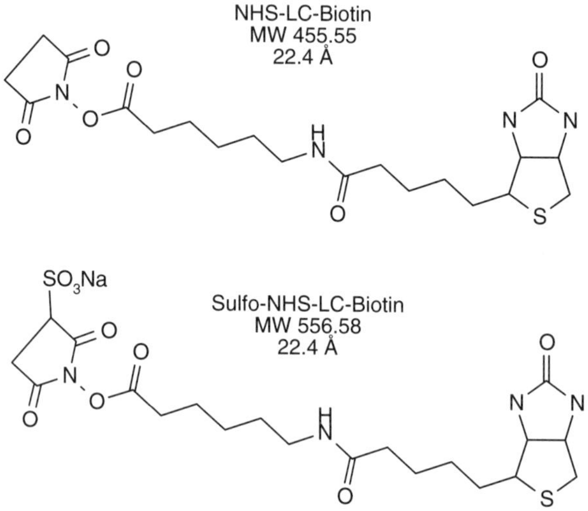

NHS-LC-Biotin
MW 455.55
22.4 Å

SO$_3$Na

Sulfo-NHS-LC-Biotin
MW 556.58
22.4 Å

The NHS ester end of NHS-LC-biotin reacts with amine groups in proteins and other molecules to form stable amide-bond derivatives (Fig. 248). Optimal reaction conditions are at a pH of 7–9. Avoid amine-containing buffers, which may compete in the acylation reaction. NHS-LC-biotin is insoluble in aqueous reaction conditions and must be solubilized in organic solvent prior to the addition of a small quantity to a

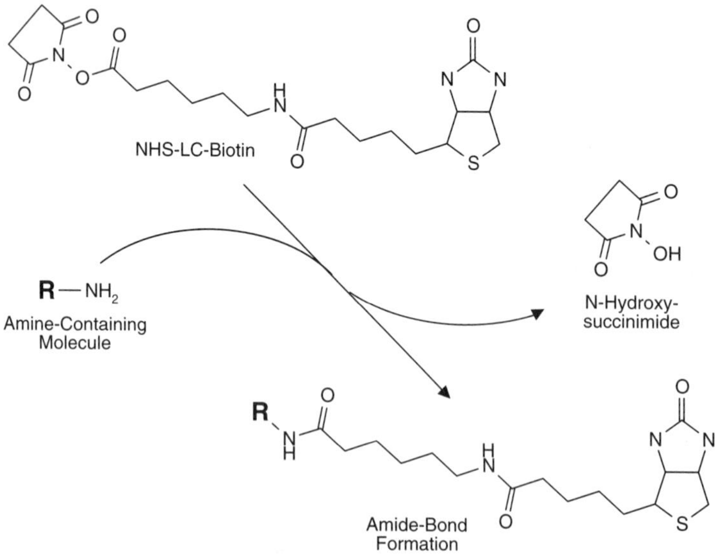

NHS-LC-Biotin

R—NH$_2$
Amine-Containing
Molecule

N-Hydroxy-
succinimide

Amide-Bond
Formation

Figure 248 NHS-LC-biotin provides an extended spacer arm to allow greater distance between the biotin rings and a modified molecule. Reaction with amines forms amide linkages.

buffered reaction. Preparation of concentrated stock solutions may be done in DMF or DMSO. Nonaqueous reactions also may be done with this reagent for the modification of molecules insoluble in water. The molar ratio of NHS-LC-biotin to a protein in a reaction can be from about 2:1 to about 50:1, with higher levels resulting in better incorporation yields (Gretch *et al.*, 1987).

In a study comparing NHS-LC-biotin with two other derivatives of biotin, NHS-SS-biotin (Section 3.1) and biotin hydrazide (Section 3.3), it was found that modification through amines on monoclonal antibodies resulted in 2.5 times more activity in binding a streptavidin–agarose affinity column than when modification of carbohydrate residues using hydrazide chemistry was done (Gretch *et al.*, 1987). This was probably due to the greater abundance of amino groups over polysaccharide residues on these antibodies.

NHS-LC-biotin can be used to add a biotin tag to monoclonal antibodies directed at certain tumor antigens. The biotinylated monoclonals are allowed to bind to the tumor cell surfaces *in vivo*, and subsequent administration of an avidin or a streptavidin conjugate can form the basis for inducing cytotoxic effects or creating traceable complexes for use in imaging techniques (Hnatowich *et al.*, 1987).

The reagent also has been used in a unique tRNA-mediated method of labeling proteins with biotin for nonradioactive detection of cell-free translation products (Kurzchalia *et al.*, 1988), in creating one- and two-step noncompetitive avidin–biotin immunoassays (Vilja, 1991), for immobilizing streptavidin onto solid surfaces using biotinylated carriers with subsequent use in a protein avidin-biotin capture system (Suter and Butler, 1986), and for the detection of DNA on nitrocellulose blots (Leary *et al.*, 1983).

Sulfo-NHS-LC-biotin, a water-soluble analog of NHS-LC-biotin, also is available (Pierce), and contains a negatively charged sulfonate group on its NHS ring structure. The presence of the negative charge creates enough polarity within the molecule to allow direct solubility in aqueous reaction mediums. All other properties of the sulfonated version of the reagent are the same as those of NHS-LC-biotin.

The following protocol is a suggested method for the biotinylation of proteins with either NHS-LC-biotin or sulfo-NHS-LC-biotin.

Protocol

1. Dissolve the protein to be biotinylated in 0.1 M sodium phosphate, 0.15 M NaCl, pH 7.2, at a concentration of 10 mg/ml.
2. Dissolve NHS-LC-biotin (Pierce) in DMF at a concentration of 40 mg/ml. This stock solution is stable for reasonable periods, although long-term storage is not recommended. For use of the water-soluble sulfo-NHS-LC-biotin, a stock solution may be prepared in water, or the solid reagent may be directly added to the reaction mixture. If a solution in water is made to facilitate the addition of a small quantity of reagent to a reaction, then the solution should be prepared quickly and used immediately to prevent hydrolysis of the NHS ester. Sulfo-NHS-LC-biotin may be dissolved in water at a concentration of 20 mg/ml.
3. Add 50 µl of the NHS-LC-biotin solution in DMF to each milliliter of the protein solution in two aliquots apportioned 10 min apart. Alternatively, add a quantity of the sulfo-NHS-biotin solution prepared in water to the protein solution to obtain a 12- to 20-fold molar excess of biotinylation reagent over the quantity of

protein present. For instance, for an immunoglobulin (MW 150,000) at a concentration of 10 mg/ml, 20 μl of the sulfo-NHS-biotin solution (8×10^{-4} mmol) should be added per milliliter of antibody solution to obtain a 12-fold molar excess.

4. React for a total of 30–60 min at room temperature or several hours at 4°C.
5. Remove unreacted biotinylation reagent and reaction by-products by gel filtration using a column of Sephadex G-25 or dialysis against PBS.
6. Assay the level of biotin incorporation using the HABA dye procedure (Chapter 13, Section 7).

NHS-Iminobiotin

NHS-iminobiotin is N-hydroxysuccinimido-2-iminobiotin, the guanidino analog of NHS-biotin that has a lower affinity constant for binding avidin or streptavidin. Iminobiotin replaces the 2-oxo-imidazole upper ring structure of D-biotin with a 2-imino-imidazole structure, causing moderated interaction with the avidin or streptavidin binding sites. This biotin analog can be used in situations requiring mild dissociation of the avidin–biotin complex. Normally, breaking the avidin–biotin interaction requires 6–8 M guanidine hydrochloride at a pH of 1.5, an environment too severe for most proteins to maintain native structure or recover activity. Iminobiotin, by contrast, can be bound to avidin or streptavidin at a pH wherein the guanidino group is unprotonated and thus uncharged. Binding occurs at pH values above 9.5 (typically done at pH 11), and elution can be accomplished simply by changing the pH to 4—an environment that protonates the 2-imino group and creates a positive charge—effectively dissociating the interaction (Fig. 249).

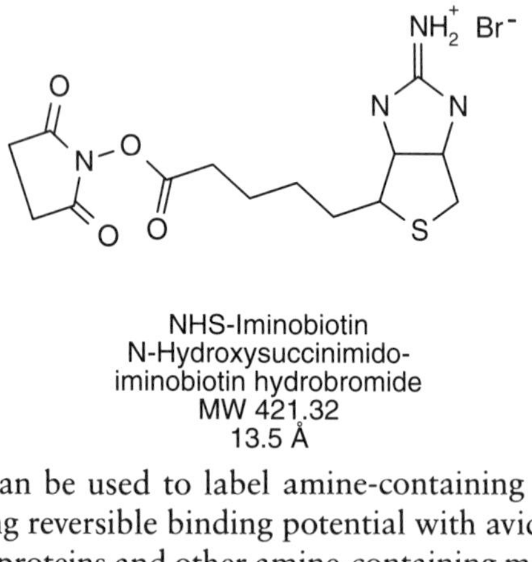

NHS-Iminobiotin
N-Hydroxysuccinimido-
iminobiotin hydrobromide
MW 421.32
13.5 Å

NHS-iminobiotin can be used to label amine-containing molecules with an iminobiotin tag, providing reversible binding potential with avidin or streptavidin. The NHS ester reacts with proteins and other amine-containing molecules to create stable amide-bond derivatives (Fig. 250). An iminobiotinylated molecule then can be used to target and purify other components in biological samples. For instance, a targeting molecule, such as an antibody, can be iminobiotinylated and allowed to bind its target in complex mixtures (such as tissue sections, cell extracts, or homogenates). The antibody–antigen complex subsequently can be purified using an affinity column of immobilized avidin with binding at pH 10–11 and simple elution at pH 4 (Orr, 1981; Zeheb et al., 1983). The relatively mild elution conditions allows recovery of the bound antigen without exposure to severe denaturing conditions.

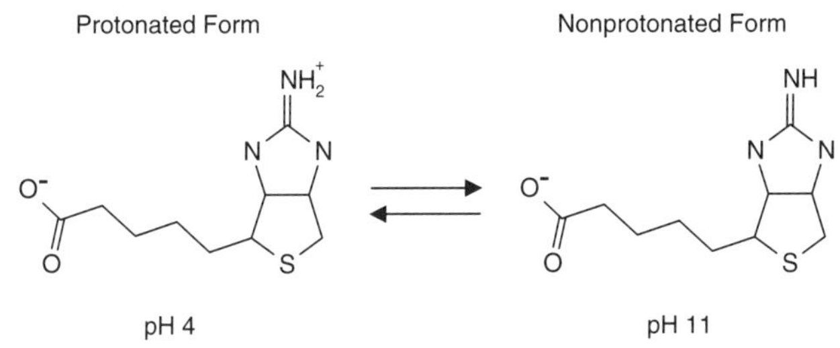

Figure 249 At pH 4, the protonated form of iminobiotin does not interact with the binding sites on avidin or streptavidin. At pH 11, the imino group is unprotonated and regains binding capability toward these proteins.

The iminobiotin–avidin interaction also can be utilized in the opposite approach. Immobilized iminobiotin affinity columns can be used to purify avidin- or streptavidin-containing complexes under mild elution conditions (Hofmann *et al.*, 1980).

NHS-iminobiotin is insoluble in aqueous solution. It can be dissolved in organic solvent (DMF) prior to addition of a small aliquot to a buffered reaction medium. Do not exceed 10% DMF in the reaction, to avoid protein precipitation problems. Optimal conditions for protein derivatization include non-amine-containing buffers at a pH of 7–9. The following protocol is a suggested method for labeling antibodies with NHS-iminobiotin. Some optimization may have to be done for particular derivatization needs.

Protocol

1. Dissolve the antibody to be modified in 50 mM sodium borate, pH 8, at a concentration of 5 mg/ml.

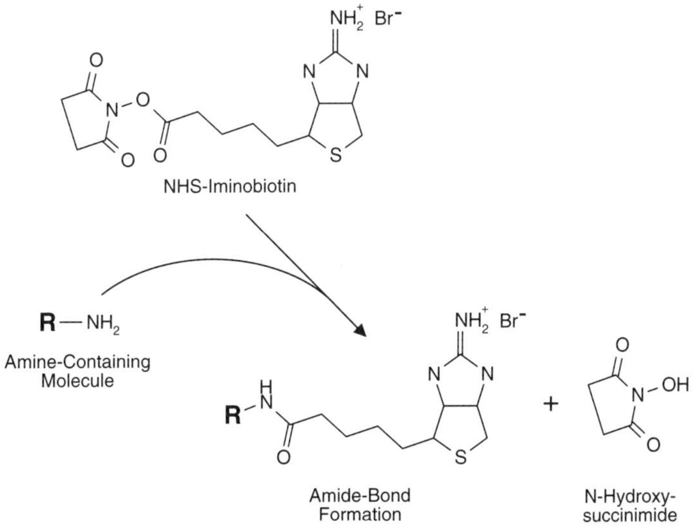

Figure 250 NHS-iminobiotin can be used to label amine-containing molecules, creating amide linkages.

2. Dissolve NHS-iminobiotin in DMF at a concentration of 1 mg/ml. Prepare fresh.

3. Add 100 μl of the NHS-iminobiotin solution to each milliliter of the antibody solution. Mix well to dissolve. Note: some turbidity may be present in the reaction due to incomplete dissolution of the NHS-iminobiotin. The solution may look cloudy or have a microparticulate suspension present. This is normal for many water-insoluble reagents when added to an aqueous solution in an organic solvent. As the reaction takes place, the NHS-iminobiotin will be driven into solution, both by coupling to the protein and by hydrolysis of the NHS ester.

4. React for 30–60 min at room temperature or for 3 h at 4°C.

5. Remove unreacted NHS-iminobiotin and reaction by-products by dialysis or gel filtration using a column of Sephadex G-25.

Sulfo-NHS-SS-Biotin

Sulfo-NHS-SS-biotin (also known as NHS-SS-biotin) is sulfosuccinimidyl-2-(biotinamido)ethyl-1,3-dithiopropionate, a long-chain cleavable biotinylation reagent that can be used to modify amine-containing proteins and other molecules (Pierce). The cross-bridge of the compound provides a 24.3-Å spacer arm that creates plenty of distance between the modified molecule and the biotin end. Using a long-chain biotinylation reagent can increase the efficiency of biotinylated molecules to bind avidin or streptavidin conjugates, thus enhancing the potential sensitivity of assay systems.

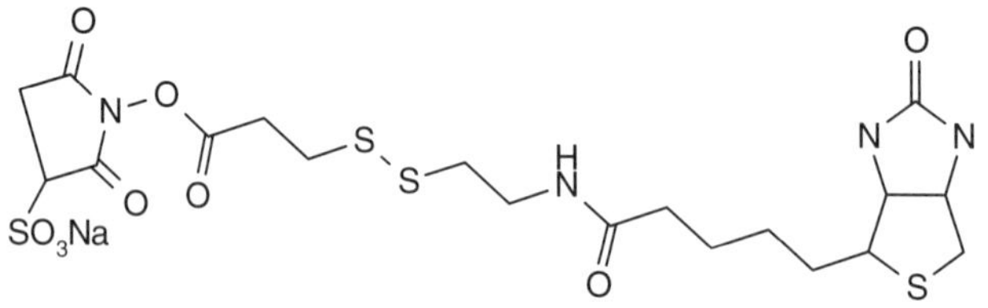

Sulfo-NHS-SS-Biotin
Sulfosuccinimidyl-2-(biotinamido)-
ethyl-1,3-dithiopropionate
MW 606.7
24.3 Å

After sulfo-NHS-SS-biotin-modified molecules are allowed to interact with avidin or streptavidin probes, the formed complexes can be cleaved at the disulfide bridge by treatment with 50 mM DTT. Reduction releases the biotinylated component from the avidin or streptavidin detection reagent. The use of disulfide biotinylation reagents thus provides much gentler conditions to break the complex than would be required if the avidin-biotin interaction itself were disrupted (which dissociates only at 6–8 M guanidine, pH 1.5).

The use of a cleavable biotinylation reagent also provides a means to purify targeted molecules using affinity chromatography on a column of immobilized avidin or streptavidin. For instance, an antibody modified with sulfo-NHS-SS-biotin can be allowed bind its target in complex mixtures (such as tissue sections, cell extracts, or homogenates). The antibody–antigen complex subsequently can be isolated using an affinity column of immobilized avidin or streptavidin. Elution from the column with DTT breaks the disulfide bonds, releasing the antibody and its bound antigen. The isolation of Herpes virus proteins (Gretch *et al.*, 1987) and the recovery of DNA binding proteins (Shimkus *et al.*, 1985) were both done using this approach.

Due to the presence of the negatively charged sulfonate group, sulfo-NHS-SS-biotin is a water-soluble biotinylation reagent that may be added directly to aqueous reactions without prior dissolution in organic solvent. For the addition of small quantities of reagent, the compound may be dissolved in water, and an aliquot transferred to the reaction medium. If an aqueous stock solution of sulfo-NHS-SS-biotin is prepared, it must be dissolved rapidly and used immediately to prevent hydrolysis of the active ester. The NHS ester reaction forms stable amide-bond linkages with amine-containing proteins and other molecules (Fig. 251). Optimal conditions for the NHS ester reaction include a pH of 7–9, avoidance of any amine-containing buffers or other components that may compete in the reaction (including imidazole buffers, which catalyze hydrolysis of these esters), and avoidance of reducing agents that could cleave the disulfide bridge.

The following protocol is a suggested method for biotinylating antibody molecules with sulfo-NHS-SS-biotin. Some optimization may have to be done with each application to ensure good biotin incorporation with retention of antigen binding activity. Other proteins and amine-containing molecules may be biotinylated using similar conditions.

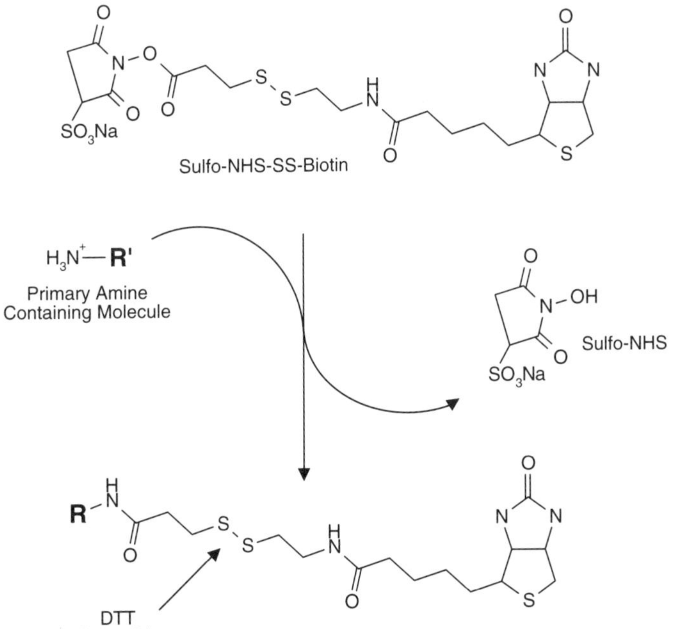

Figure 251   Sulfo-NHS-SS-biotin reacts with amine groups to form amide bonds. The biotin group can be later cleaved off the modified molecule by reduction of its internal disulfide linkage.

Protocol

1. Dissolve the antibody to be biotinylated in 50 mM sodium bicarbonate, pH 8.5, at a concentration of 10 mg/ml. Other buffers and pH conditions between pH 7 and 9 can be used as long as no amine-containing buffers like Tris are present. Avoid also the presence of disulfide-reducing agents that can cleave the disulfide group of the biotinylation reagent.

2. Add 0.3 mg of sulfo-NHS-SS-biotin (Pierce) to each milliliter of the antibody solution. To measure small amounts of the biotinylation reagent, it may be first dissolved in water at a concentration of at least 1 mg/ml. Immediately transfer the appropriate amount to the antibody solution. This level of sulfo-NHS-SS-biotin addition represents about an eight-fold molar excess over the amount of antibody present. This should result in a molar incorporation of approximately two to four biotins per immunoglobulin molecule.

3. React for 30–60 min at room temperature or for 2–4 h at 4°C.

4. Remove unreacted biotinylation reagent and reaction by-products by dialysis or gel filtration using a column of Sephadex G-25.

3.2. Sulfhydryl-Reactive Biotinylation Agents

Sulfhydryl-reactive biotinylation reagents allow modification at cysteine —SH groups or at sites of specific thiolation within proteins and other molecules. Targeting sulfhydryls for modification, as opposed to amines, usually results in more limited derivatization, often away from active centers or binding sites. Directed coupling of biotin in this manner can aid in preserving activity. For instance, antibodies may be cleaved by reduction at their disulfide groups in the hinge region, forming free sulfhydryls removed from the antigen-binding site (Chapter 10, Section 1.1). Biotinylation at these sites produces a derivative that can bind efficiently both antigen and avidin or streptavidin probes without steric hindrance.

Sulfhydryl groups also can be added to 5′-phosphate end of DNA probes (Chapter 17, Section 2.2). Biotinylation at these sites avoids disruption of base-pairing with complementary DNA targets, since the point of modification is restricted to a single end position on the oligonucleotide.

The following sections discuss three sulfhydryl-reactive biotinylation reagents that utilize maleimide, pyridyl disulfide, and iodoacetyl reactive groups, respectively. The maleimide and iodoacetyl options produce nonreversible, covalent thioether linkages with target —SH groups. The pyridyl disulfide chemistry results in disulfide bonds that are reversible through cleavage with a reducing agent.

Biotin-BMCC

Biotin-BMCC is 1-biotinamido-4-[4′-(maleimidomethyl)cyclohexane-carboxamido]butane, a biotinylation reagent containing a maleimide group at the end of an extended spacer arm (Pierce). The maleimide end reacts with sulfhydryl groups in proteins and other molecules to form stable thioether linkages (Fig. 252). The reaction is highly specific for —SH groups in the range of pH 6.5 to 7.5. The long spacer arm (32.6 Å) provides more than enough distance between modified molecules and the bicyclic biotin end to allow efficient binding of avidin or streptavidin probes.

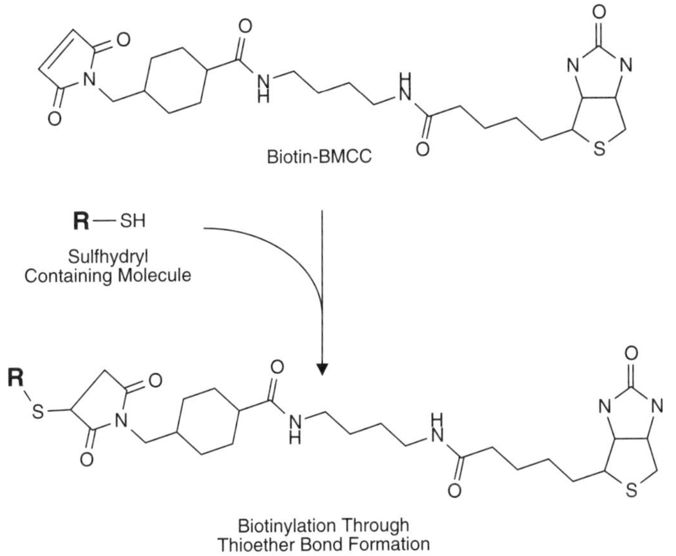

Biotin-BMCC

R—SH

Sulfhydryl
Containing Molecule

Biotinylation Through
Thioether Bond Formation

Figure 252 Biotin-BMCC provides sulfhydryl reactivity through its terminal maleimide group. The reaction creates a stable thioether linkage.

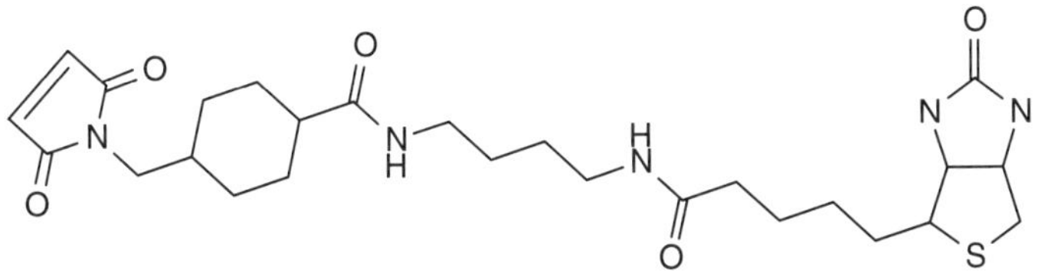

Biotin-BMCC
1-Biotinamido-4-[4'-(maleimidomethyl)-
cyclohexane-carboxamido]butane
MW 533.69
32.6 Å

The reagent is similar to another maleimide-containing biotinylation reagent, 3-(N-maleimidopropionyl) biocytin, a compound used to detect sulfhydryl-containing molecules on nitrocellulose blots after SDS–electrophoresis separation (Bayer et al., 1987b). Biotin-BMCC should be useful in similar detection procedures.

Biotin-BMCC is insoluble in water and must be dissolved in an organic solvent prior to addition to an aqueous reaction mixture. Preparing a concentrated stock solution in DMF or DMSO allows transfer of a small aliquot to a buffer reaction. The upper limit of biotin-BMCC solubility in DMSO is approximately 33 mM or 17 mg/ml. In DMF, it is only soluble to a level of about 7 mM (4 mg/ml). On addition of an organic solution of the reagent to an aqueous environment (do not exceed 10% organic solvent in the aqueous medium to prevent protein precipitation), biotin-BMCC may form a microemulsion. This is normal and during the course of the reaction the remainder of the compound will be driven into solution as it couples or hydrolyzes.

The required sulfhydryl groups for biotin-BMCC modification may be indigenous in molecules, formed through reduction of disulfides, or created by the use of thiolation reagents (Chapter 1, Section 4.1). At physiological pH values the rate of the maleimide reaction toward sulfhydryls is almost 1000-fold faster than its reaction toward amines. However, at higher pH values the maleimide will couple to amines quite readily (Ishi and Lehrer, 1986; Wu *et al.*, 1976). Maleimides also can undergo a ring-opening hydrolysis reaction that increases in rate with pH, effectively inactivating the functional group.

The following protocol is a suggested method for modifying sulfhydryl-containing proteins with biotin-BMCC. Some optimization of biotinylation levels may have to be done for particular applications.

Protocol

1. Dissolve the protein to be biotinylated (containing one or more free sulfhydryls) in 0.1 *M* sodium phosphate, 0.15 *M* NaCl, 10 m*M* EDTA, pH 7.2, at a concentration of 2.5 mg/ml.
2. Dissolve biotin-BMCC (Pierce) in DMSO at a concentration of 5 mg/ml.
3. Add 100 µl of the biotin-BMCC solution to each milliliter of the protein solution. Mix well.
4. React for at least 2 h at room temperature.
5. Remove excess biotinylation reagent and reaction by-products by dialysis or gel filtration using a column of Sephadex G-25 (Pharmacia).

Biotin-HPDP

Biotin-HPDP is *N*-[6-(biotinamido)hexyl]-3'-(2'-pyridyldithio)propionamide (Pierce). The reagent contains a 1,6-diaminohexane spacer group that is attached to biotin's valeric acid side chain. The terminal amino group of the spacer is further modified via an amide linkage with the acid of SPDP (Chapter 5, Section 1.1) to create a terminal, sulfhydryl-reactive functional group. The pyridyl disulfide end of biotin-HPDP can react with free —SH groups in proteins and other molecules to from a disulfide bond with loss of pyridine-2-thione (Fig. 253). This leaving group may be monitored by its characteristic absorbance at 343 nm to assess the level of biotinylation. However, since its extinction coefficient is rather low (about $8 \times 10^3 \, M^{-1} \mathrm{cm}^{-1}$), small-scale biotinylations may not be quantifiable using this technique.

Biotin–HPDP
N-[6-(Biotinamido)hexyl]-3'-
(2'-pyridyldithio)propionamide
MW 539.77
29.2 Å

Modifications done with biotin-HPDP produce biotinylated compounds with long spacer arms (29.2 Å), ensuring good binding efficiency with avidin or streptavidin

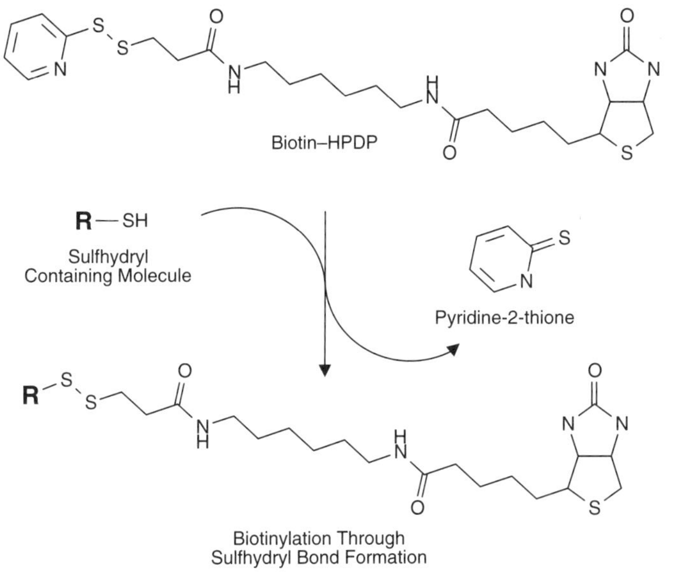

Figure 253 Biotin-HPDP reacts with sulfhydryl-containing molecules through its pyridyl disulfide group, forming reversible disulfide bonds. The biotin group may be released from modified molecules by reduction with DTT.

probes. After coupling to sulfhydryl-containing molecules, the biotin-HPDP component can be cleaved by treatment with disulfide-reducing agents, such as DTT. Breaking this bond releases the biotin modifications and regenerates the original sulfhydryl-containing molecule. This cleavability also provides a means of recovering target complexes after purification of the biotinylated molecules by affinity chromatography on immobilized avidin or streptavidin. Thus biotin-HPDP-modified antibodies directed against some specific cellular antigen can aid in the isolation of targeted components using affinity chromatography followed by elution with a disulfide reductant.

Using a similar approach, Clq has been modified with biotin-HPDP and allowed to interact with its specific receptor. Subsequent purification of the Clq receptor was accomplished through cleavage of the disulfide bridge of the biotinylation reagent (Ghebrehiwet *et al.*, 1988).

Biotin-HPDP is water-insoluble and therefore must be dissolved in an organic solvent prior to addition to an aqueous reaction medium. Suitable solvents include DMSO and DMF. Concentrated stock solutions may be prepared in DMSO, and a small aliquot transferred to a buffered reaction solution. Do not add more than 10% organic solvent to the aqueous reaction to prevent precipitation or denaturation of biological molecules. After addition, a microemulsion may result. This is normal for many water-insoluble reagents. The solution usually will become clear during the course of the reaction. Optimal conditions for the disulfide interchange reaction include a pH range of 6–9 in buffer systems that do not contain any extraneous sulfhydryl compounds. If reducing agents such as DTT or 2-mercaptoethanol are used to create sulfhydryls in the protein to be biotinylated, these must be completely removed by dialysis or gel filtration before reacting with biotin-HPDP.

A suggested protocol for the use of biotin-HPDP in the modification of sulfhydryl-containing proteins follows. Similar procedures may be used when biotinylating other molecules.

Protocol

1. Dissolve the sulfhydryl-containing protein to be biotinylated in 0.1 M sodium phosphate, 0.15 M NaCl, 10 mM EDTA, pH 7.2, at a concentration of at least 2 mg/ml.
2. Dissolve biotin-HPDP (Pierce) in DMSO at a concentration of 4 mM (2.1 mg/ml).
3. Add 100 μl of the biotin-HPDP stock solution to each milliliter of the protein solution. Mix well.
4. React for 90 min at room temperature.
5. Purify the biotinylated protein by dialysis or gel filtration using a column of Sephadex G-25. The PBS/EDTA buffer described in step 1 is suitable for either operation.

Iodoacetyl-LC-Biotin

Iodoacetyl-LC-biotin is *N*-iodoacetyl-*N*-biotinylhexylenediamine, a sulfhydryl-reactive biotinylation agent (Pierce). The reagent contains a 1,6-diaminohexane spacer group that is attached to biotin's valeric acid side chain. The terminal amino group of the spacer is further modified via an amide linkage with an iodoacetyl group to provide the sulfhydryl reactivity. Coupling to sulfhydryl-containing proteins or other molecules creates nonreversible thioether bonds (Fig. 254). Modifications done with iodoacetyl-LC-biotin produce biotinylated compounds with sufficiently long spacer arms (27.1 Å) to ensure excellent binding potential with avidin or streptavidin probes.

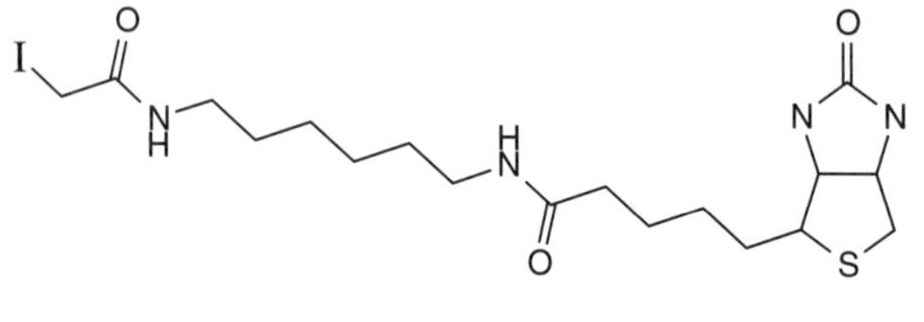

Iodoacetyl-LC-Biotin
N-Iodoacetyl-N-biotinylhexylenediamine
MW 510.42
27.1 Å

Iodoacetyl-LC-biotin is water-insoluble and therefore must be dissolved in an organic solvent prior to addition to an aqueous reaction medium. Suitable solvents include DMSO and DMF. Concentrated stock solutions may be prepared in DMSO, and a small aliquot transferred to a buffered reaction solution. Do not add more than 10% organic solvent to the aqueous reaction to prevent precipitation or denaturation of biological molecules. After addition, a microemulsion may result. This is normal for many water-insoluble reagents. The solution usually will become clear during the course of the reaction. Optimal conditions for coupling using iodoacetyl-containing reagents include a pH range of 7.5–8.5 in buffer systems that do not contain any

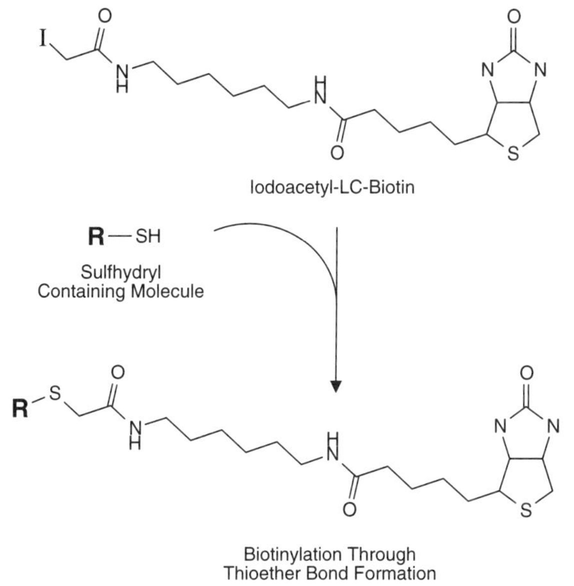

Iodoacetyl-LC-Biotin

R—SH

Sulfhydryl
Containing Molecule

Biotinylation Through
Thioether Bond Formation

Figure 254 This biotinylation reagent reacts with sulfhydryl groups through its iodoacetamide end to form thioether bonds.

extraneous sulfhydryl compounds. In addition, protect all solutions containing iodoacetyl-LC-biotin from light, since photolysis may cause liberation of iodine, degrading the activity of the compound and possibly causing modification of tyrosine or histidine residues.

Iodoacetyl-LC-biotin has been used to localize the SH_1 thiol of myosin by use of an avidin–biotin complex visualized by electron microscopy (Sutoh *et al.*, 1984) and to determine the spatial relationship between SH_1 and the actin binding site on the myosin subfragment-1 surface (Yamamoto *et al.*, 1984).

The following protocol is a suggested method for biotinylating sulfhydryl-containing proteins using iodoacetyl-LC-biotin. The required sulfhydryl groups may be provided through reductive cleavage of disulfide bonds or by the use of thiolation reagents (Chapter 1, Section 4.1). Other molecules may be modified with iodoacetyl-LC-biotin using similar techniques.

Protocol

1. Dissolve the sulfhydryl-containing protein to be biotinylated in 50 m*M* Tris, 0.15 *M* NaCl, 10 m*M* EDTA, pH 8.3, at a concentration of 4 mg/ml.
2. Dissolve iodoacetyl-LC-biotin (Pierce) in DMF at a concentration of 4 m*M* (2 mg/ml). Protect from light.
3. Add 50 μl of the iodoacetyl-LC-biotin solution to each milliliter of the protein solution. Mix well. This level of addition represents a 3.28-fold molar excess of biotinylation reagent over the quantity of protein present if the protein has a molecular weight of 67,000 and possesses one sulfhydryl. Adjustments to the amount of reagent addition may have to be made to be appropriate for other

proteins of different molecular weight. Consideration of the number of sulf-hydryls present per protein molecule also should be done. React the biotinyla-tion reagent at no more than a three- to five-fold molar excess over the amount of sulfhydryls present to ensure specificity of the iodoacetyl group for only —SH groups. Higher ratios of reagent to protein may cause reaction with amine groups present on the protein.

4. React for 90 min in the dark at room temperature.
5. Remove excess reactants and reaction by-products by dialysis or gel filtration using a column of Sephadex G-25.

3.3. Carbonyl- or Carboxyl-Reactive Biotinylation Agents

Hydrazide- or amine-containing biotinylation compounds can be used to modify carbonyl or carboxyl groups on other molecules. Hydrazides spontaneously react with aldehydes and ketones to give hydrazone linkages. The hydrazones may be further stabilized by reduction with sodium cyanoborohydride. The amine-containing biotinylation reagents (or the hydrazide ones) may be coupled to carboxylate groups using a carbodiimide reaction (Chapter 3, Section 1). In addition, amine- or hydrazide-containing biotinylation reagents may be coupled to cytosine residues in DNA or RNA by transamination catalyzed by bisulfite (Chapter 17, Section 2.3).

Biotin-Hydrazide and Biotin-LC-Hydrazide

Biotin-hydrazide is *cis*-tetrahydro-2-oxothieno[3,4-*d*]-imidazoline-4-valeric acid hy-drazide, the hydrazine derivative of *D*-biotin off its valeric acid carboxylate (Pierce). The hydrazide functional group reacts with aldehyde and ketone groups to give hy-drazone linkages. Although formyl groups are not common in biological molecules, they may be created by oxidation of diols with sodium periodate (Chapter 1, Section 4.4). Thus, glycoconjugates may be targeted specifically at their sugar residues. Bio-tinylation of these oxidized carbohydrates with biotin-hydrazide produces modifica-tions that may be away from active centers or binding sites (Fig. 255). Particularly, immunoglobulins may be biotinylated with this reagent at their polysaccharide groups, which are present in the Fc region of the IgG molecule. Directed modification in this manner avoids the antigen binding sites at the ends of the heavy and light chains, thus preserving antibody activity and allowing avidin or streptavidin probes to dock without blocking or interfering with antigen binding.

Biotin-hydrazide also may be used to couple with carboxylate-containing mole-cules. Hydrazides can be coupled with carboxylic acid groups by using the car-bodiimide reaction (Chapter 3, Section 1). The carbodiimide activates a carboxylate to an *o*-acylisourea intermediate. Biotin-hydrazide can react with this intermediate via nucleophilic addition to form a stable covalent bond.

Biotin-hydrazide has been used to biotinylate antibodies at their oxidized carbohy-drate residues (O'Shannessy *et al.*, 1984, 1987; Hoffman and O'Shannessy, 1988), to modify the low-density lipoprotein (LDL) receptor (Wade *et al.*, 1985), to biotinylate nerve growth factor (NGF) (Rosenberg *et al.*, 1986), and to modify cytosine groups in oligonucleotides to produce probes suitable for hybridization assays (Reisfeld *et al.*, 1987) (Chapter 17, Section 2.3).

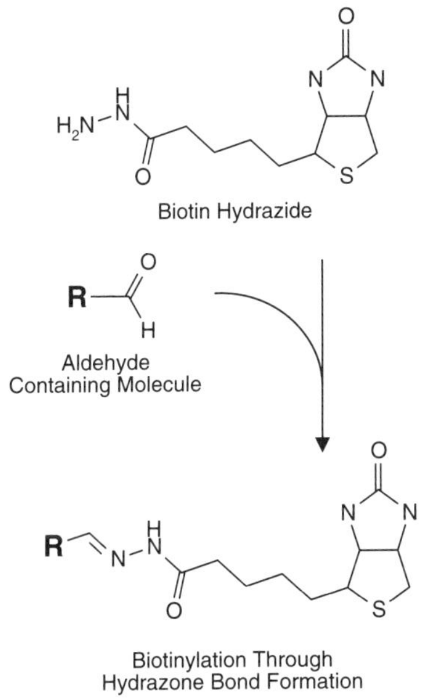

Figure 255 Biotin-hydrazide can be used to label aldehyde-containing molecules, creating hydrazone bonds.

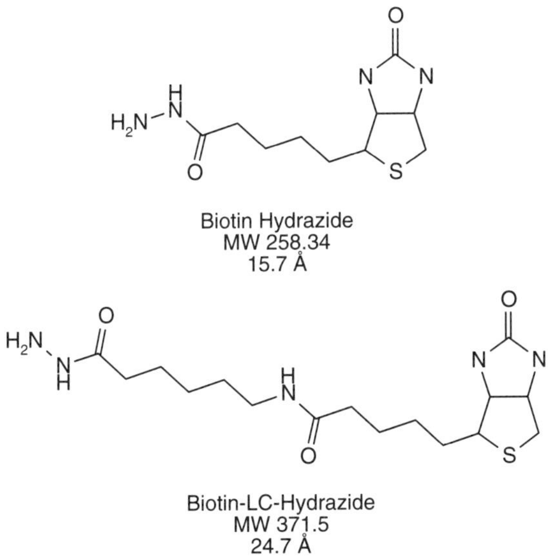

Biotin Hydrazide
MW 258.34
15.7 Å

Biotin-LC-Hydrazide
MW 371.5
24.7 Å

An analog of this biotinylation reagent with a longer spacer arm also exists. Biotin-LC-hydrazide contains a 6-aminocaproic acid extension off its valeric acid group (Pierce). The increased length of this spacer (24.7 Å) provides more efficient interaction potential with avidin or streptavidin probes, possibly increasing the sensitivity of assay systems. The reactions of biotin-LC-hydrazide are identical to those of biotin-hydrazide.

The following protocol describes the use of biotin-hydrazide to label glycosylated proteins at their carbohydrate residues. Control of the periodate oxidation level can result in specific labeling of sialic acid groups or general sugar residues (Chapter 1, Section 4.4).

Protocol

1. Dissolve a periodate-oxidized glycoprotein (i.e, an antibody—see Chapter 10, Section 1.3) in 0.1 M sodium phosphate, 0.15 M NaCl, pH 7.4, at a concentration of 2 mg/ml. Note: The buffer, 0.1 M sodium acetate, pH 5.5, is typical of literature references for reacting a hydrazide compound with an aldehyde-containing molecule to form a hydrazone linkage. Alternative buffer conditions using higher pH values also work well. Physiological pH conditions with the use of a reducing agent such as sodium cyanoborohydride (step 4) produce the most efficient labeling conditions when using hydrazide-containing reagents.
2. Add biotin-hydrazide or biotin-LC-hydrazide to a final concentration of 5 mM.
3. React for 2 h at room temperature.
4. To reduce the hydrazone bonds to more stable linkages, cool the solution to 0°C and add an equal volume of 30 mM sodium cyanoborohydride in PBS. Incubate for 40 min. Note: if the presence of a reducing agent is detrimental to protein activity, eliminate this step. In most cases, the hydrazone linkage is stable enough for avidin–biotin labeling experiments.
5. Remove excess reactants by dialysis or gel filtration using a column of Sephadex G-25.

Biocytin Hydrazide

Another biotinylation reagent that can spontaneously couple with aldehyde- or ketone-containing molecules is biocytin hydrazide (Pierce). Produced by forming the hydrazine derivative of biocytin—a lysine–biotin complex often found naturally in serum (Section 3.1)—the compound has better solubility in aqueous solutions than either biotin-hydrazide or biotin-LC-hydrazide, discussed previously. The solubility enhancement of biocytin-hydrazide is due to the presence of lysine's α-amino group, which is protonated and positively charged at physiological pH. The reagent can be used to label carbohydrate-containing molecules, such as glycoproteins, after they have been oxidized to contain reactive aldehydes (Chapter 1, Section 4.4). The hydrazide group forms a hydrazone linkage with the aldehydes, thus directing the biotinylation reaction toward the polysaccharide regions of glycoconjugates (Fig. 256).

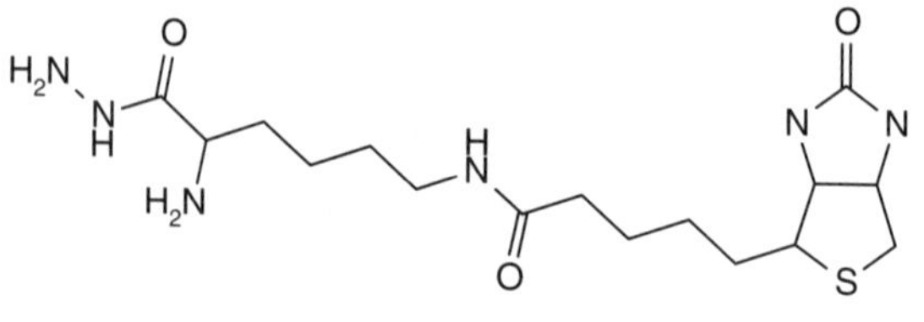

Biocytin Hydrazide
MW 386.51

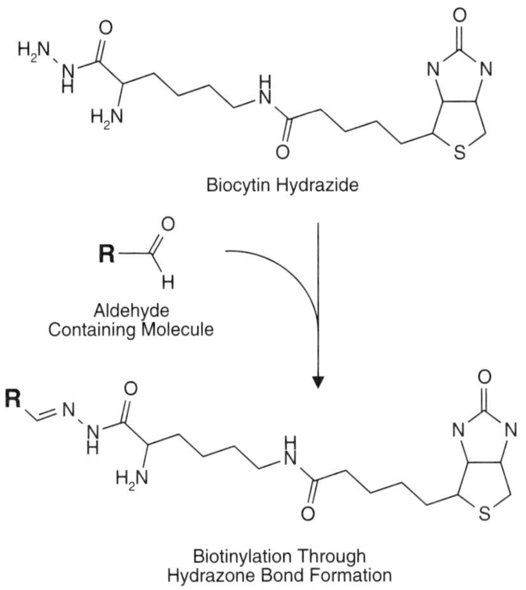

Biocytin Hydrazide

Aldehyde
Containing Molecule

Biotinylation Through
Hydrazone Bond Formation

Figure 256 Biocytin-hydrazide reacts with aldehyde-containing molecules to form hydrazone bonds.

Biocytin hydrazide was used to label specifically sialic acid residues and galactose residues, and for general sugar modification (Bayer *et al.*, 1988). The galactose residues were oxidized using galactose oxidase after treatment with neuraminidase (Chapter 1, Section 4.4). The use of this approach for labeling glycoproteins *in situ* was found to be optimal, due to the other potential side reactions that may occur when using sodium periodate.

The reactivity and use of biocytin-hydrazide is similar to that described for biotin-hydrazide in Section 3.3. The following protocol for labeling glycoproteins at oxidized carbohydrate (galactose) sites is from Bayer and Wilchek (1992).

Protocol

1. Dissolve the glycoprotein to be labeled in 0.1 M sodium phosphate, 0.15 M NaCl, pH 7.4, containing 1 mM $CaCl_2$ and 1 mM $MgCl_2$ (labeling buffer), at a concentration of 1 mg/ml.
2. Dissolve biocytin-hydrazide (Pierce) in 0.1 M sodium phosphate, 0.15 M NaCl, pH 7.4 (PBS), at a concentration of 20 mg/ml.
3. To each milliliter of glycoprotein solution, add 30 μl of neuraminidase (1 unit/ml as supplied by Behringwerke AF), then 30 μl of galactose oxidase (previously dissolved at 100 units/ml in the labeling buffer of step 1), and finally 100 μl of the biocytin hydrazide solution.
4. React for 2 h at 37°C.
5. Remove unreacted reagents by dialysis or gel filtration.

5-(Biotinamido)pentylamine

The derivative 5-(biotinamido)pentylamine contains a 5-carbon cadaverine spacer group attached to the valeric acid side chain of biotin (Pierce). The compound can be used in a carbodiimide reaction process to label carboxylate groups in proteins and other molecules, forming amide bond linkages (Chapter 3, Section 1). However, the main use of this biotinylation reagent is in the determination of factor XIIIa or transglutaminase enzymes in plasma, cell, or tissue extracts.

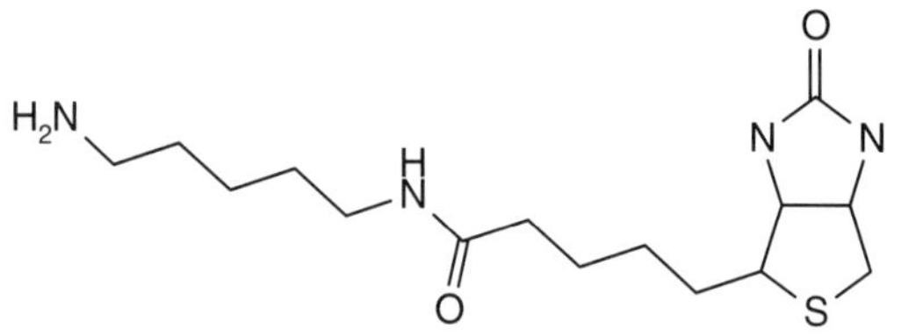

5-(Biotinamido)pentylamine
MW 328.48

Factor XIII, also known as plasma transglutaminase, is an enzyme of the blood coagulation cascade. It is activated by thrombin and calcium to factor XIIIa, at which point it catalyzes covalent cross-links between the ε-amine group of lysine side chains and the γ-glutamyl side chain of glutamine residues. Abnormal levels of factor XIII in plasma are clinically important, being associated with cancer, liver or renal disfunction, or various bleeding disorders. The assay of transglutaminase activity therefore is important for investigating the activity and function of this enzyme as it relates to post-translational protein modification as well as various disease states.

5-(Biotinamido)pentylamine is able to participate in the acyltransferase reaction, becoming covalently attached to protein substrates at their glutamine residues (Fig. 257). Lee et al., (1988) used this biotinylation reagent to quantify factor XIII in plasma. Transglutaminase activity resulted in the modification of an N,N'-dimethylcasein substrate, which was subsequently detected by an avidin–biotin assay procedure. The assay may be done in microplates using wells coated with the substrate protein and quantifying the enzyme activity with streptavidin–alkaline phosphatase (Slaughter et al., 1992). Jeon et al., (1989) subsequently applied the assay to the measurement of transglutaminase activity in cells. Components biotinylated in cellular systems also can be isolated by use of affinity chromatography on immobilized avidin (Lee et al., 1992).

3.4. Photoreactive: Photobiotin

Biotin derivatives containing a photoreactive group provide nonselective biotinylation potential at certain reactive hydrogen sites or nucleophilic groups. They can be used to incorporate an avidin-binding, biotin group into molecules that do not possess amines, sulfhydryls, or other easily modifiable functional groups. Most of these photoreactive derivatives utilize the phenyl azide type of photosensitive group, which can be activated by exposure to UV light to an intermediate nitrene or the nucleophile-reactive dehydroazepine (Chapter 2, Section 7.1, and Chapter 5, Section 3).

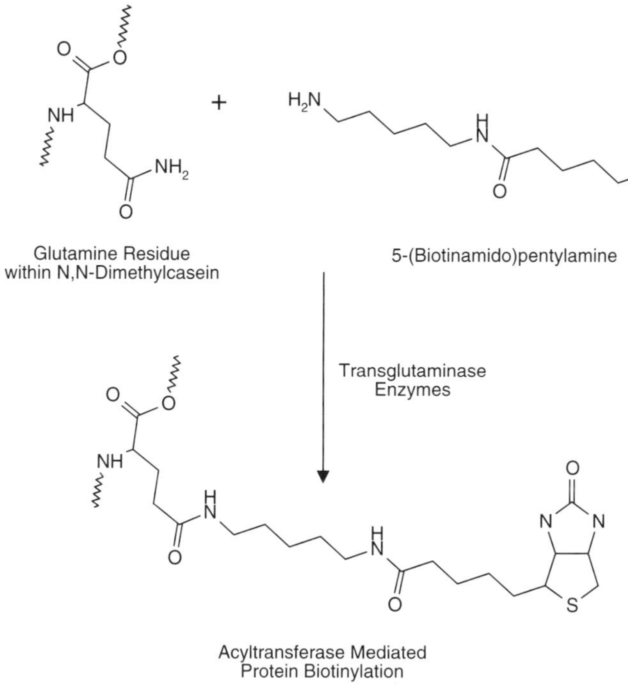

Figure 257   5-(Biotinamido)pentylamine can be used to label glutamine residues in proteins by enzymatic action of transglutaminase.

Perhaps the most common photoreactive biotin derivative is N-(4-azido-2-nitrophenyl)-aminopropyl-N'-(N-d-biotinyl-3-aminopropyl)-N'-methyl-1,3-propanediamine, simply called photoactivatable biotin or photobiotin (Forster et al., 1985) (Pierce). The compound contains a 9-atom diamine spacer group on the biotin valeric acid side chain at one end, while the other end of the spacer terminates in the aryl azide functional group. The presence of a nitro group on the phenyl azide ring allows for photoactivation at higher wavelengths approaching the visible region of the spectrum, thus avoiding potential breakdown of biological molecules through UV exposure. Photolyzing with light at a wavelength of 350 nm causes rapid activation with nitrene formation. The nitrene can couple to replaceable hydrogen sites or add to double bonds within Van der Waals distance or undergo ring expansion to the dehydroazepine. If ring expansion occurs, the principal target group for coupling is a nucleophile, such as a primary amine (Fig. 258).

Photobiotin has been used to biotinylate numerous macromolecules, including proteins and nucleic acids. The biotinylation of alkaline phosphatase was done with complete retention of activity (Forster et al., 1985). Tubulin was labeled with photobiotin and detected on dot blots down to a level of 10 pg of sample using an avidin–enzyme conjugate (Lacey and Grant, 1987). DNA and RNA were labeled for use in hybridization assays (Forster et al., 1985; Keller et al., 1989). For instance, photobiotin-modified probes have been used to detect flavivirum RNA in infected cells (Khan and Wright, 1987), to detect single-copy genes and low-abundance mRNA

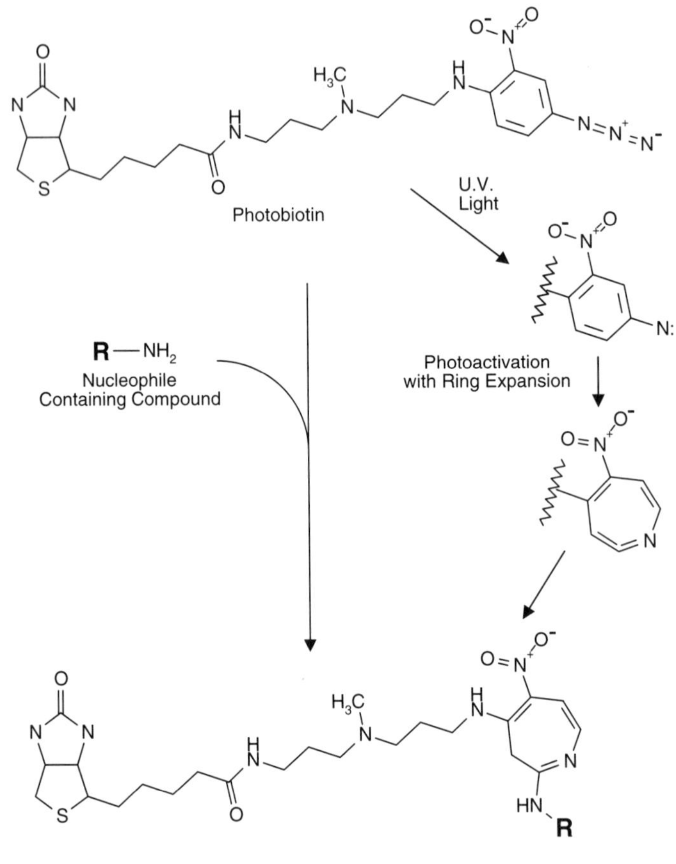

Figure 258 Photobiotin can be made to couple spontaneously with nucleophiles by exposure to UV light. The phenyl azide ring undergoes ring expansion to a highly reactive dehydroazepine intermediate, which can react with amines.

(McInnes *et al.*, 1987), for the diagnosis of barley yellow dwarf virus (Habili *et al.*, 1987), to assay luteinizing hormone β mRNA in individual gonadotropes (Childs *et al.*, 1987), and to perform DNA mapping using a cross-hybridization technique (Chetrit *et al.*, 1989).

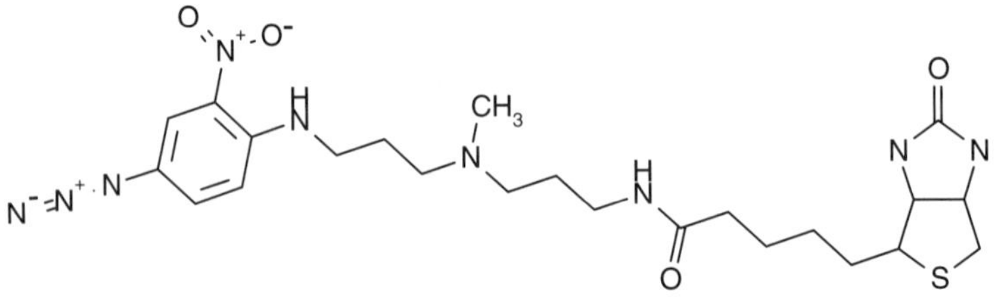

Photobiotin
N-(4-azido-2-nitrophenyl)-aminopropyl-
N'-(N-d-biotinyl-3-aminopropyl)-N'-methyl-1,3-propanediamine
MW 533.65
30.0 Å

Photobiotin can be dissolved in water or buffer at a concentration of 1 mg/ml and stored in the dark at −20°C until needed. As long as no exposure to light is permitted, the compound is stable for at least 1 year under these conditions.

The protocol for modifying DNA probes with photobiotin can be found in Chapter 17, Section 2.3. It is based on the method of Forster *et al.* (1985). The following method is a suggested protocol for the modification of proteins using a photoreactive biotin derivative. Some optimization may be necessary to obtain the best incorporation levels.

Protocol for Labeling Proteins with Photobiotin

1. Dissolve the protein to be biotinylated at a concentration of at least 1 mg/ml in water or dilute buffer at neutral pH.
2. In subdued light, dissolve photobiotin (Pierce) in water at a concentration of 1 mg/ml.
3. Add a quantity of photobiotin solution to the protein solution to give at least a five-fold molar excess of biotinylation reagent.
4. Place in an ice bath and irradiate from above (about 10 cm away) for 15 min using a sunlamp (such as Philips Ultrapnil MLU 300 W, General Electric Sunlamp RSM 275 W, or National Self-Ballasted BHRF 240–250 V 250 W W-P lamp).
5. Remove excess photobiotin by dialysis or gel filtration using a column of Sephadex G-25 (Pharmacia).

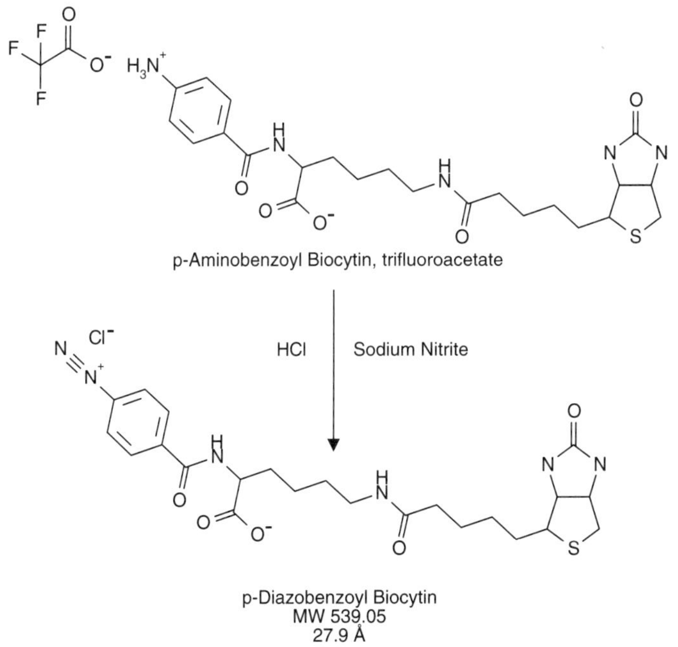

p-Aminobenzoyl Biocytin, trifluoroacetate

HCl   Sodium Nitrite

p-Diazobenzoyl Biocytin
MW 539.05
27.9 Å

Figure 259   The aminophenyl group of this biotin derivative can be transformed into a diazonium functional group by treatment with sodium nitrite in dilute HCl.

3.5. Active-Hydrogen-Reactive: *p*-Aminobenzoyl Biocytin, Diazotized

p-Aminobenzoyl biocytin contains a 4-aminobenzoic acid amide derivative off the α-amino group of biocytin's (Section 3.1) lysine residue (Pierce). The aromatic amine can be treated with sodium nitrite in dilute HCl to form a highly reactive diazonium group (Fig. 259), which is able to couple with active hydrogen-containing compounds. A diazonium reacts rapidly with histidine or tyrosine residues within proteins, forming covalent diazo bonds (Wilchek *et al.*, 1986) (Fig. 260). It also can react with guanidine residues within DNA at position 8 of the base (Rothenberg and Wilchek, 1988) (Fig. 261). Biotinylation via diazo linkages is reversible by treatment with a 10-fold molar excess of $Na_2S_2O_4$ (sodium dithionite) in 50 m*M* Tris, pH 8.5 (Gorecki *et al.*, 1971) (Chapter 2, Section 6.1, and Chapter 4, Section 9).

The procedure for creating the diazonium derivative of *p*-aminobenzoyl biocytin and coupling to a protein or a nucleic acid is as follows.

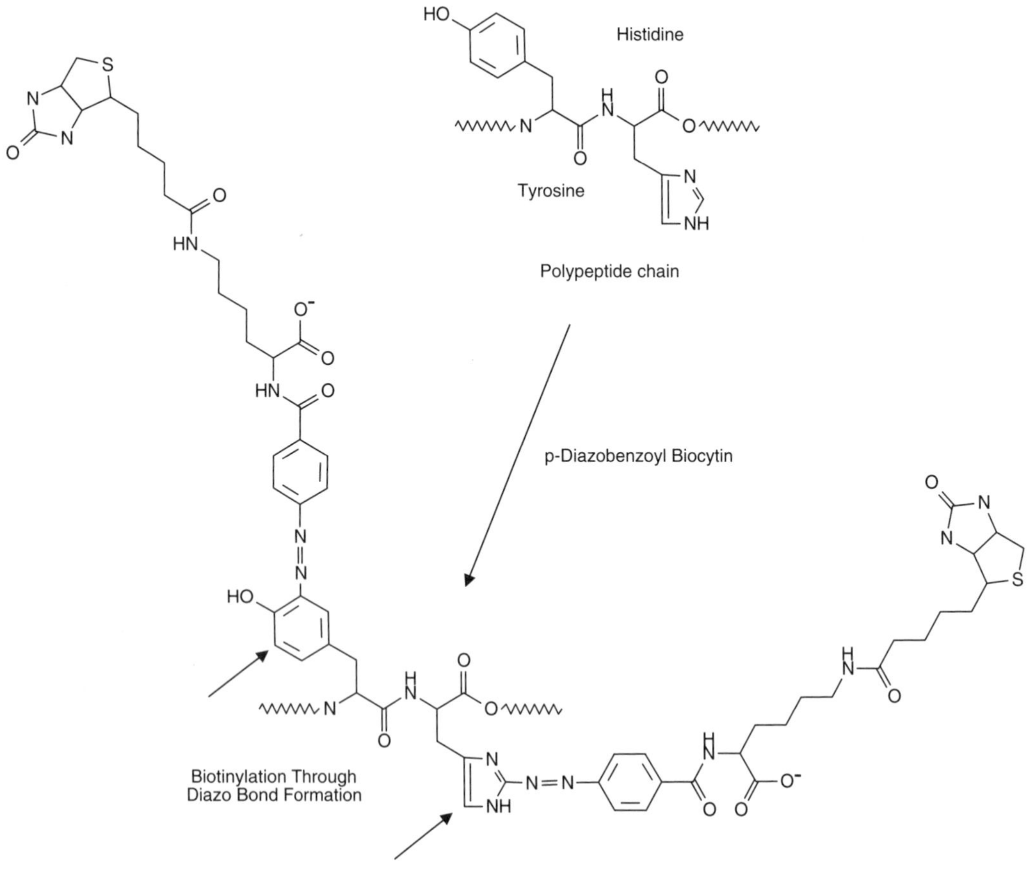

Figure 260 The diazonium group of *p*-diazobenzoylbiocytin can react with tyrosine or histidine residues in proteins to form diazo bonds.

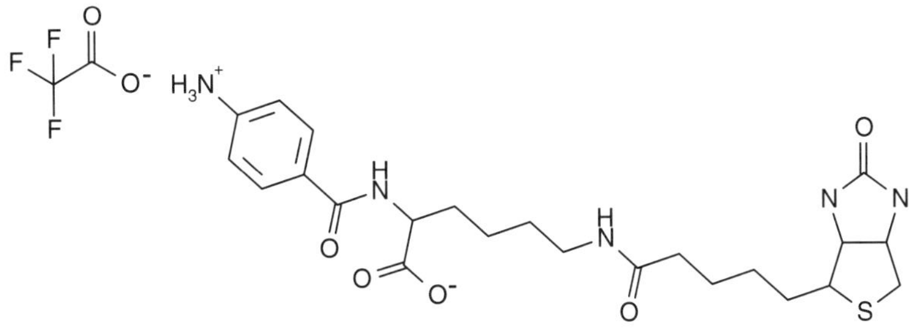

p-Aminobenzoyl Biocytin,
trifluoroacetate salt
MW 605.63

Protocol

Formation of the Diazonium Derivative

1. Dissolve 2 mg of *p*-aminobenzoyl biocytin (Pierce) in 40 μl of 1 N HCl (concentration of 50 mg/ml). Cool the solution in ice.

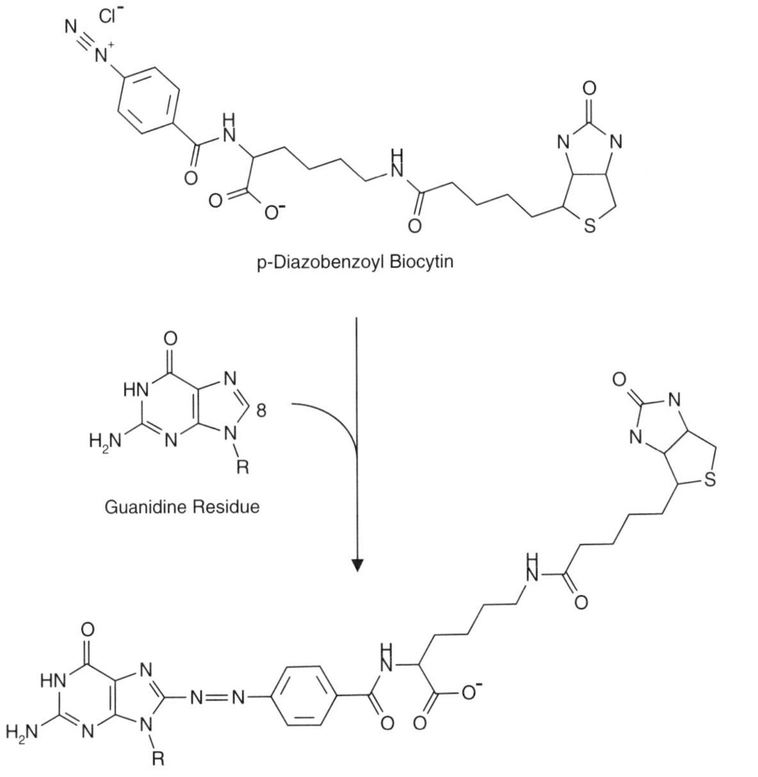

p-Diazobenzoyl Biocytin

Guanidine Residue

Biotinylation Through
Diazo Bond Formation

Figure 261   The diazonium group of *p*-diazobenzoylbiocytin can couple to the C-8 position of guanidine bases in nucleic acids, forming diazo bonds.

2. Dissolve 7.7 mg of sodium nitrite in 1 ml of ice-cold water. Prepare fresh.
3. Mix 40 μl of the *p*-aminobenzoyl biocytin solution with 40 μl of the sodium nitrite solution.
4. React for 5 min on ice to create the diazonium derivative.
5. Stop the reaction by the addition of 35 μl of 1 *N* NaOH. Use immediately for biotinylation.

Biotinylation of Proteins on Blots Using the Diazonium Derivative of *p*-Aminobenzoyl Biocytin

1. Dilute the diazonium derivative of *p*-aminobenzoyl biocytin prepared in preceding part with 0.2 *M* sodium borate, pH 8.4, to a concentration of 10 μg/ml.
2. Transfer proteins onto a nitrocellulose membrane using any appropriate procedure, including dot blotting the protein solution onto the surface.
3. Incubate the membrane with the biotin derivative at a ratio of 1 ml/cc² of membrane.
4. React for 1 h at room temperature.
5. Wash the membrane thoroughly with 0.1 *M* Tris, 0.15 *M* NaCl, pH 7.5.
6. Block nonspecific sites on the membrane with an appropriate blocking component (such as BSA) and detect the biotinylated proteins using an avidin or streptavidin conjugate.

4. Iodination Reagents

Modification of proteins and other molecules with a radioactive element provides a means of detection that can be extremely sensitive for assay, localization, and imaging applications. Among the most common radiolabels for biological studies are ^{14}C, ^{32}P, ^{35}S, and ^{3}H, and the isotopes of iodine, ^{125}I and ^{131}I. The unstable isotopes of carbon, phosphorus, sulfur, and hydrogen are all β emitters, releasing particulate radiation consisting of either positrons or electrons. To measure labeled molecules containing β emitters often necessitates tedious sample manipulation including tissue homogenization and mixing with scintillation cocktails for subsequent counting.

The radioactive isotopes of iodine, by contrast, are both γ emitters, providing a much easier route to measurement than β-particle-emitting radioisotopes. High-energy electromagnetic radiation can be detected directly without the need for intermediate scintillation cocktails. Iodine-131 was the first unstable iodine isotope to be used for labeling protein molecules (Li, 1945; Pressman and Keighley, 1948). The ^{131}I isotope decays by both β (electron) and γ emission. The specific activity of this element can be as high as 6550 Ci/mmol, providing extraordinary sensitivity for detecting labeled molecules.

Iodine-125 decays by electron capture followed by γ emission. However, the maximum energy of ^{125}I electromagnetic energy emission can be as little as one-tenth to one-third that of ^{131}I (Wilbur, 1992; Powsner, 1994). The greater energy intensity of ^{131}I emission actually can be a disadvantage, since γ rays emanating from it are more penetrating, requiring increased precautions and greater protective equipment. In addition, the relatively short half-life of ^{131}I (8.1 days) compared to that of ^{125}I (60 days) necessitates that labeled compounds be prepared more often, since activity losses will be severe on storage. Because ^{125}I is not a particulate emitter, its use *in vivo* for imaging applications limits radiation damage to surrounding proteins, cells, and tissues.

These factors make ^{125}I the iodine label of choice for radiolabeling biological molecules. Its commercial availability from a number of suppliers at relatively low cost further adds to its popularity. Even though it has lower specific activity than ^{131}I, iodine-125 still provides much greater sensitivity than ^{14}C, ^{32}P, ^{35}S, or ^{3}H in labeling biomolecules. In fact, the use of a radioactive iodine label can create probes that have 150-fold more sensitivity than tritiated molecules and as much as 35,000 times the detectability of ^{14}C-labeled molecules (Bolton and Hunter, 1986).

Radioiodination is the process of chemically modifying a molecule to contain one or more atoms of radioactive iodine. Early studies on protein modification determined that iodine in aqueous solution formed a reactive ion, H_2OI^+ (Fig. 262), that is capable of modifying tyrosine side chains and the imidazole groups of histidine, and either modifying sulfhydryl groups or catalyzing their oxidation to disulfides (Fig. 263). More modern methods utilize a chemical agent to create the reactive iodine species, thus driving the reaction at much greater rates.

There are two main methods of radioiodination commonly employed to modify proteins and other molecules: (1) direct labeling of the desired protein or other target molecule in the presence of an oxidizing agent or (2) indirect labeling of the desired molecule by first labeling an intermediate compound, which is then used to perform the final modification. Direct labeling methods are by far the most common, and the chemical reactions used in this process have been reviewed (Regoeczi, 1984).

The prevailing procedures for direct coupling of ^{125}I to a protein or other molecule is through the use of an oxidizing agent. The *in situ* preparation of an electrophilic radioiodine species is fundamental to the ability to modify certain reactive sites within the desired molecules. The most common oxidizing compounds are *N*-haloamine derivatives, such as *N*-chlorotoluenesulfonamide (chloramine-T) or 1,3,4,6-tetrachloro-3α,6α-diphenylglycouril (IODO-GEN). In most instances, such compounds do not harm the proteins being labeled, although careful control over reaction times should be done to prevent overlabeling or oxidative damage. A secondary method of producing an oxidative effect is to use an enzyme-driven system. The glucose oxidase/lactoperoxidase reaction creates reactive iodine through the production of hydrogen peroxide from glucose with the subsequent action of peroxidase to form I_2 from I^-.

Formation of the electrophilic halogen species leads to the potential for rapid reaction with compounds containing strongly activating groups, such as activated aryl compounds. Particularly, substances containing aromatic ring structures that have

$$I_3^- \rightleftharpoons I_2 + I^-$$

$$I_2 + H_2O \rightleftharpoons H_2OI^+ + I^-$$

Figure 262 Iodide anion in aqueous solution undergoes an equilibrium reaction process to form the reactive H_2OI^+ species.

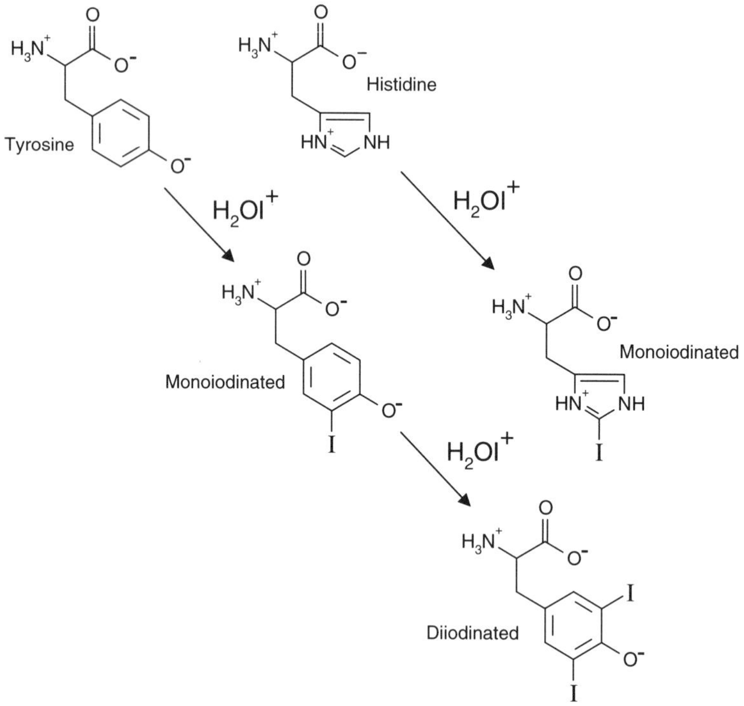

Figure 263 The iodination of tyrosine or histidine residues in proteins by H_2OI^+.

substituents on the ring that are electron-donating can sufficiently activate the carbons on the ring to undergo electrophilic substitution reactions. Therefore, phenols, aniline derivatives, or alkyl anilines that contain OH, NH_2, or NHR constituents, respectively, are very susceptible to being iodinated. In proteins, this translates into tyrosine side-chain phenolic groups and histidine side-chain imidazole groups. Cross-linking compounds or modification reagents containing ring-activated groups also are capable of being iodinated.

The addition of a radioactive iodine atom to a protein molecule typically has little effect on the resultant protein activity, unless the active center is modified in the process. The size of an iodine atom is relatively small and does not result in many steric problems with large molecules. The sites of potential protein modification are tyrosine and histidine side chains. Tyrosine may be modified with a total of two iodine atoms per phenolate group, whereas histidine can incorporate one iodine. Sulfhydryl modification at cysteine residues is typically unstable.

The result of iodination at tyrosine groups can alter the spectral characteristics of the protein in solution (Hughes and Straessle, 1950). The typical protein absorbency at 280 nm can shift to a maximum at about 305–315 nm due to the addition of iodine atoms to the phenolate ring of tyrosine. The degree of absorbance shift is dependent on how many iodine atoms are incorporated into the protein and whether they result in mainly mono-iodotyrosine or di-iodotyrosine formation. In addition, as the level of iodination increases, the solubility of a protein in aqueous solutions can dramatically decrease until complete insolubility results in proteins with high numbers of tyrosine sites.

Thus, controlling the degree of iodination is an important consideration both in choosing the oxidant used and in controlling the time of reaction. Typically, most radiohalogenations are done in a time period of 30 s to as long as 30 min. Optimization may have to be done to determine the correct time to use for a particular modification reaction. Termination of the iodination reaction may be done through addition of a reducing agent, such as sodium metabisulfite. Bisulfite reduces the electrophilic iodine species to unreactive iodide, effectively stopping the modification process.

The following sections discuss the major radioiodination reagents available for direct labeling as well as the main cross-linkers or modification reagents used for indirect labeling techniques.

4.1. Chloramine-T

Chloramine-T, or *N*-chlorotoluenesulfonamide, has been one of the most popular oxidizing reagents used for radioiodination techniques since its introduction by Greenwood *et al.* in 1963 (Sigma). It has strong oxidizing properties that readily lead to the formation of the required electrophilic halogen species that results in iodine incorporation into target molecules. The reactions of chloramine-T are well documented, being suitable for both macromolecular protein iodination and small-molecule modification (Wilbur, 1992). It also can be used to modify molecules with other radioactive halogen elements, such as isotopes of bromine and astatine (Hadi *et al.*, 1979; Mazaitis *et al.*, 1981).

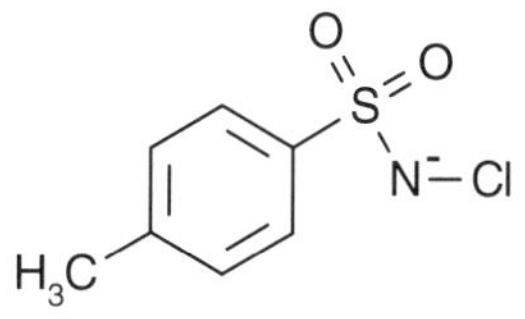

Chloramine-T
N-chlorotoluenesulfonamide

The reaction of chloramine-T with iodide ion in solution results in oxidation with subsequent formation of a reactive, mixed halogen species, ICl (Fig. 264). Either ^{125}I or ^{131}I can be used in this reaction. The ICl then rapidly reacts with any sites within target molecules that can undergo electrophilic substitution reactions. Within proteins, any tyrosine and histidine side-chain groups can be modified with iodine within 30 s to 30 min. Since chloramine-T is a water-soluble reagent and the reaction is done completely in the solution phase, higher incorporation of radioactive iodine can be obtained than using insoluble or immobilized oxidants (see subsequent sections). However, a greater yield of specific radioactivity does not always translate into a better radiolabeled probe. Chloramine-T, being a strong oxidant with rapid reaction rates can easily overlabel a target molecule or cause oxidative damage to sensitive proteins (Lee and Griffiths, 1984). The reaction may be quenched with a reductant, usually

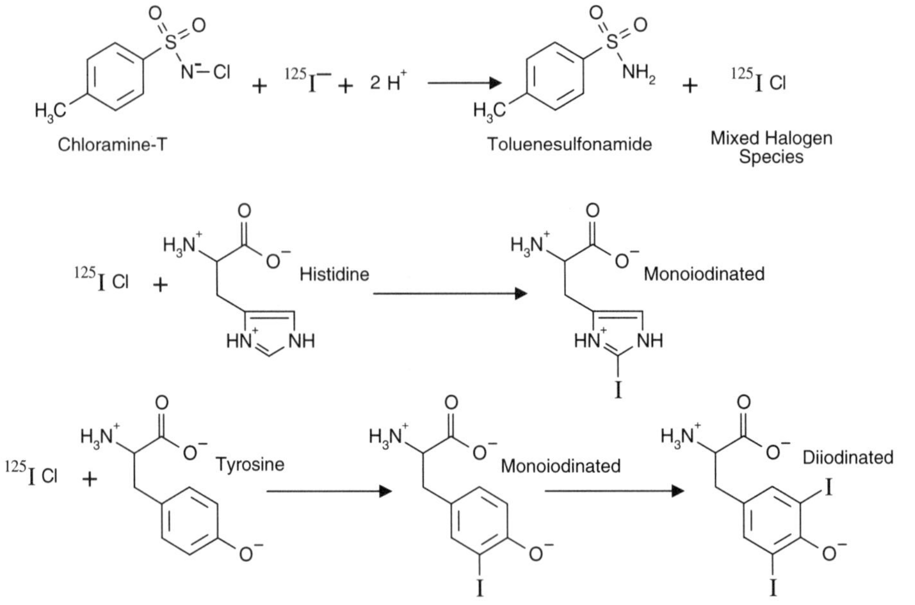

Figure 264 The strong oxidant Chloramine-T can react with iodide anion in aqueous solution to form a highly reactive mixed halogen species. ^{125}ICl then can modify tyrosine and histidine groups in proteins to form radiolabeled products.

done by addition of sodium metabisulfite. Although chloramine-T is still widely used, alternative iodination reagents that are insoluble or immobilized on insoluble supports provide milder reaction conditions and are more controllable.

The following protocol is representative of those found in the literature for iodination of protein molecules using chloramine-T.

Protocol

Caution: Handle all radioactive substances according to the radiation safety regulations instituted at each facility approved to handle such materials. Use adequate precautions to protect personal safety and the environment. Dispose of radioactive waste only by following approved guidelines.

1. Dissolve chloramine-T in 50 m*M* sodium phosphate, pH 7, at a concentration of 4 mg/ml. Prepare fresh. Approximately 25 µl of this solution (100 µg) is required to iodinate 5 µg of a protein.
2. Dissolve sodium metabisulfite in 50 m*M* sodium phosphate, pH 7, at a concentration of 12.6 m*M* (240 µg/100 µl). Prepare fresh. Approximately 100 µl of this solution is required for a 5-µg protein iodination.
3. Obtain fresh Na^{125}I and adjust its concentration to approximately 0.5 mCi/µl. Two microliters of this solution is required to iodinate 5 µg of protein.
4. Add to a suitable reaction vial 25 µl of a solution consisting of 5 µg of a protein dissolved in 50 m*M* sodium phosphate, pH 7. Mix using a small magnetic stirring chip.

5. Add 2 µl of the Na^{125}I solution (about 1 mCi) to the protein in the vial. Seal the vial using a screw-cap septum that can be penetrated with a syringe.

6. Using a syringe, add 25 µl of the chloramine-T solution and continue to mix for at least 30 s. Longer reaction times can be done, but the solution-phase iodination usually proceeds very rapidly.

7. Add 100 µl of the sodium metabisulfite solution to the iodination reaction to stop it. Stir for 10 s.

8. Purify the iodinated protein from excess reactants by gel filtration using a column of Sephadex G-25 or G-50 (Pharmacia). The column may be pretreated by passing a solution of BSA through it to eliminate nonspecific binding sites that could cause significant protein loss in small-sample applications.

4.2. IODO-BEADS

IODO-BEADS (Pierce Chemical) is an immobilized preparation of a chloramine-T analog, consisting of nonporous, polystyrene beads of diameter $\frac{1}{8}''$ that have been derivatized to contain N-chlorobenzenesulfonamide groups (as the sodium salt) (refer to U.S. Patents 4,448,764 and 4,436,718). During the manufacturing process, the hydrophobic nature of the polystyrene is changed to a rather hydrophilic surface due to the charged chlorosulfonamide groups. The surface character results in excellent protein recoveries (typically greater than 90%). The oxidizing capability of IODO-BEADS is limited to surface reactions on the outer shell of the polystyrene ball. The effect is to reduce the rate of iodine incorporation into macromolecules from the extremely rapid 30-s reaction of soluble chloramine-T to a more relaxed pace of about 2–15 min. This also creates a milder oxidizing environment, thus minimizing the potential for protein degradation or activity loss. A slower iodination process allows more control over the level of iodine derivatization, as well. Often, tyrosine iodination can be limited to mono-iodo forms, avoiding the detrimental effects on solubility or activity that excessive modification can cause.

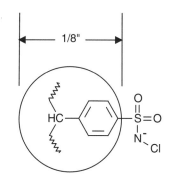

IODO-BEADS Containing
N-Chlorobenzenesulfonamide
Groups on a Polystyrene Backbone

Markwell (1982) reported that the reaction mechanism for creating the electrophilic iodine species may be somewhat different for IODO-BEADS than other oxidizing agents. She demonstrated that the active component remained at or near the surface of the beads during the course of the iodination process. Markwell speculated that an intermediate reactive species, N-iodobenzenesulfonamide, is formed from substitution of the chlorine atoms on the bead (Fig. 265). It is possible that this intermediate is involved in the direct iodination of target molecules that approach the bead surface.

IODO-BEADS may be used in a variety of buffer salts and in the presence of detergents or denaturants without affecting iodination. The reagent, however, is susceptible to inactivation by reducing agents such as disulfide reductants, and it can be inactivated by moisture on storage. Also avoid organic solvents that can dissolve or affect the surface characteristics of polystyrene, such as DMF or DMSO.

To determine the optimal reaction time for a particular radioiodination, 5-μl aliquots of the reaction medium can be removed every 30 s and diluted 1:20,000 with 20 mM Tris, 1 mM EDTA, pH 7.4, Containing 0.5 mg/ml BSA as a carrier protein. Finally, precipitate a small amount of the diluted aliquot with trichloroacetic acid (60%), centrifuge to recover the pellet, wash the pellet once with TCA, and measure the amount of radioactivity in the pellet and supernatant using a gamma counter. The reaction period representing optimal radiolabel incorporation should be used for subsequent radioiodinations.

Directing the iodination reaction toward histidine residues in proteins, as opposed to principally tyrosine modification, is possible simply by increasing the pH of the

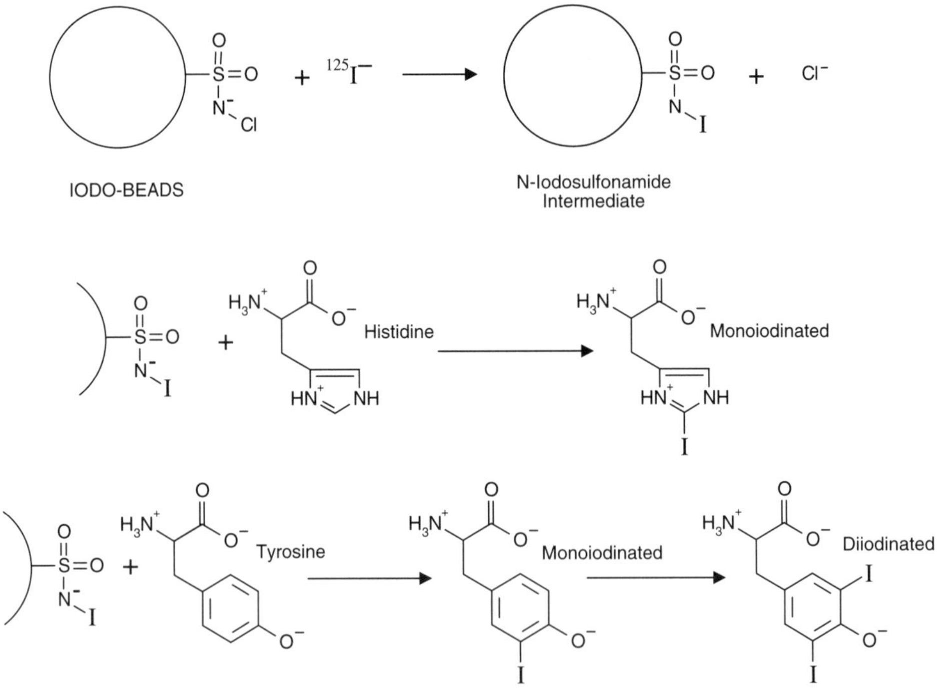

Figure 265 IODO-BEADS contains immobilized Chloramine-T functional groups that can react with radioactive iodide in aqueous solution to form a highly reactive intermediate. The active species may be an iodosulfonamide derivative, which then can iodinate tyrosine or histidine residues in proteins.

IODO-BEADS reaction from the manufacturer's recommended pH 7.0 to 8.2 (Tsomides *et al.*, 1991). No reducing agent is required to stop the iodination reaction, as is the case with chloramine-T and other methods. Simple removal of the bead(s) from the reaction is enough to eliminate the iodination process. The mild nature of the IODO-BEADS iodination reaction can result in better recovery of active protein than using soluble oxidants (Lee and Griffiths, 1984).

Each bead can iodinate up to 500 μg of tyrosine-containing protein or peptide. This translates into an oxidative capacity of about 0.55 μmol per bead. The rate of reaction can be controlled by changing the number of beads used and altering the sodium iodide concentration added to the reaction. Reaction volumes of 100–1000 μl are possible per bead. The following protocol is suggested for iodinating proteins. Optimization should be done to determine the best incorporation level to obtain good radiolabel incorporation with retention of protein activity.

Protocol

Caution: Handle all radioactive substances according to the radiation safety regulations instituted at each facility approved to handle such materials. Use adequate precautions to protect personal safety and the environment from contamination. Dispose of radioactive waste by following approved guidelines.

1. Wash one or more IODO-BEADS (Pierce) with the iodination buffer of choice. Buffers containing 0.1 *M* sodium phosphate or 0.1 *M* Tris at slightly acidic to slightly alkaline pH work well. A buffer consisting of 0.1 *M* sodium phosphate, pH 6.5, will give the highest possible reaction rates and yields.
2. Add the desired number of beads to a solution of carrier-free Na^{125}I in iodination buffer at a concentration level of about 1 mCi/100 μg of protein to be modified. The total reaction volume should be 100–1000 μl per bead.
3. Add from 5 to 500 μg of a tyrosine-containing peptide or protein dissolved in iodination buffer to the reaction mixture.
4. React for 2–15 min at room temperature. Reactions done at 4°C are possible, but will result in slightly lower incorporation of iodine.
5. Stop the reaction by removing the solution from the beads. This can be done by simply pipetting the solution away from the beads or by physically removing the beads. The beads may be washed once with iodination buffer to ensure complete recovery of protein. Exact timing of the reaction is important to obtain reproducible results.
6. Remove excess ^{125}I from the iodinated protein by gel filtration using a column of Sephadex G-25 or G-50 (Pharmacia).

4.3. IODO-GEN

IODO-GEN (Pierce), first described by Fraker and Speck in 1978, is 1,3,4,6-tetrachloro-3α,6α-diphenylglycouril, an *N*-haloamine derivative with oxidizing properties similar to those of IODO-BEADS and chloramine-T. The compound is insoluble in aqueous solution, therefore making it a type of solid-phase radioiodination reagent. However, unlike IODO-BEADS, wherein the oxidizing group is immobilized on another support material, IODO-GEN must be plated out on the surface of a reaction vessel prior to iodination. Due to the reagent's stability, the plated reaction vessels can be prepared well in advance and stored in a desiccator until needed (Markwell and Fox, 1978).

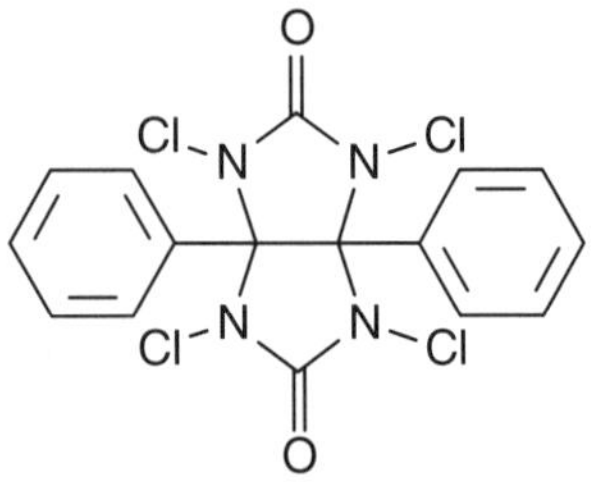

IODO–GEN
1,3,4,6-Tetrachloro-
3α,6α-diphenylglycouril
MW 432.09

The reaction of IODO-GEN with iodide ion in solution results in oxidation with subsequent formation of a reactive, mixed halogen species, ICl (Fig. 266). Either ^{125}I or ^{131}I can be used in this reaction. The ICl then rapidly reacts with any sites within target molecules that can undergo electrophilic substitution reactions. Within proteins, any tyrosine and histidine side-chain groups can be modified with iodine within

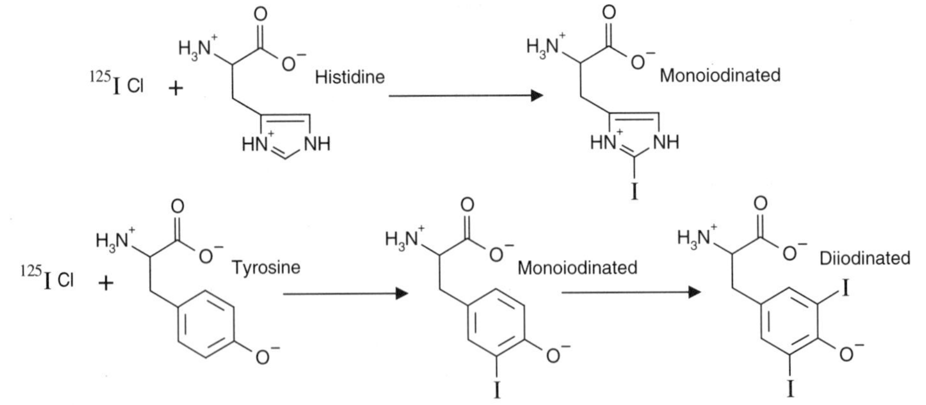

Figure 266 IODO-GEN is a water-insoluble oxidizing agent that can react with $^{125}I^-$ to form a highly reactive mixed halogen species, ^{125}ICl. This intermediate can add radioactive iodine atoms to tyrosine or histidine side-chain rings.

30 s to 30 min. In addition, cross-linking or modification reagents possessing phenyl rings with activating groups present (e.g., electron-donating constituents: —OH, —NH$_2$, etc.) can be iodinated using IODO-GEN. The incidence of side reactions appears to be negligible.

Since IODO-GEN is insoluble and plated on the surface of the vessel during the iodination reaction, it is possible to stop the reaction by simply removing the aqueous phase. The plating technique is important for the successful use of this reagent. Failure to properly plate IODO-GEN on the surface of the reaction vessel may cause the oxidizing agent to become suspended in the reaction medium. However, even with well-plated vessels, there is some potential that a portion of the iodinating reagent can break off in small pieces and contaminate the aqueous phase. For this reason, it is not advisable to stop the iodination reaction only by removing the supernatant, as in the case of IODO-BEADS (Section 4.2). To be certain that the iodination has stopped, an aliquot of sodium metabisulfite can be added to ensure complete cessation of the oxidative process. Alternatively, immediate separation of the iodinated protein from the reactants by gel filtration can be used to stop the reaction (any suspended particles of IODO-GEN will be filtered out on the top of the gel).

Specific radioactivity of 1×10^5 cpm of ^{125}I per microgram of protein easily can be obtained using IODO-GEN. Iodination efficiencies are typically 60% or better and may be controlled by regulating the amount of I$^-$ concentration added to the reaction.

When iodinating intact cells, IODO-GEN can be used to radiolabel the outer cell surface proteins or be directed more toward the inner membrane areas simply by modulating the reaction conditions. Membrane proteins in hydrophobic regions can be labeled to a greater extent by including a small excess of carrier iodide, using high salt conditions, or by employing detergents to disrupt the membrane integrity. Cell surface hydrophilic proteins may be preferentially labeled by not including components that increase cell permeability, by the use of carrier-free iodide, and using short reaction times (Markwell and Fox, 1978).

The following protocol describes the use of IODO-GEN for the radioiodination of proteins and peptides.

Protocol

Caution: Handle all radioactive substances according to the radiation safety regulations instituted at each facility approved to handle such materials. Use adequate precautions to protect personal safety and the environment from contamination. Dispose of radioactive waste by following approved guidelines.

1. In a fume hood, dissolve 10–100 μg of IODO-GEN (Pierce) in 100–500 μl of chloroform, methylene chloride, or DMSO. The use of 10 μg of IODO-GEN per 100 μg of protein or 10^7 cells to be iodinated will result in good incorporation yields.
2. Add the IODO-GEN solution to a clean, dry, glass reaction vessel in an amount needed for the quantity of protein to be labeled. Slowly evaporate the solvent in the vessel using a stream of dry nitrogen or other inert gas. Do not use a strong gas jet, since rapid evaporation or turbulence in the solvent solution will cause uneven IODO-GEN distribution with possible clumping. Do not merely leave the vessel to dry in a hood, since contaminants or moisture may get into the

reagent film. If done properly, the plating process should leave a film of IODO-GEN on the inner surface of the vessel that is difficult to see—looking like a slight clouding of the glass. After solvent evaporation, seal the container with nitrogen and store in a desiccator until needed.

3. Dissolve the protein to be iodinated in a buffer compatible with its known biological stability. Conditions ranging from pH 4.4 to 9 and temperatures from 0 to 37°C can be used with good results. The amount of protein to be labeled should be contained in a volume of 100 μl or less. The sample buffer should not contain reducing agents, antioxidants, 2-mercaptoethanol, DTT, cysteine, glycerol, high detergent concentrations, or anything that may interfere with the iodination reaction or dislodge the plated IODO-GEN.

4. Rinse the plated reaction vessel once with sample buffer to remove any loose particles of IODO-GEN that may not be strongly adhered to the surface of the glass.

5. Add carrier-free Na^{125}I to the reaction vessel in a ratio of about 500 μCi per 100 μg protein.

6. React for 10–15 min at room temperature. Optimization of the reaction time and the amount of ^{125}I added to the reaction may have to be done to obtain the best radioactivity incorporation and retention of protein activity.

7. Remove the sample from the reaction vessel. This process should terminate the iodination reaction, unless small IODO-GEN particles break off from the sides of the vessel. To ensure safe handling, carrier NaI may be added to the reaction mixture to a final concentration of 1 mM.

8. Remove excess reactants by gel filtration using a column of Sephadex G-25.

4.4. Lactoperoxidase-Catalyzed Iodination

An enzyme-catalyzed process also may be used to form reactive iodine species capable of iodinating proteins and other molecules (Marcholonis, 1969; Morrison and Bayse, 1970). The enzymatic approach utilizes lactoperoxidase in the presence of H_2O_2 to oxidize $^{125-}$ to I_2. The iodine thus formed may react with tyrosine or histidine sites within proteins, forming radiolabeled complexes. Unlike the use of chemical oxidants for iodination, the enzymatic reaction is very pH-dependent—the optimum being between pH 6 and 7. If H_2O_2 is directly added to the reaction medium, it must be highly pure with no stabilizing agents such as metals, since they inhibit the oxidation process.

An alternative to direct addition of H_2O_2 is to form it *in situ* through the use of a second enzymatic reaction. Enzymobeads (originally from BioRad, but no longer commercially available) used immobilized lactoperoxidase along with immobilized glucose oxidase to create the necessary oxidative environment. The glucose oxidase reaction transformed added glucose in the iodination medium to the required H_2O_2. As it was formed, the lactoperoxidase (coupled in tandem to the same beads) would catalyze the formation of I_2 (Fig. 267). The immobilized enzymes create an iodination environment that is more oxidatively gentle than direct addition of a soluble chemical oxidant like chloramine-T.

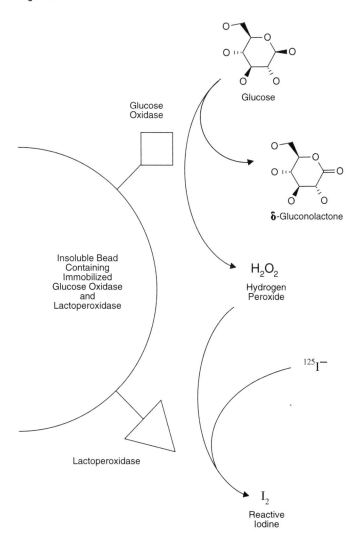

Figure 267   The immobilized glucose oxidase/lactoperoxidase system radioiodinates proteins through the intermediate formation of hydrogen peroxide from the oxidation of glucose. H_2O_2 then reacts with iodide anions to form reactive iodine (I_2). This efficiently drives the formation of the highly reactive H_2OI^+ species that is capable of iodinating tyrosine or histidine residues (See Fig. 262).

4.5. Iodinatable Modification and Cross-linking Agents

Radioiodination can be done by an indirect approach that utilizes a radiolabeled cross-linking or modification reagent which is then used to label the target molecule. The advantage of indirect labeling over direct modification of tyrosine or histidine residues in proteins is to be able to control the iodination to occur with functional groups other than just activated phenyl or imidazole rings. A second major advantage is to eliminate the potential for oxidative damage to sensitive biological molecules, as may occur when an oxidant is used in direct iodination procedures.

For instance, an amine-reactive modification reagent can be radiolabeled and subsequently used to couple with ε- and N-terminal amines on a protein molecule. The protein is not exposed to oxidative conditions, and the level of radiolabeling can be discretely controlled by the molar ratio of modification reagent addition. The use of iodinatable cross-linking reagents can similarly provide radioactive tags incorporated into conjugates at the time of formation. In addition, iodinatable bioconjugation reagents that react with groups such as sulfhydryls, aldehydes, or other functional groups of limited occurrence in proteins or other macromolecules can be used to direct the point of radiolabeling to areas away from active centers or binding sites, thus better preserving biological activity. Finally, some photoreactive cross-linking agents can be iodinated, used to label a targeting molecule, and photolyzed at the point of binding to its target, and the cross-bridge of the resulting complex chemically cleaved, resulting in the transfer the radiolabel to the targeted component. This process can be used to follow the targeted molecule *in vivo* or in cellular systems.

Direct iodination of proteins and other molecules does not provide the range of experimental options available through indirect labeling. The disadvantage of the indirect labeling process is the additional steps needed to prepare the radiolabeled cross-linker or modification reagent before iodination of the desired molecule. The following sections discuss some of the major indirect iodination methods, including the reagents available for doing such procedures.

Bolton–Hunter Reagent

Bolton and Hunter (1973) developed the reagent N-succinimidyl-3-)(4-hydroxyphenyl) propionate (SHPP) for the indirect radioiodination of proteins and other macromolecules (Pierce). The NHS ester end of the molecule reacts with amine groups in target molecules to form stable amide bond derivatives (Chapter 2, Section 1.4). The other end of the reagent contains a phenolic group that is ideally suited for modification with ^{125}I. Iodination of the phenol group occurs *ortho* to the hydroxyl, thus accommodating either one or two iodine substitutions per molecule (Fig. 268).

The use of the Bolton–Hunter reagent to incorporate radioactivity into proteins results in at least as good incorporation of ^{125}I as direct labeling procedures using an oxidant. In many cases, the degree of radioiodine labeling can be much greater than that possible by direct labeling of tyrosine residues, because the total number of amines in a protein (from N-terminal and lysine side chains) is typically significantly more than the number of tyrosines present. The major advantage of the indirect approach, however, it that non-tyrosine-containing proteins also may be iodinated. In addition, substances sensitive to oxidant exposure can be labeled without loss of activity or structural degradation. Ultimately, any molecule containing an available amino group can be radioiodinated with SHPP, even if it does not contain a strongly activated aromatic ring system to allow direct iodine substitution. For reviews of protein modification using radiolabeled Bolton–Hunter reagent, see Langone (1980, 1981) and Wilbur (1992). For its use in labeling cellular components, see Katz *et al.*, (1982) and Davies and Palek (1981).

SHPP is relatively insoluble in aqueous environments and must be dissolved in

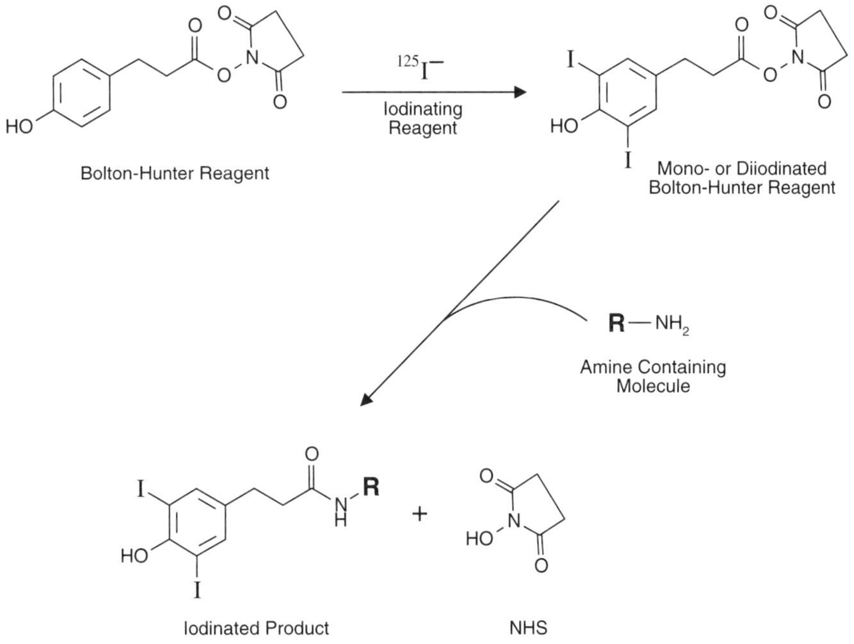

Figure 268 The Bolton–Hunter reagent may be radioiodinated at its phenolic ring structure prior to reaction with an amine-containing molecule to form an amide bond modification.

an organic solvent prior to addition to a reaction medium. Suggested solvents include dioxane and DMSO that are low in water content to avoid hydrolysis of the NHS ester.

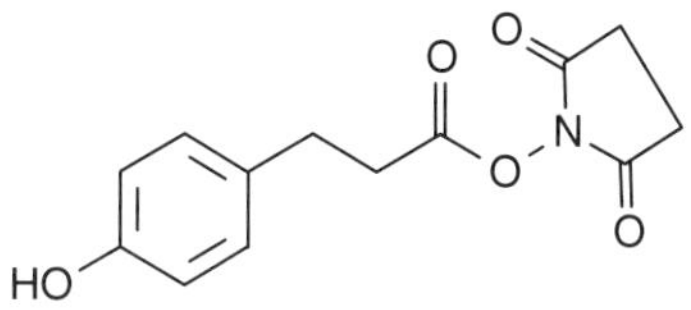

Bolton-Hunter Reagent
N-Succinimidyl-3-
(4-hydroxyphenyl)propionate
MW 263.26

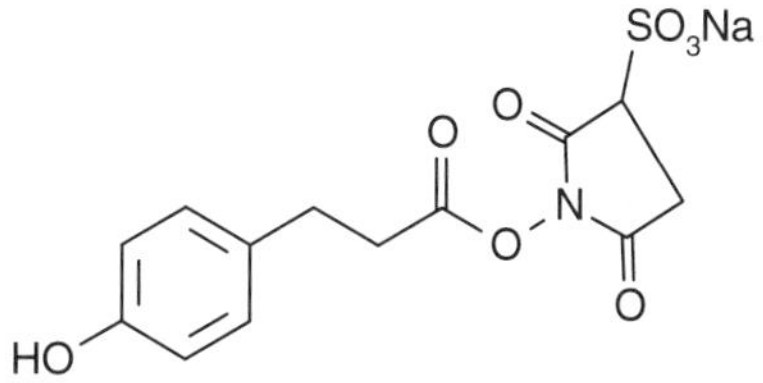

Water-soluble Bolton-Hunter Reagent
N-Sulfosuccinimidyl-3-
(4-hydroxyphenyl)propionate
MW 365.30

A water-soluble version of the original Bolton–Hunter reagent also is available, called sulfo-SHPP or sulfosuccinimidyl-3-(4-hydroxyphenyl)propionate (Pierce) This compound contains a sulfonate group on the NHS ring structure, the negative charge of which provides enough hydrophilicity to allow direct addition to aqueous reaction mediums.

The following procedure describes the iodination process for the Bolton–Hunter reagent and its subsequent use for the radiolabeling of protein molecules. Modification of other macromolecules can be done using the same general method. For particular labeling applications, optimization of the level of iodine incorporation may have to be done to obtain the best specific radioactivity with retention of biological activity.

Protocol

Caution: Handle all radioactive substances according to the radiation safety regulations instituted at each facility approved to handle such materials. Use adequate precautions to protect personal safety and the environment from contamination. Dispose of radioactive waste by following approved guidelines.

1. Dissolve SHPP (Pierce) in dry dioxane or DMSO at a concentration of 0.5 mg/ml. Prepare fresh. If sulfo-SHPP is used dissolution in organic solvent is unnecessary.
2. Dissolve chloramine-T (Sigma) in 50 mM sodium phosphate, pH 7.5 (reaction buffer), at a concentration of 100 μg per 25–50 μl of buffer. Prepare fresh.
3. Add 2 μl of the SHPP solution 50 μl of the reaction buffer containing 0.5–1 mCi Na^{125}I. Alternatively, dissolve solid sulfo-SHPP into the reaction buffer to give a final concentration of 1 μg/50 μl. To better facilitate measuring out sulfo-SHPP, a more concentrated solution may be prepared in a greater volume and then immediately diluted to this concentration. Once the Bolton–Hunter reagent is in an aqueous environment the NHS ester end of the compound will hydrolyze. Therefore, all aqueous handling of the reagent from this point on should be done quickly to preserve enough amine-coupling activity to label the protein after the iodination reaction.
4. Immediately add the chloramine-T to the SHPP solution. Mix well.
5. React for 15 s with mixing.
6. Add to the iodination reaction 5 μl of DMF and 100 μl of benzene. Mix to extract the iodinated Bolton–Hunter reagent into the organic phase.
7. Remove the aqueous phase and transfer the organic phase into a clean glass vial or tube.
8. Remove the organic solvent by evaporation using a steady, but gentle, stream of nitrogen.
9. Dissolve the equivalent of 250 ng of a protein to be labeled in 2–2.5 μl of ice-cold 50 mM sodium borate, pH 8.5.
10. Add the protein solution to the dried, iodinated Bolton–Hunter reagent.
11. React for 2 h on ice. An overnight reaction may be done at 4°C.
12. Remove excess labeled Bolton–Hunter reagent by gel filtration or dialysis.

The Bolton–Hunter reagent also may be used to modify a molecule prior to the iodination reaction. In this case, an amine-containing protein or other molecule is coupled via the NHS ester end of the reagent to form an amide-bond derivative. This derivative is then iodinated using any of the iodination reagents discussed in this section. This approach can be useful in preparing stable Bolton–Hunter derivatives that can be stored for extended periods until requiring iodination, eliminating the relatively short half-life of ^{125}I-labeled probes.

Iodinatable Bifunctional Cross-linking Agents

Bifunctional cross-linking agents containing an activated aromatic ring system may be radioiodinated using procedures similar to those described for the Bolton–Hunter reagent (Section 4.5). Certain conjugation compounds have been designed with the potential for radiolabeling in mind. For instance, there are a number of photoreactive phenyl azide cross-linkers that possess an activating hydroxyl group on their phenyl rings. The phenolic group provides sites of facile iodination *ortho* and *para* to the hydroxyl.

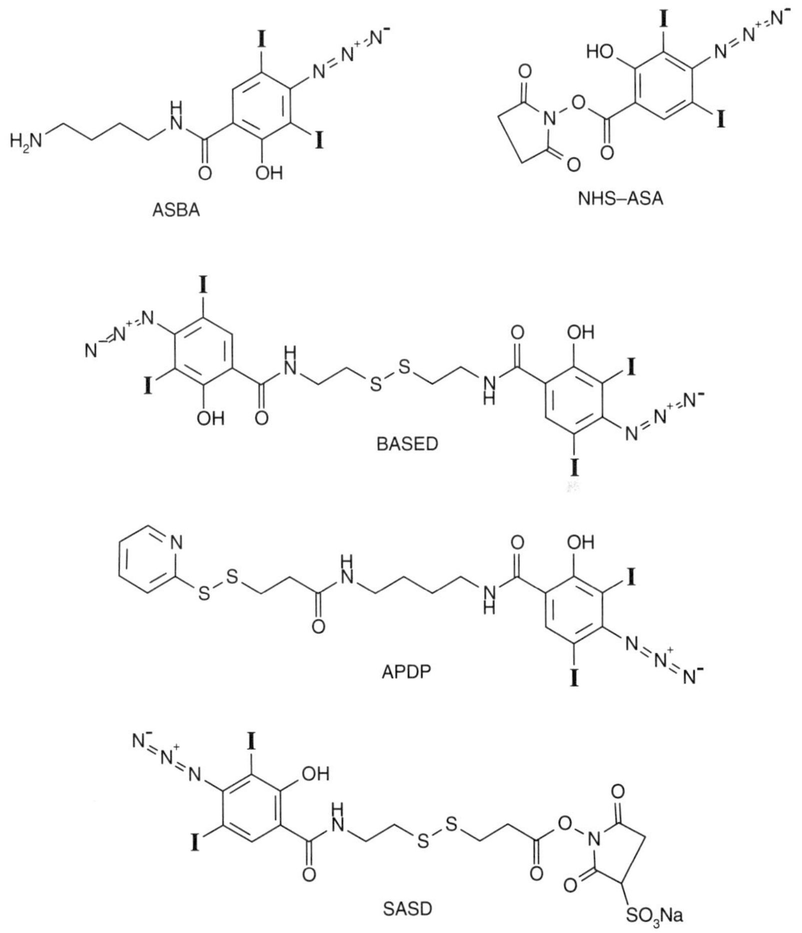

Figure 269 Some common cross-linking agents that are capable of being radioiodinated. The sites of iodination are shown in bold.

The heterobifunctional cross-linker ASBA (Chapter 5, Section 6.1) is an example of this type of iodinatable photoreactive reagent. The phenyl azide group may be radio-labeled using any standard iodination process (described previously) before coupling of its primary amine end to a carbonyl group on a macromolecule. After allowing the modified molecule to bind to a target, the complex may be photolyzed and the covalent conjugate localized by its radioactivity.

The homobifunctional photoreactive BASED (Chapter 4, Section 5.1) has two photoreactive phenyl azide groups, each of which contains an activating hydroxyl. Radioiodination of the cross-linker can yield one or two iodine atoms on each ring, creating an intensely radioactive compound. Cross-links formed between two interacting molecules are reversible by disulfide reduction, thus allowing tractability of both components of the conjugate.

An extremely versatile iodinatable heterobifunctional is APDP (Chapter 5, Section 4.2). One end of the cross-linker can couple with sulfhydryl-containing molecules, while the other end is a nonselective photoreactive phenyl azide. Again, the phenyl ring contains an activating hydroxyl group, providing radioiodination capability. Modification of a sulfhydryl-containing molecule may be done after iodination of the reagent. After the labeled molecule is allowed to interact with a target molecule, the photoreactive process can be initiated to form a covalent conjugate. Subsequently, the cross-links may be cleaved using a disulfide reducing agent, thus transferring the radiolabel to the second molecule.

SASD (Chapter 5, Section 3.2) behaves in a similar manner, except it contains an amine-reactive end that can be coupled to proteins and other molecules. Its photoreactive end can be iodinated using any of the radioiodination reagents discussed previously. Just as in the case of APDP, SASD cross-links can be cleaved by a disulfide reductant to transfer the radioactive component to a second molecule.

Finally, the small amine-reactive and photoreactive cross-linker NHS-ASA (Chapter 5, Section 3.1) can be iodinated to provide a noncleavable radioactive conjugate.

Figure 269 shows the iodination products resulting from labeling these reagents with ^{125}I. Any of the iodination reagents described previously can be used to radio-label these compounds prior to their incorporation into target molecules. However, the insoluble iodination reagents are probably the best choice, since the separation of radiolabeled compound from excess oxidant simply involves removing the solution.

There are many other compounds that have been investigated for their use in indirect radiolabeling of proteins. For an excellent overview of these chemical reactions, see Wilbur (1992).

PART III

Bioconjugate Applications

The technology of bioconjugation has affected every conceivable discipline in the life sciences. The application of the available chemical reactions and reagent systems for creating novel complexes with unique activities has made possible the assay of minute quantities of substances, the *in vivo* targeting of molecules, and the modulation of specific biological processes. Modified or conjugated molecules also have been used for purification, detection, or location of specific substances, and in the treatment of disease.

Cross-linking and modifying agents can be applied to alter the native state and function of peptides and proteins, sugars and polysaccharides, nucleic acids and oligonucleotides, lipids, and almost any other molecule imaginable that can be chemically derivatized. Through careful modification or conjugation strategies, the structure and function of proteins can be investigated, active site conformation discovered, or receptor–ligand interactions revealed. Some of these techniques are so well characterized and standardized that general protocols can be used with broad application and with excellent prospects for success. The following chapters describe how to prepare modified or conjugated biological macromolecules for use in specific applications. The chosen applications represent the most popular uses of these reagent systems, but are by no means exhaustive.

9

Preparation of Hapten–Carrier
Immunogen Conjugates

This chapter describes the design, preparation, and use of hapten–carrier conjugates used to elicit an immune response toward a coupled hapten. The chemical reactions discussed for these conjugations are useful for coupling peptides, proteins, carbohydrates, oligonucleotides, and other small organic molecules to various carrier macromolecules. The resultant conjugates are important in antibody production, in immune response research, and in the creation of vaccines.

1. The Basis of Immunity

The essence of adaptive immunity is the ability of an organism to react to the presence of foreign substances and produce components (antibodies and cells) capable of specifically interacting with and protecting the host from their invasion. An "*antigen*" or "*immunogen*" is the name given for a substance that is able to elicit this type of immune response and also is capable of interacting with the sensitized cells and antibodies that are manufactured against it.

The immune system has two basic components that respond to a challenge of a foreign substance: a cellular response mediated by T lymphocytes and a humoral response mediated by secreted proteins called antibodies produced by B lymphocytes, also called plasma cells. The B lymphocytes recognize antigens through cell-surface immunoglobulins that bind to discrete chemical and structural epitopes on the antigen molecule. Each B cell possesses surface immunoglobulin of a single type (i.e., is monoclonal) and has a binding capacity that is directed against a discrete epitopic target.

Antigen binding by a complementary immunoglobulin molecule on the surface of B cells starts a process of cellular internalization of the foreign substance by pinocytosis. Once internalized by endosomes, systematic processing of the antigen takes place that breaks it down into smaller components.

At this point, the endosome may fuse with vesicles containing newly synthesized or recycling major histocompatibility complex (MHC) antigens. Some of the partially degraded antigenic fragments may form a complex with the MHC and be transported back to the cell surface. There they are "presented" to the circulating T-helper (T_h)

cells, which contain receptors able to bind specifically to particular structural and chemical characteristics of the degraded antigen–MHC complex. If a T_h cell recognizes and binds to the presented antigen on the surface of the APC, the T_h cell proliferates and begins to produce various lymphokines. Finally, the recognition and binding of the presented antigen by the T_h cells, coupled with the release of lymphokines, stimulate the associated B cells to proliferate and produce antibodies that recognize the intact antigen (Germain, 1986).

Antigens usually are macromolecules that contain distinct antigenic sites or "*epitopes*" that are recognized and interact with the various components of the immune system. They can exist as individual molecules composed of synthetic organic chemicals, proteins, lipoproteins, glycoproteins, RNA, DNA, or polysaccharides—or they may be parts of cellular structures (bacteria or fungi) or viruses (Harlow and Lane, 1988a,b,c; Male *et al.*, 1987).

Small molecules like short peptides, although normally able to interact with the products of an immune response, often cannot cause a response on their own. These "*haptens*," as they are called, actually are incomplete antigens, and although not able by themselves to cause immunogenicity or to elicit antibody production, they can be made immunogenic by coupling them to a suitable carrier molecule (Fig. 270). Carriers typically are antigens of higher molecular weight that are able to cause an immunological response when administered *in vivo*.

In an immune response, antibodies are produced and secreted by the B lymphocytes in conjunction with the T_h cells. In the majority of hapten–carrier systems, the B cells end up producing antibodies that are specific for both the hapten and the carrier. In these cases, the T lymphocytes will have specific binding domains on the carrier, but will not recognize the hapten alone. In a kind of synergism, the B and T cells cooperate to induce a hapten-specific antibody response. After such an immune response has taken place, if the host is subsequently challenged with only the hapten, usually it will respond by producing hapten-specific antibodies from memory cells formed after the initial immunization.

Synthetic haptens mimicking some critical epitopic structures on larger macromolecules are often conjugated to carriers to create an immune response to the larger "parent" molecule. For instance, short peptide segments can be synthesized from the known sequence of a viral coat protein and coupled to a carrier to induce immu-

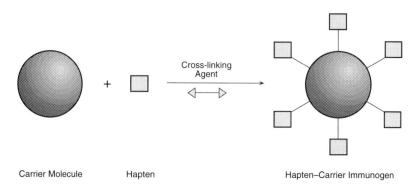

Carrier Molecule   Hapten   Hapten–Carrier Immunogen

Figure 270 Immunogens are made by the cross-linking of a hapten molecule with a carrier using a conjugation reagent.

nogenicity toward the native virus. This type of synthetic approach to immunogen production has become the basis of much of the current research into the creation of vaccines.

The complete picture of the immune system is much more complex than this brief discussion can justly describe. In many instances, merely creating a B-cell response by using synthetic peptide–carrier conjugates, however well designed, will not always guarantee complete protective immunity toward an intact antigen. The immune response generated by a short peptide epitope from, say, a larger viral particle or bacterial cell may only be sufficient to generate memory at the B cell level. In these cases it is generally now accepted that a cytotoxic T-cell response is a more important indicator of protective immunity. Designing peptide immunogens with the proper epitopic binding sites for both B-cell and T-cell recognition is one of the most challenging research areas in immunology today.

Hapten–carrier conjugates also are being used to produce highly specific monoclonal antibodies that can recognize discrete chemical epitopes on the coupled hapten. The resulting monoclonals often are used to investigate the epitopic structure and interactions between native proteins. In many cases, the haptens used to generate these monoclonals are again small peptide segments representing crucial antigenic sites on the surface of larger proteins. Monoclonals developed from known peptide sequences will interact in highly defined ways with the protein from which the sequence originated. These antibodies then can be used, for example, as competitors to the natural interactions between a receptor and its ligand. Thus, using antibodies generated from hapten–carrier conjugates, information can be obtained about the precise sites of binding between macromolecules.

2. Types of Immunogen Carriers

The most commonly used carriers are all highly immunogenic, large molecules that are capable of imparting immunogenicity to covalently coupled haptens. Some of the more useful ones are proteins, but other carriers may be composed of lipid bilayers (liposomes), synthetic or natural polymers (dextran, agarose, poly-L-lysine), or synthetically designed organic molecules. The criteria for a successful carrier molecule are the potential for immunogenicity, the presence of suitable functional groups for conjugation with a hapten, reasonable solubility properties even after derivatization—although this is not an absolute requirement, since precipitated molecules can be highly immunogenic—and lack of toxicity *in vivo*.

Some synthetic carriers actually are designed to have low immunogenicity on their own to minimize the potential for antibody production against themselves. When a hapten is coupled to these molecules, the immune response is directed principally toward the modification, not at the carrier. This design approach guides most of the immune response toward the desired target and minimizes the production of carrier-specific antibodies.

2.1. Protein Carriers

The first carrier molecules used for immunogen conjugation were proteins. A foreign protein administered *in vivo* by any one of a number of potential routes nearly ensured the elicitation of an immune response. In addition, protein carriers could be chosen to be highly soluble and possessed of abundant functional groups that could facilitate easy conjugation with a hapten molecule. When proteins are used as carriers in immunogen formation, the conjugates can be injected in any animal except the animal of origin for the carrier protein itself. In other words, the use of BSA would not be suitable for administration into cows, since self-proteins would not be expected to elicit good immune responses, even when attached with hapten molecules.

The most common carrier proteins in use today are keyhole limpet hemocyanin (KLH; MW 4.5×10^5 to 1.3×10^7), bovine serum albumin (BSA; MW 67,000), aminoethylated (or cationized) BSA (cBSA), thyroglobulin (MW 660,000), ovalbumin (OVA; MW 43,000), and various toxoid proteins, including tetanus toxoid and diphtheria toxoid. Other proteins occasionally used include myoglobin, rabbit serum albumin, immunoglobulin molecules (particularly IgG) from bovine or mouse sera, tuberculin-purified protein derivative, and synthetic polypeptides such as poly-L-lysine and poly-L-glutamic acid.

KLH

Perhaps the most popular carrier protein is KLH. The hemocyanin from keyhole limpets (the mollusk *Megathura crenulata*) is the oxygen-carrying protein of these primitive sea creatures. KLH is an extremely large, multisubunit protein that contains chelated copper of non-heme origin. In concentrated solutions above pH 7, it displays a characteristic opalescent blue color that betrays its near insolubility and copper prosthetic groups. In acidic solutions the blue color changes to green. At physiological pH the protein exists in various subunit aggregate states of large molecular weight. For instance, in Tris buffer at pH 7.4 it is known to associate in five different aggregate forms (Senozan *et al.*, 1981). In highly alkaline or acidic environments, KLH disassociates into subunits (Hersckovits, 1988). The protein exhibits increased immunogenicity when it is disassociated into subunits, probably due to exposure of additional epitopic sites to the immune system (Bartel and Campbell, 1959). The intact protein usually creates considerable light-scattering or iridescent effects due to its size and almost colloidal nature in aqueous solutions.

Since keyhole limpets are marine creatures existing in a high-salt environment, KLH maintains its best stability and solubility in buffers containing at least 0.9 *M* NaCl (not 0.9%). As the concentration of NaCl is decreased below about 0.6 *M*, the protein begins to precipitate and denature. Conjugation reactions using KLH, therefore, should be done under high-salt conditions to preserve the solubility of the hapten–carrier complex.

KLH also should not be frozen. Freeze–thaw effects cause extensive denaturation and result in considerable amounts of insolubles. Commercial preparations of KLH are typically freeze-dried solids that no longer fully dissolve in aqueous buffers and do not display the protein's typical blue color due to loss of chelated copper. The partial denatured state of these products often makes conjugation reactions difficult. Pierce Chemical is the only commercial source of KLH that includes special (proprietary) stabilizers to provide the protein in a lyophilized form that is almost completely soluble upon reconstitution and with its blue copper-binding characteristics still intact. Reconstitution of the Pierce product with water yields a buffered solution ready for conjugation reactions.

KLH contains an abundance of functional groups available for conjugation with hapten molecules. On a per-mole basis (using an average MW of 5,000,000), KLH has over 2000 amines from lysine residues, over 700 sulfhydryls from cysteine groups, and over 1900 tyrosines. Activation of the protein with SMCC (Section 5) typically results in 300–600 maleimide groups per molecule for coupling to sulfhydryl-containing haptens.

The preparation of immunogen conjugates often requires the coupling of a sparingly soluble hapten to a carrier molecule. Predissolving the hapten in an organic solvent and adding an aliquot of this solution to an aqueous reaction mixture typically is done to maintain at least some solubility of the hapten in the conjugation solution. DMSO may be used for this purpose with KLH while maintaining very good solubility characteristics of the protein as well as the hapten. KLH is completely soluble in 50% (v/v) DMSO, becomes cloudy at a level of 60%, and definitely precipitates at 67%. Therefore, conjugation reactions may be done by adding a volume of aqueous KLH to an equal volume of hapten dissolved in DMSO. Care should be taken, however, to avoid buffer salt precipitation on addition of organic solvent.

BSA and cBSA

BSA (MW 67,000) and cationized BSA (cBSA) are highly soluble proteins containing numerous functional groups suitable for conjugation. Even after extensive modification with hapten molecules these carriers usually retain their solubilities. The exception to this statement is when hydrophobic peptides or other sparingly soluble molecules are conjugated to the proteins. Modification of any carrier with hydrophobic haptens may cause enough masking of the hydrophilic surface to result in precipitation. Depending on the degree of precipitation, such conjugates often are useful in generating an immune response. To limit the production of insoluble complexes, however, the conjugation reaction can be sealed back to reduce the level of carrier modification.

BSA possesses a total of 59 lysine ε-amine groups (with only 30–35 of these typically available for derivatization), 1 free cysteine sulfhydryl (with 17 disulfides buried within its three-dimensional structure), 19 tyrosine phenolate residues, and 17 histidine imidazolides. The presence of numerous carboxylate groups gives BSA its net negative charge (pI 5.1).

Cationized BSA is prepared by modification of its carboxylate groups with ethylene diamine (Chapter 1, Section 4.3) (Fig. 271). Controlled aminoethylation using the water-soluble carbodiimide EDC results in blocking many of BSA's aspartic and glutamic acid side chains (and possibly the C-terminal carboxylate), forming an amide bond with an alkyl spacer containing a terminal primary amine group. Since the negative charge contributions of the native carboxylates are masked and positively charged amines are created in their place, the result of this process is a significant rise in the protein's pI. Cationization performed according to published procedures alters the net charge of BSA from a pI of about 5.1 (Cohn *et al.*, 1947) to over pI 11 (Muckerheide *et al.*, 1987a).

The highly positive charge of cBSA dramatically increases its immunogenicity. The positive character of the molecule aids in its binding to antigen presenting cells (APC) *in vivo*, the first step in antibody production. The protein thus gets incorporated into the APCs faster than molecules having lower pH values. It also gets processed at an

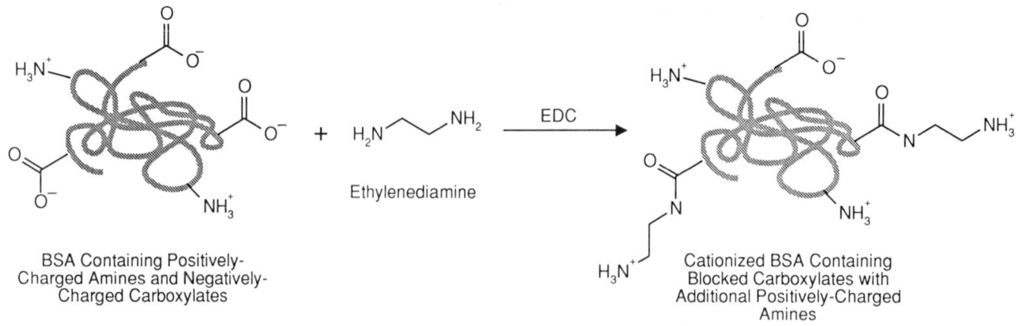

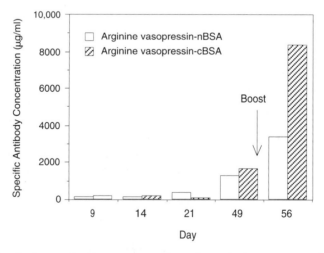

Figure 271 Cationized BSA is formed by the reaction of ethylene diamine with bovine serum albumin using the water-soluble carbodiimide EDC. Blocking of the carboxylate groups on the protein combined with the addition of terminal primary amines raises the pI of the molecule to highly basic values.

accelerated rate, producing a quicker immune response, and one that occurs with greater concentrations of specific antibody (Muckerheide *et al.*, 1987b; Domen *et al.*, 1987; Apple *et al.*, 1988; Domen and Hermanson, 1992).

Cationized BSA used as a carrier protein also induces a similar increase in the production of antibody against any attached hapten molecules. Even when haptens are coupled through cBSA's amine residues, the overall charge of the molecule remains basic enough to augment the immune response beyond that usually obtained using other carriers. This augmentation occurs even when the attached molecule is not merely a hapten, but a larger antigen macromolecule. Conjugation of a complete antigen (a molecule able to generate an immune response on its own) to cBSA causes an increased immune response against the antigen beyond that normally obtainable

Figure 272 The effectiveness of cBSA as an immunogen can be seen by the comparison of specific antibody response in mice to arginine vasopressin coupled to both native and cationized BSA. The quantity injected was standardized according to the amount of arginine vasopressin present. The cationized carrier results in higher concentrations of antibody produced against the peptide than the immunogen made with native BSA.

for the native antigen administered in unconjugated form (Domen and Hermanson, 1992).

The effectiveness of cBSA as a carrier for peptides was investigated using arginine vasopressin (AV) as the hapten. Figure 272 shows the antibody concentration resulting after injection of the AV–cBSA conjugate intraperitoneally (ip) into BDF_1 female mice. As a control, native BSA was similarly conjugated with AV and administered in a second set of mice under identical conditions. The antibody concentration in the sera were monitored periodically by ELISA. The antibody response resulting from a set of mice injected with unconjugated peptide was subtracted in all cases. All injections were done using 100 μg of conjugate mixed with an equal volume of alum (22.5 mg/ml aluminum hydroxide) as adjuvant.

After the boost, the group of mice receiving the AV–cBSA conjugate generated over twice the antibody response as the group receiving the peptide conjugated to native BSA.

In a similar study, OVA conjugated to cBSA was compared to the same protein conjugated to native BSA (nBSA) and also OVA administered in an unconjugated form in mice. Figure 273 shows that before and after the boost, the OVA–cBSA conjugate resulted in much higher antibody concentrations than either the OVA–nBSA conjugate or OVA injected in an unconjugated form. Similar results were obtained for a conjugate of human IgG with cBSA (Fig. 274).

A corollary to the use of cBSA as a carrier protein is that its increased immune response often abrogates the use of complete Freund's adjuvant, now a source of concern because of its potential side effects in animals. A relatively innocuous mixture with alum is usually all that is required as adjuvant to result in good antibody production.

As mentioned previously for KLH, DMSO may be used to solubilize hapten molecules that are rather insoluble in aqueous environments. Conjugation reactions may be done in solvent/aqueous phase mixtures to maintain some solubility of the hapten

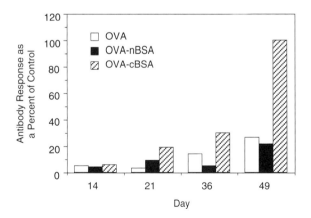

Figure 273 Cationized BSA even can increase the specific antibody response to large proteins coupled to it. This graph shows a comparison of the relative antibody response in mice to injections of ovalbumin, either in an unconjugated form or conjugated to native or cationized BSA. The quantity injected was standardized according to the amount of ovalbumin present. The highly basic cBSA molecule modulates the immune response to enhance the production of antibodies toward even proteins conjugated with it.

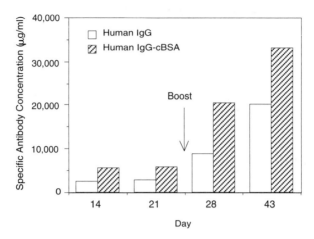

Figure 274 Human IgG was injected in mice either in an unconjugated form or cross-linked with cBSA. The quantity injected was standardized according to the amount of IgG present. A greater antibody response was obtained using the cBSA conjugate.

once it is added to a buffered solution. BSA remains soluble in the presence of up to 35% DMSO, becomes slightly cloudy at 40%, and precipitates at 45% (v/v).

Thyroglobulin and OVA

Thyroglobulin and ovalbumin (OVA) are used less often as carriers, but they are particularly valuable as nonrelevant carriers in ELISA tests designed to measure the antibody response after injection of an immunogen conjugate. Since an antibody response would be directed against both the carrier and the attached hapten, an ELISA done to quantify a specific antibody that only interacts with the hapten must not utilize the same carrier in the conjugate coated on the microplates. If the carrier conjugate used for the ELISA is identical to that used in the original immunization, the test results will be skewed by the contribution of carrier-specific antibodies. For this reason, a nonrelevant carrier—one that is not recognized by the products of the immune response—must be coupled with hapten and used for the ELISA test.

Since OVA and BSA possess some immunologically similar epitopes, a population of the antibodies produced against one often will cross-react against the other. Therefore, OVA cannot function as a nonrelevant carrier for BSA and vice versa. Either OVA or BSA, however, may be used as nonrelevant carriers for KLH, thyroglobulin, or the various toxoid proteins used as immunogen conjugates.

OVA makes up about 75% of the total protein in hen egg whites. The protein contains 20 lysine residues and 14 aspartic acid and 33 glutamic acid groups. This gives a total of 20 ε-amines, 1 N-terminal amine, 47 side-chain carboxylates, and 1 C-terminal carboxylate for conjugation reactions. The majority of acidic groups gives the protein a pI of 4.63. Additional sites of modification include 4-sulfhydryl groups, 10 tyrosines, and 7 histidine residues. OVA is sensitive to temperature (above 56°C), electric fields, and vigorous mixing. Care should be taken in handling the protein to prevent denaturation and subsequent precipitation.

One advantage of OVA is its extreme solubility characteristics in the presence of DMSO. A sparingly soluble hapten molecule may be dissolved in this solvent and added to an aqueous OVA reaction mixture to maintain solubility of the molecule during conjugation. OVA is soluble at up to 70% DMSO, becomes cloudy at 75%, and precipitates at 80% (v/v).

Thyroglobulin is a prohormone protein that is synthesized and stored in the thyroid gland. Specific proteolytic action on the protein *in vivo* causes the release of tri-iodothyronine and thyroxine, low-molecular-weight amino acid derivatives that affect metabolic rate and oxygen consumption. Thyroglobulin is a large, multisubunit protein composed of several polypeptide chains (MW 670,000). Its acidic pI (4.7) reflects the abundance of carboxylate groups. Thyroglobulin is also glycosylated, containing about 8–10% carbohydrate. Its use as an immunogen carrier protein is less frequent than that of KLH or BSA.

Tetanus and Diphtheria Toxoids

Toxoid proteins are biologically inactivated forms of native toxins. The most often used toxoid is tetanus toxoid, but diphtheria-derived toxoids and other proteins also are used occasionally (Anderson *et al.*, 1989). Tetanus toxoid (MW 150,000) has 106 amine groups, 10 sulfhydryls, 81 tyrosine residues, and 14 histidines that may participate in conjugation reactions with hapten molecules (Bizzini *et al.*, 1970). Diphtheria toxoid is derived from a protein secreted by certain strains of *Corynebacterium diphtheriae*. Its molecular weight is approximately 63,000 (Collier and Kandel, 1971). Both protein toxoids can be used to couple haptens through any of the chemical reactions described in this chapter. They generate strong immunological responses *in vivo*.

2.2. Liposome Carriers

Liposomes are artificial structures composed of phospholipid bilayers exhibiting amphiphilic properties (chapter 12). In complex liposome morphologies, concentric spheres or sheets of lipid bilayers are usually separated by aqueous regions that are sequestered or compartmentalized from the surrounding solution. The phospholipid constituents of liposomes consist of hydrophobic lipid tails connected to a head constructed of various glycerylphosphate derivatives. The hydrophobic interaction between the fatty acid tails is the primary driving force for creating liposomal bilayers in aqueous solutions.

The morphology of a liposome may be classified according to the compartmentalization of aqueous regions between bilayer shells. if the aqueous regions are sequestered by only one bilayer each, the liposomes are called unilamellar vesicles (ULV). If there is more than one bilayer surrounding each aqueous compartment, the liposomes are termed multilamellar vesicles (MLV). ULV forms are further classified according to their relative size, although rather crudely. Thus, there can be small unilamellar vesicles (SUV; usually less than 100 nm in diameter) and large unilamellar vesicles (LUV; usually greater than 100 nm in diameter). With regard to MLV, however, the bilayer structures cannot be easily classified due to the almost infinite number of ways each bilayer sheet can be associated and interconnected with the next one. MLVs typically form large complex honeycomb structures that are difficult to classify or reproduce.

The overall composition of a liposome—its morphology, composition (including a variety of potential phospholipids and the degree of its cholesterol content), charge, and any attached functional groups—can affect the antigenicity of the vesicle *in vivo* (Allison and Gregoriadis, 1974; Alving, 1987; Therien and Shahum, 1989). When liposomes are used as carriers for immunization purposes, the haptens or antigens usually are attached covalently to the head groups using various phospholipid derivatives and cross-linking chemistries (Derksen and Scherphof, 1985). Most often these derivatization reactions are done off of phosphatidylethanolamine constituents within the liposomal mixture. The primary amine modification off the glycerylphosphate head group of phosphatidylethanolamine provides an ideal functional group for activation and subsequent coupling of hapten molecules (Shek and Heath, 1983). Stock preparations of activated liposomes may be prepared and lyophilized to be used as needed in coupling hapten molecules (Friede *et al.*, 1993). All of the amine-reactive conjugation methods discussed in this section may be used with phosphatidylethanolamine-containing liposomes; however, see chapter 12 for a more complete discussion of the unique considerations associated with conjugation of molecules to liposomes.

2.3. Synthetic Carriers

Synthetic molecules may be used as carriers if they are designed with the appropriate functional groups to couple hapten molecules. These carriers may consist of simple polymers such as poly-L-lysine, poly-L-glutamic acid, Ficoll, dextran, or polyethylene glycol (Lee *et al.*, 1980; Fok *et al.*, 1982; Boyle *et al.*, 1983; Hopp, 1984; Wheat *et al.*, 1985). Coupling of hapten molecules to the principal functional groups of these polymers can produce immunogenic conjugates that may be injected in animals to generate a specific antibody response.

Poly-L-lysine may be coupled to carboxylate-containing molecules using the carbodiimide conjugation procedure to yield amide linkages (Chapter 3, Section 1.1, and Chapter 9, Section 3). Homobifunctional or heterobifunctional cross-linking agents also may be used with poly-L-lysine, such as in the use of sulfo-SMCC (Chapter 9, Section 5). The polymer can be used as well for coupling hapten molecules and subsequent coating of microplates for ELISA procedures (Gegg and Etzler, 1993). Conversely, poly-L-glutamic acid may be coupled to amine-containing haptens by the same carbodiimide protocol. Ficoll and dextran carriers may be activated by mild sodium periodate oxidation to generate reactive aldehyde groups (Chapter 1, Section 4.4, and Chapter 15, Section 2.1). Coupling to amine-containing haptens then may be done by reductive amination (Chapter 3, Section 4). Polyethylene glycol chemical reactions involve alternate activation and coupling schemes that are addressed in Chapter 15, Section 1.

A unique synthetic molecule that can be used as a carrier is the so-called multiple antigenic peptide (MAP) (Tam, 1988; Posnett *et al.*, 1988). The MAP core structure is composed of a scaffolding of sequential levels of poly-L-lysine. The matrix is constructed from a divalent lysine compound to which two additional levels of lysine are attached. The final MAP compound consists of a symmetrical, octavalent primary amine core to which hapten molecules may be attached. Coupling of up to eight peptide haptens to the MAP core yields a highly immunogenic complex having a molecular weight of typically greater than 10,000. The nature of the MAP carrier makes it ideal for remarkably defined conjugates useful in vaccine development.

One particularly novel carrier was reported to consist of 50- to 70-nm colloidal gold particles of the type often used in cytochemical labeling techniques for microscopy (Pow and Crook, 1993) (Chapter 14). Adsorption of peptide antigens onto gold and subsequent injection of the complex into rabbits in an adjuvant mixture resulted in rapid production of antibody of extremely high titer. The resultant antibodies could be used in immunocytochemistry at dilutions from 1 in 250,000 down to 1 in 1,000,000, which is orders of magnitude beyond the dilutions typically used with lower-titer antibodies.

3. Carbodiimide-Mediated Hapten–Carrier Conjugation

The coupling chemistry used to prepare an immunogen from a hapten and carrier protein is an important consideration for the successful production and correct specificity of the resultant antibodies. The choice of cross-linking methodology is governed by the functional groups present on both the carrier and the hapten as well as the orientation of the hapten desired for appropriate presentation to the immune system. An associated concern is the potential for antibody recognition and cross-reactivity toward the cross-linking reagent used to effect the conjugation. If antibodies are generated against the cross-linker bridge, then this may dilute the desired antibody response against the hapten. The use of a zero-length cross-linking procedure mediated by the water-soluble carbodiimide EDC eliminates this problem, since no bridging molecule is introduced between the hapten and the carrier.

The reactions involved in an EDC-mediated conjugation are discussed in Chapter 3, Section 1.1 [Note: EDC is 1-ethyl-3-(3-dimethylaminopropyl) carbodiimide hydrochloride; MW 191.7]. The carbodiimide first reacts with available carboxylic groups on either the carrier or hapten to form a highly active O-acylisourea intermediate. The activated carboxylic group then can react with a primary amine to form an amide bond, with release of the EDC mediator as a soluble isourea derivative. The reaction is quite efficient with no more than 2 h required for it to go to completion and form a conjugated immunogen.

Since most peptide haptens contain either amines or carboxylic groups available for coupling, EDC-mediated immunogen formation may be the simplest method for the majority of hapten–carrier protein conjugations. Figure 275 shows the coupling of a carrier protein to a short peptide molecule through its amine terminus. It should be kept in mind, however, that this type of conjugation may occur at either the C- or the N-terminal of the peptide or at any carboxyl- or amine-containing side chains. Therefore, this method probably should be avoided if a particularly interesting part of the peptide contains groups that may be blocked or undergo coupling using the carbodiimide reaction. Also, when using peptides rich in Lys, Glu, Arg, His, or Asp, unacceptable crosslinking of the hapten may occur on conjugation, and thus change the antigenic structure of the resulting immunogen. However, some cross-linking or polymerization of the peptide on the surface of the carrier actually may be beneficial to the immunogenicity of the peptide, and thus create an even greater antibody response.

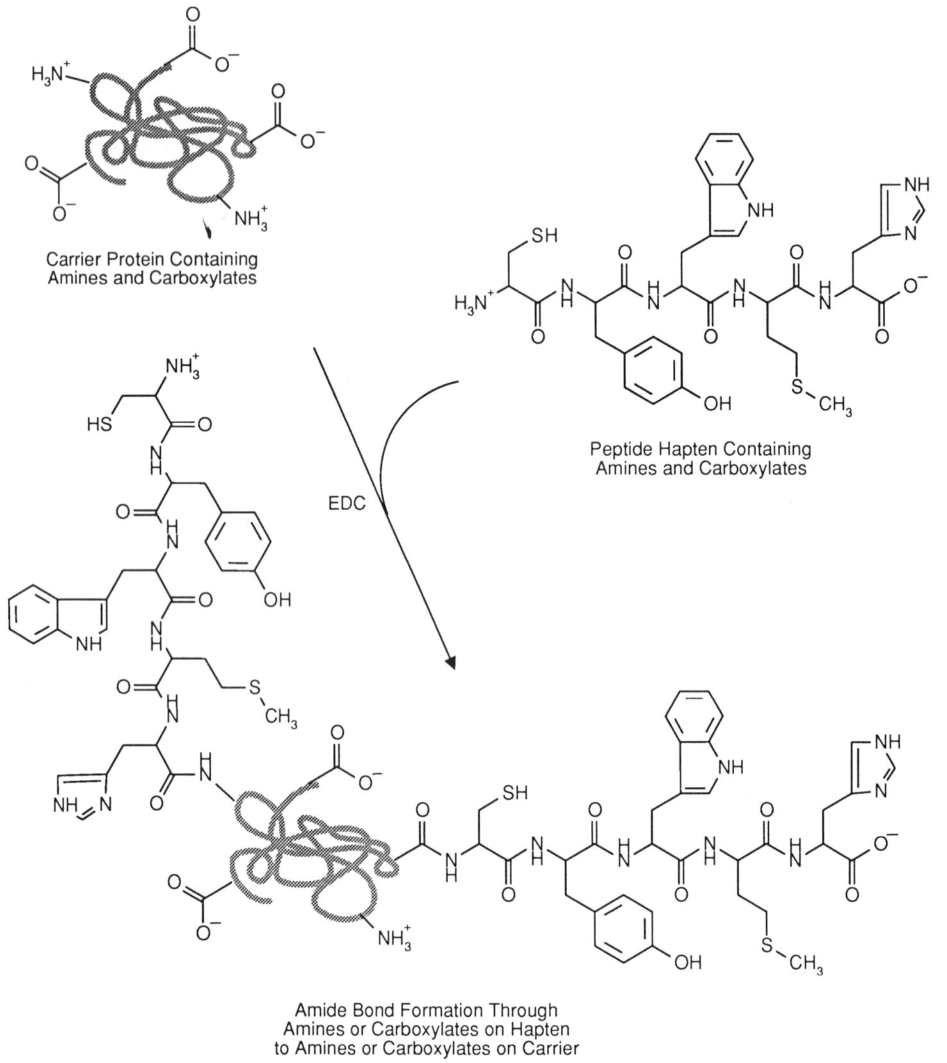

Carrier Protein Containing
Amines and Carboxylates

Peptide Hapten Containing
Amines and Carboxylates

EDC

Amide Bond Formation Through
Amines or Carboxylates on Hapten
to Amines or Carboxylates on Carrier

Figure 275   Peptide haptens are easily conjugated to carrier proteins using the water-soluble carbodiimide EDC.

Some investigators even advocate using no carrier protein when a peptide hapten is involved: merely polymerizing the peptide in the presence of EDC may result in a complex of high enough molecular weight to be immunogenic by itself. In general, EDC coupling is a very efficient, one-step method for forming a wide variety of peptide–carrier protein immunogens.

Figure 276 shows the results of an EDC conjugation study comparing a reaction done at pH 4.7 (A) to one done at pH 7.3 (B and C), with and without added sulfo-NHS (see Chapter 3, Section 1.2). The graphs show the elution profiles of a gel filtration separation after conjugation. In each case, a blank run done without the addition of EDC illustrates the separation of the protein carrier (the first peak) from

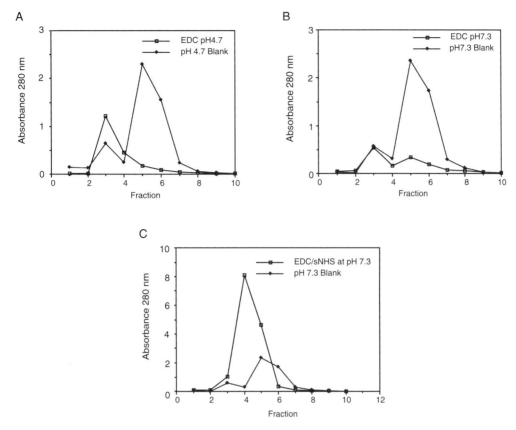

Figure 276 To assess the effectiveness of an EDC conjugation reaction of a peptide with a carrier protein, glycyl-tyrosine was coupled to BSA using various conditions. The graphs show a gel filtration profile on Sephadex G-25 after completion of the conjugation reaction. The first peak eluting off the column is the higher molecular weight carrier, while the second peak is excess peptide. The elution profiles demonstrate that the carbodiimide reaction proceeds with nearly equal efficiency at pH 4.7 (A) or pH 7.3 (B). In each graph, a comparison is shown between the separation of peptide and carrier without addition of EDC and the same mixture after reaction with EDC. Depletion of the peptide peak in the EDC-containing elution profiles indicates uptake of glycyl-tyrosine in the carrier conjugate. Some polymerization of peptide also is possible using this method, as evidence by movement of the peptide peak toward higher molecular weight elution points. Addition of sulfo-NHS to the reaction caused precipitation problems as well as obscuring the separation due to the absorbance of excess sulfo-NHS (C).

the lower molecular weight peptide and reagent peak (the second peak). Decrease in the peptide peak is indicative of successful conjugation. Complete recovery of the total absorbance at 280 nm usually does not occur, presumably due to a decrease in the peptide's absorptivity as it is conjugated or polymerized. Staros' method of adding sulfo-NHS to form an intermediate active ester that subsequently reacts with an amine to form the amide bond does not work as well due to excessive conjugation (causing precipitation in most cases) and interference from the eluting sulfo-NHS peak. The reaction proceeds with similar yields at either acid or neutral pH. Thus, the efficiency of an EDC conjugation reaction is approximately the same from pH 4.7 to physiological conditions.

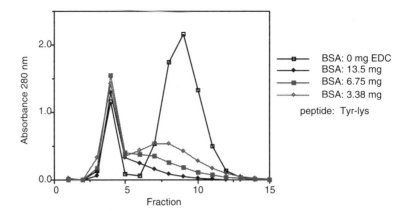

Figure 277 To study the conjugation of peptides to carriers using different levels of EDC, tyrosyl-lysine was conjugated to BSA and separated after the reaction by chromatography on a Sephadex G-25 column. As the EDC level was increased in the reaction, more peptide reacted and the peptide peak (the second peak) was depleted. The absorbance of the carrier peak (the first one) increases as more peptide is conjugated.

Figure 277 shows the result of the conjugation of the dipeptide tyrosyl-lysine to BSA using various concentrations of EDC. Again, the elution profile shows the gel filtration pattern resulting after the reaction. The first peak is the protein carrier while the second is the peptide. Progressive decrease in the peptide peak with increasing amounts of EDC added to the reaction mixture correlates to increased conjugation yields. A side reaction to EDC conjugation of haptens that contain both an amine and a carboxylate group is hapten polymerization. This is revealed in the movement of the peptide peak toward higher molecular weights (e.g., decreased time of elution) with increasing amounts of EDC added to the reaction.

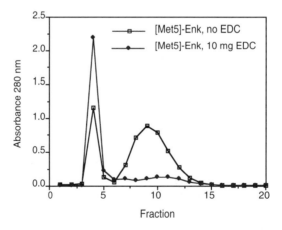

Figure 278 Conjugation of the biological peptide [Met5]-enkephalin to BSA using EDC. The graph shows the gel filtration profile (on Sephadex G-25) after completion of the conjugation reaction. A blank run with no added EDC was done to illustrate the peak absorbances that would be obtained if no conjugation took place. With addition of 10 mg of EDC to a reaction mixture consisting of 2 mg of BSA plus 2 mg of peptide, nearly complete conjugate formation was obtained.

Figure 278 illustrates the conjugation of [Met5]-enkephalin with BSA using EDC. The gel filtration profile after cross-linking reveals that the peptide peak effectively disappears upon complete conjugation with the carrier protein. With nicely soluble peptides such as this one, the immunogen remains freely soluble even at high modification levels. For less soluble peptides or haptens, reducing the amount of EDC addition may be necessary to maintain solubility in the conjugate.

To illustrate the similarity of an EDC conjugation reaction using a different carrier protein, but the same peptide, Fig. 279 shows the gel filtration separation after conjugation of [Met5]-enkephalin to OVA. The uptake of peptide on addition of EDC is almost identical to that observed when conjugating to BSA. This is logical, since on a per mass basis, there is very little difference between these proteins in the amount of amines or carboxylates available for conjugation.

Figure 280 shows the conjugation of tyrosyl-lysine to KLH using various concentrations of EDC. The elution profile shows the gel filtration pattern resulting after the reaction. Progressive decrease in the peptide peak (peak 2) with increasing amounts of EDC correlates to increased conjugation (or polymerization) yields. Despite the extremely high molecular weight of KLH compared to the other commonly used carriers, the conjugation reaction using EDC again proceeds with results virtually identical to those of the similar study shown in Fig. 277 using BSA as the carrier. In fact, superimposing the two studies on the same graph demonstrates the reproducibility of an EDC-facilitated reaction (Fig. 281).

Due to the high molecular weight of KLH and its solubility characteristics, the conjugation of this protein to some haptens can result in precipitation of the complex. This is especially true if the level of EDC addition is similar to the EDC concentrations used with lower molecular weight carriers such as BSA or OVA. Figure 282 shows the elution profile resulting from the gel filtration separation of KLH and the peptide [Met5]-enkephalin after an EDC reaction. To result in a soluble immunogen, only 0.1 to 0.2 times the amount of EDC was added, compared to similar BSA or OVA conjuga-

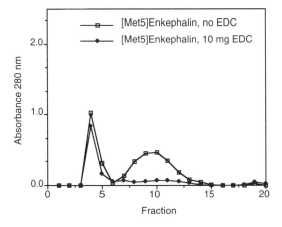

Figure 279 To illustrate the consistency of an EDC-mediated reaction, [Met5]-enkephalin was conjugated to ovalbumin using conditions identical to those described for BSA in Fig. 278. Note the similarity in the degree of conjugate formation.

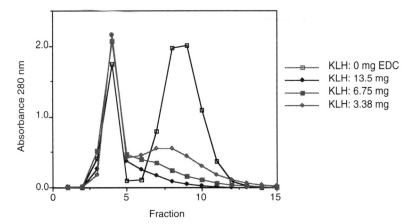

Figure 280 The conjugation of tyrosyl-lysine to KLH is illustrated by the gel filtration pattern on Sephadex G-25 after the reaction. The first peak is the carrier protein and the second peak is the peptide. A blank containing no EDC is also shown to provide baseline peak heights that would be obtained if no cross-linking occurred. When more EDC was added, more peptide was conjugated, as evidenced by peptide peak depletion.

tion reactions. Even at levels this low, however, the coupling of peptide to the carrier is very efficient and results in an excellent immunogen.

These studies using EDC-facilitated conjugations were done to develop an optimal protocol for the preparation of immunogens by carbodiimide cross-linking. For haptens (i.e., peptides) that display good solubility in aqueous solution, the level of reagent addition should result in a soluble immunogen conjugate. When using haptens that are sparingly soluble or insoluble in aqueous environments, the conjugation reaction may result in a precipitated complex. Precipitation often can be controlled by scaling back

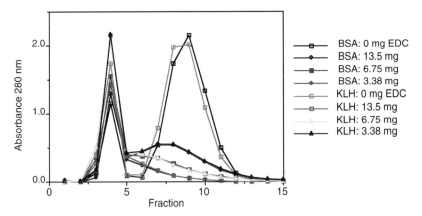

Figure 281 EDC conjugation reactions can be extraordinarily consistent using the same peptide crosslinked to two carrier proteins. This figure shows the gel filtration pattern on Sephadex G-25 after completion of the cross-linking reaction. Conjugation of tyrosyl-lysine to BSA and KLH are shown. The first peaks represent eluting carrier, while the second peaks are the excess peptide. Note the consistency of conjugation using the same levels of EDC addition.

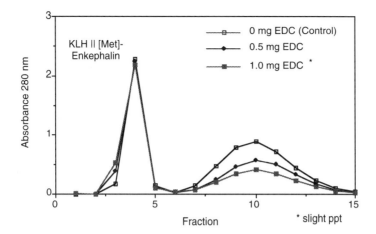

Figure 282 Conjugation to KLH often can cause precipitation due to the high molecular weight of the carrier protein. The conjugation of [Met5]-enkephalin to KLH yields a soluble immunogen if the level of EDC addition is about 0.1 times that typically used with BSA as a carrier. The figure shows the gel filtration pattern on Sephadex G-25 after completion of the cross-linking reaction. The first peak is KLH and the second peak is excess peptide. Depletion of the peptide peak correlates to hapten–carrier conjugation.

the level of EDC addition or limiting the time of the reaction. If a precipitated immunogen is not a problem (most precipitated, high-molecular-weight conjugates are very immunogenic), then the following protocol is applicable to the great majority of peptide–carrier protein conjugations. Pierce Chemical offers a kit containing all the reagents necessary for an EDC-mediated hapten–carrier conjugation.

Protocol

1. Dissolve the carrier protein in 0.1 *M* MES, 0.15 *M* NaCl, pH 4.7, at a concentration of 10 mg/ml. If using KLH, increase the NaCl concentration of all buffers to 0.9 *M* (yes, 0.9 *M*, not 0.9%) to maintain solubility of the protein. For neutral pH conjugations, substitute 0.1 *M* sodium phosphate, 0.15 *M* NaCl, pH 7.2, for the MES buffer.
2. Dissolve up to 4 mg of the peptide or hapten to be coupled in 1 ml of the reaction buffer chosen in step 1. If the peptide to be coupled is already in solution, it may be used directly if it is in a buffer containing no other amines or carboxylic acids and is at a pH between 4.7 and 7.2. Note: If an assessment of the degree of peptide coupling is desired, measure the absorbance at 280 nm of the 1 ml peptide solution before proceeding to step 3. In some cases, a dilution of the peptide solution may be necessary to keep the absorbance on scale for the spectrophotometer. If the peptide is sparingly soluble in aqueous solution, it may be dissolved in DMSO and an aliquot added to the carrier solution. See the previous discussion on carrier proteins to determine the levels of DMSO compatible with carrier protein solubility.
3. Add 500 μl of the peptide solution to 200 μl of carrier protein. For greater reaction volumes, keep the ratio of peptide/carrier addition the same and proportionally scale up the amount of EDC added in the next step. If the peptide is

initially dissolved in DMSO, much less peptide volume compared to protein volume should be used to maintain solubility (see discussion in step 2).

4. For conjugations using relatively low-molecular-weight proteins, such as BSA or OVA, add the peptide/carrier solution to a vial containing 10 mg of EDC (Pierce) and gently mix to dissolve. For high-molecular-weight KLH immunogens, first dissolve one vial containing 10 mg of EDC in 1 ml of deionized water, and immediately transfer 50 μl of this solution to the carrier/peptide solution. Gently mix.

5. Allow the reaction to continue at room temperature for 2 h.

Note: Although the conjugation protocols have been optimized by preparing a number of different peptide–carrier conjugates, some peptide sequences or other haptens may cause precipitation of the carrier upon coupling. This may occur as a result of changing the carrier's solubility characteristics through surface modification or due to polymerization. A small amount of precipitation is not a problem and can easily be removed by centrifugation before the gel filtration step. If severe precipitation occurs, however, the amount of EDC added to the reaction may have to be scaled back to eliminate or reduce it. With BSA or OVA conjugates, this may mean using as little as 1–3 mg of EDC instead of the recommended 10 mg. With KLH as a carrier, reducing the EDC levels to 0.1 mg may be necessary.

6. Purify the hapten–carrier conjugate by gel filtration or dialysis.

4. NHS Ester-Mediated Hapten–Carrier Conjugation

Hapten–carrier conjugation may be accomplished by the use of homobifunctional reagents containing NHS ester groups on both ends. The active esters are highly reactive toward amine functional groups on proteins and other molecules to form stable amide bonds. Cross-linking agents of various lengths may be used for this conjugation strategy, including the sulfo-NHS ester analogs, which are more water-soluble than the NHS esters without a sulfonic acid group (Chapter 4, Section 1).

Using homobifunctional NHS esters, amine-containing haptens may be conjugated to amine-containing carriers in a single step (Fig. 283). The carrier is dissolved in a buffer having a pH of 7–9 (0.1 M sodium phosphate, pH 7.2 works well). The hapten molecule is added to this solution at a suitable molar excess to assure multipoint attachment of the hapten to the carrier. A molar excess of 20–30 times that of the carrier concentration is a good starting point. Next, the NHS ester cross-linker is added to the solution to provide at least a threefold molar excess over that of the hapten. For cross-linkers insoluble in aqueous solution, first solubilize them in DMF or DMSO at higher concentration, and then add an aliquot of this stock solution to the hapten–carrier solution. The conjugation reaction is complete within 2 h at room temperature. Some adjustment of the level of hapten and cross-linker addition may be necessary to avoid extensive precipitation of the conjugate, especially when using rather hydrophobic hapten molecules.

Another method of NHS ester mediated hapten–carrier conjugation is to create reactive sulfo-NHS esters directly on the carboxylates of the carrier protein using the EDC/sulfo-NHS reaction described in Chapter 3, Section 1.2. A carbodiimide reaction in the presence of sulfo-NHS activates the carboxylate groups on the carrier protein to form amine-reactive sulfo-NHS esters. The activation reaction is done at pH

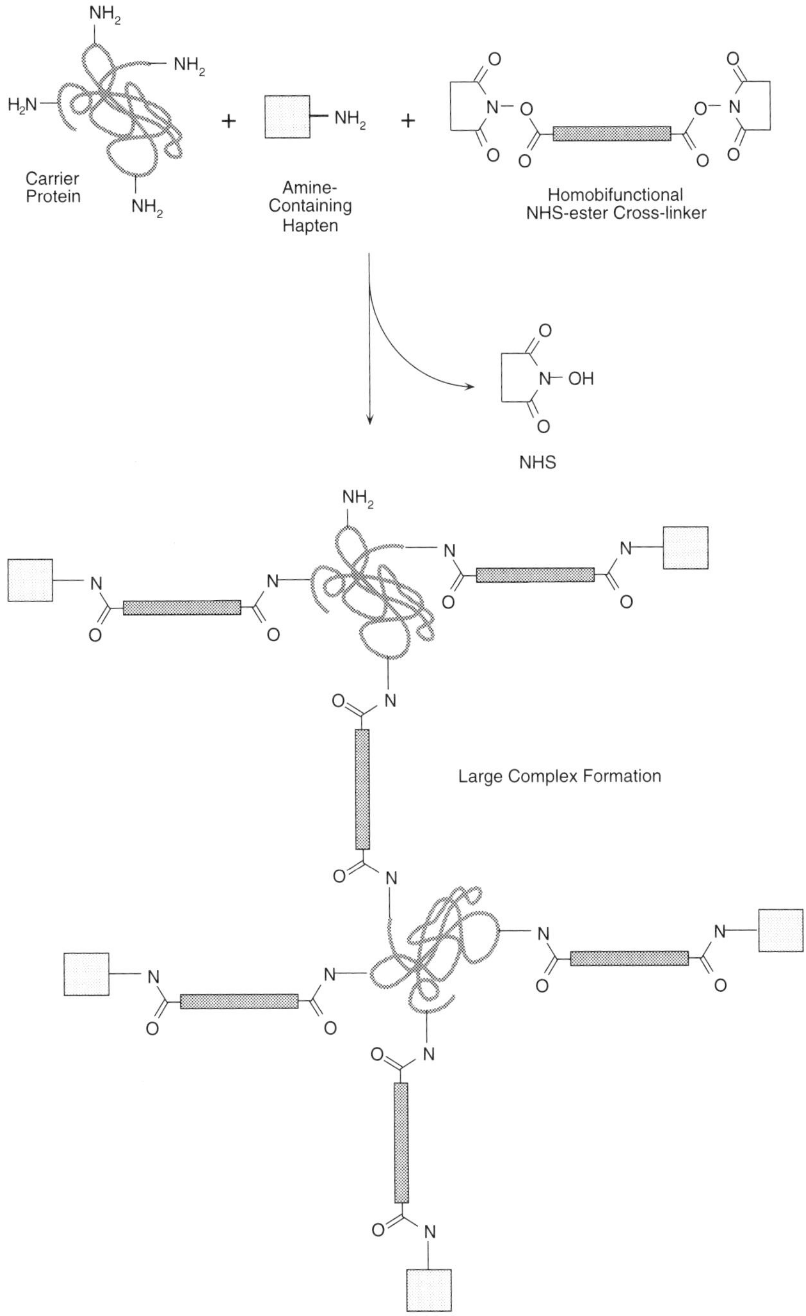

Figure 283 Hapten–carrier immunogen conjugates can be formed using homobifunctional NHS ester cross-linkers. The reaction may create large polymeric complexes, some of which could precipitate.

6, since the amines on the protein will be protonated and therefore be less reactive toward the sulfo-NHS esters that are formed. In addition, the hydrolysis rate of the esters is dramatically slower at acid pH. Thus, the active species may be isolated in a reasonable time frame without significant loss in conjugation potential. To quench unreacted EDC, 2-mercaptoethanol is added to form a stable complex with the remaining carbodiimide, according to Carraway and Triplett (1970). In the following protocol, a modification of the Grabarek and Gergely (1990) two-step method, sulfo-NHS is used instead of NHS so that active ester hydrolysis is slowed even more (Thelen and Deuticke, 1988; Anjaneyulu and Staros, 1987). Subsequent conjugation with amine-containing hapten molecules yields hapten–carrier conjugates created by amide bond formation (Fig. 284).

Protocol

1. Dissolve the carrier protein to be activated in 0.05 *M* MES, 0.5 *M* NaCl, pH 6 (reaction buffer), at a concentration of 1 mg/ml.
2. Add to the solution in step 1 a quantity of EDC and sulfo-NHS (both from Pierce) to obtain a concentration of 2 m*M* EDC and 5 m*M* sulfo-NHS. To aid in aliquoting the correct amount of these reagents, they may be quickly dissolved in

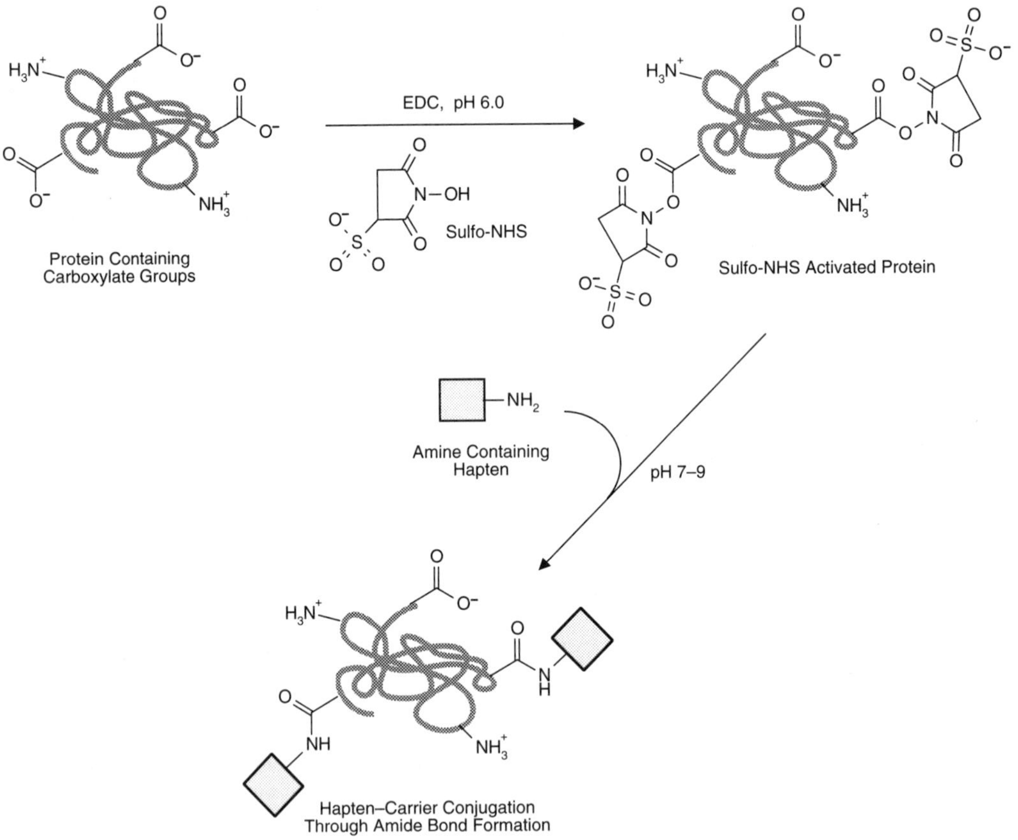

Figure 284 The carbodiimide EDC can be used in the presence of sulfo-NHS to create reactive sulfo-NHS ester groups on a carrier protein. Subsequent coupling with an amine-containing hapten can be done to create amide bond linkages.

water at a higher concentration, and then immediately a volume pipetted into the protein solution to obtain the proper molar quantities.

3. Mix and react for 15 min at room temperature to form the sulfo-NHS esters.

4. Add 2-mercaptoethanol to the reaction solution to obtain a final concentration of 20 mM. Mix and incubate for 10 min at room temperature. Note: if the protein being activated is sensitive to this level of 2-mercaptoethanol, instead of quenching the reaction chemically, the activation may be terminated by desalting (see step 5).

5. If the reaction was quenched by the addition of 2-mercaptoethanol, the activated protein may be added directly to an amine-containing hapten molecule for conjugation. Alternatively, or if no 2-mercaptoethanol was added, the activated protein may be purified from reaction by-products by gel filtration using Sephadex G-25 or equivalent. The desalting operation should be done rapidly to minimize hydrolysis and recover as much of the active ester functional group as possible. The use of centrifugal spin columns of some sort may afford the greatest speed in separation. After purification, add the activated protein to the hapten for conjugation. The hapten molecule should be dissolved in 0.1 M sodium phosphate, pH 7.5.

6. React for at least 2 h at room temperature.

7. Remove excess reactants by gel filtration or dialysis.

5. NHS Ester–Maleimide Heterobifunctional Cross-linker-Mediated Hapten–Carrier Conjugation

A common method for coupling haptens to carrier proteins involves the use of a heterobifunctional cross-linker containing an NHS ester and a maleimide group. This type of cross-linker allows better control over the coupling process than homobifunctional or zero-length conjugation methods by incorporating a two- or three-step reaction strategy directed against two different functional targets. In this approach, the carrier protein first is activated with the cross-linker through its amine groups, purified to remove excess reactants, and then cross-linked to a hapten molecule containing a sulfhydryl group. One of the most useful reagents for this conjugation approach is sulfo-SMCC.

The reactions associated with a sulfo-SMCC conjugation are shown in Fig. 285 [note: sulfo-SMCC is sulfosuccinimidyl-4-(N-maleimidomethyl) cyclohexane-1-carboxylate; MW, 436.37] (see Chapter 5, Section 1.3). This crosslinking reagent mediates the conjugation of a carrier protein through its primary amine groups to a peptide or other hapten through sulfhydryl groups. The active N-hydroxysulfosuccinimide ester (sulfo-NHS) end of sulfo-SMCC first is reacted with available primary amine groups on the carrier protein. This reaction results in the formation of an amide bond between the protein and the cross-linker with the release of sulfo-NHS as a by-product. The carrier protein is then isolated by gel filtration to remove excess reagents. At this stage, the purified carrier possesses modifications generated by the cross-linker resulting in a number of reactive maleimide groups projecting from its surface. The maleimide portion of sulfo-SMCC is a thiol-reactive group that can be used in a secondary step to conjugate with a free sulfhydryl (i.e., a cysteine residue) on a peptide or other hapten, resulting in a stable thioether bond.

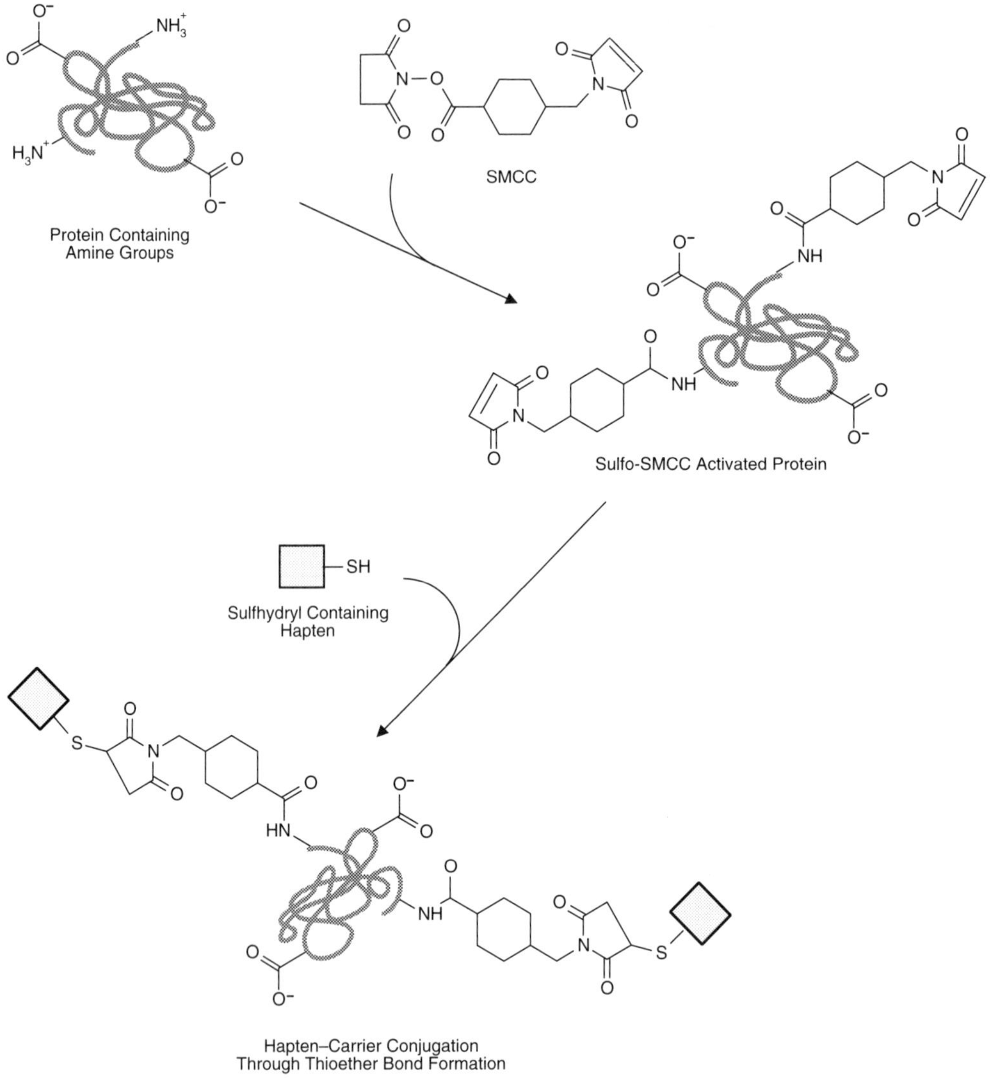

Figure 285   A common way of conjugating sulfhydryl-containing haptens to carrier proteins is to activate the carrier with sulfo-SMCC to create an intermediate maleimide derivative. The maleimide groups then can be coupled to thiols to form thioether bonds.

The use of sulfo-SMCC over the other common maleimide-containing cross-linkers such as MBS or SMPB provides the advantages of water solubility during the activation step and increased stability of the maleimide group prior to conjugation with a peptide. The improved stability ensures that the majority of the maleimide groups substituted on the carrier will survive the subsequent purification process without degradation. The relatively good stability of the maleimide group of sulfo-SMCC is probably due to the neighboring steric effects of its cyclohexane ring. The faster hydrolysis rates of other maleimide-type cross-linkers can be a significant problem, since they readily break down to the maleamic acid form, which is no longer reactive toward sulfhydryls (Fig. 286).

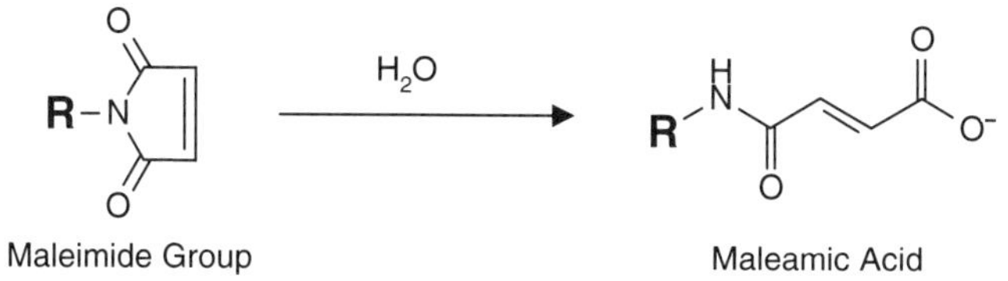

Figure 286 A maleimide group may hydrolyze in aqueous solution to an open maleamic acid form that is unreactive with sulfhydryls.

A disadvantage to using SMCC or other NHS–maleimide-type cross-linkers with hindered ring structures (such as MBS) is the relatively high immunogenicity of the cross-bridge. Studies have shown that a hapten–carrier complex formed from such cross-linkers generates significant antibody response against the spacer group itself, not just the hapten and carrier. To minimize the antibody population directed against the cross-bridge of the conjugate, the use of aliphatic straight-chain spacers will exhibit the lowest immunogenicity (Peeters *et al.*, 1989). Although SMCC (or sulfo-SMCC) is used in the following protocol, substitution of GMBS (or sulfo-GMBS) (Chapter 5, Section 1.7) will limit the immune response to the cross-linker.

Since many peptides do not naturally contain cysteine residues with free sulfhydryls, a terminal cysteine may be incorporated during peptide synthesis, or where appropriate, disulfide groups may be reduced to generate them. Alternatively, a thiolating reagent such as 2-iminothiolane (Traut's reagent) can be used to modify existing amino groups and introduce a sulfhydryl (see Chapter 1, Section 4.1). Caution must be taken when using this last technique, however, because multiple sites of modification may alter the immunogenic structure of the hapten.

If a terminal cysteine residue is added to a peptide during its synthesis, its sulfhydryl group provides a highly specific conjugation site for reacting with a sulfo-SMCC-activated carrier. All peptide molecules coupled using this approach will display the same basic conformation after conjugation. In other words, they will have a known and predictable orientation, leaving the majority of the molecule free to interact with the immune system. This method therefore can preserve the major epitopes on a peptide while still enhancing the immune response to the hapten by being covalently linked to a larger carrier protein. In addition, the well-known chemical reactivity of a sulfo-SMCC-mediated immunogen preparation permits covalent conjugation in a controllable fashion that can be highly defined for quality assurance purposes.

The process of carrier activation by sulfo-SMCC may be followed by performing a simple purification step after the reaction using Sephadex G-25. Figure 287 shows the gel filtration profiles for the separation of sulfo-SMCC-activated BSA and OVA. The first peak of both separations represents the elution point for the carrier protein, while the absorbance due to reaction by-products of the crosslinker is contained in the second peak (shown only as its leading edge). Activated proteins exhibit an increase in their absorbance at 280 nm over an identical sample with no added sulfo-SMCC due to their covalently attached maleimide groups. After isolation, the activated protein may be frozen and lyophilized to preserve maleimide coupling activity toward sulfhydryl-containing haptens. Pierce Chemical sells a number of maleimide-activated carrier proteins in lyophilized form for easy hapten conjugation.

After a carrier protein has been activated with sulfo-SMCC it is often useful to

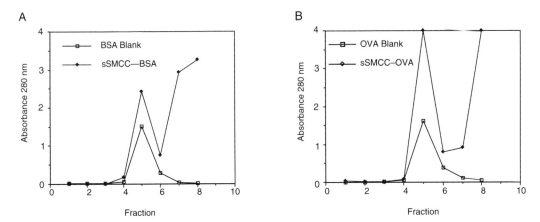

Figure 287   Carrier proteins may be activated with sulfo-SMCC to produce maleimide derivatives reactive with sulfhydryl-containing molecules. The graphs show the gel filtration separation on Sephadex G-25 of maleimide-activated BSA (A) and OVA (B) after reaction with sulfo-SMCC. The first peak is the protein and the second peak is excess cross-linker. The maleimide groups create increased absorbance at 280 nm in the activated proteins.

measure the degree of maleimide incorporation prior to coupling an expensive hapten. Ellman's reagent may be used in an indirect method to assess the level of maleimide activity of sulfo-SMCC-activated proteins and other carriers. First, a sulfhydryl-containing compound such as 2-mercaptoethanol or cysteine is reacted in excess with the activated protein. The amount of unreacted sulfhydryls remaining in solution then is determined using the Ellman's reaction (Chapter 1, Section 4.1). Comparison of the response of the sample to a blank reaction using the native, nonactivated protein at

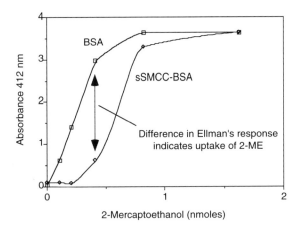

Figure 288   An Ellman's assay may be done to determine the maleimide activation level of SMCC-derivatized proteins. Reaction of the activated carrier with different amounts of 2-mercaptoethanol results in various levels of sulfhydryls remaining after the reaction. Detection of the remaining thiols using an Ellman's assay indirectly indicates the amount of sulfhydryl uptake into the activated carrier. Comparison of the Ellman's response to the same quantity of 2-mercaptoethanol plus an unactivated carrier indicates the absolute amount of sulfhydryl that reacted. Calculation of the maleimide activation level then can be done.

the same concentration and a series of standards made from a serial dilution of the sulfhydryl compound employed in the assay gives the amount of sulfhydryl compound conjugated and thus an estimate of the original maleimide activity.

Figure 288 shows a plot of the results of such an assay done to determine the maleimide content of activated BSA. This particular assay used 2-mercaptoethanol, which is relatively unaffected by metal-catalyzed oxidation. For the use of cysteine or cysteine-containing peptides in the assay, however, the addition of EDTA is required to prevent disulfide formation. Without the presence of EDTA at 0.1 M, the metal contamination of some proteins (especially serum proteins such as BSA) is so great that disulfide formation proceeds preferential to maleimide coupling. Figure 289 shows a similar assay for maleimide-activated BSA using the more innocuous cysteine as the sulfhydryl-containing compound.

Using this type of cysteine-uptake assay, it is possible to determine the percentage of maleimides that reacted over time. Thus, an indication of the reaction efficiency of a sulfhydryl-containing compound coupling with a maleimide-activated protein may be determined. Figure 290 shows the reaction rate for the coupling of cysteine to maleimide-activated BSA. Note that maximal coupling is obtained in less than 2 h, with over 80% yield in under 30 min.

The following protocol describes the activation of a carrier molecule with sulfo-SMCC and its subsequent conjugation with a hapten. The preactivated carriers containing maleimide groups ready for coupling to a sulfhydryl-containing compound are now commercially available in a stable freeze-dried form (Pierce Chemical). Substitution of GMBS (Chapter 5, Section 1.7) in the following protocol will provide a straight-chain aliphatic spacer with less immunogenicity than the ring structure of SMCC's cross-bridge.

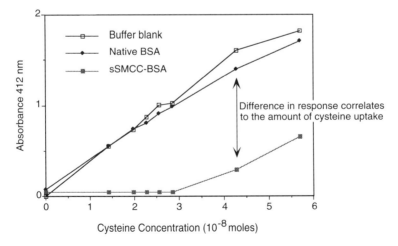

Figure 289 Cysteine also may be used in an Ellman's assay to determine the maleimide activation level of SMCC-derivatized proteins. Reaction of the activated carrier with different amounts of cysteine results in various levels of sulfhydryls remaining after the reaction. The coupling must be done in the presence of EDTA to prevent metal-catalyzed oxidation of sulfhydryls. Detection of the remaining thiols using an Ellman's assay indirectly indicates the amount of sulfhydryl uptake into the activated carrier. Comparison of the Ellman's response to the same quantity of cysteine plus an unactivated carrier indicates the absolute amount of sulfhydryl that reacted. Calculation of the maleimide activation level then can be done.

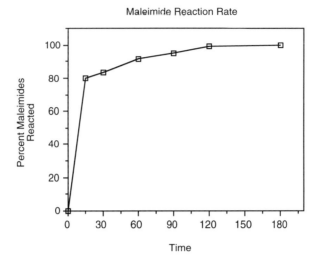

Figure 290   The rate of reaction of cysteine with maleimide-activated BSA was determined using an Ellman's assay for remaining sulfhydryl groups after the reaction, according to Fig. 289. Nearly all of the available maleimides are coupled with sulfhydryls within 2 h.

Protocol

1. Dissolve the carrier of choice at a concentration of 10 mg/ml in 0.1 M sodium phosphate, 0.15 M NaCl, pH 7.2 (activation buffer). Note: For use of KLH, increase the NaCl concentration to 0.9 M.
2. Dissolve sulfo-SMCC (Pierce) at a concentration of 10 mg/ml in the activation buffer. Immediately transfer the appropriate amount of this cross-linker solution to the vial containing the dissolved carrier protein.

Note: The amount of cross-linker solution to be transferred is dependent on the level of activation desired. Suitable activation levels can be obtained for the following proteins by adding the indicated quantities of the sulfo-SMCC solution. The degree of sulfo-SMCC modification often determines whether the carrier will maintain solubility after activation and coupling to a hapten. KLH, in particular, is sensitive to the amount of cross-linker addition. KLH usually retains solubility at about 0.1–0.2 times the mass of cross-linker added to BSA. This level of addition still results in excellent activation yields, since KLH is significantly larger than most of the other protein carriers.

Add the following quantities of sulfo-SMCC solution to each ml of carrier protein solution:

    a. BSA,   500 μl
    b. cBSA, 200 μl
    c. OVA,   500 μl
    d. KLH,   100 μl

Carriers having molecular weights similar to that of BSA or OVA may be activated at the same level with good success. Cationized BSA requires less cross-linker addition due to its greater quantity of amines present.

3. React for 1 h at room temperature.

4. Immediately purify the activated carrier protein by gel filtration using a volume of Sephadex G-25 equal to 15 times the volume of the activation reaction. To perform the chromatography use 0.1 M sodium phosphate, 0.15 M NaCl (0.9 M for KLH), 0.1 M EDTA, pH 7.2 (conjugation buffer). The EDTA is present to prevent metal-catalyzed sulfhydryl oxidation to disulfides. This is a particular problem when using BSA due to contaminating iron from hemolysis. Concentrations less than 0.1 M EDTA will not fully inhibit the oxidation reaction, especially if a cysteine containing peptide is to be conjugated to the activated carrier. Apply the sSMCC/carrier reaction mixture to the column while collecting 0.5- to 1-ml fractions. Pool the fractions containing the activated carrier (the first peak to elute from the column) and discard the fractions containing excess sulfo-SMCC (the second peak). The activated carrier should be used immediately or freeze-dried to maintain maleimide stability.

5. Dissolve a sulfhydryl-containing hapten or peptide to be conjugated at a concentration of 10 mg/ml in the conjugation buffer. Other hapten concentrations may be used, depending on its solubility. If an excess of the peptide solution is made at this time, an estimate of the degree of conjugation may be determined later (see section below). Add this solution to the pooled fractions containing the activated carrier at an equivalent mass ratio (1 mg hapten/mg of the carrier). Alternatively, the peptide may be added in solid form directly to the activated carrier solution if it is known to be freely soluble and can be weighed out in the appropriate quantity.

6. Allow the conjugation reaction to proceed for 2 h at room temperature.

7. The hapten–carrier conjugate now may be used for injection purposes without further purification.

An estimate of the degree of conjugation may be made by assaying the amount of sulfhydryl present before and after the coupling reaction. A portion of the peptide solution before mixing with the activated carrier should be saved to compare with the reaction mixture after the conjugation is complete. The comparison is made using a solution of Ellman's reagent (5,5′-dithiobis-(2-nitrobenzoic acid), which reacts with sulfhydryls to form a highly colored chromophore having an absorbance maximum at 412 nm ($\varepsilon_{412nm} = 1.36 \times 10^4$ cm^{-1} M^{-1}) (Chapter 1, Section 4.1). A generalized procedure is presented here. Modifications to this guideline may have to be made for each individual peptide to obtain the appropriate response to the Ellman's reagent.

1. Using a microtiter plate (96-well) dispense 200 μl of 0.1 M sodium phosphate, 0.15 M NaCl, 0.1 M EDTA, pH 7.2 (conjugation buffer), into each well to be used.

2. Add 10 μl of the peptide solution before conjugation to the appropriate wells in duplicate.

3. Add 10 μl of the reaction mixture after the conjugation reaction is complete to another set of wells in duplicate.

4. Add 20 μl of Ellman's reagent (1 mg/ml dissolved in the gel filtration purification buffer) to each well including one containing only buffer (220 μl) to use as a blank.

5. Incubate for 15 min at room temperature.

6. Measure the absorbance of all wells using a microplate reader with a filter set at 410 nm.

A comparison of the blank corrected values before and after conjugation should give an indication of the percentage of peptide coupled. To be more quantitative, a standard curve must be run to focus in on the linear response range of the peptide–Ellman's reaction. Using cysteine as a representative sulfhydryl compound (similar in Ellman's response to a peptide having one free sulfhydryl) it is possible to obtain very accurate determinations of the amount that coupled to the activated carrier. Figure 289, discussed previously in this section, shows the results of this type of assay.

6. Active-Hydrogen-Mediated Hapten–Carrier Conjugation

Conjugation chemistry for the coupling of haptens to carrier molecules is fairly well defined for compounds having common functional groups to facilitate such attachment. The types of functional groups generally useful for this operation include easily reactive components such as primary amines, carboxylic acids, aldehydes, or sulfhydryls.

However, for hapten molecules containing no easily reactive functional groups, conjugation can be difficult or impossible using current techniques. To solve this problem, demanding organic synthesis is frequently required to modify the hapten molecule to contain a suitable reactive portion. Particularly, certain drugs, steroidal compounds, dyes, or other organic molecules often have structures that contain no available "handles" for convenient cross-linking.

Frequently, these difficult to conjugate compounds do have certain sufficiently active hydrogens that can be reacted with a carrier molecule using specialized reactions designed for this purpose. This section describes two choices for this conjugation problem, the diazonium procedure and the Mannich reaction. Both of them are able to cross-link haptens through any available active hydrogen to carrier molecules, resulting in immunogens suitable for injection.

6.1. Diazonium Conjugation

Diazonium coupling procedures have been used for many years in organic synthesis and for cross-linking or immobilization of active hydrogen-containing compounds (Inman and Dintzis, 1969; Cuatrecasas, 1970). Diazonium derivatives can couple with haptens containing available phenolic or, to a lesser extent, imidazole groups in an electrophilic substitution reaction (Riordan and Vallee, 1972). They also may undergo minor secondary reactions with sulfhydryl groups and primary amines (Glazer *et al.*, 1975).

The most important reaction of a diazonium group, however, is with available tyrosine and histidine residues within peptide haptens, rapidly creating diazo linkages. This method of conjugation is especially useful for site-directed cross-linking of tyrosine-containing peptides. Since tyrosine usually is present only in limited quantities in a given peptide, use of diazonium conjugation can cross-link and orient all peptide molecules in an identical fashion on the carrier. The result is excellent reproducibility in preparation of the immunogen, and a consistent presentation of the peptide on the surface of the carrier to the immune system for antibody production.

Derivatives of carbohydrate antigens also have been coupled to carrier proteins through the use of an intermediate diazonium group (McBroom *et al.*, 1976). In this case, an aminophenyl glycoside was prepared by reaction of the reducing end of the oligosaccharide with β-(*p*-aminophenyl)ethylamine and then formation of the diazotized derivative with sodium nitrite (Zopf *et al.*, 1978a). On mixing with carrier proteins containing tyrosine residues, the carbohydrate derivative is coupled via a diazo bond.

Creation of a diazonium group on phenolic compounds or tyrosine side-chain groups is possible by forming an intermediate nitrophenol derivative. Reaction of tyrosine-containing proteins and peptides with tetranitromethane effectively nitrates the ring in the *ortho* position (Vincent *et al.*, 1970). Reduction of the nitro group to an amine then is done using sodium dithionite (sodium hydrosulfite; $Na_2S_2O_4$) (Sokolovsky *et al.*, 1967; Chapter 1, Section 4.3). The aminophenol derivative finally is reacted with sodium nitrite in acidic conditions to form the highly reactive diazonium group (Fig. 291). Once treated, the diazonium compound must be added immediately to the conjugation reaction, since the species is extremely unstable in aqueous environments.

The active diazonium typically is a colored compound, sometimes orange, dark brown, or even black in concentrated solutions. The conjugated immunogen therefore usually is deeply colored as well. The coupling reaction is performed at alkaline pH, optimally at pH 8 for histidinyl residues and pH 9–10 for tyrosine groups. In practice, however, it is not possible to target a histidine group in the presence of a tyrosine group.

Diazo linkages are reversible bonds that may be cleaved by addition of 0.1 *M*

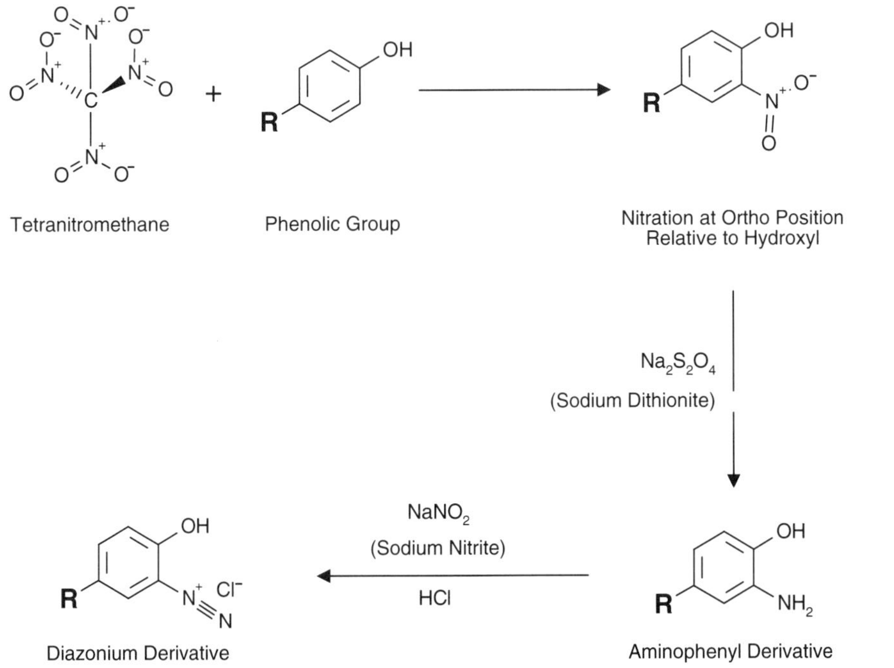

Figure 291 Phenolic compounds may be derivatized to contain reactive diazonium groups by nitration with tetranitromethane followed by reduction with sodium dithionite and diazotization with sodium nitrite in dilute HCl.

sodium dithionite in 0.2 M sodium borate, pH 9. Release of the cross-links can be followed by loss of the diazo bond color.

A simple, one-step conjugation reaction is possible with diazonium chemistry if a *bis*-aminophenyl compound is used as a homobifunctional cross-linking agent. Activation of the aminophenyl groups with sodium nitrite creates the requisite *bis*-diazonium derivative that can couple with active hydrogen-containing haptens and carriers. In this way, tyrosine-containing peptides can be conjugated with tyrosine-containing carrier proteins in a single step. Compounds useful for this procedure include *o*-tolidine and benzidine (Chapter 4, Section 9), both of which contain aromatic amines that easily can be diazotized (Fig. 292).

From a practical perspective, however, any of the conjugation methods utilizing

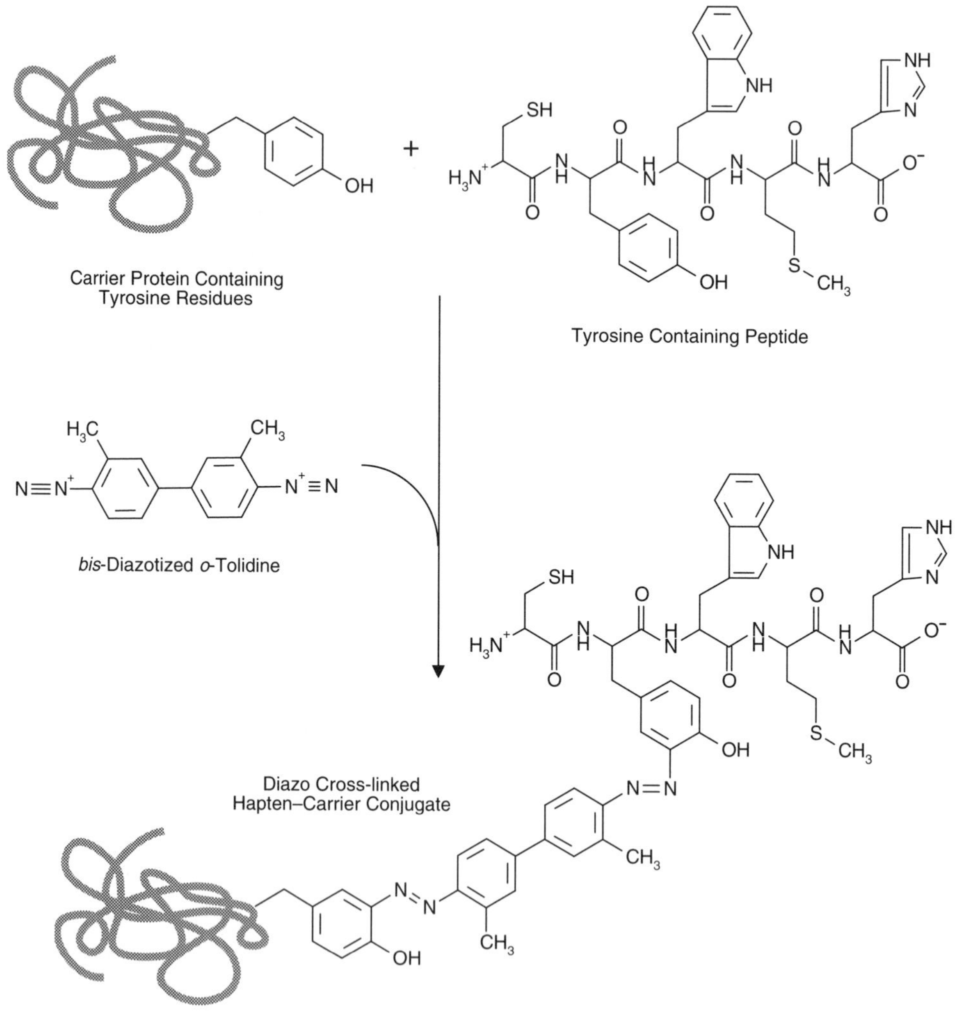

Figure 292   The conjugation of a tyrosine-containing carrier protein and a tyrosine-containing peptide may be done using *bis*-diazotized tolidine to form diazo cross-links.

diazonium chemistry can be brimming with problems. The rate of reaction of the diazonium species is so rapid that much of the total coupling potential can be lost through intramolecular cross-linking. As the diazonium groups are formed they may immediately cross-link to the active hydrogens present on the aminophenyl precursor molecules, even before addition of a second molecule to be conjugated. Even without addition of a second active-hydrogen-containing compound, the diazonium-activated molecule will turn brown to black within an hour, indicating formation of diazo bonds and self-conjugation. For this reason, the reproducibility of conjugation reactions using this method usually is poor.

The following protocol describes the use of diazotized *o*-tolidine for the cross-linking of active-hydrogen-containing haptens to active-hydrogen-containing carriers. Using a *bis*-diazonium compound is perhaps the simplest method of conjugation, but as in many one-step cross-linking procedures, it often results in some precipitation of the final product. Reaction conditions may have to be adjusted to prevent severe precipitation; however, even an insoluble immunogen can be useful in generating an antibody response.

Caution: Both *o*-tolidine and benzidine are potential carcinogens. Protective clothing, gloves, and the use of a fume hood are recommended. Avoid all contact of the compounds with skin or clothing and do not inhale vapors or dust.

Protocol

1. Diazotization of *o*-tolidine: Weigh out 25 mg of *o*-tolidine and place in a small test tube or vial. Add 4.5 ml of 0.2 N HCl and mix to dissolve. Chill the solution on ice. Dissolve 17.5 mg of sodium nitrite into 0.5 ml of ice-cold deionized water, and add it to the vial containing the *o*-tolidine. The solution should begin to turn an orange color, progressively getting darker as the reaction continues. React for 1 h on ice, mixing periodically. At the completion of the diazotization reaction, aliquots of the solution may be stored at −20°C.
2. Dissolve 10 mg of carrier protein into 0.5 ml 0.15 M sodium borate, 0.15 M NaCl, pH 9.
3. Dissolve 5–10 mg of a peptide hapten containing at least one tyrosine residue per milliliter of 0.15 M sodium borate, 0.15 M NaCl, pH 9.
4. Mix 0.5 ml of the peptide solution with 0.5 ml of the carrier protein solution. Chill on ice. Add 0.4 ml of the *bis*-diazotized tolidine solution. There should be a color change from orange to red almost immediately. Continue the reaction for 2 h on ice in the dark.
5. Purify the conjugate by gel filtration or dialysis using PBS, pH 7.4. The preparation is now ready for immunization purposes.

6.2. Mannich Condensation

Another approach for cross-linking haptens to carriers when the hapten has no available common functional groups (amines, carboxylates, sulfhydryls, etc.), but does possess active hydrogens, is to use the Mannich reaction. Using this strategy an active hydrogen-containing compound can be condensed with formaldehyde and an amine in the Mannich reaction, resulting in a stable alkylamine linkage. Particularly, compounds containing replaceable hydrogens provided by the presence of certain activating chemical constituents can be aminoalkylated using this reaction (see Chapter 2, Section 5.4, and Chapter 4, Section 6.1, for additional information on active hydrogens).

Formally, the Mannich reaction consists of the condensation of formaldehyde (or sometimes another aldehyde) with ammonia, in the form of its salt, and another compound containing an active hydrogen. Instead of using ammonia, however, this reaction can be done with primary or secondary amines, or even with amides. An example is illustrated in the condensation of acetophenone, formaldehyde, and a secondary amine salt (the active hydrogens are shown underlined):

$$C_6H_5COCH_3 + CH_2O + R_2NH{\cdot}HCl \rightarrow C_6H_5COCH_2CH_2NR_2{\cdot}HCl + H_2O$$

The Mannich reaction provides a viable alternative to the diazonium conjugation method (discussed previously), because of the disadvantages inherent in the instability of both the diazonium group and the resultant diazo linkage. By contrast, conjugations done through Mannich condensations result in stable covalent bonds.

The cross-linking scheme using this method can make use of the native ε- and N-terminal amines on carrier proteins as the source of primary amine for the condensation reaction. Added to the conjugation reaction is formaldehyde and the desired hapten to be coupled containing an appropriately active hydrogen.

To increase the yield of conjugated hapten using this procedure, cationized BSA is used as the carrier protein in the method described below (Section 2.1). The greater density of amine groups on cBSA available for participation in the Mannich reaction over that available on native proteins provides better results in coupling active hydrogen-containing haptens.

One note of caution should be realized when using the Mannich reaction. The hapten to be coupled should not contain any amine groups or hapten polymerization may occur preferential to conjugation to the carrier. For instance, when performing site-directed coupling of tyrosine-containing peptides through their phenolic side chain, the diazonium reaction should be used instead of the Mannich procedure, otherwise peptide-to-peptide coupling may occur.

Protocol

1. In a vial or test tube are placed and mixed:

    a. 200 μl of a solution containing 10 mg/ml cationized BSA (cBSA, Pierce) in 0.1 M MES, 0.15 M NaCl, pH 4.7 (coupling buffer). The acidic conditions of this coupling buffer are optimal for the Mannich reaction.

    b. 200 μl of a solution consisting of 10 mg/ml of a hapten containing an active hydrogen. The solution can be made up in absolute ethanol in the case of water insoluble haptens and is made up in coupling buffer in the case of water-soluble haptens.

    c. 50 μl of additional absolute ethanol in the case of water-insoluble haptens.

    d. 50 μl of 37% formaldehyde (Sigma) solution. Caution: Use a fume hood and avoid contact or inhalation of vapors.

2. Incubate the reaction mixture in a water bath or oven at a temperature of 37–57°C for a period of 3–24 h.

3. To separate unconjugated haptens and formaldehyde from the synthesized conjugate, apply the entire volume of reactants to a Sephadex G-25 desalting column containing a bed volume of at least 10 times the volume of reactants. PBS, pH 7.2, can be used for the desalting step. The purified conjugate is recovered in the void volume.

The yield of conjugation using the Mannich reaction is dependent on the reactivity of active hydrogens within the hapten molecule. It is often difficult to predict the relative reactivity of any given compound in this reaction. Thus, trial and error may be necessary to determine the suitability of the Mannich procedure.

Figure 293 shows the conjugation reaction of the dye phenol red to cBSA using the Mannich reaction. The active hydrogens which participate in the conjugation are *ortho* to the hydroxyl group on the phenol ring. After purification of the conjugate by gel filtration to remove any unconjugated dye and formaldehyde, a wavelength scan was done to assess the degree of conjugate formation. Figure 294 shows the results of this scan. The protein solution appeared red after conjugation and desalting, indicating successful cross-linking had occurred.

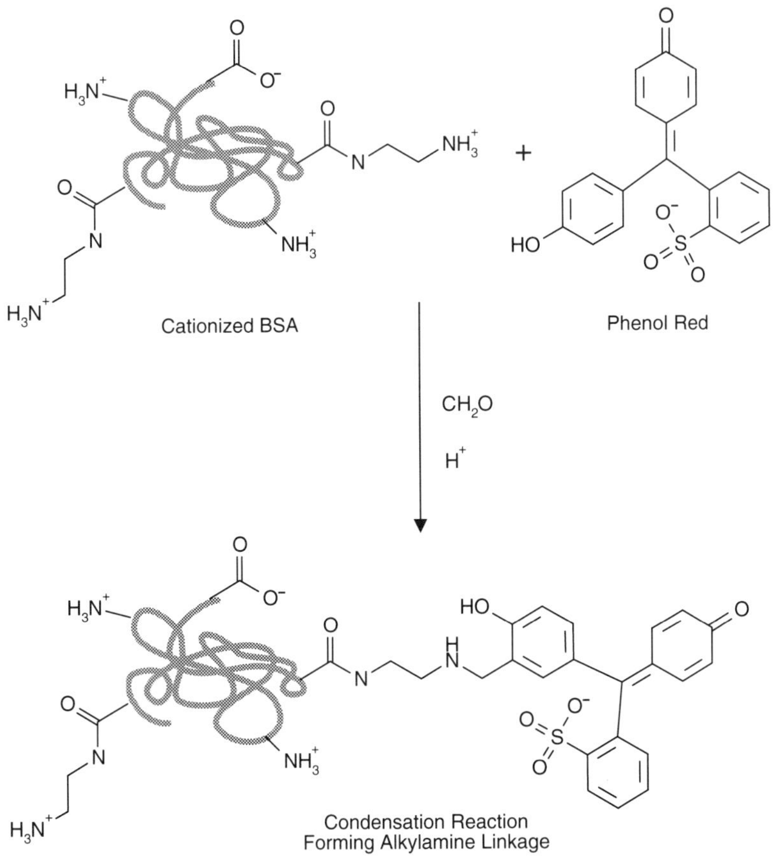

Figure 293 The conjugation of phenol red to cationized BSA using the Mannich reaction.

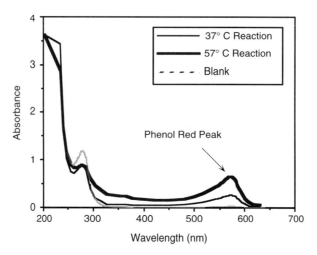

Figure 294   Absorbance scan comparing unconjugated cBSA with the same carrier that had been coupled with phenol red using the Mannich reaction. Two different reaction times are compared, indicating that extended reactions yield increased conjugate formation.

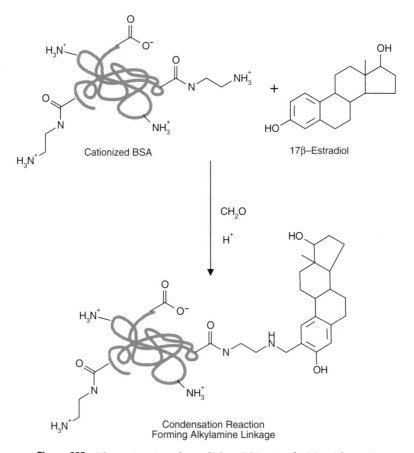

Figure 295   The conjugation of estradiol to cBSA using the Mannich reaction.

The steroidal compound 17β-estradiol was also conjugated to cBSA using the Mannich reaction. Similar to phenol red, conjugation with estradiol occurs *ortho* to the hydroxyl group on its phenolic ring (Fig. 295). After purification of the conjugate by gel filtration, it was injected in mice intraperitoneally using alum as adjuvant. Antibodies were successfully produced against the coupled estradiol. Controls consisting of unconjugated estradiol with and without mixed carrier molecules also were injected, but resulted in no antibody production.

7. Glutaraldehyde-Mediated Hapten–Carrier Conjugation

The homobifunctional cross-linking reagent glutaraldehyde can be used in a one- or two-step conjugation protocol to prepare hapten–carrier conjugates. Glutaraldehyde can react with primary amine groups to create Schiff bases or double-bond (Michael-type) addition products (Chapter 4, Section 6.2). The Schiff base intermediate may form resonance-stabilized products with the α,β-unsaturated aldehydes of the glutaraldehyde polymers predominating at basic pH values (Monsan *et al.*, 1975; Peters and Richards, 1977; Korn *et al.*, 1972). One such product, a quaternary pyridinium complex, can form as a cross-link between two lysine residues (Chapter 1, Section 4.4). Reduction of the Schiff bases with sodium borohydride or sodium cyanoborohydride yields stable secondary amine linkages.

The reaction of glutaraldehyde with protein carriers and peptide haptens involves mainly lysine ε-amine and N-terminal α-amine groups. The conjugates formed are usually of high molecular weight and may cause precipitation products. In addition, the orientation of the hapten on the carrier is indiscriminate with oligomers of the peptide predominating. However, despite the disadvantages of using glutaraldehyde-mediated cross-linking, it still remains one of the most popular techniques for creating bioconjugates.

There are several different protocols commonly used in the literature to form glutaraldehyde conjugates. Some methods utilize a neutral pH environment in phosphate buffer (pH 6.8–7.5) while others use more alkaline conditions in carbonate buffer (pH 8–9) (Price *et al.*, 1993). In general, the higher pH conditions will more effectively form Schiff base intermediates and result in greater conjugation yields, but also higher molecular weight conjugates. The concentration of glutaraldehyde in the reaction medium generally varies from 0.20 to 1% (Avrameas, 1969; Avrameas and Ternynck, 1969; Jeanson *et al.*, 1988; Ford *et al.*, 1978) with occasional use of very dilute solutions (0.05%). The lower concentrations of glutaraldehyde generate lower yields of conjugation and result in less stable conjugates (Briand *et al.*, 1985).

The following procedure utilizes the one-step glutaraldehyde method. A two-step method may be used to limit somewhat polymerization of the conjugate (Chapter 10, Section 1.2). Varying the pH and the amount of glutaraldehyde added to the reaction can control the yield and molecular weight of the conjugates formed.

Protocol

1. Dissolve the carrier protein (or another carrier that contains amine groups) in 0.1 *M* sodium carbonate, 0.15 *M* NaCl, pH 8.5, at a concentration of 2 mg/ml.
2. Add peptide hapten to the carrier solution to obtain a concentration of about 2

mg/ml. Alternatively, determine the molar ratio of peptide to carrier. Ratios of 20:1 to 40:1 (peptide:carrier) usually result in good immunogens.

3. Add fresh glutaraldehyde to the peptide/carrier solution to obtain a 1% final concentration. Mix well. Caution: Use of a fume hood is recommended when working with glutaraldehyde. Avoid contact with skin and clothing. Do not breathe vapors.

4. React for 2–4 h at 4°C. Periodically mix the solution or use a gentle rocker.

5. The conjugate may be stabilized by addition of a reductant such as sodium borohydride or sodium cyanoborohydride. Usually sodium cyanoborohydride is recommended for specific reduction of Schiff bases, but since the conjugate has already formed at this point, the use of sodium borohydride will both reduce the associated Schiff bases and eliminate any remaining aldehyde groups. Add sodium borohydride to a final concentration of 10 mg/ml. Continue to react for 1 h at 4°C.

6. Purify the conjugate by gel filtration using Sephadex G-25 or dialysis to remove excess reagents. The presence of high-molecular-weight conjugates may cause some precipitation in the final product. If turbidity is evident, instead of gel filtering, dialyze against PBS, pH 7.4.

8. Reductive-Amination-Mediated Hapten–Carrier Conjugation

Hapten molecules containing aldehyde residues may be cross-linked to carrier molecules by use of reductive amination (chapter 3, Section 4). At alkaline pH values, the aldehyde groups form intermediate Schiff bases with available amine groups on the carrier. Reduction of the resultant Schiff bases with sodium cyanoborohydride or sodium borohydride creates a stable conjugate held together by secondary amine bonds.

Oligosaccharide haptens are especially amenable for coupling to carriers by reductive amination. Carbohydrate molecules may contain reducing ends that can be utilized for this purpose (Chapter 1, Section 2.1) (Gray, 1978), or aldehyde residues may be specifically created from other functional groups (Chapter 1, Section 4.4). Often, mild oxidation using sodium periodate can be used to cleave adjacent diols on sugar residues, forming reactive aldehyde groups (Anderson et al., 1989).

If the reducing ends of oligosaccharide molecules are used for this technique, then the reaction time necessary to obtain good yields of hapten–carrier conjugates may be from several days to several weeks. The extended reaction period is due to the limited time reducing sugars are in their open, aldehydic form (usually far less than 1% of the available saccharide at any given time). By contrast, if periodate oxidized carbohydrate is used, then the reaction time is reduced to only hours. It should be noted, however, that extensive periodate oxidation could modify antigenic determinants and no longer reflect the native structure and characteristics of the carbohydrate.

Protocol

1. Dissolve the carrier protein at a concentration of 10 mg/ml in 0.1 M sodium phosphate, 0.15 M NaCl, pH 8.

2. Add the aldehyde-containing oligosaccharide to the carrier solution at a concentration sufficient to obtain at least a 20-fold molar excess of hapten to carrier.

Adding a much greater molar excess of oligosaccharide to couple through reducing ends (i.e., up to 200-fold excess) will help to drive the conjugation reaction to completion.

3. Add sodium cyanoborohydride (Aldrich) to a concentration of 20 mg/ml. *Caution:* Highly toxic! Use a fume hood and avoid inhalation of dust or vapors. Seal the reaction vessel with parafilm. Do not use a rigid sealing cap, since cyanoborohydride will liberate hydrogen gas bubbles over time and may rupture the vessel.

4. React at room temperature with periodic mixing. Reaction times can vary significantly depending on the reactivity of the aldehyde group. For coupling of the reducing ends of polysaccharide molecules, continue the reaction for at least 100 h. High-density derivatization through the reducing ends may take up to 2 weeks. For coupling of periodate oxidized carbohydrate, where the aldehyde residues are more accessible, the reaction is complete within 4 h.

5. Purify the hapten–carrier conjugate to remove excess reductant by gel filtration or dialysis using a PBS, pH 7.4, buffer. Removal of unconjugated carbohydrate may be more difficult. If the oligosaccharide was of high molecular weight so that the unconjugated carbohydrate cannot be easily separated from the conjugate using typical desalting gels or small-porosity dialysis tubing, then a gel filtration matrix possessing greater exclusion limits may be used. However, often it is not necessary to remove unconjugated hapten from such preparations.

10

Antibody Modification and Conjugation

The use of antibody molecules in immunoassay, targeting, or detection techniques encompasses a broad variety of applications affecting nearly every field of the life sciences. The availability of relatively inexpensive polyclonal and monoclonal antibodies of exacting specificity has made possible the design of reagent systems that can interact in high affinity with virtually any conceivable analyte. The directed specificity of purified immunoglobulins provides powerful tools for constructing immunological reagents. Using a number of conjugation and modification techniques, these specific antibodies can be modified to allow easy tracking in complex mixtures. For instance, an antibody molecule labeled with an enzyme, a fluorescent compound, or biotin provides a detectable complex able to be quantified or visualized through its tag.

To maintain specificity in antibody conjugates derived from polyclonal antisera, only affinity-purified immunoglobulins should be used. Such purified preparations are isolated from antisera by affinity chromatography using the corresponding immobilized antigen. These preparations thus contain only that population of antibody molecules with the desired antigenic specificity. Modification or conjugation of whole immunoglobulin fractions should be avoided, since other antibody populations will be present and cause considerable nonspecificity in the resultant activity of the reagent.

Monoclonal antibodies also should be purified by affinity chromatography prior to undergoing bioconjugation. This can be accomplished using an immobilized antigen or, if the antigen is not available in large enough quantities, an immobilized immunoglobulin binding protein (such as protein A) may be employed. Most monoclonals that can be successfully purified while maintaining activity also will be stable enough to withstand the rigors of chemical modification. Occasionally, however, a particular monoclonal will be partially or completely inactivated through the modification reaction. Sometimes this activity loss is caused by physically blocking the antigen binding sites during conjugation. In other cases, conformational changes in the complimentarity-determining regions are the cause of the problem. If the antigen binding site is merely being blocked, then choosing the appropriate site-directed chemical reaction may solve the problem. On the other hand, some monoclonals are too labile to undergo modification chemistries, regardless of the coupling method. Trial and error is often necessary when working with monoclonals to determine whether modification will severely affect activity.

The unique structural characteristics of antibody molecules supply a number of choices for modification and conjugation schemes (Roitt, 1977; Goding, 1986a; Harlow and Lane, 1988a,b,c). The chemical reaction used to effect complex formation should be chosen to yield the best possible retention of antigen binding activity. A detailed illustration of antibody structure is shown in Fig. 296. The most basic immunoglobulin molecule is composed of two light and two heavy chains, held together by noncovalent interactions as well as a number of disulfide bonds. The light chains are disulfide-bonded to the heavy chains in the C_L and C_H^1 regions, respectively. The heavy chains are in turn disulfide-bonded to each other in the hinge region.

The heavy chains of each immunoglobulin molecule are identical. Depending on the class of immunoglobulin, the molecular weight of these subunits ranges from about 50,000 to around 75,000. Similarly, the two light chains of an antibody are identical and have a molecular weight of about 25,000. For IgG molecules the intact molecular weight representing all four subunits is in the range 150,000 to 160,000.

There are two forms of light chains that may be found in antibodies. A single antibody will have light chain subunits of either lambda (λ) or kappa (κ) variety, but not both types in the same molecule. The immunoglobulin class, however, is determined by an antibody's heavy chain variety. A single antibody also will possess only one type of heavy chain (designated as γ, μ, α, ε, or δ). Thus, there are five major classes of antibody molecules, each determined from their heavy chain type, and designated as IgG, IgM, IgA, IgE, or IgD. Three of these class forms, IgG, IgE, and IgD, consist of the basic Ig monomeric structure containing two light and two heavy chains. By contrast, IgA molecules can exist as a singlet, doublet, or triplet of this basic Ig monomeric structure, while IgM molecules are large pentameric constructs (Fig. 297). Both IgA and IgM contain an additional subunit, called the J chain—a very acidic polypeptide

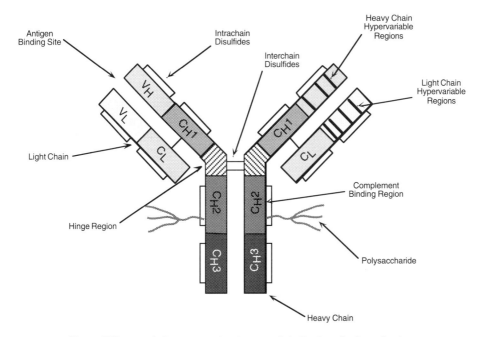

Figure 296   Detailed structure of an immunoglobulin G antibody molecule.

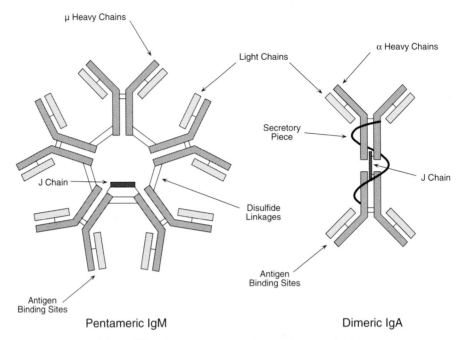

Figure 297   General structures of IgA and IgM antibodies.

of molecular weight 15,000 that is very rich in carbohydrate. The heavy chains of immunoglobulin molecules also are glycosylated, typically in the C_H2 domain within the Fc fragment region.

There are two antigen binding sites on each of the basic Ig-type monomeric structures, formed by the heavy–light chain proximity in the N-terminal, hypervariable region at the tips of the "y" structure. The unique tertiary structure created by these subunit pairings produces the conformation necessary to interact with a complementary antigen molecule. The points of interaction on the immunoglobulin molecule with an antigen involve noncovalent forces that may encompass numerous nonsequential amino acids within the heavy and light chains. In other words, the binding site is formed not strictly from the linear sequence of amino acids on each chain, but from the unique orientation of these groups in three-dimensional space. The binding site thus has affinity for a particular antigen molecule due to both structural complementarity and the combination of van der Waals, ionic, hydrophobic, and hydrogen bonding forces that may be created at each point of contact.

Useful enzymatic derivatives of antibody molecules that still retain the antigen binding sites may be prepared. Two principal digested forms of IgG antibodies are useful for creating immunological reagents. Enzymatic digestion with papain produces two small fragments of the immunoglobulin molecule, each containing an antigen binding site (called Fab fragments), and one larger fragment containing only the lower portions of the two heavy chains (called Fc, for "fragment crystallizable") (Section 1.4) (Coulter and Harris, 1983). Alternatively, pepsin cleavage produces one large fragment containing two antigen binding sites [called F(ab′)$_2$] and many smaller fragments formed from extensive degradation of the Fc region (Rousseaux *et al.,*

1983). The F(ab')$_2$ fragment is held together by retention of the disulfide bonds in the hinge region. Specific reduction of these disulfides using 2-mercaptoethylamine (Chapter 1, Section 4.1) produces two Fab' fragments, each of which has one antigen binding site.

Antibody molecules possess a number of functional groups suitable for modification or conjugation purposes. Cross-linking reagents may be used to target lysine ε-amine and N-terminal α-amine groups. Carboxylate groups also may be coupled to another molecule using C-terminal, aspartic acid, and glutamic acid residues. Although both amine and carboxylate groups are as plentiful in antibodies as they are in most proteins, the distribution of these functional groups within the three-dimensional structure of an immunoglobulin is nearly uniform throughout the surface topology. For this reason, conjugation procedures that utilize these groups will randomly cross-link to nearly all parts of the antibody molecule. This in turn leads to a random orientation of the antibody within the conjugate structure, often blocking the antigen binding sites against the surface of another coupled protein or molecule. Obscuring the binding sites in this manner results in decreased antigen binding activity in the conjugate compared to that observed for the unconjugated antibody.

Conjugation chemical reactions done with antibody molecules generally are more successful at preserving activity if the functional groups utilized are present in limiting quantities and only at discrete sites on the molecule. Such "site-directed conjugation" schemes make use of cross-linking reagents that can specifically react with residues that are only in certain positions on the immunoglobulin surface—usually chosen to be well removed from the antigen binding sites. By proper selection of the conjugation chemical reaction and a knowledge of antibody structure, the immunoglobulin molecule can be oriented so that its bivalent binding potential for antigen remains available.

Two site-directed chemical reactions are especially useful in this regard. The disulfides in the hinge region that hold the heavy chains together can be selectively cleaved with the reductant 2-mercaptoethylamine to create two half-antibody molecules, each containing an antigen binding site (Palmer and Nissonoff, 1963) (Chapter 1, Section 4.1). Alternatively, smaller antigen binding fragments may be made from pepsin digestion [F(ab')$_2$] and similarly reduced to form Fab' molecules. Both of these preparations contain exposed sulfhydryl groups that can be targeted for conjugation using sulfhydryl-reactive probes or cross-linkers. Conjugations done using hinge-area —SH groups will orient the attached protein or other molecule away from the antigen binding regions, thus preventing blockage of these sites and preserving activity.

The second method of site-directed conjugation of antibody molecules takes advantage of the carbohydrate chains attached to the C$_H$2 domain within the Fc region. Mild oxidation of the polysaccharide sugar residues with sodium periodate will generate aldehyde groups. A cross-linking or modification reagent containing a hydrazide functional group then can be used to target these formyl groups specifically for coupling to another molecule. Directed conjugation through antibody carbohydrate chains thus avoids the antigen binding regions while allowing for use of intact antibody molecules. This method often results in the highest retention of antigen binding activity within the ensuing conjugate.

The only limitation to the use of this strategy is the necessity for the antibody molecule to be glycosylated. Antibodies of polyclonal origin (from antisera) are usually glycosylated and work well in this procedure, but other antibody preparations may not possess polysaccharide. In particular, some monoclonals may not be post-translationally modified with carbohydrate after hybridoma synthesis. Recombinant antibodies grown in bacteria also may be devoid of carbohydrate. Before attempting to use a conjugation method that couples through polysaccharide regions, it is best to test the antibody to see whether it contains carbohydrate—especially if the immunoglobulin is of hybridoma or recombinant origin.

1. Preparation of Antibody–Enzyme Conjugates

The most extensive application of antibody conjugation using cross-linking reagents occurs in the preparation of antibody–enzyme conjugates. Since the development of enzyme-linked immunosorbent assay (ELISA) systems, the ability to make conjugates of specific antibodies with enzymes has provided the means to quantify or detect hundreds, if not thousands, of important analytes. The use of enzymes as labels in immunoassay procedures surpassed radioactive tags as the means of detection, primarily due to the long-term stability potential of an enzyme system and the hazards and waste problems associated with radioisotopes. Designed properly, an antibody–enzyme conjugate assay system can be just as sensitive as a radiolabeled antibody system.

The development of viable methods for cross-linking antibody and enzyme molecules—methods that retain high antigen binding activity coupled with high enzymatic activity—has formed the basis for much of today's diagnostic industries, literally a multibillion dollar enterprise with enormous impact on world health. The conjugation chemical reactions that make this possible are designed around a knowledge of both antibody and enzyme structure. The best methods make use of definitive site-directed chemical reactions that target both molecules in regions removed from their respective active centers.

The major enzymes used in ELISA technology include horseradish peroxidase (HRP), alkaline phosphatase (AP), β-galactosidase (β-gal), and glucose oxidase (GO). See Chapter 16 for a detailed description of enzyme properties and activities. HRP is by far the most popular enzyme used in antibody–enzyme conjugates. One survey of enzyme use stated that HRP is incorporated in about 80% of all antibody conjugates, most of them utilized in diagnostic assay systems. AP is the second most popular choice for antibody–enzyme conjugation, being used in almost 20% of all commercial enzyme-linked assays. Although β-gal and GO are used frequently in research and cited numerous times in the literature, their utilization for commercial ELISA applications represents less than 1% of the total assays available.

Conjugation methods for attaching these enzymes to antibody molecules vary according to the functional groups available. HRP is a glycoprotein and easily can be periodate-oxidized for coupling via reductive amination to the amino groups on immunoglobulins. β-Gal contains abundant free sulfhydryl groups in its native state. The thiols can be utilized for coupling to the sulfhydryl-reactive end of heterobifunctional cross-linkers such as SMCC (Chapter 5, Section 1.3). Any of the enzymes can be conjugated through their amine groups using cross-linking agents such as glutaraldehyde or various heterobifunctional agents. The catalytic properties and activation methods often used with these enzymes are discussed in detail in Chapter 16.

The following sections describe the most common chemical reactions used to create antibody–enzyme conjugates.

1.1. NHS Ester–Maleimide-Mediated Conjugation

Heterobifunctional reagents containing an amine-reactive NHS ester on one end and a sulfhydryl-reactive maleimide group on the other end generally have great utility for producing antibody–enzyme conjugates (see Chapter 5, Section 1). Cross-linking reagents possessing these functional groups can be used in highly controlled, multistep procedures that yield conjugates of defined composition and high activity. Among the most popular of these NHS ester–maleimide cross-linkers are SMCC (Chapter 5, Section 1.3), MBS (Chapter 5, Section 1.4), and GMBS (Chapter 5, Section 1.7). The use of any one of these cross-linkers in the following protocol can result in useful conjugates. However, SMCC and its water-soluble analog sulfo-SMCC possess the most stable maleimide functional groups and are probably the most widely used cross-linkers of this type. This increased stability to hydrolysis of SMCC's hindered maleimide group allows activation of either enzyme or antibody via the amine-reactive NHS ester end, resulting in a maleimide-activated intermediate. The intermediate species then can be purified away from excess cross-linker and reaction by-products before mixing with the second protein to be conjugated. The multistep nature of this process limits polymerization of the conjugated proteins and provides control over the extent and sites of cross-linking.

In protocols involving enzyme activation with SMCC and subsequent conjugation with an antibody molecule (the most common method of producing antibody–enzyme conjugates with this cross-linker), the antibody usually has to be prepared for coupling to the maleimide groups on the enzyme by introduction of sulfhydryl residues. Since antibodies typically do not contain free sulfhydryls accessible for conjugation, they must be fabricated by chemical means. Two main options are available for creating sulfhydryl functions on immunoglobulin molecules. The disulfide residues in the hinge region of the IgG structure may be reduced with 2-mercaptoethylamine to cleave the immunoglobulin into two half-antibody molecules each possessing one antigen binding site and the requisite sulfhydryls. Alternatively, a thiolation reagent may be used to modify the intact antibody to contain sulfhydryls (Chapter 1, Section 4.1). Both options are described below. Although there are numerous thiolation reagents from which to choose, only SATA and Traut's reagent are discussed in this section, since they are the most popular.

Activation of Enzymes with NHS Ester–Maleimide Cross-linkers

The first step in conjugation of antibody molecules and enzymes using NHS ester–maleimide cross-linkers usually is modification of the enzyme with the NHS ester end of the reagent to produce a maleimide-activated derivative (Fig. 298). The protocol described here uses sulfo-SMCC as the cross-linking agent due to the enhanced stability of its maleimide group and the water solubility afforded by the negatively charged sulfonate on its NHS ring. Other NHS ester–maleimide cross-linkers may be substituted without difficulty; however, water-insoluble varieties should be solubilized in DMSO or DMF prior to addition to the aqueous reaction mixture.

One note should be mentioned before proceeding: when conjugating antibody molecules with β-galactosidase, the antibody usually is activated with sulfo-SMCC first to take advantage of the indigenous sulfhydryl groups on the enzyme. Therefore, if β-gal is being used, substitute the antibody for the enzyme mentioned in this protocol,

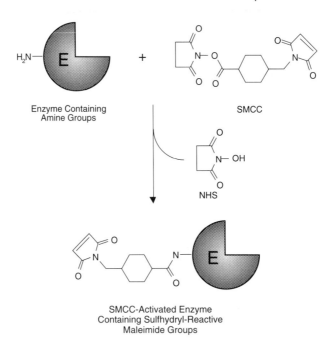

SMCC-Activated Enzyme
Containing Sulfhydryl-Reactive
Maleimide Groups

Figure 298   The reaction of SMCC with the amine groups on enzyme molecules yields a maleimide-activated derivative capable of coupling with sulfhydryl-containing antibody molecules.

and then after the purification step, add the enzyme in the desired molar excess to produce the final conjugation.

The following protocol describes the activation of HRP with sulfo-SMCC. Activation of other enzymes is done similarly, with the appropriate adjustments in the mass of enzyme added to the reaction to account for differences in molecular weight.

The gel filtration column described in step 3 should be packed and equilibrated prior to starting the modification reaction. Enzymes preactivated with sulfo-SMCC are available from Pierce.

Protocol

1. Dissolve 18 mg of HRP in 0.1 *M* sodium phosphate, 0.15 *M* NaCl, pH 7.2, at a concentration of 20–30 mg/ml. The more highly concentrated the enzyme solution, the more efficient will be the modification reaction.

2. Add 6 mg of sulfo-SMCC (Pierce) to the HRP solution. Mix to dissolve and react for 30 min at room temperature. Alternatively, two 3-mg additions of cross-linker may be done—the second one after 15 min of incubation—to obtain even more efficient modification.

3. Immediately purify the maleimide-activated HRP away from excess cross-linker and reaction by-products by gel filtration on a desalting column (Sephadex G-25 or the equivalent). Use 0.1 *M* sodium phosphate, 0.15 *M* NaCl, pH 7.2, as the chromatography buffer. HRP can be observed visually as it flows through the column due to the color of its heme ring. Pool the fractions containing the HRP peak. After elution, adjust the HRP concentration to 10 mg/ml for the conjuga-

tion reaction. At this point, the maleimide-activated enzyme may be frozen and lyophilized to preserve its maleimide activity. The modified enzyme is stable for at least one year in a freeze-dried state. If kept in solution, the maleimide-activated HRP should be used immediately to conjugate with an antibody following one of the three options outlined below.

Conjugation with Reduced Antibodies

One method of introducing sulfhydryl residues into antibody molecules for conjugation with maleimide-activated enzymes is to reduce indigenous disulfide groups in the hinge region of the immunoglobulin structure. Reduction with 2-mercaptoethylamine (MEA) will cleave principally the disulfide bonds holding the heavy chains together, but leave the disulfides between the heavy and light chains alone. Similar reduction can be done with $F(ab')_2$ fragments produced from pepsin digestion of IgG molecules. Either of these reduction steps creates half-antibody fragments, each containing one heavy and one light chain and one antigen binding site (Fig. 299). The sulfhydryl groups produced by this reduction are able to couple with maleimide-activated enzymes without blocking the antigen binding area.

Antibody reduction with MEA usually is done in the presence of EDTA to prevent reoxidation of the sulfhydryls by metal catalysis. In phosphate buffer at pH 6–7 and 4°C, one report stated that the number of available thiols decreased only by about 7% in the presence of EDTA over a 40-h time span. In the absence of EDTA, this sulfhydryl loss increased to 63–90% in the same period (Yoshitake et al., 1979).

In the following protocol, the most critical aspects are the concentration of MEA and EDTA in the reaction mixture. Good reduction of IgG will take place in 50–100 mM MEA and 1–100 mM EDTA. The pH of the reaction can vary from pH 6 to 9, with about pH 8 being optimal. The absolute concentration of antibody can vary and still yield acceptable results. With some monoclonals, however, MEA reduction may not be completely efficient in cleaving the antibody between the heavy-chain pairs. Particularly, some subclasses of immunoglobulins contain structures with unusually high numbers of disulfides in the hinge region, and a few of them may not be reduced except under much more severe conditions. Polyclonal populations typically work well in this procedure.

A final consideration is to provide adequate desalting of the reduced antibody molecule from excess MEA. If even a small amount of reductant remains, subsequent conjugation with a maleimide-activated enzyme will be inhibited.

Protocol

1. Dissolve the IgG to be reduced at a concentration of 1–10 mg/ml in 0.1 M sodium phosphate, 0.15 M NaCl, pH 7.2, containing 10 mM EDTA.
2. Add 6 mg of 2-mercaptoethylamine (Pierce) to each milliliter of antibody solution. Mix to dissolve.
3. Incubate for 90 min at 37°C.
4. Purify the reduced IgG by gel filtration using Sephadex G-25 or the equivalent. Perform the chromatography using 0.1 M sodium phosphate, 0.15 M NaCl, pH 7.2, containing 10 mM EDTA as the buffer. To obtain efficient separation between the reduced antibody and excess reductant, the sample size applied to the

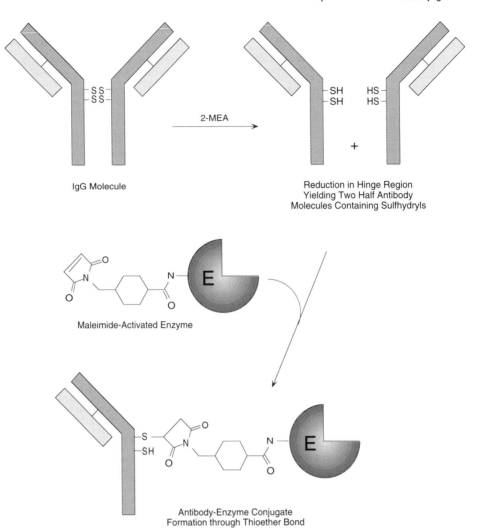

Figure 299   Reduction of the disulfide bonds within the hinge region of an IgG molecule produces half-antibody molecules containing thiol groups. Reaction of these reduced antibodies with a maleimide-activated enzyme creates a conjugate through thioether bond formation.

column should be at a ratio of no more than 5% sample volume to column volume. Collect 0.5-ml fractions and monitor for protein at 280 nm. Since MEA has no absorbance at 280 nm, the elution profile also may be monitored by use of the BCA protein assay method (Pierce). The BCA–copper reagent reacts with both the protein fractions and the sulfhydryl groups of MEA to produce a colored product. EDTA in the chromatography buffer will inhibit the BCA method somewhat, but a color response to the MEA peak will still be obtained. A micromethod for monitoring each fraction is as follows:

a. Take 5 μl from each fraction collected and place in a separate well of a microtiter plate.

b. Add 200 μl of BCA working reagent.

c. Incubate at room temperature of 37°C for 15–30 min or until color develops. The color response may be measured visually or by absorbance at 562 nm. A microplate reader with a filter close to this wavelength is sufficient. To ensure good separation between the antibody peak and excess MEA, at least one fraction of little or no color should separate the two peaks.

5. Pool the fractions containing antibody and immediately mix with an amount of maleimide-activated enzyme to obtain the desired molar ratio of antibody-to-enzyme in the conjugate. Use of a 4:1 (enzyme:antibody) molar ratio in the conjugation reaction usually results in high-activity conjugates suitable for use in many enzyme-linked immunoassay procedures.

6. React for 30–60 min at 37°C or 2 h at room temperature. The conjugation reaction also may be done at 4°C overnight.

7. The conjugate may be further purified away from unconjugated enzyme by the procedures described in Chapter 10, Section 1.5. For storage, the conjugate should be kept frozen, lyophilized, or sterile-filtered and kept at 4°C. Stability studies may have to be done to determine the optimal method of long-term storage for a particular conjugate.

Conjugation with 2-Iminothiolane-Modified Antibodies

Traut's reagent, or 2-iminothiolane, is described in Chapter 1, Section 4.1. The reagent reacts with amine groups in proteins or other molecules in a ring-opening reaction to result in permanent modifications containing terminal sulfhydryl residues (Fig. 300). Antibodies may be modified with Traut's reagent to create the requisite sulfhydryls necessary for conjugation with a maleimide-activated enzyme. Unlike the MEA reduction method described in the previous section, this protocol retains the divalent nature of the antibody molecule. However, since amine modification of antibodies can take place at virtually any available lysine ε-amine location, the resultant sulfhydryls are distributed almost randomly over the immunoglobulin structure. Conjugation through these —SH groups may result in a certain population of antibodies that have their antigen binding sites obscured or blocked by enzyme molecules. Typically, enough free antigen binding sites are available in the conjugate to result in high-activity complexes useful in ELISA procedures.

The number of sulfhydryls created on the immunoglobulin using thiolation procedures such as this one is more critical to the yield of conjugated enzyme molecules than the molar excess of maleimide-activated enzyme used in the conjugation reaction. Therefore, it is important to use a sufficient excess of Traut's reagent to obtain a high density of available sulfhydryls.

Protocol

1. Dissolve the antibody to be modified at a concentration of 1–10 mg/ml in 0.1 M sodium phosphate, 0.15 M NaCl, pH 7.2, containing 10 mM EDTA. High levels of EDTA often are required to completely stop metal-catalyzed oxidation of sulfhydryl groups when working with serum proteins—especially polyclonal antibodies purified from antisera. Presumably, carryover of iron from partially hemolyzed blood is the contaminating culprit.

2. Add 2-iminothiolane (Pierce) to this solution to give a molar excess of 20–40×

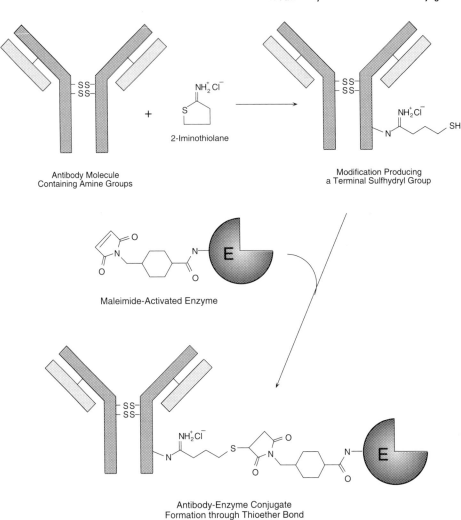

Antibody Molecule
Containing Amine Groups

2-Iminothiolane

Modification Producing
a Terminal Sulfhydryl Group

Maleimide-Activated Enzyme

Antibody-Enzyme Conjugate
Formation through Thioether Bond

Figure 300   Antibodies may be modified with 2-iminothiolane at their amine groups to create sulfhydryls for conjugation with SMCC-activated enzymes. The maleimide groups on the derivatized enzyme react with the thiols on the antibody to form thioether bonds.

over the amount of antibody present (MW of Traut's reagent is 137.63). The greater the molar excess of Traut's reagent added, the higher level of thiolation and the greater the incorporation of maleimide-activated enzyme will be during the conjugation reaction. Addition of solid 2-iminothiolane may be done despite the fact that the compound is relatively insoluble in aqueous solution. As the reagent reacts, it will be completely drawn into solution. Alternatively, a stock solution of Traut's may be made in DMF and an aliquot added to the antibody solution (not to exceed 10% DMF in the final solution).

3. React for 30 min at 37°C or 1 h at room temperature.
4. Purify the thiolated antibody by gel filtration using a desalting matrix such as Sephadex G-25. Perform the chromatography using 0.1 M sodium phosphate,

0.15 M NaCl, pH 7.2, containing 10 mM EDTA as the buffer. To obtain efficient separation between the reduced antibody and excess reductant, the sample size applied to the column should be at a ratio of no more than 5% sample volume to the total column volume. Collect 0.5-ml fractions and monitor for protein at 280 nm. To monitor the separation of the second peak (excess Traut's reagent), the BCA protein assay reagent (Pierce) may be used according to the procedure described in the previous section, protocol step 4.

5. Pool the fractions containing antibody and immediately mix with an amount of maleimide-activated enzyme to obtain the desired molar ratio of antibody-to-enzyme in the conjugate. Use of a 4:1 (enzyme:antibody) to 15:1 molar ratio in the conjugation reaction usually results in high-activity conjugates suitable for use in many enzyme-linked immunoassay procedures.

6. React for 30–60 min at 37°C or 2 h at room temperature. The conjugation reaction also may be done at 4°C overnight.

7. The conjugate may be further purified away from unconjugated enzyme by the procedures described in Section 1.5. For storage, the conjugate should be kept frozen, lyophilized, or sterile-filtered and kept at 4°C. Stability studies may have to be done to determine the optimal method of long-term storage for a particular conjugate.

Conjugation with SATA-Modified Antibodies

N-Succinimidyl-*S*-acetylthioacetate (SATA) is a thiolation reagent described in detail in Chapter 1, Section 4.1. The compound reacts with primary amines via its NHS ester end to form stable amide linkages. The acetylated sulfhydryl group is stable until deacetylated with hydroxylamine. Thus, antibody molecules may be thiolated with SATA to create the sulfhydryl target functional groups necessary to couple with a maleimide-activated enzyme (Fig. 301). Using this reagent, stock preparations of SATA-modified antibodies may be prepared and deacetylated as needed. Unlike thiolation procedures that immediately form a free sulfhydryl residue, the protected sulfhydryl group of SATA-modified proteins is stable to long-term storage without degradation.

Although amine-reactive protocols, such as SATA thiolation, result in nearly random attachment over the surface of the antibody structure, it has been shown that modification with up to 6 SATA/molecules per antibody molecule typically results in no decrease in antigen binding activity (Duncan *et al.*, 1983). Even higher ratios of SATA to antibody are possible with excellent retention of activity.

The following protocol should be compared to the method described for SATA thiolation in Chapter 1, Section 4.1. Although the procedures are slightly dissimilar, the differences indicate the flexibility inherent in the chemistry. For convenience, the buffer composition indicated here was chosen to be consistent throughout this section on enzyme–antibody conjugation using SMCC. Other buffers and alternate protocols can be found in the literature.

Protocol

1. Dissolve the antibody to be modified in 0.1 M sodium phosphate, 0.15 M NaCl, pH 7.2, at a concentration of 1–5 mg/ml. Note: phosphate buffers at various pH

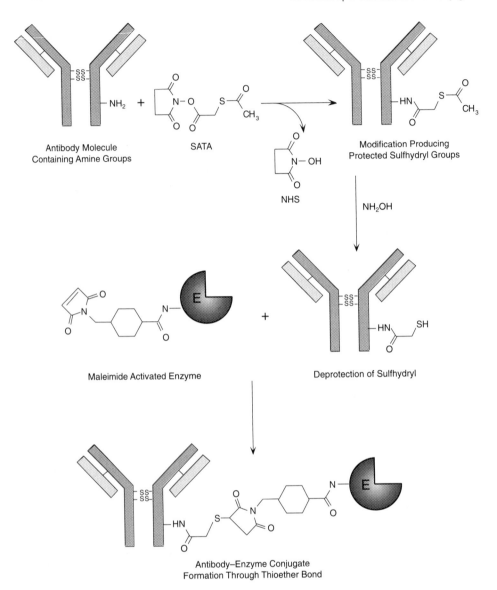

Figure 301 Available amine groups on an antibody molecule may be modified with the NHS ester end of SATA to produce amide bond derivatives containing terminal protected sulfhydryls. The acetylated thiols may be deprotected by treatment with hydroxylamine at alkaline pH. Reaction of the thiolated antibody with a maleimide-activated enzyme results in thioether cross-links.

values between 7.0 and 7.6 have been used successfully with this protocol. Other mildly alkaline buffers may be substituted for phosphate in this reaction, providing they do not contain extraneous amines (e.g., Tris) or promote hydrolysis of SATA's NHS ester (e.g., imidazole).

2. Prepare a stock solution of SATA (Pierce) by dissolving it in DMF or DMSO at a concentration of 8 mg/ml. Use a fume hood to handle the organic solvents.

3. Add 10–40 µl of the SATA stock solution per milliliter of 1 mg/ml antibody solution. This will result in a molar excess of approximately 12- to 50-fold of SATA over the antibody concentration (for an initial antibody concentration of 1 mg/ml). A 12-fold molar excess works well, but higher levels of SATA incorporation will potentially result in more maleimide-activated enzyme molecules able to couple to each thiolated antibody molecule. For higher concentrations of antibody in the reaction medium, proportionally increase the amount of SATA addition; however, do not exceed 10% DMF in the aqueous reaction medium.

4. React for 30 min at room temperature.

5. To purify the SATA-modified antibody perform a gel filtration separation using Sephadex G-25 or by dialysis against 0.1 M sodium phosphate, 0.15 M NaCl, pH 7.2, containing 10 mM EDTA. Purification is not absolutely required, since the following deprotection step is done using hydroxylamine at a significant molar excess over the initial amount of SATA added. Whether a purification step is done or not, at this point, the derivative is stable and may be stored under conditions that favor long-term antibody activity (i.e., sterile filtered at 4°C, frozen, or lyophilized).

6. Deprotect the acetylated sulfhydryl groups on the SATA-modified antibody according to the following protocol.

   a. Prepare a 0.5 M hydroxylamine (Pierce) solution in 0.1 M sodium phosphate, pH 7.2, containing 10 mM EDTA.

   b. Add 100 µl of the hydroxylamine stock solution to each milliliter of the SATA-modified antibody. Final concentration of hydroxylamine in the antibody solution is 50 mM.

   c. React for 2 hours at room temperature.

   d. Purify the thiolated antibody by gel filtration on Sephadex G-25 using 0.1 M sodium phosphate, 0.1 M NaCl, pH 7.2, containing 10 mM EDTA as the chromatography buffer. To obtain efficient separation between the thiolated antibody and excess hydroxylamine and reaction by-products, the sample size applied to the column should be at a ratio of no more than 5% sample volume to the total column volume. Collect 0.5-ml fractions. Pool the fractions containing protein by measuring the absorbance of each fraction at 280 nm.

7. Immediately mix the thiolated antibody with an amount of maleimide-activated enzyme to obtain the desired molar ratio of antibody-to-enzyme in the conjugate. Use of a 4:1 (enzyme:antibody) to 15:1 molar ratio in the conjugation reaction usually results in high-activity conjugates suitable for use in many enzyme-linked immunoassay procedures.

8. React for 30–60 min at 37°C or 2 h at room temperature. The conjugation reaction also may be done at 4°C overnight.

9. The conjugate may be further purified away from unconjugated enzyme by the procedures described in Section 1.5. For storage, the conjugate should be kept frozen, lyophilized, or sterile filtered and kept at 4°C. Stability studies may have to be done to determine the optimal method of long-term storage for a particular conjugate.

1.2. Glutaraldehyde-Mediated Conjugation

Glutaraldehyde was one of the first and still is one of the most commonly used cross-linking agents available for creating antibody–enzyme conjugates. The cross-linking process using glutaraldehyde is believed to proceed by a number of mechanisms, including Schiff base formation with possible rearrangement to a stable product or through a Michael-type addition reaction that takes place at points of double-bond unsaturation created by polymerization of the reagent in solution (Chapter 1, Section 4.4, and Chapter 4, Section 6.2) (Avrameas, 1969). Reduction of Schiff base intermediates also is possible using sodium cyanoborohydride to form stable secondary amine linkages.

The problem of indeterminate reaction products is a deficiency that plagues all conjugations done using glutaraldehyde. Part of this difficulty is due to the reagent's homobifunctional nature, but a significant part of the problem is also due to the ambiguous nature of the commercial product. In aqueous solutions at alkaline pH, glutaraldehyde can undergo aldol condensation reactions with itself to form large polymer structures containing α,β-unsaturated aldehydes (Hardy et al., 1969, 1976). Another disadvantage of the reagent is the tendency to form high-molecular-weight conjugates due to uncontrollable polymerization during the cross-linking process. The resultant conjugates often have a significant amount of insoluble polymer, which causes yield and activity losses in the preparation of antibody–enzyme conjugates. This is especially true when the conjugation is done using the one-step method where glutaraldehyde is simply added to a solution containing the two proteins to be cross-linked (Fig. 302). Enzymatic activity yields using this process can be as little as 10% in the final antibody–enzyme conjugate. To somewhat overcome the polymerization problem, a two-step procedure was developed that involves first activating one of the proteins with glutaraldehyde, purifying the intermediate from excess reagent, and then adding the second protein to effect the final conjugation. Unfortunately, even the two-step method results in significant formation of large-molecular-weight species that may precipitate out of solution. The only enzyme that the two-step method seems to work well with is HRP, since it only contains a limited number of available lysine amine groups.

Despite these deficiencies, antibody–enzyme conjugates are still being made using glutaraldehyde—particularly for many commercial diagnostic ELISA kits, which were developed before the advent of more controllable, heterobifunctional cross-linking procedures. Today, choosing another method of producing antibody–enzyme conjugates will result in much better conjugates of high activity and high yield.

One-Step Glutaraldehyde Protocol

1. Prepare a solution containing 2 mg/ml antibody and 5 mg/ml enzyme in 0.01 M sodium phosphate, 0.15 M NaCl, pH 7.4, chilled to 4°C.
2. In a fume hood, add 10 μl of 25% glutaraldehyde (Sigma) per milliliter of antibody/enzyme solution. Mix well.
3. React for 2 h at 4°C.
4. To reduce the resultant Schiff bases and any excess aldehydes, add sodium borohydride (Aldrich) to a final concentration of 10 mg/ml.

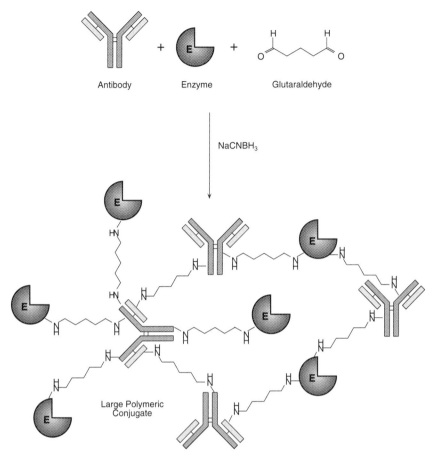

Figure 302   Glutaraldehyde antibody–enzyme cross-linking procedures usually produce a wide range of high-molecular-weight complexes, some of which may precipitate from solution.

Note: Some protocols do not call for a reduction step. As an alternative to reduction, add 50 μl of 0.2 M lysine in 0.5 M sodium carbonate, pH 9.5, to each milliliter of the conjugation reaction to block excess reactive sites. Block for 2 h at room temperature. Other amine-containing small molecules may be substituted for lysine—such as glycine, Tris buffer, or ethanolamine.

5. Reduce for 1 h at 4°C.
6. To remove any insoluble polymers that may have formed, centrifuge the conjugate or filter it through a 0.45-μm filter. Purify the conjugate by gel filtration or dialysis using PBS, pH 7.4.

Two-Step Glutaraldehyde Protocol

1. Dissolve the enzyme at a concentration of 10 mg/ml in 0.1 M sodium phosphate, 0.15 M NaCl, pH 6.8.

2. Add glutaraldehyde to a final concentration of 1.25%.
3. React overnight at room temperature.
4. Purify the activated enzyme from excess glutaraldehyde by gel filtration (using Sephadex G-25) or by dialysis against PBS, pH 6.8.
5. Dissolve the antibody to be conjugated at a concentration of 10 mg/ml in 0.5 M sodium carbonate, pH 9.5. Mix the activated enzyme with the antibody at the desired molar ratio to effect the conjugation. Mixing the equivalent of 4 mg of enzyme per milligram of antibody usually results in acceptable conjugates.
6. React overnight at 4°C.
7. To reduce the resultant Schiff bases and any excess aldehydes, add sodium borohydride to a final concentration of 10 mg/ml.

Note: Some protocols avoid a reduction step. As an alternative to reduction, add 50 μl of 0.2 M lysine in 0.5 M sodium carbonate, pH 9.5, to each milliliter of the conjugation reaction to block excess reactive sites. Block for 2 h at room temperature. Other amine-containing small molecules may be substituted for lysine—such as glycine, Tris buffer, or ethanolamine.

8. Reduce for 1 h at 4°C.
9. To remove any insoluble polymers that may have formed, centrifuge the conjugate or filter it through a 0.45-μm filter. Purify the conjugate by gel filtration or dialysis using PBS, pH 7.4.

1.3. Reductive-Amination-Mediated Conjugation

Oxidation of polysaccharide residues in glycoproteins with sodium periodate provides a mild and efficient way of generating reactive aldehyde groups for subsequent conjugation with amine- or hydrazide-containing molecules via reductive amination (Chapter 1, Section 4.4, and Chapter 2, Section 5.3). Some selectivity of monosaccharide oxidation may be accomplished by regulating the concentration of periodate in the reaction medium. In the presence of 1 mM sodium periodate sialic acid groups will be specifically oxidized at their adjacent hydroxyl residues on the Nos. 7, 8, and 9 carbon atoms, cleaving off two molecules of formaldehyde and leaving one aldehyde group on the No. 7 carbon. At higher concentrations of sodium periodate (10 mM or greater) other sugar residues will be oxidized at points where adjacent carbon atoms contain hydroxyl groups.

Thus, glycoproteins such as horseradish peroxidase, glucose oxidase, or most antibody molecules can be activated for conjugation by brief treatment with periodate. Cross-linking with an amine-containing protein takes place under alkaline pH conditions through the formation of Schiff base intermediates. These relatively labile intermediates can be stabilized by reduction to a secondary amine linkage with sodium cyanoborohydride (Fig. 303).

The use of periodate coupling chemistry for HRP was first introduced by Nakane and Kawaoi (1974; see also Nakane, 1975). In the first step of their protocol, the few amine groups on HRP were initially blocked with 2,4-dinitrofluorobenzene (DNFB) before periodate oxidation. The blocking step was designed to eliminate the possibility of self-conjugation of enzyme molecules during reductive amination with an immunoglobulin. However, Boorsma and Streefkerk (1976a,b) determined that HRP can

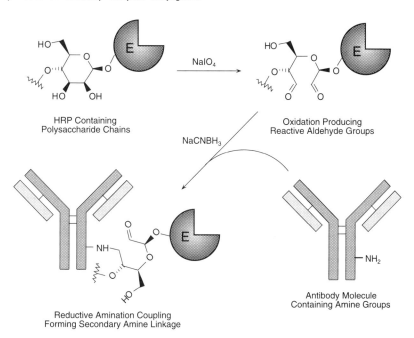

Figure 303 Enzymes that are glycoproteins like HRP may be oxidized with sodium periodate to produce reactive aldehyde residues. Conjugation with an antibody then may be done by reductive animation using sodium cyanoborohydride.

dimerize even after DNFB blocking, perhaps by a mechanism similar to Mannich condensation (Chapter 2, Section 6.2) or through aldol formation. In fact, amine-blocked, periodate-oxidized HRP will form insoluble complexes during storage after just weeks in solution at room temperature or 4°C, indicating that another route of conjugation is taking place.

Reductive amination cross-linking has been done using sodium borohydride or sodium cyanoborohydride; however, cyanoborohydride is the better choice since it is more specific for reducing Schiff bases and will not reduce aldehydes. Small blocking agents such as lysine, glycine, ethanolamine, or Tris can be added after conjugation to quench any unreacted aldehyde sites (Mannik and Downey, 1973; Mattiasson and Nilsson, 1977; Barbour, 1976). Ethanolamine and Tris are the best choices for blocking agents, since they contain hydrophilic hydroxyl groups with no charged functional groups.

The pH of the reductive amination reaction can be controlled to affect the efficiency of the cross-linking process and the size of the resultant antibody–enzyme complexes formed. At physiological pH, the initial Schiff base formation is less efficient and conjugates of lower molecular weight result. At more alkaline pH (i.e., pH 9–10), Schiff base formation occurs rapidly and with high efficiency, resulting in conjugates of higher molecular weight and greater incorporation of enzyme (when oxidized HRP is reacted in excess).

The ability to select the relative size of the antibody–enzyme complex is important, depending on the assay application. Low-molecular-weight conjugates may be more optimal for immunohistochemical staining or blotting techniques where penetration of the complex through membrane barriers is an important consideration. Washing steps also more effectively remove excess reagent if the conjugate is of low molecular weight, thus maintaining a low background signal in an assay. By contrast, conjugates of high molecular weight are more appropriate for ELISA procedures in a microplate format, where high sensitivity is important, but washing off excess conjugate is not a problem.

The protocols appearing in the literature vary according to the amount of periodate used during polysaccharide oxidation, the type of reductant and blocking agent employed for reductive amination, and the pH at which the various reactions are done. This variability indicates considerable flexibility in the protocols, all of which yield usable antibody–enzyme conjugates. There are, however, several conclusions that can be drawn from these studies: Investigations done using HRP indicate the optimal concentration of sodium periodate during oxidation to be approximately 4–8 mM (Tussen and Kurstak, 1984). This reaction should be performed in the dark to prevent periodate breakdown and for a limited period of time (15–30 min) to avoid loss of enzymatic activity. The conjugation reaction should be done at alkaline pH (7.2–9.5) in the presence of a reducing agent to stabilize the Schiff base intermediates. If sodium cyanoborohydride is used as the reductant, a blocking agent should be added at the completion of the conjugation reaction to cap excess aldehyde sites. The following protocol follows these general guidelines and works well especially in the preparation of HRP–antibody conjugates.

Activation of Enzymes with Sodium Periodate

Enzymes that are glycosylated (i.e., HRP and glucose oxidase) may be oxidized according to the following method to produce aldehyde groups for reductive amination coupling to an antibody molecule.

Protocol

1. Dissolve the enzyme to be oxidized in water or 0.01 M sodium phosphate, 0.15 M NaCl, pH 7.2, at a concentration of 10–20 mg/ml.
2. Dissolve sodium periodate in water at a concentration of 0.088 M. Protect from light.
3. Immediately add 100 μl of the sodium periodate solution to each milliliter of the enzyme solution. This results in a 8 mM periodate concentration in the reaction mixture. Mix to dissolve. Protect from light.
4. React in the dark for 20 min at room temperature. If HRP is the enzyme being oxidized, a color change will be apparent as the reaction proceeds—changing the brownish/gold color of concentrated HRP to green.
5. Immediately purify the oxidized enzyme by gel filtration using a column of Sephadex G-25. The chromatography buffer is 0.01 M sodium phosphate, 0.15 M NaCl, pH 7.2. To obtain efficient separation between the oxidized enzyme and excess periodate, the sample size applied to the column should be at a ratio of no more than 5% sample volume to the total column volume. Collect 0.5-ml fractions and monitor for protein at 280 nm. HRP also may be detected by its absorbance at 403 nm. In oxidizing large quantities of HRP, the fraction collection process may be done visually—just pooling the colored HRP peak as it comes off the column.

6. Pool the fractions containing protein. Adjust the enzyme concentration to 10 mg/ml for the conjugation step (see next section). The periodate-activated enzyme may be stored frozen or freeze-dried for extended periods without loss of activity. Do not store the preparation in solution at room temperature or 4°C, since precipitation will occur over time due to self-polymerization.

Activation of Antibodies with Sodium Periodate

Many immunoglobulin molecules are glycoproteins that can be periodate-oxidized to contain reactive aldehyde residues. Polyclonal IgG molecules contain carbohydrate in the Fc portion of the molecule. This is sufficiently removed from the antigen binding sites to allow conjugation to take place through the polysaccharide chains without compromising activity. Although antibody–enzyme conjugation by reductive amination typically is done by oxidation of the enzyme with subsequent cross-linking to an amine-containing antibody, oxidation of the antibody with subsequent conjugation to an amine- or hydrazide-containing molecule also is possible. It should be noted, however, that many monoclonal antibodies are not glycosylated and therefore can not be used in this protocol. A given monoclonal should be checked to verify the presence of carbohydrate before attempting to use a periodate-mediated conjugation protocol.

Protocol

1. Dissolve the antibody to be periodate-oxidized at a concentration of 10 mg/ml in 0.01 M sodium phosphate, 0.15 M NaCl, pH 7.2.
2. Dissolve sodium periodate in water to a final concentration of 0.1 M. Protect from light.
3. Immediately add 100 μl of the sodium periodate solution to each milliliter of the antibody solution. Mix to dissolve. Protect from light.
4. React in the dark for 30 min at room temperature.
5. Immediately purify the oxidized antibody by gel filtration using a column of Sephadex G-25. The chromatography buffer is 0.1 M sodium phosphate, 0.15 M NaCl, pH 7.2. To obtain efficient separation between the oxidized antibody and excess periodate, the sample size applied to the column should be at a ratio of no more than 5% sample volume to the total column volume. Collect 0.5-ml fractions and monitor for protein at 280 nm.
6. Pool the fractions containing protein. Adjust the antibody concentration to 10 mg/ml for the conjugation step. The oxidized antibody should be used immediately.

Conjugation of Periodate-Oxidized HRP to Antibodies by Reductive Amination

The following protocol assumes that HRP has already been periodate-oxidized by the method of Section 1.3, above.

Protocol

1. Dissolve the IgG to be conjugated at a concentration of 10 mg/ml in 0.2 M sodium bicarbonate, pH 9.6, at room temperature. The high-pH buffer will

result in very efficient conjugation with the highest possible incorporation of enzyme molecules per antibody molecule. To produce lower molecular weight conjugates, dissolve the IgG at a concentration of 10 mg/ml in 0.1 M sodium phosphate, 0.15 M NaCl, pH 7.2.

2. The periodate-oxidized enzyme (HRP) prepared in Section 1.3 was finally purified using 0.01 M sodium phosphate, 0.15 M NaCl, pH 7.2. For conjugation using the lower pH buffered environment, this HRP preparation can be used directly at 10 mg/ml concentration. For conjugation using the higher pH carbonate buffer, dialyze the HRP solution against 0.2 M sodium carbonate, pH 9.6, for 2 h at room temperature prior to use. A volume of HRP solution equal to the volume of antibody solution will be required.

3. Mix the antibody solution with the enzyme solution at a ratio of 1:1 (v/v). Since an equal mass of antibody and enzyme is present in the final solution, this will result in a 3.75 molar excess of HRP over the amount of IgG. For conjugates consisting of greater enzyme-to-antibody ratios, proportionally increase the amount of enzyme solution as required. Typically, molar ratios of 4:1 to 15:1 (enzyme:antibody) give acceptable conjugates useful in a variety of ELISA techniques.

4. React for 2 h at room temperature.

5. In a fume hood, add 10 μl of 5 M sodium cyanoborohydride (Sigma) per milliliter of reaction solution. Caution: cyanoborohydride is extremely toxic. All operations should be done with care in a fume hood. Also, avoid any contact with the reagent, as the 5 M solution is prepared in 1 N NaOH.

6. React for 30 min at room temperature (in a fume hood).

7. Block unreacted aldehyde sites by addition of 50 μl of 1 M ethanolamine, pH 9.6, per milliliter of conjugation solution. Approximately a 1 M ethanolamine solution may be prepared by addition of 300 μl ethanolamine to 5 ml of deionized water. Adjust the pH of the ethanolamine solution by addition of concentrated HCl, keeping the solution cool on ice.

8. React for 30 min at room temperature.

9. Purify the conjugate from excess reactants by dialysis or gel filtration using Sephadex G-25. Use 0.01 M sodium phosphate, 0.15 M NaCl, pH 7, as the buffer for either operation. The conjugate may be further purified by removal of unconjugated enzyme by one of the methods of Section 1.5.

Conjugation of Periodate-Oxidized Antibodies with Amine or Hydrazide Derivatives

The following protocol assumes that the antibody has already been periodate-oxidized by the method of Section 1.3 (above) to create reactive aldehyde groups suitable for coupling with amine-containing or hydrazide-containing molecules. This is an excellent method for directing the antibody modification reaction away from the antigen binding sites. For instance, biotinylation of intact antibodies can be done after mild periodate treatment using biotin-hydrazide (Chapter 8, Section 3.3) (Fig. 304). It should be noted, however, that periodate-oxidized antibodies can self-conjugate through their own amines if high-pH reductive amination is used. Conjugation with periodate-oxidized antibodies works best if the receiving molecule is modified to

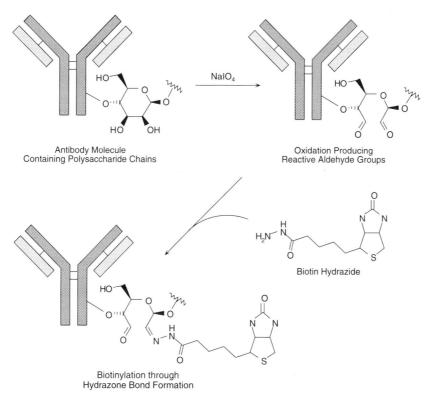

Figure 304 Polysaccharide groups on antibody molecules may be oxidized with periodate to create aldehydes. Modification with biotin–hydrazide results in hydrazone linkages. The sites of modification using this technique are away from the antibody–antigen binding regions, thus preserving antibody activity.

contain hydrazide functional groups and the reaction is done at more moderate pH values.

1. For conjugation to hydrazide-containing proteins, dissolve the periodate-oxidized antibody at a concentration of 10 mg/ml in 0.1 M sodium phosphate, 0.15 M NaCl, pH 7.2. This is the final solution makeup from the oxidation method described previously. For conjugation to amine-containing molecules and proteins, dissolve the oxidized antibody at 10 mg/ml in 0.2 M sodium carbonate, pH 9.6.

2. Dissolve a hydrazide-containing enzyme or other protein at a concentration of 10 mg/ml in 0.1 M sodium phosphate, 0.15 M NaCl, pH 7.2. For the preparation of a hydrazide-activated enzyme see Chapter 16, Section 2.4. For modification with a hydrazide-containing probe, such as biotin-hydrazide, use a concentration of 5 mM in the phosphate buffer. For conjugation through the amine groups of a secondary molecule, dissolve the amine-containing protein at 10 mg/ml in 0.2 M sodium carbonate, pH 9.6.

3. Mix the antibody solution from step 1 with the protein solution from step 2 in amounts necessary to obtain the desired molar ratio for conjugation. Often, the

secondary molecule is reacted in approximately a 4- to 15-fold molar excess over the amount of antibody present.

4. React for 2 h at room temperature.

5. In a fume hood, add 10 μl of 5 M sodium cyanoborohydride (Sigma) per milliliter of reaction solution. Caution: cyanoborohydride is extremely toxic. All operations should be done with care in a fume hood. Also, avoid any contact with the reagent, as the 5 M solution is prepared in 1 N NaOH. The addition of a reductant is necessary for stabilization of the Schiff bases formed between an amine-containing protein and the aldehydes on the antibody. For coupling to a hydrazide-activated protein, however, most protocols do not include a reduction step. Even so, hydrazone linkages may be further stabilized by cyanoborohydride reduction. The addition of a reductant during hydrazide/aldehyde reactions also increases the efficiency and yield of the reaction.

6. React for 30 min at room temperature (in a fume hood).

7. Block unreacted aldehyde sites by addition of 50 μl of 1 M ethanolamine, pH 9.6, per milliliter of conjugation solution. Approximately a 1 M ethanolamine solution may be prepared by addition of 300 μl ethanolamine to 5 ml of deionized water. Adjust the pH of the ethanolamine solution by addition of concentrated HCl, while keeping the solution cool on ice.

8. React for 30 min at room temperature.

9. Purify the conjugate from excess reactants by dialysis or gel filtration using Sephadex G-25. Use 0.01 M sodium phosphate, 0.15 M NaCl, pH 7, as the buffer for either operation. The conjugate may be further purified by removal of unconjugated enzyme by one of the methods of Section 1.5.

1.4. Conjugation Using Antibody Fragments

It is often advantageous to use antibody fragments in the preparation of antibody–enzyme conjugates. Selected fragmentation carried out by enzymatic digestion of intact immunoglobulins can yield lower molecular weight molecules still able to recognize and bind antigen. Conjugation of these fragments with enzyme molecules can result in ELISA reagents that possess better characteristics than corresponding conjugates prepared with intact antibody. Such antibody fragment conjugates display less interference with various Fc binding proteins and also less immunogenicity (due to lack of the Fc region), more facile membrane penetration for immunohistochemical staining techniques (due to lower overall conjugate molecular weight) (Farr and Nakane, 1981; Wilson and Nakane, 1978), and lower nonspecific binding to solid materials (resulting in increased signal-to-noise ratios) (Hamaguchi et al., 1979; Ishikawa et al., 1981a,b).

Enzymatic digests of IgG can result in two particularly useful fragments called Fab and F(ab')$_2$, prepared by the action of papain and pepsin, respectively. Most specific enzymatic cleavages of IgG occur in relatively unfolded regions between the major domains. Papain and pepsin, and similar enzymes including bromelain, ficin, and trypsin, cleave immunoglobulin molecules in the hinge region of the heavy-chain pairs (Liener and Friedenson, 1970; Murachi, 1976). Depending on the location of cleavage, the disulfide groups holding the heavy chains together may or may not remain attached to the antigen binding fragments that result. If the disulfide-bonded region does remain with the antigen binding fragment, as in pepsin digestion, then a divalent molecule is produced [F(ab')$_2$] that differs from the intact antibody by lack of an extended Fc portion. If the disulfide region is below the point of digestion, then the two heavy–light chain complexes that produce the two antigen binding sites of an antibody are cleaved and released, forming individual dimeric fragments (Fab) containing one antigen binding site each (see Fig. 299, discussed previously).

Methods for producing immobilized papain or pepsin for antibody fragmentation can be found in Hermanson *et al.* (1992). The following protocol describes the use of pepsin to cleave IgG molecules at the C-terminal side of the inter-heavy-chain disulfides in the hinge region, producing a bivalent antigen binding fragment, F(ab')$_2$, with a molecular weight of about 105,000 (Fig. 305). Using this enzyme, most of the Fc fragments undergo extensive degradation and cannot be recovered intact.

Preparation of F(ab')$_2$ Fragments Using Pepsin

1. Equilibrate by washing 0.25 ml of immobilized pepsin (Pierce) with 4 × 1 ml of 20 mM sodium acetate, pH 4.5 (digestion buffer). Finally, suspend the gel in 1 ml of digestion buffer.
2. Dissolve 1–10 mg of IgG in 1 ml digestion buffer and add it to the gel suspension.
3. Mix the reaction slurry in a shaker at 37°C for 2–48 h. The optimal time for complete digestion varies depending on the IgG subclass and species of origin. Mouse IgG1 antibodies are usually digested within 24 h and human antibodies are fragmented in 12 h, whereas some minor subclasses (e.g., mouse IgG2a) require a full 48-h digestion period.
4. After the digestion is complete, add 3 ml of 10 mM Tris–HCl, pH 8, to the gel suspension. Separate the gel from the antibody solution by using a serum separator tube or by centrifugation.

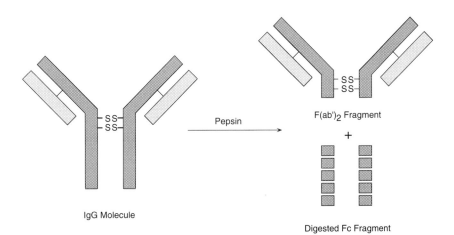

Figure 305 Digestion of IgG class antibodies with pepsin results in heavy chain cleavage below the disulfide groups in the hinge region. The bivalent fragments that are formed are called F(ab')$_2$. The remaining Fc region is normally severely degraded into smaller peptide fragments.

5. Apply the fragmented IgG solution to an immobilized protein A column containing 2 ml of gel (Pierce) that was previously equilibrated with 10 mM Tris–HCl, pH 8.
6. After the sample has entered the gel, wash the column with 10 mM Tris–HCl, pH 8, while collecting 2-ml fractions. The fractions may be monitored for protein by measuring absorbance at 280 nm. The protein peak eluting unretarded from the column is F(ab')$_2$.
7. Bound Fc or Fc fragments and any undigested IgG may be eluted from the column with 0.1 M glycine, pH 2.8.

Similarly, immobilized papain may be used to generate Fab fragments from immunoglobulin molecules. Papain is a sulfhydryl protease that is activated by the presence of a reducing agent. Cleavage of IgG occurs above the disulfides in the hinge region, creating two types of fragments, two identical Fab portions and one intact Fc fragment (Fig. 306). For preparation of the immobilized papain gel used in the following protocol see Hermanson *et al.* (1992). The gel also is commercially available from Pierce Chemical.

Preparation of Fab Fragments Using Papain

1. Wash 0.5 ml of immobilized papain (Pierce) with 4 × 2 ml of 20 mM sodium phosphate, 20 mM cysteine–HCl, 10 mM EDTA, pH 6.2 (digestion buffer), and finally suspend the gel in 1 ml of digestion buffer.
2. Dissolve 10 mg of human IgG solution in 1 ml of digestion buffer and add it to the immobilized papain gel suspension.
3. Mix the gel suspension in a shaker at 37°C for 4–48 h. Maintain the gel in suspension during mixing. The optimal time for complete digestion varies depending on the IgG subclass and species of origin. Mouse IgG1 antibodies are usually digested within 27 h, whereas other mouse subclasses require only 4 h; human antibodies are fragmented in 4 h (IgG1 and IgG3), 24 h (IgG4), or 48 h

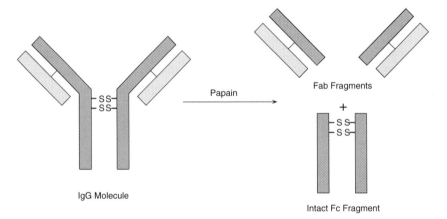

Figure 306 Papain digestion of IgG antibodies primarily results in cleavage in the hinge region above the interchain disulfides. This produces two heavy–light chain pairs, called Fab fragments, containing one antigen binding site each. The Fc region normally can be recovered intact.

(IgG2), whereas bovine, sheep, and horse antibodies are somewhat resistant to digestion and require a full 48 h.

4. After the required time of digestion, add 3.0 ml 10 mM Tris–HCl buffer, pH 8, to the gel suspension, mix, and then separate the digest solution from the gel by use of a serum separator tube or by centrifugation at 2000 g for 5 min.

5. Apply the supernatant liquid to an immobilized protein A column (2 ml gel; Pierce) that was previously equilibrated by washing with 20 ml of 10 mM Tris–HCl buffer, pH 8.

6. After the sample has entered the gel bed, wash the column with 15 ml of 10 mM Tris–HCl buffer, pH 8, while 2-ml fractions are collected. Monitor the fractions for protein by their absorbance at 280 nm. The protein eluted unretarded from the column is purified Fab.

7. Elute Fc and undigested IgG bound to the immobilized protein A column with 0.1 M glycine–HCl buffer, pH 2.8.

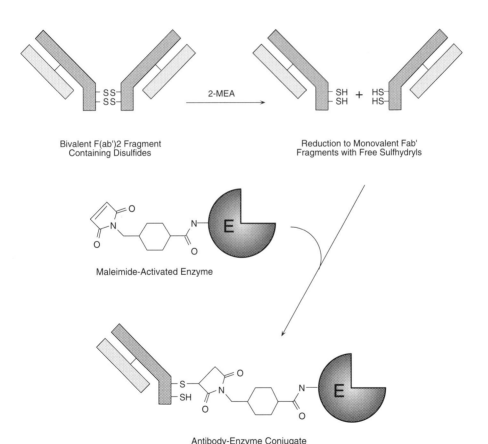

Figure 307 F(ab′)$_2$ fragments produced by pepsin digestion of IgG can be reduced at their heavy chain disulfides using a reducing agent, such as 2-mercaptoethylamine. Conjugation then can be done with a maleimide-activated enzyme to produce low-molecular-weight complexes linked by thioether bonds.

Conjugation of these fragments with enzymes is done using methods similar to those previously discussed for intact antibody molecules. F(ab′)$_2$ fragments may be selectively reduced in the hinge region with 2-mercaptoethylamine using the protocols identical to those outlined for whole antibody molecules (Chapter 1, Section 4.1, and Chapter 10, Section 1.1). Mild reduction results in cleaving the disulfides holding the heavy chain pairs together at the central portion of the fragment, thus creating two F(ab′) fragments each containing one antigen binding site (Fig. 307).

The amine groups on these fragments also may be modified with thiolating agents, such as SATA or 2-iminothiolane, to create sulfhydryl residues suitable for coupling to

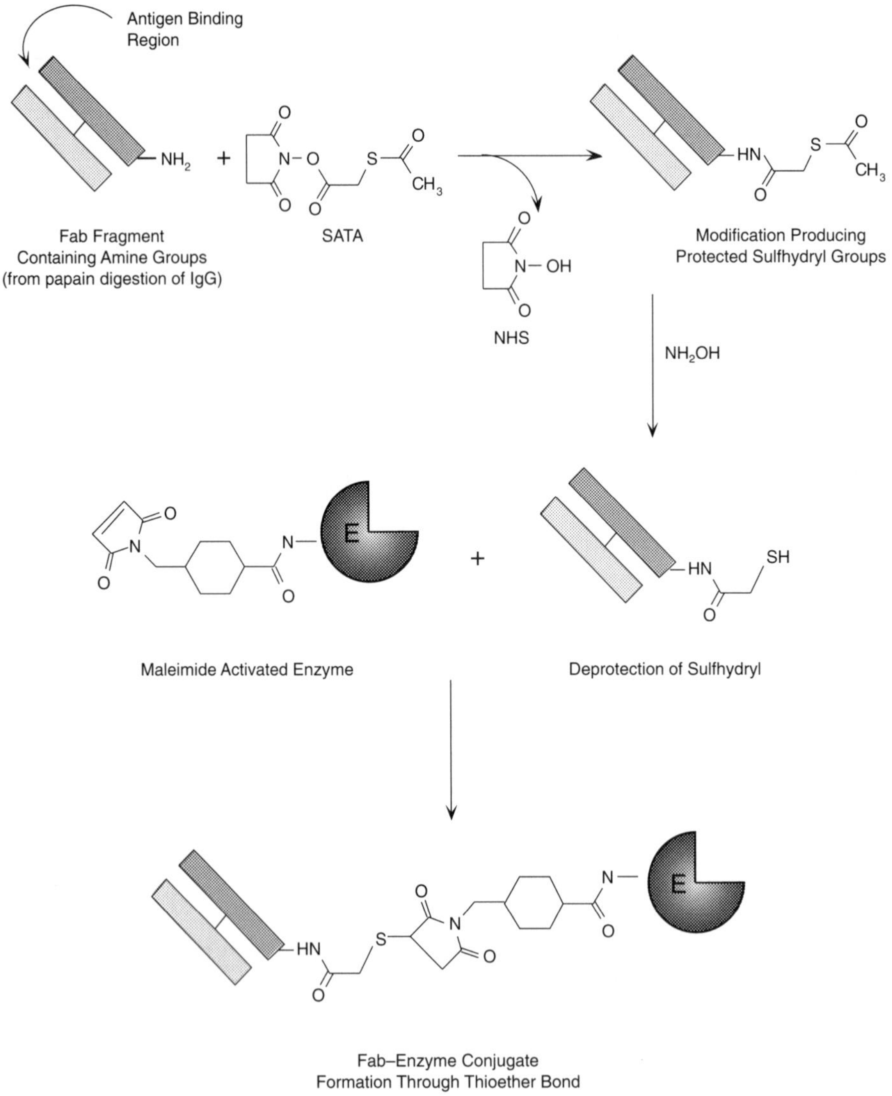

Figure 308   The thiolation reagent SATA can be used to create sulfhydryl groups on Fab fragments. After deprotection of the acetylated thiol of SATA with hydroxylamine, conjugation with a maleimide-activated enzyme may take place, producing thioether linkages.

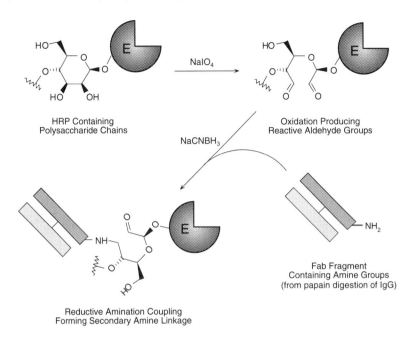

Figure 309 Periodate oxidation of HRP creates aldehyde groups on the carbohydrate chains of the enzyme. Reaction with a Fab fragment then may be done using reductive amination to produce a low-molecular-weight complex.

a maleimide-activated enzymes (Section 1.1) (Fig. 308). Amine groups further may be utilized in reductive amination coupling to periodate-oxidized glycoproteins, such as in the protocol outlines for HRP conjugation, previously (Section 1.3) (Fig. 309). Successful periodate oxidation of the fragments themselves, however, may not be possible since immunoglobulin molecules contain carbohydrate mainly in the C_H2 domain, most of which usually is cleaved off and lost during digestion of the antibody. Finally, glutaraldehyde-mediated conjugation techniques will work with antibody fragments, but are not recommended due to the reasons discussed in Section 1.2.

The primary goal of any of these conjugation schemes using antibody fragments is to maintain the activity of the antigen binding site while limiting the size of the final complex with a second molecule. The use of heterobifunctional cross-linkers such as SMCC or reductive amination techniques allows sufficient control over the process to reach these goals.

1.5. Removal of Unconjugated Enzyme from Antibody–Enzyme Conjugates

Conjugates of antibodies and enzymes are essential components in immunoassay and detection systems. In the preparation of such conjugates, a molar excess of enzyme typically is cross-linked to a specific antibody to obtain a complex of high activity. The result of this ratio is excess enzyme left unconjugated after completion of the reaction. The unconjugated enzyme confers nothing to the utility of the final product and can be detrimental by contributing to increased backgrounds in assay procedures. The removal of this free enzyme component may be advantageous to improving the resultant signal-to-noise ratio in some immunoassays. Commercial preparations of antibody–enzyme conjugates usually are not purified to remove unconjugated enzyme. Frequently, the major proteinaceous part of these products is not active conjugate, but leftover enzyme that contributes nothing to the immunochemical activity of what was purchased.

Boorsma and Kalsbeek (1976) state that unconjugated HRP must be removed from antibody–enzyme conjugates to obtain optimal staining in immunoassay procedures. This is especially true in blotting techniques and cytochemical staining where free enzyme may become entrapped nonspecifically within the membrane or cellular structures. The presence of this unconjugated enzyme leads to diffuse substrate noise that can obscure the immunospecific signal.

Several methods may be used to purify an antibody–enzyme conjugate and remove unconjugated enzyme. For instances where the enzyme molecular weight is significantly different than the conjugate molecular weight, separation may be achieved by gel filtration chromatography. Using the proper support with an exclusion limit and separation range able to accommodate all the proteins in the sample, the conjugate peak will elute before the enzyme peak, thus providing an efficient way of removing free enzyme. However, gel filtration procedures can be time consuming and of relatively low capacity for the amount of gel required. In addition, separation of higher molecular weight enzymes from antibody conjugates, such as in the case of alkaline phosphatase (MW 140,000), is considerably less efficient or impossible. Gel filtration separation also becomes a problem if the conjugate itself consists of a broad range of molecular weights, as is often true when glutaraldehyde is used as the cross-linking agent.

The most effective methods of removing unconjugated enzyme all make use of affinity chromatography systems using specific ligands that can interact with the antibody portion of the conjugate. Thus, the supports retain any unconjugated antibody (usually in very low percentage when the enzyme is reacted in excess) as well as the antibody–enzyme conjugate produced from the cross-linking reaction. The unconjugated enzyme, however, passes through such affinity columns unretarded. Two main methods are discussed below: (1) affinity chromatography, which makes use of immobilized immunoglobulin binding proteins or immobilized antigen molecules having specificity for the antibody used in the conjugate, and (2) nickel-chelate affinity chromatography, which binds the Fc region of antibody molecules.

The use of immunoaffinity techniques (whether antigen specific or immunoglobulin binding proteins such as protein A) allows strong binding of the antibody conjugate, but has the significant disadvantage of having elution conditions that are often too severe for maintaining activity of the antibody or enzyme components. By contrast, nickel-chelate affinity techniques give excellent binding of the conjugate while allowing free enzyme to pass through the gel unretarded. It also has the significant advantage of having mild elution conditions, which preserve the activity of the conjugate.

Immunoaffinity Chromatography

Immunoaffinity chromatography makes use of immobilized antigen molecules to bind and separate specific antibody from a complex mixture. After the preparation of an antibody–enzyme conjugate, the antibody binding capability of the cross-linked complex toward its complementary antigen ideally remains intact. This highly specific interaction can be used to purify the conjugate from excess enzyme if the antibody and enzyme can survive the conditions necessary for binding and elution from such a column. Binding conditions typically are mild physiological pH conditions that cause no difficulty. However, many elution conditions require acidic or basic conditions or the presence of a chaotropic agent to deform the antigen binding site. Sometimes these conditions can irreversibly damage the antigen binding recognition capability of the antibody or denature the active site of the enzyme, thus diminishing enzymatic activity. Activity losses for both the antibody and the enzyme can be severe under such conditions.

Another potential disadvantage of an immunoaffinity separation is the assumed abundance of the purified antigen in sufficient quantities to immobilize on a chromatography support. Protein antigens should be immobilized at densities of at least 2–3 mg/ml of affinity gel to produce supports of acceptable capacity for binding antibody. Often, the antigen is too expensive or scarce for the amounts needed to be obtained.

However, if the antigen is abundant and inexpensive and the antibody–enzyme complex will survive the associated elution conditions, then immunoaffinity chromatography can provide a very efficient method of purifying a conjugate from excess enzyme. This method also ensures that the recovered antibody still retains its ability to bind specific target molecules (i.e., the antigen binding sites were not blocked during conjugation). The preparation of immunoaffinity supports can be found in Hermanson *et al.* (1992). A suggested method for performing immunoaffinity chromatography follows:

Protocol

1. Equilibrate the immunoaffinity column with 50 mM Tris, 0.15 M NaCl, pH 8 (binding buffer). Wash with at least 5 column volumes of buffer. The amount of gel used should be based on the total binding capacity of the support. A determination of binding capacity can be done by overloading a small-scale column, eluting, and measuring the amount of conjugate that bound. Such an experiment may be coupled with a determination of conjugate viability for using immunoaffinity as the purification method. The final column size should represent an amount of gel capable of binding at least 1.5 times more than the amount of conjugate that will be applied.

2. Apply the conjugate to the column in the binding buffer while taking 2-ml fractions.

3. Wash with binding buffer until the absorbance at 280 nm decreases back to baseline. The unbound protein flowing through the column will consist of mainly unconjugated enzyme. Some conjugate may flow through also if some of the conjugate is inactive or the column is overloaded.

4. Elute the bound conjugate with 0.1 M glycine, 0.15 M NaCl, pH 2.8, or another suitable elution buffer. A neutral pH alternative to this buffer is the Gentle Elution Buffer from Pierce Chemical. If acid pH conditions are used, immediately neutralize the fractions eluting from the column by the addition of 0.5 ml of 1 M Tris, pH 8, per fraction.

Nickel-Chelate Affinity Chromatography

Metal chelate affinity chromatography is a powerful purification technique whereby proteins or other molecules can be separated based upon their ability to form coordination complexes with immobilized metal ions (Porath *et al.*, 1975; Porath and Belew, 1983; Porath and Olin, 1983; Lonnerdal and Keen, 1982; Sulkowski, 1985; Kagedal, 1989). The metal ions are stabilized on a matrix through the use of chelation compounds, which usually have multivalent points of interaction with the metal atoms. To form useful affinity supports, these metal ion complexes must have some free or weakly associated and exchangeable coordination sites. These exchangeable sites can then form complexes with coordination sites on proteins or other molecules. Substances that are able to interact with the immobilized metals will bind and be retained on the column. Elution is typically accomplished by one or a combination of the following options: (1) lowering of pH, (2) raising the salt strength, and/or (3) the inclusion of chelation agents such as EDTA in the buffer.

Sorensen (1993) reported that a nickel chelate column will specifically bind IgG class immunoglobulins while allowing certain enzymes to pass through the gel unretarded (Pierce). This phenomenon allows the separation of antibody–enzyme complexes containing, in particular, HRP or alkaline phosphatase conjugated to common polyclonal or monoclonal antibodies. The nickel chelate column binds the conjugate through the Fc region of the associated antibody, even if enzyme molecules are covalently attached. Any unconjugated enzyme will pass through the affinity column unretarded (Fig. 310).

Elution of the bound antibody–enzyme conjugate occurs by only a slight shift in pH to acidic conditions or through the inclusion of a metal chelating agent like EDTA in the binding buffer. Either method of elution is mild compared to most immunoaffinity separation techniques (discussed in previous section). Thus, purification of the antibody–enzyme complex can be done without damage to the activity of either component.

One limitation to this method should be noted. If the antibody–enzyme conjugate is prepared using antibody fragments such as Fab or F(ab′)$_2$, then nickel-chelate affinity chromatography will not work, since the requisite Fc portion of the antibody necessary for complexing with the metal is not present.

The preparation of a metal chelate affinity support containing iminodiacetic acid functional groups may be found in Hermanson *et al.* (1992), or it can be purchased from a commercial source. The following protocol is adapted from the instructions accompanying the nickel chelate support. Pierce offers a kit based on this technology for the purpose of removing unconjugated enzyme from antibody–enzyme conjugates (called the FreeZyme conjugate purification kit).

Protocol

1. Pack a column containing an immobilized iminodiacetic acid support (Pierce). The column size should be no less than 1.5 times that required to bind the anticipated amount of conjugate to be applied. The maximal capacity of such a column for binding antibody can be up to 50 mg/ml gel; however, best results are obtained if no more than 10–20 mg/ml of conjugate is applied.
2. Dissolve 50 mg of nickel ammonium sulfate per milliliter of deionized water. Apply 1 ml of nickel solution per milliliter of gel to the column.

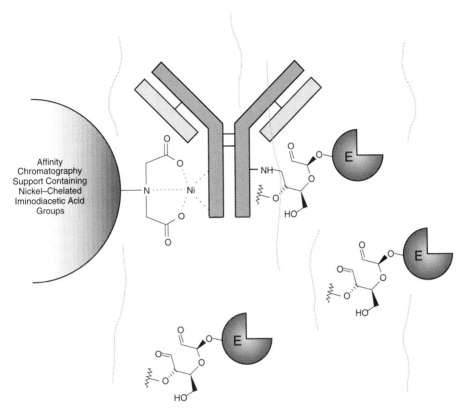

Figure 310 An affinity chromatography support containing iminodiacetic acid groups chelated with nickel may be used to remove excess enzyme after reactions to produce antibody–enzyme conjugates. The nickel chelate binds to the antibody in the Fc region, retaining the conjugate while allowing free enzyme to pass through the gel unretarded.

3. Wash the column with 10 vol of water, then equilibrate the support with 2 vol of 10 mM sodium phosphate, 0.15 M NaCl, pH 7 (binding buffer).
4. Dissolve or dialyze the conjugate into binding buffer. Apply the conjugate solution to the column while collecting 2-ml fractions.
5. Continue to wash the gel with 0.15 M NaCl (saline solution) until the absorbance at 280 nm is down to baseline. The protein eluting from the column at this point is unconjugated enzyme.
6. Elute the bound conjugate with 0.1 M sodium acetate, 0.5 M NaCl, pH 5. Pool the fractions containing protein and dialyze the conjugate into 10 mM sodium phosphate, 0.15 M NaCl, pH 7, or other suitable storage buffers.

2. Preparation of Labeled Antibodies

In addition to labeling immunoglobulins with enzymes to provide detectability through their catalytic action on a substrate, antibody molecules also can be labeled or tagged with small compounds that can provide indigenous traceable properties. The specificity of the antibody then can be used to bind unique antigenic determinants, while the attached tag supplies the chemical visibility necessary for detection. Such small chemical labels typically are one of two types: intense fluorophores or unstable, radioactive isotopes.

Radiolabeling antibodies with ^{125}I forms the basis for highly sensitive radioimmunoassays (RIA) that were first developed in the early days of immunoglobulin-mediated testing. The use of radioisotopes in tagging antibodies is used less often today for *in vitro* immunoassays due to the hazards associated with handling and disposal of radioactive compounds. However, isotopes other than ^{125}I are becoming very important as monoclonal labels for use in *in vivo* diagnostic or therapeutic procedures for cancer therapy. In addition, a radiolabel has distinct advantages over other chemical tags. It is not influenced by conformational changes within the antibody molecule or by changes in its chemical environment as enzymes or labels with unique spectral characteristics can be. Thus, radiolabels still can provide a means of detection equal to or exceeding the most sensitive and reliable tags now available.

Another form of label often used to tag antibody molecules is chemical modification with a reagent terminating in a biotin group. Biotinylation (Chapter 13, Section 6) creates a handle on the immunoglobulin with the ability to bind avidin or streptavidin strongly in one of the most tightly held noncovalent interactions known. With a dissociation constant (K_d) on the order of 1.3×10^{-15}, the avidin–biotin interaction can be used to detect biotinylated molecules with extreme sensitivity. In this type of system, instead of the antibody being labeled, the avidin (or streptavidin) molecules are modified to contain the detection complex—consisting of enzyme, fluorophore, or radiolabel. Interaction of the biotinylated antibody with its targeted antigen is amplified and detected by addition of such labeled avidin or streptavidin reagents.

The following three sections describe the preparation and properties of fluorescent, radiolabeled, and biotinylated antibodies.

2.1. Fluorescently Labeled Antibodies

Antibody molecules can be labeled with any one of more than a dozen different fluorescent probes currently available from commercial sources. Each probe option has its own characteristic spectral signals of excitation (or absorption) and emission (or fluorescence). Many derivatives of these fluorescent probes possess reactive functional groups convenient for covalently linking to antibodies and other molecules. Each of the main fluorophore families contains at least a few different choices in coupling chemistry to direct the modification reaction to selected functional groups on the molecule to be labeled. These choices include amine-reactive, sulfhydryl-reactive, and carbonyl-reactive. Examples of some of the more popular varieties can be found in Chapter 8, Section 1.

In addition to the wide range of commercial probes obtainable, many other fluorescent molecules have been synthesized and described in the literature. Only a handful, however, are generally used to label antibody molecules. Perhaps the most common fluorescent tags with application to immunoglobulin assays are reflected in the main derivatives produced by the prominent antibody-manufacturing companies. These include derivatives of fluorescein, rhodamine, Texas red, aminomethylcoumarin (AMCA), and phycoerythrin. Figure 311 shows the reaction of fluorescein isothiocyanate, one of the most common fluorescent probes, with an antibody molecule.

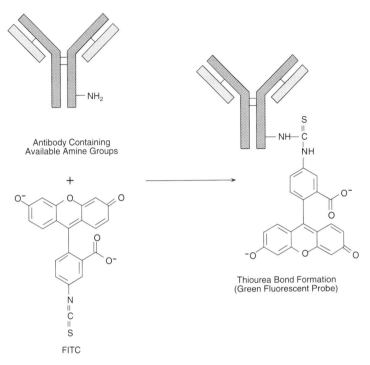

Figure 311 FITC may be used to label amine groups on antibody molecules, forming isothiourea bonds.

Industrial standardization has occurred for the use of these five probes due to the large literature documentation available on their successful application to antibody-based assays. As a result of this, instrumentation has become widely available for measuring any of these probes' fluorescent response, including standard filter selections that match their emission patterns. Such fluorescently labeled antibodies can be used in immunohistochemical staining (Osborn and Weber, 1982), in flow cytometry or cell sorting techniques (Ormerod, 1990; Watson, 1991), for tracking and localization of antigens, and in various double-staining methods (Kawamura, 1977).

In choosing a fluorescent tag, the most important factors to consider are good absorption, stable excitation, and efficient, high-quantum yield of fluorescence. Some fluorophores, such as fluorescein, exhibit fluorescent quenching, which lowers the quantum yield over time. Up to 50% of the fluorescent intensity observed on a fluorescein-stained slide can be lost within 1 month in storage. AMCA, by contrast, has much better stability, but all fluorophores lose some intensity upon exposure to light or upon storage.

In some cases, the preparation of a fluorescently labeled antibody is not even necessary. Particularly, if indirect methods are used to detect antibody binding to antigen, then preparing a fluorescently labeled primary antibody is not needed. Instead, the selection from a commercial source of a labeled secondary antibody having specificity for the species and class of primary antibody to be used is all that is required. However, if the primary antibody needs to be labeled and it is not manufactured commercially, then a custom labeling procedure will have to be done.

Generalized protocols for the attachment of these fluorophores to protein molecules, including antibodies, can be found in Chapter 8, Section 1. The main consideration for the modification of immunoglobulins is to couple these probes at an optimal level to allow good detectability without high backgrounds. Too low a substitution level and the response of the fluorophore will yield low signal strength and poor sensitivity. Too high a substitution level and the fluorophore may decrease the antibody's ability to bind target molecules by blocking the antigen binding sites or cause nonspecific interactions resulting in high background or noise levels. In some cases, trial and error will be required to optimize this process.

For other examples of antibody labeling protocols see Goding (1976) and Harlow and Lane (1988a,b,c).

2.2. Radiolabeled Antibodies

The attachment of a radioactive label onto an antibody molecule provides a powerful means of detection in immunoassay procedures, tracking of analytes, for *in vivo* diagnostic procedures, and, more recently, for the detection or therapy of numerous types of malignancies. Originally, radiolabeling of antibodies merely meant modifying tyrosine residues with [125]I. Now, a number of different radioactive elements are being attached, both covalently and through specialized chelating compounds, to provide imaging capabilities for the detection of primary tumors and metasticies (Order, 1989).

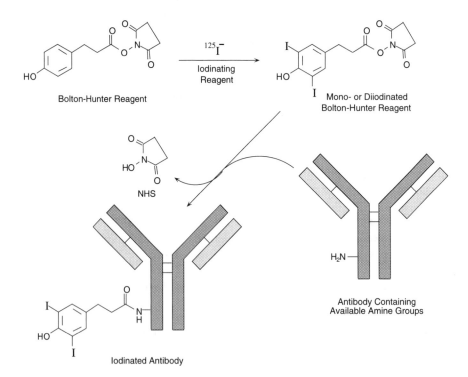

Figure 312 Bolton–Hunter reagent may be used to add radioactive iodine labels to antibody molecules by modification of amines.

Radioiodination can be done using any one of a number of techniques. Most of the procedures utilize ^{125}I as the unstable isotope of choice for *in vitro* use due to its easy availability, comparably long 60-day half-life, and relatively low-energy photon emissions. Radioactive ^{125}I usually is supplied as its sodium salt and must be oxidized to create an electrophilic species capable of modifying molecules. Commonly used oxidizing agents include chloramine-T, IODO-GEN, and IODO-BEADS (Chapter 8, Section 4). When used in direct labeling techniques with antibodies and other proteins, these oxidants cause an iodination reaction to occur at available tyrosine residues within the polypeptide chain. If tyrosine is important to antibody activity and cannot be labeled, then certain cross-linking or modification reagents containing an activated aromatic ring also may be iodinated to label at other functional sites within the protein molecule. An example of this technique is to use the Bolton–Hunter reagent (Chapter 8, Section 4.5) labeled with radioactive iodine to modify the primary amines within the antibody. This reagent also can be used to add an iodinatable site to molecules containing no tyrosine residues (Fig. 312).

Reagent options and protocols for the radioiodination of antibodies and other molecules may be found in Chapter 8, Section 4.

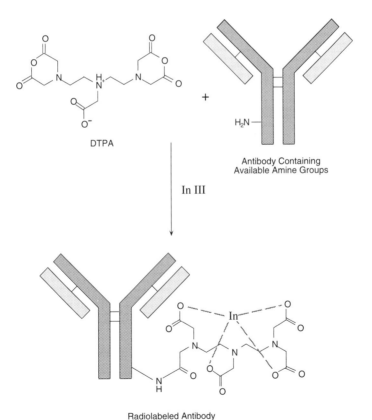

Figure 313 The bifunctional chelating reagent DTPA may be used to modify amine groups on antibody molecules, forming amide bond linkages. Radioactive In III then may be complexed to create a radiolabeled targeting reagent.

Another method of adding a radioactive tag to antibodies is to use a chelating compound capable of complexing metal isotopes. One of the most frequently used chelating reagents is DTPA (Chapter 8, Section 2.1). The reagent contains two anhydride groups that can be used to modify primary amines in proteins and other molecules. The reaction process involves ring opening and the formation of an amide bond. Ring opening also creates up to four free carboxylate groups that are able to form coordination complexes with radionuclides such as indium-III (Fig. 313). Monoclonals labeled with chelated radioactive isotopes of such metals can be used in targeting tumor cells *in vivo*. The detection sensitivity of radiolabeled antibodies has led to effective diagnostic procedures to monitor primary and secondary cancer growths. In addition, the intensity of radioactivity at the tumor site when labeled monoclonals are targeted therapeutically can be great enough to cause tumor cell death and remission.

2.3. Biotinylated Antibodies

Another popular tag for use with immunoglobulins is biotin. Modification reagents that can add a functional biotin group to proteins, nucleic acids, and other molecules now come in many shapes and reactivities (Chapter 8, Section 3). Depending on the functional group present on the biotinylation reagent, specific reactive groups on antibodies may be modified to create an avidin (or streptavidin) binding site. Amines, carboxylates, sulfhydryls, and carbohydrate groups can be specifically targeted for biotinylation through the appropriate choice of biotin derivative. Figure 314 shows the biotinylation of an antibody with NHS-LC-biotin, one of the most common biotinylation reagents.

The presence of biotin labels on an antibody molecule provides multiple sites for the binding of avidin or streptavidin. If the biotin binding protein is in turn labeled with an enzyme, fluorophore, etc., then a very sensitive detection system is created. The potential for more than one labeled avidin to become attached to each antibody through its multiple biotinylation sites provides an increase in detectability over antibodies directly labeled with a detectable tag.

Several assay designs that use the enhanced sensitivity afforded through biotinylated antibodies have been developed. Most of these systems use conjugates of avidin or streptavidin with enzymes (such as HRP or alkaline phosphatase), although other labels (such as fluorophores) can be used as well. In the simplest assay design, called the labeled avidin–biotin system (LAB), a biotinylated antibody is allowed to incubate and bind with its target antigen. Next, an avidin–enzyme conjugate is introduced and allowed to interact with the available biotin sites on the bound antibody. Substrate development then provides the chemical detectability necessary to quantify the antigen.

In a slightly more complex design, the bridged avidin–biotin system (BRAB) uses avidin's multiple biotin binding sites to create an assay potentially of higher sensitivity than that of the LAB assay. Again the biotinylated antibody is allowed to bind its target, but next an unmodified avidin is introduced to bind with the biotin binding sites on the antibody. Finally, a biotinylated enzyme is added to provide a detection vehicle. Since the bound avidin still has additional biotin-binding sites available, the potential exists for more than one biotinylated enzyme to interact with each bound

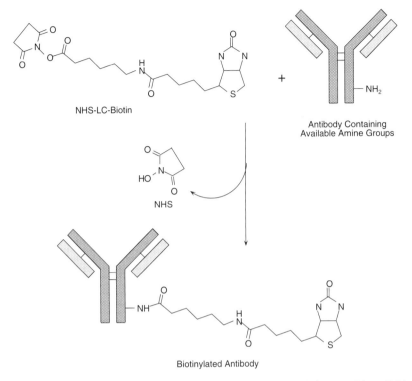

Figure 314 Biotinylated antibodies can be formed by reacting NHS-LC–biotin with available amine groups to create amide bonds.

avidin. In some cases, sensitivity can be increased over that of the LAB technique by using this bridging ability of avidin or streptavidin (Chapter 13, Section 2).

A modification on this theme can be used to produce one of the most sensitive enzyme-linked assay systems known. The ABC system (for avidin–biotin complex) increases the detectability of antigen beyond that possible with either the LAB or the BRAB designs by forming a polymer of biotinylated enzyme and avidin before addition to an antigen-bound, biotinylated antibody. When avidin and a biotinylated enzyme are mixed together in solution in the proper proportion, the multiple binding sites on avidin create a linking matrix to form a high-molecular-weight complex. If the biotinylated enzyme is not in large enough excess to block all the binding sites on avidin, then additional sites will still be available on this complex to bind a biotinylated antibody bound to its complementary antigen. The large complex provides multiple enzyme molecules to enhance the sensitivity of detecting antigen. Thus, the ABC procedure is currently among the highest-sensitivity methods available for immunoassay work.

More detailed discussions of avidin–biotin systems and the process of adding a biotin handle to proteins, nucleic acids, and other biomolecules can be found in Chapter 8, Section 3 and Chapter 13.

11

Immunotoxin Conjugation Techniques

Monoclonal antibodies directed against tumor antigens may be used as targeting agents for conducting certain cytotoxic substances to malignant cells for selective killing. Numerous cell surface markers are known to proliferate in human solid tumors (Boyer *et al.*, 1988). The ability to raise monospecific antibodies to these markers creates the capacity to target discretely the tumor, causing cell death while leaving healthy cells alone (Salmon, 1989).

The design of such cytotoxic antibodies is conceptually simple: attach a toxic substance or a mediator of toxicity to the appropriate monoclonal and you have a "magic bullet" that can find and eliminate the one-in-a-billion cells that have the requisite marker (Fig. 315). The antibody provides the recognition and binding capacity, while the associated toxic component effects cellular alterations leading to cell death.

The approach to constructing antibody immunoconjugates for cancer has taken a number of forms (Vogel, 1987). One of the first designs used conjugates of monoclonal antibodies with toxins that were able to block protein synthesis at the ribosome level inside the cell (Lord *et al.*, 1988). Other conjugate forms used radioactive labels that killed cells by overexposure to radiation in the proximity of where antibody docking occurred (Order, 1989). Drug conjugates were also constructed that combined the known benefits of chemotherapy with the targeting capability of monoclonals (Reisfeld *et al.*, 1989; Willner *et al.*, 1993). The expected result was the effective concentration of the chemotherapeutic agent at the location of the tumor—hopefully eliminating or minimizing the side effects of traditional chemotherapy. In addition, conjugates of monoclonals with certain biological modulators such as lymphokines or growth factors were tried to affect malignant cell viability (Obrist *et al.*, 1988).

Some immunoconjugates utilize intermediate carrier systems consisting of polymeric molecules such as polysaccharides, particularly dextran (Hurwitz *et al.*, 1978, 1980, 1983a,b, 1985; Sela and Hurwitz, 1987; Manabe *et al.*, 1983). The activated dextran is cross-linked to both the monoclonal and the cytotoxic agent, providing multivalent conjugation sites to create larger complexes (Section 2.3 and Chapter 15, Section 2). Liposomes can be used in similar fashion by anchoring the antibody to its outer surface and charging the vesicle with cytotoxic compounds (Gregoriadis, 1984; Matthay *et al.*, 1986; Singhal and Gupta, 1986; Ho *et al.*, 1986). See Chapter 12 for a survey of liposome conjugation techniques.

494

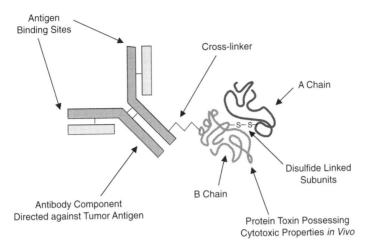

Figure 315 The basic design of an immunotoxin conjugate consists of an antibody targeting component cross-linked to a toxin molecule. The complexation typically includes a disulfide bond between the antibody portion and the cytotoxic component of the conjugate to allow release of the toxin intracellularly. In this illustration, an intact A–B toxin protein provides the requisite disulfide, but the linkage also may be designed into the cross-linker itself.

Other systems were designed to use two-stage approaches where an antibody was conjugated with an intermediate agent, which, when combined with another factor, could elicit cytotoxicity. For instance, enzyme conjugates with monoclonals have been used that could transform an inactive prodrug into a chemotactic agent *in vivo* (Senter *et al.*, 1988). Monoclonals also could be tagged with a biotin label and a secondary, avidin–toxin conjugate used to target the antibody once it bound the tumor cell. Two-stage radiative treatment also has been tried through the use of boron neutron capture therapy (Holmberg and Meurling, 1993). In this case, a carrier molecule is modified to contain ^{10}B. After administration *in vivo* to target tumor cells, neutron bombardment yields an unstable intermediate ^{11}B, which immediately undergoes a fission reaction to yield ^{7}Li and ^{4}He. The induced radiation then kills the cells.

Tumor-targeting conjugates that use biospecific agents other than monoclonal antibodies have been developed as well. The targeting component in these systems consists of any molecule that can function as a ligand having specific affinity for some receptor molecule on the surface of the tumor cells. Such an affinity molecule might be a hormone (typically called hormonotoxins; Singh *et al.*, 1989, 1993), a growth factor, an antigen specific for binding particular antibodies projecting from B cell surfaces, transferrin, α_2-macroglobulin, or anything else able to specifically interact with the targeted tumor cells.

It became apparent very early in the development of such agents that their conception and design were much easier to imagine than to implement successfully. Monoclonal antibodies used *in vitro* could easily detect antigen molecules in complex mixtures with very little nonspecificity or cross-reactivity. Very quickly following the invention of hybridoma technology, monoclonals were employed in numerous diagnostic assays. Their use as therapeutic agents *in vivo*, however, was complicated by the body's natural immune response designed to prevent the invasion of foreign substances.

Injection of immunotoxin conjugates prepared from mouse monoclonals usually results in antibody production against the foreign protein. Sometimes allergic reactions can further complicate the side effects, making continued therapy infeasible. Most often, however, induced host immunoglobulins will quickly bind the immunoconjugate and remove it from the circulatory system. Instead of finding the targeted tumor cells, the immunotoxin ends up sequestered and degraded in the liver or removed by the kidneys. Since the common culprit in this scenario is the monoclonal, the acronym HAMA, for *human anti-mouse antibody*, is given to the response.

In an attempt to overcome the HAMA problem, "humanized" mouse monoclonals were designed where large portions of the murine antibody are substituted for their human counterparts. For instance, replacing a mouse Fc portion with the corresponding human ones can significantly decrease the immune response against such conjugates. Replacing everything but the hypervariable regions that code for antigen binding has been accomplished, too. Unfortunately, regardless of how much "humanization" is done, the remaining murine part still has the potential of causing immunological reactions. In addition, there is the potential for generating an immune response from the toxic component of the immunoconjugate, but this is unavoidable unless immune modulators are administered to reduce the overall immune system responsiveness.

Modification of immunotoxin conjugates with synthetic polymers has been used to mask the complex from the host immune response. Particularly, polyethylene glycol (PEG; Chapter 15, Section 1) has been found to be quite successful in reducing or eliminating the HAMA reaction (Roffler and Tseng, 1994). Modification of an antibody conjugate with two to four PEG molecules increases the serum half-life and improves tumor localization of the targeted reagent.

Other innovations in preparing targeted conjugates for cancer utilize recombinant DNA techniques to create antibody molecules that are entirely of human origin (Huse *et al.*, 1989; Sastry *et al.*, 1989; Orlandi *et al.*, 1989). A completely human antibody molecule eliminates the immunological problems associated with mouse monoclonals. Intact antibodies, Fab fragments, small Fv fragments held together by synthetically designed amino acid segments, and short peptides representing the antigen binding site have all been developed by recombinant means. Although frequently the word "antibody" is used to describe these engineered proteins, many of the molecules are far removed from the traditional picture of an antibody molecule. The terms "single-chain antibody" or "single-chain Fv protein" are commonly used and more closely describe these new targeting molecules.

With the great diversity of targeted toxic agents being developed for cancer therapy, it would be difficult to characterize this section strictly as antibody conjugation. While many, if not most, studies utilize monoclonal antibodies of one form or another as the biospecific targeting component, the complete picture involves the cross-linking of a wide variety of molecules together to create the final conjugate. This section presents some of the most common methods of immunotoxin preparation. For the preparation of other unique targeted toxin conjugates, the methods found throughout this book for linking one particular functional group to another can be followed with an excellent probability for success.

1. Properties and Use of Immunotoxin Conjugates

Conjugates of monoclonal antibodies and protein toxins are undergoing extensive research for their usefulness in the treatment of cancer. Toxins of many different types can be used to create effective immunotoxin conjugates, including the proteins ricin from castor beans (*Ricinus communis*), abrin from *Abrus precatorius,* modeccin, gelonin from *Gelonium multiflorum* seeds, diphtheria toxin produced by *Corynebacterium diphtheriae,* pokeweed antiviral proteins (PAPs; three types: PAP, PAP II, and PAP-S) from *Phytolacca americana* seeds, cobra venom factor (CVF), *Pseudomonas* exotoxin, restrictocin from *Aspergillus restrictus,* momordin from *Momordica charantia* seeds, and saporin from *Saponaria officinalis* seeds, as well as other ribosome-inactivating proteins (RIPs).

By far the most popular choices for the toxin component of immunotoxins are ricin, abrin, modeccin, and diphtheria toxin. The three plant toxins have lectin binding activity toward terminal β-galactosyl residues and can be inhibited by the presence of simple sugars like galactose and lactose. They are able to bind to cell-surface polysaccharide receptors with high affinity (K_a in the range of 10^7–10^8). Ricin, abrin, and modeccin consist of two subunits with remarkably similar structures and activities. The intact proteins have molecular weights of approximately 63,000–65,000 with each subunit of about equal size. The subunits are joined by disulfide linkages that are important reversible bonds in the mechanism of cytotoxicity. The A chain is called the effectomer and possesses ribosomal-inactivating properties. The B chain contains the carbohydrate binding site and is termed the haptomer. Although the intact toxin molecules have potent cytotoxic effects on cells, they exhibit no ribosomal-inactivating activity on ribosomes in a cell-free system. By contrast, reduction of the toxins with a disulfide-reducing agent creates the opposite effects. Reduced, dissociated toxin subunits inhibit ribosomal activity in cell-free systems, but they have no effect on intact cells.

The reason for these properties is due to the toxins' mode of action. Toxin molecules bind through saccharide-recognition sites on the B chain to particular β-galactosyl-containing glycoprotein or glycolipid components on the surface of cell membranes. In animals sensitive to these toxins, the necessary polysaccharide ligands are present in large quantities on virtually all cell types (Cumber *et al.*, 1985). On binding of the dimer to the cell, the A chain enters the cell either by active transport into endocytic vesicles or through some mechanism of its own. Once inside the cell membrane, the A chain enters the cytoplasmic space, binding to and enzymatically inactivating the 60S subunit of ribosomes (Olsnes and Pihl, 1976, 1982a). The result is cessation of protein synthesis and eventual cell death. Because the A chain's action is through enzymatic means, as little as one active toxin molecule is enough to disrupt seriously protein synthesis operations and probably sufficient to kill a target cell (Eiklid *et al.*, 1980). The turnover rate of one A chain molecule is about 1500 ribosomes inactivated per minute (Olsnes, 1978).

Diphtheria toxin also is a two-subunit protein, but it is initially synthesized by certain strains of *C. diphtheriae* as a single polypeptide chain of molecular weight 63,000. Proteolytic processing results in the formation of a "nicked toxin" that is enzymatically inactive, but consists of two subunits bonded together by an interchain cystine disulfide. On reduction of the disulfide, the A chain (MW 24,000) is released and manifests enzymatic activity toward ribosomal proteins (Sandvig and Olsnes, 1981; Collier and Cole, 1969). Its mode of action is different than the plant toxins. The A-chain fragment of diphtheria toxin catalyzes the ADP-ribosylation of eukaryotic aminoacyl-transferase II (EF2) using NAD^+ (Honjo *et al.*, 1968; Gill *et al.*, 1969). The B chain, by contrast, possesses no enzymatic activity, but evidence points to the fact that a binding site on it recognizes certain cell surface receptors. As in the action of ricin, abrin, and modeccin, the B chain of diphtheria toxin is necessary for cytotoxicity (Colombatti *et al.*, 1986). Recent work demonstrates the role of the C-terminus cysteine residue of the B chain in cell penetration (Dell'Arciprete *et al.*, 1988).

Figure 316 illustrates the basic structure of these common two-subunit toxins, showing schematically their major characteristics.

Due to the extraordinary toxicity of intact ribosome-inactivating toxins like ricin, abrin, and modeccin, purification and handling of these proteins must be done with extreme care. Even dust from crude seed powders or lyophilized proteins should be considered dangerous. During the height of the Cold War days, a Soviet KGB agent killed a man from the West by injecting at most only milligram quantities of ricin into his leg using a modified umbrella tip. There even have been instances of worker deaths at companies that routinely purify these proteins. For this reason, all handling operations of intact toxin dimers and purified subunits should be done in fume or laminar-flow hoods. Avoid, also, the use of laboratory tools that could lead to puncture wounds causing contaminating toxin injection.

Some potentially cytotoxic proteins contain only a single polypeptide chain, such as gelonin and PAPs. Such toxins manifest enzymatic ribosome-inactivating properties similar to those of the multisubunit proteins like ricin, but do not possess the cell-recognition capacity that the B-chain subunit contains. The result is the inability of these toxins to bind or affect intact cells. However, they do maintain the typical ribosome-inactivating properties in a cell-free system that the A chain of two-subunit toxins possesses. If these proteins are conjugated with a cell targeting agent, such as the B chain of ricin or a specific antibody that recognizes cell-surface epitopes, full cytotoxicity results.

Gelonin and PAPs are much more convenient to work with than ricin and the other two-subunit toxins. Most importantly, they are relatively nontoxic to cells unless conjugated with something that can facilitate cell binding and internalization (Stirpe *et al.*, 1980; Irvin, 1983). The pI of these toxins is in the basic range, and they each have a molecular weight of about 30,000 (Barbieri and Stirpe, 1982). These single-subunit

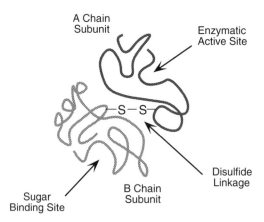

Figure 316 Conceptualized construction of an A–B subunit protein toxin. The B chain contains a binding region for docking onto cell surfaces, while the A chain contains a catalytic site that produces cytotoxic affects intracellularly. The two subunits are joined by a disulfide bond that is somehow reductively cleaved at the cellular level to allow the A subunit to affect cell death.

proteins are very stable, especially to purification techniques, but also to most modification and cross-linking steps associated with preparing immunotoxins. Studies
have shown (Lambert *et al.*, 1985) that modification of gelonin or PAPs can be done
with 2-iminothiolane (Chapter 1, Section 4.1) to create sulfhydryl groups without loss
of activity. Previous studies, however, have determined that the use of SPDP to modify
gelonin resulted in a 90% inactivation (Thorpe *et al.*, 1981). The difference in these
results may be due to the retention of positive charge characteristics on the modified
amine when using Traut's reagent, but neutralization of that charge when using SPDP.

Immunotoxin conjugates consist of an antibody covalently cross-linked to a toxin
molecule in a way that maintains the unique properties of both proteins. The antibody
component consists of a monoclonal having specificity for an antigenic determinant
on the surface of a particular cell type. Most often, the targeted cells are tumors that
express a unique cell-surface marker that can be recognized by the monoclonal. The
role of the antibody, therefore, is to function as a passive taxi, carrying the toxin
component to the targeted cells. Once at the tumor location, the toxin component
effects its intended ribosome-inhibiting action, ultimately causing cell death and
tumor destruction.

Since immunotoxin conjugates are destined to be used *in vivo,* their preparation
involves more critical consideration of cross-linking methods than most of the other
conjugation protocols described in this book. The following sections discuss the problems associated with toxin conjugates and the main cross-linking methods currently in
use for preparing them.

2. Preparation of Immunotoxin Conjugates

It has become apparent that the method of cross-linking can dramatically affect the
activity of an immunotoxin *in vivo.* Not only is this true with regard to possible direct
blocking by the cross-linker of the enzymatic active site that is responsible for inactivation of ribosomes, but also the chemical reaction of conjugation is an important factor
in proper binding and entry of the conjugate into the cell. Preparation of the conjugate
should maintain the antigen binding character of the attached antibody and at the
same time not block the ribosome-inactivating activity of the toxin component.

Studies have been done to investigate the importance of using a cleavable linker
between the antibody and the toxin. This configuration in the immunoconjugate
would mimic the natural state of two-subunit toxins like ricin that are held together by
disulfide bonds. There is evidence that disulfide reduction and cleavage of the A chain
from the B chain is necessary for cytotoxicity in native toxins (Olsnes, 1978). There is
similar evidence that the creation of cytotoxic immunotoxins using only A-chain
subunits requires that the conjugation be done with a monoclonal using a cross-linker
that possesses a disulfide bond in its cross-bridge or creates a disulfide linkage upon
coupling (Masuho *et al.*, 1982). Using disulfide-cleavable cross-linkers in the preparation of immunotoxins results in the antibody taking on the role of the B chain in
recognizing and binding to antigenic determinants on the surface of cells. After binding, some mechanism internalizes the conjugate where the two components then are
separated by disulfide reduction. The A-chain subunit is then freed to enter the cytoplasmic space where enzymatic degradation of the ribosomal proteins occurs.

Other investigators, however, have demonstrated that conjugations of antibody with intact, two-subunit toxins can be done using noncleavable cross-linkers such as NHS ester–maleimide heterobifunctionals (Chapter 5, Section 1) (Myers *et al.*, 1989). Presumably, the toxin is still able to release the A chain after the antibody has bound to the cell, since the conjugation process does not permanently attach the two toxin subunits together.

Thus, two main strategies can be used in making immunotoxin conjugates (Fig. 317). In the most often used method, the isolated A chain of two-subunit toxins (or the intact polypeptide of single-subunit toxins like gelonin) is conjugated to a monoclonal using a cross-linker that can introduce a disulfide bond. When using only purified A chain, it is common (but not absolutely required) to couple through the sulfhydryl that is freed during A–B chain cleavage by disulfide reduction. The single-chain toxins like gelonin, however, have no free sulfhydryls, so a thiolating agent such as 2-iminothiolane (Chapter 1, Section 4.1) may be used to create them (Lambert *et al.*, 1985).

When using ricin A chains, it has been found that chemical deglycosylation of the subunit prevents its nonspecific binding to receptors for mannose on certain cells of the reticuloendothelial system (Vitetta and Thorpe, 1985; Ghetie *et al.*, 1988, 1991).

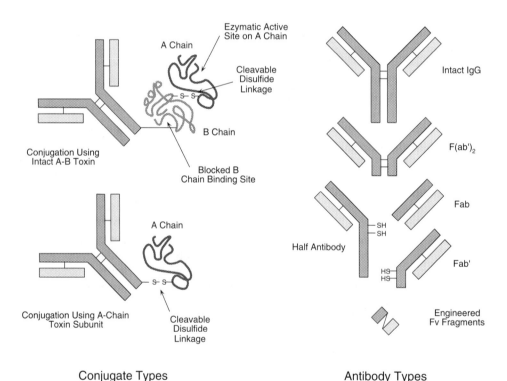

Conjugate Types   Antibody Types

Figure 317   The strategies involved in creating an immunotoxin conjugate are numerous. Intact antibody molecules or enzymatic fragments may be used as the targeting component. Even small recombinant Fv fragments that are genetically engineered to limit host immune response can be employed. Conjugates can include two-subunit toxins or purified A-chain components. If intact toxins are used, the B-chain binding site must be blocked to prevent nonspecific cell death. If A-chain subunits are used, to induce cytotoxic effects in the conjugate the cross-linking agent must generate a disulfide bond that can allow toxin release.

Thus, immunotoxin conjugates consisting of deglycosylated ricin A chain (dgA) have been shown to survive longer *in vivo* and are more efficient at reaching their intended target cells. In addition, if the antibody component does not contain Fc region, but consists of only F(ab')$_2$, Fab', Fab, or smaller Fv fragments, then nonspecific binding of the immunotoxin *in vivo* will be reduced to a minimum. One study found that constructing immunotoxin conjugates with molar ratios of two dgA per antibody molecule resulted in a sevenfold increase in cytotoxicity over a 1:1 conjugate ratio (Ghetie *et al.*, 1993).

A-chain immunotoxins, however, may not be quite as cytotoxic as conjugates formed from intact toxin molecules (Manske *et al.*, 1989). In an alternative approach to A chain use, the intact toxin of two-subunit proteins is directly conjugated to a monoclonal without isolation of the A chain. Conjugation of an antibody with intact A–B chain toxins can be done without a cleavable linker, as long as the A chain can still separate from the B chain once it is internalized. Therefore, it is important to avoid intramolecular cross-linking during the conjugation process, which can prevent release of the A–B complex. In addition, since the B chain possesses a recognition site for most cell surfaces, it still has the ability to nonselectively bind and kill nontumor cells. To maintain antibody specificity in intact toxin conjugates toward only one cell type (and thus prevent nonspecific cell death), all cell binding capability within the toxin itself must be removed. Fortunately, a large proportion of the binding sites on the B chains usually are blocked during the conjugation process, and the galactose binding potential is significantly impaired. Further purification to remove conjugates that have remaining galactose binding potential can be done on an acid-treated Sepharose column (which contains galactose residues) or on a column of asialofetuin bound to Sepharose (Cumber *et al.*, 1985). Conjugate fractions that do not bind to both affinity gels contain no nonspecific binding potential toward nontargeted cells.

More elaborate methods of blocking or eliminating the B-chain galactose binding site also can be done. For instance, the cross-linking agent may have a lactose molecule designed into it that can block B-chain activity. The lactose portion possesses natural affinity for the B-chain binding site, and thus it occupies that area while a nearby reactive group covalently attaches the cross-linker to neighboring functional groups on the protein.

Moroney *et al.* (1987) used this approach in creating a ricin conjugate. Lactose was modified at its reducing end with cystamine via reductive amination (Chapter 2, Section 5.3). The cystamine was reduced with DTT and immobilized on an affinity gel that was activated with a pyridyl dithiol group (Chapter 2, Section 2.6). The coupled lactose then was modified with a chlorotriazine derivative through the secondary amine created by the reductive amination process. Next, the ricin molecule was immobilized by the additional reactive group on the chlorotriazine ring. Since this reactive group was immediately adjacent to the lactose residue, the ricin bound the sugar at its B-chain binding site and was covalently coupled through a nearby amine. This process effectively blocked and eliminated all nonspecific binding potential in the ricin dimer. After removal of the blocked ricin from the support by DTT, the free sulfhydryl group on the protein was conjugated to an SMCC-modified antibody, forming the final conjugate (Fig. 318). Although this process worked in preparing an effective immunotoxin conjugate, most conjugation schemes are less elaborate.

Regardless of their method of preparation, the required and ideal characteristics of immunotoxin conjugates can be summarized in the following points:

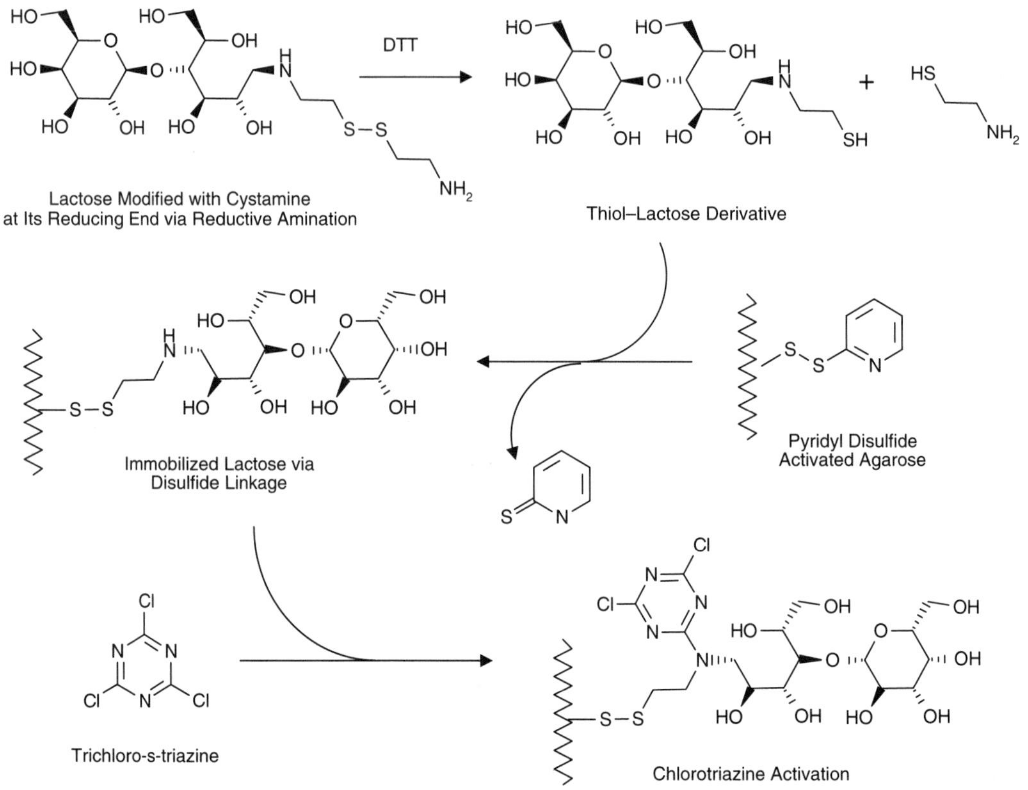

Figure 318   In an elaborate strategy to block the B-chain binding site in the construction of immunotoxins using intact A–B subunit toxins, cystamine was first coupled to the reducing end of lactose by reductive amination. DTT was then used to reduce the cystamine disulfide group, revealing the free thiol. A pyridyldisulfide-activated agarose gel then was used to couple the lactose derivative through its sulfhydryl. Next, trichloro-s-triazine was reacted with the support to modify the secondary amine on the cystamine component, forming a reactive gel. Finally, addition of an intact toxin to the affinity support caused binding with the lactose group at the B-chain binding site. Since the B chain was now in proximity to the chlorotriazine ring, covalent coupling occurred with available amines on the protein toxin, thus permanently blocking the binding site. After removal of the modified toxin from the gel using a disulfide reducing agent, the free thiol of the cystamine group was used to conjugate with an SMCC-activated immunoglobulin.

1. The conjugation process must leave the antigen binding sites on the antibody component free to interact with its intended target. Cross-linker modification or blockage of these binding sites by the attached toxin must be kept to a minimum.

2. The toxin component of the conjugate must be able to elicit cytotoxicity by ribosomal damage as it could in its native state. This means that the cell penetration and enzymatic properties of the toxin remain unaltered, although an antibody molecule is conjugated to it.

3. The nature of toxin binding to cells through the B chain must be eliminated in the conjugate to prevent nonspecific binding and nonselective cell death. This may be accomplished by using only the A-chain subunit or by blocking the B-chain binding site in the intact toxin conjugate.

4. Avoid covalently linking the A and B chains together during the cross-linking process. This can be done by using heterobifunctional cross-linkers that are more controllable in their reactivity than homobifunctional reagents.
5. The cross-linking process must minimize polymerization of either the antibody or the toxin. Low-molecular-weight, 1:1 or 1:2 conjugates of antibody-to-toxin are best.
6. The cross-linker used to form the bond between the antibody and toxin must be able to survive *in vivo* and not be cleaved by enzymatic or reductive means before reaching the targeted cells.
7. The conjugate must reach its intended target without being intercepted, bound, and destroyed by the host immune system.
8. Administration of the immunotoxin should result in cell death and complete elimination of all target cells.

The last two points are the most difficult to realize. The following methods describing the conjugation chemical reactions used to prepare immunotoxins work well in creating complexes containing active antibody and toxin. The majority of research today is not so much concerned with further optimization of the cross-linking process, but primarily is directed at overcoming the host immune system and making the conjugates more effective in accomplishing complete tumor destruction.

2.1. Preparation of Immunotoxin Conjugates via Disulfide Exchange Reactions

Since the cytotoxic potential of most common toxins relies on their subunit disulfide cleavability with subsequent release of associated A chains, most successful conjugation techniques for preparing immunotoxins involve the use of disulfide exchange reactions. Heterobifunctional cross-linking agents containing an amine-reactive group at one end and a disulfide bond with a good leaving group on the other end are common choices for making these conjugates (Chapter 5, Section 1). The leaving group on the disulfide portion of the cross-linker permits efficient disulfide interchange with a free sulfhydryl on the antibody or toxin. The resultant covalent bond thus is a cleavable disulfide that mimics the native cleavability inherent in the toxin dimer.

Pyridyldisulfide Reagents

The most common reactive group for initiating disulfide interchange reactions is a pyridyldisulfide. Attack of a nucleophilic thiolate anion dissociates the pyridine-2-thione leaving group from this functional group, forming a new disulfide bond with the incoming sulfhydryl compound. Several cross-linking reagents containing these groups are popular choices for producing antibody–toxin conjugates.

SPDP

N-succinimidyl 3-(2-pyridyldithio)propionate (SPDP) is by far the most popular heterobifunctional cross-linking agent used for immunotoxin conjugation (Chapter 5, Section 1.1). The activated NHS ester end of SPDP reacts with amine groups in one of the two proteins to form an amide linkage. The 2-pyridyldithiol group at the other end reacts with sulfhydryl groups in the other protein to form a disulfide linkage (Carlsson *et al.*, 1978). The result is a cross-linked antibody–toxin conjugate containing cleavable disulfide bonds that can emulate the activity of native two-subunit toxin molecules.

LC-SPDP (Chapter 5, Section 1.1) is an analog of SPDP containing a hexanoate spacer arm within its internal cross-bridge. The increased length of the extended cross-linker is important in some conjugations to avoid steric problems associated with closely-linked macromolecules. However, for the preparation of immunotoxins, no advantages were observed for LC-SPDP over SPDP (Singh *et al.*, 1993).

SPDP is also useful in creating sulfhydryls in one of the two proteins being conjugated (Chapter 1, Section 4.1). Once modified with SPDP, the protein can be treated with DTT (or other disulfide reducing agents) to release the pyridine-2-thione leaving group and form the free sulfhydryl. The terminal —SH group then can be used to conjugate with any cross-linking agents containing sulfhydryl-reactive groups, such as maleimide or iodoacetyl (for covalent conjugation) or 2-pyridyldithiol groups (for reversible conjugation).

In the preparation of immunotoxins, some procedures call for the modification of the antibody with SPDP to introduce reactive thiols (Cumber *et al.*, 1985). The NHS ester end of the cross-linker is reacted at slightly alkaline pH with the primary amines on the antibody. After removal of excess reagent by gel filtration, the pyridyldisulfide groups are reduced by DTT. The reductant causes the removal of pyridine-2-thione groups and the creation of sulfhydryl groups on the immunoglobulin. The reason the antibody is thiolated in this manner and not the toxin is to avoid exposing the intact toxin to reducing conditions that could disassociate the A and B subunits.

To activate the toxin, SPDP again can be used to modify the intact A–B component. After purification of the modified toxin from excess cross-linker, the SPDP–toxin is mixed with the thiolated antibody to effect the final conjugate (Fig. 319).

This multistep cross-linking method employing SPDP on both molecules has been used to prepare a number of immunotoxin conjugates (Thorpe *et al.*, 1982; Edwards *et al.*, 1982; Colombatti *et al.*, 1983; Wiels *et al.*, 1984; Vogel, 1987; Reiter and Fishelson, 1989). Although this method has worked well for many different toxins, its main potential disadvantage is exposure of the antibody to reducing conditions that potentially could cleave the disulfide bonds holding its heavy and light chains together. Alternative methods using SPDP in a nonreducing environment may result in better conjugates.

For instance, if toxin A chain–antibody conjugates are to be prepared, the antibody can be similarly activated with SPDP, but in this case not treated with reductant. After removal of excess cross-linker, the activated antibody can be directly mixed with isolated A chain to create the conjugate (Fig. 320). This procedure makes use of the indigenous sulfhydryl residues produced during reductive separation of the A and B chains and therefore does not require cross-linker thiolation of one of the proteins.

Another way of utilizing SPDP is to again activate the antibody to create the pyridyldisulfide derivative, but this time thiolate the toxin component using 2-iminothiolane (Chapter 1, Section 4.1). 2-Iminothiolane reacts with primary amines in a ring-opening reaction that creates a terminal sulfhydryl group without reduction. Intact A–B toxins and toxins containing only one subunit, like gelonin, PAPs, and *Pseudomonas* toxin A, can be coupled to antibodies using this procedure (Lambert *et al.*, 1985, 1988; Bjorn *et al.*, 1986; Scott *et al.*, 1987; Ozawa *et al.*, 1989). Mixing the

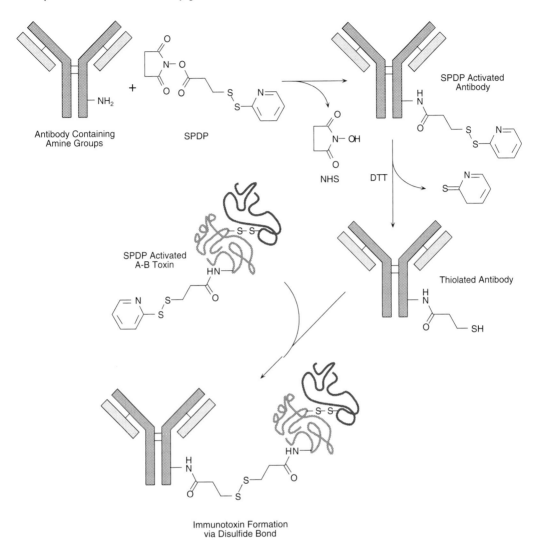

Figure 319 SPDP can be used to modify both an antibody and a toxin molecule for conjugation purposes. In this case, the antibody is thiolated to contain a sulfhydryl group by modification with SPDP followed by reduction with DTT. A toxin molecule is then activated with SPDP and reacted with the thiolated antibody to effect the final conjugate through a disulfide bond.

SPDP-activated antibody with the thiolated toxin effects the final conjugation (Fig. 321).

SPDP also can be used to conjugate other targeting molecules to toxins, such as transferrin, epidermal growth factor, α_2-macroglobulin, and human chorionic gonadotropin (Fizgerald *et al.*, 1980; Helenius *et al.*, 1980; Keen *et al.*, 1982; Oeltmann, 1985). To create these conjugates, one of the two components must be activated with SPDP to generate the sulfhydryl-reactive pyridyldisulfide groups, while the other component must be modified to contain the —SH functional group. Mixing these modified molecules together forms the toxin conjugate.

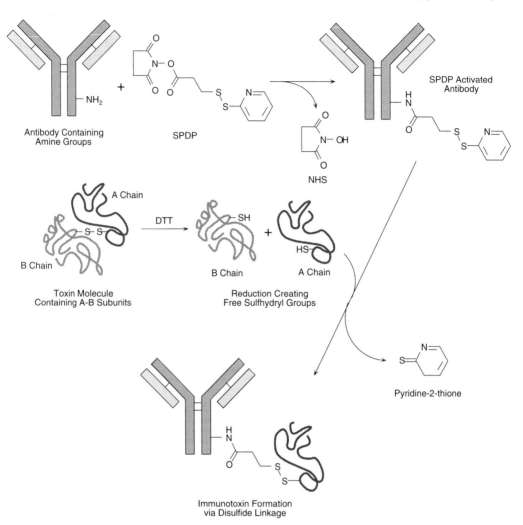

Figure 320 SPDP can be used to activate an antibody molecule through its available amine groups to form a sulfhydryl-reactive derivative. Toxin molecules containing disulfide-linked A–B chains may be reduced with DTT to isolate the A-chain component containing a free thiol. The SPDP-activated antibody is then mixed with the reduced A chain to effect the final conjugate by disulfide bond formation.

The following methods are generalized to provide an overview of how SPDP can be used in these conjugation techniques. The appropriate optimization for a particular toxin conjugate should be done. *Caution! Toxins are highly toxic even in very low amounts. Handle all toxin molecules and their isolated subunits with extreme care.*

Protocol for Thiolation of Antibody with SPDP and Conjugation to an SPDP-Activated Toxin
Caution: Toxin molecules are dangerously toxic even in small amounts. Use extreme care in handling.

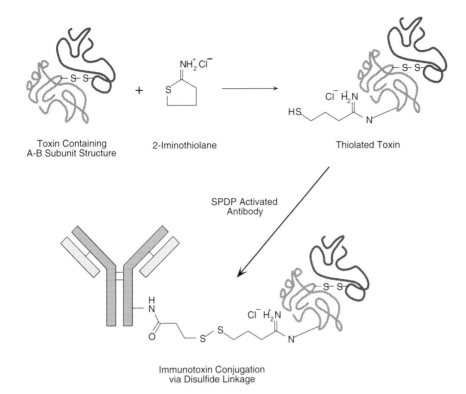

Toxin Containing
A-B Subunit Structure

2-Iminothiolane

Thiolated Toxin

SPDP Activated
Antibody

Immunotoxin Conjugation
via Disulfide Linkage

Figure 321 An intact A–B subunit toxin molecule may be activated with 2-iminothiolane with good retention of cytotoxic activity. The thiolated toxin then may be conjugated with SPDP-activated antibody to generate the immunotoxin conjugate through a disulfide bond.

Note: In this protocol, for every mg of toxin employed, 2.5 mg of antibody is required to obtain the correct molar ratios in the final conjugate.

Activation of Toxin with SPDP

1. Dissolve the toxin to be conjugated in 0.1 M sodium phosphate, 0.15 M NaCl, pH 7.5, at a concentration of 10 mg/ml. Some protocols use as an SPDP reaction buffer, 50 mM sodium borate, 0.3 M NaCl, 0.5% n-butanol, pH 9. Both buffer systems work well for the NHS ester modification reaction, although the pH 9 buffer is at the higher end of effective derivatization with active esters, since the hydrolysis rate is dramatically increased at this level of alkalinity.
2. Dissolve SPDP (Pierce) in DMF at a concentration of 3 mg/ml. Add 20 µl of this solution to each milliliter of the toxin solution. Gently mix to effect dissolution. Retain the SPDP stock solution for use in the antibody modification step, below.
3. React for 30 min at room temperature.
4. Purify the SPDP-activated toxin from excess reagents and reaction by-products by gel filtration using Sephadex G-25 or its equivalent. For the chromatography use 0.1 M sodium phosphate, 0.15 M NaCl, pH 7.5, containing 10 mM EDTA.

5. Concentrate the toxin to 10 mg/ml using a centrifugal concentrator with a molecular weight cutoff of 10,000 (Amicon Centricon 10 concentrators work well for this purpose). Retain this solution for the conjugation reaction.

Thiolation of Antibody with SPDP

1. Dissolve the antibody to be conjugated in 0.1 M sodium phosphate, 0.15 M NaCl, pH 7.5, at a concentration 10 mg/ml. Note: Some protocols use the borate buffer system described in step A. Use 2.5 mg of antibody per milligram of toxin to be conjugated.
2. Dissolve SPDP in DMF at a concentration of 3 mg/ml. Add 24 µl of this solution to each milliliter of the antibody solution with gentle mixing to effect complete dissolution.
3. React for 30 min at room temperature.
4. Remove excess cross-linker by gel filtration using a column of Sephadex G-25. Perform the chromatography using 0.1 M sodium phosphate, 0.15 M NaCl, 10 mM EDTA, pH 7.5. The buffer should be degassed under vacuum and nitrogen bubbled through it to remove oxygen. The presence of EDTA stabilizes the free sulfhydryls formed in the following steps against metal-catalyzed oxidation.
5. Concentrate the fractions containing protein from the gel filtration step to 10 mg/ml using a centrifugal concentrator (MW cutoff of 10,000).
6. To reduce the pyridyl dithiol groups and create reactive sulfhydryls, dissolve DTT in water at a concentration of 17.2 mg/ml and immediately add 500 µl of this solution to each milliliter of concentrated antibody solution. Mix to dissolve and react for 30 min at room temperature.
7. Remove excess DTT by gel filtration using the same buffer as in step 4. Pool the fractions containing protein and concentrate to 10 mg/ml.

Conjugation of SPDP-Activated Toxin with Thiolated Antibody

1. Immediately mix the concentrated, thiolated antibody solution with the SPDP-activated toxin.
2. React for 18 hours at room temperature to form the final conjugate. isolation of the ideal 1:1 or 1:2 antibody-toxin conjugate can be done through gel filtration separation using a column of Sephacryl S-300.

Protocol for Activation of Antibody with SPDP
and Conjugation to a Toxin A Chain
Caution: Toxin molecules are dangerously toxic even in small amounts. Use extreme care in handling.

Since the A chain of toxin molecules contains a free sulfhydryl group, there is no need in this conjugation strategy to thiolate one of the molecules. The following protocol calls for 1.73 mg of antibody per milligram of toxin A chain to produce a conjugate possessing the correct molar ratio of components. Best results for creating a highly cytotoxic immunotoxin will be obtained if deglycosylated ricin A chain is used.

1. Dissolve the antibody to be conjugated in 0.1 M sodium phosphate, 0.15 M NaCl, pH 7.5, at a concentration 10 mg/ml.
2. Dissolve SPDP (Pierce) in DMF at a concentration of 3.0 mg/ml. Add 30 µl of

this solution to each milliliter of the antibody solution with gentle mixing to effect complete dissolution.

3. React for 30 min at room temperature.
4. Remove excess cross-linker by gel filtration using a column of Sephadex G-25. Perform the chromatography using 0.1 M sodium phosphate, 0.15 M NaCl, 10 mM EDTA, pH 7.5.
5. Concentrate the fractions containing SPDP-activated antibody from the gel filtration step to 10 mg/ml using a centrifugal concentrator (MW cutoff of 10,000) or the equivalent.
6. Mix the activated antibody solution with a solution of deglycosylated toxin A chain (dgA) dissolved in 0.1 M sodium phosphate, 0.15 M NaCl, pH 7.5, containing 10 mM EDTA. The ratio of mixing should equal 1.73 mg of antibody per milligram of A chain or 580 μl of A-chain solution at 10 mg/ml per milliliter of activated antibody solution at 10 mg/ml. The A-chain solution must not contain any reducing agents left over from the disassociation of the toxin subunits during the A-subunit isolation. Reductants will compete for the SPDP activation sites on the antibody molecule.
7. React for 18 h at room temperature. Isolation of the 1:1 or 1:2 antibody–toxin conjugate can be done through a gel filtration separation using a column of Sephacryl S-200. Isolation of conjugates containing molar ratios of 1:2 antibody:dgA have resulted in greater cytotoxicity behavior *in vivo* (Ghetie *et al.*, 1993).

Protocol for the Conjugation of SPDP-Activated Antibodies with 2-Iminothiolane-Modified Toxins

Caution: Toxin molecules are dangerously toxic even in small amounts. Use extreme care in handling.

A third option for immunotoxin preparation is to again activate the antibody with SPDP, while this time thiolating a single-chain toxin molecule to conjugate with it. This method works especially well using 2-iminothiolane (Chapter 1, Section 4.1) to create sulfhydryls on gelonin or PAPs. Gelonin is a single-polypeptide toxin containing no free sulfhydryls. A number of options are available for thiolation; however, the use of SPDP to add sulfhydryl groups inactivates the toxin, while 2-iminothiolane preserves its activity, perhaps by maintaining the positive charge on the amines that are being modified (Lambert *et al.*, 1985).

Activation of Antibody with SPDP

1. Dissolve the antibody to be conjugated in 0.1 M sodium phosphate, 0.15 M NaCl, pH 7.5, at a concentration of 10 mg/ml.
2. Dissolve SPDP (Pierce) in DMF at a concentration of 3.0 mg/ml. Add 30 μl of this solution to each milliliter of the antibody solution with gentle mixing to effect dissolution.
3. React for 30 min at room temperature.
4. Remove excess cross-linker by gel filtration using a column of Sephadex G-25. Perform the chromatography using 0.1 M sodium phosphate, 0.15 M NaCl, 10 mM EDTA, pH 7.5.
5. Concentrate the fractions containing SPDP-activated antibody from the gel

filtration step to 10 mg/ml using a centrifugal concentrator (MW cutoff of 10,000) or the equivalent.

Thiolation of Gelonin (or other single-polypeptide toxins)
with 2-Iminothiolane (Traut's Reagent)

1. Dissolve gelonin at a concentration of 10 mg/ml in 50 mM triethanolamine hydrochloride, pH 8, containing 10 mM EDTA. The buffer should be degassed under vacuum and bubbled with nitrogen to remove oxygen that may cause sulfhydryl oxidation after thiolation.
2. Dissolve 2-iminothiolane (Pierce) in degassed, nitrogen-bubbled deionized water at a concentration of 20 mg/ml (makes a 0.14 M stock solution). The solution should be used immediately. Add 70 µl of the 2-iminothiolane solution to each milliliter of the gelonin solution (final concentration is about 10 mM).
3. React for 1 h at 0°C (or on ice) under a nitrogen blanket.
4. Purify the thiolated protein from unreacted Traut's reagent by gel filtration using 0.1 M sodium phosphate, 0.15 M NaCl, pH 7.5, containing 10 mM EDTA. The presence of EDTA in this buffer helps to prevent oxidation of the sulfhydryl groups and resultant disulfide formation. The degree of —SH modification in the purified protein may be determined using the Ellman's assay (Chapter 1, Section 4.1).
5. Concentrate the thiolated toxin to 10 mg/ml using centrifugal concentrators.

Conjugation of SPDP-Activated Antibody with Thiolated Gelonin

1. Mix SPDP-activated antibody with thiolated gelonin in equal mass quantities (or equal volumes if they are at the same concentration). This ratio results in about a fivefold molar excess of toxin over the amount of antibody.
2. React for 20 h at 4°C under a nitrogen blanket.
3. To block unreacted sulfhydryl groups, add iodoacetamide to the solution to a final concentration of 2 mM.
4. React for an additional 1 h at room temperature.
5. Remove unconjugated gelonin by passage of the conjugate solution over a column of immobilized protein A (Pierce). Use 2 ml of the protein A column for each 10 mg of conjugate to be purified. Equilibrate the column with 50 mM sodium phosphate, 0.15 M NaCl, pH 7.5. Apply the conjugate sample and allow it to enter the gel. Continue to wash the column with equilibration buffer while taking 2-ml fractions until baseline is reached (monitored at an absorbance of 280 nm). Unconjugated gelonin will pass through the column unretarded. Elute bound conjugate with 0.1 M acetic acid, 0.15 M NaCl. Immediately add 0.1 ml of 1 M potassium phosphate, pH 7.5, to each bound fraction for neutralization. Alternatively, gel filtration may be used to isolate the conjugate from lower molecular weight antibody and gelonin. A column of Sephacryl S-200 works well for this purpose.

SMPT

Succinimidyloxycarbonyl-α-methyl-α-(2-pyridyldithio)toluene (SMPT) is a heterobifunctional cross-linking agent similar to SPDP that contains an amine-reactive NHS ester on one end and a sulfhydryl-reactive pyridyl disulfide group on the other (Chapter 5, Section 1.2). Reaction with a sulfhydryl-containing protein results in a cleavable disulfide linkage, important for immunotoxin activity. SMPT is an analog of SPDP that differs only in its cross-bridge, which contains an aromatic ring and a hindered disulfide group (Thorpe *et al.*, 1987; Ghetie *et al.*, 1990). The spacer arm of SMPT is slightly longer than that of SPDP, but the presence of the benzene ring and α-methyl group adjacent to the disulfide sterically hinders the structure sufficiently to provide increased half-life of immunotoxin conjugates *in vivo*.

SMPT increasingly is used in place of SPDP for the preparation of immunotoxin conjugates. The hindered disulfide of SMPT has distinct advantages in this regard. Thorpe *et al.* (1987) showed that SMPT conjugates had approximately twice the half-life *in vivo* as SPDP conjugates. Antibody–toxin conjugates prepared with SMPT possess a half-life *in vivo* of up to 22 h, presumably due to the decreased susceptibility of the hindered disulfide toward reductive cleavage.

Ghetie *et al.* (1991) developed a large-scale preparation procedure for antibody–deglycosylated ricin A chain (dgA) conjugates utilizing this cross-linker. The following procedure describes a generalized method for using SMPT to prepare dgA–antibody conjugates. It is based on the Ghetie protocol, but using smaller quantities of reagents. Figure 322 illustrates the reactions involved in using SMPT.

Protocol

Caution: Toxin molecules are dangerously toxic even in small amounts. Use extreme care in handling.

The following method calls for mixing activated antibody with ricin A chain at a ratio of 2 mg antibody per milligram of A chain. Adjustments to the amount of antibody and A chain initially dissolved in the reaction buffers should be done to anticipate this ratio.

1. Dissolve the antibody to be conjugated in 0.1 *M* sodium phosphate, 0.15 *M* NaCl, pH 7.5, at a concentration of 10 mg/ml. If the antibody contains oligomers (as evidenced by nondenaturing electrophoresis or HPLC gel filtration analysis), then the monomeric IgG form should be isolated by gel filtration using a column of Sephacryl S-200HR. If no oligomers are present, then omit the chromatographic purification.

2. Dissolve SMPT (Pierce) in DMF at a concentration of 4.8 mg/ml. Add 27 μl of this solution to each milliliter of the antibody solution. Mix gently. The final concentration of SMPT in the reaction mixture is 0.13 mg/ml, which translates into about a 4.8-fold molar excess of cross-linker over the amount of antibody present.

3. React for 1 h at room temperature.

4. Remove unreacted SMPT and reaction by-products by gel filtration on Sephadex G-25. Pool fractions containing SMPT-activated antibody (the first peak eluting from the column) and concentrate to 10 mg/ml using centrifugal concentrators with a molecular weight cut-off of 10,000.

5. Dissolve deglycosylated ricin A chain (dgA) in 0.1 *M* sodium phosphate, 0.15 *M* NaCl, 10 m*M* EDTA, pH 7.5, at a concentration of 10 mg/ml. The buffer should be degassed under vacuum and nitrogen bubbled through it to remove oxygen. Prepare half the amount of A-chain solution as the amount of antibody prepared

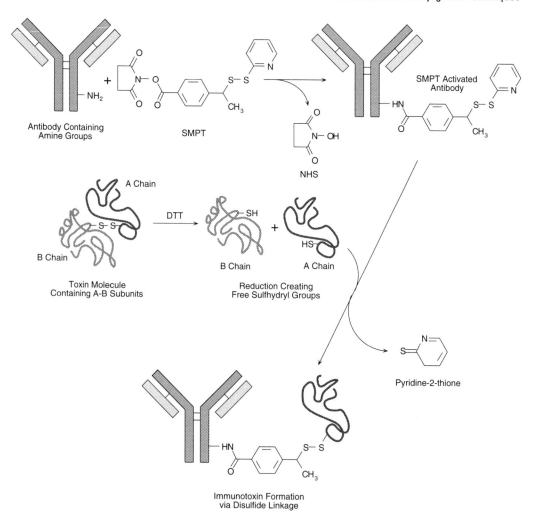

Figure 322 SMPT may be used to form immunotoxin conjugates by activation of the antibody component to form a thiol-reactive derivative. Reduction of an A–B toxin molecule with DTT can facilitate subsequent isolation of the A chain containing a free thiol. Mixing the A chain containing a sulfhydryl group with the SMPT-activated antibody causes immunotoxin formation through disulfide bond linkage. The hindered disulfide of SMPT cross-links has been found to survive *in vivo* for longer periods than conjugates formed with SPDP.

in step 1. If the A-chain preparation is done in bulk quantities or if the protein has been stored for lengthy periods, it may be necessary to reduce the sulfhydryls with DTT prior to proceeding with the cross-linking reaction. If A-chain sulfhydryl oxidation is suspected, add 2.5 mg of DTT per milliliter of A-chain solution. React for 1 h at room temperature. Purify the reduced ricin A chain by gel filtration on a Sephadex G-25 column using the PBS–EDTA buffer. Apply no greater volume of sample to the gel than is represented by 5% of the column volume to ensure good removal of excess DTT. Collect the protein and concentrate to 10 mg/ml using centrifugal concentrators.

6. Mix the reduced A-chain solution with activated antibody solution at a ratio of 2 mg of antibody per milligram of A chain. Sterile filter the solution through a 0.22-μm membrane, and react at room temperature under nitrogen for 18 h.

7. To block excess pyridyl disulfide-active sites on the antibody, add cysteine to a final concentration of 25 μg/ml. React for an additional 6 h at room temperature.

8. To isolate the conjugate, apply the immunotoxin solution to a column of Sephacryl S-200HR. Collect the peaks with molecular weights between 150,000 and 210,000. Further purification to remove excess unconjugated antibody can be done on a column of immobilized Cibracron Blue [available commercially (Pierce) or for column preparation, see Hermanson *et al.*, 1992]. Equilibration of the column with 50 mM sodium borate, 1 mM EDTA, pH 9, will bind the conjugate, but not free antibody. Elution of purified immunotoxin conjugate can be done with 50 mM sodium borate, 1 mM EDTA, 0.5 M NaCl, pH 9 (see Ghetie *et al.*, 1991).

3-(2-Pyridyldithio)propionate

A lesser-used reagent to introduce sulfhydryl-reactive pyridyl disulfide groups is 3-(2-pyridyldithio)propionate (PDTP), the acid precursor of SPDP containing no NHS ester group on the carboxylate. Sulfhydryl interchange reaction at the pyridyl dithiol end results in the formation of a disulfide linkage with —SH-containing molecules. The carboxylate end is not further derivatized to contain a reactive species, but may be coupled to amines by the carbodiimide reaction (Chapter 3, Section 1). Reaction of PDTP with an antibody molecule in the presence of 1-ethyl-3-(3-dimethylaminopropyl) carbodiimide (EDC) results in the formation of amide linkages with the active pyridyldisulfide groups still available for coupling to sulfhydryl-containing toxins (Fig. 323). Mixing the PDTP antibody with purified ricin A chain results in disulfide cross-links identical to those obtained using SPDP as the cross-linker (Jansen *et al.*, 1980; Gros *et al.*, 1985). PDTP also has been used to activate transferrin to contain reactive pyridyldithiol groups for conjugation to ricin A-chain molecules (Raso and Basala, 1984, 1985).

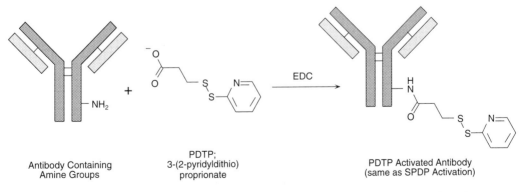

PDTP;
3-(2-pyridyldithio)
proprionate

Antibody Containing
Amine Groups

PDTP Activated Antibody
(same as SPDP Activation)

Figure 323   PDTP may be used to modify antibody molecules using a carbodiimide reaction with EDC. The derivative is the same as that obtained using SPDP activation and is highly reactive toward sulfhydryls.

Since activated molecules and cross-links formed between two species are identical to those formed using SPDP, it is of little advantage to use PDTP. Furthermore, an EDC-mediated reaction of the carboxylate end of the cross-linker with amine groups on proteins can cause concomitant zero-length cross-linking and polymerization of protein molecules. For these reasons, SPDP is the better choice for preparing conjugates.

Use of Cystamine, Ellman's Reagent, or S-Sulfonate

Other reagent systems can be used to form disulfide linkages between antibody and toxin molecules in immunotoxin conjugates. Cystamine can be incorporated into

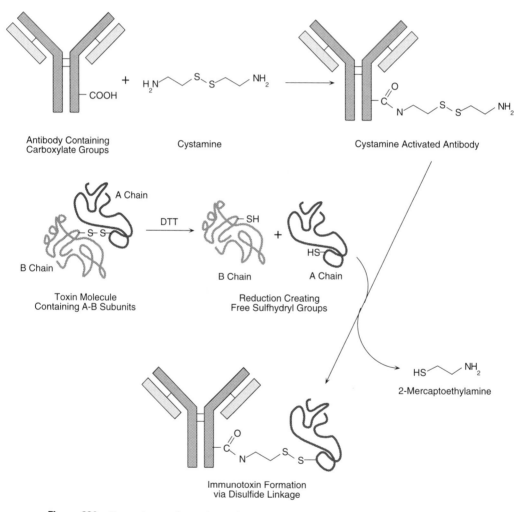

Figure 324 Cystamine may be used to make immunotoxin conjugates by a disulfide interchange reaction. Modification of antibody molecules using an EDC-mediated reaction creates a sulfhydryl-reactive derivative. A-chain toxin subunits containing a free thiol can be coupled to the cystamine-activated antibody to form disulfide cross-links.

proteins by reaction of its terminal amines with the carboxylates on the proteins via the carbodiimide reaction (Chapter 3, Section 1). The resultant modifications contain disulfide linkages that can undergo disulfide interchange reactions with other sulfhydryl-containing molecules (Chapter 1, Section 4.1). For instance, a cystamine-modified targeting component, such as an antibody, can be mixed with the reduced A chain of a toxin molecule to cause conjugate formation through the creation of a disulfide bond (Fig. 324) (Oeltmann and Forbes, 1981). Epidermal growth factor was modified with cystamine and coupled with reduced diphtheria toxin using this approach (Shimisu *et al.*, 1980).

Similarly, Ellman's reagent [5,5'-dithiobis(2-nitrobenzoic acid)] can be used to activate one thiol-containing molecule by disulfide exchange and subsequently used to couple to a second sulfhydryl-containing molecule by the same mechanism (Chapter 1, Section 5.2) (Fig. 325). The disulfide of Ellman's reagent readily undergoes disulfide

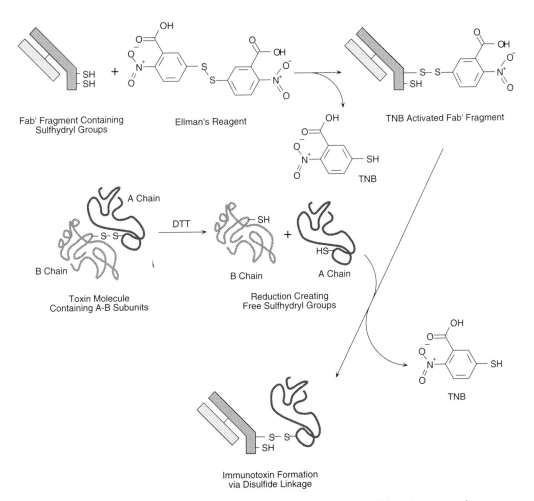

Figure 325 Fab' antibody fragments containing free thiols can be activated with Ellman's reagent to form a sulfhydryl-reactive derivative. A-chain toxin subunits containing a free thiol group may be coupled to the activated Fab' molecule to produce an immunotoxin complex.

exchange with a free sulfhydryl to form a mixed disulfide with concomitant release of one molecule of the highly chromogenic 5-sulfido-2-nitrobenzoate, also called 5-thio-2-nitrobenzoic acid (TNB). The intense yellow color produced by the TNB anion can be measured by its absorbance at 412 nm. Thus, the efficiency of conjugation can be determined spectrophotometrically using this procedure (Pirker *et al.*, 1986; Fitzgerald *et al.*, 1988). A method for the large-scale conjugation of Fab′ fragments containing an available sulfhydryl group and deglycosylated ricin A chain (also containing an —SH group) was developed using Ellman's reagent as the cross-linker (Ghetie *et al.*, 1988).

A final method of forming disulfide cross-links between toxins and targeting molecules is the use of *S*-sulfonate formation using sodium sulfite (Na_2SO_3) in the presence of sodium tetrathionate ($Na_2S_4O_6$). Tetrathionate reacts with sulfhydryls to form sulfenylthiosulfate intermediates (Chapter 1, Section 5.2). These derivatives are reactive toward other thiols to create disulfide linkages rapidly. Sulfite ions react with disulfides to form S-substituted thiosulfates, also known as *S*-sulfonates, and a thiol. The combination of these reagents results in the transformation of available thiols and disulfides into reactive *S*-sulfonates that can be used to cross-link with sulfhydryl-containing molecules. *S*-Sulfonate conjugation can be used to conjugate the A chain of toxin molecules with sulfhydryl-containing Fab′ fragments with good efficiency (Masuho *et al.*, 1979).

Although the use of these alternative disulfide generating agents has proven successful in some applications, pyridyldisulfide-containing cross-linkers, as discussed previously, by far are more common for producing immunotoxin conjugates.

2.2. Preparation of Immunotoxin Conjugates via Amine- and Sulfhydryl-Reactive Heterobifunctional Cross-linkers

Other forms of heterobifunctional cross-linkers that can be used for this purpose are the amine- and sulfhydryl-reactive agents that produce a thioether bond with —SH-containing molecules (Chapter 5, Section 1). The amine-reactive end of these cross-linkers is usually an NHS ester group that can form a stable amide bond with amine-containing proteins. One of two main reactive groups usually are used on the sulfhydryl-reactive end: an iodoacetyl group that couples to sulfhydryls with loss of HI or a maleimide group that undergoes a double-bond addition reaction with —SH groups (see Chapter 2, Sections 2.1 and 2.2).

Since this type of cross-linker forms noncleavable thioether bonds between toxin molecules and the targeting component of the conjugates, they are not appropriate for use with A-chain or single-chain toxins. This is because the cross-linker will now allow the conjugated A chain to break free of the antibody by disulfide reduction after docking at the cellular target. Since release of the A chain is a prerequisite to ribosomal inactivation, such conjugates will prove to be ineffective cytotoxic agents. One report found a 1000-fold increase in cytotoxicity when an immunotoxin containing PAP or gelonin was prepared using a cleavable disulfide linker as opposed to a noncleavable thioether linkage (Lambert *et al.*, 1985).

To make effective immunotoxin conjugates using the following cross-linkers, it is necessary to cross-link intact A–B toxins to antibodies, not single-chain or A-chain toxins. Using intact two-subunit toxins allows the A chain to break free of the complex and perform its cytotoxic duties upon entering the target cell. Two main criteria are especially important when using A–B toxin conjugates: first, the B-chain binding site must be blocked or inoperative in the final immunotoxin complex to prevent non-specific cell death, and second, the two subunits of the toxin must not be covalently cross-linked by the conjugation procedure, precluding them from being separated *in vivo*.

Fortunately, satisfying these criteria is not difficult. The heterobifunctional cross-linkers described in this section are sufficiently controllable to prevent A–B chain cross-linking. In addition, during the conjugation process, the B-chain binding site often becomes inactivated or physically blocked by the attached antibody molecule. Subsequent cleanup of the conjugate using affinity chromatography over a column containing an immobilized sugar can completely eliminate any potential non-specificity contributed by the B chain in the final preparation.

It should be noted that the use of the following cross-linkers to create other forms of toxic conjugates for cancer therapy is not restricted by the disulfide bridge requirement (Willner *et al.*, 1993; Trail *et al.*, 1993). Drug–toxin conjugates, hormono-toxins, lymphokine–toxin, or growth factor–toxin conjugates all can be made without difficulty using nonreversible thioether linkages.

SIAB

N-Succinimidyl(4-iodoacetyl)aminobenzoate (SIAB) is a heterobifunctional cross-linker containing amine-reactive and sulfhydryl-reactive ends (Chapter 5, Section 1.5). The NHS ester on one end of the reagent can be used to couple with primary amine-containing molecules, forming stable amide linkages (Chapter 2, Section 1.4). The other end contains an iodoacetyl group that is specific for coupling to sulfhydryl residues, potentially creating stable thioether bonds (Chapter 2, Section 2.1).

Conjugations using SIAB to create immunotoxins can be done by first reacting the NHS ester end of the cross-linker with available amine groups on the antibody and then coupling to a thiolated toxin dimer—or by first reacting it with the toxin and coupling to a thiolated antibody (Fig. 326). Thiolation of the secondary component is usually done with SPDP or 2-iminothiolane. Other cross-linkers containing an io-doacetyl group can be used in a similar fashion.

Conjugations with iodoacetyl cross-linkers have been done using ricin and cobra venom factor (Myers *et al.*, 1989; Vogel, 1987; Thorpe *et al.*, 1984). The following generalized protocol for using SIAB is based on the method of Cumber *et al.* (1985).

Protocol
Caution: Toxin molecules are dangerously toxic even in small amounts. Use extreme care in handling.

To prepare an antibody–ricin conjugate using this protocol, 2.25 mg of antibody is needed for every milligram of toxin.

Activation of Toxin with SIAB

1. Dissolve intact ricin in 0.1 *M* sodium phosphate, 0.15 *M* NaCl, pH 7.5, at a concentration of 10 mg/ml.
2. Dissolve SIAB (Pierce) in DMSO at a concentration of 1.4 mg/ml. Prepare fresh and protect from light to avoid breakdown of the active halogen group.

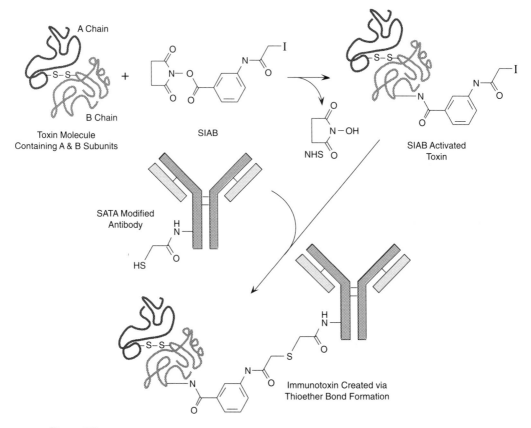

Figure 326 SIAB can be used to activate toxin molecules for coupling with sulfhydryl-containing anti-bodies. In this case, the antibody molecule was thiolated using SATA and deprotected to reveal the free sulfhydryl. Reaction with the SIAB-activated toxin forms the final conjugate by thioether bond formation.

3. Add 160 μl (225 μg) of the SIAB solution to each milliliter of the ricin solution.
4. React for 30 min at room temperature in the dark.
5. Remove excess cross-linker from the activated ricin by gel filtration using a column of Sephadex G-25.
6. Concentrate the purified, SIAB-activated toxin to 10 mg/ml using centrifugal concentrators with a molecular weight cut-off of 10,000. Protect the activated toxin from light.

Thiolation of Specific Antibody Molecule with SPDP

1. Dissolve the antibody to be conjugated in 0.1 *M* sodium phosphate, 0.15 *M* NaCl, pH 7.5, at a concentration 10 mg/ml. Use 2.25 mg of antibody per milligram of toxin to be conjugated.
2. Dissolve SPDP (Pierce) in DMF at a concentration of 3 mg/ml. Add 24 μl of this solution to each milliliter of the antibody solution with gentle mixing to effect complete dissolution.
3. React for 30 min at room temperature.

4. Remove excess cross-linker by gel filtration using a column of Sephadex G-25. Perform the chromatography using 0.1 M sodium phosphate, 0.15 M NaCl, 10 mM EDTA, pH 7.5. The buffer should be degassed under vacuum and nitrogen bubbled through it to remove oxygen. The presence of EDTA stabilizes the free sulfhydryls formed in the following steps against metal-catalyzed oxidation.

5. Concentrate the fractions containing protein from the gel filtration step to 10 mg/ml using a centrifugal concentrator (MW cut-off of 10,000).

6. To reduce the pyridyldithiol groups and create reactive sulfhydryls, dissolve DTT (Pierce) in water at a concentration of 17.2 mg/ml and immediately add 500 µl of this solution to each milliliter of concentrated antibody solution. Mix to dissolve and react for 30 min at room temperature.

7. Remove excess DTT by gel filtration using the same buffer as in step 4. Pool the fractions containing protein and concentrate to 10 mg/ml.

Conjugation of SIAB-Activated Toxin with Thiolated Antibody

1. Mix activated toxin from part A with thiolated antibody from part B at a ratio of 2.25 mg of antibody per milligram of toxin. Protect the solution from light.

2. React for 18 h at room temperature in the dark.

3. To block unreacted sulfhydryl groups, add iodoacetamide to the solution to a final concentration of 2 mM. React for an additional 1 h at room temperature.

4. Isolation of the ideal 1:1 or 1:2 antibody–toxin conjugate can be done by gel filtration separation using a column of Sephacryl S-300.

SMCC

Succinimidyl-4-(N-maleimidomethyl)cyclohexane-1-carboxylate (SMCC) is a cross-linker with significant utility in protein conjugation (Chapter 5, Section 1.3). It is a popular choice among heterobifunctional reagents, especially for the preparation of antibody–enzyme and hapten–carrier conjugates (Hashida and Ishikawa, 1985; Dewey *et al.*, 1987). The NHS ester end of the reagent can react with primary amine groups on proteins to form stable amide bonds. The maleimide end of SMCC is specific for coupling to sulfhydryls when the reaction pH is in the range of 6.5–7.5 (Smyth *et al.*, 1964). The nature of the reactive groups of SMCC allow for highly controlled cross-linking procedures to be performed, wherein the resulting products can be closely limited to a 1:1 ratio in the final complex. Thus, low-molecular-weight conjugates that make ideal reagents for *in vivo* purposes can be made.

However, since SMCC forms nonreversible thioether linkages with sulfhydryl groups, it only can be used in the preparation of immunotoxins if intact A–B toxins are employed in the conjugate. In such conjugates, the A chain still has the potential for reductive release from the B-chain subunit after cellular docking and internalization. Immunotoxins prepared with A-chain or single-subunit toxins will not display cytotoxicity if cross-linked with SMCC, since the cross-linker does not create cleavable disulfide bonds upon conjugation.

SMCC has been used to prepare immunotoxins with cobra venom factor (Vogel, 1987) and was compared to other cross-linkers in the preparation of gelonin and PAP conjugates (Lambert *et al.*, 1985).

The following protocol is a suggested method for the conjugation of SMCC-activated antibodies with 2-iminothiolane modified, intact toxin molecules (Fig. 327). It utilizes the water-soluble analog of SMCC, sulfo-SMCC, which contains a negatively charged sulfonate group on its NHS ring.

Protocol
Caution: Toxin molecules are dangerously toxic even in small amounts. Use extreme care in handling.

Note: This protocol requires mixing activated antibody with thiolated toxin at a ratio of 2.25 mg of antibody per milligram of toxin. This ratio should be taken into account before starting the reactions.

Activation of Antibody with Sulfo-SMCC

1. Dissolve 10 mg of specific antibody in 1 ml of 0.1 M sodium phosphate, 0.15 M NaCl, pH 7.2.

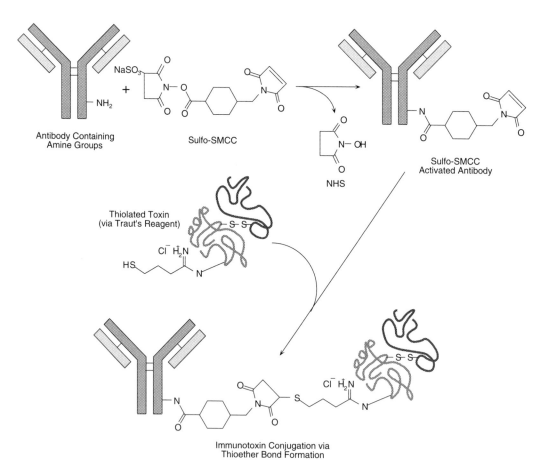

Immunotoxin Conjugation via
Thioether Bond Formation

Figure 327 Sulfo-SMCC may be used to activate antibody molecules for coupling to thiolated toxin components. An intact A–B toxin molecule can be modified to contain sulfhydryls by treatment with 2-iminothiolane. Thiolation with this reagent retains the cytotoxic properties of the toxin while generating a sulfhydryl for conjugation. Reaction of the thiolated toxin with the maleimide activated antibody creates the immunotoxin through thioether bond formation.

2. Weigh out 2 mg of sulfo-SMCC (Pierce) and add it to the above solution. Mix gently to dissolve. To aid in measuring the exact quantity of cross-linker, a concentrated stock solution may be made in water and an aliquot equal to 2 mg transferred to the reaction solution. If a stock solution is made, it should be dissolved rapidly and used immediately to prevent extensive hydrolysis of the active ester.
3. React for 1 h at room temperature.
4. Immediately purify the maleimide-activated protein by applying the reaction mixture to a desalting column packed with Sephadex G-25 or the equivalent. Do not dialyze the solution, since the maleimide activity will be lost over the time course required to complete the operation. To obtain good separation between the protein peak (eluting first) and the peak representing excess reagent and reaction by-products (eluting second), the applied sample size should be no more than 8% of the column bed volume.
5. Collect the peak containing the activated antibody and concentrate to 10 mg/ml using centrifugal concentrators. Use immediately for conjugating to a thiolated toxin.

Thiolation of Intact A–B Toxin

1. Dissolve the toxin (e.g., intact ricin) at a concentration of 10 mg/ml in 50 mM triethanolamine hydrochloride, pH 8, containing 10 mM EDTA. The buffer should be degassed under vacuum and bubbled with nitrogen to remove oxygen that may cause sulfhydryl oxidation after thiolation.
2. Dissolve 2-iminothiolane in degassed, nitrogen-bubbled deionized water at a concentration of 20 mg/ml (makes a 0.14 M stock solution). The solution should be used immediately. Add 70 μl of the 2-iminothiolane solution to each milliliter of the toxin solution (final concentration is about 10 mM).
3. React for 1 h at 0°C (on ice) under a nitrogen blanket.
4. Purify the thiolated toxin from unreacted Traut's reagent by gel filtration using 0.1 M sodium phosphate, 0.15 M NaCl, pH 7.5, containing 10 mM EDTA. The presence of EDTA in this buffer helps to prevent oxidation of the sulfhydryl groups with resultant disulfide formation. The degree of —SH modification in the purified protein may be determined using the Ellman's assay (Chapter 1, Section 4.1).
5. Concentrate the thiolated toxin to 10 mg/ml using centrifugal concentrators.

Conjugation of SMCC-Activated Antibody with Thiolated Toxin

1. Mix the thiolated toxin with SMCC-activated antibody at a ratio of 2.25 mg of antibody per milligram of toxin. Protect the solution from light.
2. React for 18 h at room temperature.
3. To block unreacted sulfhydryl groups, add iodoacetamide to the solution to a final concentration of 2 mM. React for an additional 1 h at room temperature.
4. Isolation of the ideal 1:1 antibody–toxin conjugate can be done by gel filtration separation using a column of Sephacryl S-300.
5. Removal of nonspecific binding potential in the B chain must be done before using an A–B intact toxin conjugate *in vivo*. A large proportion of the binding

sites on the B chains usually are blocked during the above conjugation process, and the galactose binding potential is significantly impaired. Further purification to remove conjugates that have galactose binding potential can be done on an acid-treated Sepharose column (which contains galactose residues) or on a column of asialofetuin bound to Sepharose (Cumber *et al.*, 1985). Conjugate fractions that do not bind to both affinity gels contain no nonspecific binding potential toward nontargeted cells.

MBS

m-Maleimidobenzoyl-*N*-hydroxysuccinimide ester (MBS) is a heterobifunctional cross-linking agent containing an NHS ester and a maleimide group. The NHS ester can react with primary amines in proteins and other molecules to form stable amide bonds, while the maleimide end reacts with sulfhydryl groups to create stable thioether linkages (Chapter 5, Section 1.4). The reagent can be used in many different conjugation protocols to cross-link amine-containing proteins with sulfhydryl-containing proteins. Since the thioether bond formed at the maleimide end is nonreversible, MBS can be used for immunotoxin preparation only if the conjugate involves cross-linking intact A–B toxins with antibody molecules. Using intact toxins (as opposed to single-chain or A-chain isolates), the A chain still is able to release from the complex after cellular docking and inactivate ribosomal activity (Youle and Nevelle, 1980; Myers *et al.*, 1989; Dell'Arciprete *et al.*, 1988).

MBS contains a benzoic acid derivative as its cross-bridge. In many applications involving NHS–maleimide cross-linkers, nonaromatic cross-bridges are considered superior to aromatic ones. This is reflected in the stability of the maleimide group to hydrolysis prior to conjugating with a sulfhydryl group. For immunotoxin preparation, however, aromatic maleimides resulted in better conjugate yield and more potent cytotoxic effects when compared to aliphatic ones (Myers *et al.*, 1989). MBS, therefore, may be a cross-linker of choice when making conjugates with intact toxin molecules.

The following protocol is adapted from Myers *et al.* (1989). It involves activation of ricin with MBS and conjugation with a partially reduced antibody (Fig. 328).

Protocol
Caution: Toxin molecules are dangerously toxic even in small amounts. Use extreme care in handling.

This method uses a molar ratio of 15:1 for ricin:antibody. This requires 6.24 mg of ricin per milligram of antibody. This ratio should be considered when determining how much starting materials to use for each step.

Activation of Ricin with MBS

1. Dissolve ricin at a concentration of 10 mg/ml in 0.1 *M* sodium phosphate, 0.15 *M* NaCl, pH 7.5.
2. Dissolve MBS (Pierce) in DMF at a concentration of 2 mg/ml.
3. Add 76 μl of the MBS solution to each milliliter of the ricin solution. This represents a 3:1 molar ratio of cross-linker to protein.
4. React for 30 min at room temperature.

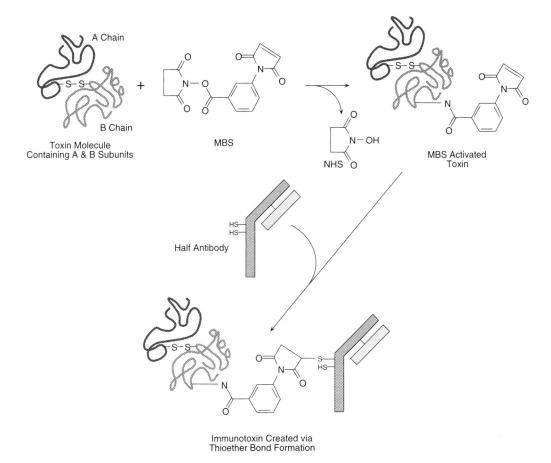

Figure 328 Activation of an intact A–B toxin molecule with MBS with subsequent conjugation with a reduced antibody fragment to produce an immunotoxin.

5. Immediately purify the MBS-activated toxin by gel filtration using a column of Sephadex G-25. Apply no more sample than represents 5–8% of the gel volume. Isolate the protein peak by its absorbance at 280 nm and concentrate to 10 mg/ml using centrifugal concentrators with a molecular weight cut-off of 10,000.

Partial Reduction of Antibody with DTT

1. Dissolve the antibody in 0.1 M sodium phosphate, 0.15 M NaCl, 10 mM EDTA, pH 7.5, at a concentration of 10 mg/ml.
2. Add DTT to a final concentration of 50 mM.
3. Reduce for 30 min at room temperature.
4. Purify the reduced antibody using gel filtration on a column of Sephadex G-25. Concentrate the protein to 10 mg/ml using centrifugal concentrators.

Conjugation of MBS Activated Ricin
with Partially Reduced Antibody

1. Mix the MBS-activated ricin with the partially reduced antibody in a molar ratio of 15:1 (or 6.24 mg activated ricin per milligram of reduced antibody). This represents a volume ratio (at 10 mg/ml for both proteins) of 1 ml ricin solution mixed with 160 μl antibody solution.
2. React for 18 h at room temperature.
3. Purification of the immunotoxin conjugate from unconjugated ricin can be done using a column of TSK3000 SW (Toya Soda, Japan) according to the method of Myers *et al.* (1989).
4. Removal of nonspecific binding potential in the B chain must be done before using an A–B intact toxin conjugate *in vivo*. See step 5 of the SMCC conjugation protocol discussed previous to this section.

SMPB

Succinimidyl-4-(*p*-maleimidophenyl)butyrate (SMPB) is a heterobifunctional analog of MBS containing an extended cross-bridge (Chapter 5, Section 1.6). The cross-linker has an amine-reactive NHS ester on one end and a sulfhydryl-reactive maleimide group on the other. Conjugates formed using SMPB thus are linked by stable amide and thioether bonds.

As in the case of MBS, discussed previously, SMPB was found to be more effective than aliphatic cross-linkers in producing immunotoxin conjugates with ricin that have high yields of cytotoxicity (Myers *et al.*, 1989). This was attributed to the reagent's aromatic ring structure. A comparison with SPDP-produced immunotoxin conjugates concluded that SMPB formed more stable complexes that survive in serum for longer periods (Martin and Papahadjopoulos, 1982).

The method for the preparation of immunotoxins with SMPB is identical to that used for MBS (above). Since the thioether bonds formed with sulfhydryl-containing molecules are noncleavable, A-chain isolates or single-chain toxin molecules cannot be conjugated with antibodies with retention of cytotoxicity. Only intact A–B toxin molecules may be used with this cross-linker, since the A chain still is capable of being reductively released from the complex.

2.3. Preparation of Immunotoxin Conjugates via Reductive Amination

Conjugations involving aldehyde groups and amine-containing molecules can be done through Schiff base formation with subsequent reduction using sodium cyanoborohydride to form stable secondary amine linkages (Chapter 2, Section 5.3). Carbohydrates, glycoproteins, and other polysaccharide-containing molecules can be oxidized to contain aldehyde residues by sodium periodate or specific oxidases (Chapter 1, Section 4.4). Some antibodies and toxin molecules are glycoproteins and contain sufficient carbohydrate to be utilized for reductive amination cross-linking.

A second method of immunotoxin preparation by reductive amination involves the use a polysaccharide spacer. Soluble dextran may be oxidized with periodate to form a multifunctional cross-linking polymer. Reaction with antibodies and cytotoxic molecules in the presence of a reducing agent forms multivalent immunotoxin conjugates. The following sections discuss these options.

Periodate Oxidation of Glycoproteins Followed by Reductive Conjugation

Antibody molecules usually contain carbohydrate in their Fc regions. Similarly, many toxins, such as ricin and abrin, are glycoproteins that contain abundant polysaccharide. These carbohydrate residues can be oxidized with 10 mM sodium periodate to form reactive aldehyde groups capable of being conjugated with primary amines (Chapter 1, Section 4.4). Mixing an aldehyde-containing glycoprotein with another amine-containing molecule in the presence of sodium borohydride or sodium cyanoborohydride reduces the intermediate Schiff bases that are formed to stable secondary amine bonds. Since functional groups on the antibody and the toxin components are the only ones necessary for this type of conjugation strategy, it is often referred to as a zero-length cross-linking procedure (Chapter 3). In other words, no additional cross-linking reagents are introduced into the site of the cross-link. This method of conjugation is used with great success in the formation of antibody–enzyme conjugates, especially using the glycosylated enzyme, HRP (Chapter 10, Section 1.3).

The disadvantage of this type of conjugation approach for producing immunotoxins is that many of the monoclonal antibodies or antibody fragments used for immunotoxin conjugation are devoid of carbohydrate. Especially when using small Fv fragments or single-chain antibodies produced by recombinant techniques, there are typically no polysaccharide portions attached to them. In this case, creation of aldehydes on the targeting component is not possible. In addition, not all toxin molecules contain carbohydrate. Ricin, abrin, and cobra venom factor are glycoproteins and can be oxidized and coupled to antibodies without difficulty (Olsnes and Pihl, 1982b,c; Vogel and Muller-Eberhard, 1984). However, it is not well known whether immunotoxin conjugates formed by this procedure retain their ability to inhibit ribosomal activity.

Suggested procedures for using reductive amination techniques may be found in Chapter 1, Section 4.4, and Chapter 3, Section 4.

Periodate-Oxidized Dextran as Cross-linking Agent

Dextran polymers consist of glucose residues bound together predominantly in α-1,6 linkages. The main repeating unit is an isomaltose group. Most preparations of dextran contain some branching, mainly incorporating 1,2, 1,3, and 1,4 linkages. The degree of branching is characteristic of its source—the strain and species of yeast or bacteria from which the dextran originated. The terminating monosaccharide in a dextran polymer is often a fructose group. Dextrans of molecular weight 10,000–40,000 provide long, hydrophilic arms that can accommodate multiple attachment points for macromolecules along their length. Soluble dextrans can be oxidized in aqueous solution to create numerous aldehyde residues suitable for use in reductive amination techniques (Hurwitz *et al.*, 1978, 1985; Sela and Hurwitz, 1987; Manabe *et al.*, 1983). Periodate oxidation results in the cleavage of the carbon—carbon bonds between the Nos. 2 and 3 carbons within each monosaccharide unit of the chain, transforming the associated hydroxyl groups into aldehydes (Chapter 1, Section 4.4).

Periodate oxidized dextran can be used as a protein modification or cross-linking

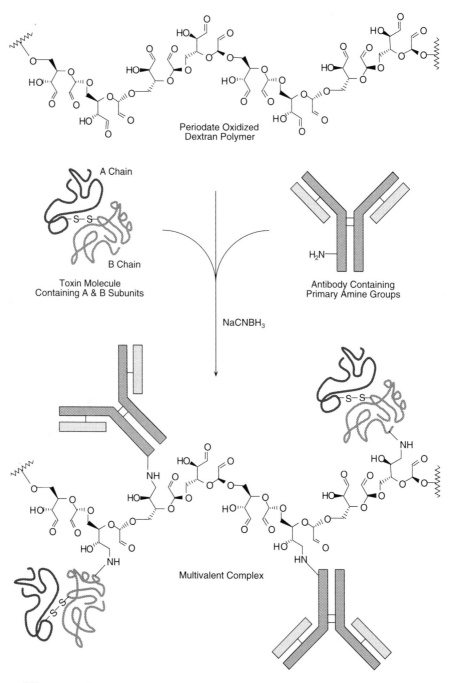

Figure 329   A periodate-oxidized dextran polymer may be reacted with both an antibody and an intact toxin component using reductive amination to form a multivalent immunotoxin complex.

agent (Chapter 15, Section 2). Conjugation of antibody molecules to toxins can be done with dextran to produce immunotoxins suitable for *in vivo* administration. Mixing of the antibody and toxin together with the oxidized dextran under alkaline conditions results in the formation of Schiff base interactions with the amines on both proteins. Reduction of these Schiff base linkages with sodium borohydride or sodium cyanoborohydride results in stable amide bonds, covalently attaching multiple antibody and toxin molecules along the length of the polysaccharide chain (Fig. 329).

Chemoimmunoconjugates consisting of drugs attached to antibody targeting molecules also can be formed using oxidized dextran carriers. Cancer therapeutic agents such as adriamycin, bleomycin, and daunomycin can be coupled to the oxidized dextran through their amine groups. After formation of Schiff base linkages between these drugs and the carrier, the antibody is added and a reducing agent used to create the final amide bond linkages (Sela and Hurwitz, 1987). The dextran backbone provides many more drug molecules associated with each antibody than could be accomplished by direct conjugation to the antibody itself.

Although dextran can be a versatile cross-linking agent for the preparation of many forms of macromolecular conjugates, immunotoxin conjugation may be impeded by the nonreversibility of the multiple amide bond linkages formed during reductive amination. Certainly, only intact A–B toxins have a chance of succeeding with this method, since A-chain or single-subunit toxins would not be capable of release from the complex after cellular docking. Even intact two-subunit toxins, however, may not be capable of releasing an A-chain unit, due to the multivalent nature of the oxidized dextran linker. For this reason, activated dextran may be more useful for constructing antibody conjugates consisting of some cytotoxic component other than protein toxins—for example, drug, hormone or radioactive complexes.

Methods for using oxidized dextran, including reductive amination techniques, can be found in Chapter 1, Section 4.4, Chapter 3, Section 4, and especially Chapter 15, Section 2).

12

Preparation of Liposome Conjugates and Derivatives

One of the fastest growing fields that heavily depends on bioconjugate technology involves the use of liposomes. At one time, liposomes were studied only for their interesting structural characteristics in solution. Their physicochemical properties were investigated extensively as models of membrane morphology. Today, they are being put to use as macromolecular carriers for nearly every application of bioconjugate chemistry imaginable. They are used as delivery devices to encapsulate cosmetics, drugs, and fluorescent detection reagents and as vehicles to transport nucleic acids, peptides, and proteins to cellular sites *in vivo*. Targeting components such as antibodies can be attached to liposomal surfaces and used to create large antigen-specific complexes. In this sense, liposomal derivatives are being used to target cancer cells *in vivo*, to enhance detectability in immunoassay systems, and as multivalent cross-bridges in avidin–biotin-based assays. Covalent attachment of antigens to the surface of liposomes provides excellent immunogen complexes for the generation of specific antibodies or as vaccine carriers to elicit protective immunity.

The end-products of liposome technology are used in retail markets, for the diagnosis of disease, as therapeutic agents, as vaccines, and as important components in assays designed to either detect or quantify certain analytes.

The following sections discuss the properties and applications of liposome technology as well as the most common methods of preparing conjugates of them with proteins and other molecules.

1. Properties and Use of Liposomes

1.1. Liposome Morphology

Liposomes are artificial structures primarily composed of phospholipid bilayers exhibiting amphiphilic properties. Other molecules, such as cholesterol or fatty acids also may be included in the bilayer construction. In complex liposome morphologies, concentric spheres or sheets of lipid bilayers are usually separated by aqueous regions that are sequestered or compartmentalized from the surrounding solution. The phospholipid constituents of liposomes consist of hydrophobic lipid "tails" connected to a "head" constructed of various glycerylphosphate derivatives. The hydrophobic interaction between the fatty acid tails is the primary driving force for creating liposomal bilayers in aqueous solution.

However, the organization of liposomes in aqueous solution may be highly complicated. The nature of the lipid constituents, the composition of the medium, and the temperature of the solution all affect the association and morphology of liposomal construction. Small "monomers" or groupings of lipid molecules may assemble to create larger structures having several main forms (Fig. 330). Aggregation of these monomers may fuse them into spherical micelles, wherein the polar head groups are all facing outward toward the surrounding aqueous medium and the hydrophobic tails are all pointing inward, excluding water. In addition, aggregation may result in bilayer construction. In this case, sheets of lipid molecules, all with their head groups facing one direction and their tails facing the other way, are fused with another lipid sheet having their tails and heads facing the opposite direction. Thus, the inside of the bilayer contains only hydrophobic tails from both sheets, while the outside contains the hydrophilic heads facing the outer aqueous environment.

The configuration the bilayers can assume also can be complex. A bilayer may be in a spherical form, having one layer of hydrophilic head groups pointing outward toward the surrounding solution and the second hydrophilic layer pointing inward toward a compartment of aqueous solution sequestered within the sphere. The morphology of a liposome may be classified according to the compartmentalization of aqueous regions between bilayer shells. If the aqueous regions are segregated by only one bilayer each, the liposomes are called unilamellar vesicles (ULV) (Fig. 331). If there

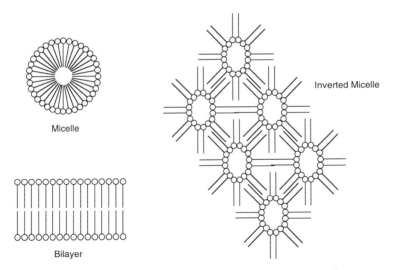

Micelle

Inverted Micelle

Bilayer

Figure 330 The amphiphilic nature of phospholipids in solution drives the formation of complex structures. Spherical micelles may form in aqueous solution, wherein the hydrophilic head groups all point out toward the surrounding water environment and the hydrophobic tails point inward to the exclusion of water. Larger lipid bilayers may form by similar forces, creating sheets, spheres, and other highly complex morphologies. In nonaqueous solution, inverted micelles may form, wherein the tails all point toward the outer hydrophobic region and the heads point inward forming hexagonal shapes.

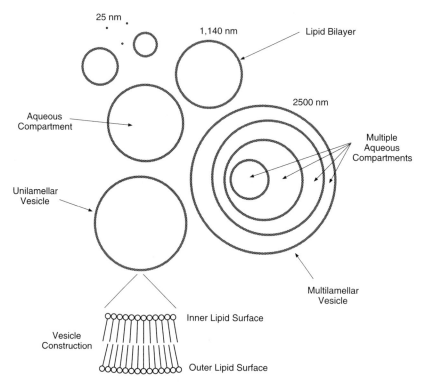

Figure 331   The highly varied morphologies of lipid bilayer construction.

is more than one bilayer surrounding each aqueous compartment, the liposomes are termed multilamellar vesicles (MLV). ULV forms are further classified according to their relative size, although rather crudely. Thus, there can be small unilamellar vesicles (SUV; usually thought of as less than 100 nm in diameter, with a minimum of about 25 nm) and large unilamellar vesicles (LUV; usually greater than 100 nm in diameter, with a maximal size of about 2500 nm). With regard to MLV, however, the bilayer structures cannot be as easily classified due to the almost infinite number of ways each bilayer can be associated and interconnected with the next one. MLVs typically form large complex honeycomb structures that are difficult to categorize or exactly reproduce. However, MLVs are the simplest to prepare, the most stable, and the easiest to scale up to large production levels.

Small lipid groupings or monomers also may fuse into bigger, inverted micelles, wherein their hydrophobic tails point outward toward other inverted micelle lipid tails. The individual structures are usually hexagonal in shape, but typically they exist as large groupings of inverted micelles, the outer edge of which contains partial inverted micelles, exposing their inner hydrophilic heads to the surrounding aqueous environment.

The most useful form of liposomes for bioconjugate applications consists of small, spherical ULVs that possess layers of hydrophilic head groups on their inner and outer surfaces. The inside of each vesicle can contain hydrophilic molecules that are protected from the outer environment by the lipid shell. The outside surface can be derivatized to contain covalently attached molecules designed to target the liposome for specific interactions.

1.2. Preparation of Liposomes

Mixtures of phospholipids in aqueous solution will spontaneously associate to form liposomal structures. To prepare liposomes having morphologies useful for bioconjugate or delivery techniques, it is necessary to control this assemblage to create vesicles of the proper size and shape. Many methods are available to accomplish this goal; however, all of them have at least several steps in common: (1) dissolving the lipid mixture in organic solvent, (2) dispersion in an aqueous phase, and (3) fractionation to isolate the correct liposomal population.

In the first stage, the desired mix of lipid components is dissolved in organic solvent [usually chloroform:methanol (2:1 by volume)] to create a homogeneous mixture. This mixture will include any phospholipid derivatized to contain reactive groups as well as other lipids used to form and stabilize the bulk of the liposomal structure. During all handling procedures using lipids or their derivatives, it is essential that the solutions be protected from oxidation and excessive exposure to light, especially the sun. Organic solvents should be maintained under a nitrogen or argon atmosphere to prevent introduction of oxygen. Water and buffers should be degassed using a vacuum and bubbled with inert gas before lipid components are introduced.

The correct ratio of lipid constituents is important to form stable liposomes. For instance, a reliable liposomal composition for encapsulating aqueous substances may contain molar ratios of lecithin:cholesterol:negatively charged phospholipid (e.g., phosphatidyl glycerol) of 0.9:1:0.1. A composition that is typical when an activated phosphoethanolamine is included may contain molar ratios of phosphatidyl choline:cholesterol:phosphatidyl glycerol:derivatized phosphatidyl ethanolamine of 8:10:1:1. Another typical composition using a maleimide derivative of phosphoethanolamine (PE) without phosphatidyl glycerol is phosphatidyl choline:maleimide-PE:cholesterol of 85:15:50 (Friede *et al.*, 1993). In general, to maintain membrane stability, the PE derivative should not exceed a concentration ratio of about 1–10 mol PE per 100 mol of total lipid.

An example of a lipid mixture preparation based on mass would be to dissolve 100 mg of phosphatidyl choline, 40 mg of cholesterol, and 10 mg of phosphatidyl glycerol in 5 ml of chloroform/methanol solution. When using activated PE components, inclusion of 10 mg of the PE derivative to this recipe will result in a stable liposome preparation.

Once the desired mixture of lipid components is dissolved and homogenized in organic solvent, one of several techniques may be used to disperse the liposomes in aqueous solution. These methods may be broadly classified as (1) mechanical dispersion, (2) detergent-assisted solubilization, and (3) solvent-mediated dispersion.

Probably the most popular option is mechanical dispersion, simply because the greatest number of methods that utilize it have been developed. When using mechanical means to form vesicles, the lipid solution first is dried to remove all traces of organic solvent prior to dispersion in an aqueous media. The dispersion process is the key to producing liposomal membranes of the correct morphology. This method uses mechanical energy to break up large lipid agglomerates into smaller vesicles having the optimal size and shape characteristics necessary for encapsulation or bioconjugation.

Mechanical dispersion methods involve adding an aqueous solution (which may contain substances to be encapsulated) to the dried, homogeneous lipid mixture and manipulating it to effect dispersion. Major methods of mechanical dispersion include simple shaking (Bangham *et al.*, 1965), nonshaken aqueous contact (Reeves and Dowben, 1969), high-pressure emulsification (Mayhew *et al.*, 1984), sonication (Huang, 1969), extrusion through small-pore membranes (Olson *et al.*, 1980), and various freeze–thaw techniques (Pick, 1981). Some devices are available commercially that automate the mechanical dispersion process, usually by high-pressure emulsification or sonication (Branson Ultrasonics Corp.).

Most of these methods result in a population of vesicles ranging from SUVs of only 25 nm diameter to very large MLVs. Classification of the desired liposomal morphology may be done by chromatographic means using columns of Sepharose 2B or Sepharose 4B, by density-gradient centrifugation using Ficoll or metrizamide gradients, or by dialysis.

Liposome formation by detergent-assisted solubilization utilizes the amphipathic nature of detergent molecules to bring more effectively the lipid components into the aqueous phase for dispersion. The detergent molecules presumably bind and mask the hydrophobic tails of lipids from the surrounding water molecules. Detergent treatment may take place from a dried lipid mixture or after formation of small vesicles. Usually, nonionic detergents such as the Triton X family, alkyl glycosides, or bile salts such as sodium deoxycholate are employed for this procedure. The immediate structures that form as the detergent molecules solubilize the lipids from a dried state are small micelles. On removal of the detergent from the solution, the lipid micelles aggregate to create larger liposome structures. Liposomes of up to 1000 Å containing a single bilayer may be formed using detergent-assisted methods (Enoch and Strittmatter, 1979). Unfortunately, some detergent-removal processes also may remove other molecules that were to be entrapped in the liposomes during formation.

Solvent-mediated dispersion techniques used to create liposomes first involve dissolving the lipid mixture in an organic solvent to create a homogeneous solution, and then introducing this solution into an aqueous phase. The solvent may or may not be soluble in the aqueous phase to effect this process. There also may be components dissolved in the aqueous phase to be encapsulated in the developing liposomes.

Perhaps the simplest solvent dispersion method is that developed by Batzri and Korn (1973). Phospholipids and other lipids to be a part of the liposomal membrane are first dissolved in ethanol. This ethanolic solution is then rapidly injected into an aqueous solution of 0.16 M KCl using a Hamilton syringe, resulting in a maximum concentration of no more than 7.5% ethanol. Using this method, single bilayer liposomes of about 25-nm diameter can be created that are indistinguishable from those formed by mechanical sonication techniques. The main disadvantages of ethanolic injection are the limited solubility of some lipids in the solvent (about 40 mM for phosphatidyl choline) and the dilute nature of the resultant liposome suspension. However, for the preparation of small quantities of SUVs, this method may be one of the best available.

Other solvent dispersion methods utilize solvents that are insoluble in the aqueous phase. The key to the production of liposomes by this procedure involves the formation of a "water-in-oil" emulsion. To create the proper reverse-phase emulsion, a small quantity of aqueous phase must be introduced into a large quantity of organic phase containing the dissolved liposomes. The result is a milky dispersion containing the "homogenized" liposomes. A number of techniques have been developed to perform this procedure (Kim and Martin, 1981; Kim *et al.*, 1983; Pidgeon *et al.*, 1986; Szoka *et al.*, 1980). The emulsification process in each of these solvent-dispersion techniques involves the use of mechanical means (shaking, stirring, or sonication) to effect the formation of small droplets of aqueous solution uniformly dispersed in the lipid-organic phase.

For the preparation of large quantities of liposomes, mechanical dispersion using a commercially available emulsifier is probably the best route. For limited quantities, the use of simple shaking or ethanolic dispersion techniques work well.

Regardless of their method of fabrication, most liposome preparations need to be further classified and purified before use. To remove excess aqueous components that were not encapsulated during the vesicle formation process, gel filtration using a column of Sephadex G-50 or dialysis can be employed. To fractionate the liposome population according to size, gel filtration using a column of Sepharose 2B or 4B should be done.

Small liposome vesicles often aggregate on standing to form larger, more complex structures. Therefore long-term storage in aqueous solution is usually not possible without major transformations in liposome morphology. Freezing also fractures the liposomal membrane, releasing any entrapped substances. The inclusion of cryoprotectants such as sugars or polyhydroxylic containing compounds can overcome the structural degradation problems on freezing (Harrigan *et al.*, 1990; Talsma *et al.*, 1991; Park and Huang, 1992). Presumably, the hydroxyl groups in cryoprotectants can take the place of water in hydrogen bonding activities, thus providing structural support even under conditions in which water is removed. A procedure by Friede *et al.* (1993) allows freezing and lyophilization of SUVs in the presence of 4% sorbitol with complete retention of liposome integrity on reconstitution. Thus freeze-drying may be the best method for the long-term storage of intact liposomes.

1.3. Chemical Constituents of Liposomes

The overall composition of a liposome—its morphology, chemical constituents (including a large variety of phospholipids and other lipids or fatty acids), charge, and any attached functional groups—can affect the properties of the vesicle both *in vitro* and *in vivo* (Allison and Gregoriadis, 1974; Alving, 1987; Therien and Shahum, 1989). Although there are literally dozens of lipid components that potentially can be included in a liposomal recipe, only a handful are commonly used.

Phospholipids are the most important of these liposomal constituents. The major component of cell membranes, phospholipids are composed of a hydrophobic, fatty acid tail and a hydrophilic head group. The amphipathic nature of these molecules is the primary force that drives the spontaneous formation of bilayers in aqueous solution and holds the spherical vesicles together.

Naturally occurring phospholipids can be isolated from a variety of sources. One of the most common phospholipid raw materials is egg yolk. However, since the composition of egg phospholipid is from a biological source and can vary considerably depending on age of the eggs, the diet of the chickens, and the method of processing, newer enzymatic and synthetic chemical methods now are being employed to manufacture the required phospholipid derivatives in higher purity and yield.

Two main forms of lipid derivatives exist biologically: molecules containing a glycerol backbone and those containing a sphingosine backbone. The most important type for liposomal construction is a phosphodiglyceride derivative, which consists of a glycerol backbone that links two fatty acid molecules with a polar head group (Fig. 332). The fatty acids are acyl bonded in ester linkages to the Nos. 1 and 2 carbon hydroxyls of the glycerol bridge. The No. 3 carbon hydroxyl of the glycerol group is phosphorylated and possesses a negative charge at physiological pH. This basic phosphodiglyceride construct of two fatty acids and one glycerylphosphate group is called phosphatidic acid. This is the simplest form of phospholipid available.

The fundamental phosphatidyl group also can be further derivatized at the phosphate to contain an additional polar constituent. Several common derivatives of phosphodiglycerides are naturally occurring, including phosphatidyl choline (PC; commonly called lecithin), phosphatidyl ethanolamine (PE), phosphatidyl serine (PS), phosphatidyl glycerol (PG), and phosphatidyl inositol (PI) (Fig. 333). All of these phospholipids have polar groups that are linked to the phosphatidyl moiety in a phosphate ester bond. The most abundant of these derivatives in biological cell membranes is PC—the trimethyl derivative of PE, possessing a positive charge at physiological pH. Some or all of these phosphodiglyceride derivatives can be mixed to create a particular liposomal recipe.

The fatty acid constituents of phosphodiglycerides can vary considerably in nature among a number of different chain lengths and points of unsaturation. A given isolated phosphatidyl derivative from a biological source usually possesses a range of fatty acid

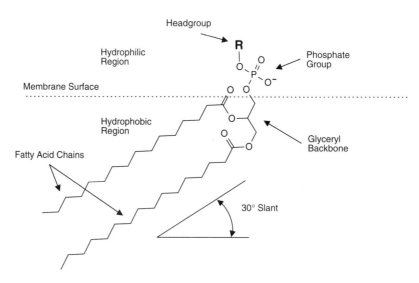

Figure 332   The basic construction of phosphodiglyceride molecules within lipid bilayers. The fatty acid chains are embedded in the hydrophobic inner region of the membrane, oriented at an angle to the plane of the membrane surface. The hydrophilic head group, including the phosphate portion, points out toward the hydrophilic aqueous environment.

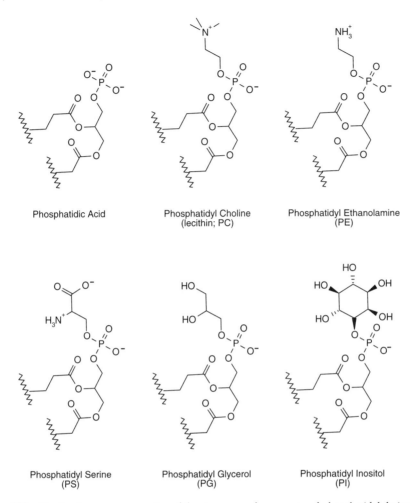

Phosphatidic Acid

Phosphatidyl Choline
(lecithin; PC)

Phosphatidyl Ethanolamine
(PE)

Phosphatidyl Serine
(PS)

Phosphatidyl Glycerol
(PG)

Phosphatidyl Inositol
(PI)

Figure 333 The head-group construction of the six commonly encountered phosphatidyl derivatives.

components, varying in chain length from C16 to about C24. Some of the fatty acids also may contain points of unsaturation—one or more double bonds between certain carbon atoms within the chain (Matreya and Avanti Polar Lipids, suppliers). For instance, egg lecithin is not a single compound, but contains a mixture of phosphatidyl cholines containing about 31% saturated fatty acid having a chain length of 16 carbons, 16% saturated fatty acid with 18 carbons, about 48% also with 18 carbons but having at least 1–2 points of unsaturation, and the rest a variety of other fatty acid constituents. Naturally occurring, unsaturated fatty acids typically are of the *cis* conformation, not *trans*. The existence of unsaturation within a fatty acid is usually abbreviated as the chain length followed by a colon and the number of double bonds. For instance, *cis*-9-hexadecenoic acid (palmitoleic acid) has one double bond at carbon 9 and is abbreviated as C16:1.

By contrast, a given synthetic preparation of a major phospholipid possesses fatty acid constituents all of identical chain length and unsaturation. A synthetic PC derivative can be purchased that contains only, for instance, 1,2-dimyristoyl (C14) fatty acid substitutions on its glyceryl backbone (Genzyme). The use of synthetic rather than natural phospholipids for making liposomes thus produces reagents of known chemical purity—very important for regulatory requirements surrounding the introduction of products used topically or *in vivo*.

Despite the large variety of potential fatty acid components in natural-occurring phosphodiglycerides, only three major fatty acid derivatives of synthetic phospholipids are commonly used in liposome preparation: (1) myristic acid (*n*-tetradecanoic acid; containing 14 carbons), (2) palmitic acid (*n*-hexadecanoic acid; containing 16 carbons), and (3) stearic acid (*n*-octadecanoic acid; containing 18 carbons) (Fig. 334).

The nomenclature for associating individual fatty acid groups with particular phosphodiglyceride derivatives is straightforward. For instance, a phosphatidic acid derivative that contains two myristic acid chains is commonly called dimyristoyl phosphatidic acid (DMPA). Likewise, a PC derivative containing two palmitate chains is called dipalmitoyl phosphatidyl choline (DPPC). Other phosphodiglyceride derivatives are similarly named.

The second form of lipid derivative that occurs naturally in membrane structures is derived from sphingosine. Unlike the phosphodiglyceride derivatives discussed above, sphingolipids contain no glycerol backbone. Instead, these lipids are constructed from a derivative of 4-sphingenine, containing an N-acyl-linked fatty acid group and possibly other constituents off the No. 1 carbon hydroxyl group (Fig. 335). Sphingolipids are highly similar in their construction to glyceryl lipids, in that there are two hydrophobic tails present on a 3-carbon backbone (one of them contributed from 4-sphingenine itself and the other from the attached fatty acid), and there also exists a hydrophilic head group. This creates the typical amphipathic properties common to all lipid membrane components.

The simplest form of sphingolipid, ceramide, contains a fatty acid group, but no additional components on the No. 1 hydroxyl. Major derivatives of ceramide at the 1-hydroxyl position include a positively charged phosphocholine compound, called sphingomyelin, a glucose derivative, called glucosylcerebroside, and other complex carbohydrate derivatives, termed gangliosides (Fig. 336). Gangliosides are involved in

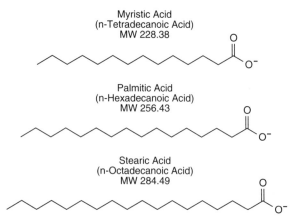

Figure 334   The three fatty acid components commonly used in liposome construction.

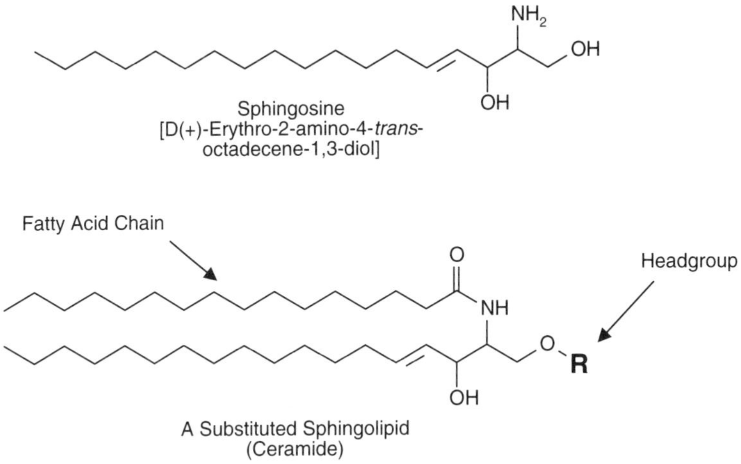

Figure 335 Sphingolipids are constructed of sphingosine derivatives containing an acylated fatty acid and a head group attached to the hydroxyl.

various cellular recognition phenomena, including being part of the blood group determinants A, B, and O in humans.

The use of sphingolipids in liposome formation is possible due to the natural amphipathic properties of the molecules. Some sphingolipids can lend structural advantages to the integrity of liposomal membranes. Sphingomyelin, for example, is capable of hydrogen bonding with adjacent glyceryl lipids, thus increasing the order and stability of the vesicle construction. This stability may translate into a lower potential for passage of molecules through the membrane bilayer, forming vesicles that are better able to retain their contents than more fluid liposome constructions. The temperature of phase transition in sphingolipid-containing membranes is often greater than membranes constructed of only phosphodiglyceride derivatives. Liposomes containing sphingomyelin or gangliosides also have prolonged lifetimes *in vivo* (Allen and Chonn, 1987; Gregoriadis and Senior, 1980) and may be advantageous for creating liposome immunogen complexes.

The main disadvantage of incorporating sphingolipids in liposomes is their high cost. Purified phosphodiglyceride derivatives may be obtained in bulk quantities and in highly defined synthetic preparations, whereas sphingolipid derivatives are not so readily available in similar purity.

Another significant component of many liposome preparations is cholesterol. In natural cell membranes, cholesterol makes up about 10–50% of the total lipid on a molar basis. For liposome preparation, it is typical to include a molar ratio of about 50% cholesterol in the total lipid recipe. The addition of cholesterol to phospholipid bilayers alters the properties of the resultant membrane in important ways. As it dissolves in the membrane, cholesterol orients itself with its polar hydroxyl group pointed toward the aqueous outer environment, approximately even, in a three-dimensional sense, with the glyceryl backbone of the bilayer's phosphodiglyceride components (Fig. 337). Structurally, cholesterol is a rigid component in membrane construction, not having the same freedom of movement that the fatty acid tails of

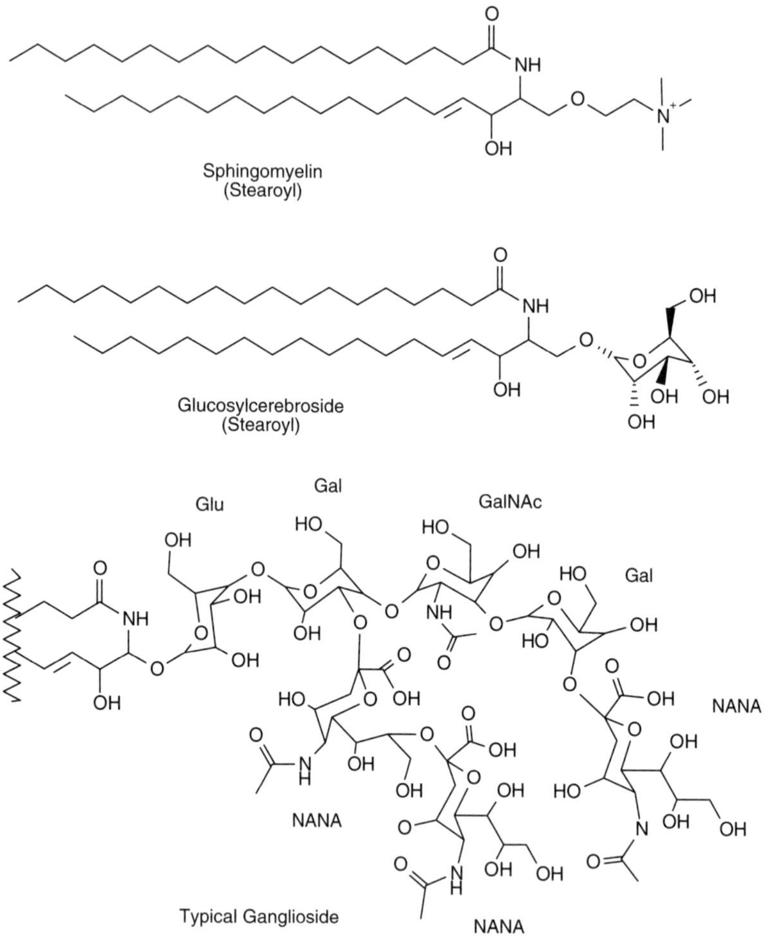
Sphingomyelin
(Stearoyl)
Glucosylcerebroside
(Stearoyl)
Glu
Gal
GalNAc
Gal
NANA
NANA
NANA
Typical Ganglioside

Figure 336 Common sphingolipid derivatives include small and highly complex head groups.

phosphodiglycerides possess. Adjacent phospholipid molecules are restricted in their freedom of movement throughout the length of their fatty acid chains that are abutting the cholesterol molecules. However, since the cholesterol components have the effect of creating spaces in the uniform hydrophobic morphology of the bilayer, the portion of the fatty acid chains below the abutted regions are increased in their freedom of movement.

Cholesterol's presence in liposome membranes has the effect of decreasing or even abolishing (at high cholesterol concentrations) the phase transition from the gel state to the fluid or liquid crystal state that occurs with increasing temperature. It also can modulate the permeability and fluidity of the associated membrane—increasing both parameters at temperatures below the phase transition point and decreasing both above the phase transition temperature. Most liposomal recipes include cholesterol as an integral component in membrane construction.

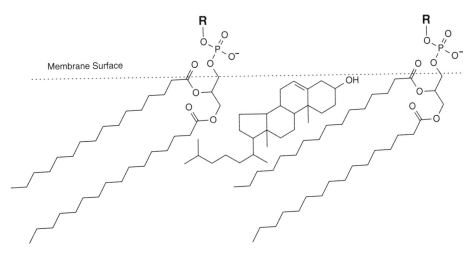

Figure 337 The orientation of cholesterol in phospholipid bilayers.

1.4. Functional Groups of Phospholipids

For the production of liposomal conjugates, lipid derivatives must be incorporated into the bilayer construction that contains available functional groups able to be chemically cross-linked or modified. A number of phosphodiglyceride compounds can be employed for conjugation purposes. Each of these components contains a head group that can be directly derivatized or chemically modified to contain a reactive functional group. For instance, several lipid derivatives contain amine groups that can be utilized in nucleophilic reactions with cross-linkers or other modification reagents. These include phosphatidyl ethanolamine (PE), phosphatidyl serine (PS), and stearylamine. Carboxyl-containing molecules include all the individual fatty acids as well as PS. These can be coupled to amine-containing molecules by the use of the carbodiimide reaction (Chapter 3, Section 1). Hydroxyl-containing lipids include phosphatidyl glycerol (PG), fatty acid alcohols, phosphatidyl inositol (PI), and various gangliosides and cerebrosides of sphingolipid derivation. Lipids possessing hydroxyl groups on adjacent carbon atoms, such as those containing sugar constituents, may be oxidized with sodium periodate to produce reactive aldehyde residues (Chapter 1, Section 4.4). Coupling aldehydes to amine-containing molecules can be accomplished by reductive amination (Chapter 3, Section 4). Finally, the phosphate groups of phosphatidic acid (PA) residues may be used to conjugate with amine-containing molecules—similar to modification of the 5'-phosphates of DNA probes—through the use of the carbodiimide reaction with EDC (Chapter 17, Section 2.1). Figure 338 shows the structures and reactive sites of these lipid functional groups.

When liposomes are used as part of a conjugate system, the targeting molecules usually are attached covalently to these head group functional groups using standard cross-linking chemistries (Derksen and Scherphof, 1985). Although all of the above-mentioned functional groups on lipid molecules can be used for the conjugation

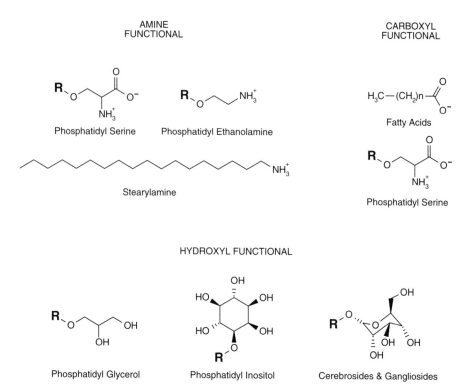

Figure 338 The major functional groups of lipids that may participate in bioconjugate chemistries include amines, carboxylates, and hydroxyls.

process, most often the derivatization reaction is done off the phosphatidylethanol-amine constituents within the liposomal mixture. The primary amine modification off the glycerylphosphate head of phosphatidylethanolamine provides an ideal functional group for activation and subsequent coupling of targeting or detection molecules (Shek and Heath, 1983). Liposomes may be constructed with active groups already prepared on their PE constituents, all set to be conjugated with selected molecules having the correct chemical functional group. Stock preparations of activated lipo-somes may even be prepared and lyophilized to be used as needed in coupling macro-molecules (Friede *et al.*, 1993). All of the amine-reactive conjugation methods dis-cussed in this section may be used with phosphatidylethanolamine-containing liposomes.

2. Derivatization and Activation of Lipid Components

Two approaches for the activation of lipid components may be used to create reactive groups in liposomes. A purified lipid component may be activated prior to incorpora-tion into the bilayer construction or the activation step may occur after formation of the intact liposome. Either way, the goal of the activation process is to provide a reactive species that can be used to couple with selected target functional groups on proteins or other molecules. Although numerous cross-linking methods can be used with lipid functional groups, three main strategies commonly are used to conjugate proteins with liposomes: (1) reductive amination to couple aldehyde residues with amines, (2) carbodiimide-mediated coupling of an amine to a carboxylate or an amine to a phosphate group, and (3) multistep, heterobifunctional cross-linker-mediated conjugation. Both reductive amination and heterobifunctional processes involve activation of particular lipid components.

2.1. Periodate Oxidation of Liposome Components

Reductive amination-mediated conjugation can be done by periodate-oxidizing carbohydrate or glycerol groups on lipid components and using them to couple with amine-containing molecules. It also may be accomplished by using amines on liposomes (i.e., by the incorporation of PE or SA residues) and coupling them to aldehydes present on proteins or other molecules. Using the first approach, liposomes containing PG or glycosphingolipid residues are oxidized by sodium periodate, purified, and then used to conjugate with protein molecules in the presence of sodium cyanoborohydride (Fig. 339) (Chapter 3, Section 4). A protocol for the formation of aldehyde groups on liposomes can be found in the method of Heath *et al.* (1981).

Protocol

1. Prepare a 5 mg/ml liposome construction in 20 mM sodium phosphate, 0.15 M NaCl, pH 7.4, containing, on a molar ratio basis, a mixture of PC:cholesterol:PG:other glycolipids of 8:10:1:2. The other glycolipids that can be incorporated include phosphatidyl inositol, lactosylceramide, galactose cerebroside, or various gangliosides. Other liposome compositions may be used, for example, recipes without cholesterol, as long as a periodate-oxidizable component

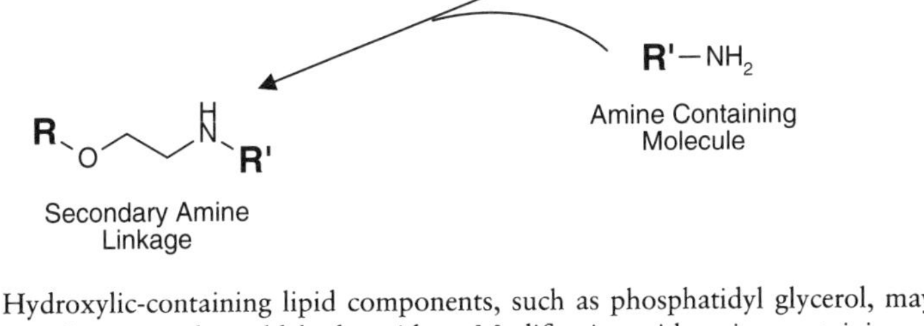

Figure 339 Hydroxylic-containing lipid components, such as phosphatidyl glycerol, may be oxidized with sodium periodate to produce aldehyde residues. Modification with amine-containing molecules then may take place using reductive amination.

containing vicinal hydroxyls is present. Any method of liposome formation may be used.

2. Dissolve sodium periodate to a concentration of 0.6 M by adding 128 mg per milliliter of water. Add 200 μl of this stock periodate solution to each milliliter of the liposome suspension with stirring.
3. React for 30 min at room temperature in the dark.
4. Dialyze the oxidized liposomes against 20 mM sodium borate, 0.15 M NaCl, pH 8.4, to remove unreacted periodate. This buffer is ideal for the subsequent coupling reaction. Chromatographic purification using a column of Sephadex G-50 also can be done.

The periodate-oxidized liposomes may be used immediately to couple with amine-containing molecules such as proteins (see Section 7.6, below), or they may be stored in a lyophilized state in the presence of sorbitol (Friede *et al.*, 1993) for later use.

2.2. Activation of PE Residues with Heterobifunctional Cross-linkers

The most common type of heterobifunctional reagent used for the activation of lipid components includes the amine- and sulfhydryl-reactive cross-linkers containing an NHS ester group on one end and a maleimide, iodoacetyl, or pyridyl disulfide group on the other end (Chapter 5, Section 1). Principle reagents used to effect this activation process include SMCC (Chapter 5, Section 1.3), MBS (Chapter 5, Section 1.4), SMPB (Chapter 5, Section 1.6), SIAB (Chapter 5, Section 1.5), and SPDP (Chapter 5, Section 1.1).

Activation of PE residues with these cross-linkers can proceed by one of two routes: the purified PE phospholipid may be modified in organic solvent prior to incorporation into a liposome, or an intact liposome containing PE may be activated while suspended in aqueous solution. Most often, the PE derivative is prepared before the liposome is constructed. In this way, a stable, stock preparation of modified PE may be made and used in a number of different liposomal recipes to determine the best formulation for the intended application. However, it may be desirable to modify PE after formation of the liposomal structures to ensure that only the outer half of the lipid bilayer be altered. This may be particularly important if substances to be entrapped within the liposome are sensitive or react with the PE derivatives.

Cross-linkers used to activate PE should be of the longest spacer variety available. The length of the spacer is important in providing enough distance from the liposome surface to accommodate the binding of another macromolecule. Short activating reagents often do not allow a protein to approach close enough to react with the functional groups on the bilayer surface. For instance, direct modification of PE with iodoacetate results in little or no sulfhydryl-modified IgG binding to the associated liposomes. When an aminoethylthioacetyl spacer is used to move the iodoacetyl group further away from the bilayer surface, good IgG coupling occurs (Hashimoto *et al.*, 1986). However, this concept does not apply to the coupling of low-molecular-weight molecules, which can access the surface chemistry more readily than macromolecules.

For the activation of PE prior to liposome formation, it is best to employ a highly purified form of the molecule. Although egg PE is abundantly available, it consists of a range of fatty acid derivatives—many of which are unsaturated—and is highly susceptible to oxidation. By contrast, synthetic PE, which has a discrete fatty acid composition and is much more stable to oxidative degradation, can be obtained.

The following suggested protocols are modifications of those described by Martin and Papahadjopoulos (1982), Martin *et al.* (1990), and Hutchinson *et al.* (1989). Although the methods were developed for use with SMPB, SPDP, and MBS, the same basic principles can be used to activate PE with any of the heterobifunctional cross-linkers mentioned above. The reaction sequence for activation and coupling using SMPB is shown in Fig. 340. The PE employed should be of a synthetic variety having fatty acid constituents of dimyristoyl (DMPE), dipalmitoyl (DPPE), or distearoyl (DSPE) forms. For activation of pure PE, the heterobifunctional reagents should not be of the sulfo-NHS variety, since they are best used in aqueous reaction mediums and PE is activated in nonaqueous conditions. For activation of intact liposomes in aqueous suspension, the sulfo-NHS variety of the cross-linkers may be the best choice, since they are incapable of penetrating membranes, and thus only the outer surfaces of vesicles will be modified.

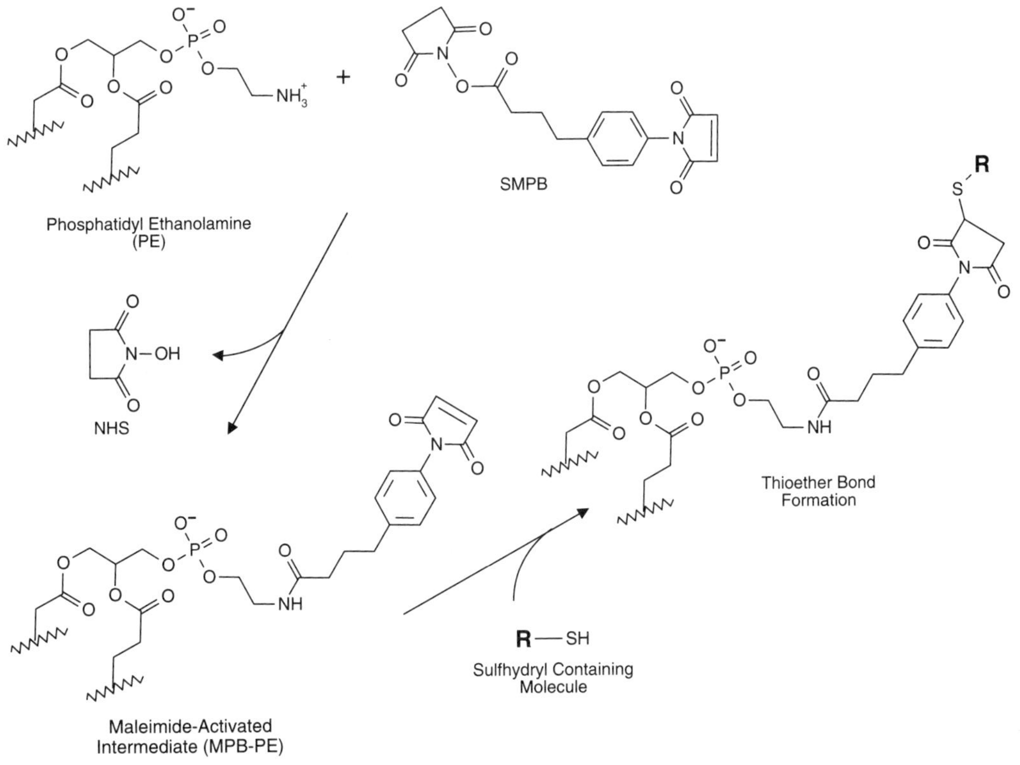

Figure 340 A sulfhydryl-reactive lipid derivative may be prepared through the reaction of SMPB with phosphatidyl ethanolamine to produce a maleimide-containing intermediate. Sulfhydryl-containing molecules then may be coupled to the phospholipid via stable thioether linkages.

Protocol for the Activation of DPPE with SMPB

1. Dissolve 100 μmol of PE in 5 ml of argon-purged, anhydrous methanol containing 100 μmol of triethylamine (TEA). Maintain the solution over an argon or nitrogen atmosphere. The reaction also may be done in dry chloroform. Note: methanol or chloroform and TEA should be handled in a fume hood.

2. Add 50 mg of SMPB (Pierce) to the PE solution. Mix well to dissolve.

3. React for 2 h at room temperature, while maintaining the solution under an argon or nitrogen atmosphere. Reaction progress may be determined by thin-layer chromatography (TLC) using silica gel 60-F_{254} plates (Merck) and developed with a 65:25:4 (by volume) mixture of chloroform:methanol:water. The activated PE derivative will develop faster on TLC ($R_F = 0.52$ for MPB–PE) than the unmodified PE.

4. Remove the methanol from the reaction solution by rotary evaporation and redissolve the solids in chloroform (5 ml).

5. Extract the water-soluble reaction by-products from the chloroform with an equal volume of 1% NaCl. Extract twice.

6. Purify the MPB–PE derivative by chromatography on a column of silicic acid (Martin *et al.*, 1981). The following description is from Martin *et al.* (1990). Dissolve 2 g silicic acid in 10 ml of chloroform and pour the solution into a syringe barrel containing a plug of glass wool at the bottom. Apply the chloroform-dissolved lipids on the silicic acid column. Wash with 4 ml of chloroform, then elute with 4 ml each of the following series of chloroform:methanol mixtures: 4:0.25, 4:0.5, 4:0.75, and 4:1. During the chromatography collect 2-ml fractions. Monitor for the presence of purified MPB–PE by TLC according to step 3.

7. Remove chloroform from the MBP–PE by rotary evaporation. Store the derivative at −20°C under a nitrogen atmosphere until use.

SPDP also may be used to activate pure PE lipids in a manner similar to that for SMPB. The result will be a derivative containing pyridyldisulfide groups rather than maleimide groups (Fig. 341). Pyridyldisulfides react with sulfhydryls to form disulfide linkages. Either the standard SPDP or the long-chain version, LC-SPDP, may be employed in the following protocol.

Protocol for Activation of PE with SPDP

1. Dissolve 20 μmol of PE (15 mg) in 2 ml of argon-purged, anhydrous methanol containing 20 μmol of triethylamine (TEA; 2 mg). Maintain the solution over an argon or nitrogen atmosphere. The reaction also may be done in dry chloroform. Note: methanol or chloroform and TEA should be handled in a fume hood.

2. Add 30 μmol (10 mg) of SPDP (Pierce) to the PE solution. Mix well to dissolve.

3. React for 2 h at room temperature, while maintaining the solution under an argon atmosphere. Reaction progress may be determined by thin-layer chromatography (TLC) using silica gel plates developed with a 65:25:4 (by volume) mixture of chloroform:methanol:water. The activated PE derivative (PDP–PE) will develop faster on TLC than the unmodified PE.

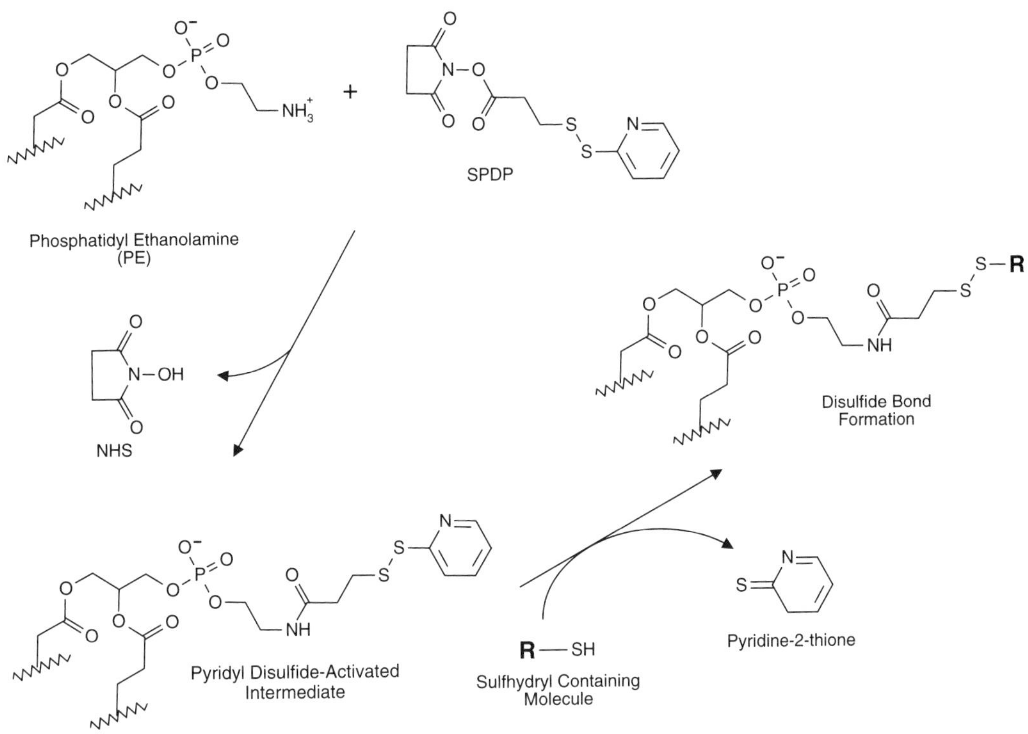

Figure 341 The reaction of SPDP with phosphatidylethanolamine creates a maleimide derivative capable of coupling thiols. Reaction with a sulfhydryl-containing molecule forms a conjugate through a thioether linkage.

4. Remove the methanol from the reaction solution by rotary evaporation and redissolve the solids in chloroform (5 ml).
5. Extract the water-soluble reaction by-products from the chloroform with an equal volume of 1% NaCl. Extract twice.
6. Purify the PDP–PE derivative by chromatography on a column of silicic acid (Martin *et al.*, 1981). The following description is from Martin *et al.* (1990). Dissolve 2 g of silicic acid in 10 ml of chloroform and pour the solution into a syringe barrel containing a plug of glass wool at the bottom. Apply the chloroform-dissolved lipids on the silicic acid column. Wash with 4 ml of chloroform, then elute with 4 ml each of the following series of chloroform:methanol mixtures: 4:0.25, 4:0.5, 4:0.75, and 4:1. During the chromatography collect 2-ml fractions. Monitor for the presence of purified PDP–PE by TLC according to step 3.
7. Remove chloroform from the PDP–PE by rotary evaporation. Store the derivative at −20°C under a nitrogen atmosphere until use.

Other heterobifunctional reagents containing an NHS ester end can be used to activate PE in a similar manner to that in the protocols described above. A somewhat abbreviated protocol (eliminating the silicic acid chromatography step) for the activation of DPPE with MBS (Chapter 5, Section 1.4), as adapted from Hutchinson *et al.* (1989), follows. The NHS ester of MBS reacts with PE's free amine group to create an amide bond. The maleimide end of the cross-linker then remains available for subsequent conjugation with a sulfhydryl-containing molecule after liposome formation (Fig. 342).

Protocol for the Activation of DPPE with MBS

1. Dissolve 40 mg of DPPE in a mixture of 16 ml dry chloroform and 2 ml dry methanol containing 20 mg triethylamine. Maintain under nitrogen to prevent lipid oxidation.
2. Add 20 mg of MBS to the lipid solution and mix to dissolve.
3. React for 24 h at room temperature under nitrogen.
4. Wash the organic phase three times with PBS, pH 7.3, to extract excess cross-linker and reaction by-products.
5. Remove the organic solvents by rotary evaporation under vacuum.
6. Analyze the MBS–DPPE derivative by TLC using a silica plate and developing with a solvent mix containing chloroform:methanol:glacial acetic acid in the volume ratio of 65:25:13. The R_f value of underivatized DPPE is 0.56, while that of the MBS–DPPE product is 0.78.

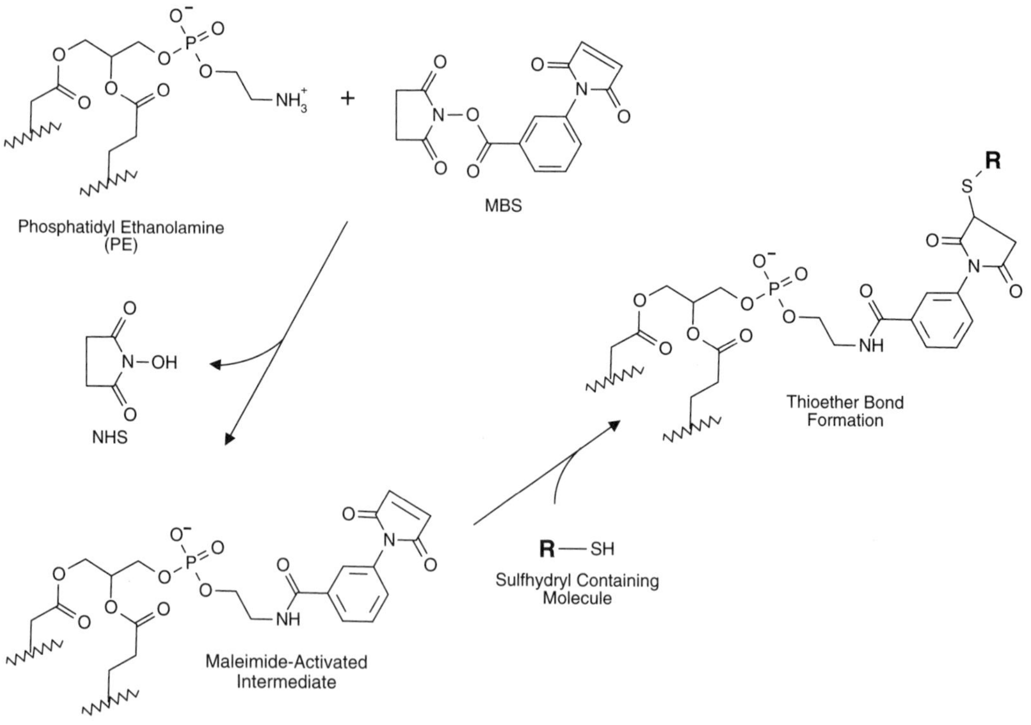

Figure 342 MBS reacted with phosphatidylethanolamine produces a maleimide derivative that can couple to thiol compounds through a stable thioether bond.

The MBS–DPPE derivative can be stored dry under a nitrogen blanket at 4°C or dissolved in chloroform:methanol (9:1, v/v) under the same conditions.

If intact liposomes containing PE are to be activated with these cross-linkers, the methods employed are similar to those used to modify proteins and other macromolecules in aqueous solution. The following protocol is a generalized version for the activation of liposomes containing PE with SPDP (Fig. 343).

Protocol for the Activation of Liposomes with SPDP

1. Prepare a 5 mg/ml liposome suspension in 0.1 M sodium phosphate, 0.15 M NaCl, pH 7.5, containing, for example, a mixture of PC:cholesterol:PG:PE at molar ratios of 8:10:1:1. Other lipid recipes may be used as long as they contain about this percentage of PE. In addition, if this level of cholesterol is maintained in the liposome, then the integrity of the bilayer will be stable up to a level of organic solvent addition of about 5%. This factor is important for adding an aliquot of the cross-linker to the liposome suspension as a concentrated stock solution in an organic solvent. Dispersion of the liposomes to the desired size and morphology may be done by any common method (see Section 1.2).

2. Dissolve SPDP (Pierce) at a concentration of 6.2 mg/ml in DMF (Makes a 20 mM stock solution). Alternatively, LC-SPDP may be used and dissolved at a concentration of 8.5 mg/ml in DMF (also makes a 20 mM solution). If the water-

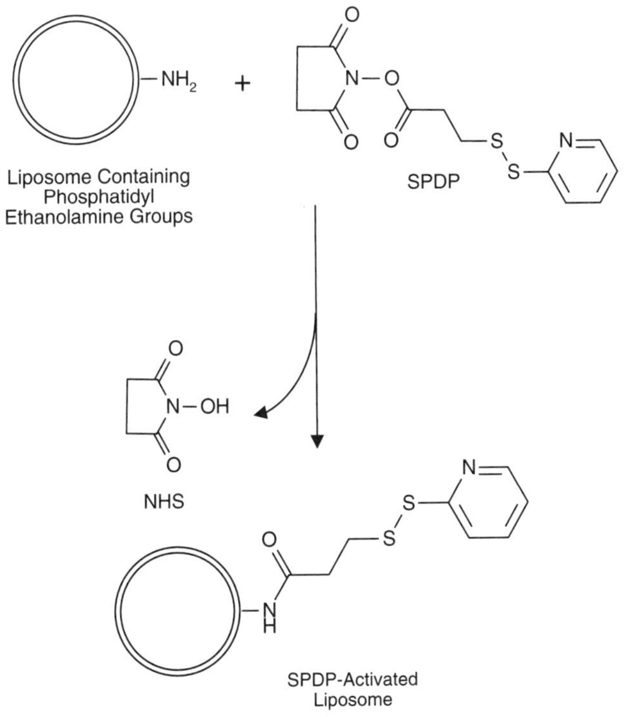

Figure 343 Intact liposomes containing phosphatidylethanolamine components may be modified with SPDP to produce thiol reactive derivatives.

soluble Sulfo-LC-SPDP is used, a stock solution in water may be prepared just prior to addition of an aliquot to the reaction. The sulfo-NHS form of the crosslinker contains a negatively charged sulfonate group that prevents the reagent from penetrating lipid bilayers. Thus, only the outer surfaces of the liposomes can be activated using Sulfo-LC-SPDP. If this is desirable, prepare a 10 mM solution of Sulfo-LC-SPDP by dissolving 5.2 mg/ml in water. Since an aqueous solution of the crosslinker will degrade by hydrolysis of the sulfo-NHS ester, it should be used quickly to prevent significant loss of activity. If a sufficiently large amount of liposomes will be modified, the solid Sulfo-LC-SPDP may be added directly to the reaction mixture without preparing a stock solution in water.

3. Add 25–50 µl of the stock solution of either SPDP or LC-SPDP in DMF to each milliliter of the liposome suspension to be modified. If Sulfo-LC-SPDP is used, add 50–100 µl of the stock solution in water to each milliliter of liposome suspension.

4. Mix and react for at least 30 min at room temperature. Longer reaction times, even overnight, will not adversely affect the modification.

5. Purify the modified liposomes from reaction by-products by dialysis or gel filtration using a column of Sephadex G-50.

The SPDP-activated liposomes may be used immediately to couple with sulfhydryl-containing molecules such as proteins (see Section 7.7, below), or they may be stored in a lyophilized state in the presence of sorbitol (Friede *et al.*, 1993) for later use.

3. Use of Glycolipids and Lectins to Effect Specific Conjugations

Glycolipids are carbohydrate-containing molecules, usually of sphingosine derivation, possessing a hydrophobic, fatty acid tail that embeds them into membrane bilayers. The hydrophilic carbohydrate ends of these amphipathic molecules orient toward the outer aqueous phase, protruding from the bilayer surface, and thus having the capability to interact with molecules dissolved in the surrounding environment. Sphingosine glycolipids may consist of the simple glucosylcerebroside molecules (containing a single glucose residue), lactosylceramide (containing up to four glucose and galactose residues), or complex gangliosides (containing elaborate oligosaccharides that may approach the complexity of those carbohydrate "trees" found on glycoproteins). In membranes of biological origin, glycoconjugates provide sites of cellular recognition for the binding or attachment of proteins and other molecules that possess binding sites able to interact with the particular saccharides present. Such proteins, called lectins, recognize unique sugars or polysaccharide sequences within the carbohydrates of glycoconjugates.

Liposomes partially constructed of glycolipids of known carbohydrate content may be targeted by lectin molecules possessing the requisite binding properties. The liposome may be labeled in this manner with a lectin conjugate, wherein the lectin possesses another molecule covalently attached to it having secondary detection or recognition properties. For instance, a liposome containing a glycolipid may be modified by a lectin–antibody complex, producing a conjugated antibody for specific antigen-targeting applications (Fig. 344). In addition, since lectins typically are multivalent in character, having more than one binding site for a particular carbohydrate type, they

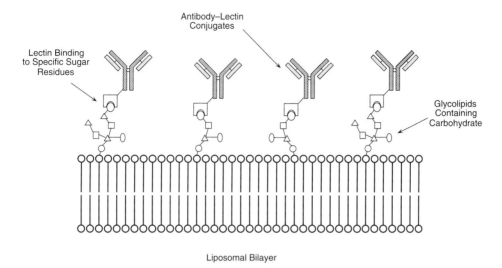

Figure 344   Glycolipid components included in liposome construction may be used to couple antibody molecules by using conjugates of lectins with the proper specificity for binding the sugar groups.

can act as multifunctional cross-linking agents to agglutinate cells or liposomes containing the proper saccharide receptors.

The advantage of this approach to liposome conjugation is that the linkage between the lectin complex and the membrane bilayer is noncovalent and reversible. The addition of a saccharide containing the proper sequence or sugar type recognized by the lectin breaks the binding and releases the attached molecules. This property can be a significant disadvantage, as well, if stable linkages are desired.

Carbohydrate residues on the surface of liposomes may be used to bind selected receptor molecules on cell surfaces. This approach can provide a targeting ability to the vesicles *in vivo,* delivering drugs or toxic agents to intended cellular destinations (for review see Leserman and Machy, 1987). Lectins also may be covalently conjugated to liposome surfaces to provide targeting capability toward cells or molecules containing the complementary carbohydrate needed for binding.

4. Antigen or Hapten Conjugation to Liposomes

Liposomes exert an adjuvant effect *in vivo,* and thus they may be used as carrier systems in the generation of a specific immune response directed against associated antigen or hapten molecules (Heath *et al.,* 1976). Since the main targets of liposomes are the reticulo-endothelial system, particularly macrophages, they naturally associate with the very cells important for mediating humoral immunity. The antigenicity of a liposomal vesicle to a large degree is determined by its overall composition. Its morphology, phospholipid composition, charge, and any attached functional groups all affect the resultant antibody response (Allison and Gregoriadis, 1974; Alving, 1987; Therien and Shahum, 1989).

For instance, incorporation of beef sphingomyelin instead of egg lecithin into liposomal antigen–carrier systems can increase the antibody response to the associated antigen (Yasuda *et al.*, 1977). Liposomal recipes using immune modulators such as lipid A (0.8 nmol of lipid A per micromole of phospholipid) or attaching muramyl dipeptide to the surface of the bilayer also can stimulate the immune response (Daemen *et al.*, 1989). In addition, the fatty acid composition of the phospholipid components can dramatically affect liposome immunogenicity. In general, the higher the transition temperature of the associated phospholipids, the greater the immune response to the liposome. Thus, the relative order of immunogenicity related to fatty acid composition is distearoyl > dipalmitoyl > dimyristoyl. It is also possible that the presence of a positive charge on the bilayer surface may increase the resultant immune response (Muckerheide *et al.*, 1987b; Domen *et al.*, 1987; Apple *et al.*, 1988; Domen and Hermanson, 1992).

Two methods may be used to make immunogenic antigen–liposome or hapten–liposome complexes: (1) the molecule may be dissolved in solution and encapsulated within the vesicle construction or (2) it may be covalently coupled to the phospholipid constituents using standard cross-linking chemical reactions (Shek and Sabiston, 1982a,b). If the antigen molecules are not chemically coupled to the liposome, then they must be entrapped within them to effect an enhancement of the antibody response. If antigen is simply mixed with preformed liposomal vesicles, then there is no beneficial modulation of immunogenicity (Therien and Shahum, 1989). Encapsulation of soluble or particulate vaccines into giant liposomes provides a means of extending the half-life of the vaccine molecules *in vivo* and potentiating the immune response toward the vaccine (Antimisiaris *et al*, 1993).

When haptens or antigens are covalently attached to liposomes, it is typically done through the head groups using various phospholipid derivatives and cross-linking chemical reactions, as described previously (Derksen and Scherphof, 1985). Usually, these derivatization reactions are done using PE constituents within the liposomal mixture. The primary amine on PE molecules provides an ideal functional group for activation and subsequent coupling of hapten molecules (Shek and Heath, 1983). All of the amine-reactive activation methods discussed in this section using heterobifunctional cross-linkers may be used with PE-containing liposomes to prepare immunogen conjugates. In addition, the cross-linking methods in Chapter 9, dealing specifically with hapten–carrier conjugation, should be consulted for potential use with liposomes.

The following protocols provide suggested cross-linking strategies for producing an antigen or hapten complex with liposomes. The first procedure is a simple encapsulation of antigen molecules. The second method involves the coupling of a sulfhydryl-containing peptide hapten to a liposome that had been previously activated with SMPB (Fig. 345) (see Section 2). They are based on the methods of Van Regenmortel *et al.* (1988).

Protocol for the Encapsulation of Antigen into Liposomal Vesicles

1. Prepare a homogeneous lipid mixture by dissolving the desired components in chloroform. A suggested recipe may be to use a mixture of DPPC:DPPG:cholesterol at a molar ratio of 7.5:2.5:5. Evaporate the chloroform using a rotary

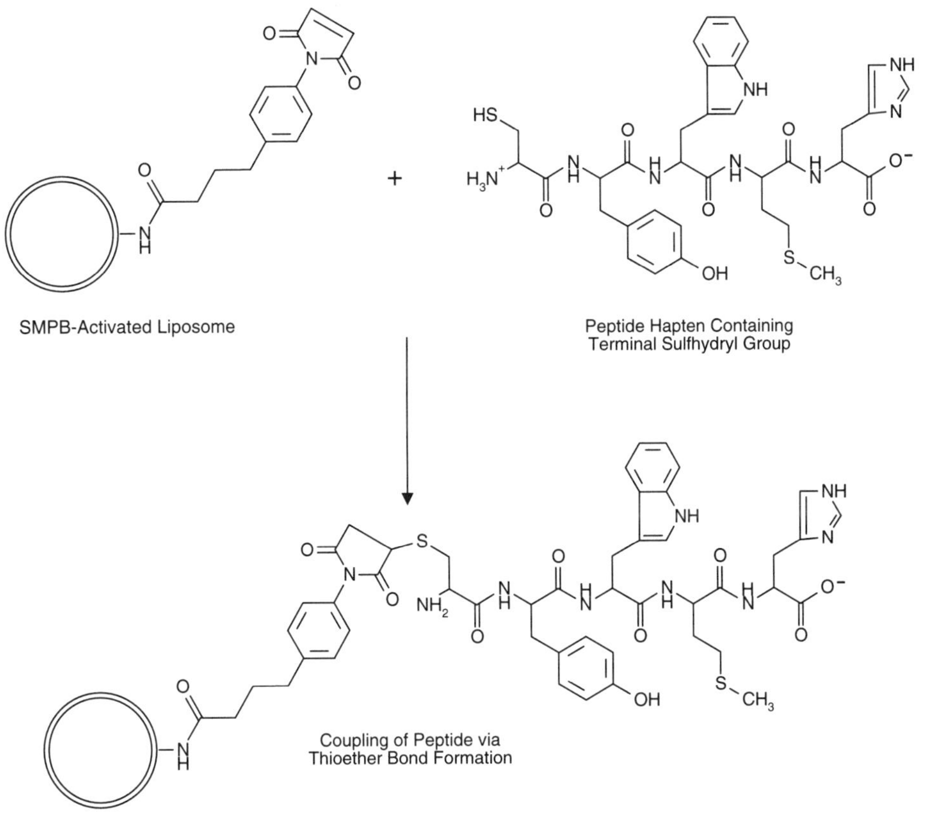

Figure 345 SMPB-activated liposomes may be modified with peptide hapten molecules containing cysteine thiol groups. The resultant immunogen may be used for immunization purposes to generate an antibody response against the coupled peptide.

evaporator under vacuum. Maintain lipids under nitrogen or argon to prevent air oxidation.

2. Dissolve the hapten or antigen to be encapsulated at a concentration of 21 μmol/ml in degassed, nitrogen-purged 10 mM Hepes, 0.15 M NaCl, pH 6.5.

3. Create liposomal vesicles using any established method (see Section 1) by mixing the antigen solution with the lipid mixture to obtain a final concentration of 5 mg/ml lipid in the aqueous buffer. A suggested procedure may be to redissolve the lipids in a minimum quantity of diethylether, and then mix the buffer with the ether phase at a ratio necessary to give a 5 mg/ml concentration of lipid in the buffer. Sonicate to emulsify the liposomal preparation. Remove diethylether by vacuum evaporation. Periodically mix by vortexing to maintain a homogeneous suspension of liposomes.

4. Remove free antigen from encapsulated antigen by gel filtration using a column of Sephadex G-75 or by dialysis using 10 mM Hepes, 0.15 M NaCl, pH 6.5. Store under an inert gas at 4°C in the dark until use.

Protocol for the Coupling of Peptide Haptens Containing
Sulfhydryl Groups to Liposomal Vesicles

1. Prepare a homogeneous lipid mixture by dissolving in chloroform a mixture of
   DPPC:DPPG:cholesterol:MPB-DPPE at a molar ratio of 6.3:2.12:4.25:1.5.
   Preparation of MPB-activated DPPE can be found in Section 2. Maintain all
   solutions under nitrogen or argon. The lipid derivative provides reactive maleimide groups for the coupling of sulfhydryl-containing molecules. Evaporate the
   chloroform using a rotary evaporator under vacuum.

2. Create liposomal vesicles using any established method (see Section 1) by mixing the lipid mixture with degassed, nitrogen-purged 10 mM Hepes, 0.15 M
   NaCl, pH 7, to obtain a final concentration of 5 mg/ml lipid in the aqueous
   buffer. Sonicate to emulsify the liposomal preparation. Remove diethylether by
   vacuum evaporation. Periodically mix by vortexing to maintain a homogeneous
   suspension of liposomes.

3. Dissolve a sulfhydryl-containing peptide hapten at a concentration of 25
   μmol/ml in degassed, nitrogen-purged 10 mM Hepes, 0.15 M NaCl, pH 7. Add
   the peptide solution to the liposome suspension at a molar ratio necessary to
   obtain at least a 5:1 excess of thiol groups to the amount of maleimide groups
   present (as MPB–DPPE).

4. React overnight at room temperature. Maintain an inert gas blanket over the
   vessel to prevent lipid oxidation.

5. Purify the derivatized liposomes from excess peptide by gel filtration using a
   column of Sephadex G-75 or by dialysis. Store the immunogenic vesicles at 4°C
   under nitrogen or argon and protected from light until use.

5. Preparation of Antibody–Liposome Conjugates

Covalent attachment of antibody molecules to liposomes can provide a targeting
capacity to the vesicle that can modulate its binding to specific antigenic determinants
on cells or to molecules in solution. Antibody-bearing liposomes may possess encapsulated components that can be used for detection or therapy (Fig. 346). For instance,
fluorescent molecules encapsulated within antibody-targeted vesicles can be used as
imaging tools or in flow cytometry (Truneh *et al.*, 1987). Specific antibodies coupled to
the vesicle surface can improve diagnostic assays involving agglutination of latex
particles (Kung *et al.*, 1985). Liposomes possessing antibodies directed against tumor
cell antigens can deliver encapsulated toxins or drugs to the associated cancer cells,
effecting toxicity and cell death (Straubinger *et al.*, 1988; Matthay *et al.*, 1984; Heath
et al., 1983, 1984).

Encapsulation of chemotherapeutic agents within lipid bilayers reduces systemic
toxicity and local irritation often caused by anticancer drugs (Gabizon *et al.*, 1986).
The liposomal membrane acts as a slow-release agent so that cytotoxic components do
not come into contact with nontumor cells. Liposome binding to cells causes internalization and release of the encapsulated drugs. Antibody targeting can increase the
likelihood of vesicle binding to the desired tumor cells.

However, there are problems associated with the use of antibody–liposome conjugates for drug delivery *in vivo*. Particularly, since lipid vesicles are huge compared to

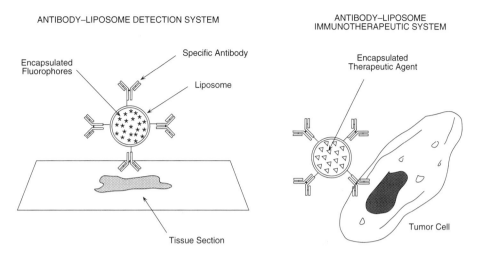

Figure 346 Antibody–liposome conjugates may be used as targeting reagents for detection or therapeutic applications. The liposome may be constructed to contain fluorescent molecules for detection purposes or bioactive agents for therapy. The antibody component targets the complex for binding to specific antigenic determinants.

similar immunotoxin conjugates (Chapter 11), their passage to particular tissue destinations may be difficult or impossible. Liposomes are almost entirely limited to the reticuloendothelial system. Their ability to pass through tissue barriers to target cells in other parts of the body is limited by their size. If liposomal conjugates can reach their intended destination, their contents are delivered to the cells by endocytosis. Endocytic vesicles arising from the surface of cells have diameters in the range 1000–1500 Å. This limits the size of liposomes that can be used to small vesicles that can bind to the surface of a cell and be internalized efficiently. Large liposomes, by contrast, will not be internalized and therefore not be able to deliver their contents (Leserman and Machy, 1987).

The methods for coupling antibody molecules to liposomal surfaces are not unlike those described for general protein coupling (Section 7), below. Antibodies may be coupled through sulfhydryl residues using liposomes containing PE groups derivatized with heterobifunctional cross-linkers such as SMCC (Chapter 5, Section 1.3), MBS (Chapter 5, Section 1.4), SMPB (Chapter 5, Section 1.6), SIAB (Chapter 5, Section 1.5), and SPDP (Chapter 5, Section 1.1). They also may be coupled through their amine groups using reductive amination to periodate-oxidized glycolipids (Sections 2 and 7.6).

6. Preparation of Biotinylated or Avidin-Conjugated Liposomes

Liposome conjugates may be used in various immunoassay procedures. The lipid vesicle can provide a multivalent surface to accommodate numerous antigen–antibody interactions and thus increase the sensitivity of an assay. At the same time, it can function as a vessel to carry encapsulated detection components needed for the assay system. This type of ELISA is called a liposome immunosorbent assay or LISA.

One method of using liposomes in an immunoassay is to modify the surface so that it can interact to form biotin–avidin or biotin–streptavidin complexes. The avidin–biotin interaction can be used to increase detectability or sensitivity in immunoassay tests (Chapter 13) (Savage *et al.*, 1992).

Liposomes containing biotinylated phospholipid components can be used in a bridging assay system with avidin and a biotinylated antibody molecule, creating large multivalent complexes able to bind antigen (Plant *et al.*, 1989) (Fig. 347). The inside of the vesicles may contain fluorescent detection reagents that can be used to localize or quantify target analytes. One small liposome provides up to 10^5 molecules of fluorophore to allow excellent detectability of a binding event. LISA systems using biotinylated liposomes to detect antigen molecules can increase the sensitivity of an immunoassay up to 100-fold over that obtainable using traditional antibody–enzyme ELISAs.

Biotinylated liposomes usually are created by modification of PE components with an amine-reactive biotin derivative, for example, NHS-LC-Biotin (Chapter 8, Section 3.1). The NHS ester reacts with the primary amine of PE residues, forming an amide bond linkage (Fig. 348). Since the modification occurs at the hydrophilic end of the phospholipid molecule, after vesicle formation the biotin component protrudes from the liposomal surface. In this configuration, the surface-immobilized biotins are able to bind avidin molecules present in the outer aqueous medium.

Biotinylation may be done before or after liposome formation, but having a stock supply of biotin-modified PE is an advantage, since it can then be used to test a number

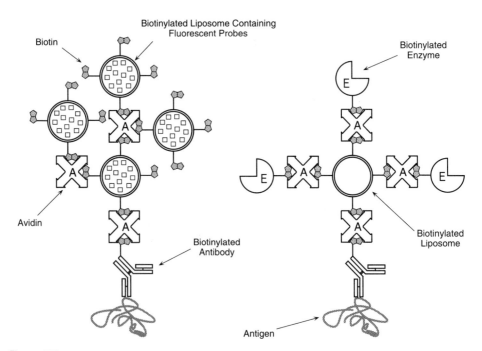

Figure 347   Biotinylated liposomes may be used in immunoassay systems to enhance the signal for detection or measurement of specific analytes. The liposome components may be constructed to include fluorescent molecules to facilitate detection of antigens within tissue sections.

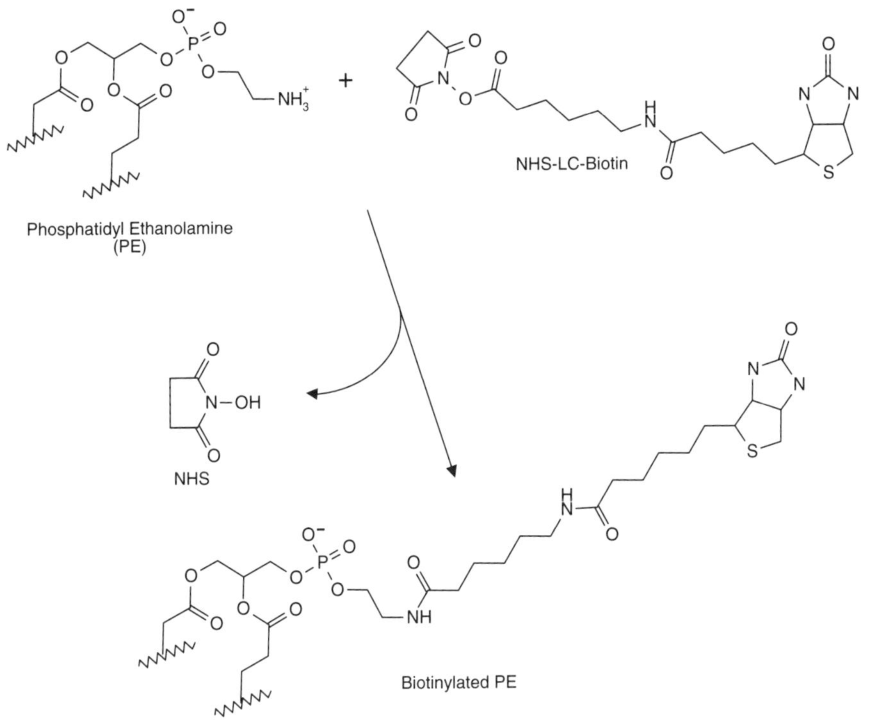

Figure 348 Biotinylated liposomes may be formed using biotinylated phosphatidyl ethanolamine. Reaction of NHS-LC–biotin with PE results in amide bond linkages and a long spacer arm terminating in a biotin group.

of liposomal recipes. In addition, only a very small percentage of the total lipid should be biotinylated to prevent avidin-induced aggregation in the absence of antigen. It is difficult to control finely the biotin content if direct biotinylation of intact liposomes is done. Using pure biotinylated phospholipid allows incorporation of discrete amounts of biotin binding sites into the final liposomal membrane. The preparation of biotinylated PE (B-PE) can be done using a method similar to that described for activation of PE with SMPB (Section 2) or it may be obtained commercially (Pierce, Molecular Probes).

The following method for the formation of a biotinylated liposome is adapted from Plant *et al.* (1989). It assumes prior production of B-PE.

Protocol

1. Prepare a biotinylated liposome construction by first dissolving in chloroform the lipids DMPC:cholesterol:dicetylphosphate (Sigma) at molar ratios of 5:4:1 and adding to this solution 0.1 mol% B-PE. Larger molar ratios of B-PE to total lipid may result in nonspecific aggregation of liposomes in the presence of avidin. Maintain all lipids under an inert atmosphere to prevent oxidation.
2. Evaporate 2 μmol of total homogenized lipid in chloroform using a stream of nitrogen or a rotary vacuum evaporator.

3. Redissolve the dried lipid in 50 μl of dry isopropanol.
4. Take the lipid solution up into a syringe and inject it into 1 ml of degassed, nitrogen-purged 20 mM Tris, 0.15 M NaCl, pH 7.4, which is being vigorously stirred using a vortex mixer. To encapsulate a fluorescent dye using this procedure, include 100 mM 5,6-carboxyfluorescein (or another suitable fluorophore) in the buffer solution before adding the lipids. The fluorescent probe used in the encapsulation procedure should be chemically nonreactive so that no lipid components are covalently modified during the process.

The biotinylated liposomes prepared by this procedure may be stored under an inert gas atmosphere at 4°C for long periods without degradation.

7. Conjugation of Proteins to Liposomes

Covalent attachment of proteins to the surface of liposomal bilayers is done through reactive functional groups on the head groups of phospholipids with the intermediary use of a cross-linker or other activating agent. The lipid functional groups described in Section 1 are modified according to the methods discussed in Section 2 to be reactive toward specific target groups in proteins. Conjugation of liposomes with proteins may be done with homobifunctional or heterobifunctional cross-linking reagents, with carbodiimides, by reductive amination, by NHS ester activation of carboxylates, or through the noncovalent use of the avidin–biotin interaction.

Characterizing the resultant complex for the amount of protein per liposome is somewhat more difficult than in other protein conjugation applications. The resultant protein–liposome composition is highly dependent on the size of each liposomal particle, the amount of protein charged to the reaction, and the molar quantity of reactive lipid present in the bilayer construction. An approach to solving this problem is presented by Hutchinson *et al.* (1989). In analyzing at least 17 different protein–liposome preparations, the ratio of protein:lipid content (μg protein/μg lipid) in most of the complexes ranged from a low of about 4 to as much as 675. In some instances, however, up to 6000 molecules of a particular protein could be incorporated into each liposome.

Coupling of protein molecules to liposomes occasionally may induce vesicle aggregation. This may be due to the unique properties or concentration of the protein used, or it may be a result of liposome-to-liposome cross-linking during the conjugation process. Adjusting the amount of protein charged to the reaction as well as the relative amounts of cross-linking reagents employed may have to be done to solve an aggregation problem.

The following sections present suggested protocols for creating protein-bearing liposomes. Each method utilizes specific lipid functional groups and targets amines, sulfhydryls, aldehydes, or carboxylates on the protein molecules.

7.1. Coupling via the NHS Ester of Palmitic Acid

Huang *et al.* (1980) coupled monoclonal antibodies to liposomes using an NHS ester modification of palmitic acid incorporated into the bilayer construction (Lapidot *et al.*,

1967). The NHS ester reacts with amine groups on the protein molecule, producing stable amide bond linkages (Fig. 349). The specificity of the antibody-bearing liposomes for mouse L-929 cells was documented, illustrating the preservation of antibody-binding activity.

Protocol
Preparation of NHS-palmitate

1. Dissolve 3.45 mg of N-hydroxysuccinimide (NHS) in 30 ml dry ethyl acetate.
2. Add 30 mmol of palmitic acid to the NHS solution. Maintain a nitrogen blanket over the solution.
3. Dissolve 6.18 g of dicyclohexyl carbodiimide (DCC: Chapter 3, Section 1.4) in 10 ml of ethyl acetate and add it to the NHS/palmitic acid solution.
4. React overnight at room temperature under a nitrogen blanket.
5. Remove the insoluble dicyclohexyl urea (DCU) by filtration using a glass fiber filter pad and vacuum.
6. Remove solvent from the filtered solution by using a rotary evaporator under vacuum.
7. The NHS-palmitate may be purified by recrystallization using ethanol. Dissolve the activated fatty acid in a minimum quantity of hot ethanol. Immediately on dissolving, filter it through a filter funnel containing a fluted glass fiber filter pad, both of which have been warmed to the same temperature as the ethanol solution. Allow the NHS-palmitate to recrystallize overnight at room temperature. Remove solvent from the recrystallized solid by filtration. Dry under vacuum in a desiccator.

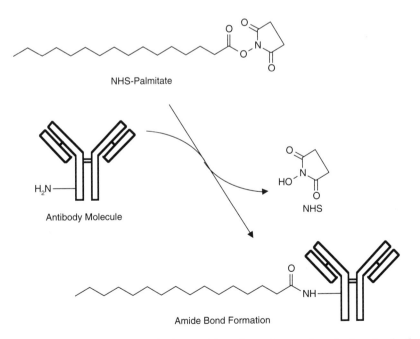

Figure 349 The NHS ester derivative of palmitic acid may be used to couple antibody molecules through amide bonds. These complexes then may be incorporated into liposomes.

8. Analyze the NHS-palmitate for purity using TLC on silica plates. Develop using a solvent mixture of chloroform:petroleum diethyl ether (bp 40–60°C) of 8:2. Excess NHS and NHS-palmitate may be detected by staining with 10% hydroxylamine in 0.1 N NaOH, followed after 2 min by a 5% solution of $FeCl_3$ in 1.2 N HCl (creates red-colored spots).

Coupling of Protein to NHS-palmitate

1. Add 2 mg of protein to 44 μg NHS-palmitate in 20 mM sodium phosphate, 0.15 M NaCl, pH 7.4, containing 2% deoxycholate.
2. Incubate at 37°C for 10 h.
3. Remove excess palmitic acid by chromatography on a column of Sephadex G-75. Use PBS, pH 7.4, containing 0.15% deoxycholate to perform the gel filtration. Collect the fractions containing derivatized protein, as monitored by absorbance at 280 nm.

Addition of Protein–Palmitate Conjugate
to Liposomal Membranes
Since the protein–palmitate derivative cannot be dissolved in organic solvent during homogenization of lipid to form liposomal membranes, it must be inserted into intact liposomes by detergent dialysis.

1. Construct a liposome by dissolving the desired lipids in chloroform to homogenize fully the mixture, drying them to remove solvent, and using any established method of forming bilayer vesicles in aqueous solution (i.e., sonication; see Section 1).
2. Add protein–palmitate conjugate to the formed liposomes in a ratio of 20:1 (w/w). Add concentrated deoxycholate to give a final concentration of 0.7%. Mix thoroughly using a vortex mixer.
3. Dialyze the liposome preparation against PBS, pH 7.4.
4. The liposome vesicles may be characterized for size by chromatography on a column of Sepharose 4B.

7.2. Coupling via Biotinylated Phosphatidylethanolamine Lipid Derivatives

Biotinylated PE (B-PE) incorporated into liposomal membranes can be used to interact noncovalently with avidin–protein conjugates or with other biotinylated proteins using avidin as a bridging molecule (Plant *et al.,* 1989). It is important that a long-chain spacer be used in constructing the B-PE derivative to allow enough spatial separation from the bilayer surface to accommodate avidin docking (Hashimoto *et al.,* 1986). Thus, any biotinylated protein can be coupled to the liposome surface through the strength of the avidin–biotin interaction. Section 6 describes the preparation of B-PE derivatives and their addition into vesicle construction. Incubation of avidin conjugates or avidin plus a biotinylated protein with the biotinylated liposome in PBS, pH 7.4, will form essentially nonreversible complexes, immobilizing the proteins to the outer surface of the bilayers. This method can be used to couple biotin-modified antibody molecules to liposomes (Fig. 350). Removal of noncomplexed protein may be done using gel filtration chromatography on a column of Sephadex G-50 or G-75.

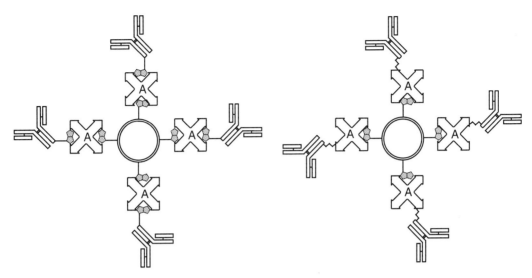

Figure 350   Antibodies may be conjugated to liposomes using an indirect approach incorporating an avidin–biotin system. Biotinylated liposomes may be complexed with biotinylated antibodies using avidin as a bridging molecule or may be complexed with an antibody–avidin conjugate.

7.3. Conjugation via Carbodiimide Coupling to Phosphatidylethanolamine Lipid Derivatives

Underivatized PE in liposomal membranes contains an amine group that can participate in the carbodiimide reaction with carboxylate groups on proteins (Dunnick *et al.*, 1975). The water-soluble carbodiimide EDC (Chapter 3, Section 1.1) activates carboxylate groups to form active-ester intermediates that can react with PE to form an amide linkage. Unfortunately, EDC coupling of proteins to surfaces often results in considerable protein-to-protein cross-linking, since proteins contain an abundance of both amines and carboxylates. There also is potential for vesicle aggregation by proteins coupling to more than one liposome. Martin *et al.* (1990) suggest avoiding this polymerization problem by first blocking the amine groups of the protein with citraconic acid, which has been used successfully with antibodies (Jansons and Mallett, 1980).

However, even with the potential for protein–protein conjugation, carbodiimide coupling of peptides and proteins to liposomes can be done with EDC without blocking polypeptide amines. The approach is similar to that described for the EDC conjugation of hapten molecules to carrier proteins to form immunogen complexes (Chapter 9, Section 3). Thus, this method may be used to prepare peptide hapten–liposome conjugates for immunization purposes (Fig. 351). The procedure also works particularly well for coupling molecules containing only carboxylates to the amines on the liposomes.

Protocol

1. Prepare liposomes containing PE by any desired method. For instance, the common recipe mentioned in Section 1 mixing PC:cholesterol:PG:PE in a molar

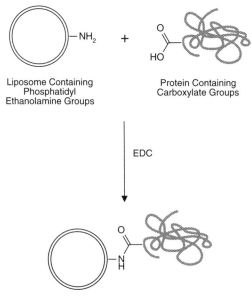

Figure 351   A protein may be conjugated with a liposome containing PE groups using a carbodiimide reaction with EDC.

ratio of 8:10:1:1 may be used. Thoroughly emulsify the liposome construction to obtain a good population of SUVs. The final liposome suspension should be in 10 mM sodium phosphate, 0.15M NaCl, pH 7.2. Adjust the concentration to about 5 mg lipid/ml buffer.

2. Add the protein or peptide to be conjugated to the liposome suspension. The protein may be dissolved first in PBS, pH 7.2, and an aliquot added to the reaction lipid mixture. The amount of protein to be added can vary considerably, depending on the abundance of the protein and the desired final density required. From 1 mg protein per milliliter liposome suspension to about 20 mg protein/ml can be reacted.

3. Add 10 mg EDC per milliliter of lipid/protein mixture. Solubilize the carbodiimide using a vortex mixer.

4. React for 2 h at room temperature. If liposome aggregation or protein precipitation occurs during the cross-linking process, scale back the amount of EDC added to the reaction.

5. Purify the conjugate by gel filtration using a column of Sephadex G-75.

7.4. Conjugation via Glutaraldehyde Coupling to Phosphatidylethanolamine Lipid Derivatives

Glutaraldehyde is among the earliest homobifunctional cross-linkers employed for protein conjugation (Chapter 4, Section 6.2). It reacts with amine groups through several routes, including the formation of Schiff base linkages, which can be reduced with borohydride or cyanoborohydride to create stable secondary amine bonds. Although very efficient in reacting with proteins, glutaraldehyde typically causes extensive polymerization accompanied by precipitation of high-molecular-weight oligomers. Even with this significant disadvantage, the reagent is still used routinely in protein conjugation techniques.

Liposomes containing PE residues can be reacted with glutaraldehyde to form an activated surface possessing reactive aldehyde groups. A two-step glutaraldehyde reaction strategy is probably best when working with liposomes, since precipitated protein would be difficult to remove from a vesicle suspension.

The following protocol describes the two-step method wherein the liposome is glutaraldehyde-activated, purified away from excess cross-linker, and then coupled to a protein by reductive amination (Fig. 352).

Protocol

1. Prepare a liposome suspension, containing PE, at a total lipid concentration of 5 mg/ml in 0.1 M sodium phosphate, 0.15 M NaCl, pH 6.8. Maintain all lipid-containing solutions under an inert gas atmosphere. Degas all buffers and bubble them with nitrogen or argon prior to use.
2. Add glutaraldehyde to this suspension to obtain a final concentration of 1.25%.
3. React overnight at room temperature under a nitrogen blanket.
4. Purify the activated liposomes from excess glutaraldehyde by gel filtration (using Sephadex G-50) or by dialysis against PBS, pH 6.8.

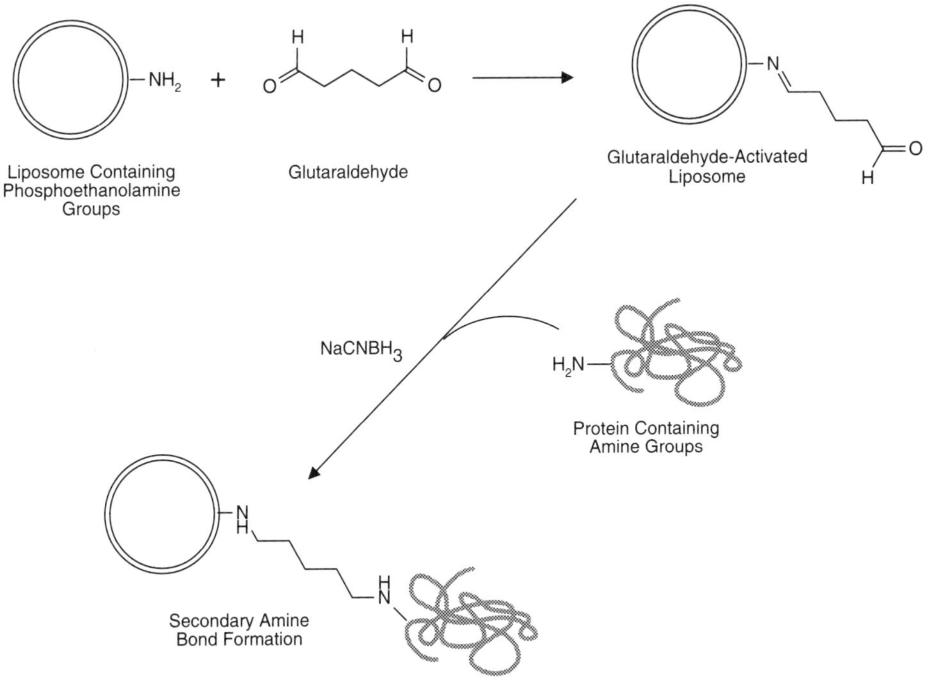

Figure 352 Glutaraldehyde activation of PE-containing liposomes may be used to couple protein molecules.

5. Dissolve the protein or peptide to be conjugated at a concentration of 10 mg/ml in 0.5 *M* sodium carbonate, pH 9.5. Mix the activated liposome suspension with the polypeptide solution at the desired molar ratio to effect the conjugation. Mixing the equivalent of 4 mg of protein per milligram of total lipid usually results in acceptable conjugates.
6. React overnight at 4°C under an atmosphere of nitrogen.
7. To reduce the resultant Schiff bases and any excess aldehydes, add sodium borohydride to a final concentration of 10 mg/ml.

7.5. Conjugation via DMS Cross-linking to Phosphatidylethanolamine Lipid Derivatives

Dimethyl suberimidate (DMS) is a homobifunctional cross-linking agent containing amine-reactive imidoester groups on both ends. The compound is reactive toward the ε-amine groups of lysine residues and *N*-terminal α-amines in the pH range of 7–10 (pH 8–9 is optimal). The resulting amidine linkages are positively charged at physiological pH, thus maintaining the positive charge contribution of the original amine.

Bis-imidoesters like DMS may be used to couple proteins to PE-containing liposomes by cross-linking with the amines on both molecules (Fig. 353). However, single-step cross-linking procedures using homobifunctional reagents are particularly subject to uncontrollable polymerization of protein in solution. Polymerization is possible because the procedure is done with the liposomes, protein, and cross-linker all in solution at the same time.

The reaction is carried out in 0.2 *M* triethanolamine, pH 8.2. DMS should be the limiting reagent in the reaction to avoid blocking all amines on both molecules with only one end of the cross-linker, thus eliminating any conjugation. The amounts of total lipid and protein in solution may have to be adjusted to optimize each conjugation reaction and avoid precipitation of protein or aggregation of liposomes.

7.6. Coupling via Periodate Oxidation Followed by Reductive Amination

Periodate-oxidized liposomes that contain glycolipid moieties may be used to couple proteins and other amine-containing molecules by reductive amination. Section 2 describes the oxidative procedure, which results in the formation of reactive aldehyde groups on the liposomal surface. Amine-containing polypeptides form Schiff base linkages with the aldehyde groups under alkaline conditions. The addition of a reducing agent such a borohydride or cyanoborohydride reduces the labile Schiff bases to form stable secondary amine bonds (Fig. 354).

The following generalized method is based on the procedure described by Heath *et al.* (1981) for the coupling of immunoglobulins to liposomes containing glycosphingolipids.

Protocol

1. Periodate-oxidize a liposome suspension containing glycolipid components according to Section 2. Adjust the concentration of total lipid to about 5 mg/ml.

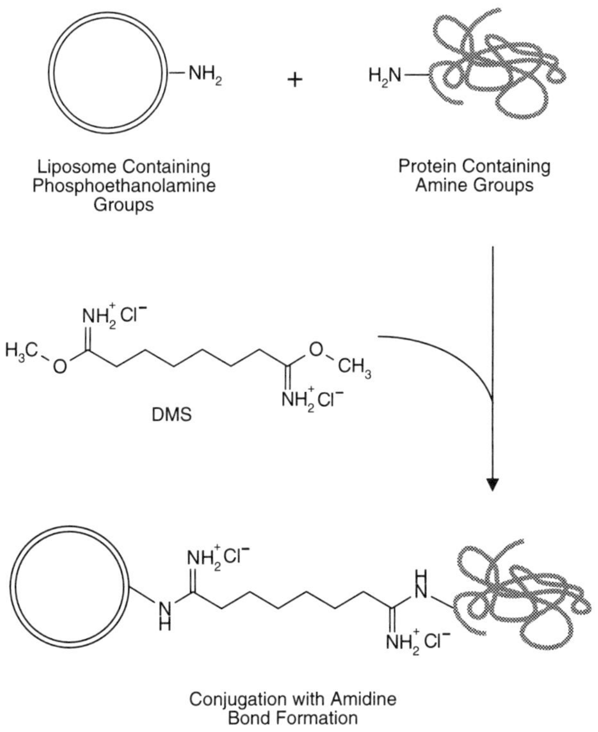

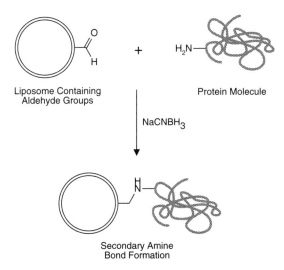

Figure 353 The homobifunctional cross-linker DMS may be used to conjugate PE-containing liposomes with proteins via amidine bond formation.

Figure 354 Glycolipids incorporated into liposomes may be oxidized with periodate to produce aldehydes suitable for coupling proteins via reductive amination.

2. Dissolve the protein to be coupled in 20 mM sodium borate, 0.15 M NaCl, pH 8.4, at a concentration of at least 10 mg/ml.

3. Add 0.5 ml of protein solution to each milliliter of liposome suspension with stirring.

4. Incubate for 2 h at room temperature to form Schiff base interactions between the aldehydes on the vesicles and the amines on the protein molecules.

5. In a fume hood, dissolve 125 mg of sodium cyanoborohydride in 1 ml water (makes a 2 M solution). *Caution: Highly toxic compound; handle with care.* This solution may be allowed to sit for 30 min to eliminate most of the hydrogen-bubble evolution that could affect the vesicle suspension.

6. Add 10 μl of the cyanoborohydride solution to each milliliter of the liposome reaction.

7. React overnight at 4°C.

8. Remove unconjugated protein and excess cyanoborohydride by gel filtration using a column of Sephadex G-50 or G-75.

7.7. Coupling via SPDP-Modified Phosphatidylethanolamine Lipid Derivatives

N-Succinimidyl 3-(2-pyridyldithio)propionate (SPDP) may be the most popular heterobifunctional cross-linking agent available, especially for protein conjugation (Chapter 5, Section 1.1). The activated NHS ester end of SPDP reacts with amine groups in proteins and other molecules to form an amide linkage (Fig. 355). The

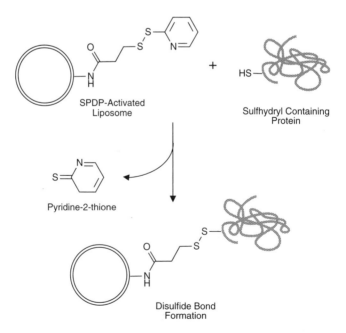

Figure 355   SPDP-activated liposomes can be used to couple sulfhydryl-containing proteins, forming disulfide linkages.

2-pyridyldithiol group at the other end reacts with sulfhydryl residues to form a disulfide linkage with sulfhydryl-containing molecules (Carlsson *et al.*, 1978).

SPDP also is a popular choice for coupling sulfhydryl-containing molecules to liposomes. PE residues in vesicles may be activated with this cross-linker to form pyridyldisulfide derivatives that can react with sulfhydryls to form disulfide linkages. Unlike the iodoacetyl- and maleimide-based cross-linkers discussed previously, the linkage formed with SPDP is reversible by simple disulfide reduction. Pure PE may be activated with SPDP prior to its incorporation into a liposome, or intact liposomes containing PE may be activated using the methods described in Section 2. Stearylamine also can be activated with SPDP to be used in liposome conjugation (Goundalkar *et al.*, 1983). If the long-chain version, Sulfo-LC-SPDP, is used with intact vesicles, the cross-linker will be water-soluble and may be added directly to the buffered suspension without prior organic-solvent dissolution. The negatively charged sulfonate group on its NHS ring prevents the reagent from penetrating the hydrophobic region of the lipid bilayer. Thus, only the outer surface of the liposomes will be modified. Using preactivated PE, both inner and outer surfaces end up containing reactive pyridyl disulfide groups. If sequestered components will react with this functional group or are sensitive, activation of intact liposomes with Sulfo-LC-SPDP may be a better tactic.

The following protocol is a suggested method for coupling sulfhydryl-containing proteins to SPDP-activated vesicles.

1. Prepare a 5 mg/ml liposome suspension containing a mixture of PC:cholesterol:PG:PDP-PE in molar ratios of 8:10:1:1. The emulsification may be done by any established method (Section 1). Suspend the vesicles in 50 mM sodium phosphate, 0.15 M NaCl, 10 mM EDTA, pH 7.2.

2. Add at least 5 mg/ml of a sulfhydryl-containing protein or other molecule to the SPDP-modified vesicles to effect the conjugation reaction. Molecules lacking available sulfhydryl groups may be modified to contain them by a number of methods (Chapter 1, Section 4.1). The conjugation reaction should be done in the presence of at least 10 mM EDTA to prevent metal-catalyzed sulfhydryl oxidation.

3. React overnight with stirring at room temperature. Maintain the suspension in a nitrogen or argon atmosphere to prevent lipid oxidation.

4. The modified liposomes may be separated from excess protein by gel filtration using Sephadex G-75 or by centrifugal floatation in a polymer gradient (Derksen and Scherphof, 1985).

7.8. Coupling via SMPB-Modified Phosphatidylethanolamine Lipid Derivatives

Succinimidyl-4-(*p*-maleimidophenyl)butyrate (SMPB) (Chapter 5, Section 1.6) is a heterobifunctional cross-linking agent that has an amine-reactive NHS ester on one end and a sulfhydryl-reactive maleimide group on the other. Conjugates formed using SMPB are linked by stable amide and thioether bonds.

SMPB can be used to activate PE residues to contain sulfhydryl-reactive maleimide groups (Section 2). Lipid vesicles formed with MPB—PE components thus can couple proteins through available —SH groups, forming thioether linkages (Derksen and Scherphof, 1985) (Fig. 356). A comparison with SPDP-produced conjugates con-

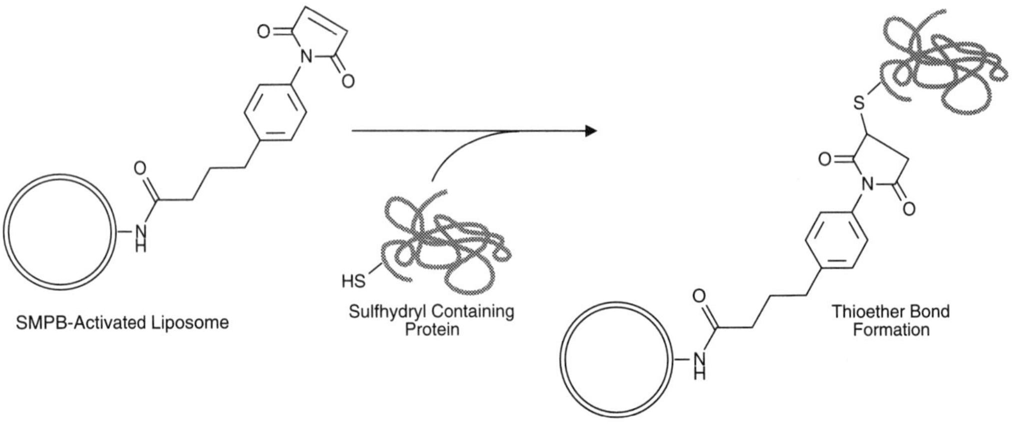

Figure 356   SMPB-activated liposomes may be used to couple thiol-containing protein molecules, forming stable thioether linkages.

cluded that SMPB formed more stable complexes that survive in serum for longer periods (Martin and Papahadjopoulos, 1982). The following protocol is a generalized method for the conjugation of proteins to SMPB-activated liposomes.

1. Prepare a liposome suspension, containing MPB–PE, at a total lipid concentration of 5 mg/ml in 0.05 M sodium phosphate, 0.15 M NaCl, pH 7.2. Activation of DPPE with SMPB is described in Section 2. A suggested lipid composition for vesicle formation is PC:cholesterol:PG:MPB–PE mixed at a molar ratio of 8:10:1:1. The presence of relatively high levels of cholesterol in the liposomal recipe dramatically enhances the conjugation efficiency of the component MPB–PE groups (Martin *et al.*, 1990). Any method of emulsification to create liposomes of the desired size and morphology may be used (Section 1).
2. Dissolve a sulfhydryl-containing protein at a concentration of at least 5 mg/ml in 0.05 M sodium phosphate, 0.15 M NaCl, 10 mM EDTA, pH 7.2. The sulfhydryl groups on the protein molecule may be indigenous or created by any of the methods described in Chapter 1, Section 4.1.
3. Mix the protein solution with the liposome suspension in equal volume amounts.
4. React overnight at room temperature with stirring. Maintain an atmosphere of nitrogen over the reaction to prevent lipid oxidation.
5. Separate unreacted protein from modified liposomes by gel filtration using a column of Sephadex G-75 or by centrifugal floatation in a polymer gradient (Derksen and Scherphof, 1985).

7.9. Coupling via SMCC-Modified Phosphatidylethanolamine Lipid Derivatives

SMCC is a heterobifunctional cross-linker with significant utility in cross-linking proteins, particularly in the preparation of antibody–enzyme (Chapter 10) and hapten–carrier (Chapter 9) conjugates (Hashida and Ishikawa, 1985; Dewey *et al.*,

1987). It is normally used in a two-step cross-linking procedure, wherein the NHS
ester end of the reagent first is reacted with primary amine groups on proteins or other
molecules to form stable amide bonds. This creates a reactive intermediate containing
terminal maleimide groups on the modified molecule. The maleimide end is specific
for coupling to sulfhydryls when the reaction pH is in the range of 6.5–7.5 (Smyth *et
al.*, 1964). Addition of a sulfhydryl-containing protein forms a stable thioether linkage
with the SMCC-activated molecule.

In a similar manner, PE may be activated through its head-group primary amine to
possess reactive maleimide groups capable of coupling sulfhydryl-containing proteins
to liposomes (Fig. 357). The method of derivatizing DPPE with SMCC is essentially
the same as that described for SMPB (Section 2). SMCC, however, contains a more
stable maleimide functional group toward hydrolysis in aqueous reaction environments, due to the proximity of an aliphatic cyclohexane ring rather than the aromatic
phenyl group of SMPB. In protein conjugation to liposomes, this stability may translate into higher activity and more efficient cross-linking. A general protocol for the
coupling of sulfhydryl-containing proteins to liposomes containing SMCC-PE is essentially the same as that described previously for SMPB (Section 7.8).

7.10. Coupling via Iodoacetate-Modified Phosphatidylethanolamine Lipid Derivatives

Iodoacetate derivatives have been used for decades to block or cross-link sulfhydryl
groups in proteins and other molecules (Chapter 1, Section 5.2). At mildly alkaline pH

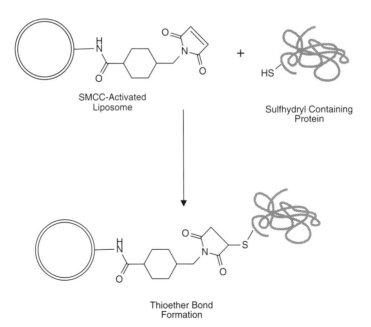

Figure 357   The reaction of an SMCC-activated liposome with a sulfhydryl-containing protein forms
stable thioether bonds.

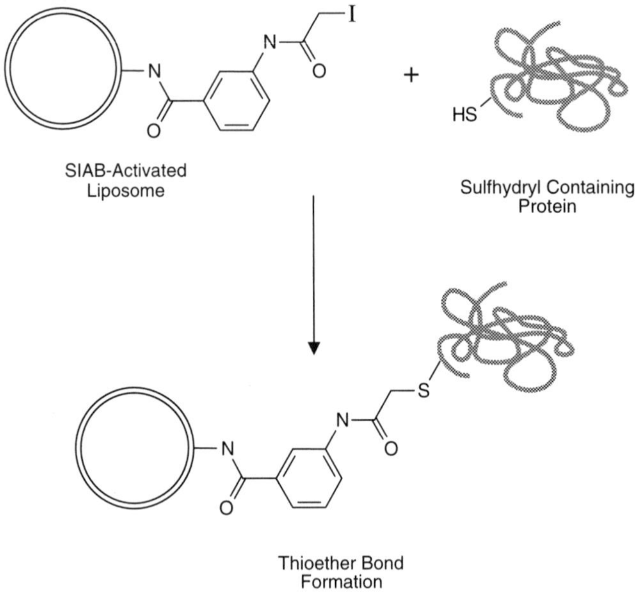

Figure 358 SIAB-activated liposomes can couple sulfhydryl-containing proteins to produce thioether linkages.

values (pH 8–8.5), iodoacetyl derivatives are almost entirely selective toward the cysteine —SH groups in proteins. Disulfide reduction or thiolation reagents can be used to create the required sulfhydryl groups on proteins containing no free sulfhydryls.

Cross-linking reagents, containing an amine-reactive NHS ester on one end and an iodoacetyl group on the other end are particularly useful for two-step protein conjugation. Heterobifunctional reagents like SIAB (Section 2.3.1.5), SIAX (Section 2.3.1.8), or SIAC (Section 2.3.1.9) can be used to modify amine-containing molecules, resulting in iodoacetyl derivatives capable of coupling to sulfhydryl-containing molecules.

Liposomes containing PE lipid components may be activated with these cross-linkers to contain iodoacetyl derivatives on their surface (Fig. 358). The reaction conditions described in Chapter 5, Section 1.5 may be used, substituting a liposome suspension for the initial protein being modified in that protocol. The derivatives are stable enough in aqueous solution to allow purification of the modified vesicles from excess reagent (by dialysis or gel filtration) without significant loss of activity. The only consideration is to protect the iodoacetyl derivative from light, which may generate iodine and reduce the activity of the intermediate. Finally, the modified liposome can be mixed with a sulfhydryl-containing molecule to effect the conjugation through a thioether bond.

Alternatively, pure PE may be derivatized to contain iodoacetyl groups prior to vesicle formation. This may be done using heterobifunctional cross-linkers or through the use of iodoacetic anhydride according to Hashimoto *et al.* (1986). However, a single iodoacetyl group on PE was found not to be sufficiently extended from the vesicle surface to allow efficient protein coupling. Only after creating a longer spacer

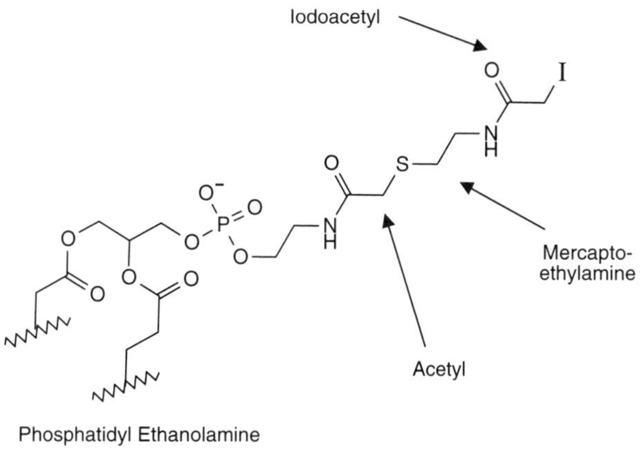

Figure 359 An iodoacetamide derivative of PE containing an extended spacer arm can be constructed through a carbodiimide coupling of iodoacetic acid to PE, followed by reaction with 2-mercaptoethylamine, and finally another reaction with iodoacetate.

by reacting 2-mercaptoethylamine with the initial iodoacetamide derivative and then reacting a second iodoacetic anhydride to form an extended arm did the active derivative possess enough length to give it conjugation capability (Fig. 359). This example illustrates the importance of a long spacer in avoiding steric problems during conjugation to the vesicle surface.

13

Avidin–Biotin Systems

One of the most popular methods of noncovalent conjugation is to make use of the natural strong binding of avidin for the small molecule biotin. The strength of the avidin–biotin interaction has made it a useful tool in specific targeting applications and assay design. Since each avidin molecule contains a maximum of four biotin binding sites, the interaction can be used to enhance the signal strength in immunoassay systems.

Modification reagents that can add a functional biotin group to proteins, nucleic acids, and other molecules now come in many shapes and reactivities (Chapter 8, Section 3). Depending on the functional group present on the biotinylation compound, specific reactive groups on antibodies or other proteins may be modified to create an avidin (or streptavidin) binding site. Amines, carboxylates, sulfhydryls, and carbohydrate groups can be specifically targeted for biotinylation through the appropriate choice of biotin derivative. In addition, photoreactive biotinylation reagents (Chapter 8, Section 3.4) are used to add nonselectively a biotin group to molecules containing no convenient functional groups for modification. In this manner, oligonucleotide probes often are modified for detection with avidin or streptavidin conjugates (Chapter 19, Section 2.3).

The following sections discuss the concept and use of the avidin–biotin interaction in bioconjugate techniques. Preparation of biotinylated molecules and avidin conjugates also are reviewed with suggested protocols. For a discussion of the major biotinylation reagents, see Chapter 8, Section 3.

1. The Avidin–Biotin Interaction

Avidin is a glycoprotein found in egg whites that contains four identical subunits of 16,400 daltons each, giving an intact molecular weight of approximately 66,000 daltons (Green, 1975). Each subunit contains one binding site for biotin, or vitamin H, and one oligosaccharide modification (Asn-linked). The tetrameric protein is highly basic, having a pI of about 10. The biotin interaction with avidin is among the strongest non-covalent affinities known, exhibiting a dissociation constant of about 1.3×10^{-15} M. Tryptophan and lysine residues in each subunit are known to be involved in forming the binding pocket (Gitlin *et al.*, 1987, 1988).

The tetrameric native structure of avidin is resistant to denaturation under extreme chaotropic conditions. Even in 8 M urea or 3 M guanidine hydrochloride the protein maintains structural integrity and activity (Green, 1963). When biotin is bound to avidin, the interaction promotes even greater stability to the complex. An avidin–biotin complex is resistant to breakdown in the presence of up to 8 M guanidine at pH 5.2. A minimum of 6–8 M guanidine at pH 1.5 is required for inducing complete dissociation of the avidin–biotin bond (Cuatrecasas and Wilchek, 1968; Bodanszky and Bodanszky, 1970). Since the subunits in avidin are not held together by disulfide bonds, conditions that cause denaturation also result in subunit disassociation.

The strength of the noncovalent avidin–biotin interaction along with its resistance to breakdown makes it extraordinarily useful in bioconjugate chemistry. Biotinylated molecules and avidin conjugates can "find" each other under the most extreme conditions to bind and complex. The biospecificity of the interaction is similar to antibody–antigen or receptor–ligand recognition, but on a much higher level with respect to affinity constants. Variations in buffer salts and pH, the presence of denaturants or detergents, and extremes of temperature will not prevent the interaction from occurring (Ross *et al.*, 1986).

The only disadvantage to the use of avidin is its tendency to bind nonspecifically to components other than biotin due to its high pI and carbohydrate content. The strong positive charge on the protein causes ionic interactions with more negatively charged molecules, especially cell surfaces. In addition, carbohydrate binding proteins on cells can interact with the polysaccharide portions on the avidin molecule to bind them in regions devoid of targeted biotinylated molecules. These nonspecific interactions can lead to elevated background signals in some assays, preventing the full potential of the avidin–biotin amplification process to be realized.

Streptavidin is another biotin binding protein isolated from *Streptomyces avidinii* that can overcome some of the nonspecificities of avidin (Chaiet and Wolf, 1964). Similar to avidin, streptavidin contains four subunits, each with a single biotin binding site. After some postsecretory modifications, the intact tetrameric protein has a molecular mass of about 60,000 D, slightly less than that of avidin (Bayer *et al.*, 1986, 1989).

The primary structure of steptavidin is considerably different than that of avidin, despite the fact that they both bind biotin with similar avidity. This variation in the amino acid sequence results in a much lower isoelectric point for streptavidin (pI 5–6) than the highly basic pI of 10 for avidin. Moderation in the overall charge of the protein substantially reduces the amount of nonspecific binding due to ionic interaction with other molecules. Of additional significance is the fact that streptavidin is not a glycoprotein; thus there is no potential for binding to carbohydrate receptors. These factors lead to better signal-to-noise ratios in assays using streptavidin–biotin interactions than those employing avidin–biotin.

Both avidin and streptavidin can be conjugated to other proteins or labeled with various detection reagents without loss of biotin binding activity. Streptavidin is slightly less soluble in water than avidin, but both are extremely robust proteins that can tolerate a wide range of buffer conditions, pH values, and chemical modification processes. Bioconjugate techniques can utilize the ε- or N-terminal amines on these proteins for direct conjugation or employ modification reagents, such as thiolation compounds, to transform their existing functional groups into other reactive groups (Chapter 1, Section 4).

2. Use of Avidin–Biotin in Assay Systems

The specificity of biotin binding to avidin or streptavidin provides the basis for developing assay systems to detect or quantify analytes. Biotinylated molecules can be targeted in complex mixtures by using the appropriate avidin or streptavidin conjugates. If the biotinylated component has affinity for binding a particular antigen, then the antigen can be located through the use of an avidin conjugate containing a detectable molecule. A series of avidin–biotin interactions can be built upon each other—utilizing the multivalent nature of each tetrameric avidin molecule—to enhance further the detection capability for the target.

A common application for avidin–biotin chemistry is in immunoassays. The specificity of antibody molecules provides the targeting capability to recognize and bind particular antigen molecules. If there are biotin labels on the antibody molecule, it creates multiple sites for the binding of avidin or streptavidin. If avidin or streptavidin is in turn labeled with an enzyme, fluorophore, etc., then a very sensitive antigen-detection system is created. The potential for more than one labeled avidin to become attached to each antibody through its multiple biotinylation sites is the key to dramatic increases in assay sensitivity over that obtained through the use of antibodies directly labeled with a detectable tag.

There are several basic immunoassay designs that make use of the enhanced sensitivity afforded by the avidin–biotin interaction. Most of these assays use conjugates of avidin or streptavidin with enzymes (such as HRP or alkaline phosphatase), although other labels (such as fluorophores) can be used as well. In the simplest assay design, called the labeled avidin–biotin system (LAB) (Fig. 360), a biotinylated antibody is allowed to incubate and bind with its target antigen. Next, an avidin–enzyme conjugate is introduced and allowed to interact with the available biotin sites on the bound antibody. Just as in other ELISA tests, substrate development then provides the chemical detectability necessary to quantify the antigen (Guesdon *et al.*, 1979).

In a slightly more complex design, the bridged avidin–biotin system (BRAB) uses avidin's multiple biotin binding sites to create an assay of potentially higher sensitivity than that of the LAB assay. Again the biotinylated antibody is allowed to bind to its target, but in the next step an unmodified avidin is introduced to bind with the biotin binding sites on the antibody. Finally, a biotinylated enzyme is added to provide a detection vehicle (Fig. 361). Since the bound avidin still has additional biotin binding sites available, the potential exists for more than one biotinylated enzyme to interact with each bound avidin. In some cases, sensitivity can be increased over that of the LAB technique by using this bridging ability of avidin or streptavidin.

A modification on this theme can be used to produce one of the most sensitive enzyme-linked assay systems known. The ABC system (for avidin–biotin complex) increases the detectability of antigen beyond that possible with either the LAB or BRAB designs by forming a polymer of biotinylated enzyme and avidin before its addition to an antigen-bound, biotinylated antibody (Bayer *et al.*, 1988). When avidin and a biotinylated enzyme are mixed together in solution in the proper proportion, the multiple binding sites on avidin create a linking matrix to form a high-molecular-weight complex. If the biotinylated enzyme is not in large enough excess to block all the binding sites on avidin, then additional sites will still be available on this complex to bind a biotinylated antibody that is bound to its complementary antigen. The large complex provides multiple enzyme molecules to enhance the sensitivity of detecting

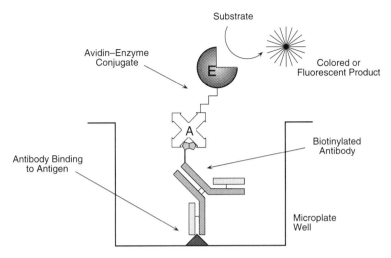

Figure 360   The basic design of the labeled avidin–biotin (LAB) assay system.

antigen (Fig. 362). Thus, the ABC procedure is currently among the highest sensitivity methods available for immunoassay work.

Similar techniques can be used to devise avidin–biotin assay systems for detection of nucleic acid hybridization. DNA probes labeled with biotin can be detected after they bind their complementary DNA target through the use of avidin-labeled complexes (Bugawan *et al.*, 1990; Lloyd *et al.*, 1990). Direct detection of hybridized probes can be accomplished, in a manner similar to that for LAB, by incubating with an avidin–enzyme conjugate followed by substrate development. BRAB-like and ABC-like assays also can be utilized to further enhance a DNA probe signal (Chapter 17, Section 2.3).

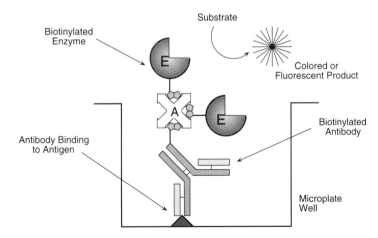

Figure 361   The basic design of the bridged avidin–biotin (BRAB) assay system.

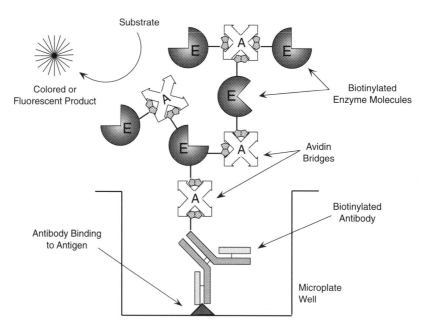

Figure 362   The assay design of the avidin–biotin complex (ABC) system.

Nonenzyme assay systems can be designed with the avidin–biotin interaction, as well. Fluorescently labeled avidin molecules can be used to detect a biotinylated molecule after it has bound its target. In fact, a single preparation of a fluorescent avidin derivative can be used as a universal detection reagent for any biotinylated targeting molecule. The main application of this technique is in cytochemical staining wherein the fluorescence signal is used to localize antigen or receptor molecules in cells and tissue sections.

Other tags or probes (Chapter 8) can be coupled to avidin and used in a similar fashion. For instance, radiolabeled avidin can be employed as a universal detection reagent in radioimmunoassay designs (Wojchowski and Sytkowski, 1986). Avidin labeled with ^{125}I can be used to localize biotinylated monoclonal antibodies directed against tumor cells *in vivo* for imaging purposes (Paganelli *et al.*, 1988). Chemical tags such as in hydrazide–avidin (or streptavidin) derivatives can be made to site-direct avidin's interaction toward oxidized carbohydrate residues for specific detection of glycoconjugates (Section 5). Colloidal gold labeled avidin or streptavidin can be used as highly sensitive detection reagents for microscopy techniques (Cubie and Norval, 1989) (Section 3.6). Finally, cytotoxic substances coupled to avidin can be used to direct cell-killing activity toward a tumor cell-bound, biotinylated monoclonal antibody (or other targeting molecule) for cancer therapy (Hashimoto *et al.*, 1984) (Section 3.3).

Universal detection reagents also can be constructed through biotinylation techniques. Modification of immunoglobulin binding proteins with biotin tags, for instance, creates a reagent useful for the general assay of antibody molecules. In this sense, biotinylated protein A or biotinylated protein G can be used to detect the binding of any primary IgG to its antigen target (provided there are no other antibody molecules present to cause nonspecific binding of the protein A component). Subsequent addition of a labeled avidin molecule binds to the biotinylated protein A, completing the formation of a detection complex (Jagannath and Sehgal, 1989).

To develop assay systems using the avidin–biotin interaction, it is first necessary to produce the associated avidin conjugates and/or biotinylated components. When the LAB technique is employed, the avidin conjugate is made using cross-linking agents, not biotinylation reagents, in order to maintain the binding capacity of the avidin tetramer toward other biotinylated molecules. In the BRAB assay system, avidin is left unconjugated and acts merely as the multivalent bridging molecule, while both the targeting molecule and the detection molecule are biotinylated. The components for the ABC assay are identical to those of the BRAB system.

The following sections discuss the main techniques used to make avidin (or streptavidin) conjugates and various biotinylated components. Chapter 8, Section 3, should be consulted for a complete overview of the biotinylation reagents currently available.

3. Preparation of Avidin or Streptavidin Conjugates

Conjugates of avidin or streptavidin with other protein molecules must be prepared to design systems using the LAB assay technique. Suitable protein molecules attached to avidin or streptavidin either possess indigenous detectability, such as in the case of ferritin or phycobiliproteins, or possess catalytic activity (enzymatic) that can be utilized to produce a detectable substrate product. The majority of conjugation procedures for making avidin–protein or streptavidin–protein conjugates use the amines, sulfhydryls, and carbohydrates on each protein as functional groups for cross-linking.

Perhaps the most common conjugates of avidin or streptavidin involve attaching enzyme molecules for use in ELISA systems. As in the case of antibody–enzyme conjugation schemes (Chapter 10), by far the most commonly used enzymes for this purpose are horseradish peroxidase and alkaline phosphatase. Other enzymes such as β-galactosidase and glucose oxidase are used less often, especially with regard to assay tests for clinically important analytes (Chapter 16).

Other proteins commonly cross-linked to avidin or streptavidin are chromogenic or fluorescent molecules, such as ferritin or phycobiliproteins (Chapter 8, Section 1.7). These conjugates can be used in microscopy techniques to stain and localize certain antigens or receptors in cells or tissue sections.

The following sections discuss three main methods for preparing these types of avidin- or streptavidin–protein conjugates. They involve using an NHS ester–maleimide heterobifunctional cross-linker, making use of the carbohydrate on glycoproteins for reductive amination coupling, and employing the old technique of homobifunctional cross-linking with glutaraldehyde.

3.1. NHS Ester–Maleimide-Mediated Conjugation Protocols

Heterobifunctional cross-linking agents can be used to control the degree of protein conjugation, thus limiting polymerization and controlling the molar ratio of each component in the final complex (Chapter 5). Particularly useful heterobifunctionals include the amine- and sulfhydryl-reactive NHS ester–maleimide cross-linkers discussed in Chapter 5, Section 1. Chief among these is SMCC or sulfo-SMCC (Chapter 5, Section 1.3), which contains a reasonably long spacer and an extremely stable maleimide group due to the adjacent cyclohexane ring in its cross-bridge.

Conjugations done with SMCC usually involve up to 3 steps. In the first stage, one of the proteins is modified at its amine groups via the NHS ester end of the cross-linker to form amide linkages terminating in active maleimide groups. If the other protein to be conjugated does not contain sulfhydryl residues necessary to react with the maleimide-activated protein, it must be modified to contain them (Chapter 1, Section 4.1). Finally, the two reactive components are mixed together in the proper ratio to effect the conjugation reaction.

For the preparation of avidin–enzyme conjugates, either protein may be first modified with SMCC and the other one modified to contain —SH groups. Since avidin or streptavidin does not possess any free sulfhydryls—and the disulfides present in avidin are inaccessible to reduction—these proteins must be modified with either the cross-linker or with a thiolating agent before conjugation. If the enzyme employed contains free sulfhydryls in its native state, such as β-galactosidase, then it is convenient to activate avidin with SMCC and simply add the sulfhydryl-containing protein to it for conjugation. If the enzyme does not contain free sulfhydryls (as is the case with alkaline phosphatase or horseradish peroxidase), then the choice of which component gets maleimide-activated and which gets thiolated is up to the individual.

The following protocol describes the activation of avidin or streptavidin with sulfo-SMCC and its subsequent conjugation with an enzyme modified to contain sulfhydryls using SATA (Chapter 1, Section 4.1). A method for the opposite approach, wherein the enzyme is activated with SMCC and the avidin component is thiolated, is presented immediately after this protocol. This strategy may be the most common approach to forming these conjugates (Fig. 363). In addition, since there are enzymes commercially available that are preactivated with SMCC (Pierce), their use may be the easiest solution.

Protocol for the Conjugation of SMCC-Activated Avidin or Streptavidin with Thiolated Enzyme
Activation of Avidin or Streptavidin with SMCC

1. Dissolve avidin or streptavidin (Pierce) in 0.1 M sodium phosphate, 0.15 M NaCl, pH 7.2, at a concentration of 10 mg/ml.
2. Add 1 mg of sulfo-SMCC (Pierce) to each milliliter of avidin or streptavidin solution. Mix to dissolve.
3. React for 30–60 min at room temperature. Since maleimide groups are labile in aqueous solution, extended reaction times should be avoided.
4. Immediately purify the maleimide-activated avidin or streptavidin away from excess cross-linker and reaction by-products by gel filtration on a desalting column (Sephadex G-25 or the equivalent). Use 0.1 M sodium phosphate, 0.15 M NaCl, pH 7.2, as the chromatography buffer. Pool the fractions containing protein (the first peak eluting from the column). After elution, adjust the protein concentration to 10 mg/ml for the conjugation reaction (centrifugal concentrators work well for this step). At this point, the maleimide-activated avidin may be frozen and lyophilized to preserve its maleimide activity. The modified pro-

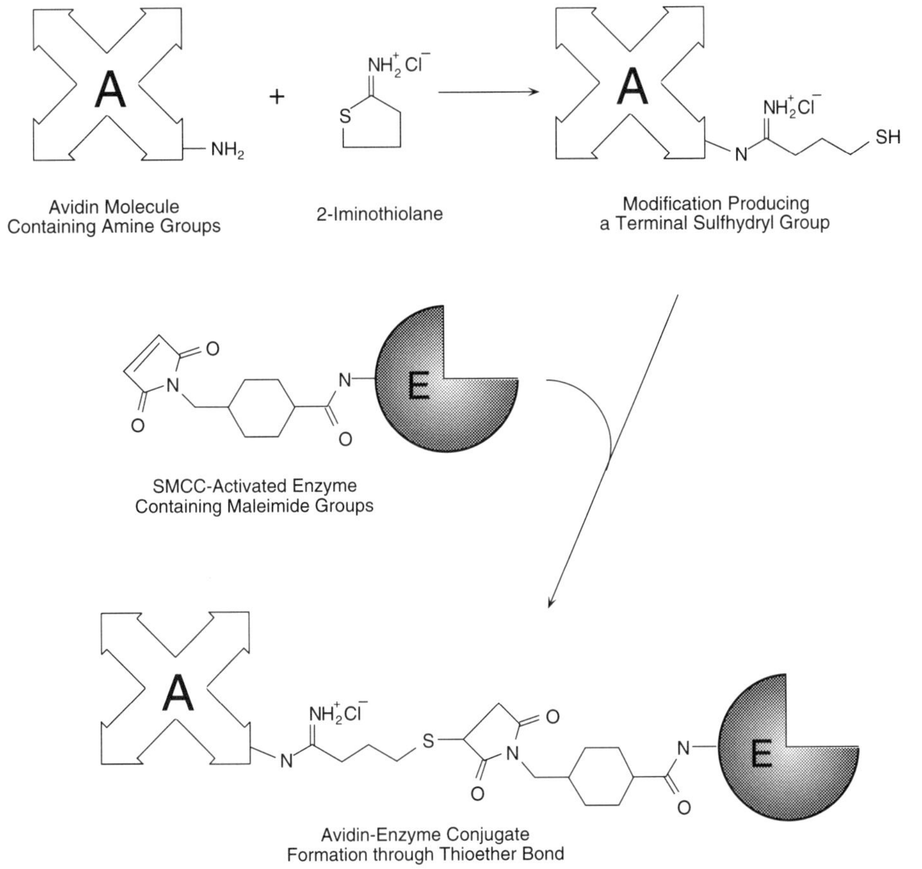

Figure 363 Avidin may be modified with 2-iminothiolane to produce sulfhydryl groups. Subsequent reaction with a maleimide-activated enzyme produces a thioether-linked conjugate.

tein is stable for at least 1 year in a freeze-dried state. If kept in solution, the maleimide-activated avidin is labile and should be used immediately to conjugate with a thiolated enzyme following the procedure described below.

Modification of Enzyme with SATA

If β-galactosidase is used to conjugate with an SMCC-activated avidin or streptavidin, then there is no need to thiolate the enzyme, since it contains sulfhydryls in its native state (Sivakoff and Janes, 1988; Fujiwara *et al.*, 1988). For conjugations using horseradish peroxidase, alkaline phosphatase, or glucose oxidase, however, thiolation is necessary to add the requisite sulfhydryls.

1. Dissolve the enzyme to be modified in 0.1 *M* sodium phosphate, 0.15 *M* NaCl, pH 7.2, at a concentration of 10 mg/ml.
2. Prepare a stock solution of SATA (Pierce) by dissolving it in DMSO at a concentration of 13 mg/ml. Use a fume hood to handle the organic solvent.
3. Add 25 μl of the SATA stock solution to each milliliter of 10 mg/ml enzyme

solution. For different concentrations of enzyme in the reaction medium, proportionally adjust the amount of SATA addition; however, do not exceed 10% DMSO in the aqueous reaction medium.

4. React for 30 min at room temperature.
5. To purify the SATA-modified enzyme perform a gel filtration separation using Sephadex G-25 or dialyze against 0.1 M sodium phosphate, 0.15 M NaCl, pH 7.2, containing 10 mM EDTA. Purification is not absolutely required, since the following deprotection step is done using hydroxylamine at a significant molar excess over the initial amount of SATA added. Whether a purification step is done or not, at this point, the derivative is stable and may be stored under conditions that favor long-term enzyme activity.
6. Deprotect the acetylated sulfhydryl groups on the SATA-modified enzyme according to the following protocol:

   a. Prepare a 0.5 M hydroxylamine solution in 0.1 M sodium phosphate, pH 7.2, containing 10 mM EDTA.
   b. Add 100 μl of the hydroxylamine stock solution to each milliliter of the SATA-modified enzyme. Final concentration of hydroxylamine in the enzyme solution is 50 mM.
   c. React for 2 h at room temperature.
   d. Purify the thiolated enzyme by gel filtration on Sephadex G-25 using 0.1 M sodium phosphate, 0.1 M NaCl, pH 7.2, containing 10 mM EDTA as the chromatography buffer. To obtain efficient separation between the thiolated protein and excess hydroxylamine and reaction by-products, the sample size applied to the column should be at a ratio of no more than 5% sample volume to the total column volume. Collect 0.5-ml fractions. Pool the fractions containing protein by measuring the absorbance of each fraction at 280 nm.

Production of Conjugate

1. Immediately mix the thiolated enzyme with an amount of maleimide-activated avidin (or streptavidin) to obtain the desired molar ratio of enzyme to avidin in the conjugate. Use of a 4:1 (enzyme:avidin) molar ratio in the conjugation reaction usually results in high-activity conjugates suitable for use in many enzyme-linked immunoassay procedures employing the LAB approach.
2. React for 30–60 min at 37°C or 2 h at room temperature. The conjugation reaction also may be done at 4°C overnight.

A variation of the above method can be used, wherein the enzyme is first activated with SMCC and conjugated to a thiolated avidin or streptavidin molecule. This approach probably is the most common way of preparing avidin–enzyme conjugates, and since the preactivated enzymes are readily available (Pierce), it also may be the easiest.

Protocol for the Conjugation of SMCC-Activated Enzymes with Thiolated Avidin or Streptavidin
Activation of Enzyme with Sulfo-SMCC
The following protocol describes the activation of HRP with sulfo-SMCC. Other enzymes may be activated in a similar manner. The activated enzyme possesses male-

imide groups that are relatively unstable in aqueous solution. Therefore, the thiolation reaction (part B) should be coordinated with the activation process so that the conjugation (part C) can be done immediately. Note that if preactivated enzymes are obtained, this step may be eliminated.

1. Dissolve HRP in 0.1 M sodium phosphate, 0.15 M NaCl, pH 7.2, at a concentration of 10 mg/ml.
2. Add 3.3 mg of sulfo-SMCC (Pierce) to each milliliter of the HRP solution. Mix to dissolve and react for 30 min at room temperature. Alternatively, two equal additions of cross-linker may be done—the second one after 15 min of incubation—to obtain even more efficient modification.
3. Immediately purify the maleimide-activated HRP away from excess cross-linker and reaction by-products by gel filtration on a desalting column (Sephadex G-25 or the equivalent). Use 0.1 M sodium phosphate, 0.15 M NaCl, pH 7.2, as the chromatography buffer. HRP can be observed visually as it flows through the column due to the color of its heme ring. Pool the fractions containing the HRP peak. After elution, adjust the HRP concentration to 10 mg/ml for the conjugation reaction. At this point, the maleimide-activated enzyme may be frozen and lyophilized to preserve its maleimide activity. The modified enzyme is stable for at least 1 year in a freeze-dried state. If kept in solution, the maleimide-activated HRP should be used immediately to conjugate with thiolated avidin or streptavidin following the protocols outlined below.

Thiolation of Avidin or Streptavidin

1. Dissolve avidin or streptavidin in 0.1 M sodium phosphate, 0.15 M NaCl, pH 7.2, at a concentration of 10 mg/ml.
2. Prepare a stock solution of SATA by dissolving it in DMSO at a concentration of 13 mg/ml. Use a fume hood to handle the organic solvent.
3. Add 25 μl of the SATA stock solution to each milliliter of 10 mg/ml avidin or streptavidin solution. For different concentrations of protein in the reaction medium, proportionally adjust the amount of SATA addition; however, do not exceed 10% DMSO in the aqueous reaction medium.
4. React for 30 min at room temperature.
5. To purify the SATA-modified avidin or streptavidin use gel filtration on a column of Sephadex G-25 or dialyze against 0.1 M sodium phosphate, 0.15 M NaCl, pH 7.2, containing 10 mM EDTA. At this point, the derivative is stable and may be stored under conditions that favor long-term avidin activity.
6. Deprotect the acetylated sulfhydryl groups on the SATA-modified protein according to the following protocol:
   a. Prepare a 0.5 M hydroxylamine solution in 0.1 M sodium phosphate, pH 7.2, containing 10 mM EDTA.
   b. Add 100 μl of the hydroxylamine stock solution to each milliliter of the SATA-modified avidin or streptavidin. Final concentration of hydroxylamine in the solution is 50 mM.
   c. React for 2 h at room temperature.
   d. Purify the thiolated protein by gel filtration on Sephadex G-25 using 0.1 M sodium phosphate, 0.1 M NaCl, pH 7.2, containing 10 mM EDTA as the chromatography buffer.

*Conjugation of SMCC-Activated Enzyme with Thiolated Avidin
or Streptavidin*

1. Immediately mix the SMCC-activated enzyme with an amount of thiolated avidin (or streptavidin) to obtain the desired molar ratio of enzyme to avidin in the conjugate. Use of a 4:1 (enzyme:avidin) molar ratio in the conjugation reaction usually results in high-activity conjugates suitable for use in many enzyme-linked immunoassay procedures employing the LAB approach.
2. React for 30–60 min at 37°C or 2 h at room temperature. The conjugation reaction also may be done at 4°C overnight.

3.2. Periodate Oxidation/Reductive Amination Conjugation Protocols

Glycoproteins may be conjugated with another amine-containing protein through the process of periodate oxidation and reductive amination. Periodate oxidation of polysaccharide components on the glycoprotein results in the formation of reactive aldehyde residues by cleavage of carbon—carbon bonds and oxidation of the associated adjacent hydroxyls (Chapter 1, Section 4.4). Conjugation with another protein may be done by reacting the aldehydes with amines to form intermediate Schiff bases with subsequent reduction using sodium cyanoborohydride to create stable secondary amine bonds.

This method of conjugation is particularly well suited to coupling HRP or ferritin with avidin or streptavidin. Both HRP and avidin are glycoproteins that can be oxidized with sodium periodate to generate aldehydes. Thus, HRP–avidin, HRP–streptavidin, and ferritin–avidin may be prepared by reductive amination. Ferritin is a large, complex protein of molecular weight 750,000. Its structure is made of a protein shell of diameter approximately 12 nm that surrounds a micelle core consisting of ferric hydroxide of about 6 nm in diameter. This core contains more than 2000 iron atoms, making the protein extremely electron dense and thus perfect for electron microscopy applications. The properties of HRP are described in Chapter 16, Section 1.

The following protocol is adapted from Bayer *et al.* (1976).

*Protocol for the Conjugation of Avidin with Ferritin Using
Reductive Amination*

1. Dissolve avidin (Pierce) in 0.1 *M* sodium acetate, 0.15 *M* NaCl, pH 4.5, at a concentration of 3 mg/ml.
2. Dissolve ferritin (Sigma) in 0.1 *M* sodium acetate, 0.15 *M* NaCl, pH 4.5, at a concentration of 100 mg/ml.
3. Add 1 ml of ferritin solution to every 5 ml of avidin solution. Chill on ice.
4. Dissolve sodium periodate in water at a concentration of 100 m*M*. Prepare fresh and protect from light.
5. Add 110 µl of sodium periodate solution to each milliliter of avidin/ferritin solution.
6. React for 3 h on ice with periodic mixing. Protect from light.
7. Remove excess periodate by gel filtration on a column of Sephadex G-25 or by overnight dialysis against 50 m*M* sodium borate, 0.15 *M* NaCl, pH 8.5.

8. Dissolve 10 mg of sodium borohydride in 1 ml of 10 mM NaOH. Prepare fresh. Add 83 μl of this reducing solution to each ml of avidin/ferritin solution.
9. React for 1 h on ice.
10. Remove excess reductant by gel filtration using a column of Sephadex G-25 or by extensive dialysis against 20 mM sodium phosphate, 0.15 M NaCl, pH 7.4.

Conjugation of HRP by reductive amination can be done by oxidizing the carbohydrate on the enzyme and subsequently coupling to the amines on avidin or streptavidin (Fig. 364).

Protocol for the Preparation of Avidin–HRP or Streptavidin–HRP by Reductive Amination
Oxidation of HRP with Sodium Periodate

1. Dissolve HRP in water or 0.01 M sodium phosphate, 0.15 M NaCl, pH 7.2, at a concentration of 10–20 mg/ml.
2. Dissolve sodium periodate in water at a concentration of 0.088 M. Protect from light.
3. Immediately add 100 μl of the sodium periodate solution to each ml of the HRP solution. This results in an 8 mM periodate concentration in the reaction mixture. Mix to dissolve. Protect from light.
4. React in the dark for 20 min at room temperature. A color change will be

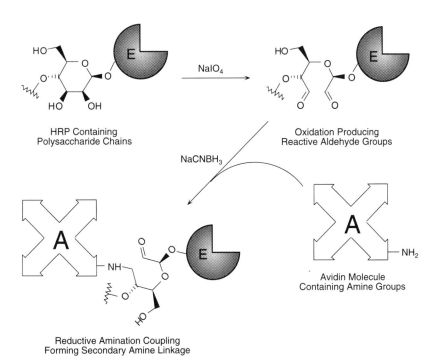

HRP Containing Polysaccharide Chains — NaIO$_4$ → **Oxidation Producing Reactive Aldehyde Groups**

NaCNBH$_3$

Avidin Molecule Containing Amine Groups

Reductive Amination Coupling Forming Secondary Amine Linkage

Figure 364 Oxidation of the polysaccharide components of HRP produces reactive aldehyde groups. Conjugation to avidin then may be done by reductive amination.

apparent as the reaction proceeds—changing from the brownish/gold color of concentrated HRP to green.

5. Immediately purify the oxidized enzyme by gel filtration using a column of Sephadex G-25. The chromatography buffer is 0.01 M sodium phosphate, 0.15 M NaCl, pH 7.2. Collect 0.5-ml fractions and monitor for protein at 280 nm. HRP also may be detected by its absorbance at 403 nm. In oxidizing large quantities of HRP, the fraction collection process may be done visually—just pooling the colored HRP peak as it comes off the column.

6. Pool the fractions containing protein. Adjust the enzyme concentration to 10 mg/ml for the conjugation step. The periodate-activated HRP may be stored frozen or freeze-dried for extended periods without loss of activity. However, do not store the preparation in solution at room temperature or 4°C, since precipitation will occur over time due to self-polymerization.

Conjugation of Periodate-Oxidized HRP
with Avidin or Streptavidin

1. Dissolve avidin or streptavidin at a concentration of 10 mg/ml in 0.2 M sodium bicarbonate, pH 9.6, at room temperature. The high-pH buffer will result in very efficient Schiff base formation and conjugation with the highest possible incorporation of enzyme molecules per avidin or streptavidin molecule. To produce lower molecular weight conjugates (using less efficient Schiff base formation conditions), dissolve the proteins at a concentration of 10 mg/ml in 0.1 M sodium phosphate, 0.15 M NaCl, pH 7.2.

2. The periodate-oxidized HRP (prepared above) was finally purified using 0.01 M sodium phosphate, 0.15 M NaCl, pH 7.2. For conjugation using the lower pH buffered environment, this HRP preparation can be used directly at 10 mg/ml concentration. For conjugation using the higher pH carbonate buffer, dialyze the HRP solution against 0.2 M sodium carbonate, pH 9.6, for 2 h at room temperature prior to use.

3. Mix the avidin or streptavidin solution with the enzyme solution at a ratio of 1:6.6 (v/v). Since avidin has a molecular weight of about 66,000 and the molecular weight of HRP is 40,000, this ratio of volumes will result in a molar ratio of HRP:avidin equal to 4:1. For conjugates consisting of greater enzyme-to-avidin ratios, proportionally increase the amount of enzyme solution as required. Typically, molar ratios of 2:1 to 10:1 (enzyme:avidin) give acceptable conjugates useful in a variety of ELISA techniques.

4. React for 2 h at room temperature to form the initial Schiff base interactions.

5. In a fume hood, add 10 μl of 5 M sodium cyanoborohydride (Sigma) per milliliter of reaction solution. Caution: cyanoborohydride is extremely toxic. All operations should be done with care in a fume hood. Also, avoid any contact with the reagent, as the 5 M solution is prepared in 1 N NaOH.

6. React for 30 min at room temperature (in a fume hood).

7. Block unreacted aldehyde sites by addition of 50 μl of 1 M ethanolamine, pH 9.6, per milliliter of conjugation solution. Approximately a 1 M ethanolamine solution may be prepared by addition of 300 μl ethanolamine to 5 ml of deionized water. Adjust the pH of the ethanolamine solution by addition of concentrated HCl, keeping the solution cool on ice.

8. React for 30 min at room temperature.
9. Purify the conjugate from excess reactants by dialysis or gel filtration using Sephadex G-25. Use 0.01 M sodium phosphate, 0.15 M NaCl, pH 7, as the buffer for either operation. Use a fume hood, since cyanoborohydride will be present in some of the fractions.

3.3. Glutaraldehyde Conjugation Protocol

Glutaraldehyde is one of the oldest homobifunctional reagents used for protein conjugation. It reacts with amine groups to create cross-links by one of several routes (Chapter 4, Section 6.2). Under reducing conditions, the aldehydes on both ends of glutaraldehyde will couple with amines to form secondary amine linkages. The reagent is highly efficient at protein conjugation, but has a tendency to form high-molecular-weight polymers due to its homobifunctional nature. Single-step protocols using glutaraldehyde are particularly notorious at resulting in some degree of insoluble protein oligomers (Porstmann *et al.*, 1985). Two-step methods somewhat alleviate this problem, but the potential for conjugate precipitation is still present.

Preparation of avidin or streptavidin conjugates with other proteins can be accomplished using either a one- or a two-step glutaraldehyde procedure. Both methods may result in some degree of oligomer formation; however, the two-step protocol may keep insolubles to a minimum. Although the following procedures are described using particular proteins, they may be used as a general guide for coupling enzymes, ferritin, phycobiliproteins, or other detectable proteins to avidin or streptavidin. Some optimization may be necessary to obtain the best yield of active conjugate.

Protocol for the One-Step Glutaraldehyde Conjugation
of Ferritin to Avidin or Streptavidin
This protocol is adapted from Bayer and Wilchek (1980).

1. Prepare a solution containing 5 mg/ml avidin (or streptavidin) and 25 mg/ml ferritin in 0.02 M sodium phosphate, 0.15 M NaCl, pH 7.4, at room temperature. Note: for the coupling of other proteins to avidin or streptavidin, their concentration may be reduced from the 25 mg/ml stated for ferritin.
2. In a fume hood, add 10 μl of 25% glutaraldehyde per milliliter of avidin/ferritin solution. Mix well.
3. React for 1 h at room temperature.
4. To reduce the resultant Schiff bases and any excess aldehydes, add sodium borohydride to a final concentration of 10 mg/ml.

Note: Some protocols do not call for a reduction step. As an alternative to reduction, add 50 μl of 0.2 M lysine in 0.5 M sodium carbonate, pH 9.5, to each milliliter of the conjugation reaction to block excess reactive sites. Block for 2 h at room temperature. Other amine-containing small molecules may be substituted for lysine—such as glycine, Tris buffer, or ethanolamine.

5. Reduce for 1 h at 4°C.
6. To remove any insoluble polymers that may have formed, centrifuge the conjugate or filter it through a 0.45-μm filter. Purify the conjugate by gel filtration or dialysis using PBS, pH 7.4.

A two-step glutaraldehyde protocol may result in lower molecular weight conjugates, thus limiting the degree of insoluble material formed during the cross-linking process. The following protocol is adapted from Avrameas (1969).

Protocol for the Two-Step Glutaraldehyde Conjugation of Enzymes to Avidin or Streptavidin

1. Dissolve the enzyme at a concentration of 10 mg/ml in 0.1 M sodium phosphate, 0.15 M NaCl, pH 6.8.
2. Add glutaraldehyde to a final concentration of 1.25%.
3. React overnight at room temperature.
4. Purify the activated enzyme from excess glutaraldehyde by gel filtration (using Sephadex G-25) or by dialysis against PBS, pH 6.8.
5. Dissolve avidin or streptavidin at a concentration of 10 mg/ml in 0.5 M sodium carbonate, pH 9.5. Mix the activated enzyme with the avidin or streptavidin solution at the desired molar ratio to effect the conjugation. Mixing the equivalent of 1–2 mol of enzyme per mole of avidin usually results in acceptable conjugates.
6. React overnight at 4°C.
7. To reduce the resultant Schiff bases and any excess aldehydes, add sodium borohydride to a final concentration of 10 mg/ml.

Note: Some protocols avoid a reduction step. As an alternative to reduction, add 50 μl of 0.2 M lysine in 0.5 M sodium carbonate, pH 9.5, to each milliliter of the conjugation reaction to block excess reactive sites. Block for 2 h at room temperature. Other amine-containing small molecules may be substituted for lysine—such as glycine, Tris buffer, or ethanolamine.

8. Reduce for 1 h at 4°C.
9. To remove any insoluble polymers that may have formed, centrifuge the conjugate or filter it through a 0.45-μm filter. Purify the conjugate by gel filtration or dialysis using PBS, pH 7.4.

4. Preparation of Fluorescently Labeled Avidin or Streptavidin

Fluorophore modification of avidin or streptavidin creates a reagent system that can be used to detect and localize biotinylated targeting molecules. The application of such reagents in immunohistochemical staining techniques is significant (Bonnard *et al.*, 1984). A biotinylated antibody directed against a particular tissue antigen can be allowed to bind its target *in situ*, and then a fluorescently tagged avidin or streptavidin may be added to bind and visualize the antibody-bound antigenic sites by luminescence. Individual cellular structures can be labeled in similar assay strategies and detected by fluorescent microscopy or cell sorting techniques (Sternberger, 1986; Abou-Samra *et al.*, 1990). Biotinylated targeting molecules like antibodies usually possess low nonspecific binding potential despite the presence of a biotin tag. Thus, background fluorescence of tissue sections can be kept to a minimum using avidin–biotin detection. The multivalent nature of avidin or streptavidin combined with the potential of more than one biotin tag per antibody creates a system of much greater potential sensitivity than when using fluorescently modified antibodies directly. The complex formed from the avidin–biotin interaction amplifies the fluorescent signal beyond that capable in standard labeled antibody techniques.

Double-labeling systems also can be developed using the avidin–biotin interaction. If two primary antibodies directed against separate antigenic determinants are labeled, one with biotin and the other with another detection component (such as a fluorophore, enzyme, and gold particles), then both may be used to localize simultaneously different antigens in tissue sections. The biotinylated antibody may be subsequently detected by the addition of a fluorescently labeled avidin or streptavidin reagent. An example of a double-labeling avidin–biotin detection system is that of Feller *et al.* (1983). A pair of tonsil antigens were visualized using two monoclonal antibodies, one fluorescein-labeled and the other biotinylated. The biotinylated antibody was detected by using a phycoerythrin-labeled avidin conjugate (Chapter 13, Section 4.4, and Chapter 8, 1.7). Even triple-labeling systems may be developed using this strategy (van Dongen *et al.*, 1985).

The following sections present suggested protocols for labeling avidin or streptavidin with selected fluorophores. Other fluorescent probes may be constructed using the reagents and methods discussed in Chapter 8, Section 1.

4.1. Modification with FITC

Fluorescein isothiocyanate (FITC) is the most common fluorescent label used to modify proteins and other biomolecules (Chapter 8, section 1.1). The isothiocyanate group reacts with amines in protein molecules to form a stable thiourea linkage (Fig. 365). Avidin or streptavidin may be tagged with this reagent to yield highly fluorescent

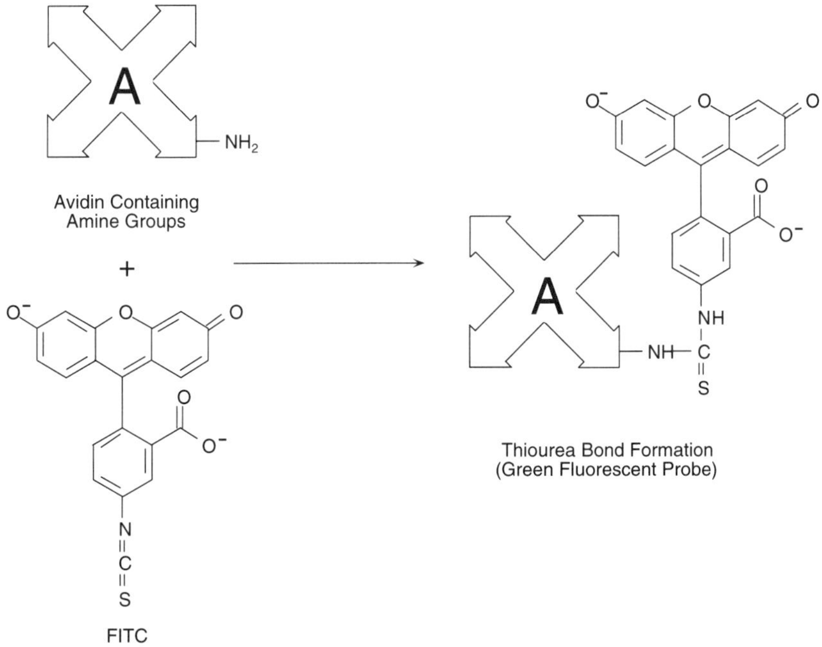

Avidin Containing
Amine Groups

+

FITC

Thiourea Bond Formation
(Green Fluorescent Probe)

Figure 365 The reaction of FITC with avidin produces a fluorescent probe via isothiourea bonds.

derivatives useful both in single-staining and double-staining techniques (Bayer and Wilchek, 1980; Szabo *et al.*, 1989; Bakkus *et al.*, 1989). Optimal modification levels for fluorescein are in the range 3–8 fluorophores per avidin or streptavidin molecule. Lower incorporation levels than these result in low luminescence and poor sensitivity. Higher levels may cause fluorescein–fluorescein quenching effects, resulting in decreased fluorescence. Too high a modification level also may result in nonspecific binding of the derivatized proteins to nontargeted components in assay systems.

Protocol

1. Dissolve avidin or streptavidin (Pierce) in 0.1 M sodium carbonate, pH 9.5, at a concentration of 2–4 mg/ml.
2. Dissolve FITC (Pierce) in DMF at a concentration of 2 mg/ml. Protect from light.
3. Add 50–100 μl of the FITC solution to each milliliter of the avidin or streptavidin solution.
4. React overnight at 4°C in the dark.
5. Remove excess fluorescein by gel filtration using a column of Sephadex G-25.

4.2. Modification with Lissamine Rhodamine B Sulfonyl Chloride

Rhodamine derivatives are popular probes to use in tandem with fluorescein labels. The Lissamine derivatives of rhodamine (Chapter 8, Section 1.2) are intensely fluorescent, strongly emitting in the red region of the spectrum. The red luminescence of Lissamine rhodamine contrasts sharply with the green emission of fluorescein. Lissamine rhodamine B sulfonyl chloride can be used to modify proteins at their ε- and N-terminal amine functional groups. The resultant derivatives are linked through stable sulfonamide bonds, resulting in rhodamine's intensely fluorescent character being incorporated into the modified molecules. Avidin and streptavidin derivatives of this fluorophore are particularly popular for use in fluorescent assay systems (Fig. 366).

Protocol

1. Dissolve avidin or streptavidin (Pierce) in 0.1 M sodium carbonate/bicarbonate buffer, pH 9, at a concentration of 1–5 mg/ml.
2. Dissolve Lissamine rhodamine B sulfonyl chloride (Molecular Probes) in DMF at a concentration of 1–2 mg/ml. Protect from light and use immediately. Do not use DMSO as the solvent, as sulfonyl chlorides react with it.
3. In a darkened lab and with gentle mixing, slowly add 50–100 μl of the fluorophore solution to each milliliter of the avidin or streptavidin solution.
4. React for 1 h at room temperature in the dark.
5. Remove excess fluorophore by gel filtration using a column of Sephadex G-25 or by dialysis.

Modification of avidin or streptavidin with Texas Red sulfonyl chloride may be done similarly, except the fluorophore is first dissolved in acetonitrile prior to addition to the aqueous reaction mixture.

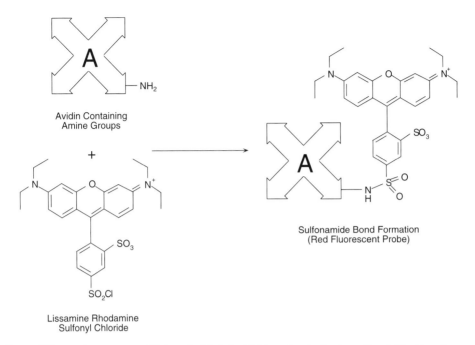

Figure 366 Avidin (or streptavidin) can be labeled with Lissamine rhodamine sulfonyl chloride to form a fluorescent probe.

4.3. Modification with AMCA–NHS

AMCA derivatives possess intense fluorescent properties within the blue region of the visible spectrum (Chapter 8, Section 1.3). Their emission range is well removed from other common fluorophores, making them excellent choices for use in double-labeling techniques, for example, with fluorescein-labeled molecules. Coumarin-based fluorescent probes are very good donors for excited-state energy transfer to fluoresceins. AMCA–NHS reacts with amine-containing molecules to result in stable amide-bond derivatives (Fig. 367). Avidin and streptavidin may be labeled with this reagent to give probes useful for immunohistochemical staining of biotinylated targeting molecules. AMCA labeled proteins are fairly stable to photoquenching and exhibit large Stokes shifts, allowing sensitive measurements to be made without interference from scattered excitation light.

Protocol

1. Dissolve avidin or streptavidin (Pierce) in 50 mM sodium borate, pH 8.5, at a concentration of 10 mg/ml. Other buffers may be used for an NHS ester reaction, including 0.1 M sodium phosphate, pH 7.5 (Chapter 2, Section 1.4).
2. Dissolve AMCA–NHS (Pierce) in DMSO at a concentration of 2.6 mg/ml. Protect from light.
3. In subdued lighting conditions, slowly add 50–100 μl of the AMCA–NHS stock solution to each milliliter of the avidin or streptavidin solution, with gentle mixing.

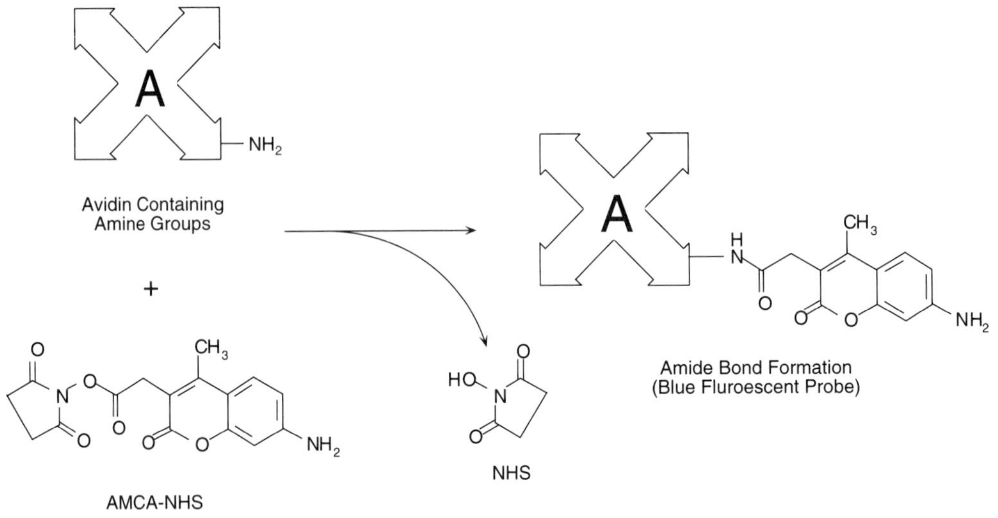

Figure 367 AMCA–NHS reacts with the amine groups of avidin (or streptavidin) to produce amide bonds.

4. React for 1 h at room temperature in the dark.
5. Remove excess reagent and reaction by-products by gel filtration using a column of Sephadex G-25 or by dialysis.

4.4. Conjugation with Phycobiliproteins

Phycobiliproteins are incredibly fluorescent due to their multiple chromophoric bilin prosthetic groups, conferring extremely high absorbance coefficients to each protein molecule (Chapter 8, Section 1.7). Conjugates of these biliproteins with targeting molecules form extraordinarily luminescent probes. Labeling with phycobiliprotein derivatives can provide absorption coefficients 30-fold higher than labeling with small, synthetic fluorophores. Their ability to be monitored by fluorescing in the red region of the spectrum decreases potential interferences from indigenous biological fluorescence. Phycoerythrin-labeled avidin or streptavidin probes can be used in double-staining procedures with a fluorescein-labeled antibody, detecting two antigens in the same tissue section simultaneously (Feller *et al.*, 1983).

The bilin content of these fluorescent proteins ranges from a low of 4 prosthetic groups in C-phycocyanin to the 34 groups of B- and R-phycoerythrin. Phycoerythrin derivatives, therefore, can be used to create the most intensely fluorescent probes possible using these proteins. Streptavidin–phycoerythrin conjugates, for example, have been used to detect as little as 100 biotinylated antibodies bound to receptor proteins per cell (Zola *et al.*, 1990).

Conjugates of avidin or streptavidin with these fluorescent probes may be prepared by activation of the phycobiliprotein with SPDP to create a sulfhydryl-reactive derivative, followed by modification of avidin or streptavidin with 2-iminothiolane or SATA (Chapter 1, Section 4.1) to create the free sulfhydryl groups necessary for conjugation. The protocol for SATA modification of avidin or streptavidin can be found in Section

3.1. The procedure for SPDP activation of phycobiliproteins can be found in Chapter 8, Section 1.7. Reacting the SPDP-activated phycobiliprotein with SH-labeled avidin at a molar ratio of 2:1 will result in highly fluorescent biotin-binding probes.

5. Preparation of Hydrazide-Activated Avidin or Streptavidin

Hydrazide groups can react with aldehydes or ketones to form hydrazone linkages (Chapter 2, Section 5.1). Proteins may be labeled with hydrazide residues by reaction of their indigenous carboxylate groups with *bis*-hydrazine compounds such as adipic acid dihydrazide or carbohydrazide (Chapter 4, Section 8). A carbodiimide-mediated reaction between the protein and the bis-hydrazine reagent forms diimide bond derivatives terminating in hydrazide groups (Fig. 368). Avidin or streptavidin labeled with adipic acid dihydrazide can form the basis of a carbohydrate detection system using the avidin–biotin interaction (Bayer and Wilchek, 1990; Bayer *et al.*, 1987a, 1990). Glycoconjugates in tissue sections, cells, or blots may be treated with sodium periodate or galactose oxidase to create aldehyde groups on the associated sugar components. Introduction of hydrazide-activated avidin or streptavidin causes hydrazone bonds to form between the hydrazides and aldehydes, thus specifically targeting glycoproteins and other carbohydrate-containing molecules. Subsequent detection with a biotinylated enzyme allows precise localization of glycoconjugates. Detection in a single-step using this strategy is possible using preformed complexes of hydrazide-activated streptavidin and a biotinylated enzyme (Fig. 369).

   The activation of avidin or streptavidin with adipic dihydrazide may be done using the method of Bayer *et al.* (1987a). A summary of this protocol is given below.

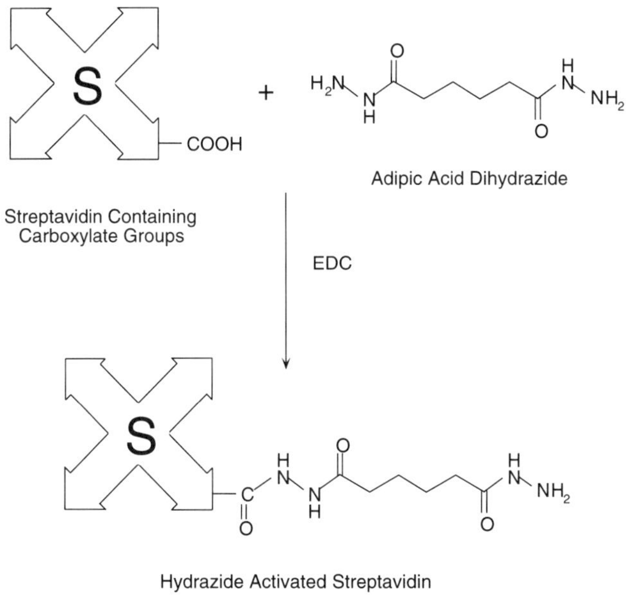

Streptavidin Containing
Carboxylate Groups

Adipic Acid Dihydrazide

EDC

Hydrazide Activated Streptavidin

Figure 368 Reaction of adipic acid dihydrazide with avidin (or streptavidin) produces a hydrazide derivative highly reactive toward periodate-oxidized polysaccharides.

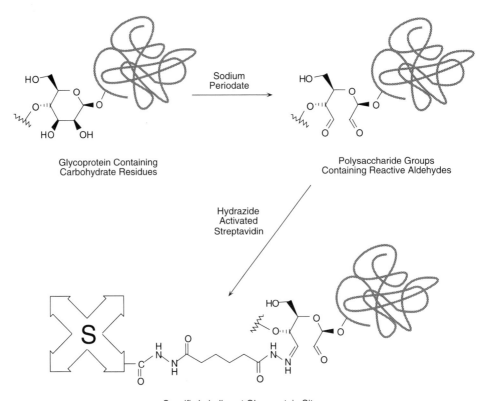

Figure 369 Glycoproteins may be oxidized with sodium periodate to generate aldehyde residues. These may be specifically labeled using a hydrazide–streptavidin derivative through hydrazone bond formation. Subsequent detection may be done using biotinylated enzymes.

Protocol

1. In a test tube, dissolve 160 mg of adipic acid dihydrazide (Aldrich) in 5 ml of 0.1 M sodium phosphate, pH 6. Some heating of the tube under a hot-water tap may be required to help solubilize the compound. Cool to room temperature.
2. Dissolve 50 mg of avidin or streptavidin (Pierce) in the adipic acid dihydrazide solution.
3. Add 160 mg of the water-soluble carbodiimide EDC (Pierce) (Chapter 3, Section 1.1) to the solution, and mix to dissolve.
4. React for 4 h at room temperature.
5. Dialyze against PBS, pH 7.2 to remove excess reagent and reaction by-products.

Hydrazide-activated avidin or streptavidin may be stored as a freeze-dried preparation without loss of activity.

6. Biotinylation Techniques

In addition to preparing the avidin or streptavidin conjugates necessary to develop avidin–biotin-based systems, the process of modifying targeting molecules with a biotin tag is just as critical and forms the other key component of the interacting complex. Since biotin is a relatively small molecule (MW 244.31), coupling it to macromolecules usually can be done without disturbing the activity or binding capability of either the targeting molecule or the biotin handle. Proteins, carbohydrates, lipid molecules, and nucleic acids can be modified to contain one or more biotins able to interact strongly with avidin or streptavidin. The technique of biotinylation is made easy through the commercial availability of a range of different biotin derivatives having a number of important reactivity and property characteristics useful in avidin–biotin chemistry.

Chapter 8, section 3, describes the major biotinylation compounds and their properties. Also provided in that section are suggested protocols for reacting each of these reagents with specific functional groups on macromolecules.

7. Determination of the Level of Biotinylation

It is often important to determine the extent of biotin modification after a biotinylation reaction is complete. Measuring biotin incorporation into macromolecules can aid in optimizing a particular avidin–biotin assay system, and it also can be used to ensure reproducibility in the biotinylation process. The most common method of measuring the degree of biotinylation makes use of the 4′-hydroxyazobenzene-2-carboxylic acid (HABA) dye assay (Green, 1965). In the absence of biotin, the dye is capable of specifically forming noncovalent complexes with avidin at its biotin-binding sites. On binding to avidin in aqueous solution, HABA exhibits a characteristic absorption band at 500 nm ($\varepsilon = 35,500\ M^{-1}cm^{-1}$, expressed as per mole of HABA bound). The addition of biotin to this complex results in displacement of HABA from the binding site, since the affinity constant of the avidin–biotin interaction ($1.3 \times 10^{15}\ M^{-1}$) is much greater than that for avidin–HABA ($6 \times 10^{6}\ M^{-1}$). As HABA is displaced, the absorbance of the complex decreases proportionally. Thus, the amount of biotin present in the solution can be determined by plotting the avidin–HABA absorbance at 500 nm versus the absorbance modulation with increasing concentrations of added biotin. Comparing an unknown biotin-containing sample to this standard response curve can result in the determination of the biotin concentration in the sample.

Since a biotinylated molecule is able to interact with avidin at its biotin binding sites just as strongly as biotin in solution, the degree of biotinylation may be determined using the HABA method as well. Comparison of the response of a biotinylated protein, for example, with a standard curve of various biotin concentrations allows calculation of the molar ratio of biotin incorporation.

Two variations of HABA dye assay for biotinylated proteins are possible. In one approach, the biotinylated protein is digested using the enzyme pronase prior to doing the assay. The digestion process breaks the protein into small fragments, some of which possess biotin modifications. The digestion is done to eliminate any sterically hindered biotinylation sites from not being able to interact with avidin. The second approach merely uses the intact biotinylated protein in the assay, assuming that the HABA assay results then will provide a truer picture of the level of *accessible* biotin sites on the molecule. Pronase addition is obviously not necessary for assessing biotinylated molecules that are not proteins.

The following protocol describes both of these HABA-based tests for determining the level of biotinylation.

Protocol

1. Dissolve avidin in 0.05 M sodium phosphate, 0.15 M NaCl, pH 6, at a concentration of 0.5 mg/ml. A total of 3 ml of the avidin solution is required to create a standard curve using known concentrations of biotin and an additional 3 ml is needed for each sample determination.

2. Dissolve the HABA dye (Sigma) in 10 mM NaOH at a concentration of 2.42 mg/ml (10 mM). Prepare about 100 μl of the HABA solution for each 3-ml portion of avidin solution required.

3. Dissolve the biotinylated protein to be measured in 0.05 M sodium phosphate, 0.15 M NaCl, pH 6, at a concentration of 10–20 mg/ml. The amount required is about 100 μl of sample per determination.

4. Dissolve D-biotin (Pierce) in 0.05 M sodium phosphate, 0.15 M NaCl, pH 6, at a concentration of 0.5 mM.

5. For the proteolytic digestion procedure, dissolve pronase in water at a concentration of 1% (w/v).

6. If pronase digestion of the biotinylated protein is to be done, heat 100 μl of the sample at 56°C for 10 min, then add 10 μl of the pronase solution. Allow the sample to enzymatically digest at room temperature overnight. If no pronase digestion is desired, simply use the biotinylated protein solution prepared in step 3 without further treatment.

7. To construct a standard curve of various biotin concentrations, first zero a spectrophotometer at an absorbance setting of 500 nm with both sample and reference cuvettes filled with 0.05 M sodium phosphate, 0.15 M NaCl, pH 6. Remove the buffer solution from the sample cuvette and add 3 ml of the avidin solution plus 75 μl of the HABA dye solution. Mix well and measure the absorbance of the solution at 500 nm. Next add 2-μl aliquots of the biotin solution to this avidin–HABA solution, mix well after each addition, and measure and record the resultant absorbance at 500 nm. With each addition of biotin, the absorbance of the avidin–HABA complex at 500 nm decreases. The absorbance readings are plotted against the amount of biotin added to construct the standard curve.

8. To measure the response of the biotinylated protein sample, add 3 ml of the avidin solution plus 75 μl of the HABA dye to a cuvette. Mix well and measure the absorbance of the solution at 500 nm. Next add a small amount of sample to this solution and mix. Record the absorbance at 500 nm. If the change in absorbance due to sample addition was not sufficient to obtain a significant difference from the initial avidin–HABA solution, add another portion of sample and measure again. Determine the amount of biotin present in the protein sample by using the standard curve. The number of moles of biotin divided by the moles of protein present gives the number of biotin modifications on each protein molecule.

14

Preparation of Colloidal-Gold-Labeled Proteins

As early as the first decade of this century colloidal gold sols containing particles of less than 10 nm were produced by chemical means (Zsigmondy, 1905). However, the application of these inorganic suspensions to protein labeling did not occur until 1971, when Faulk and Taylor invented the immunogold staining procedure. Since that time, the labeling of targeting molecules, especially proteins, with gold nanoparticles has revolutionized the visualization of cellular or tissue components by electron microscopy (Horisberger et al., 1975; Horisberger, 1979). The silver enhancement technique further broadened the application of gold labeling to include light microscopy (Holgate et al., 1983). The electron-dense and visually dense nature of gold labels also provided excellent detection qualities for such techniques as blotting, flow cytometry, and hybridization assays (Jackson et al., 1990; Gee et al., 1991). Double- or triple-labeling systems have been constructed using immunogold methods in tandem with immunoenzymatic techniques to detect more than one antigen at the same time (Gillitzer et al., 1990).

This chapter discusses the properties of gold particles as well as the common methods of labeling proteins and other biomolecules with them. The cited references should be consulted to obtain protocols for using these protein–gold complexes in assay and detection systems.

1. Properties and Use of Gold Conjugates

Colloidal gold suspensions consist of small granules of this transition metal in a stable, uniform dispersion. Viewed under the light or electron microscope, they appear as solid spheres of dense material. In electron microscopy the gold particles are visible as dense, dark markers usually black in appearance. In light microscopy, they can appear as light dots on a darker background due to the high reflectance of the particles, or as an orange-red coating where they are localized in large conglomerates on cells or tissues. Colloidal gold particles act as efficient nuclei for deposition of silver, thus markedly enhancing their detection under light microscopy (Danscher and Rytter-

Norgaard, 1983). The same silver–gold combination also provides increased sensitivity in blotting applications (Moeremans *et al.*, 1984).

Most preparations of colloidal gold consist of particles varying in diameter from about 5 nm to around 150 nm. The methods of forming small-particle gold suspensions of known diameter are discussed in Section 2.

The labeling of macromolecules with gold particles proceeds through a number of rather poorly understood processes. Preparing stable protein–gold complexes depends on several interactions: (*a*) the electronic attraction between the negatively charged gold particles and the abundant positively charged sites on the protein molecule, (*b*) an adsorption phenomena involving hydrophobic pockets on the protein binding to the metal surface, and (*c*) the potential for covalent binding of gold to free sulfhydryl groups, if present (dative binding) (Fig. 370).

Deryagin and Landau (1941) and Verwey and Overbeek (1948) working independently developed a theory of the behavior of colloidal systems that aids in understanding macromolecular labeling with gold particles. Called the DLVO theory from the initials of the four authors, it views the particles in a sol as consisting of two components producing opposite effects in aqueous suspension. The overlap of the electrical double layer of each particle causes a negative charge on the surface, leading to gold–gold repulsion and stabilizing the sol from aggregation. The other phenomenon is electromagnetic in nature and leads to the potential for Van der Waals attraction between the metal surface and other molecules.

In the colloidal suspension, there exists a balance between the negative-charge repulsion and the attractive forces that could cause coagulation. As particles approach each other, an energy barrier must be traversed to overcome the repulsive character and enter the region of Van der Waals attraction. This barrier can be breached by the addition of electrolytes to the solution that can mask the negative surface charge on each particle. At a certain concentration of electrolytes, the colloid will begin to collapse as the gold particles adsorb onto one another, forming large aggregates and ultimately falling out of suspension.

Electrolyte-mediated coagulation forms the basis for creating all gold conjugates with other molecules. If macromolecules such as proteins are present in the colloidal suspension as the electrolyte concentration is raised to surpass the negative repulsion effects, then adsorption will occur with the protein molecules instead of with other

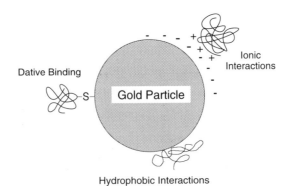

Figure 370   Protein binding to gold particles can occur through several types of interactions.

gold particles. Thus, in place of aggregation and collapse of the suspension, labeling occurs.

The most common electrolyte additions in protein–gold labeling are NaCl or buffer salts. If no macromolecules are present, the addition of NaCl would itself cause gold particle coagulation. The aggregation is accompanied by a color change from orange-red to red-violet or blue (Roth and Binder, 1978), and it may be quantified spectrophotometrically by the change in absorbance at 580 nm (Horisberger *et al,* 1975).

In practice, the addition of a protein to a gold sol will result in spontaneous adsorption on the surface of the gold particles due to electrostatic, hydrophobic, and Van der Waals interactions. To prepare labeled proteins, initially the gold suspension is rapidly mixed while a quantity of protein is added. As the gold is bound to the protein molecules, a decrease in the absorbance at 580 nm occurs as the gold particles become stabilized and less coagulated. To check for the completeness of the adsorption process and to determine whether the gold particles are totally blocked, a portion of the sol can be removed and an aliquot of NaCl added. If coagulation occurs upon addition of salt (increase in A_{580nm}), more protein should be added to stabilize the sol completely. Finally, many protocols further stabilize the colloidal suspension after protein binding by the addition of PEG or an immunochemical blocking agent, such as BSA or a solution of dried milk. These blocking agents completely mask any remaining sites of potential gold–gold or gold–protein interactions, thus preventing aggregation or non-specific binding during assays.

To produce acceptable gold probes, it is often common practice to add the minimum quantity of protein needed to prevent NaCl-induced aggregation plus about 10–20% excess (Horisberger and Rosset, 1977; De Mey *et al.,* 1981). Other investigators have reported that the addition of large excesses of protein to the amount of gold present yields conjugates of higher specific activity (Tokuyasu, 1983; Tinglu *et al.,* 1984). However, there is some evidence that overloading may cause leaching of loosely bound protein (Horisberger and Clerc, 1985).

As in any conjugation procedure, optimization of the ratios of reactants must be done to obtain the best probes. In labeling proteins with gold particles, several parameters should be considered: (a) the pI of the protein, (b) the pH of the adsorption process, and (c) the quantity of protein charged to the labeling reaction. It is generally believed that most proteins can be made to adsorb maximally at or near their isoelectric point (Norde, 1986). This is the pH of net electrical neutrality for a protein, wherein any electrically induced repulsive or attractive forces are balanced. For many proteins, especially antiserum-derived immunoglobulins, the average pI is a broad band encompassing a range of pH values. Thus, a polyclonal antibody preparation may possess an average pI much different than that of a particular purified monoclonal.

Geoghegan (1988) determined that as the pH of the adsorption reaction increased beyond the pI range, the percentage of IgG bound to the gold particles decreased. However, for high-pI immunoglobulins, coupling at basic pH values increased the coupling yield. Geoghegan also noted that the more immunoglobulin that was charged to the adsorption process, the more ended up being coupled, although the percentage bound would decrease.

Thus, while definite standards for the ratio of protein to gold are not universally agreed upon, the efficiency of the process can be improved by following these general guidelines: (a) perform the adsorption reaction at a pH within the range of the pI of the protein being modified or at slightly higher pH, (b) charge an amount of protein to the gold particles that is slightly more (by about 10%) than necessary to maintain colloidal stability upon addition of NaCl, (c) avoid high overloads of protein, since this may promote subsequent leaching of bound material, (d) evaluate the degree of adsorption and the relative coagulation of the gold particles by measuring the absorbance of the solution at 580 nm, and (e) each protein–gold conjugate should be optimized with regard to colloidal stability and retention of activity.

An approximation of the correct amount of protein to be added to a gold sol to maintain stability of the colloid can be done using the following protocol (Slot and Geuze, 1984).

Protocol

1. Add 0.25 ml of the gold suspension to separate tubes containing 25 μl of different concentrations of the protein to be adsorbed. The amount of protein required to stabilize 1 ml of most gold sols is in the microgram range. The protein concentrations should be from about 10 μg/100 μl to about 150 μg/100μl. Mix well.
2. After about 1 min, add 0.25 ml of 10% NaCl to the gold/protein suspension. Mix well.
3. Monitor the stability of the gold sol by its color or by the absorbance of the mixture at 580 nm. As long as the colloid continues to turn blue, and thus forms gold aggregates, with addition of electrolyte, the amount of protein added is not sufficient to stabilize the suspension. This condition translates into a decrease in the absorbance at 580 nm. When the concentration of protein added is enough to stabilize the colloidal suspension the solution no longer changes color (or the absorbance at 580 nm no longer decreases).
4. The amount of protein added at the stabilization point plus 10% should be used to produce the final protein–gold conjugate.

The use of gold probes in detection systems has a number of advantages. The ability to label macromolecules with a range of gold particle sizes makes it possible to visualize the probe under a variety of microscopic conditions. Gold avoids all the disadvantages of radioactive labels, while being much more stable to quenching or fading than fluorescent probes or enzymatically developed substrate chromophores. A gold-labeled tissue, cell, or blot will maintain its record of staining on a permanent basis. Under sufficient magnification, an assessment of the degree of antigen labeling can be made simply by counting the number of gold particles present per unit area of cell or tissue mass. This cannot be done with other labeling systems, since chemical stains develop an amorphous quality that does not allow differentiation of individual molecules. Finally, gold probes are essentially nontoxic and relatively inexpensive to use.

A variety of biological molecules can be labeled with gold particles. Proteins are perhaps the most common gold probes; toxins, antibodies, immunoglobulin binding proteins such as protein A, enzymes, lectins, avidin and streptavidin, lipoproteins, and glycoproteins all have been labeled with colloidal gold to form highly sensitive reagents. In addition, polymers, hormones, carbohydrates, and lipids have been gold-labeled for various applications. Small hapten molecules coadsorbed with adjuvant peptides to gold particles make extraordinary immunogen complexes, producing polyclonal antibody responses having very high titers (Pow and Crook, 1993).

Very small gold particles can even be derivatized to contain specific chemical reactive groups for covalent coupling to macromolecules. For instance, an NHS ester containing gold particle of 1.4 nm is manufactured by Nanoprobes (Stony Brook, NY). Presumably, such derivatives are formed by adsorption of chemically reactive polymers or by dative binding with a sulfhydryl-containing modification reagent.

The following sections discuss the preparation of colloidal gold suspensions of various particle sizes and their use in labeling proteins for detection purposes. Gold-labeled molecules and proteins are available from a number of manufacturers (Janssen, E-Y Labs, and Nanoprobes).

2. Preparation of Mono-disperse Gold Suspensions for Protein Labeling

Mono-disperse colloidal gold suspensions useful for labeling macromolecules can be produced by a variety of chemical methods. Three main procedures have become common for making particles which fall into predictable particle-size ranges. All of them use reductive processes on chloroauric acid ($HAuCl_4$) to create the spheroidal gold particles. In general, the greater the power and concentration of the reducing agent, the smaller the resultant particles.

To create large-particle colloidal gold dispersions, chloroauric acid normally is treated with sodium citrate. The result is a particle range of about 15–150 nm, depending on the concentration of citrate utilized (Horisberger, 1979; Horisberger and Rosset, 1977; Pow and Morris, 1991). Medium-sized gold particles of diameter between 6 and 15 nm (average 12 nm) are formed by treatment with sodium ascorbate as the reductant (although some procedures use trisodium citrate at concentrations higher than that of the sodium citrate used for making large particles) (Horisberger and Tacchini-Vonlanthen, 1983; Albrecht et al., 1989). The smallest gold particles (<5 nm diameter) are created by reduction with either yellow or white phosphorus (Zsigmondy, 1905; Faulk and Taylor, 1971; Horisberger and Rosset, 1977; Pawley and Albrecht, 1988). Particles as small as 2 nm may be created by reduction with sodium borohydride (Bonnard et al., 1984).

The following protocols for creating colloidal gold sols are adaptations from the above-cited articles. To obtain reproducible preparations, extreme care should be taken in making each batch to maintain the same reagent concentrations, temperatures, and times for the reactions. In each preparation, a color change is noted as the chloroauric acid is reduced from its initial state to the final gold sol. The initial color is typically a brown, purple-red, or dark blue, depending on the reductant used and other conditions. The final color of the mono-disperse colloidal gold preparation is typically red.

2.1. Preparation of 2-nm Gold Particle Sols

1. Prepare 1 ml of a 4% $HAuCl_4$ solution in deionized water.
2. Add 375 μl of the chloroauric acid solution plus 500 μl of 0.2 M K_2CO_3 to 100 ml deionized water, cooled on ice to 4°C. Mix well.
3. Dissolve sodium borohydride ($NaBH_4$) (Aldrich) in 5 ml of water at a concentration of 0.5 mg/ml. Prepare fresh.
4. Add five 1-ml aliquots of the sodium borohydride solution to the chloroauric

acid/carbonate suspension with rapid stirring. A color change from bluish-purple to reddish-orange will be noted as the additions take place.

5. Stir for 5 min on ice after the completion of sodium borohydride addition.

2.2. Preparation of 5-nm Gold Particle Sols

1. Prepare 7 ml of a 1% $HAuCl_4$ solution in deionized water.
2. Add 6.25 ml of the chloroauric acid solution plus 5.8 ml of 0.1 M K_2CO_3 to 500 ml deionized water. Mix well.
3. Prepare a saturated solution of white phosphorus in diethylether, then dilute 1 part of the saturated phosphorus solution with 4 parts of diethylether.
4. Add 4.16 ml of the diluted phosphorus solution to the chloroauric acid/carbonate solution with mixing.
5. React at room temperature for 15 min.
6. Bring the mixture to a boil and reflux until the color of the suspension turns from brownish to red. This should take no more than about 5 min.
7. Cool the sol to room temperature.
8. The pH of the suspension will be around 6. Adjustments to more alkaline conditions for adsorbing macromolecules of higher pI may be done by addition of 0.1 M K_2CO_3 with stirring. Monitor pH of the sol using a gel-filled electrode (Orion Research, No. 9115, Cambridge, MA) (Geoghegan et al., 1980). After pH adjustment, the gold should be used immediately for complexing with a protein or other macromolecule.

2.3. Preparation of 12-nm Gold Particle Sols

1. Prepare 5 ml of a 1% $HAuCl_4$ solution in deionized water.
2. Add 4 ml of the chloroauric acid solution plus 4 ml of 0.1 M K_2CO_3 to 100 ml deionized water. Mix well and cool the solution on ice.
3. With rapid mixing of the chloroauric acid/carbonate solution, quickly add 1 ml of a 7% sodium ascorbate solution prepared in water. Maintain the solution cooling in an ice bath. Higher temperatures will create larger particle sizes. The color of the solution at this point will turn to a purple-red.
4. Adjust the volume of the reaction to 400 ml with deionized water.
5. Bring the mixture to a boil and reflux until the color of the suspension turns from purple-red to red.
6. Cool the sol to room temperature.
7. The pH of the suspension will be around 6. Adjustments to more alkaline conditions for adsorbing macromolecules of higher pI may be done by addition of 0.1 M K_2CO_3 with stirring. Monitor pH of the sol using a gel-filled electrode (Geoghegan et al., 1980). After pH adjustment, the gold should be used immediately for complexing with a protein or other macromolecule.

2.4. Preparation of 30-nm Gold Particle Sols

1. Prepare 1 ml of a 4% $HAuCl_4$ solution in deionized water.
2. Add 0.5 ml of the chloroauric acid solution to 200 ml of deionized water and bring to a boil while mixing.

3. Add to the boiling, rapidly mixing solution of chloroauric acid 3 ml of a 1% sodium citrate solution.
4. Reflux for 30 min. The color of the suspension will change from a dark blue to a red as the monodisperse colloidal gold particles are formed.
5. Cool to room temperature.

Any of the particle sols prepared above may be used to adsorb macromolecules to create gold probes. To concentrate the suspensions, the solutions may be filtered through a small-pore filter. Centrifugation also may be done. Each protein–gold complexation should be optimized for the proper amount of protein to add to maintain stability of the colloid. This can be done according to the method described in Section 1.

3. Preparation of Protein A–Gold Complexes

Protein A–gold probes (as well as other immunoglobulin binding proteins adsorbed to gold) have been used to visualize antibody binding to antigenic sites in tissue sections, cells, and blots (Jemmerson and Agre, 1987; Yokota, 1988; Bendayan and Garzon, 1988; Hearn, 1987; Lethias *et al.*, 1987; Herbener, 1989; Roth *et al.*, 1989; Bendayan, 1989; Stump *et al.*, 1989). Gold labeling of immunoglobulin binding proteins provides "universal" probes for detection of any antibody–antigen interaction (Fig. 371). Thus, only one gold-labeled reagent need be prepared to visualize many different immunochemical procedures. This avoids the need to make antibody–gold probes for each specific immunoglobulin used.

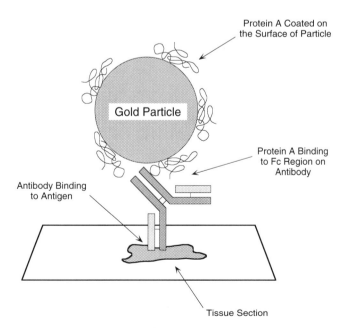

Figure 371 Antigens may be detected in cells or tissue sections through the use of protein A-coated gold particles. The binding of a specific antibody to its target antigen can be localized by the immunoglobulin binding capability of protein A, which occurs in the Fc region of the antibody.

Protocol

1. Determine the minimum amount of protein A required to stabilize the colloidal gold sol being used. The colloidal suspension should be adjusted, if needed, with 0.1 M K_2CO_3 to pH 6–7. Measure the pH of the sol using a gel-filled electrode. Determining the stabilization amount of protein A can be done according to the method described in Section 1.

2. Mix a stabilizing amount of protein A plus an additional 10% with the appropriate volume of colloidal gold. For example, Herbener (1989) mixed 10 ml of a 14-nm gold particle sol at pH 6.9 with 0.3 mg of protein A dissolved in 0.2 ml water. Mix well.

3. After 1 min, add 250 μl of 1% PEG (molecular weight 20,000) per 10 ml of gold sol used. The PEG helps to further stabilize the sol against aggregation.

4. Stir for an additional 5 min.

5. To remove excess protein A, centrifuge the preparation at a minimum of 50,000 g for 30 min to several hours (4°C), depending on the size of the particles and the amount of solution. Discard the supernatant, and resuspend the protein A–gold pellet in 0.01 M sodium phosphate, pH 7.4, containing 1% PEG.

4. Preparation of Antibody–Gold Complexes

Immunocytochemical staining with antibody–gold probes is a powerful way to detect, localize, and quantify antigen molecules in tissue sections and cells (Fig. 372). Metabolic processes can be followed, epitope mapping of the structural characteristics of macromolecules can be done, and detection of pathogens or other foreign substances within cells can be accomplished using gold-labeled antibodies (van den Brink *et al.*, 1990; Cramer *et al.*, 1989; De Waele *et al.*, 1989; Nielsen *et al.*, 1989; Martinez-Ramon *et al.*, 1990; Albrecht *et al.*, 1989; Ellis *et al.*, 1988).

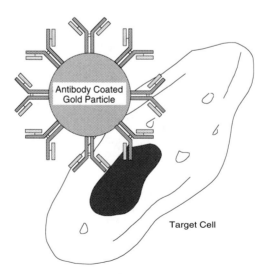

Figure 372 Antibodies coated on colloidal gold particles can be used to detect specific antigens in cells.

The optimal coupling pH for an antibody should be determined by measurement of the relative pI range of the immunoglobulin. Many antibodies, however, adsorb best at a pH of 8–9. The optimal level of protein addition to the gold sol to prevent aggregation should be determined according to the method of Section 1. In addition, BSA is often added instead of PEG (see the protein A coupling procedure, described previously) to further stabilize the antibody–gold suspension.

Protocol

1. Determine the minimum amount of antibody required to stabilize the colloidal gold sol being used. The colloidal suspension should be adjusted, if needed, with 0.1 M K_2CO_3 or NaOH to pH 8–9. Measure the pH of the sol using a gel-filled electrode. Determining the stabilization amount of antibody can be done according to the method described in Section 1.
2. Mix a stabilizing amount of antibody plus an additional 10% with the appropriate volume of colloidal gold. For example, Geoghegan (1988) found that an addition of 10–14 µg of antibody per milliliter of gold colloid resulted in stable preparations. Mix well after addition of antibody to the gold suspension.
3. After 1 min, add a quantity of 10% BSA to bring the concentration to 0.25% in the antibody–gold suspension. The BSA helps to further stabilize the sol against aggregation and also blocks nonspecific binding sites. Alternatively, PEG may be added according to step 3 of Section 3.
4. Stir for an additional 5 min.
5. To remove excess IgG, centrifuge the preparation at a minimum of 50,000 g for 30 min to several hours (4°C), depending on the size of the particles and the amount of solution. Discard the supernatant and resuspend the antibody–gold pellet in 0.01 M sodium phosphate, pH 7.4, containing 0.25% BSA (or 1% PEG, as desired).

5. Preparation of Lectin–Gold Complexes

Lectins, or proteins with specific binding sites for carbohydrates, can be used as targeting molecules to localize particular glycoconjugates such as glycoproteins or glycolipids on cell surfaces (Fig. 373). Labeled with gold particles, lectins are important probes for detection of cell surface components and intracellular receptors and in immunological or biochemical assay procedures (Bog-Hansen *et al.*, 1978; Kimura *et al.*, 1979; Nicolson, 1978; Roth, 1983; Benhamou *et al.*, 1988; Nakajima *et al.*, 1988).

The following generalized protocol is an adaptation for the labeling of 15-nm gold particles with *Aplysia* gonad lectin, as described by Benhamou *et al.* (1988). Each lectin–gold preparation will have its own unique pH optimum and ratio of lectin-to-gold for the absorption process.

Protocol

1. Determine the minimum amount of lectin required to stabilize the colloidal gold sol being used. The colloidal suspension should be adjusted, if needed, with 0.1 M K_2CO_3 or NaOH to a pH equal to or slightly above the pI of the lectin being

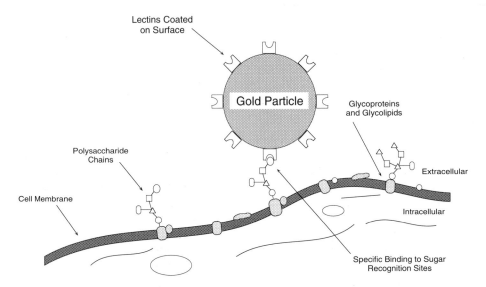

Figure 373 Lectins coated on gold particles can be used to detect specific carbohydrate sequences in cell surface glycoconjugates.

used. For *Aplysia* gonad lectin, the optimal pH for adsorption was determined to be 9.5. Nakajima *et al.* (1988) include pI conditions for a number of different lectins. Measure the pH of the sol using a gel-filled electrode. Determining the stabilization amount of lectin can be done according to the method described in Section 1.

2. Mix a stabilizing amount of lectin plus an additional 10% with the appropriate volume of colloidal gold. For example, Benhamou *et al.* (1988) found that an addition of 5 μg of lectin per milliliter of gold colloid resulted in stable preparations. However, in their final lectin–gold preparation a fivefold increase in this ratio (25 μg lectin/ml gold) was used to stabilize fully the sol. Mix well after addition of lectin to the gold suspension.

3. After 1 min, add 250 μl of 1% polyethylene glycol (PEG; molecular weight 20,000) per 10 ml of gold sol used. The PEG helps to further stabilize the sol against aggregation.

4. Stir for an additional 5 min.

5. To remove excess lectin (particularly important if the fivefold excess ratio is used), centrifuge the preparation at a minimum of 50,000 *g* for 30 min to several hours (4°C), depending on the size of the particles and the amount of solution. Discard the supernatant and resuspend the lectin–gold pellet in 0.01 *M* sodium phosphate, pH 7.4, containing 1% PEG.

6. Preparation of Avidin–Gold or Streptavidin–Gold Complexes

Avidin–gold or streptavidin–gold conjugates can be used to detect, localize, or quantify the binding of biotinylated molecules in cells, tissue sections, or blots (Morris and Saelinger, 1984; Bonnard *et al.*, 1984; Gillitzer *et al.*, 1990; Bronckers *et al.*, 1987)

(Fig. 374). Use of these reagents is similar to the use of protein A–gold complexes in detecting immunoglobulins (Section 3) in that they are "universal" for detecting any biotin-labeled molecules. Thus, targeting molecules need not be directly modified with gold, only biotinylated so that they are able to interact with avidin–gold or streptavidin–gold conjugates. See Chapter 8, Section 3, for biotinylation reagents and protocols that can be used to add a biotin tag to macromolecules. Also see Chapter 13 for additional information on avidin–biotin techniques, including conjugation protocols.

The following protocol is based on the method of Morris and Saelinger (1984) for the labeling of succinylated avidin with gold particles of 5.2-nm diameter. Succinylated avidin was used to reduce the pI of the protein, thus eliminating nonspecific binding due to the strong positive charge of the native tetramer.

Protocol

1. Prepare a 200 ml gold sol by using white phosphorus reduction as described in Section 2.
2. Prepare 5 ml of a 1 mg/ml succinylated avidin solution by dissolving the protein in 50 mM sodium phosphate, pH 7.5.
3. With stirring, add the succinylated avidin solution to the colloidal gold suspension at room temperature.
4. React for 30 min with constant mixing.
5. Remove excess protein by centrifugation at a minimum of 50,000 g for several hours.

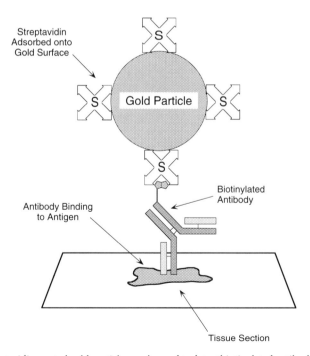

Figure 374   Streptavidin-coated gold particles can be used to detect biotinylated antibodies that are bound to specific antigenic determinants.

6. Suspend the succinylated avidin–gold pellet in 50 mM Tris, 0.15 M NaCl, pH 7.5, containing 0.5 mg/ml PEG (MW 20,000).
7. Centrifuge again under the same conditions to ensure complete removal of nonadsorbed protein.
8. Resuspend the pellet in Tris buffer containing PEG and store at 4°C.

A similar protocol has been used by Bonnard *et al.* (1984) in the preparation of streptavidin–gold probes.

Protocol

1. Prepare 20 ml of a gold sol by the white phosphorus method described in Section 2.
2. Dissolve streptavidin in 0.1 M sodium phosphate buffer, pH 7.4, at a concentration of 1 mg/ml.
3. With stirring add the streptavidin solution to the gold suspension. Immediately add 200 μl of 1 M sodium bicarbonate.
4. React for 10 min at room temperature.
5. To further stabilize the colloid, add 200 μl of 2% PEG 6000.
6. Centrifuge the streptavidin–gold suspension at a minimum of 50,000 g for several hours to remove excess protein solution.
7. Resuspend the pellet in 0.1 M sodium phosphate, 0.02% PEG, pH 7.4, and store at 4°C.

15
___

Modification with Synthetic Polymers

Modification or attachment of proteins or other molecules with synthetic polymers can provide many benefits for both *in vivo* and *in vitro* applications. Covalent coupling of polymers to large macromolecules can alter their surface and solubility properties, creating increased water solubility or even organic solvent solubility for molecules normally sparingly miscible in such environments. Polymer modification of foreign molecules can provide increased biocompatibility, reducing the immune response, increasing *in vivo* stability, and delaying clearance by the reticuloendothelial system. Modification of enzymes with polymers can dramatically enhance their stability in solution. Polymer attachment can provide cryoprotection for proteins sensitive to freezing. Polymers with multivalent reactive sites can be used to couple numerous small molecules for creating pharmacologically active agents that possess long half-lives in biological systems. Similar complexes can be formed to create highly potent immunogens consisting of hapten–polymer conjugates for induction of an antibody response toward the hapten. Polymer modification of surfaces can effectively mask the intrinsic character of the surface and thus prevent nonspecific protein adsorption. Finally, multifunctional polymers can serve as extended cross-linking agents for the conjugation of more than one molecule of one protein to multiple numbers of a second molecule, creating large complexes with increased sensitivity or activity in detecting or acting upon target analytes.

Many polymers have been studied for their usefulness in producing pharmacologically active complexes with proteins or drugs. Synthetic and natural polymers such as polysaccharides, poly(L-lysine) and other poly(amino acids), poly(vinyl alcohols), polyvinylpyrrolidinones, poly(acrylic acid) derivatives, various polyurethanes, and polyphosphazenes have been coupled to with a diversity of substances to explore their properties (Duncan and Kopecek, 1984; Braatz *et al.*, 1993). Copolymer preparations of two monomers also have been tried (Nathan *et al.*, 1993).

The two polymers most often used in these applications are dextran and PEG. Both polymers consist of repeating units of a single monomer—glucose in the case of dextran and an ethylene oxide basic unit in the case of PEG. The polymers may be composed of linear strands (PEG or dextran) or branched constructs (dextran). An additional similarity is that both of them possess hydroxyl and ether linkages, lending hydrophilicity and water solubility to the molecules. Dextran and PEG can be activated through their hydroxyl groups by a number of chemical methods to allow efficient coupling of other molecules. Dextran can be activated at multiple sites throughout its chain, since each monomer contains hydroxyl resides. PEG, by contrast, only has hydroxyls at the termini of each linear strand. Derivatives of both polymers are commercially available.

The following sections discuss the major properties and conjugation chemistries associated with the use of these polymers in modifying or conjugating proteins and other molecules.

1. Protein Modification with Activated Polyethylene Glycols

Since the first reports by Abuchowski and co-workers in 1977 (Abuchowski *et al.*, 1977b) concerning the alteration of immunological properties toward BSA that had been modified with PEG, the interest in polymer modification of biological molecules has grown almost exponentially. PEG coupled to other molecules can be used for altering solubility characteristics in aqueous or organic solvents (Inada *et al.*, 1986), for modulation of the immune response (Delgado *et al.*, 1992), to increase the stability of proteins in solution (Berger and Pizzo, 1988), to enhance the half-life of substances *in vivo* (Knauf *et al.*, 1988), to aid in penetrating cell membranes, to alter pharmacological properties (Dunn and Ottenbrite, 1991), to increase biocompatibility, especially toward implanted foreign substances, and to prevent protein adsorption to surfaces.

PEG consists of repeating units of ethylene oxide that terminate in hydroxyl groups on either end of a linear chain. It is made from the anionic polymerization of ethylene oxide, resulting in the formation of polymer strands of various potential molecular weights, depending on the polymerization conditions. Most forms of PEG useful in bioconjugate applications have molecular weights less than 20,000 and are soluble both in aqueous solution and in many organic solvents.

PEG
Poly(ethylene glycol)

mPEG
Monomethoxy-
Poly(ethylene glycol)

Since the polymer backbone of PEG is not of biological origin, it is not readily degraded by mammalian enzymes (although some bacterial enzymes will break it down). This property results in only slow degradation of the polymer when used *in vivo*, thus extending the half-life of modified substances. PEG modification serves to mask any molecule to which it is coupled—the "pegylated" molecule being protected from immediate breakdown or from being complexed and inactivated by immunoglobulins in the bloodstream.

The properties of PEG in solution are especially unusual, frequently displaying amphiphilic tendencies, having the ability to solubilize both in aqueous layers and in hydrophobic membranes or organic phases. The partitioning quality of PEG across membranes is important in aiding the formation of hybridomas in the production monoclonal antibodies (Goding, 1986b). The partitioning characteristics of PEG also create the ability to use it in aqueous two-phase systems for the purification of biological molecules (Johansson, 1992; Tjerneld, 1992).

PEG in solution is a highly mobile molecule that creates a large exclusion volume for its molecular weight, much larger in fact than proteins of comparable size. Whether in solution or attached to other insoluble supports or surfaces, PEG has a tendency to exclude other polymers. This property forms a protein-rejecting region that is effective in preventing nonspecific protein binding (Bergström *et al.*, 1992). Conjugation with PEG can create the same exclusion effects surrounding a macromolecule, preventing interaction between a ligand and its target (Klibanov *et al.*, 1991), an enzyme and its substrate (Berger and Pizzo, 1988), or the immune system and a foreign substance (Davis *et al*, 1979). Thus, PEG-modified molecules display low immunogenicity, have good resistance to proteolytic digestion, and survive in the bloodstream for extended periods (Abuchowski *et al.*, 1977a; Dreborg and Akerblom, 1990).

PEG can be conjugated to other molecules through its two hydroxyl groups at the ends of each linear chain. This process is typically done by the creation of a reactive electrophilic intermediate that is capable of spontaneously coupling to nucleophilic residues on a second molecule. To prevent the potential for cross-linking when using a bifunctional polymer, monofunctional PEG polymers can be used that contain one end of each chain blocked with a methyl ether group. Monomethoxypolyethylene glycol (mPEG) contains only one hydroxyl per chain, thus limiting activation and coupling to one site and preventing the cross-linking and polymerization of modified molecules.

1.1. Trichloro-*s*-triazine Activation and Coupling

The most common activation methods for PEG create amine-reactive derivatives that can form amide or secondary amine linkages with proteins and other amine-containing molecules. The oldest method of PEG activation is through the use of trichloro-*s*-triazine (TsT; cyanuric chloride) (Abuchowski *et al.*, 1977a,b). TsT is a symmetrical heterocyclic compound containing three reactive acyl-like chlorines. This reagent and its derivatives are extensively used in industrial applications to form strong covalent bonds between dye molecules and fabrics. The compound also has been used to activate affinity chromatography supports for the coupling of amine-containing ligands (Finlay *et al.*, 1978). Reaction of the TsT with PEG results in the formation of an activated derivative with an ether bond to the hydroxyl group of the polymer. If mPEG is used, TsT activation will be restricted to the one free hydroxyl, thus forming a monovalent intermediate that can be coupled to proteins without polymerization (Fig. 375).

The three reactive chlorines on TsT have dramatically different reactivities toward nucleophiles in aqueous solution. The first chlorine is reactive toward hydroxyls as well as primary and secondary amine groups at 4°C and a pH of 9 (Mumtaz and Bachhawat, 1991; Abuchowski *et al.*, 1977a). Once the first chlorine is coupled, the second one requires at least room temperature conditions at the same pH to efficiently react. If two chlorines are conjugated to nucleophilic groups, the third is even more difficult to couple, requiring at least 80°C. After activation of mPEG with TsT, it is therefore, for all practical purposes, only possible to couple one additional component to the triazine ring.

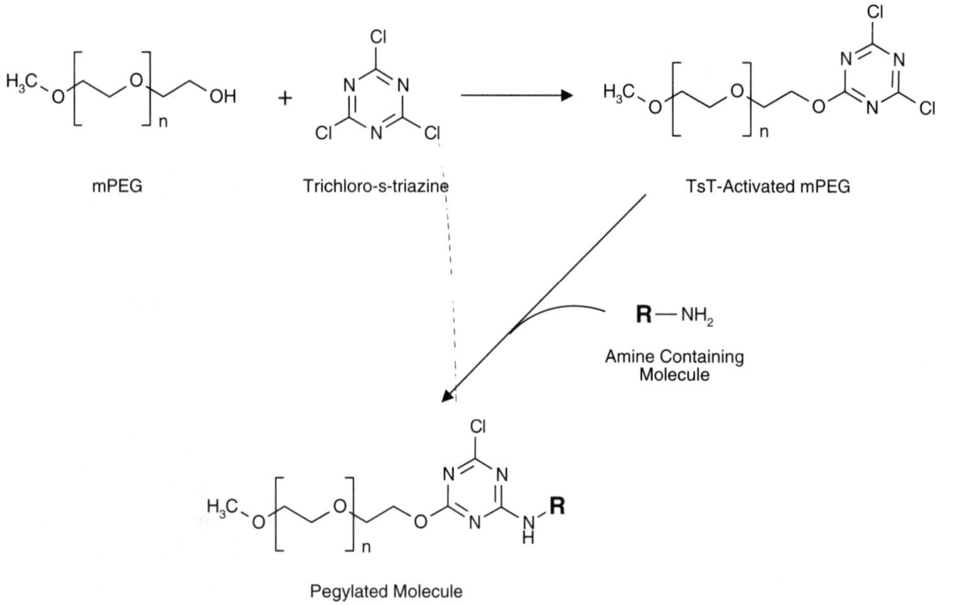

mPEG   Trichloro-s-triazine   TsT-Activated mPEG

R—NH$_2$

Amine Containing
Molecule

Pegylated Molecule

Figure 375   mPEG polymers may be activated by trichloro-*s*-triazine for the modification of amine-containing molecules.

TsT activation provides a simple route to an amine-reactive PEG derivative and has been used extensively as an activation method for modifying proteins (Wieder *et al.*, 1979; Zalipsky and Lee, 1992; Gotoh *et al.*, 1993). The modification of primary amine-containing molecules such as proteins is pH dependent. At physiological pH values, the reaction will proceed slower than in a more alkaline pH environment. Optimal derivatization efficiency is reached at conditions equal to or above pH 9. However, TsT reactivity is not exclusive toward amines. TsT–mPEG modification of proteins can result in modifying other nucleophilic groups such as sulfhydryls and the phenolate ring of tyrosine. In addition, there is potential for toxicity associated with TsT and its derivatives—an especially important consideration for *in vivo* use.

The following protocol for mPEG activation using TsT and its coupling to proteins is based on the protocols of Abuchowski *et al.* (1977b) and Gotoh *et al.* (1993).

Protocol for the Activation of mPEG with TsT

Note: All operations should be done in a fume hood. Dispose of hazardous waste according to EPA guidelines.

1. Dissolve 5.5 g of TsT in 400 ml of anhydrous benzene that contains 10 g of anhydrous sodium carbonate.
2. Add to the TsT solution, 50 g of mPEG-5000 (monomethoxypolyethylene glycol having a molecular weight of 5000). Mix well to dissolve.
3. React overnight at room temperature with stirring.
4. Filter the solution through a glass-fiber filter pad and slowly add, with stirring, 600 ml of petroleum ether (bp 35–60°C).

5. Collect the precipitated product by filtration and redissolve it in 400 ml of benzene. Repeat steps 4–5 several times to ensure complete removal of unreacted TsT. The residual TsT may be detected by HPLC using a 250 × 3.2-mm LiChrosorb (5-μm particle size) column from E. Merck. The separation is done using a mobile phase of hexane, and peaks are detected with a UV detector.
6. Remove excess solvents by rotary evaporation. The TsT–mPEG should be used immediately or stored in anhydrous conditions at 4°C.

Protocol for Coupling of TsT–mPEG to Proteins

1. Dissolve the protein to be modified with TsT–mPEG in ice-cold 0.1 M sodium borate, pH 9.4, at a concentration of 2–10 mg/ml. Other buffers at lower pH values (down to pH 7.2) can be used and still obtain modification, but the yield will be less. Avoid amine-containing buffers such as Tris or the presence of sulfhydryl-containing compounds, such as disulfide reductants.
2. Slowly add TsT–mPEG to the protein solution at a level of at least a fivefold molar excess over the desired modification level. For example, Gotoh et al. (1993) added 100 mg of TsT–mPEG-5000 to 19 mg of protein dissolved in 6 ml of buffer. Add the polymer over a period of about 15 min with stirring at 4°C.
3. React for 1 h at 4°C.
4. Remove excess TsT–mPEG by dialysis or gel filtration using a column of Sephacryl S-300.

1.2. NHS Ester and NHS Carbonate Activation and Coupling

Carboxylate groups activated with NHS esters are highly reactive toward amine nucleophiles. In the mid-1970s, NHS esters were introduced as reactive ends of cross-linking reagents (Bragg and Hou, 1975; Lomant and Fairbanks, 1976). Their excellent reactivity at physiological pH quickly established NHS esters as the major amine-coupling chemistry in bioconjugate chemistry.

NHS ester-containing compounds react with nucleophiles to release the NHS leaving group and form an acylated product (Chapter 2, Section 1.4). The reaction of such esters with sulfhydryl or hydroxyl groups is possible, but does not yield stable conjugates, forming thioesters or ester linkages. Both of these bonds typically hydrolyze in aqueous environments. Histidine side-chain nitrogens of the imidazolyl ring also may be acylated with an NHS ester reagent, but they too hydrolyze rapidly (Cuatrecasas and Parikh, 1972). Reaction with primary and secondary amines, however, creates stable amide and imide linkages, respectively, that do not readily break down. In protein molecules NHS ester groups primarily react with the α-amines at the N-terminals and the ε-amines of lysine side chains, due to their relative abundance.

PEG contains no carboxylate groups in its native state, but can be modified to possess them by reaction with anhydride compounds. Either PEG or mPEG may be acylated with anhydrides to yield ester derivatives terminating in free carboxylate groups. Modification of PEG with succinic anhydride or glutaric anhydride gives bis-modified products having carboxylates at both ends. Modification of mPEG yields the monosubstituted derivative containing a single carboxylate. Creation of the succinimidyl succinate and succinimidyl glutarate derivative of PEG was described by

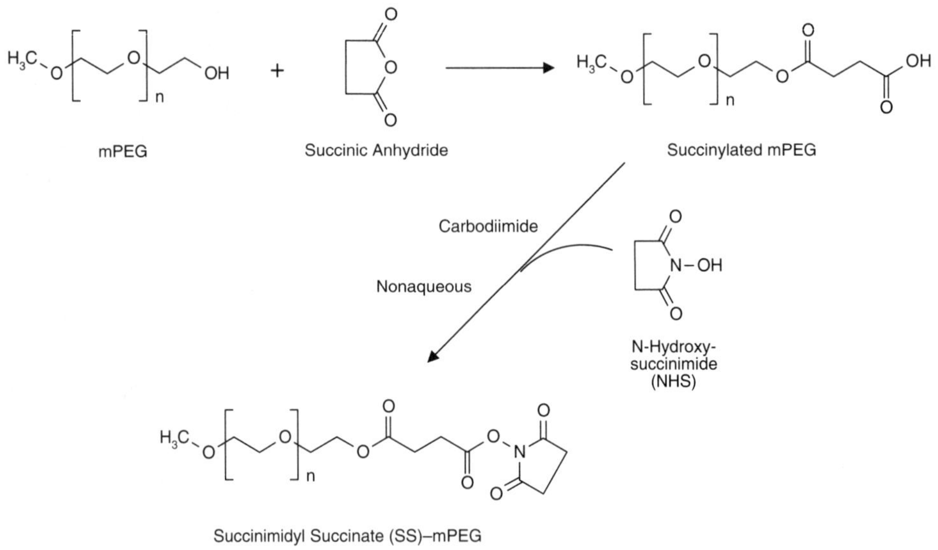

Figure 376 mPEG may be derivatized with succinic anhydride to produce a carboxylate end. A reactive NHS ester may be formed from this derivative by use of a carbodiimide-mediated reaction under nonaqueous conditions. The succinimidyl succinate–mPEG is highly reactive toward amine nucleophiles.

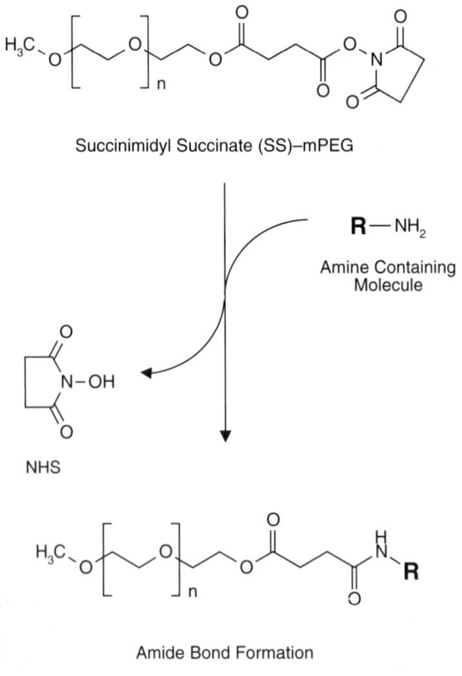

Figure 377 Succinimidyl succinate–mPEG may be used to modify amine-containing molecules to form amide bond derivatives. The ester bond of the succinylated mPEG, however, is subject to hydrolysis.

Abuchowski *et al.* (1984). A method for the succinylation of mPEG can be found in Section 1.3. Subsequent formation of the NHS ester derivatives of these acylated PEG compounds produce a highly reactive polymer that can be used to modify amine-containing molecules under mild conditions and with excellent yields (Figs. 376 and 377). The main deficiency of the succinimidyl succinate or succinimidyl glutarate activation procedures is the potential for hydrolysis of the ester bond formed by acylation of the hydroxyl groups of PEG.

A modification of the anhydride-acylation route to obtaining reactive NHS ester–PEG compounds was introduced by Zalipsky *et al.* (1991, 1992). In this approach, the terminal hydroxyl group of mPEG is treated with phosgene to give a reactive intermediate, an mPEG-chloroformate compound. Next, the addition of NHS gives the succinimidyl carbonate derivative (Fig. 378). Nucleophiles, such as the primary amino groups of proteins, can react with the succinimidyl carbonate functional groups to give stable carbamate (aliphatic urethane) bonds (Fig. 379). The linkage is identical to that obtained through CDI activation of hydroxyl groups with subsequent coupling of amines (Chapter 3, Section 3, and this chapter, Section 1.4). However, the reactivity of the succinimidyl carbonate is much greater than that of the imidazole carbamate formed as the active species in CDI activation.

Unlike the succinimidyl succinate or succinimidyl glutarate activation methods, succinimidyl carbonate chemistry does not suffer from the presence of a labile ester bond. The intermediate carbonate may hydrolyze in aqueous solution to release NHS and CO_2, essentially regenerating the underivatized PEG hydroxyl. After coupling to amine-containing molecules, however, the resultant carbamate linkage stabilizes the chemistry to the point that a modified molecule will not lose PEG by hydrolytic cleavage. For these reasons, the succinimidyl carbonate method of PEG activation and

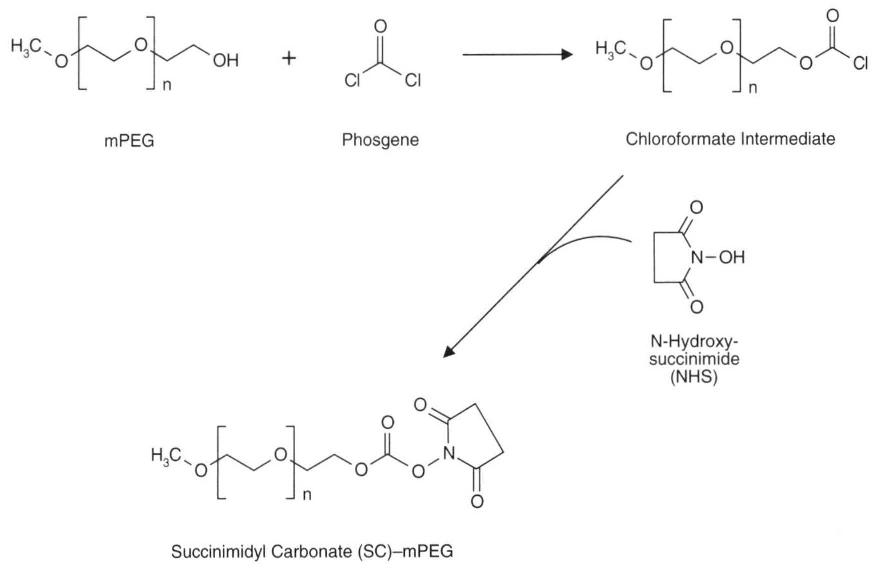

mPEG   Phosgene   Chloroformate Intermediate

N-Hydroxy-
succinimide
(NHS)

Succinimidyl Carbonate (SC)–mPEG

Figure 378 A succinimidyl carbonate derivative of mPEG was first prepared through the use of phosgene to form a chloroformate intermediate. Reaction with NHS gives the amine-reactive succinimidyl carbonate–mPEG.

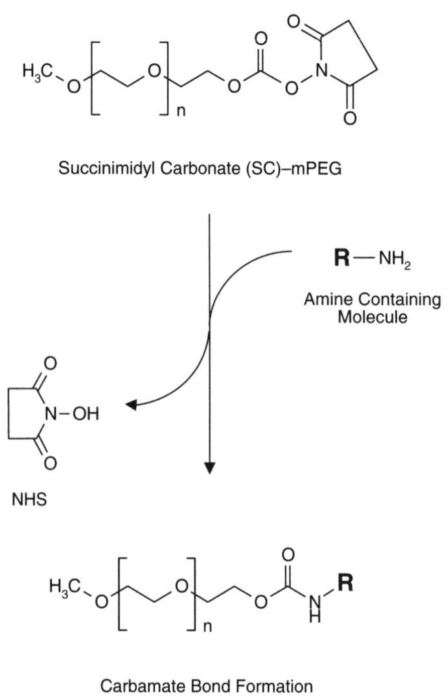

Succinimidyl Carbonate (SC)–mPEG

R—NH₂

Amine Containing
Molecule

N–OH

NHS

Carbamate Bond Formation

Figure 379   Succinimidyl carbonate–mPEG can be used to modify amine-containing molecules to form stable carbamate linkages.

coupling has become the chemical reaction of choice for attaching the polymer to amine-containing proteins and other molecules.

A modification of the Zalipsky method by Miron and Wilchek (1993) simplifies the creation of the succinimidyl carbonate-activated species. Instead of using highly toxic phosgene to form a chloroformate intermediate and then reacting with NHS, the new procedure utilizes either N-hydroxysuccinimidyl chloroformate or N,N'-disuccinimidyl carbonate (DSC; Chapter 4, Section 1.7) to produce the succinimidyl carbonate–PEG in one step (Fig. 380). Since both activation reagents are commercially available, creating an amine-reactive PEG derivative has never been easier.

The following procedure is based on the Miron and Wilchek (1993) modification of the Zalipsky method.

Protocol for the Activation of PEG with N-Succinimidyl
Chloroformate or N,N'-Disuccinimidyl Carbonate
Caution: The steps using flammable solvents, especially diethyl ether, should be done in a fume hood.

1. Dissolve 5 g PEG or mPEG (MW 5000; 1 mmol) in 25 ml of dry dioxane. Heating in a water bath may be necessary to solubilize fully the polymer. Cool to room temperature.
2. Dissolve 6 mmol of either N-succinimidyl chloroformate or N,N'-disuccinimidyl carbonate (Aldrich) in 10 ml of dry acetone.

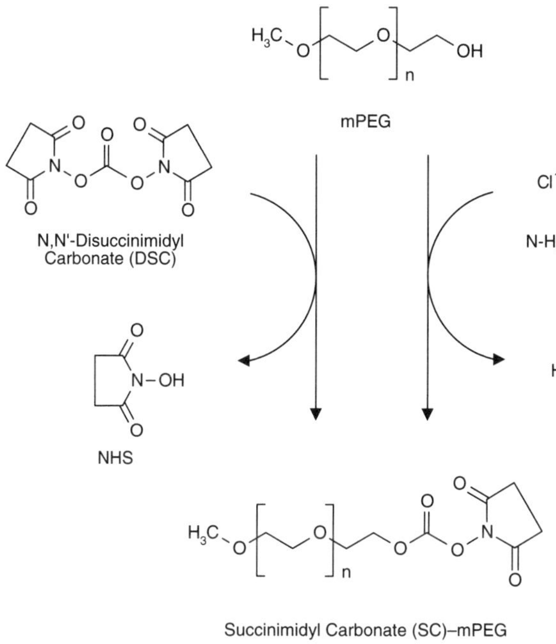

Figure 380 An alternative route to an succinimidyl carbonate derivative of mPEG can be accomplished by the reaction of the terminal hydroxyl group of the polymer with either *N,N'*-disuccinimidyl carbonate or *N*-hydroxysuccinimidyl chloroformate.

3. Dissolve 6 mmol of 4-(dimethylamino)pyridine in 10 ml of dry acetone (catalyst).
4. With stirring, add the solution prepared in step 2 to the PEG solution prepared in step 1. Next, slowly add the solution prepared in step 3.
5. React for 2 h if activating with *N*-succinimidyl chloroformate or 6 h if using *N,N'*-disuccinimidyl carbonate. Maintain stirring with a magnetic stirring bar.
6. If *N*-succinimidyl chloroformate was used, filter out the white precipitate of 4-(dimethylamino)pyridine hydrochloride using a glass-fiber filter pad. Collect the supernatant.
7. For either activation chemistry, precipitate the succinimidyl carbonate (SC)–PEG formed by addition of diethyl ether until no further precipitation is observed (typically 3–4 vol of solvent).
8. Redissolve the precipitated product in acetone and precipitate again using diethyl ether. Repeat at least once more to remove completely excess reactants.
9. Dry the SC–PEG and store at 4°C.

Protocol for the Coupling of SC–mPEG to Proteins

1. Dissolve the protein to be "pegylated" in cold 0.1 *M* sodium phosphate, pH 7.5, at a concentration of 1–10 mg/ml.
2. With stirring, add a quantity of SC–mPEG to the protein solution at the molar ratio of polymer to protein desired. The ratio of activated polymer addition typically is expressed versus the molar quantity of primary amines present on

the protein being modified. Ratios of SC–PEG to amines between 0.3:1 and 8:1 were investigated by Miron and Wilchek (1993) for the derivatization of egg white lysozyme. The greater the ratio of activated polymer to protein, the higher the molecular weight of the resultant complex. Experiments may have to be done using a number of different reaction ratios to determine the optimal pegylation level for a particular protein.

3. React overnight at 4°C.
4. Remove excess SC–PEG by dialysis or gel filtration.

1.3. Carbodiimide Coupling of Carboxylate–PEG Derivatives

PEG contains only hydroxyl functional groups in its native state that need to be activated or modified in some manner to allow efficient conjugation to other molecules. These hydroxyls can be modified to possess carboxylates by reaction with anhydride compounds. Acylation of PEG with succinic anhydride or glutaric anhydride gives *bis*-modified products having carboxylates at both ends. Modification of mPEG yields the monosubstituted derivative containing a single carboxylate. Creation of these derivatives was first described by Abuchowski *et al.* (1984). Once the carboxylate–PEG modification is formed, it can be used to couple directly to amine-containing molecules by use of the carbodiimide reaction (Chapter 3, Section 1).

A carbodiimide may be used to activate the carboxylates to highly reactive o-acylisourea intermediates. When generated in the presence of an amine-containing protein or other molecule, these active esters will react with the nucleophiles to give amide bond derivatives (Fig. 381). Atassi and Manshouri (1991) used this technique to pegylate various peptides. In this instance, the reaction was carried out in DMF due to the solubility of the peptides in this solvent. The organic-soluble carbodiimides DCC (Chapter 3, Section 1.4) or DIC (Chapter 3, Section 1.5) were used to perform the conjugation. However, aqueous phase reactions can be done using this approach just as easily as organic-based conjugations if the water-soluble reagent EDC is employed (Chapter 3, Section 1.1). The general protocols for using carbodiimides outlined in the referenced sections may be used to conjugate a carboxylate-containing PEG derivative to an amine-containing protein or other molecule. The formation of a carboxylate-containing PEG derivative can be done according to the following protocol (adapted from Atassi and Manshouri, 1991).

Protocol

1. Dissolve 1 g of mPEG (MW 5000) in 5 ml of anhydrous pyridine by heating to 50°C.

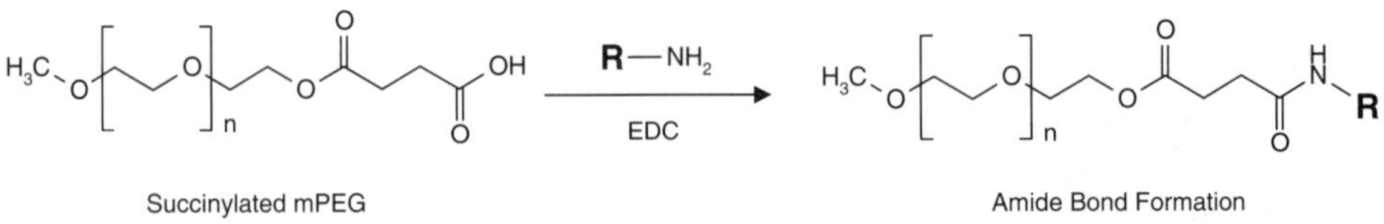

Succinylated mPEG   Amide Bond Formation

Figure 381   A succinylated mPEG derivative may be coupled to amine-containing molecules using a carbodiimide reaction to form an amide bond.

2. To the stirring mPEG solution, add several 0.5-g aliquots of solid succinic anhydride over a period of several hours.
3. React for a further 2 h at 50°C.
4. Evaporate the pyridine solvent by using a flash evaporator or a rotary evaporator under vacuum.
5. Redissolve the residue in water (the solution may have to be heated to fully dissolve) and again evaporate to dryness. Repeat until the odor of pyridine is nearly gone.
6. Remove remaining reactants by dialysis against water using a membrane having a molecular weight 1000 cut-off.

1.4. CDI Activation and Coupling

N,N'-Carbonyldiimidazole is a highly reactive carbonylating compound that was first shown to be an excellent amide bond-forming agent in peptide synthesis (Paul and Anderson, 1962). Later it was used to activate both carboxylic groups and hydroxyls in the immobilization of amine-containing ligands (Bartling et al., 1973; Hearn, 1987).

The activation of a carboxylate group with CDI proceeds to give an intermediate imide with imidazole as the active leaving group. In the presence of a primary amine-containing compound, the nucleophile attacks the electron-deficient carbonyl, displacing the imidazole and forming a stable amide bond.

For hydroxyl-containing compounds, CDI will react to form an intermediate imidazolyl carbamate that in turn can react with N-nucleophiles to give an N-alkyl carbamate linkage. Proteins normally couple through their N-terminals (α-amine) and lysine side chain (ε-amine) functional groups. The final bond is an uncharged, urethane-like derivative having excellent chemical stability (Fig. 382). The result of CDI activation of PEG and subsequent coupling to a protein or other amine-containing molecule is a linkage identical to that obtained using succinimidyl carbonate chemistry, described previously (Section 1.2).

CDI-activated PEG is stable for years in a dried state or in organic solvents devoid of water. The activated polymer also will have an excellent half-life to hydrolysis even in the coupling environment. Unlike some activation chemical reactions that degrade rapidly and have half-lives on the order of minutes, imidazole carbamates have half-lives measured in hours. For instance, an agarose chromatography support activated with CDI will take up to 30 h at pH 8.5–9 for complete loss of activity. The hydrolysis of CDI–PEG derivatives causes the release of CO_2 and imidazole. The hydrolyzed product thus reverts back to the original hydroxylic PEG compound, leaving no residual groups with the potential to cause sites for nonspecific interactions.

The optimal coupling condition for a CDI–PEG or CDI–mPEG reaction is in an alkaline pH environment, typically above pH 8.5. The coupling reaction proceeds at greatest efficiency when the target molecule is reacted at about 1 pH value above its pI or pK_a. The reaction can be done directly in an organic solvent environment if the molecule to be modified demonstrates poor solubility in aqueous systems. The advantage of an organic coupling reaction is that there is no competing hydrolysis of the active groups, so very high substitution yields of PEG can be realized.

There are a few precautions that should be noted when doing a CDI activation and

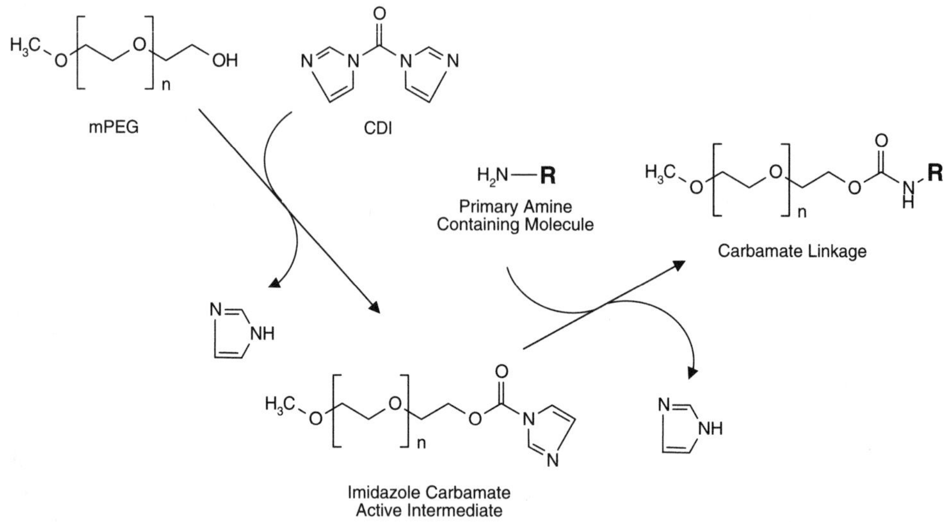

Figure 382 *N,N′*-Carbonyldiimidazole (CDI) may be used to activate the terminal hydroxyl of mPEG to an imidazole carbamate. Reaction of this intermediate with an amine-containing compound results in the formation of a stable carbamate linkage.

coupling experiment. First, CDI itself is extremely unstable to aqueous environments, much more so than the active imidazolyl carbamate that is formed after PEG activation. Therefore, the activation step must be done in a solvent that is free of water. If unacceptable amounts of water are present, CDI will be immediately broken down to CO_2 and imidazole. The evolution of bubbles upon addition of CDI to a PEG solution is the telltale sign of high water content. Only freshly obtained solvents analyzed to be extremely low in moisture or those dried over a molecular sieve should be used. A water content of less than 0.1% in the solvent is usually all right for a CDI activation.

A second precaution is to carry out the activation step in a fume hood away from sources of ignition. Most CDI activation protocols use flammable or toxic solvents and care should be taken in handling and disposing of them.

The coupling reaction using CDI–mPEG or CDI–PEG derivatives is slower than that obtained using NHS ester or succinimidyl carbonate coupling methods. Therefore, the reaction times used with CDI chemistry are typically on the order of 1–2 days at 4°C at a pH of about 8.5. Increasing the pH of the reaction to pH 9 or 10 will speed up the coupling. In addition, doing the reaction at room temperature also helps in this regard. If the molecule to be modified is stable at alkaline pH values and room temperature, then these conditions may be used to decrease the time of the suggested protocol.

The following method is adapted from Beauchamp *et al.* (1983).

Protocol for the Activation of mPEG with CDI

1. Dissolve mPEG (MW 5000) in dioxane at a concentration of 50 m*M* (0.25 gm/ml) by heating to 37°C.
2. Add solid CDI to a final concentration of 0.5 *M* (81 mg/ml).

3. React for 2 h at 37°C with stirring.
4. To remove excess CDI and reaction by-products, Beauchamp *et al.* (1983) dialyzed against water at 4°C. However, the imidazole carbamate groups on mPEG formed during the activation process are subject to hydrolysis in aqueous environments. A better method may be to precipitate the activated mPEG with diethyl ether as in the protocol described for succinimidyl carbonate activation (Section 1.2).
5. Finally, dry the isolated product by lyophilization (if the water dialysis method is used) or by use of a rotary evaporator (if the ether precipitation method is used).

Protocol for the Coupling of CDI–mPEG to Proteins

1. Dissolve the protein to be pegylated in 10 mM sodium borate, pH 8.5, at a concentration of 1–10 mg/ml. Higher pH values may be used to increase the reaction rate, for instance, 0.1 M sodium carbonate, pH 9–10.
2. Add CDI–mPEG to this solution with stirring to bring the final concentration of the activated polymer to 180 mM. Note: other ratios of polymer to protein may be used, depending on the modification level desired. Some optimization of the derivatization level may have to be done to obtain conjugates having the best amount of polymer substitution with retention of protein activity.
3. React for 48 h at 4°C. If higher pH or room temperature conditions are used, the reaction time can be decreased to 24 h.
4. Remove unconjugated mPEG and reaction by-products by dialysis or gel filtration.

1.5. Miscellaneous Coupling Reactions

PEG or mPEG may be conjugated to proteins or other molecules using other coupling chemistries in addition to the ones mentioned in the previous sections. Almost any activation method that can be built off of the terminal hydroxyl(s) of PEG may be employed to pegylate target molecules. For instance, Bergström *et al.* (1992) created an epoxy derivative of the polymer by reaction with epichlorohydrin under alkaline conditions. The reactive alkyl halogen end of epichlorohydrin is first coupled to the hydroxyls of PEG to give the terminal glycidyl ether derivative (Fig. 383). The epoxy-functionalized polymer could then be used to modify covalently a poly(ethylene imine)-coated polystyrene surface to prevent nonspecific protein adsorption. This type of derivative also could be used to modify other amine-, hydroxyl-, or sulfhydryl-containing molecules (Chapter 2, Section 4.1).

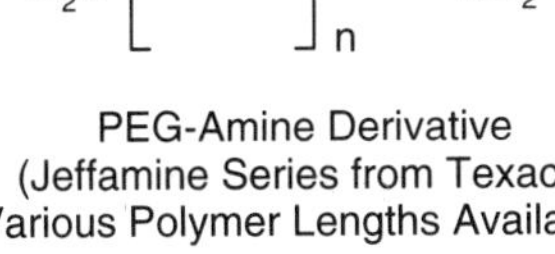

PEG-Amine Derivative
(Jeffamine Series from Texaco;
Various Polymer Lengths Available)

Creation of a sulfhydryl-reactive PEG derivative was done by Goodson and Katre (1990) by reacting a active ester–maleimide heterobifunctional cross-linker with the

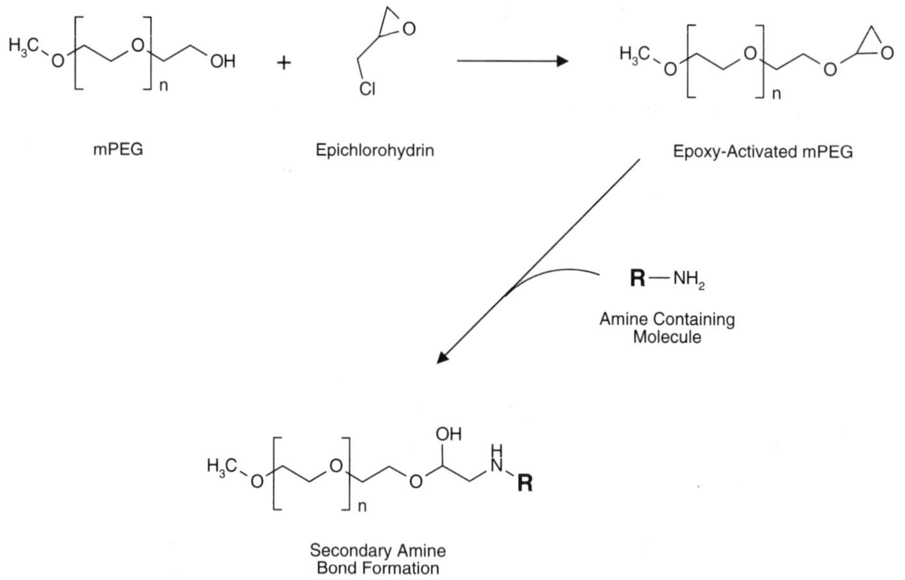

Figure 383   Epichlorohydrin may be used to activate the hydroxyl group of mPEG, creating an epoxy derivative. Reaction with amine-containing molecules yields secondary amine bonds.

amino groups of a PEG-amine polymer (Pillai and Mutter, 1980). An amine-terminal derivative of a PEG-type polymer is the Jeffamine series from Texaco Chemical. Reaction with the N-maleimido-6-aminocaproyl ester of 1-hydroxy-2-nitro-4-benzene sulfonic acid resulted in an amide bond derivative (Fig. 384). This created terminal maleimide groups on each PEG-amine molecule. Maleimide compounds can be used in a site-directed coupling procedure to pegylate specifically at the sulfhydryl groups of proteins and other molecules (Chapter 2, Section 2.2).

In another approach, Wirth *et al.* (1991) and Chamow *et al.* (1994) transformed the terminal hydroxyl group of mPEG into an aldehyde residue by the Moffatt oxidation procedure (Harris *et al.*, 1984). In this reaction, the hydroxyl is treated with acetic anhydride in DMSO containing triethylamine, converting it to the aldehyde (Fig. 385). After stirring at room temperature for 48 h, the aldehyde derivative is isolated by precipitation with ether and ethyl acetate. An aldehyde can be conjugated to proteins or other amine-containing molecules by reductive amination using sodium cyanoborohydride (Chapter 3, Section 4). The advantage of an aldehyde–PEG derivative over the other amine reactive chemistries described previously is that the active function will not hydrolyze or readily degrade before the coupling reaction is initiated. In addition, reductive amination is a reasonably mild conjugation technique that is well tolerated by most proteins.

2. Protein Modification with Activated Dextrans

Dextran is a naturally occurring polymer that is synthesized in yeasts and bacteria for energy storage. It is mainly a linear polysaccharide consisting of repeating units of

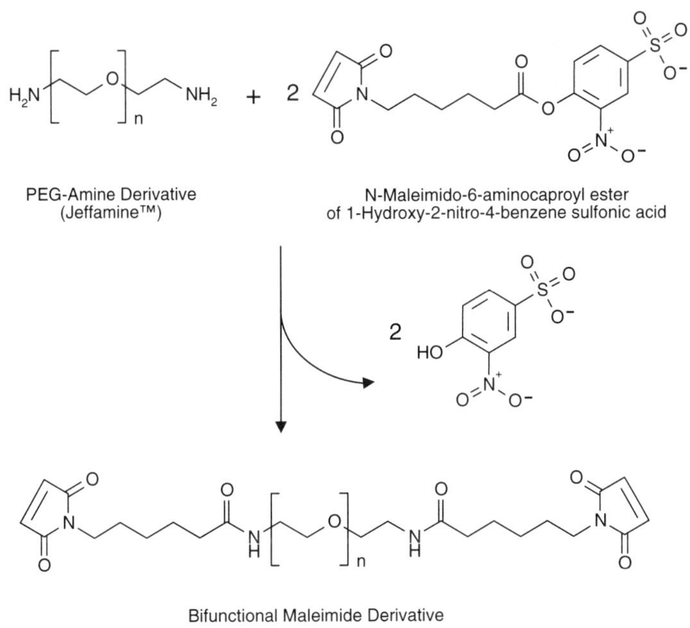

Figure 384 A PEG–amine compound, such as the Jeffamine polymers from Texaco, may be reacted with this heterobifunctional cross-linker to form amide bond derivatives terminating in maleimide groups. This results in a homobifunctional reagent capable of cross-linking thiol molecules. Subsequent reaction with sulfhydryl-containing molecules yields thioether linkages.

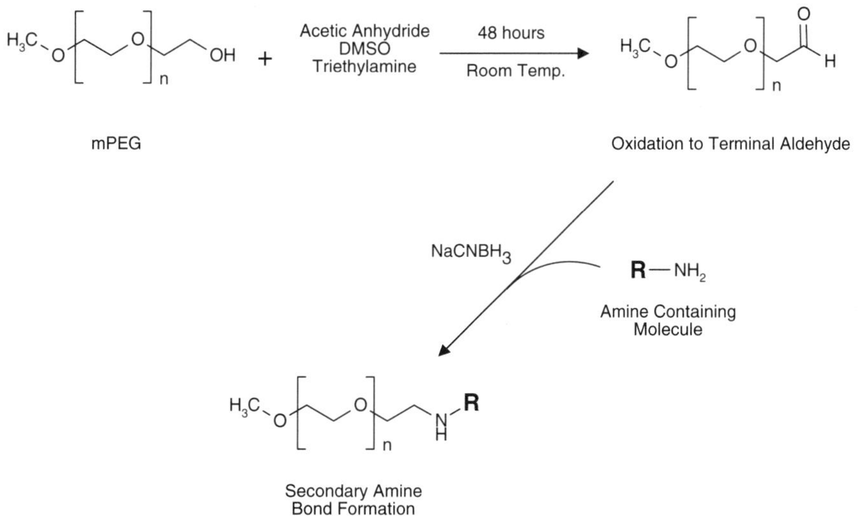

Figure 385 A terminal aldehyde function on mPEG may be formed through an oxidative process at elevated temperatures. This derivative may be used to modify amine-containing molecules by reductive amination.

D-glucose linked together in glycosidic bonds (Chapter 1, Section 2), wherein the carbon-1 of one monomer is attached to the hydroxyl group at the carbon-6 of the next residue. This configuration is the same as that found in the α-1,6-linked disaccharide isomaltose. The same disaccharide is found at the branch points of glycogen and amylopectin. Occasional branch points also may be present in a dextran polymer, occurring as α-1,2, α-1,3, or α-1,4 glycosidic linkages. The branch type and degree of branching vary by species.

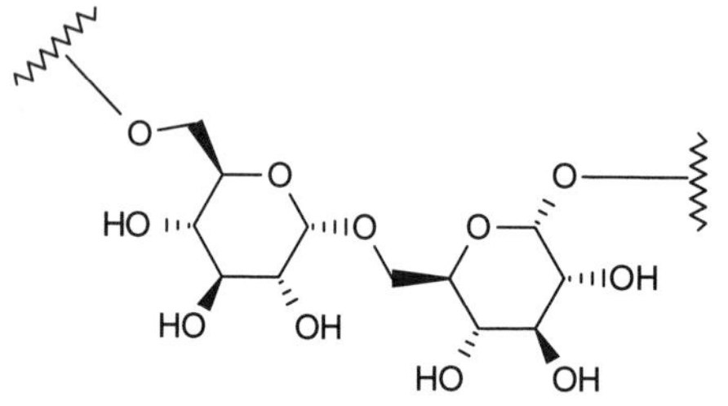

Isomaltose (a-1,6) Repeating
Unit of Dextran Polymer Chains

The hydroxylic content of the dextran sugar backbone makes the polymer very hydrophilic and easily modified for coupling to other molecules. Unlike PEG, discussed previously, which has modifiable groups only at the ends of each linear polymer, the hydroxyl functional groups of dextran are present on each monomer in the chain. The monomers contain at least three hydroxyls (four on the terminal units) that may undergo derivatization reactions. This multivalent nature of dextran allows molecules to be attached at numerous sites along the polymer chain.

Soluble dextran of molecular weight 10,000–500,000 has been used extensively as a modifying or cross-linking agent for proteins and other molecules. It has been used as a drug carrier to transport greater concentrations of antineoplastic pharmaceuticals to tumor sites *in vivo* (Bernstein *et al.*, 1978; Heindel *et al.*, 1990), as conjugated to biotin to make a sensitive anterograde tracer for neuroatomic studies (Brandt and Apkarian, 1992), as a hapten carrier to elicit an immune response against coupled molecules (Shih *et al.*, 1991; Dintzis *et al.*, 1989), as an inducer of B-cell proliferation by coupling anti-Ig antibodies (Brunswick *et al.*, 1988), as a multifunctional linker to cross-link monoclonal antibody conjugates with chemotherapeutic agents (Heindel *et al.*, 1991), and as a stabilizer of enzymes and other proteins (Zlateva *et al.*, 1988; Nakamura *et al.*, 1990). As is true of PEG conjugates with proteins, dextran modification of macromolecules provides increased circulatory half-life *in vivo*, decreased immunogenicity, and a heat and protease protective effect when coupled at sufficient density (Mumtaz and Bachhawat, 1991).

The following sections describe the major activation and coupling methods used with dextran polymers. The active derivatives may be used to couple with proteins and other molecules containing the appropriate functional groups.

2.1. Polyaldehyde Activation and Coupling

The dextran polymer contains adjacent hydroxyl groups on each glucose monomer. These diols may be oxidized with sodium periodate to cleave the associated

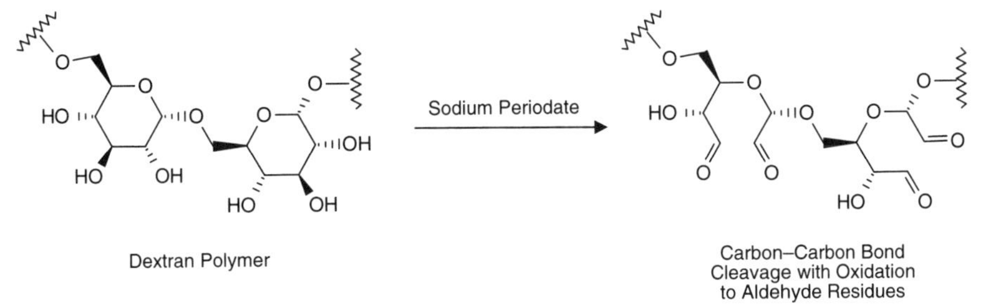

Figure 386 Dextran polymers may be oxidized with sodium periodate to create a polyaldehyde derivative.

carbon—carbon bonds and produce aldehydes (Chapter 1, Section 4.4). This procedure results in two aldehyde groups formed per glucose monomer, thus producing a highly reactive, multifunctional polymer able to couple with numerous amine-containing molecules (Bernstein *et al.*, 1978) (Fig. 386). Polyaldehyde dextran may be conjugated with amine groups by Schiff base formation followed by reductive amination to create stable secondary (or tertiary amine) linkages (section 2.1.4) (Fig. 387).

Proteins may be modified with oxidized dextran polymers under mild conditions using sodium cyanoborohydride as the reducing agent. The reaction proceeds primarily through ε-amino groups of lysine located at the surface of the protein molecules. The optimal pH for the reductive amination reaction is an alkaline environment between pH 7 and 10. The rate of reaction is greatest at pH 8–9 (Kobayashi and Ichishima, 1991), reflecting the efficiency of Schiff base formation at this pH.

Polyaldehyde dextran can be used to couple many small molecules, such as drugs, to a targeting molecule like an antibody. The multivalent nature of the oxidized dextran backbone provides more sites for conjugation than possible using direct coupling of the drug with the antibody itself. Similarly, detection molecules such as fluorescent probes can be conjugated in greater amounts using a dextran carrier than is feasible with direct modification of a protein.

The following protocol for creating the polyaldehyde dextran derivative is based on the method of Bernstein *et al.* (1978).

Protocol for Oxidizing Dextran with Sodium Periodate

1. Dissolve sodium periodate ($NaIO_4$) (Sigma) in 500 ml of deionized water at a concentration of 0.03 M (6.42 g).

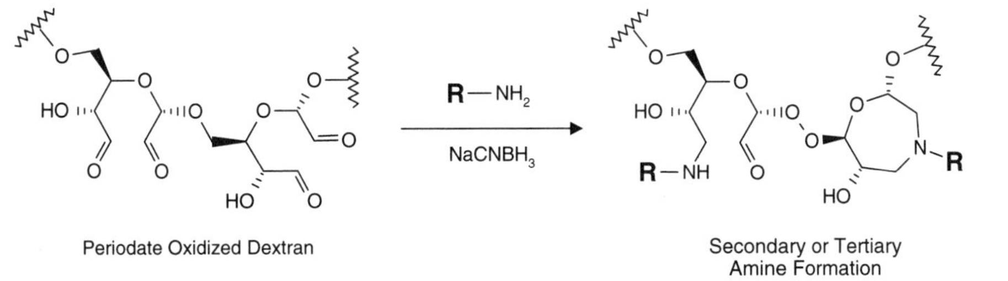

Figure 387 Polyaldehyde dextran may be used as a multifunctional cross-linking agent for the coupling of amine-containing molecules. Reductive amination creates secondary amine linkages.

2. Dissolve dextran (Polysciences) of molecular weight between 10,000 and 40,000 in the sodium periodate solution with stirring.
3. React overnight at room temperature in the dark.
4. Remove excess reactant by dialysis against water. The purified polyaldehyde dextran may be lyophilized for long-term storage.

The degree of oxidation may be assessed by measurement of the aldehydes formed. Zhao and Heindel (1991) suggest derivatizing the polyaldehyde dextran with hydroxylamine hydrochloride and measuring the amount of HCl released by titration. However, this may be tedious and time-consuming. A simpler method may be to take advantage of the fact that periodate-oxidized sugars are capable of reducing Cu^{2+} to Cu^+, which can be detected using the bicinchoninic acid (BCA) reagent (Pierce Chemical) (Smith *et al.*, 1985). The formation of Cu^+ is in direct proportion to the amount of aldehydes present in the polymer. BCA will form a purple-colored complex with Cu^+, which can be measured at 562 nm.

Protocol for Coupling Polyaldehyde Dextran to Proteins

1. Dissolve or buffer-exchange the periodate-oxidized dextran in 0.1 M sodium phosphate, 0.15 M NaCl, pH 7.2, at a concentration of 10–25 mg/ml. Other buffers having a pH range of 7–10 may be used with success, as long as they do not contain competing amines (such as Tris). A reaction environment of pH 8–9 (0.1 M sodium bicarbonate) will give the greatest yield of reductive amination coupling.
2. Add 10 mg of the protein to be coupled to the dextran solution. Other ratios of dextran to protein may be used as appropriate. For instance, if more than one protein or a protein plus a smaller molecule are both to be conjugated to the dextran backbone, the amount of protein added initially may have to be scaled back to allow the second molecule to be coupled later. Many times, a small molecule such as a drug will be coupled to the dextran polymer first, and then a targeting protein such as an antibody conjugated secondarily. The optimal ratio of components forming the dextran conjugate should be determined experimentally to obtain the best combination possible.
3. In a fume hood, add 0.2 ml of 1 M sodium cyanoborohydride (Aldrich) to each milliliter of the protein/dextran solution. Mix well. *Caution: Cyanoborohydride is extremely toxic and should be handled only in well-ventilated fume hoods. Dispose of cyanide-containing solutions according to approved guidelines.*
4. React for at least 6 h at room temperature. Overnight reactions also may be done.
5. To block excess aldehydes, add 0.2 ml of 1 M Tris, pH 8, to each milliliter of the reaction. Note: If a second molecule is to be coupled after the initial protein conjugation, do not block the remaining aldehydes until the second molecule is added.
6. React for an additional 2 h at room temperature.
7. Purify the protein–dextran conjugate from unconjugated protein and dextran by gel filtration using a column of Sephacryl S-200 or S-300. Small molecules may be removed from a dextran conjugate by dialysis.

2.2. Carboxyl, Amine, and Hydrazide Derivatives

Dextran derivatives containing carboxyl or amine-terminal spacer arms may be prepared by a number of techniques. These derivatives are useful for coupling amine- or carboxylate-containing molecules through a carbodiimide-mediated reaction to form an amide bond (Chapter 3, Section 1). Amine-terminal spacers also can be used to create secondary reactive groups by modification with a heterobifunctional cross-linking agent (Chapter 5).

This type of modification process has been used to form sulfhydryl-reactive dextran polymers by coupling amine spacers with cross-linkers containing an amine-reactive end and a thiol-reactive end (Noguchi *et al.*, 1992; Brunswick *et al.*, 1988). The result was a multivalent sulfhydryl-reactive dextran derivative that could couple numerous sulfhydryl-containing molecules per polymer chain.

Several chemical approaches may be used to form the amine- or carboxyl-terminal dextran derivative. The simplest procedure may be to prepare polyaldehyde dextran according to the procedure of Section 2.1, and then make the spacer arm derivative by reductively aminating an amine-containing organic compound onto it. For instance, short diamine compounds such as ethylene diamine or diaminodipropylamine (3,3'-imino*bis*propylamine) can be coupled in excess to polyaldehyde dextran to create an amine-terminal derivative. Carboxyl-terminal derivatives may be prepared similarly by coupling molecules such as 6-aminocaproic acid or β-alanine to polyaldehyde

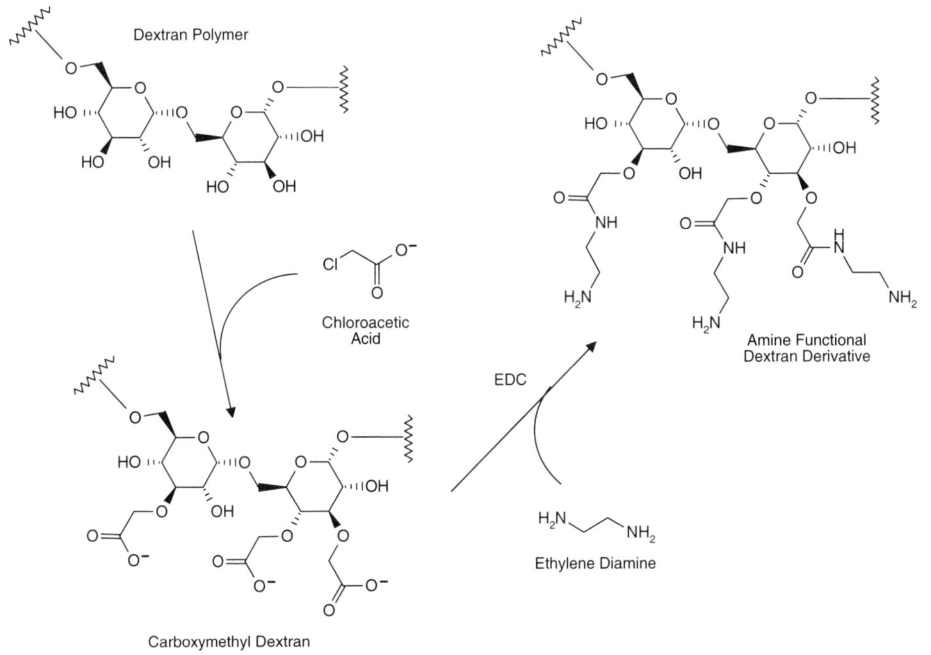

Figure 388　An amine terminal derivative of dextran may be prepared through a two-step process involving the reaction of chloroacetic acid with the hydroxyl groups of the polymer to create carboxylates. Next, ethylene diamine is coupled using a carbodiimide-mediated reaction to give the primary amine functional groups.

dextran. Alternatively, an amine-terminal spacer may be reacted with succinic anhydride to form the carboxylate derivative (Chapter 1, Section 4.2).

Another approach uses reactive alkyl halogen compounds containing a terminal carboxylate group on the other end to form spacer arms off the dextran polymer. In this manner, Brunswick *et al.* (1988) used chloroacetic acid to modify the hydroxyl groups of dextran, forming the carboxymethyl derivative. The carboxylates were then aminated with ethylene diamine to create an amine-terminal derivative (Inman, 1985). The amine was then modified with iodoacetate to form a sulfhydryl-reactive polymer (Fig. 388).

In a somewhat similar scheme, Noguchi *et al.* (1992) prepared a carboxylate spacer arm by reacting 6-bromohexanoic acid with the dextran polymer. The carboxylate was then aminated with ethylene diamine to form an amine-terminal spacer (Fig. 389). This dextran derivative was finally reacted with SPDP (Chapter 5, Section 1.1) to create the final sulfhydryl-reactive polymer (Section 2.4). The SPDP-activated polymer then could be used to prepare an immunoconjugate composed of an antibody against human colon cancer conjugated with the drug mitomycin-C.

Hydrazide derivatives of dextran also may be prepared from the periodate-oxidized polymer or from a carboxyl derivative by reaction with *bis*-hydrazide compounds (Chapter 4, Section 8). A hydrazide terminal spacer provides reactivity toward al-

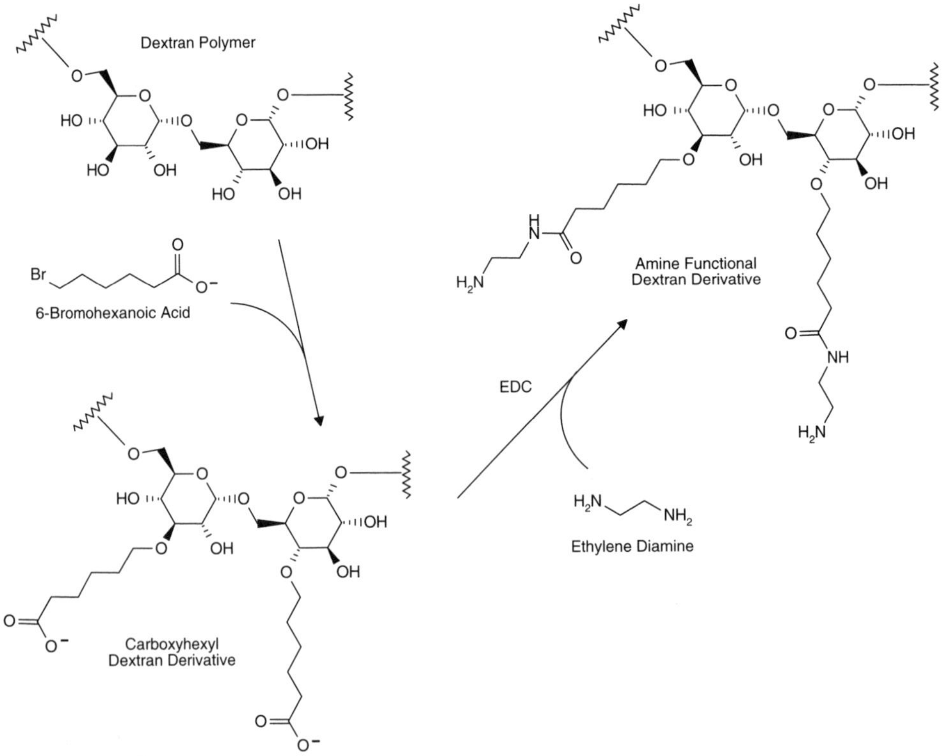

Figure 389 Amino-dextran derivatives may be prepared by the reaction of 6-bromohexanoic acid with the hydroxyl groups of the polymer followed by coupling of ethylene diamine using EDC.

dehyde- or ketone-containing molecules. Thus, the hydrazide–dextran polymer can be used to conjugate specifically glycoproteins or other polysaccharide-containing molecules after they have been oxidized with periodate to form aldehydes (Chapter 1, Section 4.4).

The following protocols may be used to create carboxyl-, amine-, or hydrazide-containing derivatives of dextran.

Protocol for the Preparation of Amine or Hydrazide Derivatives by Reductive Amination

1. Prepare polyaldehyde dextran according to the method of Section 2.1.
2. To make an amine derivative of dextran, dissolve ethylene diamine (or another suitable diamine) in 0.1 M sodium phosphate, 0.15 M NaCl, pH 7.2, at a concentration of 3 M. Note: use of the hydrochloride form of ethylene diamine is more convenient, since it avoids having to adjust the pH of the highly alkaline free-base form of the molecule. Alternatively, to prepare a hydrazide–dextran derivative, dissolve adipic acid dihydrazide (Chapter 4, Section 8.1) in the coupling buffer at a concentration of 30 mg/ml (heating under a hot water tap may be necessary to dissolve completely the hydrazide compound). Adjust the pH to 7.2 with HCl.
3. Dissolve polyaldehyde dextran in the ethylene diamine (or adipic dihydrazide) solution at a concentration of 25 mg/ml.
4. In a fume hood, add 0.2 ml of 1 M sodium cyanoborohydride to each milliliter of the diamine/dextran solution. Mix well. *Caution: Cyanoborohydride is extremely toxic and should be handled only in well-ventilated fume hoods. Dispose of cyanide-containing solutions according to approved guidelines.*
5. React for at least 6 h at room temperature. Overnight reactions also may be done.
6. Remove excess diamine and reaction by-products by dialysis.

The ethylene diamine–dextran derivative may be used to couple with carboxylate-containing molecules by the carbodiimide reaction, to couple amine-reactive probes, or to further modify using heterobifunctional cross-linkers. The hydrazide–dextran derivative may be used to cross-link aldehyde-containing molecules, such as oxidized carbohydrates or glycoproteins.

Protocol for the Modification of Dextran with Chloroacetic Acid

1. In a fume hood, prepare a solution consisting of 1 M chloroacetic acid in 3 M NaOH.
2. Immediately add dextran polymer to a final concentration of 40 mg/ml. Mix well to dissolve.
3. React for 70 min at room temperature with stirring.
4. Stop the reaction by adding 4 mg/ml of solid NaH_2PO_4 and adjusting the pH to neutral with 6 N HCl.
5. Remove excess reactants by dialysis.

The carboxymethyl–dextran derivative may be used to couple amine-containing molecules by the carbodiimide reaction. Heindel *et al.* (1994) prepared the lactone derivative of carboxymethyl–dextran by refluxing for 5 h in toluene or other anhydrous solvents. The lactone derivative is highly reactive toward amine-containing molecules, thus creating a preactivated polymer for conjugation purposes.

2.3. Epoxy Activation and Coupling

Epoxy activation of hydroxylic polymers is commonly used as a means to immobilize molecules on solid-phase chromatographic supports (Sundberg and Porath, 1974). *Bis*oxirane compounds also can be used to introduce epoxide functional groups into soluble dextran polymers in much the same manner (Böcher *et al.*, 1992; Böldicke *et al.*, 1988). The epoxide group can react with nucleophiles in a ring-opening process to form a stable covalent linkage. The reaction can take place with primary amines, sulfhydryls, or hydroxyl groups to create secondary amine, thioether, or ether bonds, respectively (Chapter 2, Section 1.7).

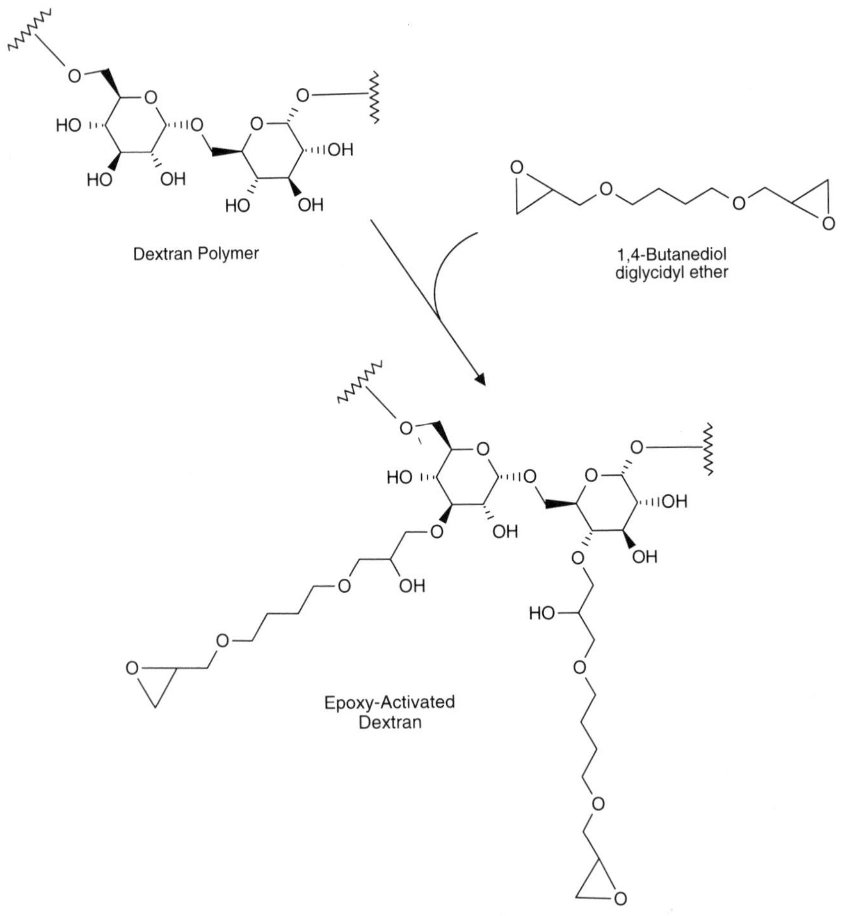

Figure 390 An epoxy-functional dextran derivative may be prepared by the reaction of 1,4-butanediol diglycidyl ether with the hydroxyl groups of the polymer.

Modification of dextran polymers with 1,4-butanediol diglycidyl ether results in ether derivatives on the dextran hydroxyl groups containing hydrophilic spacers with terminal epoxy functions (Fig. 390).

Protocol

1. In a fume hood, mix 1 part 1,4-butanediol diglycidyl ether with 1 part 0.6 N NaOH containing 2 mg/ml sodium borohydride.
2. With stirring, add 5 mg of dextran to each ml of the *bis*-epoxide solution. Mix well to dissolve.
3. React for 12 h at 25°C or 3–4 h at 37°C.

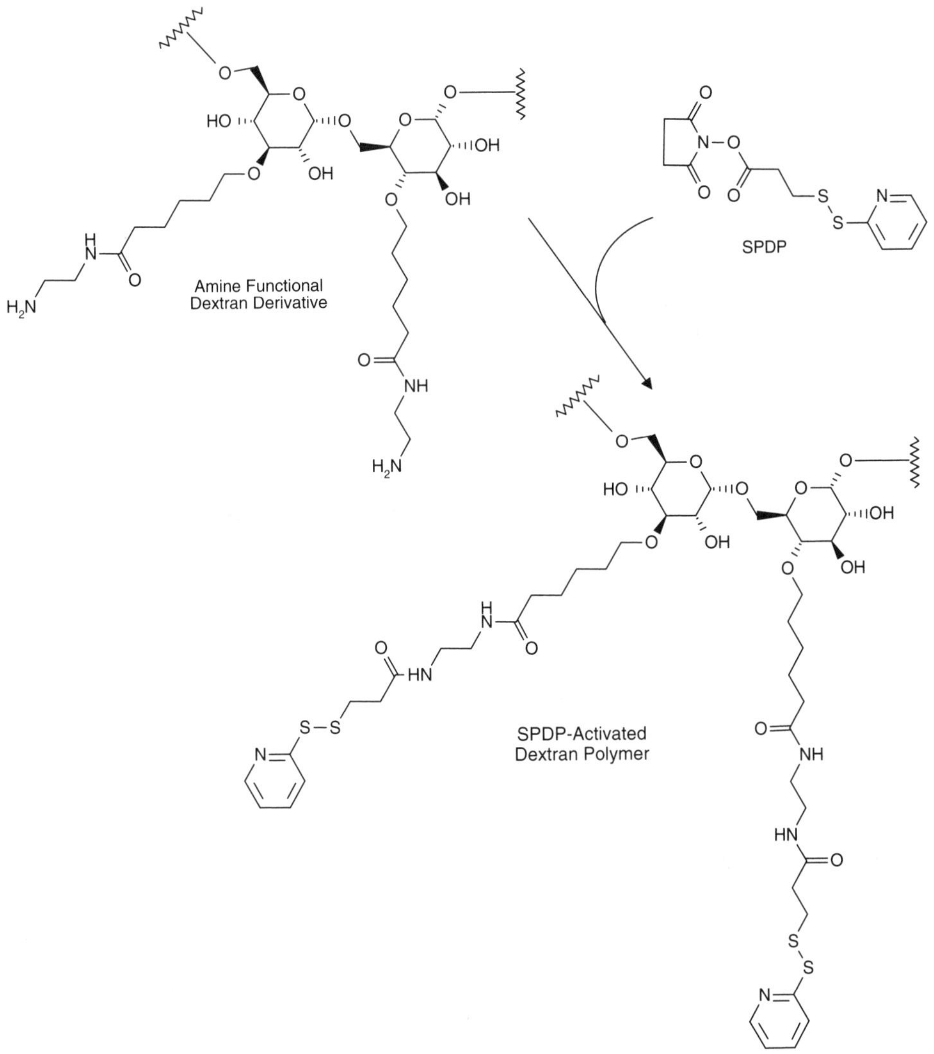

Figure 391 An amine-functionalized dextran derivative may be further reacted with SPDP to create a sulfhydryl-reactive product.

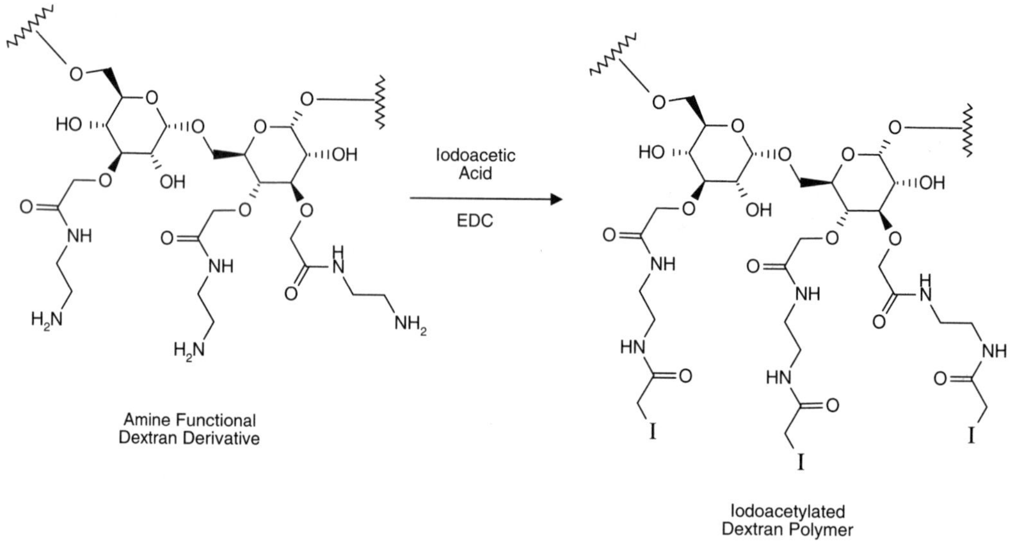

Figure 392 An amine-derivative of dextran may be coupled with iodoacetic acid using a carbodiimide reaction to produce a sulfhydryl-reactive iodoacetamide polymer.

4. Extensively dialyze the solution against water to remove excess reactants. The activated dextran may be lyophilized for long-term storage.

The epoxide-activated dextran may be used to conjugate amine-, sulfhydryl-, or hydroxyl-containing molecules. The reaction of the epoxide functional groups with hydroxyls requires high pH conditions, usually in the range pH 11–12. Amine nucleophiles react at more moderate alkaline pH values, typically needing buffer environments of at least pH 9. Sulfhydryl groups are the most highly reactive nucleophiles with epoxides, requiring a buffered system closer to the physiological range, pH 7.5–8.5, for efficient coupling.

2.4. Sulfhydryl-Reactive Derivatives

Sulfhydryl-reactive dextran derivatives may be prepared through the use of heterobifunctional cross-linking agents (Chapter 5). In particular, cross-linkers containing pyridyldisulfide, maleimide, or iodoacetyl groups on one end are quite effective in directing a conjugation reaction to thiols. Both maleimide and iodoacetyl activation procedures will yield nonreversible bonds with sulfhydryl-containing molecules. Pyridyl disulfide compounds, however, react with thiols to form cleavable disulfide bonds that can be reversed by reduction.

Noguchi *et al.* (1992) used an amine-terminal spacer arm derivative of dextran to react with SPDP (Chapter 5, Section 1.1) in the creation of a pyridyldisulfide-activated polymer (Fig. 391). Brunswick *et al.* (1988) used a different amine-terminal spacer arm derivative of dextran and subsequently coupled iodoacetate to form a sulfhydryl-reactive polymer (Fig. 392). Heindel *et al.* (1991) used a unique approach. They

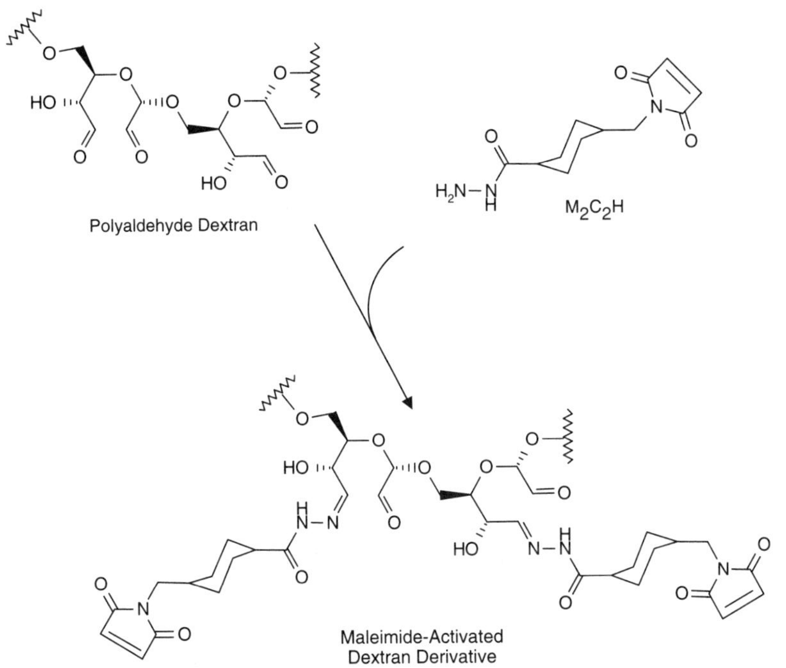

Polyaldehyde Dextran

M₂C₂H

Maleimide-Activated
Dextran Derivative

Figure 393 Polyaldehyde dextran may be modified with the hydrazide end of M_2C_2H to create a thiol-reactive polymer.

modified polyaldehyde dextran with a heterobifunctional cross-linker containing a hydrazide group on one end and a maleimide group on the other (Chapter 5, Section 2). The hydrazides reacted with the aldehyde groups to form hydrazone linkages, leaving the maleimide ends free to result in a thiol-reactive dextran derivative (Fig. 393).

16

Enzyme Modification and Conjugation

Enzymes are widely used in bioconjugate chemistry as detection components in assay systems. The catalytic activity of an enzyme can be used to turn substrate molecules into chromogenic or fluorescent products, easily detectable or quantifiable by microscopy or spectroscopy. If an enzyme is conjugated to a targeting molecule specific for some analyte of interest, then an assay system can be constructed to localize or measure the analyte. The most common targeting molecule is an antibody having antigen binding specificity for the substance to be measured. An enzyme conjugated to such an antibody can be used to visualize the presence of antigen. Due to the advantages of this simple concept, enzyme linked immunoadsorbent assays (ELISAs) have become the most important type of immunoassay system available.

The rapid turnover rate of some enzymes allows ELISAs to be designed that rival the sensitivity of radiolabeling techniques. In addition, substrates can be chosen to produce soluble products that can be accurately quantified by their absorbance or fluorescence. Alternatively, substrates are available that form insoluble, highly colored precipitates, excellent for localizing antigens in blots, cells, or tissue sections. The flexibility of enzyme-based assay systems makes the chemistry of enzyme conjugation one of the most important application areas in bioconjugate techniques.

The following sections briefly describe the principal enzymes utilized for conjugation with other protein molecules, particularly in the design of ELISA systems.

1. Properties of Common Enzymes

1.1. Horseradish Peroxidase

HRP (donor:hydrogen peroxide oxidoreductase; EC 1.11.1.7), derived from horseradish roots, is a enzyme of molecular weight 40,000 that can catalyze the reaction of hydrogen peroxide with certain organic, electron-donating substrates to yield highly colored products. The reaction of HRP with its fundamental substrate, H_2O_2, forms a stable intermediate that can dissociate in the presence of a suitable electron donor, oxidizing the donor and potentially creating a color change. The donor can consist of oxidizable molecules like ascorbate, cytochrome c, ferrocyanide, or the leuco forms of many dyes. A large variety of electron-donating dye substrates are commercially available for use as HRP detection reagents. Some of them can be used to form soluble colored products for use in spectrophotometric detection systems, while other substrates form insoluble products that are especially apropos for staining techniques. The pH optimum for HRP is 7.

HRP is a hemoprotein containing photohemin IX as its prosthetic group. The presence of the heme structure gives the enzyme its characteristic color and maximal absorptivity at 403 nm. The ratio of its absorbance in solution at 403 nm to its absorbance at 275 nm, called the RZ or Reinheitzahl ratio, can be used to approximate the purity of the enzyme. However, at least seven isoenzymes exist for HRP (Shannon et al., 1966; Kay et al., 1967; Strickland et al., 1968), and their RZ values vary from 2.50 to 4.19. Thus, unless the RZ ratio is precisely known or determined for the particular isoenzyme of HRP utilized in the preparation of a antibody–enzyme conjugate, subsequent measurement after cross-linking would yield questionable results in the determination of the amount of HRP present in the conjugate.

HRP is a glycoprotein that contains significant amounts of carbohydrate. Its polysaccharide chains are often used in cross-linking reactions. Mild oxidation of its associated sugar residues with sodium periodate generates reactive aldehyde groups that can be used for conjugation to amine-containing molecules. Reductive amination of oxidized HRP to antibody molecules in the presence of sodium cyanoborohydride is perhaps the simplest method of preparing highly active conjugates with this enzyme (Chapter 3, Section 4, and Chapter 10, Section 1.3).

Other methods of HRP conjugation include the use of the homobifunctional reagent glutaraldehyde (Chapter 4, Section 6.2) and the heterobifunctional cross-linker SMCC (Chapter 5, Section 1.3). Using glutaraldehyde, a two-step protocol usually is employed to limit the extent of oligomer formation. Nevertheless, this method often causes unacceptable amounts of precipitated conjugate. Despite this disadvantage, glutaraldehyde conjugation is still routinely used, especially in the preparation of some antibody–enzyme reagents that go into established diagnostic assays. The use of the NHS ester–maleimide cross-linker SMCC provides better control over the conjugation process. SMCC usually is reacted first with HRP to create a derivative containing sulfhydryl-reactive maleimide groups. The maleimide-activated enzyme can be purified and freeze-dried, providing a ready source of modified HRP to react with a sulfhydryl-containing antibody. Several preactivated forms of this enzyme are available from Pierce.

The size of HRP is an advantage in preparing antibody–enzyme conjugates, since the overall complex size also can be designed to be small. Relatively low-molecular-weight conjugates are able to penetrate cellular structures better than large, polymeric complexes. This is why HRP conjugates are often the best choice for immunocytochemical staining techniques. Small conjugate size means greater accessibility to antigenic structures within tissue sections.

Another distinctive advantage of HRP is its robust nature and stability, especially under the conditions employed for cross-linking. HRP is stable for years in a freeze-dried state, and the purified enzyme can be stored in solution at 4°C for many months without significant loss of activity. The enzyme also retains excellent activity after being modified with a conjugation reagent or periodate-oxidized to form aldehyde groups on its polysaccharide chains. Depending on the methods used for cross-linking, HRP conjugates can be constructed to have a high ratio of enzyme-to-antibody or a low ratio—both retaining high specific activity.

The disadvantages associated with HRP are several. The enzyme only contains two available primary ε-amine groups—extraordinarily low for most proteins—thus limiting its ability to be activated with amine-reactive heterobifunctionals. HRP is sensitive to the presence of many antibacterial agents, especially azide. It is also reversibly inhibited by cyanide and sulfide (Theorell, 1951). Finally, while the enzymatic activity of HRP is extremely high, its useful lifespan or practical substrate development time is somewhat limited. After about an hour of substrate turnover, its activity can be decreased severely.

Nevertheless, HRP is by far the most popular enzyme used in antibody–enzyme conjugates. One survey of enzyme use stated that HRP is incorporated in about 80% of all antibody conjugates, most of them utilized in diagnostic assay systems.

1.2. Alkaline Phosphatase

Alkaline phosphatases [AP, orthophosphoric-monoester phosphorylase (alkaline optimum); EC 3.1.3.1] represent a large family of almost ubiquitous isoenzymes found in organisms from bacteria to animals. In mammals there are two forms of alkaline phosphatase, one form present in a variety of tissues and another form found only in the intestines. They share common attributes in that the phosphatase activity is optimal at pH 8 to 10, is activated by the presence of divalent cations, and is inhibited by cysteines, cyanides, arsenate, various metal chelators, and phosphate ions. Most conjugates created with alkaline phosphatase utilize the form isolated from calf intestine.

AP isoenzymes can cleave associated phosphomonoester groups from a wide variety of substrates. The exact biological function of these enzymes is still uncertain. They can behave *in vivo* in their classic phosphohydrolase role at alkaline pH, but at neutral pH AP isoenzymes can act as phosphotransferases. In this sense, suitable phosphate acceptor molecules can be utilized in solution to increase the reaction rates of AP on selected substrates. Typical phosphate acceptor additives include diethanolamine, Tris, and 2-amino-2-methyl-1-propanol. The presence of these additives in substrate buffers can dramatically increase the sensitivity of AP ELISA determinations, even when the substrate reaction is done in alkaline conditions.

Calf intestinal alkaline phosphatase has a molecular weight of about 140,000. The active site of AP contains two zinc atoms and a single magnesium atom, both of which are essential for activity (Kim and Wyckoff, 1991). Substrate development with AP thus should be done in buffered environments containing small concentrations of these divalent cations to maintain optimally active site conformation. Avoid the presence of metal chelators such as EDTA, since they may extract these ions from the enzyme and inhibit activity. The pH optimum for APs can vary from 8 to 10, depending on the type of isoenzyme. Calf intestinal AP peaks in activity at the higher pH values of this range, and substrate reactions are commonly performed in diethanolamine buffer at pH 9.8. The calf intestinal enzyme has the highest catalytic rate constant yet discovered for AP isoenzymes, 3500 s^{-1}.

Purified preparations of calf intestinal AP maintained in solution are usually stored in the presence of a stabilizer, typically 3 M NaCl. The enzyme also may be lyophilized, but may experience activity loss with each freeze–thaw cycle. AP is not stable under acidic conditions. Lowering the pH of an AP solution to 4.5 reversibly inhibits the enzyme. It is recommended that all handling, storage, and use of AP be done under conditions >pH 7 to maintain the highest possible catalytic activity.

Alkaline phosphatase often is a difficult enzyme to work with when preparing enzyme conjugates. Activity losses may occur upon modification with a cross-linking agent or after coupling to an antibody molecule. Simply following established protocols for making antibody–AP conjugates does not always assure retention of activity. Sometimes activity losses can be traced to particular batches or to certain suppliers of the enzyme. Using a highly purified, high-activity AP preparation helps to maintain good resultant activity in the conjugate.

Ironically, AP is the enzyme of choice for some applications due to its stability. Since it can withstand the moderately high temperatures associated with hybridization assays, AP is the enzyme of choice for labeling oligonucleotide probes. Alkaline phosphatase also is capable of maintaining enzymatic activity for extended periods of substrate development. Increased sensitivity can be realized in ELISA procedures by extending the substrate incubation time to hours and sometimes even days. These properties make AP the second most popular choice for antibody–enzyme conjugates (behind HRP), being used in almost 20% of all commercial enzyme-linked assays.

Conjugation methods typically employed with AP include glutaraldehyde-mediated cross-linking (Chapter 4, Section 6.2) and the use of the heterobifunctional reagents SMCC (Chapter 5, Section 1.3) or SPDP (Chapter 5, Section 1.1). Heterobifunctional cross-linkers provide the best control over the cross-linking process and typically result in antibody–enzyme conjugates of high activity. Many conjugation protocols incorporate a sodium phosphate buffer system to block reversibly the AP active site during chemical modification. This prevents derivatization from occurring in the catalytic site, thus better retaining activity in the resultant conjugate.

1.3. β-Galactosidase

β-Galactosidase (β-Gal; β-D-galactoside galactohydrolase; EC 3.2.1.23; also called lactase) catalyzes the hydrolysis of β-D-galactoside in the presence of water to galactose and alcohol. This type of enzyme is found widespread in many microorganisms, plants, and animals. β-Gal can be used to determine lactose in biological fluids, is employed in food processing operations—particularly in immobilized form—and has good characteristics when conjugated to antibody molecules for use in ELISA systems (Wallenfels and Weil, 1972; Byrne and Johnson, 1975; Kato et al., 1975a,b).

β-Gal has a molecular weight of 540,000 and is composed of four identical subunits of MW 135,000, each with an independent active site (Melchers and Messer, 1973). The enzyme has divalent metals as cofactors, with chelated Mg^{2+} ions required to maintain active site conformation. The presence of NaCl or dilute solutions (5%) of low-molecular-weight alcohols (methanol, ethanol, etc.) causes enhanced substrate turnover. β-Gal contains numerous sulfhydryl groups and is glycosylated.

Commercially available β-gal usually is isolated from *Escherichia coli* and has a pH optimum at 7–7.5. By contrast, mammalian β-galactosidases usually have a pH optimum within the range 5.5–6; thus interference from endogenous β-gal during immunohistochemical staining can be avoided.

Due to the relatively high molecular weight of the enzyme, conjugates formed with antibodies and β-gal tend to be much bulkier than those associated with alkaline phosphatase or horseradish peroxidase. For this reason, antibody conjugates made with β-gal may have more difficulty penetrating tissue structures during histochemical staining techniques than those made with the other enzymes.

Although numerous research articles have been written describing the preparation and use of antibody conjugates with β-gal, the enzyme remains a minor player in ELISA procedures. Less than 1% of all commercial ELISA products utilize this enzyme.

β-Gal may be conjugated to antibody molecules using the heterobifunctional reagent SMCC. This cross-linker is reacted first with an antibody through its amine-reactive NHS ester end to form a maleimide-activated derivative. This is in contrast to most antibody–enzyme conjugation schemes utilizing SMCC, wherein the enzyme is modified first and a sulfhydryl-containing antibody is coupled secondarily. However, since β-gal already contains abundant free sulfhydryl residues that can participate in coupling to a maleimide-activated protein, conjugations with this enzyme often are done with the antibody being the first modified. This route avoids having to create sulfhydryls on the antibody molecule, by either reduction or modification with a thiolating reagent. Thus, antibody–β-gal conjugates usually are simpler to make than those using other enzymes.

1.4. Glucose Oxidase

Glucose oxidase (β-D-glucose: oxygen 1-oxidoreductase; EC 1.1.3.4; GO) is a flavoenzyme that catalyzes the oxidation of β-D-glucose to δ-D-gluconolactone. The intermediate product of the catalysis is a reduced enzyme–$FADH_2$ complex that in the presence of oxygen is oxidized back to enzyme–flavin adenine dinucleotide (FAD) with release of hydrogen peroxide. The enzyme consists of two identical subunits (MW 80,000 each) bound together by disulfide linkages (O'Malley and Weaver, 1972). GO contains two tightly bound FAD cofactors, one per subunit, which are critical to its oxidoreductase activity. Each subunit also contains one molecule of chelated iron. The intact protein consists of about 74% amino acids, 16% neutral carbohydrate, and 2% amino sugars (total molecular weight 160,000). GO operates under a relatively broad pH range of 4–7, but its pH optimum is 5.5. The commercially available preparation of GO is typically isolated from *Aspergillus niger*.

Glucose oxidase is widely used in diagnostic assays for the determination of glucose concentration in physiological fluids. Detectability of the oxidation products is done through an enzyme-coupled reaction, wherein liberated H_2O_2 is reacted with peroxidase and a suitable chromogenic substrate. The development of substrate color is thus proportional to the amount of H_2O_2 released, which is in turn related to the amount of glucose originally present. The production of hydrogen peroxide also can be quantified using a luminescence procedure with luminol to produce light in proportion to the glucose concentration (Williams *et al.*, 1976).

GO often is used in solution phase chemical reactions as well as being immobilized on "dip-sticks" and electrodes. Although its overall clinical usage is widespread, its use as conjugated to antibodies in enzyme-linked assay systems is minor compared to the popularity of other enzymes like horseradish peroxidase and alkaline phosphatase.

Of the total number of commercial diagnostic assays utilizing antibody–enzyme conjugates, GO is employed in less than 1% of clinical tests. The enzyme remains, however, an important tool in many assays developed for research use. One particular advantage to the enzyme is that there is no endogenous GO activity in mammalian tissues, making it an excellent choice for immunohistochemical staining procedures.

Antibody conjugates with GO can be made using the cross-linking agents glutaraldehyde (Chapter 10, Section 1.2) or SMCC (Chapter 10, Section 1.1). The heterobifunctional reagent SMCC provides the best control over the conjugation process and usually results in high-activity preparations.

2. Preparation of Activated Enzymes for Conjugation

Enzymes may be modified to contain functional groups useful for conjugation with other proteins. This operation may be done using homobifunctional (Chapter 4) or heterobifunctional (Chapter 5) reagents that can covalently couple to some chemical target on the enzyme and result in a terminal reactive group that can cross-link with another molecule. Enzyme activation also may take advantage of the presence of polysaccharide constituents—oxidizing them with sodium periodate to form reactive aldehydes.

Whatever the method of conjugate creation, the most important considerations are retention of activity in the complex and prevention of extensive oligomer generation, which may cause precipitation. The following sections discuss some of the more common methods for producing enzyme conjugates. The list, however, is by no means inclusive of every possible procedure used in the literature.

2.1. Glutaraldehyde-Activated Enzymes

Glutaraldehyde is a homobifunctional cross-linker containing an aldehyde residue at both ends of a 5 carbon chain. Its primary reactivity is toward amine groups, but the reaction may occur by more than one mechanism. As discussed in Chapter 4, Section 6.2, glutaraldehyde is able to form Schiff base linkages with amines that can be reduced with sodium cyanoborohydride to create stable secondary amine linkages. However, glutaraldehyde also exists as α,β-unsaturated polymers in aqueous solution. The double bonds of these polymers can undergo an addition reaction with amines that results in covalent bond formation even without a reductant being present. Fresh glutaraldehyde may contain little polymer formation. However, the older the preparation of glutaraldehyde, the more likely it is that it contains appreciable amounts of polymer. Thus, reactions with this cross-linking agent can result in indistinct conjugation products.

The high reactivity of glutaraldehyde also makes it difficult to control the conjugation process. Proteins cross-linked with this reagent often form substantial amounts of precipitated products due to polymerization. The degree of oligomer formation can be moderated somewhat by using a two-step protocol, but the first protein activated with the molecule can still form large-molecular-weight complexes.

Despite the obvious disadvantages of glutaraldehyde-mediated conjugation, the cross-linker continues to be used to form enzyme–antibody complexes and in other applications. Many diagnostic tests still utilize antibody–enzyme conjugates prepared through glutaraldehyde cross-linking procedures.

The one- and two-step procedures for enzyme activation and conjugation using glutaraldehyde can be found in Chapter 10, Section 1.2.

2.2. Periodate Oxidation Techniques

Molecules containing polysaccharide chains may be oxidized to possess reactive aldehyde residues by treatment with sodium periodate. Any adjacent carbon atoms containing hydroxyl groups will be affected, cleaving the carbon—carbon bond and transforming the hydroxyls into aldehydes. Glycoproteins may be oxidized in this manner to form reactive intermediates useful for cross-linking procedures involving reductive amination (Chapter 3, Section 4). This conjugation technique can direct the coupling process away from polypeptide active regions, thus helping to preserve catalytic activity or avoid binding sites.

Enzymes that contain carbohydrate, such as HRP or GO, may be oxidized with periodate to create active derivatives that subsequently can be used to label antibodies or other targeting molecules at their amine functional groups. The aldehyde–HRP intermediate may be stored for extended periods in a frozen or lyophilized state without loss of activity (either enzymatic or coupling potential). Avoid, however, storage in a liquid state, since polymerization may occur—resulting in precipitation and loss of activity.

The protocol for periodate oxidation of HRP and its conjugation with other proteins may be found in Chapter 10, Section 1.3.

2.3. SMCC-Activated Enzymes

The heterobifunctional cross-linker SMCC, or its water soluble analog sulfo-SMCC, can be used to activate enzymes through their amines, leaving terminal maleimide groups on the protein surface (Chapter 5, Section 1.3). The NHS ester end of the cross-linker reacts with ε-lysine or N-terminal amines to form amide bonds. The maleimide end of the reagent is stable enough in aqueous solution to allow purification of the activated enzyme prior to conjugation with a second protein. The maleimide group can react with sulfhydryl groups to create thioether linkages. A maleimide-activated enzyme may be stored in a lyophilized state for extended periods without loss of sulfhydryl-coupling capability.

The use of this type of heterobifunctional reagent allows controlled conjugations to take place, precisely regulating the exact ratio of each protein in the final complex and the size of the resultant conjugate. In addition, the second-stage conjugation through sulfhydryl groups provides directed coupling at discrete sites within a protein molecule, thus providing the potential to better avoid active centers or binding regions. For instance, antibody molecules can be coupled to enzymes in their hinge region after mild disulfide reduction to effect the cross-link in an area away from the antigen binding site.

Protocols for the activation of enzyme molecules with SMCC (or sulfo-SMCC) can be found in Chapter 10, Section 1.1. Conjugates formed using this method usually result in high-activity complexes giving excellent sensitivity for use in immunoassays or other applications.

2.4. Hydrazide-Activated Enzymes

Hydrazide functional groups can react with carbonyl groups to form stable hydrazone linkages. Derivatives of proteins formed from the reaction of their carboxylate side chains with adipic acid dihydrazide (Chapter 4, Section 8.1) and the water soluble carbodiimide EDC (Chapter 3, Section 1.1) create activated proteins that can covalently bind to formyl residues. Hydrazide-modified enzymes prepared in this manner can specifically bind to aldehyde groups formed by mild periodate oxidation of carbohydrates (Chapter 1, Section 4.4). These reagents can be used in assay systems to detect or measure glycoproteins in cells, tissue sections, or blots (Gershoni *et al.*, 1985).

Other molecules can be used in this type of assay approach. Hydrazide-modified avidin, streptavidin, lectins, biocytin, fluorescent probes, and other detectable molecules can specifically detect glycoconjugates in this manner (Wilchek and Bayer, 1987).

The activation of enzymes using adipic acid dihydrazide and EDC is identical to the procedure outlined for the modification of avidin or streptavidin (Chapter 13, Section 5).

2.5. SPDP-Activated Enzymes

SPDP is a heterobifunctional cross-linker containing an NHS ester on one end and a pyridyldisulfide group on the other end (Chapter 5, Section 1.1). The NHS ester end can be used to modify amine groups on enzymes, forming amide bonds. The result of this procedure is to create sulfhydryl-reactive pyridyldisulfide groups on the surface of each enzyme molecule that are able to complex with thiol-containing proteins and other molecules. SPDP-activated enzymes may be purified and stored for extended periods without breakdown of its coupling capacity. The reaction with a sulfhydryl group forms a reversible disulfide linkage that can be cleaved with reducing agents.

The two-step nature of SPDP cross-linking provides control over the conjugation process. Complexes of defined composition can be constructed by adjusting the ratio of enzyme to secondary molecule in the reaction as well as the amount of SPDP used in the initial activation. The use of SPDP in conjugation applications is extensive, perhaps making it the most popular cross-linker available. It is commonly used to form immunotoxins, antibody–enzyme conjugates, and enzyme-labeled DNA probes. A standard activation and coupling procedure can be found in Chapter 5, Section 1.1.

3. Preparation of Biotinylated Enzymes

Biotinylated enzymes can be used as detection reagents in avidin–biotin assay procedures. Particularly, in the bridged avidin–biotin (BRAB) approach or the ABC technique (Chapter 13, Section 2), a biotin-labeled enzyme is used as the signaling agent after the binding to an antigen of a biotinylated antibody and an avidin bridging molecule. The biotins on the surface of the enzyme can bind with extraordinary affinity to avidin–antibody complexes, providing near-covalent interaction potential with high specificity.

Adding a biotin label to an enzyme molecule is simple, given the wide variety of options available. A biotinylation reagent is chosen that has a reactive group that will couple to functional groups on the enzyme (Chapter 8, Section 3). For instance, NHS-LC–biotin can be used to modify amine groups—a good choice for most biotinylation procedures involving proteins (Chapter 8, Section 3.1). When free sulfhydryls are present, as in β-gal, a thiol-reactive biotin label may be more appropriate, such as biotin-BMCC (Chapter 8, Section 3.2). The protocols for labeling proteins with these reagents can be found in Chapter 8, Section 3.

17

Nucleic Acid and Oligonucleotide Modification and Conjugation

Molecular biology techniques utilizing nonradioactive detection methods are rapidly replacing radiolabeled probes for assaying oligonucleotide interactions. This trend mimics the movement away from radioimmunoassays, prevalent in the mid-1970s, to nonradioactive ELISA techniques using antibody–enzyme conjugates. A major factor in the development of nonradioactive systems for RNA and DNA measurement is the ability to modify a nucleic acid with a detectable component while not affecting base-pairing. The attachment of a small probe, such as a fluorescent molecule, or a large catalytic enzyme to an oligonucleotide forms the basis for constructing a sensitive hybridization reagent. Unfortunately, the methods developed to cross-link or label proteins do not always apply to nucleic acids. The major reactive sites on proteins involve primary amines, sulfhydryls, carboxylates, or phenolates— groups that are familiar and relatively easy to derivatize. RNA and DNA contain none of these functional groups. They also are unreactive with many of the common bioconjugate reagents discussed in Part II.

To modify the unique chemical groups on nucleic acids, novel methods have been developed that allow derivatization through discrete sites on the available bases, sugars, or phosphate groups (see Chapter 1, Section 3, for a discussion of RNA and DNA structure). These chemical methods can be used to add a functional group or a label to an individual nucleotide or to one or more sites in oligonucleotide probes or full-sized DNA or RNA polymers.

If an individual nucleotide is modified in the appropriate way, various enzymatic techniques can be used to polymerize the derivative into an existing oligonucleotide molecule. Alternatively, nucleotide polymers can be treated with chemical activators that can facilitate the attachment of a label at particular reactive sites. Thus, there are two main approaches to modifying DNA or RNA molecules: enzymatic or chemical. Both procedures can produce highly active conjugates for sensitive hybridization assays to quantify or localize the binding of an oligo to its complementary strand in a complex mixture.

The following sections describe the major enzymatic and chemical modification procedures used to label nucleic acids and oligonucleotides.

1. Enzymatic Labeling of DNA

Enzymatic techniques can employ a variety of DNA or RNA polymerases to add controlled amounts of modified nucleotides to an existing stand. However, the most common procedures utilize either DNA polymerase I or terminal deoxynucleotide transferase. The polymerase is used with a template to add modified nucleoside triphosphates to the end of a DNA molecule or to various sites within the middle of a sequence. The terminal transferase can add modified monomers to the 3' end of a chain without a template.

Three main procedures of enzyme labeling make use of a DNA polymerase: (a) random-primed labeling, (b) nick-translation, and (c) polymerase chain reaction (PCR). In random-primed labeling, modified nucleoside triphosphates are added to a DNA template using a random mixture of hexa-deoxynucleotides to serve as 3'-OH primers. The form of polymerase used is the Klenow fragment, which lacks the 5'–3' exonuclease activity of intact *E. coli* DNA polymerase I (Kessler *et al.*, 1990; Feinberg and Vogelstein, 1983, 1984). This method is a simple way of tagging probes prepared from a restriction digest template with randomly incorporated, labeled nucleotides.

Nick-translation labeling involves the use of a dual enzyme system acting on double-stranded DNA (Rigby *et al.*, 1977; Langer *et al.*, 1981; Höltke *et al.*, 1990). The enzymes pancreatic deoxyribonuclease I (DNase I) and *E. coli* DNA polymerase I act in tandem on a DNA helix to incorporate labeled nucleotides into the sequence. DNase I is capable of breaking phosphodiester bonds in intact DNA double-stranded molecules. If it is used in the presence of magnesium ions, it limits the hydrolysis caused by the enzyme to a single strand at a time within the DNA helix. If DNase I is further restricted in the amount added to a reaction, the number of breaks caused in the double helix can be controlled. The addition of DNA polymerase I and the appropriate labeled and unlabeled nucleotide monomers causes the breaks to be filled as quickly as they form. Since a quantity of labeled nucleoside triphosphates are present during the reaction, the labels get incorporated and the parent DNA strands are modified.

Enzymatic labeling of DNA by use of PCR techniques perhaps provides the most powerful way not only of adding a label, but also of amplifying the labeled polymer to produce numerous copies of itself. First invented by Mullis (who went on to win the Nobel Prize; see Saiki *et al.*, 1985, 1988), PCR utilizes heat-stable forms of DNA polymerase, for example, the commonly employed *Taq* polymerase isolated from thermophilic eubacterium (*Thermus aquaticus*). The stability of the enzyme allows repeated elevated-temperature denaturations of target DNA, followed by hybridization of two primers onto the single strands. *Taq* DNA polymerase then creates a complementary sequence to the two single strands by elongation of the primers. If repeated cycles of denaturation, hybridization, and elongation are done, the result is an exponential amplification of the original DNA strands (for a review of PCR, see Innis *et al.*, 1990). Labeling of these amplified strands can be accomplished by one of two routes: using labeled primers or using labeled deoxynucleoside triphosphates. Either way, the *Taq* polymerase incorporates the labels into the growing DNA copies of each PCR cycle.

Enzymatic labeling using any of these polymerase methods results in derivatized nucleoside triphosphates being incorporated at numerous locations within an oligonucleotide strand. These modifications potentially can interfere with the hybridization of a probe to a complementary sequence, especially if the level of labeling is high. Enzymatic labeling using terminal transferase is a way to avoid derivatization in the middle of a strand, and thus preserve sequence or targeting specificity. The enzyme is able to add deoxyribonucleotides to the 3'-OH ends of existing DNA probes without the need for a template. Since modification is limited to a single end of the oligonucleotide, the probe sequence is not disturbed by labeling groups that could possibly prevent hydrogen bonding interactions between base pairs.

Terminal transferase labeling was originally developed using radiolabeled (typically ^{32}P) nucleoside triphosphates (Roychoudhury *et al.*, 1979; Tu and Cohen, 1980). Later, the technique was extended to the use of nonradioactive nucleotide derivatives (Kumar *et al.*, 1988).

Regardless of the type of enzymatic labeling used, it is important that the label be incorporated into the nucleoside triphosphates or primers in a way that does not affect enzyme recognition and activity. Thus, every enzymatic labeling procedure for modifying RNA or DNA probes must start with chemical derivatization of individual nucleotides. Of the many chemical procedures that can be used to modify a nucleoside triphosphate monomer, there are only a few that will result in a derivative still able to be enzymatically added to an existing oligonucleotide strand.

Of the purine nucleosides, dATP may be derivatized at its N-6 position using a long linker arm terminating in a detectable group without losing the ability to be enzymatically incorporated into DNA probes. By contrast, if modification is done at the C-8 position of purine bases, DNA polymerase cannot be used to add the labeled monomer to an existing strand. C-8 derivatives, however, can be added at the 3' terminal using terminal transferase enzyme.

The pyrimidine nucleosides dUTP or dCTP can be modified at their C-5 position with a spacer arm containing a tag, such as a biotin group, and still remain good substrates for DNA polymerase. Enzymatic labeling with a biotin-modified pyrimidine nucleoside triphosphate is one of the most common methods of adding a detectable group to an existing DNA strand.

Figure 394 illustrates some of the common nucleoside triphosphate derivatives that can be used in enzymatic labeling processes. The following sections describe procedures for enzymatic labeling using nick-translation, random prime labeling, and 3' tailing with terminal transferase. For a review of these methods in greater detail, see Kricka (1992) or Keller and Manak (1989). See Section 2 for a discussion of the chemical methods that can be used to label individual nucleic acids for incorporation into oligonucleotides by enzymatic means.

Protocol for the Labeling of DNA by Nick-Translation

1. In a tube kept cold on ice, add 10 µl of 10× nick-translation buffer (0.5 *M* Tris, 0.1 *M* MgCl$_2$, 0.08 *M* 2-mercaptoethanol, pH 7.5, containing 0.5 mg/ml BSA), 0.5 µg of double-stranded probe DNA to be labeled, 1 µl of DNase I at a concentration of 2 ng/ml, 1 µl each of three types of unmodified deoxynucleoside triphosphates (dNTPs at 100 µ*M* concentration), 1 µl of a labeled dNTP (at 300 µ*M*), 32 µl water. Then add 1 µl of DNA polymerase containing 5–10 units of activity.

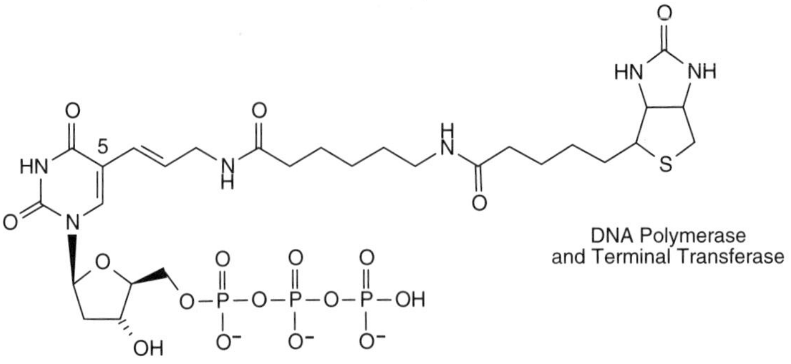

Biotin–11–dUTP
Derivatized at the C-5 Position

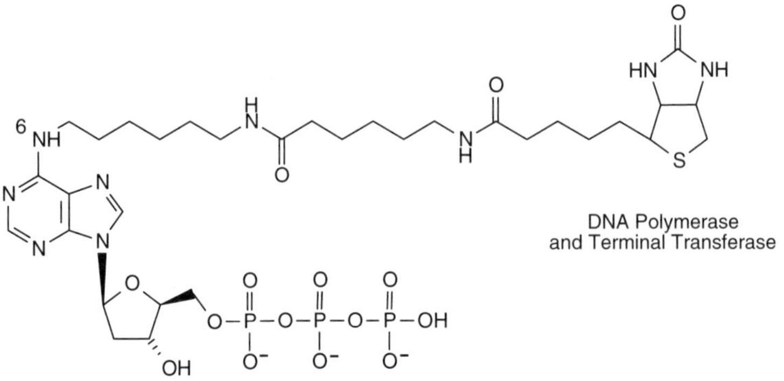

DNA Polymerase
and Terminal Transferase

Biotin–14–dATP
Derivatized at the N-6 Position

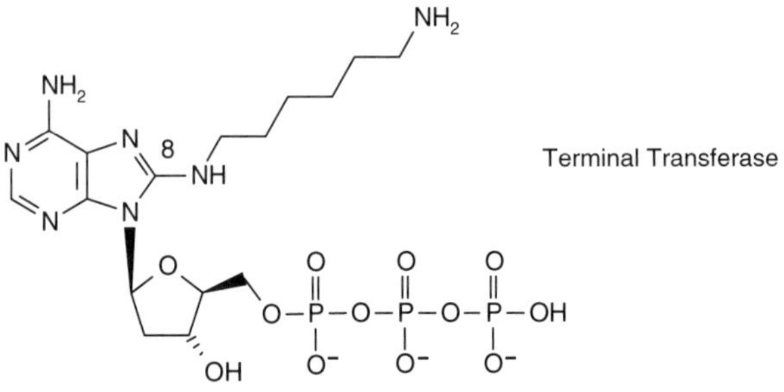

DNA Polymerase
and Terminal Transferase

8–Aminohexyl–dATP
Derivatized at the C-8 Position

Terminal Transferase

Figure 394   Three common nucleoside triphosphate derivatives that can be incorporated into oligonucleotides by enzymatic means. The first two are biotin derivatives of pyrimidine and purine bases, respectively, that can be added to an existing DNA strand using either polymerase or terminal transferase enzymes. Modification of DNA with these nucleosides results in a probe detectable with labeled avidin or streptavidin conjugates. The third nucleoside triphosphate derivative contains an amine group that can be added to DNA using terminal transferase. The modified oligonucleotide then can be labeled with amine-reactive bioconjugation reagents to create a detectable probe.

2. React for 1 h at 15°C.
3. Quench the reaction by the addition of 4 µl of 0.25 *M* EDTA, 2 µl of 10 mg/ml tRNA, and 150 µl of 10 m*M* Tris, pH 7.5.
4. Purify the labeled DNA from excess reactants by precipitation. Add 20 µl of 4 *M* LiCl and 500 µl of ethanol (chilled to −20°C). Mix well.
5. Store at −20°C for 30 min, then separate the precipitated DNA by centrifugation at 12,000 *g*.
6. Remove the supernatant and wash the pellet with 70 and 100% ethanol, centrifuging after each wash.
7. Redissolve the labeled DNA pellet in water and store at −20°C until use.

Protocol for the Labeling of DNA by Random Priming

1. Denature 1 µg of probe DNA (single stranded) with 5 µg of random hexanucleotide primers (Pharmacia) by boiling for 5 min and then rapidly chilling on ice. Incubate at least 10 min to allow the primers to hybridize to random sites within the probe DNA.
2. Add to a tube on ice 5 µl of 10× random priming buffer (0.5 *M* Tris, 0.1 *M* MgCl₂, 10 m*M* 2-mercaptoethanol, pH 6.6, containing 0.5 mg/ml BSA), 1 µl each of three types of unmodified deoxynucleoside triphosphates (dNTPs at 100 µ*M* concentration), 1 µl of a labeled dNTP (at 100 µ*M*), plus 48 µl of water. Then add 2 µl of DNA polymerase (5–10 units).
3. Combine the probe DNA:hexanucleotide preparation with the reaction solution and incubate for 2 h at 37°C.
4. Quench the reaction by the addition of 2 µl of 0.5 *M* EDTA, 2 µl of 10 mg/ml tRNA, and 150 µl of 10 m*M* Tris, pH 7.5.
5. Purify the labeled probe by ethanol precipitation according to steps 4–7 of the protocol previously described for nick-translation.

Protocol for Labeling DNA at the 3' End Using Terminal Transferase

1. Prepare 1 µg of purified DNA probe, either by restriction digestion or by synthetic means.
2. Add to the purified probe (a) 20 µl of 0.5 *M* potassium cacodylate, 5 m*M* CoCl₂, 1 m*M* DTT, pH 7, (b) 100 µ*M* of a modified deoxynucleoside triphosphate, 4 µl of 5 m*M* dCTP, and 100 µl of water. Mix.
3. Add terminal transferase to a final concentration of 50 units in the reaction mixture.
4. React for 45 min at 37°C.
5. Isolate the labeled probe by alcohol precipitation as described previously for nick-translation.

2. Chemical Modification of Nucleic Acids and Oligonucleotides

The chemical modification of nucleic acids at specific sites within individual nucleotides or within oligonucleotides allows various labels to be incorporated into DNA or RNA polymers. This labeling process can produce conjugates having sensitive detection properties for the localization or quantification of oligo binding to a complementary strand using hybridization assays.

Some form of chemical labeling process must be used regardless of whether the final oligo conjugate is created by enzymatic or strictly chemical means. If enzymatic modification is to be done, the initial label still must be incorporated into an individual nucleoside triphosphate, which then is polymerized into an existing oligonucleotide strand (Section 1). Fortunately, many useful modified nucleoside triphosphates are now available from commercial sources, often eliminating the need for custom derivatization of individual nucleotides.

Chemical modification also may be used to label directly an oligonucleotide, eliminating the enzymatic step altogether. The chemical modification of nucleic acids can encompass several strategies. The initial derivatization only may be to add a spacer arm to a particular reactive group on the nucleotide structure. The spacer typically contains a terminal functional group, such as an amine, that can be used to couple another molecule. A secondary modification might be to add a fluorescent tag to the end of the spacer, thus creating a detectable complex. The spacer also may be used to react with a cross-linking agent, such as a heterobifunctional compound (Chapter 5), that can facilitate the conjugation of a protein or another molecule to the modified nucleotide. It should be noted that if enzymatic methods are used to incorporate a small spacer into an oligonucleotide, subsequent chemical conjugation steps still will be needed to add a small label or protein tag.

In some cases, if an oligonucleotide contains the appropriate functional group, a label may be directly incorporated into it using chemical methods. For instance, certain fluorescent molecules or biotin tags can be used to modify nucleotides without going through an initial derivatization step with a spacer arm. Such labels usually contain nucleophilic groups or photoreactive portions that can couple directly to the oligo using an intermediate activating agent or by photolyzing with UV light.

Many of the chemical derivatization methods employed in these strategies involve the use of an activation step that produces a reactive intermediary. The activated species then can be used to couple a molecule containing a nucleophile, typically a primary amine. The following sections describe the chemical modification methods suitable for derivatizing individual nucleic acids as well as oligonucleotide polymers.

2.1. Diamine or Bis-hydrazide Modification of DNA

One of the more useful chemical modifications that can be done on nucleic acids or oligonucleotides is to add an amine-terminal spacer arm using a diamine compound. The resultant amine derivative can be targeted by numerous amine-reactive cross-linkers or tags to create a detectable conjugate. A similar approach modifies a DNA probe with a bis-hydrazide compound (Chapter 4, Section 8) to produce terminal hydrazide functional group. The oligonucleotide derivative then can be coupled with aldehyde-containing molecules to form conjugates. The following methods utilize activation reagents that transform a particular site on nucleic acids into an amine-reactive or hydrazide-reactive intermediate. Coupling a diamine or bis-hydrazide compound to these activated species results in the formation of an alkyl spacer arm terminating in a primary amine group or a hydrazide functional group, respectively.

Via Bisulfite Activation of Cytosine

Single-stranded DNA molecules can react with sodium bisulfite, adding a sulfonate group across the 5,6 double bond of cytidine bases and creating 6-sulfo-cytosine derivatives. The reaction catalyzes the deamination of cytosine to uracil by loss of the 4-amino group. Subsequent loss of HSO_3^- effectively forms uracil bases (Fig. 395). This reaction sequence was recognized in the early 1970s as potential evidence for the mutagenicity of bisulfite (Shapiro *et al.*, 1973, 1974). Shapiro and Weisgras (1970) demonstrated that the bisulfite reaction also can cause transamination to occur at the N-4 position of cytosine. In the presence of an amine-containing molecule, such as a diamine, sodium bisulfite will cause the exchange of the N-4 amine for another amine-containing compound, effectively forming a new covalent linkage with release of ammonium ion (Fig. 396). Draper and Gold (1980) used this reaction to produce primary amine groups on a poly(C) oligonucleotide by coupling diaminopropane to a limited number of the cytosine residues. The amine derivative subsequently could be used to couple a fluorescent probe to the polymer, allowing sensitive studies of messenger RNA.

Bisulfite-catalyzed transamination also can be used to label oligonucleotide probes for application in nonradioisotopic hybridization assays. Viscidi *et al.* (1986) described a method for derivatizing cytosine groups in DNA probes using the short spacer ethylene diamine. Other diamine molecules also may be used, such as 1,3-diaminopropane, 1,6-diaminohexane, or 3,3′-iminobispropylamine. The use of the long, hydrophilic Jeffamine molecules (Texaco Chemical Co; see Chapter 1, Section 4.3) may be especially well suited to this type of modification. Longer spacer arms may provide better steric accommodation for larger detection components without inter-

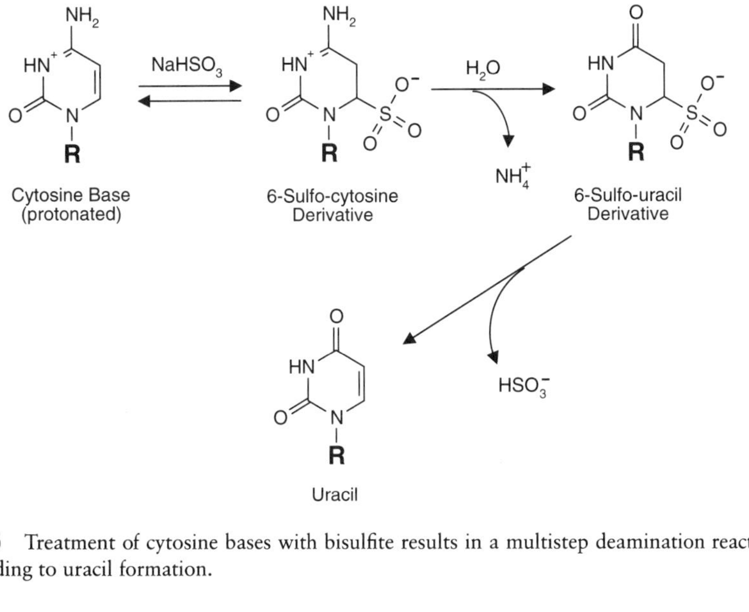

Figure 395   Treatment of cytosine bases with bisulfite results in a multistep deamination reaction, ultimately leading to uracil formation.

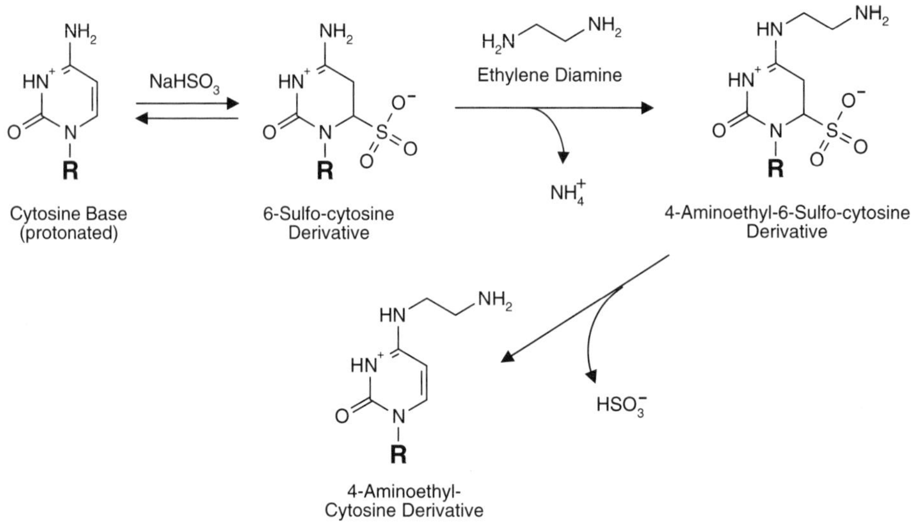

Figure 396   The reaction of cytosine with bisulfite in the presence of an excess of an amine nucleophile (such as a diamine compound) leads to transamination at the N_4 position. This process is a route to adding an amine functional group to cytosine residues in oligonucleotides.

fering substantially in the probe's ability to hybridize to a complementary DNA strand.

If an amine-containing fluorescent probe or hydrazide-containing compound is transaminated onto an oligonucleotide using bisulfite, the labeling of nucleic acids can be done in a single step. An example of this approach is the coupling of biotin hydrazide (Chapter 8, Section 3.3) to cytosine residues, resulting in a biotinylated oligonucleotide suitable for avidin- or streptavidin-based detection systems (Reisfeld *et al.*, 1987) (Section 2.3).

Since the site of modification on cytosine bases is at a hydrogen bonding position in double helix formation, the degree of bisulfite derivatization should be carefully controlled. Reaction conditions such as pH, diamine concentration, and incubation time and temperature affect the yield and type of products formed during the transamination process. At low concentrations of diamine, deamination and uracil formation dramatically exceed transamination. At high concentrations of diamine (3 M), transamination can approach 100% yield (Draper and Gold, 1980). Ideally, only about 30–40 bases should be modified per 1000 bases to ensure hybridization ability after derivatization.

Bisulfite modification of cytosine residues also can be used to add a sulfone group permanently to the C-6 position. In this scheme, the sulfone functions as a hapten recognizable by specific antisulfone antibodies. At high concentrations of bisulfite and in the presence of methylhydroxylamine, cytosines are transformed into N^4-methoxy-5,6-dihydrocytosine-6-sulfonate derivatives (Herzberg, 1984; Nur *et al.*, 1989). Labeled antibodies can then be used to detect the hybridization of such probes.

Protocol for Labeling Nucleic Acids
by Bisulfite-Catalyzed Transamination

1. Prepare single-stranded DNA (denatured) at a concentration of 1 mg/ml.
2. Prepare bisulfite modification solution consisting of 3 M concentration of a diamine (i.e., ethylenediamine), 1 M sodium bisulfite, pH 6. The use of the dihydrochloride form of the diamine avoids the necessity to adjust the pH down from the severe alkaline pH of a free-base form. Note: The optimum pH for transaminating biotin–hydrazide to cytosine residues using bisulfite is 4.5 (see Section 2.3).
3. Add 20 μl of the DNA to 180 μl of bisulfite modification solution. Mix well.
4. React for 3 h at 42°C.
5. Dialyze the solution against water overnight at 4°C to remove excess reactants.
6. The modified DNA may be recovered by alcohol precipitation according to the method in Section 1 described previously for nick-translation modification. Alternatively, dialysis or gel filtration may be done to remove excess reactants.

Via Bromine Activation of Thymine, Guanine, and Cytosine

The nucleotide bases of DNA and RNA can be activated with bromine to produce reactive intermediates capable of coupling to nucleophiles (Traincard *et al.*, 1983; Sakamoto *et al.*, 1987; Keller *et al.*, 1988). Bromination occurs at the C-8 position of guanine residues and the C-5 of cytosine, yielding reactive derivatives that can be used to couple diamine spacer molecules by nucleophilic substitution (Fig. 397). Other pyrimidine derivatives also are reactive to bromine compounds, but adenine residues are more resistant. However, even AMP can be immobilized through the introduction of an aminohexyl spacer at the C-8 position using bromination (Lowe, 1979). Either an aqueous solution of bromine or the compound *N*-bromosuccinimide can be used for this reaction. The alkaline modification proceeds rapidly, but may be too severe for RNA molecules. Coupling of amine-containing molecules is done at elevated temperatures (50°C) to ensure good incorporation. Both amine-bearing spacers or probes may be coupled using this strategy. Moreover, the sites of derivatization using bromine activation are not involved in hydrogen bonding during base-pairing, thus maintaining hybridization ability in the final conjugate.

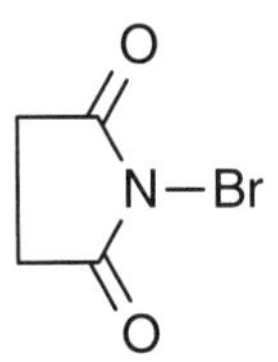

N–Bromosuccinimide
MW 177.99

Optimal bromination of a DNA probe is in the range 30–35 bases per 1000 bases, a level that can be controlled by the amount of *N*-bromosuccinimide added. Over-labeling can prevent specific interactions with target DNA, even if the point of initial modification is not a hydrogen bonding site.

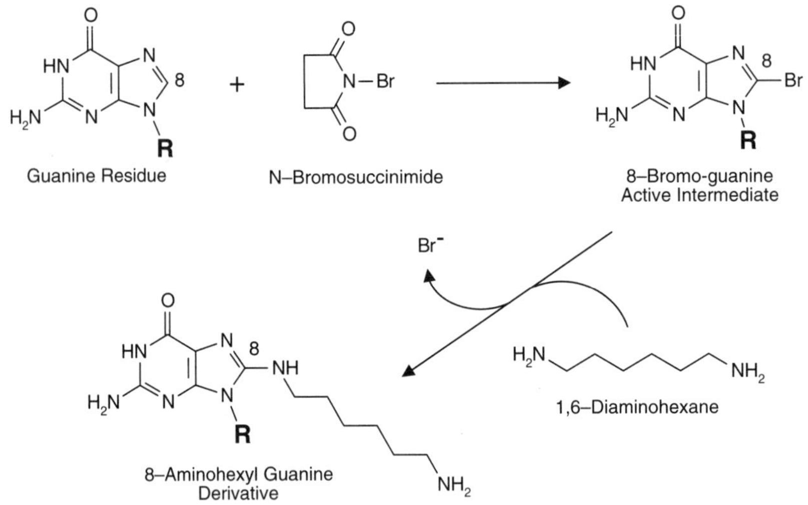

Figure 397   Reaction of guanine bases with N-bromosuccinimide causes bromination at the C-8 position of the ring. Amine nucleophiles can be coupled to this active derivative by nucleophilic displacement. Reaction of diamine compounds results in amine terminal spacers that can be further modified to contain detectable components.

The major disadvantage with bromination is the extreme toxicity of bromine. Use a fume hood for all operations. Avoid the breathing of fumes or contact with skin or eyes. Protective clothing and gloves are recommended.

Protocol for Labeling Nucleic Acids by N-Bromosuccinimide Activation
Bromination

1. Mix in a microfuge tube 20 μg of the DNA probe to be labeled, 20 μl of 1 M sodium bicarbonate, pH 9.6, and 196 μl of water. Chill on ice.
2. In a fume hood, dissolve N-bromosuccinimide (Sigma) in water at a concentration of 1.42 mg/ml.
3. Add 4 μl of the N-bromosuccinimide solution to the DNA solution (makes an 8 mM final concentration of brominating reagent). Mix well.
4. React on ice for 10 min. Use the bromine-activated DNA immediately.

Coupling a Diamine-Containing Spacer or Probe

1. Dissolve a diamine spacer (i.e., ethylene diamine or 1,6-diaminohexane—Aldrich) in water at a concentration of 80–100 mM. Caution: amine-containing molecules such as these diamines may be highly corrosive if they are in the free-base form (not the dihydrochlorides). Wear gloves and other protective clothing. The pH of an aqueous solution of free-base diamines will be >12 and may fume. The solution also may generate heat upon dissolution of the amine. Keeping it in an ice bath will help maintain a cool solution with less fuming. Using a dihydrochloride form of a diamine will avoid the problems associated with corrosiveness, heat, and fuming.

2. Add 25 μl of the diamine solution to the bromine-activated DNA solution prepared in under Bromination, above.
3. React for 1 h at 50°C.
4. The diamine-modified DNA may be isolated from excess reactants by ethanol precipitation according to steps 4–7 of the protocol described previously for nick-translation (Section 1). Alternatively, dialysis or gel filtration may be done to remove excess reactants.

Via Carbodiimide Reaction with 5′ Phosphates of DNA (Phosphoramidate Formation)

The water-soluble carbodiimide EDC (Chapter 3, Section 1.1), rapidly reacts with carboxylates or phosphates to form an active complex able to couple with primary amine-containing compounds. The carbodiimide activates an alkyl phosphate group to a highly reactive phosphodiester intermediate. Diamine spacer molecules or amine-containing probes then may react with this active species to form a stable phosphoramidate bond. Alternatively, bis-hydrazide compounds (Chapter 4, Section 8) may be coupled to DNA using this protocol to result in terminal hydrazide functional groups able to react with aldehyde-containing molecules (Ghosh et al., 1989). Specific labeling of DNA probes only at the 5′ end is possible using these technique.

Carbodiimide modification of the phosphomonoester end groups on DNA molecules was first used in Khorana'a lab to determine nucleotide sequences (Ralph et al., 1962). That early work used the water-insoluble reagent DCC (Chapter 3, Section 1.4) in an organic/aqueous solvent system to effect the conjugations.

Chu et al. (1983, 1986) and Ghosh et al. (1990) describe modified carbodiimide protocols using the water-soluble reagent EDC instead of DCC. They also incorporate a second reactive intermediate, a phosphorimidazolide, created from the reaction of the phosphomonoester at the 5′-terminus of DNA with EDC in the presence of imidazole. A reactive phosphorimidazolide will rapidly couple to amine-containing molecules to form a phosphoramidate linkage (Fig. 398). The chemical reaction had been used previously to effect the formation of phosphodiester linkages between short DNA strands (Shabarova et al., 1983).

The formation of a phosphorimidazolide intermediate provides better reactivity toward amine nucleophiles than the EDC phosphodiester intermediate if EDC is used without added imidazole. The EDC phosphodiester intermediate also is shorter-lived in aqueous conditions due to hydrolysis than the imidazolide. Although EDC alone will create nucleotide phosphoramidate conjugates with amine-containing molecules (Shabarova, 1988), the result of forming the secondary phosphorimidazolide-activated species is increased derivatization yield over that of carbodiimide-only reactions.

The downside of EDC conjugation with oligonucleotides is the potential for reaction of the carbodiimide at the guanosine N-1 site or with thymidine residues (von der Haar et al., 1971). In practice, however, this cross-reactivity appears to be low enough to maintain complete biological activity and hybridization efficiency in the final conjugate, indicating that most of the derivatization occurs at the 5′ phosphate group (Chu et al., 1983).

The following protocol describes the modification of DNA or RNA probes at their 5′-phosphate ends with a bis-hydrazide compound, such as adipic acid dihydrazide or

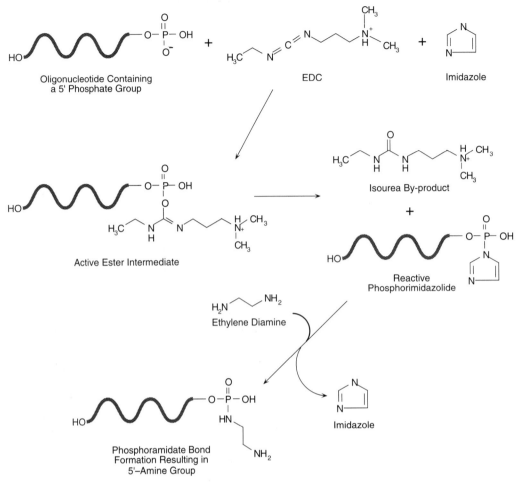

Figure 398   Oligonucleotides containing a 5′-phosphate group can be reacted with EDC in the presence of imidazole to form an active phosphorimidazolide intermediate. This derivative is highly reactive with amine nucleophiles, forming a phosphoramidate linkage. Diamines reacted with the phosphorimidazolide result in amine terminal spacers that can be modified with detectable components.

carbohydrazide. A similar procedure for coupling the diamine compound cystamine can be found in Section 2.2.

Protocol

1. Weigh out 1.25 mg of the carbodiimide EDC (1-ethyl-3-(3-dimethylamino-propyl)carbodiimide hydrochloride; Pierce) into a microfuge tube.
2. Add to the tube 7.5 μl of RNA or DNA containing a 5′ phosphate group. The concentration of the oligonucleotide should be 7.5–15 nmol or a total of about 57–115.5 μg. Also immediately add 5 μl of 0.25 M bis-hydrazide compound dissolved in 0.1 M imidazole, pH 6. Because EDC is labile in aqueous solutions, the addition of the oligo and bis-hydrazide/imidazole solutions should be done quickly.

3. Mix by vortexing, then place the tube in a microcentrifuge and spin for 5 min at maximal rpm.
4. Add an additional 20 μl of 0.1 M imidazole, pH 6. Mix and react for 30 min at room temperature.
5. Purify the hydrazide-labeled oligo by gel filtration on Sephadex G-25 using 10 mM sodium phosphate, 0.15 M NaCl, 10 mM EDTA, pH 7.2. The probe now may be used to conjugate with an aldehyde-containing molecule.

2.2. Sulfhydryl Modification of DNA

Creating a sulfhydryl group on nucleic acid probes allows conjugation reactions to be done with sulfhydryl-reactive heterobifunctional cross-linkers (Chapter 5), providing increased control over the derivatization process. Proteins can be activated with a cross-linking agent containing an amine-reactive and a sulfhydryl-reactive end, such as SPDP (Chapter 5, Section 1.1), leaving the sulfhydryl-reactive portion free to couple with the modified DNA probe. Having a sulfhydryl group on the probe directs the coupling reaction to discrete sites on the nucleotide strand, thus better preserving hybridization ability in the final conjugate. In addition, heterobifunctional cross-linkers of this type allow two- or three-step conjugation procedures to be done, which result in better yield of the desired conjugate than when using homobifunctional reagents.

Cystamine Modification of 5′ Phosphate Groups Using EDC

DNA or RNA may be modified with cystamine at the 5′ phosphate group using a carbodiimide reaction identical to that described previously (Section 2.1). In some procedures, the reaction is carried out in a two-step process by first forming a reactive phosphorylimidazolide by EDC conjugation in an imidazole buffer. Next, cystamine is reacted with the activated oligonucleotide, causing the imidazole to be replaced by the amine and creating a phosphoramidate linkage (Chu et al., 1986). An easier protocol was described by Ghosh et al. (1990) in which the oligo, cystamine, and EDC were all reacted together in an imidazole buffer. A modification of this method developed by Zanocco et al. (1993) is described below.

Reduction of the cystamine-labeled oligo using a disulfide reducing agent releases 2-mercaptoethylamine and creates a thiol group (Fig. 399). DNA probes labeled in this manner have been successfully conjugated with SPDP-activated alkaline phosphatase (Chapter 16, Section 1.5, and Section 2.4), maleimide-activated horseradish peroxidase (section 3.8.1), NHS-LC–biotin (Chapter 8, Section 3.1, and Section 2.3), and the fluorescent tag AMCA–HPDP (Chapter 8, Section 1.3, and Section 2.5).

A kit designed specifically to perform 5′-phosphate labeling on DNA probes is available from Pierce.

Protocol

1. Weigh out 1.25 mg of the carbodiimide EDC (Pierce) into a microfuge tube.
2. Add to the tube 7.5 μl of RNA or DNA containing a 5′ phosphate group. The concentration of the oligonucleotide should be 7.5–15 nmol or total of about

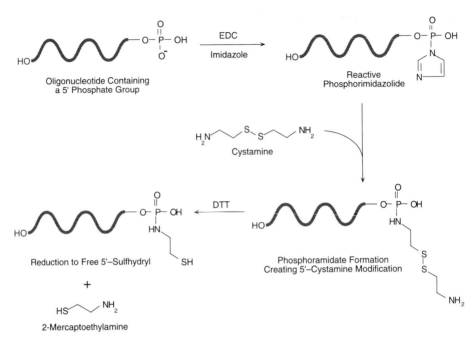

Figure 399 The 5′-phosphate group of oligonucleotides may be labeled with cystamine using the EDC/imidazole reaction. This results in the formation of an amine terminal spacer containing an internal disulfide group. Reduction of the disulfide provides a route to creating a free thiol for further derivatization.

57–115.5 μg. Also, immediately add 5 μl of 0.25 M cystamine in 0.1 M imidazole, pH 6. Because EDC is labile in aqueous solutions, the addition of the oligo and cystamine/imidazole solutions should be done quickly.

3. Mix by vortexing, then place the tube in a microcentrifuge and spin for 5 min at maximal rpm.

4. Add an additional 20 μl of 0.1 M imidazole, pH 6. Mix and react for 30 min at room temperature.

5. For reduction of the cystamine disulfides, add 20 μl of 1 M DTT and incubate at room temperature for 15 min. This will release 2-mercaptoethylamine from the cystamine modification site and create the free sulfhydryl on the 5′ terminus of the oligonucleotide.

6. Purify the SH-labeled oligo by gel filtration on Sephadex G-25 using 10 mM sodium phosphate, 0.15 M NaCl, 10 mM EDTA, pH 7.2. The probe now may be used to conjugate with an activated enzyme, biotin, fluorescent tag, or other molecules containing a sulfhydryl-reactive group.

SPDP Modification of Amines on Nucleotides

Oligonucleotide probes that have been modified with an amine-terminal spacer arm using any of the methods discussed in Sections 1 and 2, may be thiolated to contain a sulfhydryl residue. Theoretically, any of the amine-reactive thiolation reagents described in Chapter 1, Section 4.1, may be used to convert an amino group on a DNA molecule into a thiol. One of the more common choices, both for cross-linking and for thiolation reactions, is the heterobifunctional reagent SPDP (Chapter 5, Section 1.1). The NHS ester end of SPDP reacts with primary amine groups to produce stable amide bonds. The other end of the cross-linker contains a thiol-reactive pyridyldisulfide group that also can be reduced with DTT to create a free sulfhydryl.

The reaction of a 5′-diamine-modified oligonucleotide probe with SPDP proceeds under mildly alkaline conditions (optimal pH 7–9) to give the pyridyldisulfide-activated intermediate (Fig. 400). This derivative has dual functions. It can be used to couple directly with sulfhydryl-containing detection reagents or enzymes, or it may be converted into a free sulfhydryl for coupling to thiol-reactive compounds (Gaur *et al.*, 1989; Gaur, 1991). In an alternative approach, Chu and Orgel (1988) used 2,2′-dipyridyldisulfide (Chapter 1, Section 5.2) to create reactive pyridyldisulfide groups on a reduced 5′-cystamine-labeled oligonucleotide probe. This derivative then can be used to couple with sulfhydryl-containing molecules, forming a disulfide bond.

Reduction of the pyridyldisulfide end after SPDP modification releases the pyridine-2-thione leaving group and generates a terminal —SH group. This procedure allows sulfhydryl-reactive derivatives such as maleimide-activated enzymes (Chapter 16, Sec-

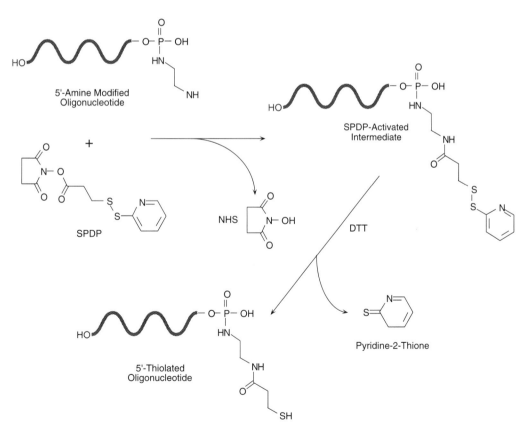

Figure 400 A oligonucleotide modified at its 5′-phosphate with a diamine compound may be reacted with SPDP and subsequently reduced to create a free sulfhydryl.

tion 1) to be conjugated with DNA probes for use in hybridization assays (Malcolm and Nicolas, 1984).

Protocol

1. Dissolve the amine-modified oligonucleotide to be thiolated in 250 μl of 50 m*M* sodium phosphate, pH 7.5.
2. Dissolve SPDP (Pierce) at a concentration of 6.2 mg/ml in DMSO (makes a 20 m*M* stock solution). Alternatively, LC-SPDP may be used and dissolved at a concentration of 8.5 mg/ml in DMSO (also makes a 20 m*M* solution). The "LC" form of the cross-linker provides a longer spacer arm that often results in better probe activity after modification. If the water-soluble Sulfo-LC-SPDP is used, a stock solution in water may be prepared just prior to addition of an aliquot to the thiolation reaction. In this case, prepare a 10 m*M* solution of Sulfo-LC-SPDP by dissolving 5.2 mg/ml in water. Since an aqueous solution of the cross-linker will degrade by hydrolysis of the sulfo-NHS ester, it should be used quickly to prevent significant loss of activity.
3. Add 50 μl of the SPDP (or LC-SPDP) solution to the oligo solution. Add 100 μl of the Sulfo-LC-SPDP solution, if the water-soluble cross-linker is used. Mix.
4. React for 1 h at room temperature.
5. Remove excess reagents from the modified oligo by gel filtration. The modified probe now may be used to conjugate with a sulfhydryl-containing molecule, or it may be reduced to create a thiol for conjugation with sulfhydryl-reactive molecules.
6. To release the pyridine-2-thione leaving group and form the free sulfhydryl, add 20 μl of 1 *M* DTT and incubate at room temperature for 15 min. If present in sufficient quantity, the release of pyridine-2-thione can be followed by its characteristic absorbance at 343 nm ($\varepsilon = 8.08 \times 10^3$ M^{-1} cm^{-1}). For many oligonucleotide modification applications, however, the leaving group will be present in too low a concentration to be detectable.
7. Purify the thiolated oligonucleotide from excess DTT by dialysis or gel filtration using 50 m*M* sodium phosphate, 1 m*M* EDTA, pH 7.2. The modified probe should be used immediately in a conjugation reaction to prevent sulfhydryl oxidation and formation of disulfide cross-links.

SATA Modification of Amines on Nucleotides

Oligonucleotides containing amine groups introduced by enzymatic or chemical means may be modified with SATA (Chapter 1, Section 4.1) to produce protected sulfhydryl derivatives. The NHS ester end of SATA reacts with a primary amine to form a stable amide bond. After modification, the acetyl protecting group can be removed as needed by treatment with hydroxylamine under mildly alkaline conditions (Fig. 401). The result is terminal sulfhydryl groups that can be used for subsequent labeling with thiol-reactive probes or activated-enzyme derivatives (Kumar and Malhotra, 1992).

The advantage of using SATA over disulfide-containing thiolation reagents such as SPDP (previous section) is that the introduction of sulfhydryl residues does not include the use of a disulfide reducing agent. Typically, the pyridyl dithiol group resulting from an SPDP thiolation must be reduced with a sulfhydryl-containing disulfide reducing

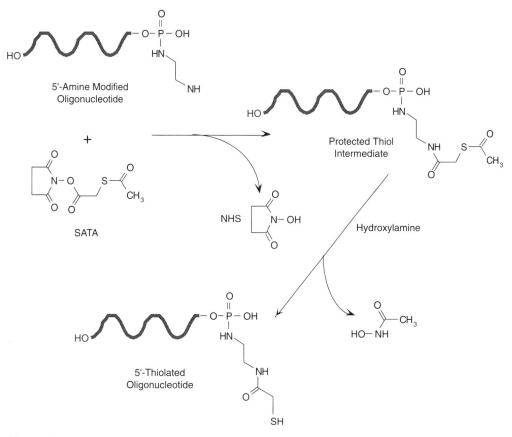

Figure 401   SATA may be used to modify a 5′-amine derivative of an oligonucleotide, forming a protected sulfhydryl. Deprotection with hydroxylamine results in generation of a free thiol.

compound like DTT to free the —SH group. With SATA, the sulfhydryl is freed by hydroxylamine cleavage, thus eliminating the need for removal of sulfhydryl reductants prior to a conjugation reaction.

Protocol

1. Dissolve the amine-modified oligonucleotide to be thiolated in 250 μl of 50 mM sodium phosphate, pH 8.
2. Dissolve SATA in DMF at a concentration of 8 mg/ml.
3. Add 250 μl of the SATA solution to the oligo solution. Mix.
4. React for 3 h at 37°C.
5. Remove excess reagents from the modified oligo by gel filtration.
6. To deprotect the thioacetyl group, add 100 μl of 50 mM hydroxylamine hydrochloride, 2.5 mM EDTA, pH 7.5.
7. React for 2 h.
8. The sulfhydryl-containing oligonucleotide may be used immediately to conjugate with a sulfhydryl-reactive label, or it can be purified from excess hydroxylamine by gel filtration.

2.3. Biotin Labeling of DNA

Biotinylation of oligonucleotides probes provides a highly specific biological recognition site for detection of DNA using avidin or streptavidin conjugates. The preparation of biotin-labeled DNA can be done by either enzymatic or chemical means. Enzyme-catalyzed reactions utilize biotinylated nucleoside triphosphates that can be incorporated into an oligonucleotide randomly or at the 3′ terminus (Section 1). Chemical derivatization methods make use of certain reactive biotin compounds that can couple to functionally modified probes or react with DNA with the use of an activating reagent. For a description of the wide range of biotinylation compounds available, see Chapter 8, Section 3. The preparation and use of avidin and streptavidin conjugates is discussed in Chapter 13.

Biotin–LC–dUTP

Perhaps the most common method of DNA biotinylation is through enzymatic incorporation with the use of a biotin-labeled deoxynucleoside triphosphate. First reported by Langer *et al.* and Leary *et al.* in 1981, the procedure is probably the most popular nonradioactive labeling technique reported for oligonucleotide probes. Although biotinylated derivatives of dCTP and dATP are reported in the literature, by far the most frequently employed derivative is biotin–dUTP prepared from the reaction of an amine-modified dUTP with an amine-reactive biotinylation reagent, such as NHS-LC–biotin (Chapter 8, Section 3.1).

Biotin–dUTP derivatives are formed by modification of the C-5 position of uridine. This location is not involved in hydrogen bonding activity with complementary DNA strands; thus hybridization efficiency is not immediately compromised. By contrast, biotin–dCTP or biotin–dATP derivatives involve modification of the bases at the N-4 position of cytosine and the N-6 position of adenine, locations directly involved in hydrogen bond formation with complementary bases. Thus, DNA biotinylation through the use of modified deoxynucleoside triphosphates to be incorporated into existing DNA strands may result in better activity of the probe if dUTP is used over dATP or dCTP.

The length of the spacer arm between the C-5 position of uridine and the biotin group is another important parameter for activity of the resulting conjugate. The spacer affects the incorporation efficiency into existing probes using DNA polymerases, and it also affects the ability of an avidin or streptavidin conjugate to bind effectively the biotinylated probe. Designation of spacer length is usually expressed as the number of atoms separating the nucleotide base from the biotin component. Thus, biotin–*n*-dUTP would describe a biotinylated deoxynucleoside triphosphate having a spacer arm *n* atoms long. The shorter the spacer arm, the better the derivative is able to be recognized and incorporated into DNA polymers using polymerase enzymes. Conversely, the longer the spacer arm, the better the biotinylated probe is able to hybridize to its target and still maintain the capacity to have a streptavidin conjugate be complexed with it. Thus, there is an optimal trade-off in spacer length between enzymatic incorporation efficiency and labeled-probe detectability. Studies have determined that this optimal range is rather broad—between 7 and 21 atoms in length. Perhaps the most common derivative is biotin–11-dUTP, wherein an 11-atom spacer is employed (Fig. 402).

General protocols for the enzymatic incorporation of biotin–11-dUTP into DNA

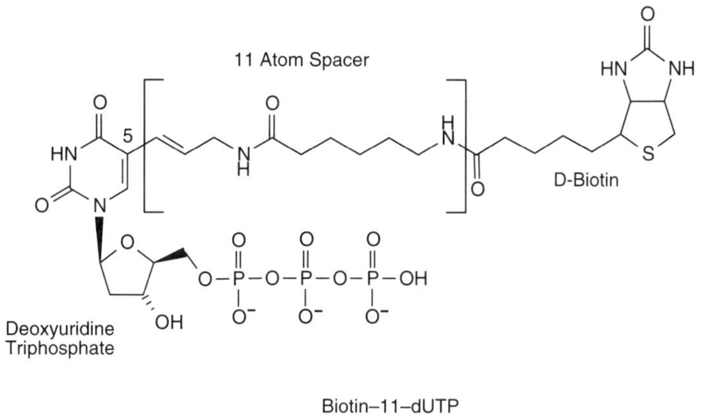

Figure 402 Biotin−11−dUTP is perhaps the most popular nucleotide derivative used for enzymatic biotinylation of oligonucleotides. The "11" designation refers to the number of atoms in its spacer arm.

probes can be found in Section 1. A particularly interesting modification of the typical enzymatic incorporation protocol for biotin is described by Didenko (1993). The single-strand template is immobilized by adsorption onto membranes before synthesis of the biotinylated probe. After polymerase incorporation of biotin−11−dUTP, the labeled probe is removed by brief heating to 90°C in water. The result is highly pure probe with no contaminating complementary DNA strands.

Photo-Biotin Modification of DNA

The photoreactive biotin derivative N-(4-azido-2-nitrophenyl)-aminopropyl-N'-(N-d-biotinyl-3-aminopropyl)-N'-methyl-1,3-propanediamine, simply called photoactivatable biotin or photobiotin (Forster *et al.*, 1985), contains a 9-atom diamine spacer group on the biotin valeric acid side chain on one end, while the other end of the spacer terminates in an aryl azide functional group. The phenyl azide group can be photolyzed with UV light (350 nm), resulting in the formation of a highly reactive nitrene intermediate. In most instances, this nitrene rapidly reacts via ring expansion to form a dehydroazepine that is reactive with nucleophiles, such as amine groups (Chapter 5, Section 3).

When photobiotin is irradiated in the presence of DNA the reaction process nonselectively couples a biotin label to every 100–200 base residues. The result is a oligonucleotide probe detectable by the use of avidin or streptavidin conjugates. The uses of photobiotin for DNA or RNA modification are summarized in Chapter 8, Section 3.4.

The following protocol is based on the method of Forster *et al.* (1985). Some optimization may be necessary to obtain the best signal and activity for particular probes in hybridization assays.

Protocol for Labeling DNA Probes with Photobiotin

1. In subdued lighting conditions, dissolve photobiotin in water at a concentration of 1 μg/μl. Protect from light.

2. Dissolve the oligonucleotide probe in water or 0.1 mM EDTA, pH 7, at a concentration of 1 μg/μl.
3. Mix an equal volume of the photobiotin solution with the DNA probe solution.
4. Place the solution in an ice bath and irradiate from above (about 10 cm away) for 15 min using a sunlamp (such as Philips Ultrapnil MLU 300 W, General Electric sunlamp RSM 275 W, or National Self-Ballasted BHRF 240–250 V 250 W W-P lamp).
5. Add 50 μl of 0.1 M Tris, pH 9, and increase the total volume of the solution to 100 μl (if it is less than this amount).
6. To extract excess photobiotin, add 100 μl of 2-butanol. Mix well and centrifuge. Discard the upper phase. Repeat this process two more times.
7. To recover the biotinylated DNA, add 75 μl of 4 M NaCl and mix.
8. Add 100 μl of ethanol and cool the sample in dry ice (CO_2) for 15 min.
9. Centrifuge to collect the precipitated, biotinylated DNA.

Reaction of NHS-LC–Biotin with Diamine-Modified DNA Probes

NHS-LC–biotin is an extended spacer arm derivative of biotin containing an amine-reactive NHS ester (Chapter 8, Section 3.1). The compound is a popular choice for biotinylating a wide range of molecules containing primary amine groups, especially proteins. Oligonucleotides modified to contain amine-terminal spacer arms also can be modified with NHS-LC–biotin to create stable amide bond derivatives. Whether an amine is incorporated into an oligo by enzymatic means or chemical derivatization, an NHS ester-containing biotinylation reagent can be used to label the derivative in high yield. If an amine group is added to the 5′ end of a DNA probe by phosphoramidate formation (Section 2.1), then biotinylation of such molecules directs the label to a region totally removed from interfering in subsequent hybridization with a target DNA strand (Fig. 403).

The following protocol assumes that the amine-containing oligo has already been synthesized by any of the methods discussed in Section 2.1.

Protocol

1. Prepare 10–20 μg of amine-containing oligonucleotide in 200 μl of water. Add to this solution 20 μl of 1 M sodium bicarbonate, pH 9.
2. Dissolve NHS-LC–biotin (Pierce) in DMSO at a concentration of 10 mg/ml. Add 50 μl of the biotinylation solution to the oligo solution. Mix well.
3. React for 2 h at room temperature.
4. Isolate the biotinylated probe by ethanol/salt precipitation as described in Section 1 for nick-translation modification of DNA probes. Alternatively, dialysis, gel filtration, or n-butanol extraction may be used to remove excess reagents.

Biotin–Diazonium Modification of DNA

Diazonium groups are able to couple at the C-8 position of adenosine or guanosine residues, forming diazo bonds. p-Aminobenzoyl biocytin can be used in this reaction to add a biotin handle to purine bases within oligonucleotides (Chapter 8, Section 3.5). This biotinylation reagent contains a 4-aminobenzoic acid amide derivative off the

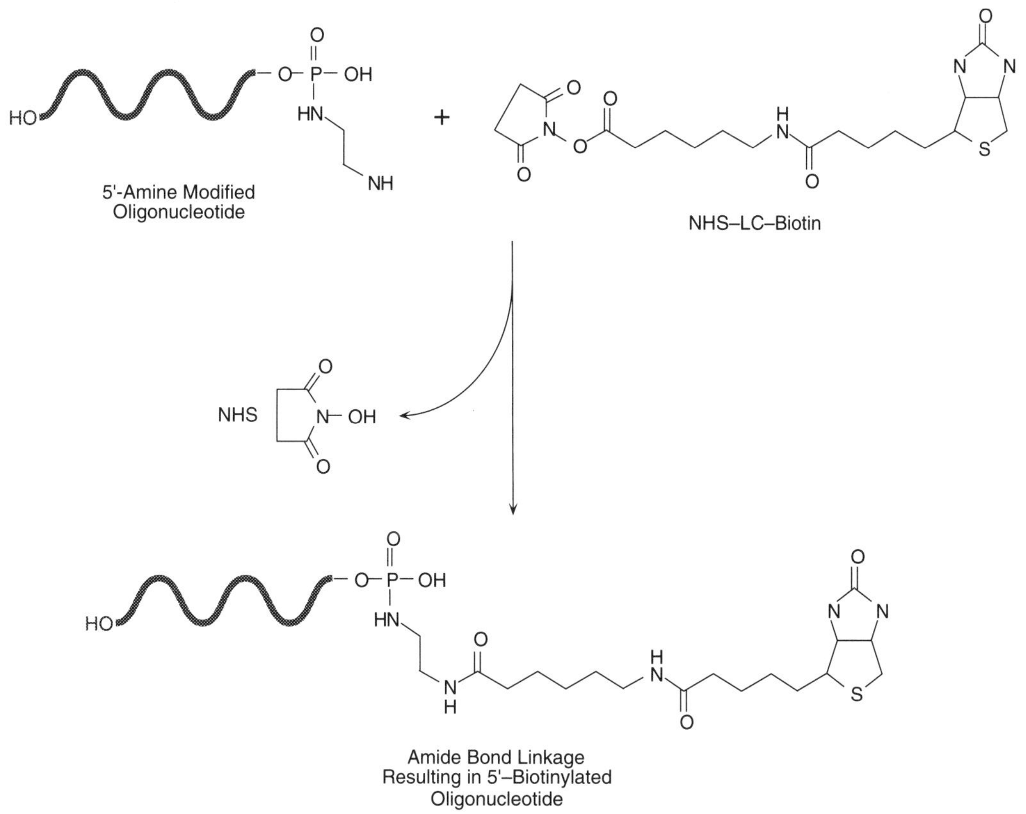

Figure 403 Biotinylation of oligonucleotides may be done at the 5'-phosphate end using a diamine derivative and reacting with NHS-LC-biotin.

α-amino group of the lysine residue of biocytin (Pierce). The aromatic amine can be treated with sodium nitrite in dilute HCl to form a highly reactive diazonium derivative, which is able to couple with active hydrogen-containing compounds. A diazonium reacts rapidly with histidine or tyrosine residues within proteins, forming covalent diazo bonds (Wilchek *et al.*, 1986). It also can react with purine residues within DNA at position 8 of the bases (Rothenberg and Wilchek, 1988; Lowe, 1979) (Fig. 404).

Protocol

1. Prepare diazotized *p*-aminobenzoyl biocytin by using the protocol outlined in Chapter 8, Section 3.5, but instead of starting with 2 mg of the biotinylation reagent dissolved in 40 μl of 1 N HCl, use 9 mg in 180 μl of 1 N HCl. Proportionally scale up the other reactant quantities used in the protocol. After the reaction is complete, immediately adjust the pH of the final solution to 9.
2. Add 1 μg of single-stranded DNA to the above solution.
3. React for 30 min at room temperature.
4. Purify the biotinylated DNA probe by ethanol precipitation, gel filtration, *n*-butanol extraction, or dialysis as discussed in other sections.

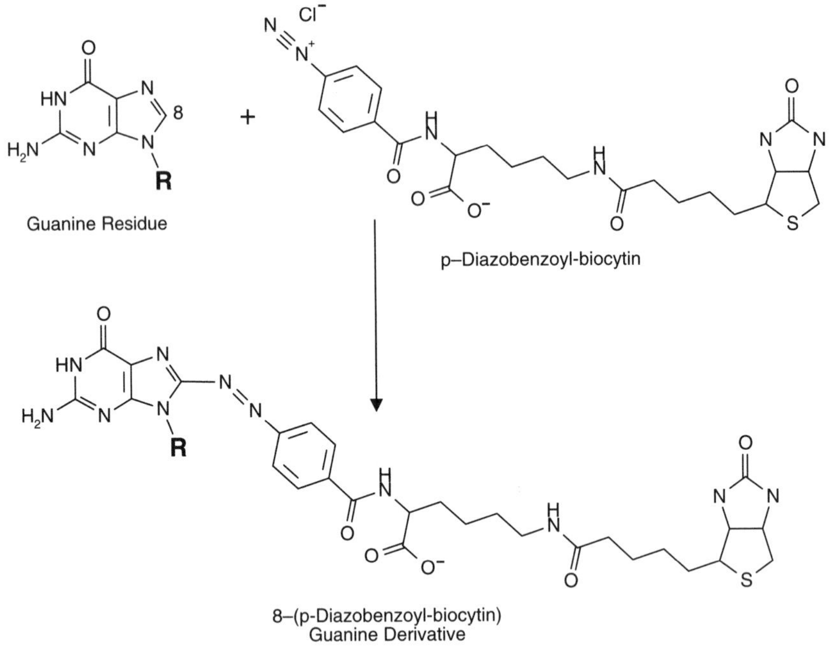

Figure 404   This diazo derivative of biocytin may be used to modify guanine bases at the C-8 position.

Reaction of Biotin–BMCC with Sulfhydryl-Modified DNA

Biotin–BMCC is a sulfhydryl-reactive biotinylation reagent containing a maleimide functional group at the end of an extended spacer arm. The long spacer (32.6 Å) provides enough distance between modified oligonucleotides and the bicyclic biotin end to allow efficient binding of avidin or streptavidin probes, even when hybridized to target sequences. The reagent may be used to add a biotin label to DNA or RNA molecules after they have been modified to contain thiol groups. For instance, cystamine labeling at the 5′ phosphate group of DNA via carbodiimide-mediated phosphoramidate formation followed by disulfide reduction (Section 2.2) can create the required sulfhydryl groups. Subsequent reaction with biotin–BMCC results in a derivative labeled only at an end of the DNA probe (Fig. 405), thus avoiding the potential for hydrogen bonding interference in hybridization assays.

Since maleimide groups are highly specific for coupling to thiols in the pH range 6.5 to 7.5, side reaction products can be avoided. The reaction is complete within 2 h at room temperature.

Protocol

1. Prepare 10–20 μg of a sulfhydryl-containing oligonucleotide in 200 μl of 50 mM sodium phosphate, 10 mM EDTA, pH 7.2 (the methods outlined in Section 2.2 can be used to form the thiol group).
2. Dissolve biotin–BMCC in DMSO at a concentration of 5 mg/ml. Prepare fresh.
3. Add 50 μl of the biotinylation solution to the oligo. Mix well.
4. React for 2 h at room temperature.

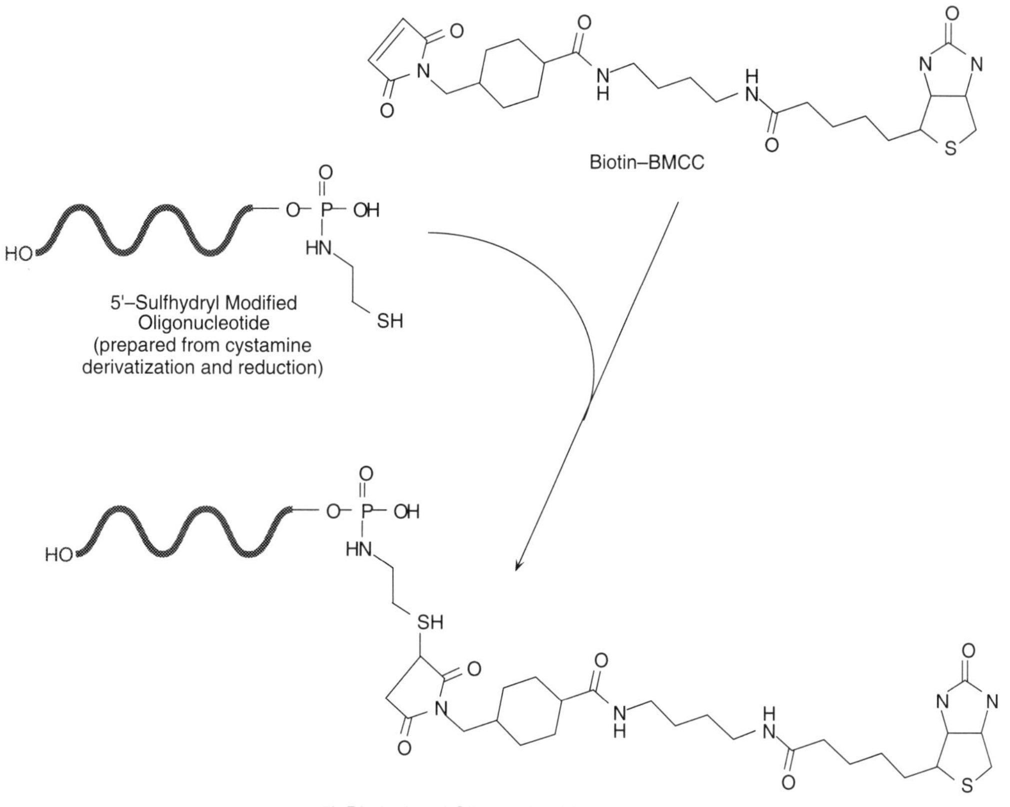

Figure 405 Biotin–BMCC may be used to modify a reduced, cystamine derivative of DNA, forming a thioether linkage.

5. Isolate the biotinylated probe by ethanol/salt precipitation as described in Section 1 for nick-translation.

Biotin–Hydrazide Modification of Bisulfite-Activated Cytosine Groups

Biotin–hydrazide is the hydrazine derivative of D-biotin prepared using its valeric acid carboxylate (Chapter 8, Section 3.3). The hydrazide functional group typically is used to react with aldehyde and ketone groups to give hydrazone linkages. However, the hydrazide compound also can undergo transamination reactions with cytosine residues via catalysis with bisulfite (Section 2.1) (Fig. 406). DNA or RNA probes containing cytosine groups may be modified to contain biotin labels using a simple, one-step procedure (Reisfeld *et al.*, 1987). The detection limit of DNA probes biotinylated using this technique can be less than 1 pg on blots, when analyzed using a streptavidin–alkaline phosphatase conjugate. A longer chain analog of biotin–hydrazide, biotin–LC-hydrazide, may be used to create an extended spacer between the oligonucleotide and the bicyclic biotin group, thus increasing the binding efficiency of avidin or streptavidin conjugates. Leary *et al.* (1983) reported that increasing the spacer arm length

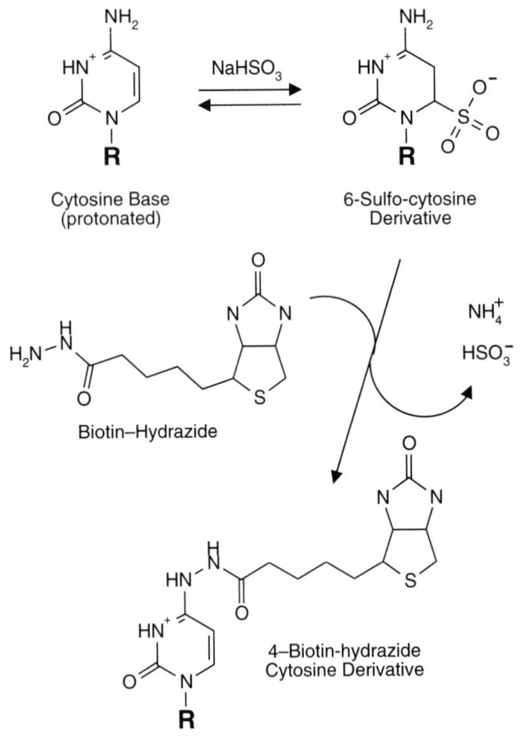

Figure 406 Biotin–hydrazide may be incorporated into cytosine bases using a bisulfite-catalyzed trans-amination reaction.

from 4 to 11 atoms when biotinylating DNA probes can increase the detectability of the target DNA approximately fourfold.

The following method is adapted from Reisfeld *et al.* (1987).

Protocol

1. Prepare 50 µg of a single-stranded DNA probe in 300 µl of 50 mM sodium acetate, pH 4.5.
2. Dissolve biotin–hydrazide in water at a concentration of 10 mg/ml.
3. Add 300 µl of the biotin–hydrazide solution to the DNA solution.
4. Add sodium bisulfite to obtain a final concentration of 1 M in the reaction medium.
5. React for 24 h at 37°C.
6. Remove excess reactants by dialysis against water at 4°C.

2.4. Enzyme Conjugation to DNA

Enzymes useful for detection purposes in ELISA techniques (Chapter 16) also can be employed in the creation of highly sensitive DNA probes for hybridization assays. The attached enzyme molecule provides detectability for the oligonucleotide through turn-over of substrates that can produce chromogenic or fluorescent products. Enzyme-based hybridization assays are perhaps the most common method of nonradioactive detection used in nucleic acid chemistry today. The sensitivity of enzyme-labeled probes can approach or equal that of radiolabeled nucleic acids, thus eliminating the need for radioactivity in most assay systems.

The conjugation reactions involved in DNA–enzyme cross-linking are not unlike the methods used to form antibody–enzyme conjugates (Chapter 10, Section 1). Bifunctional cross-linkers can be used to couple a modified oligonucleotide to an enzyme molecule using the same basic principles effective in protein–protein conjugation. The only requirement is that the DNA molecule be modified to contain one or more suitable reactive groups, such as nucleophiles like amines or sulfhydryls. The modification process used to create these functions can employ enzymatic (Section 1) or chemical (Section 2) means and can result in random incorporation of modification sites or be directed exclusively at one end of the DNA molecule, such as in 5′ phosphate coupling.

The following sections describe some of the more common procedures of preparing DNA–enzyme conjugates.

Alkaline Phosphatase Conjugation to Cystamine-Modified DNA Using Amine- and Sulfhydryl-Reactive Heterobifunctional Cross-linkers

A cystamine group added to the 5′ phosphate of DNA molecules using a carbodiimide reaction (Section 2.2) can be used in a heterobifunctional cross-linking scheme to conjugate with alkaline phosphatase. Cross-linking agents containing an amine-reactive portion and a sulfhydryl-reactive part work best in forming this type of conjugate. Perhaps the most common heterobifunctional reagent used for DNA–enzyme formation is SPDP (Chapter 5, Section 1.1). SPDP contains an NHS ester on one end able to create an amide bond linkage with amino groups on protein molecules. After modification of alkaline phosphatase with this cross-linker, the enzyme is activated to contain pyridyldisulfide groups for coupling to the sulfhydryls of a cystamine-modified DNA probe (Fig. 407). The reaction forms disulfide bonds between the oligonucleotide and the alkaline phosphatase enzyme. Since the cross-link occurs only at the 5′ end of the DNA strand, the presence of an enzyme molecule does not adversely affect the ability of base-pairing and hybridization to a target sequence.

The following protocol assumes that the labeling process used to create a sulfhydryl-modified DNA probe already has been done according to the method of Section 2.2. The modification procedure for activating alkaline phosphatase with SPDP may be done according to the protocol described in Chapter 5, Section 1.1. To obtain efficient labeling of all the alkaline phosphatase added to the reaction medium, the modified oligo is reacted in a 10-fold molar excess. Reacting the DNA probe in excess allows easy separation of not-coupled oligo from conjugated probe, thus eliminating any potential interference in hybridization assays due to unlabeled oligonucleotide. Pierce sells a kit containing SPDP-activated alkaline phosphatase to perform this conjugation procedure.

Protocol

1. Dissolve a 5′-sulfhydryl-modified oligonucleotide in water or 10 mM EDTA at a concentration of 0.05–25 μg/μl. Calculate the total nanomoles of oligo present based on its molecular weight.

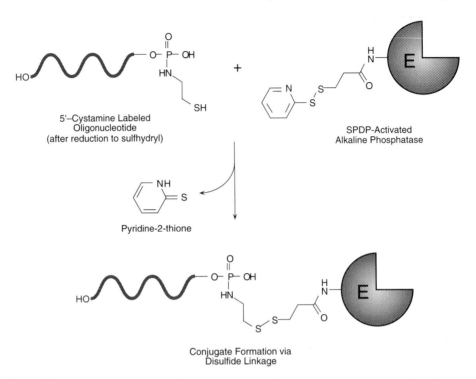

5′–Cystamine Labeled Oligonucleotide
(after reduction to sulfhydryl)

SPDP-Activated Alkaline Phosphatase

Pyridine-2-thione

Conjugate Formation via Disulfide Linkage

Figure 407   An oligonucleotide modified with cystamine and reduced to generate a free sulfhydryl may be conjugated with an SPDP-modified enzyme, forming a disulfide linkage.

2. Prepare SPDP-activated alkaline phosphatase in 50 m*M* sodium phosphate, 0.15 *M* NaCl, 10 mM EDTA, pH 7.2. Add the oligo solution to the activated enzyme in a 10-fold molar excess.

3. React at room temperature for 30 min with gentle mixing.

4. The alkaline phosphatase–DNA conjugate may be purified away from excess oligo by dialysis or gel filtration, or through the use of centrifugal concentrators. A simple way of removing unreacted oligo is to use Centricon-30 concentrators (Amicon) that have a molecular weight cutoff of 30,000. Since the enzyme molecular weight is approximately 140,000 and the conjugate is even higher, a relatively small DNA probe will pass through the membranes of these units while the conjugate will not. To purify the prepared conjugate using Centricon-30s, add 2 ml of the phosphate buffer from step 2 to one concentrator unit, then add the reaction mixture to the buffer and mix. Centrifuge at 1000 *g* for 15 min or until the retentate volume is about 50 μl. Add another 2 ml of buffer and centrifuge again until the retentate is 50 μl. Invert the Centricon-30 unit and centrifuge to collect the retentate in the collection tube provided by the manufacturer.

Alkaline Phosphatase Conjugation to Diamine-Modified DNA Using DSS

DSS is a homobifunctional cross-linker containing an amine-reactive NHS ester at both ends (Chapter 4, Section 1.2). Reaction of the reagent in excess with diamine-modified DNA probes creates an activated intermediate able to conjugate with enzyme molecules through their available amine groups (Fig. 408) (Jablonski *et al.*, 1986). The coupling reaction produces stable amide linkages under mildly alkaline conditions. Although the following protocol has been used to label oligonucleotides with success, it may be less efficient than the previous protocol at forming the desired conjugate due to the homobifunctional nature of the cross-linker. During the activation step, the modified DNA must be purified away from excess DSS. Since this is done under aqueous conditions, hydrolysis of the free NHS ester at the other end of the cross-linker takes place at the same time. Activity losses can be severe if the separation step is not done rapidly. In fact, the original protocol called for several hours of gel filtration chromatography and concentration before the conjugation reaction was done. Farmer and Castaneda (1991) made a significant improvement to this procedure by including

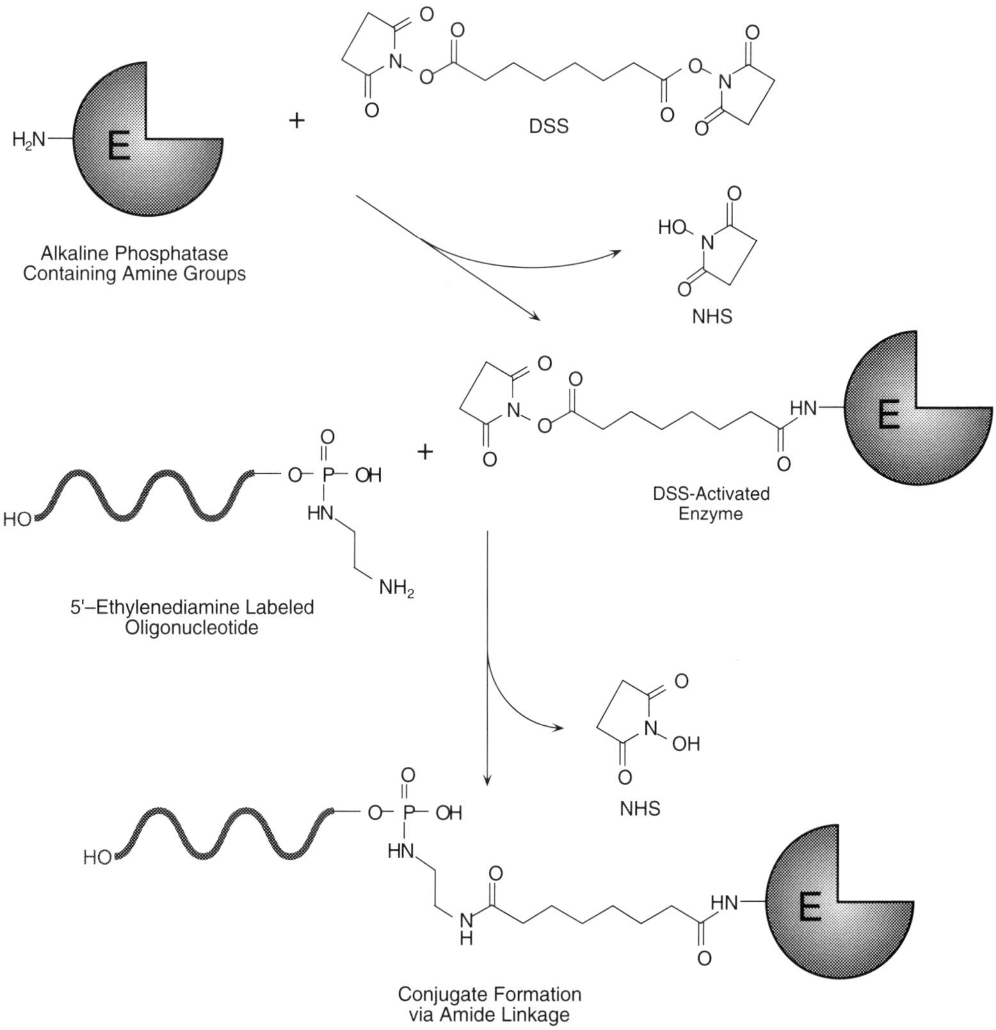

Figure 408 The homobifunctional cross-linker DSS may be used to conjugate an enzyme to a 5'-diamine modified oligonucleotide. The NHS ester groups on DSS react with the amines to form amide bonds.

a faster separation step using alcohol extraction after activation of the oligo with DSS. The purification time decreased to minutes instead of hours. This does help to limit hydrolysis, but cannot completely avoid it.

The following protocol is based on the methods of Farmer and Castaneda (1991), Kiyama *et al.* (1992), and Ruth (1993).

Protocol

1. Prepare an amine-modified oligonucleotide according to any of the protocols discussed in Section 2.1. Dissolve or buffer-exchange the oligo into 0.1 M sodium borate, 2 mM EDTA, pH 8.25, at a concentration of 9 nmol (2 $A_{260\ nm}$ units) in 15 μl.
2. Dissolve DSS in dry DMSO at a concentration of 1 mg/100 μl. Prepare fresh.
3. Add 30 μl of the DSS solution to the oligo. Mix well.
4. React for 15 min at room temperature in the dark.
5. Immediately extract excess DSS and reaction by-products by the addition of 0.5 ml of *n*-butanol. Mix vigorously by vortexing and centrifuge (1 min, 15,000 rpm) to separate the two phases. Carefully remove the upper layer and discard. Extract two more times with *n*-butanol.
6. Chill the remaining sample on dry ice and lyophilize to remove the last traces of liquid. The drying period will only take 15–30 min. The dried, DSS-activated DNA is stable.
7. Dissolve or dialyze alkaline phosphatase into 3 M NaCl, 30 mM triethanolamine, 1 mM MgCl$_2$, pH 7.6, at a concentration of 20 mg/ml.
8. Add 70 μl of the alkaline phosphatase to the dried, DSS-activated DNA. Mix gently to dissolve.
9. React overnight at 4°C in the dark.
10. Remove unconjugated oligo by using a Centricon-30 centrifugal concentrator according to step 4 of the protocol described in the previous section. Unconjugated enzyme may be removed by ion-exchange chromatography using a MonoQ-10 (0.5 × 5 cm) FPLC column (Pharmacia) or the equivalent. Binding buffer is 20 mM Tris, pH 8. Elute using a linear gradient of 0–100% 20 mM Tris, 1 M NaCl, pH 8. Free enzyme will elute before the more negatively charged oligo–enzyme conjugate.

Enzyme Conjugation to Diamine-Modified DNA Using PDITC

1,4-Phenylene diisothiocyanate (Aldrich) (PDITC) is a homobifunctional cross-linker containing two amine-reactive isothiocyanate groups on a phenyl ring. Reaction in excess with an amine-modified oligonucleotide results in the formation of a thiourea linkage, leaving the second isothiocyanate group free to couple with amine-containing enzymes or other molecules (Urdea *et al.*, 1988) (Fig. 409).

The following protocol is adapted from Keller and Manak (1989).

Protocol

1. Prepare an amine-modified oligonucleotide using any of the methods of Sections 2.1 or 1. Dissolve or buffer exchange 70 μg of the oligo into 25 μl of 0.1 M sodium borate, pH 9.3.

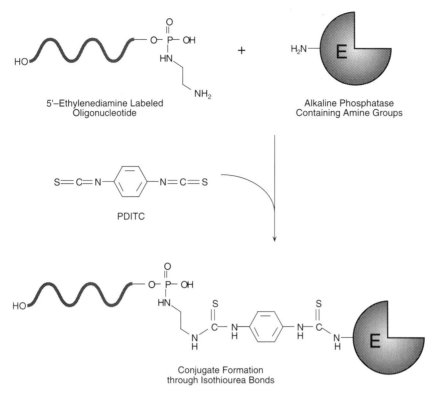

Figure 409 The homobifunctional cross-linker PDITC may be used to conjugate an enzyme to a 5′-diamine-modified oligonucleotide, creating isothiourea linkages.

2. Dissolve 10 mg of PDITC (Aldrich) in 500 μl of DMF. Add this solution to the oligo prepared in step 1.

3. React for 2 h at room temperature in the dark.

4. To extract excess reactant, add to the reaction medium, 3 ml of *n*-butanol and 3 ml of water. Mix well. Centrifuge the mixture to separate the two phases. Discard the upper yellow layer. Repeat the extraction process several times, then dry the remaining solution containing activated oligo using a lyophilizer or a rotary evaporator. The PDITC-activated DNA is stable in a dry state.

5. Dissolve HRP (Chapter 16) in 200 μl 0.1 *M* sodium borate, pH 9.3, at a concentration of 10 mg/ml. If the HRP is supplied as an ammonium sulfate suspension, all ammonium ions must be removed by dialysis prior to the conjugation reaction. Add the HRP solution to the activated oligo.

6. React overnight at room temperature in the dark.

7. Excess enzyme may be removed through isolation of the oligo–enzyme conjugate using electrophoresis separation under nondenaturing conditions. The reaction solution is applied to a 7% polyacrylamide gel using 90 m*M* Tris, 90 m*M* boric acid, 2.7 m*M* EDTA, pH 8.3, as the running buffer. The conjugate appears as a brown band in the middle of the gel.

Conjugation of SFB-Modified Alkaline Phosphatase to Bis-hydrazide-Modified Oligonucleotides

DNA probes modified to contain a 5'-terminal hydrazide functional group (Section 2.1) can be conjugated to aldehyde-containing molecules, resulting in the formation of a hydrazone bond. The cross-linking agent succinimidyl *p*-formylbenzoate (SFB) can be used to add aldehyde groups to proteins and other molecules that do not naturally contain them (Chapter 1, Section 4.4). Reaction of SFB-modified alkaline phosphatase with a hydrazide–DNA derivative can produce a conjugate having excellent sensitivity for use as a hybridization probe (Fig. 410). Other enzymes and detection molecules modified to contain aldehydes may be coupled to hydrazide–DNA probes using similar methods. Using SFB-modified HRP (or periodate-oxidized HRP) containing aldehyde groups to prepare the DNA conjugate gives about 40-fold less sensitivity in hybridization assays when compared to an alkaline phosphatase conjugate in this procedure (Ghosh *et al.*, 1989).

The following protocol assumes the prior derivatization of an oligonucleotide at the 5' end using a bis-hydrazide compound according to the protocol of Section 2.1 using a carbodiimide-mediated reaction.

Protocol

1. Dissolve alkaline phosphatase at a concentration of 10 mg/ml in 0.1 M sodium bicarbonate, 3 M NaCl, pH 8.5. Dialyze against this solution if the enzyme is already dissolved in another buffer. This protocol requires at least 0.4 ml of the enzyme solution.
2. Dissolve SFB in acetonitrile at a concentration of 50 mM (12.35 mg/ml). Make at least 100 μl.
3. Add 40 μl of the SFB solution to the 0.4-ml alkaline phosphatase solution with mixing.
4. React for 30 min at room temperature.
5. Remove excess reactants by dialysis against 50 mM Mops, 0.1 M NaCl, pH 7.5.
6. Add the aldehyde-derivatized alkaline phosphatase to 8 nmol of a 5'-hydrazide oligonucleotide preparation made according to Section 2.1 using a carbodiimide coupling protocol.
7. React overnight at room temperature.
8. Remove unconjugated oligonucleotide using gel filtration on a column of Bio-Rad P-100 (1.5 × 65 cm). Use 50 mM Tris, pH 8.5 as the chromatography buffer. Pool the enzyme fractions and apply the sample to a 1 × 7-cm column of DEAE-cellulose equilibrated with the same buffer. After washing the column with 0.1 M Tris, pH 8.5, a salt gradient from 0 to 0.2 M NaCl in the same buffer is used to remove unconjugated enzyme. The enzyme–DNA conjugate is then eluted with 0.1 M Tris, 0.5 M NaCl, pH 8.5.

2.5. Fluorescent Labeling of DNA

Oligonucleotide probes may be labeled with small fluorescent molecules for detection of hybridization by luminescence. Fluorescent probes are widely used in assay systems involving biospecific interactions. Receptors for ligands may be localized in tissues or

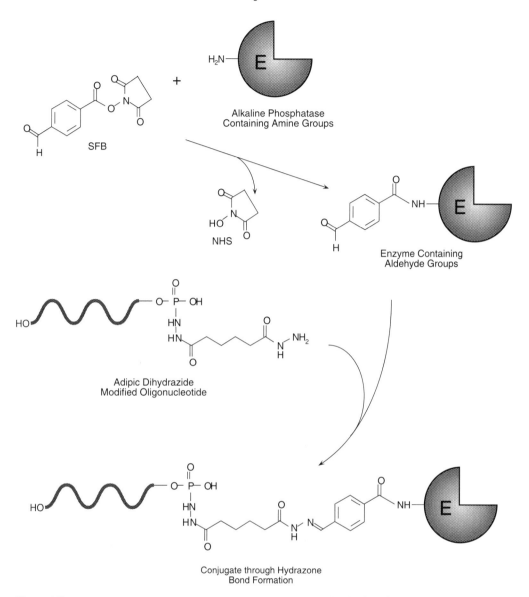

Figure 410 SFB may be used to create aldehyde groups on enzyme molecules for subsequent conjugation to a 5'-*bis*hydrazide modified oligonucleotide, forming hydrazone bonds.

cells by modification of the ligand with the appropriate fluorophore. Targeted molecules may be quantified through measurement or modulation of a fluorescent signal upon binding of a tagged ligand. The sensitivity of fluorescent assays can approach that obtained using radiolabels.

Fluorescently labeled DNA probes can be used for detection, localization, or quantification of target DNA sequences. *In situ* hybridization mapping of genomic DNA sequences can be done using fluorescent probes to target particular regions within chromosomes. Called FISH for fluorescent *in situ* hybridization, the technique is used extensively to identify marker chromosomes or chromosomal rearrangements. Since many genomic sequences are repeated, usually occurring in multiple copies within isolated regions of the chromosome, the fluorescent label on the DNA probe allows localization of targeted genes with high sensitivity. For a review of FISH, see Meyne (1993).

Fluorescently labeled DNA probes also can be used in homogeneous assay systems to detect and quantify target complementary sequences. The majority of these systems use a process of energy transfer and fluorescent quenching to detect hybridization phenomena. The principle of these assays involves the labeling of two binding components that can specifically interact with a target DNA. One or both of the labels may be a luminescent compound. The luminescent quality of the first label may consist of a chemiluminescent probe that can be excited through specific chemical processes, producing light emission. Alternatively, the label may be a fluorescent probe that can absorb light of a particular wavelength and subsequently emit light at another wavelength.

The second label also may be a fluorescent compound, but does not necessarily have to be. As long as the second label can interact with the emission of the first label and modulate its signal, binding events can be observed. Thus, the two labeled DNA probes interact with each other to produce luminescence modulation only after both have bound target DNA and are in enough proximity to initiate energy transfer. Common labels utilized in such assay techniques include the chemiluminescent probe N-(4-aminobutyl)-N-ethylisoluminol and the fluorescent compounds FITC (Chapter 8, Section 1.1), TRITC, and Texas Red sulfonyl chloride (Chapter 8, Section 1.2). For a review of these techniques, see Morrison (1992).

To prepare labeled DNA molecules for use in fluorescent assays, the oligo must be first derivatized to contain a functional group. Any of the methods of Sections 2.1 and 2.2 may be used to add an amine or sulfhydryl residue to specific regions of the DNA polymer. Once modified in this manner, the oligo may be further reacted with a fluorescent probe to create the final derivative. However, many of the fluorescent quenching formats exclusively specify either 3'- or 5'-labeled DNA molecules. This is because discrete modification at just one end of the oligo ensures that the label on a hybridized probe will be near enough to a second hybridized and labeled DNA molecule to effect the luminescent modulation necessary to make the system viable. Multiple fluorescent labels on nucleotides within the DNA probe would not be affected to the same degree by a second label attached to another oligo hybridized some distance away on the target strand. Therefore, use of the terminal transferase method of adding a modified nucleoside triphosphate to the 3' end (Section 1) or 5'-phosphate modification using a carbodiimide-mediated reaction (Section 2.1) works best for creating functionalized DNA derivatives for fluorescent modulation techniques.

The following sections describe two methods of coupling fluorescent labels to functionalized DNA probes. Other fluorophores may be attached using similar procedures with careful reference to the properties and reactivities of such labels as discussed in Chapter 8, Section 1.

Conjugation of Amine-Reactive Fluorescent Probes to Diamine-Modified DNA

DNA modified with a diamine compound to contain terminal primary amines may be coupled with amine-reactive fluorescent labels. The most common fluorophores used for oligonucleotide labeling are FITC, TRITC, and Texas Red sulfonyl chloride (Chapter 8, Sections 1.1 and 1.2). However, any of the amine-reactive labels discussed throughout Chapter 8, Section 1, are valid candidates for DNA work.

The majority of fluorescent probes are water-insoluble and must be dissolved in an organic solvent prior to addition to an aqueous reaction medium containing the DNA to be labeled. Suitable solvents are identified for each fluorophore, but mainly DMF or DMSO are used to prepare a stock solution. Some protocols utilize acetone when labeling DNA. However, avoid the use of DMSO for sulfonyl chloride compounds, as this group reacts with the solvent. For oligonucleotide labeling, the amount of solvent added to the reaction mixture should not exceed more than 20% (although at least one protocol calls for a 50% acetone addition—Nicolas et al., 1992).

The following protocol is a generalized method for labeling amine-modified oligonucleotides with a fluorescent probe, such as FITC. It is based on the method of Morrison (1992).

Protocol

1. Prepare 10 nmol of a diamine-modified DNA probe using the chemical methods discussed in Section 2.1 or through enzymatic derivatization using an amine-containing nucleoside triphosphate (Section 1). Dissolve or buffer exchange the oligo into 1 ml of a suitable coupling buffer for the type of amine-reactive fluorophore utilized (see recommended reaction conditions for the particular fluorescent label in Chapter 8, Section 1). For FITC, the appropriate buffer condition for the oligo is 0.1 M sodium carbonate, pH 9.
2. Dissolve the fluorophore in DMF or another suitable solvent at a concentration of 0.01 M. For FITC, this translates into a concentration of 3.89 mg/ml.
3. Add 50 μl of the FITC solution to the oligo solution and mix. For the use of NHS ester or sulfonyl chloride fluorescent probes, add up to 150 μl of the fluorophore solution to the DNA.
4. React overnight at room temperature.
5. Remove excess fluorophore from the labeled oligo using gel filtration on a column of Sephadex G-25, dialysis, or a Centricon centrifugal concentrator.

Conjugation of Sulfhydryl-Reactive Fluorescent Probes to Sulfhydryl-Modified DNA

Fluorescent probes containing sulfhydryl-reactive groups can be coupled to DNA molecules containing thiol modification sites. The chemical derivatization methods outlined in Section 2.2 may be used to thiolate the oligo for subsequent modification with a fluorophore. Appropriate fluorescent compounds and their reaction conditions may be found in Chapter 8, Section 1. The protocol discussed in the previous section can be used as a general guide for labeling DNA molecules.

References

Abdella, R. M., Smith, P. K., and Royer, G. P. (1979) A new cleavable reagent for cross-linking and reversible immobilization of proteins. *Biochem. Biophys. Res. Commun.* **87**, 734–742.

Abou-Samra, A. B., Freeman, M., Juppner, H., Uneno, S., and Segre, G. V. (1990) Characterization of fully active biotinylated parathyroid hormone analogs. Applications to fluorescence activated cell sorting of parathyroid hormone receptor bearing cells. *J. Biol. Chem.* **265**, 58–62.

Abuchowski, A., McCoy, J. R., Palczuk, N. C., van Es, T., and Davis F. F. (1977a) Effect of covalent attachment of polyethylene glycol on immunogenicity and circulating life of bovine liver catalase. *J. Biol. Chem.* **252**, 3582–3586.

Abuchowski, A., van Es, T., Palczuk, N. C., and Davis, F. F. (1977b) Alteration of immunological properties of bovine serum albumin by covalent attachment of polyethylene glycol. *J. Biol. Chem.* **252**, 3578–3581.

Abuchowski, A., Kazo, G. M., Verhoest, C. R., Jr., van Es, T., Kafkewitz, D., Nucci, M. L., Viau, A. T., and Davis, F. F. (1984) Cancer therapy with chemically modified enzymes. I. Antitumor properties of polyethylene glycol asparaginase conjugates. *Cancer Biochem. Biophys.* **7**, 175–186.

Adams, R., Bachman, W. E., Fieser, L. F., Johnson, J. R., and Snyder, H. R. (1942) *in* "Organic Reactions, Vol. 1, p. 303. Wiley, New York.

Adolfson, R., and Moudrianokis, E. N (1976) Molecular polymorphism and mechanisms of activation and deactivation of the hydrolytic function of the coupling factor of oxidative phosphorylation. *Biochemistry* **15**, 4164–4170.

Aguirre, R., Gonsoulin, F., and Cheung, H. C. (1986) Interaction of fluorescently labelled myosin subfragment 1 with nucleotide and actin. *Biochemistry* **25**, 6827.

Ahn, B., Rhee, S. G., and Stadtman, E. R. (1987) Use of fluorescein hydrazide and fluorescein thiosemicarbazide reagents for the fluorometric determination of protein carbonyl groups and for the detection of oxidized proteins on polyacrylamide gels. *Anal. Biochem.* **161**, 245–257.

Aime, S., Anelli, P. L., Botta, M., Fedeli, F., Grandi, M., Paoli, P., and Uggeri, F. (1992) Synthesis, characterization, and $1/T_1$ NMRD profiles of gadolinium (III) complexes of monoamide derivatives of DOTA-like ligands. X-ray structure of the 10-[2-[[2-hydroxy-1-(hydroxyl-methyl)ethyl]amino]-1-[phenylmethoxy)methyl]-2-oxo-ethyl]-1,4,7,10-tetraaza-cyclododecane-1,4,7-triacetic acid-gadolinium (III) complex. *Inorg. Chem.* **31**, 2422–2428.

Aithal, H. N., Knigge, K. M., Kartha, S., Czyewski, E. A., and Toback, F. G. (1988) An alternate method utilizing small quantities of ligand for affinity purification of monospecific antibodies, *J. Immunol. Methods* **112**, 63–70.

Ajtai, K. (1992) Stereospecific reaction of muscle fiber proteins with the 5′ or 6′ iodoacetamido derivative of tetramethylrhodamine: Only the 6′ isomer is mobile on the surface of S1. *Biophys. J.* **61**, A278, Abstract 1647.

Alagon, A. C. and King, T. P. (1980) Activation of polysaccharides with 2-iminothiolane and its uses. *Biochemistry* **19**, 4341–4345.

Albrecht, R. M., Goodman, S. L., and Simmons, S. R. (1989) Distribution and movement of membrane-associated platelet glycoproteins: Use of colloidal gold with correlative video-enhanced light microscopy, low-voltage high-resolution scanning electron microscopy, and high-voltage transmission electron microscopy. *Am. J. Anat.* **185**, 149–164.

Alexander, P. (1954) The reactions of carcinogens with macromolecules. *Adv. Cancer Res.* **2**, 1.

Aliosman, F., Caughlan, J., and Gray, G. S. (1989) Diseased DNA intrastrand cross-linking and cytotoxicity induced in human brain tumor cells by 1,3-bis(2-chloroethyl)-1-nitrosourea after *in vitro* reaction with glutathione. *Cancer Res.* **49**, 5954.

Allen, T. M., and Chonn, A. (1987) Large unilamellar liposomes with low uptake into the reticuloendothelial system. *FEBS Lett.* **223**, 42–46.

Allison, A. C., and Gregoriadis, G. (1974) Liposomes as immunological adjuvants. *Nature (London)* **252**, 252.

Allmer, K., Hilborn, J., Larsson, P. H., Hult, A., and Ranby, B. (1990) Surface modification of polymers. V. Biomaterial applications. *J. Polym. Sci.: Part A: Polym. Chem.* **28**, 173–183.

Alvear, M., Jabalquinto, A. M., and Cardemil, E. (1989) Inactivation of chicken liver mevalonate 5-diphosphate decarboxylase by sulfhydryl-directed reagents: Evidence of a functional dithiol. *Biochim. Biophys. Acta* **994**, 7–11.

Alving, C. R. (1987) Liposomes as carriers for vaccines, *in* "Liposomes from Biophysics to Therapeutics" (M. J. Ostro, ed.), pp. 195–218. Dekker, New York.

Anderson, G. W. (1958) N, N'-Carbonyldiimidazole, a new reagent for peptide synthesis. *J. Am. Chem. Soc.* **80**, 4323.

Anderson, P. W., Pichichero, M. E., Stein, E. C., Porcelli, S., Betts, R. F., Connuck, D. M., Korones, D., Insel, R. A., Zahradnik, J. M., and Eby, R. (1989) Effect of oligosaccharide chain length, exposed terminal group, and hapten loading on the antibody response of human adults and infants to vaccines consisting of *Haemophilus influenzae* type b capsular antigen uniterminally coupled to the diphtheria protein CRM197. *J. Immunol.* **142**, 2464–2468.

Anderson, W. L., and Wetlaufer, D. B. (1975) A new method for disulfide analysis of peptides. *Anal. Biochem.* **67**, 493–502.

Ando, T. (1984) Fluorescence of fluorescein attached to myosin SH_1 distinguishes the rigor state from the actin-myosin-nucleotide state. *Biochemistry* **23**, 375.

Anjaneyulu, P. S. R., and Staros, J. V. (1987) Reactions of N-hydroxysulfosuccinimide active esters. *Int. J. Pept. Protein Res.* **30**, 117–124.

Annunziato, M. E., Patel, U. S., Ranade, M., and Palumbo, P. S. (1993) p-Maleimidophenyl isocyanate: A novel heterobifunctional linker for hydroxyl to thiol coupling. *Bioconjugate Chem.* **4**, 212–218.

Antimisiaris, S. G., Jayasekera, P., and Gregoriadis, G. (1993) Liposomes are vaccine carriers: Incorporation of soluble and particulate antigens in giant vesicles. *J. Immunol. Methods* **166**, 271–280.

Apple, R. J., Domen, P. L., Muckerheide, A., and Michael, J. G. (1988) Cationization of protein antigens. IV. Increased antigen uptake by antigen-presenting cells. *J. Immunol.* **140**, 3290–3295.

Atassi, M. Z., and Manshouri, T. (1991) Synthesis of tolerogenic monomethoxypolyethylene glycol and polyvinyl alcohol conjugates of peptides. *J. Protein Chem.* **10**, 623–627.

Atha, D. H., Brew, S. A., and Ingham, K. C. (1964) Interactions and thermal stability of fluorescent labeled derivatives of thrombin and antithrombin III. *Biochim. Biophys. Acta* **785**, 1.

Avigad, E., Amaral, D., Asensio, C., and Horecker, B. L. (1962) *J. Biol. Chem.* **237**, 2736.

Avrameas, S. (1969) Coupling of enzyme to proteins with glutaraldehyde. *Immunochemistry* **6**, 43–52.

Avrameas, S., and Ternynck, T. (1969) The cross-linking of proteins with glutaraldehyde and its use for the preparation of immunosorbents. *Immunochemistry* **6**, 53–66.

Avrameas, S., and Ternynck, T. (1971) Peroxidase labelled antibody and Fab conjugates with enhanced intracellular penetration. *Immunochemistry* **8**, 1175.

Bacha, P., Murphy, J. R., and Reichlin, S. (1983) Thyrotropin-releasing hormone-diphtheria toxin-related polypeptide conjugates. *J. Biol. Chem.* **258**, 1565.

Baenziger, J. U., and Fiete, D. (1982) Photoactivatable glycopeptide reagents for site-specific labeling of lectins. *J. Biol. Chem.* **257**, 4421–4425.

Baird, B. A., and Hammes, G. G. (1976) Chemical cross-linking studies of chloroplast coupling factor 1. *J. Biol. Chem.* **251**, 6953–6962.

Baird, B. A., and Hammes, G. G. (1977) Chemical cross-linking studies of beef heart mitochondrial coupling factor 1. J. Biol. Chem, **252**, 4743–4748.

Bakkus, M. H., Brakel-van Peer, K. M., Adriaansem, H. J., Wierenga-Wolf, A. F., van den Akker, T. W., Dicke-Evinger, M. J., and Benner, R. (1989) Detection of oncogene expression by fluorescent *initu* hybridization in combination with immunofluorescent staining of cell surface markers. *Oncogene* **4**, 1255–1262.

Ballmer-Hofer, K., Schlup, V., Burn, P., and Burger, M. M. (1982) Isolation of *in situ* cross-linked ligand-receptor complexes using an anticross-linker specific antibody. *Anal. Biochem.* **126**, 246–250.

Balls, A. K., and Wood, H. N. (1956) *J. Biol. Chem.* **219**, 245.

Bangham, A. D., Standish, M. M., and Watkins, J. C. (1965) Diffusion of univalent ions across the lamellae of swollen phospholipids. *J. Mol. Biol.* **13**, 238.

Bangs, J. D., Andrews, N. W., Hart, G. W., and Englund, P. T. (1986) Post-translational modification and intracellular transport of a trypanosome variant surface glycoprotein. *J. Cell Biol.* **103**, 255–263.

Baranowska-Kortylewicz, J., and Kassis, A. I. (1993a) Labeling of sulfhydryl groups in intact mammalian cells with coumarins. *Bioconjugate Chem.* **4**, 305–307.

Baranowska-Kortylewicz, J., and Kassis, A. I. (1993b) Labeling of immunoglobulins with bifunctional, sulfhydryl-selective, and photoreactive coumarins. *Bioconjugate Chem.* **4**, 300–304.

Barany, G., and Merrifield, R. B. (1980) *in* "The Peptides" (E. Gross and J. Meienhofer, eds.), pp 1–284. Academic Press, New York.

Barbieri, L., and Stirpe, F. (1982) *Cancer Surveys* **1**, 489–520.

Barbour, H. M. (1976) Development of an enzyme immunoassay for human placental lactogen using labelled antibodies. *J. Immunol. Methods* **11**, 15.

Bartel, A., and Campbell, D. (1959) Some immunochemical differences between associated and dissociate hemocyanin. *Arch. Biochem. Biophys.* **82**, 2332.

Bartling, G. J., Brown, H. D., and Chattopadhyay, S. K. (1973) Synthesis of a matrix-supported enzyme in non-aqueous conditions. *Nature (London)* **243**, 342–344.

Baskin, L. S., and Yang, C. S. (1980a) Cross-linking studies of cytochrome P-450 and reduced nicotinamide adenine dinucleotide phosphate-cytochrome P-450 reductase. *Biochemistry* **19**, 2260–2264.

Baskin, L. S., and Yang, C. S. (1980b) *in* "Microsomes, Drug Oxidations, and Chemical Carcinogenesis," pp. 103–106. Academic Press, New York.

Baskin, L. S., and Yang, C. S. (1982) Cross-linking studies of the protein topography of rat liver microsomes. *Biochim. Biophys. Acta* **684**, 263–271.

Batzri, S., and Korn, E. D. (1973) Single bilayer liposomes prepared without sonication. *Biochim. Biophys. Acta* **298**, 1015–1019.

Baues, R. J., and Gray, G. R. (1977) Lectin purification on affinity columns containing reductively aminated disaccharides. *J. Biol. Chem.* **252**, 57.

Bayer, E. A., and Wilchek, M. (1980) The use of the avidin–biotin complex as a tool in molecular biology. *Methods Biochem. Anal.* **26**, 1–45.

Bayer, E. A., and Wilchek, M. (1990) Avidin- and streptavidin-containing probes. *in* "Methods in Enzymology" (M. Wilchek and E. A. Bayer, eds.), Vol. 184, pp. 174–185. Academic Press, San Diego.

Bayer, E. A., and Wilchek, M. (1992) Labeling and detection of proteins and glycoproteins. *in* "Nonradioactive Labeling and Detection of Biomolecules" (C. Kessler, ed.), pp. 98–99. Springer-Verlag, New York.

Bayer, E. A., Skutelsky, E., Wynne, D., and Wilchek, M. (1976) Preparation of ferritin–avidin conjugates by reductive alkylation for use in electron microscopic cytochemistry. *J. Histochem. Cytochem.* **24**, 933–939.

Bayer, E., Ben-Hur, H., Gitlin, G., and Wilchek, M. (1986) An improved method for the single-step purification of streptavidin. *J. Biochem. Biophys. Methods* **13**, 103–112.

Bayer, E. A., Ben-Hur, H., and Wilchek, M. (1987a) Enzyme-based detection of glycoproteins on blot transfers using avidin-biotin technology. *Anal. Biochem.* **161**, 123–131.

Bayer, E. A., Safars, M., and Wilchek, M. (1987b) Selective labeling of sulfhydryls and disulfides on blot transfers using avidin-biotin technology: Studies on purified proteins and erythrocyte membranes. *Anal. Biochem.* **161**, 262–271.

Bayer, E. A., Ben-Hur, H., and Wilchek, M. (1988) Biocytin hydrazide—a selective label for sialic acids, galactose, and other sugars in glycoconjugates using avidin-biotin technology. *Anal. Biochem.* **170**, 271–281.

Bayer, E., Ben-Hur, H., Hiller, T., and Wilchek, M. (1989) Postsecretory modifications of streptavidin. *Biochem. J.* **259**, 369–376.

Bayer, E. A., Ben-Hur, H., and Wilchek, M. (1990) Direct labeling of blotted glycoproteins. *in* "Methods in Enzymology" (M. Wilchek and E. A. Bayer, eds.), Vol. 184, pp. 427–429. Academic Press, San Diego.

Bayne, S., and Ottesen, M. (1977) *Carlsberg Res. Commun.* **42**, 465–474.

Beaucage, S. L., and Iyer, R. P. (1993) The functionalization of oligonucleotides via phosphoramidite derivatives. *Tetrahedron* **49**, 1925–1963.

Beauchamp, C. O., Gonias, S. L., Menapace, D. P., and Pizzo, S. V. (1983) A new procedure for the synthesis of polyethylene glycol-protein adducts; Effects on function, receptor recognition, and clearance of superoxide dismutase, lactoferrin, and $\alpha_2$macroglobulin. *Anal. Biochem.* **131**, 25–33.

Bendayan, M. (1989) Ultrastructural localization of insulin and C-peptide antigenic sites in rat pancreatic B cell obtained by applying the quantitative high-resolution protein A–gold approach. *Am. J. Anat.* **185**, 205–216.

Bendayan, M., and Garzon, S. (1988) Protein G–gold complex: Comparative evaluation with protein A–gold for high-resolution immunocytochemistry. *J. Histochem. Cytochem.* **36**, 597–607.

Benesch, R., and Benesch, R. E. (1956) Formation of peptide bonds by aminolysis of homocysteine thiolactones. *J. Am. Chem. Soc.* **78**, 1597.

Benesch, R., and Benesch, R. E. (1958) Thiolation of proteins. *Proc. Natl. Acad. Sci. U.S.A.* **44**, 848.

Benesch, R. E., and Kwong, S. (1988) Bis-pyridoxal polyphosphates: A new class of specific intramolecular cross-linking agents for hemoglobin. *Biochem. Biophys. Res. Commun.* **156**, 9.

Benhamou, N., Gilboa-Garber, N., Trudel, J., and Asselin, A. (1988) A new lectin–gold complex for ultrastructural localization of galacturonic acids. *J. Histochem. Cytochem.* **36**, 1403–1411.

Berg, H. C., Diamond, J. M., and Marfey, P. S. (1965) Erythrocyte membrane: Chemical modification. *Science* **150**, 64.

Berger, H., and Pizzo, S. V. (1988) Preparation of polyethylene glycol-tissue plasminogen activator adducts that retain functional activity: Characteristics and behavior in three different species. *Blood* **71**, 1641–1647.

Bergmann, K. E., Carlson, K. E., and Katzenellenbogen, J. A. (1994) Hexestrol diazirine photoaffinity labeling reagent for the estrogen receptor. *Bioconjugate Chem.* **5**, 141–150.

Bergström, K., Holmberg, K., Safranj, A., Hoffman, A. S., Edgell, M. J., Kozlowski, A., Hovanes, B. A., and Harris, J. M. (1992) Reduction of fibrinogen adsorption on PEG-coated polystyrene surfaces. *J. Biomed. Mater. Res.* **26**, 779–790.

Bernstein, A., Hurwitz, E., Maron, R., Arnon, R., Sela, M., and Wilchek, M. (1978) Higher antitumor efficacy of daunomycin when linked to dextran: *In vivo* and *in vitro* studies. *J. Natl. Cancer Inst.* **60**, 379–384.

Beth, A. H., Conturo, T. E., Venkataramu, S. D., and Staros, J. V. (1986) Dynamics and interactions of the anion channel in intact human erythrocytes: An electron paramagnetic resonance spectroscopic study employing a new membrane-impermeant bifunctional spin-label. *Biochemistry* **25**, 3824–3832.

Bethell, G. S., Ayers, J. S., Hancock, W. S., and Hearn, M. T. W. (1979) A novel method of activation of cross-linked agaroses with 1,1'-carbonyldiimidazole which gives a matrix for affinity chromatography devoid of additional charged groups. *J. Biol. Chem.* **254**, 2572–2574.

Beutner, E. H. (1971) *Ann. N.Y. Acad. Sci.* **177**, 506.

Bewley, T. A., and Li, C. H. (1969) *Int. J. Protein Res.* **1**, 379.

Bewley, T. A., Dixon, J. S., and Li, C. H. (1968) *Biochim. Biophys. Acta* **154**, 420.

Biermann, C. J., and McGinnis, G. D. (eds.) (1989) "Analysis of Carbohydrates by GLC and MS." CRC Press, Boca Raton, Florida.

Bigelow, D. J., and Inesi, G. (1991) Frequency-domain fluorescence spectroscopy resolves the location of maleimide-directed spectroscopic probes within the tertiary structure of the Ca-ATPase of sarcoplasmic reticulum. *Biochemistry* **30**, 2113–2125.

Binkley, R. W. (1988) "Modern Carbohydrate Chemistry." Dekker, San Diego.

Birnbaumer, M. E., Schrader, W. T., and O'Malley, B. W. (1979) Chemical cross-linking of chick oviduct progesterone-receptor subunits using a reversible bifunctional cross-linking agent. *Biochem. J.* **181**, 201–213.

Bizzini, B., Blass, J., Turpin, A., and Raynaud, M. (1970) *Eur. J. Biochem.* **17**, 100.

Bjorn, M. J., Groetsema, G., and Scalapino, L. (1986) Antibody–*Pseudomonas* exotoxin A conjugates cytotoxic to human breast cancer cells *in vitro*. *Cancer Res.* **46**, 3262.

Blass, J., Bizzini, B., and Raynaud, M. (1965) Mechanism of detoxication by formol. *Compt. Rend.* **261**, 1448.

Blattler, W. A., Kuenzi, B. S., Lambert, J. M., and Senter, P. D. (1985a) New heterobifunctional protein cross-linking reagent that forms an acid-labile link. *Biochemistry* **24**, 1517–1524.

Blattler, W. A., Kuenzi, B. S., Lambert, J. M., and Senter, P. D. (1985b) New heterobifunctional protein cross-linking reagents and their use in the preparation of antibody–toxin conjugates. *Photochem. Photobiol.* **42**, 231.

Bloxham, D. P., and Cooper, G. K. (1982) Formation of a polymethylene bis(disulfide) inter-subunit cross-link between cys-281 residues in rabbit muscle glyceraldehyde-3-phosphate dehydrogenase using octamethylene bis(methane[^{35}]thiosulfonate). *Biochemistry* **21**, 1807.

Bloxham, D. P., and Sharma, R. P. (1979) The development of S,S'-polymethylenebis-(methanethiosulfonates) as reversible cross-linking reagent for thiol groups and their use to form stable catalytically active cross-linked dimers with glyceraldehyde-3-phosphate dehydrogenase. *Biochem. J.* **181**, 355.

Boas, M. A. (1927) The effect of desiccation upon the nutritive properties of egg-white. *Biochem. J.* **21**, 712–724.

Bobbitt, J. M. (1956) Periodate oxidation of carbohydrates. *Adv. Carbohydr. Chem.* **11**, 1–41.

Böcher, M., Giersch, T., and Schmid, R. D. (1992) Dextran, a hapten carrier in immunoassays for *s*-triazines. A comparison with ELISAs based on hapten–protein conjugates. *J. Immunol. Methods* **151**, 1–8.

Bodanszky, A., and Bodanszky, M. (1970) Sepharose–avidin column for the binding of biotin or biotin-containing peptides. *Experientia* **26**, 327.

Bog-Hansen, T. C., Prahl, P., and Lowenstein, H. (1978) A set of analytical electrophoresis experiments to predict the results of affinity chromatographic separations. Fractionation of allergens from cow's hair and dander. *J. Immunol. Methods* **22**, 293.

Böldicke, T., Kindt, S., Maywald, F., Fitzlaff, G., Böcher, M., Frank, R., and Collins, J. (1988) Production of specific monoclonal antibodies against the active sites of human pancreatic secretory trypsin inhibitor variants by *in vitro* immunization with synthetic peptides. *Eur. J. Biochem.* **175**, 259–264.

Bolton, A. E., and Hunter, W. M. (1973) The labeling of proteins to high specific radioactivities by conjugation to a ^{125}I-containing acylating agent. *Biochem. J.* **133**, 529–539.

Bolton, A. E., and Hunter, W. M. (1986) Radioimmunoassay and related methods. *in* "Handbook of Experimental Immunology. Volume 1: Immunochemistry." (D. M. Weir, ed.), 4th Ed., pp. 26.1–26.56. Blackwell, London.

Bonnard, C., Papermaster, D. S., and Kraehenbuhl, J.-P. (1984) The streptavidin–biotin bridge technique: Application in light and electron microscope immunocytochemistry. *in* "Immunolabelling for Electron Microscopy" (J. M. Polak and I. M. Varndell, eds.), pp. 95–111. Elsevier, New York.

Boorsma, D. M., and Kalsbeek, G. L. (1976) A comparative study of horseradish peroxidase conjugates prepared with a one-step and a two-step method. *J. Histochem. Cytochem.* **23**, 200–207.

Boorsma, D. M., and Streefkerk, J. G. (1976a) Peroxidase-conjugate chromatography. Isolation of conjugates prepared with glutaraldehyde or periodate using polyacrylamide-agarose gel. *J. Histochem. Cytochem.* **24**, 481.

Boorsma, D. M., and Streefkerk, J. G. (1976b) Some aspects of the preparation, analysis, and use of peroxidase–antibody conjugates in immunohistochemistry. *Protides Biol. Fluids, Proc. Colloq.* **24**, 795.

Bos, F. (1981) Optimization of spectral coverage in an eight-cell oscillator-amplifier dye laser pumped at 308 nm. *Appl. Opt.* **20**, 3553.

Bouizar, Z., Fouchereau-Person, M., Taboulet, J., Moukhtar, M. S., and Milhaud, G. (1986) Purification and characterization of calcitonin receptors in rat kidney membranes by covalent cross-linking techniques. *Eur. J. Biochem.* **155**, 141–147.

Boyer, C. M., Lidor, Y., Lottich, S. C., and Bast, R. C., Jr. (1988) Antigenic cell surface markers in human solid tumors. *Antibody, Immunoconjugates, Radiopharm.* **1**, 105.

Boyer, T. D. (1986) Covalent labeling of the nonsubstrate ligand-binding site of glutathione S-transferase with bilirubin-Woodward's reagent K. *J. Biol. Chem.* **261**, 5363.

Boyle, R. (1966) The reaction of dimethyl sulfoxide and 5-dimethylaminonaphthalene-1-sulfonyl chloride. *J. Org. Chem.* **31**, 3880–3882.

Boyle, W. J., Lipsick, J. S., Reddy, E. P., and Baluda, M. A. (1983) Identification of the leukemogenic protein of avian myeloblastosis virus and of its normal cellular homologue. *Proc. Natl. Acad. Sci. U.S.A.* **80**, 2834–2838.

Braatz, J. A., Yasuda, Y., Olden, K., Yamada, K. M., and Heifetz, A. H. (1993) Functional peptide–polyurethane conjugates with extended circulatory half-lives. *Bioconjugate Chem.* **4**, 262–267.

Bragg, P. D., and Hou, C. (1975) Subunit composition, function, and spatial arrangement in the Ca^{2+}-and Mg^{2+}-activated adenosine triphosphatases of *Escherichia coli* and *Salmonella typhimurium*. *Arch. Biochem. Biophys.* **167**, 311–321.

Bragg, P. D., and Hou, C. (1980) A crosslinking study of the Ca^{+2}, Mg^{+2}-activated adenosine triphosphate of *Escherichia coli*. *Eur. J. Biochem.* **106**, 495–503.

Brandon, D. L. (1980) *Cell. Mol. Biol.* **26**, 569–573.

Brandt, H. M., and Apkarian, A. V. (1992) Biotin–dextran: A sensitive anterograde tracer for neuroatomic studies in rat and monkey. *J. Neurosci. Methods* **45**, 35–40.

Brechbiel, M. W., McMurry, T. J., and Gansow, O. A. (1993) A direct synthesis of a bifunctional chelating agent for radiolabeling proteins. *Tetrahedron Lett.* **34**, 3691–3694.

Brew, K., Shaper, J. H., Olsen, K. W., Trayer, I. P., and Hill, R. L. (1975) Cross-linking of the components of lactose synthetase with dimethylpimelimidate. *J. Biol. Chem.* **250**, 1434–1444.

Brewer, C. F., and Riehm, J. P. (1967) Evidence for possible nonspecific reactions between N-ethylmaleimide and proteins. *Anal. Biochem.* **18**, 248.

Briand, J. P., Muller, S., and Van Regenmortel, M. H. V. (1985) *J. Immunol. Methods* **78**, 59.

Bright, F. V. (1988) Bioanalytical applications of fluorescence spectroscopy. *Anal. Chem.* **60**, 1031A.

Brillhart, K. L., and Ngo, T. T. (1991) Use of microwell plates carrying hydrazide groups to enhance antibody immobilization in enzyme immunoassays. *J. Immunol. Methods* **144**, 19–25.

Brinkley, M. (1992) A brief survey of methods for preparing protein conjugates with dyes, haptens, and cross-linking reagents. *Bioconjugate Chem.* **3**, 2.

Brocklehurst, K., Carlsson, J., Kierstan, M. P. J., and Crook, E. M. (1974) Covalent chromatography by thiol–disulfide interchange. *in* "Methods of Enzymology" (W. B. Jakoby and M. Wilchek, eds.), Vol. 34, pp. 531–544. Academic Press, New York.

Bronckers, A. J. J., Gay, S., Finkelman, R. D., and Butler, W. T. (1987) Immunolocalization of Gla proteins (osteocalcin) in rat tooth germs: Comparison between indirect immunofluorescence, peroxidase–antiperoxidase, avidin–biotin–peroxidase complex, and avidin–biotin–gold complex with silver enhancement. *J. Histochem. Cytochem.* **35**, 825–830.

Brooks, B. R., and Klamerth, O. L. (1968) Interaction of DNA with bifunctional aldehydes. *Eur. J. Biochem.* **5**, 178.

Brown, D. M. (1974) Chemical reactions of polynucleotides and nucleic acids. *in* "Basic Principles in Nucleic Acid Chemistry" (P. O. P. Ts'O, ed.), Vol. 2, pp. 1–90. Academic Press, New York.

Browne, D. T., and Kent, S. B. H. (1975) Formation of nonamidine products in the reaction of primary amines with imido esters. *Biochem Biophys. Res. Commun.* **67**, 126.

Browning, J., and Ribolini, A. (1989) Studies on the differing effects of tumor necrosis factor and lymphotoxin on the growth of several human tumor lines. *J. Immunol.* **143**, 1859–1867.

Brunner, J. (1993) New photolabeling and cross-linking methods. *Annu. Rev. Biochem.* **62**, 483–514.

Brunswick, M., Finkelman, F. D., Highet, P. F., Inman, J. K., Dintzis, H. M., and Mond, J. J. (1988) Picogram quantities of anti-Ig antibodies coupled to dextran induce B cell proliferation. *J. Immunol.* **140**, 3364–3372.

Bugawan, T. L., Begovich, A. B., and Erlich, H. A. (1990) Rapid HLA-DPB typing using enzymatically amplified DNA and nonradioactive sequence-specific oligonucleotide probes. *Immunogenetics* **32**, 231–241.

Bunnett, J. F. (1963) Nucleophilic reactivity. *Annu. Rev. Phys. Chem.* **14**, 271.

Burns, J. A., Butler, J. C., Moran, J., and Whitesides, G. M. (1991) Selective reduction of disulfides by tris(2-carboxyethyl)phosphine. *J. Org. Chem.* **56**, 2648–2650.

Burtnick, L. D. (1984) Modification of actin with fluorescein isothiocyanate. *Biochim. Biophys. Acta* **791**, 57.

Butler, P. J. G., Harris, J. I., Hartley, B. S., and Leberman, R. (1967) Use of maleic anhydride for the reversible blocking of amino groups in polypeptide chains. *Biochem. J.* **103**, 78P.

Byrne, M. J., and Johnson, D. B. (1975) Studies on the immobilization of β-galactosidase. *Biochem. Soc. Trans.* **2**, 496.

Caamano, C. A., Fernandez, H. N., and Paladani, A. C. (1983) Specificity of covalently stabilized complexes of 125I-labeled human somatotropin and components of the lactogenic binding sites of rat liver. *Biochem. Biophys. Res. Commun.* **115**, 29–37.

Cabacungan, J. C., Ahmed, A. I., and Feeney, R. E. (1982) Amine boranes as alternative reducing agents for reductive alkylation of proteins. *Anal. Biochem.* **124**, 272–278.

Cai, S. X., Glenn, D. J., Gee, K. R., Yan, M., Cotter, R. E., Reddy, N. L., Weber, E., and Keana, J. F. W. (1993) Chlorinated phenyl azides as photolabeling reagents. Synthesis of an ortho, ortho-dichlorinated arylazido PCP receptor ligand. *Bioconjugate Chem.* **4**, 545–548.

Campbell, P., and Gioannini, T. L. (1979) The use of benzophenone as a photoaffinity label. Labeling in *p*-benzoylphenylacetyl chymotrypsin at unit efficiency. *Photochem. Photobiol.* **29**, 883.

Cardoza, J. D., Kleinfeld, A. M., Stallcup, K. C., and Mescher, M. F. (1984) Hairpin configuration of H-2K^k in liposomes formed by detergent dialysis. *Biochemistry* **23**, 4401–4409.

Carlsson, J., Drevin, H., and Axen, R. (1978) Protein thiolation and reversible protein–protein conjugation. *N*-Succinimidyl 3(2-pyridyldithio)propionate, a new heterobifunctional reagent. *Biochem. J.* **173**, 723–737.

Carraway, K. L., and Koshland, D. E., Jr. (1968) *Biochim. Biophys. Acta* **160**, 272–274.

Carraway, K. L., and Triplett, R. B. (1970) *Biochim. Biophys. Acta* **200**, 564–566.

Casanova, J., Horowitz, Z. D., Copp, R. P., NcIntyre, W. R., Pascual, A., and Samuels, H. H. (1984) Photoaffinity labeling of thyroid hormone nuclear receptors. *J. Biol. Chem.* **259**, 12084–12091.

Cater, C. W. (1963) The evaluation of aldehydes and other difunctional compounds as cross-linking agents for collagen. *J. Soc. Leather Trade Chem.* **47**, 259.

Caufield, M. P., Horiuchi, S., Tai, P. C., and Davis, B. D. (1984) The 64-kilodalton membrane protein of *Bacillus subtilis* is also present as a multiprotein complex on membrane-free ribosomes. *Biochemistry* **81**, 7772–7776.

Chaiet, L., and Wolf, F. J. (1964) The properties of streptavidin, a biotin-binding protein produced by *Streptomycetes*. *Arch. Biochem. Biophys.* **106**, 1–5.

Chamberlain, N. R., Deogny, L., Slaughter, C., Radolf, J. D., and Norgard, M. V. (1989) Acylation of the 47-kilodalton major membrane immunogen of *Treponema pallidum* determines its hydrophobicity. *Infect. Immun.* **57**, 2878–2885.

Chamow, S. M., Kogan, T. P., Peers, D. H., Hastings, R. C., Byrn, R. A., and Ashkenazi, A. (1992) Conjugation of soluble CD4 without loss of biological activity via a novel carbohydrate-directed cross-linking reagent. *J. Biol. Chem.* **267**, 15916–15922.

Chamow, S. M., Kogan, T. P., Venuti, M., Gadek, T., Harris, R. J., Peers, D. H., Mordenti, J., Shak, S., and Ashkenazi, A. (1994) Modification of CD4 immunoadhesin with mono-methoxypoly(ethylene glycol) aldehyde via reductive alkylation. *Bioconjugate Chem.* **5**, 133–140.

Chang, F. N., and Flaks, J. G. (1972) Specific cross-linking of *Escherichia coli* 30S ribosomal subunit. *J. Mol. Biol.* **68**, 177.

Chantler, P., and Bower, S. M. (1988) Cross-linking between translationally equivalent sites on the heads of myosin: Relationship to energy transfer results between the same pair of sites. *J. Biol. Chem.* **263**, 938.

Chase, J. W., Merrill, B. M., and Williams, K. P. (1983) F sex factor encodes a single-stranded DNA binding protein (SSB) with extensive sequence homology to *Escherichia coli* SSB. *Proc. Natl. Acad. Sci. U.S.A.* **80**, 5480–5484.

Chattopadhyay, A., James H. I., and Fair, D. S. (1992) Molecular recognition sites on factor Xa which participate in the prothrombinase complex. *J. Biol. Chem.* **267**, 12323–12329.

Chazotte, B., and Hackenbrock, C.R. (1991) Lateral diffusion of redox components in the mitochondrial inner membrane is unaffected by inner membrane folding and matrix density. *J. Biol. Chem.* **266**, 5973.

Chehab, F. F., and Kan, Y. W. (1989) Detection of specific DNA sequences by fluorescence amplification: A color complementation assay. *Proc. Natl. Acad. Sci. U.S.A.* **86**, 9178.

Chelsky, D., and Dahlquist, F. W. (1980) Chemotaxis in *Escherichia coli*: Association of protein components. *Biochemistry* **19**, 4633–4639.

Chen, K. K., Rose, C. L., and Clowes, G. H. A. (1934) *Am. J. Med. Sci.* **188**, 767.

Chen, R. F., and Knutson, J. R. (1988) Mechanism of fluorescent concentration quenching of carboxyfluorescein in liposomes: Energy transfer to nonfluorescent dimers. *Anal. Biochem.* **172**, 61.

Cheng, Y., and Dovichi, N. J. (1988) Subattomole amino acid analysis by capillary zone electrophoresis and laser-induced fluorescence. *Science* **242**, 562.

Chetrit, P., Gaudin, V., de Courcel, A., and Vedel, F. (1989) A cross-hybridization method for DNA mapping with photobiotin-labeled probes. *Anal. Biochem.* **178**, 273–275.

Childs, G. V., Lloyd, J. M., Unabia, G., Gharib, S. D., Wierman, M. E., and Chin, W. W. (1987) Detection of luteinizing hormone b messenger ribonucleic acid (RNA) in individual gonadotropes after castration: Use of a new *in situ* hybridization method with a photobiotinylated complementary RNA probe. *Mol. Endocrinol.* **1**, 926–932.

Chowdhry, V., Vaughn, R., and Westheimer, F. H. (1976) 2-Diazo-3,3,3-trifluoropropionyl chlorides: Reagent for photoaffinity labeling. *Proc. Natl. Acad. Sci. U.S.A.* **73**, 1406–1408.

Chu, B. C. F., and Orgel, L. E. (1988) Ligation of oligonucleotides to nucleic acids or proteins via disulfide bonds. *Nucleic Acids Res.* **16**, 3671.

Chu, F. S., and Ueno, I. (1977) Production of antibody against aflatoxin B$_1$. *Appl. Environ. Microbiol.* **33**, 1125–1128.

Chu, F. S., Fred Chi C., and Hinsdill, R. D. (1976) Production of antibody against ochratoxin A. *Appl. Environ. Microbiol.* **31**, 831–835.

Chu, F. S., Lau, H. P., Fan, T. S., and Zhang, G. S. (1982) Ethylenediamine modified bovine serum albumin as protein carrier in the production of antibody against mycotoxins. *J. Immunol. Methods* **55**, 73–78.

Chu, B. C. F., Wahl, G. M., and Orgel, L. E. (1983) Derivatization of unprotected polynucleotides. *Nucleic Acids Res.* **11**, 6513–6529.

Chu, B. C. F., Kramer, F. R., and Orgel, L. E. (1986) Synthesis of an amplifiable reporter RNA for bioassays. *Nucleic Acids Res.* **14**, 5591–5603.

Cimino, G. D., Gamper, H. B., Isaacs, S. T., and Hearst, J. E. (1985) Psoralens as photoactive probes of nucleic acid structure and function: Organic chemistry, photochemistry, and biochemistry. *Annu. Rev. Biochem.* **54**, 1151–1193.

Clausen, J. (1988) Immunochemical techniques for the identification and estimation of macromolecules. *in* "Laboratory Techniques in Biochemistry and Molecular Biology," (R. H. Burdon and P. H. Knippenberg, eds.), 3rd ed., Vol 1, Part 3. Elsevier, New York.

Cleland, W. W. (1964) Dithiothreitol, a new protective reagent for SH groups. *Biochemistry* **3**, 480–482.

Cocco, L., Martelli, A., Billi, A., Matteucci, A., Vitale, M., Neri, L., and Manzoli, F. (1986) Changes in nucleosome structure and histone H3 accessibility. Iodoacetamidofluorescein labeling after treatment with phosphatidylserine vesicles. *Exp. Cell Res.* **166**, 465–474.

Cohn, E. J., *et al.* (1947) Preparation and properties of serum and plasma proteins. XIII. Crystallization of serum albumins from ethanol-water mixtures. *J. Am. Chem. Soc.* **69**, 1753–1761.

Cole, R. D. (1967) *S*-Aminoethylation. *in* "Methods in Enzymology" (C. H. W. Hirs, ed.), Vol. 11, pp. 315–317. Academic Press, New York.

Cole, R. D., Stein, W. H., and Moore, S. (1958) *J. Biol. Chem.* **233**, 1359.

Coleman, P. L., Walker, M. M., Milbrath, D. S., Stauffer, D. M., Rasmussen, J. K., Krepski, L. R., and Heilmann, S. M. (1990) *J. Chromatogr.* **512**, 345.

Collier, R. J., and Cole, H. A. (1969) Diphtheria toxin subunit active *in vitro. Science* **164**, 1179.

Collier, R. J., and Kandel, J. (1971) Structure and activity of diphtheria toxin. *J. Biol. Chem.* **246**, 1496–1503.

Collioud, A., Clemence, J.-F., Sänger, M., and Sigrist, H. (1993) Oriented and covalent immobilization of target molecules to solid supports: Synthesis and application of a light-activatable and thiol-reactive cross-linking reagent. *Bioconjugate Chem.* **4**, 528–536.

Colman, R. F. (1969) The role of sulfhydryl groups in the catalytic function of isocitrate dehydrogenase. I. Reaction with 5,5'-dithiobis(2-nitrobenzoic acid). *Biochemistry* **8**, 888.

Colombatti, M., Greenfield, L., and Youle, R. J. (1986) Cloned fragment of diphtheria toxin linked to T cell-specific antibody identifies regions of B chain active in cell entry. *J. Biol. Chem.* **261**, 3030.

Colombatti, M., Nabholz, M., Gros, O., and Brown, C. (1983) Selective killing of target cells by antibody-ricin A-chain or antibody-gelonin hybrid molecules: Comparison of cytotoxic potency and use in immunoselection procedures. *J. Immunol.* **131**, 3091.

Conn, P. M., Rogers, D. C., Stewart, J. M., Niedel, J., and Sheffield, T. (1982a) Conversion of a gonadotropin-releasing hormone antagonist to an agonist. *Nature (London)* **296**, 633–655.

Conn, P. M., Rogers, D. C., and McNeil, R. (1982b) Potency enhancement of a GnRH agonist: GnRH–receptor microaggregation stimulates gonadotropin release. *Endocrinology (Baltimore)* **111**, 335–337.

Coulter, A., and Harris, R. (1983) Simplified preparation of rabbit Fab fragments. *J. Immunol. Methods* **59**, 199–203.

Cover, J. A., Lambert, J. M., Norman, C. M., and Traut, R. R. (1981) Identification of proteins at the subunit interface of the *Escherichia coli* ribosome by cross-linking with dimethyl 3,3′-dithiobis(propionimidate). *Biochemistry* **20**, 2843–2852.

Cox, J. P. L., Craig, A. S., Helps, L. M., Jandowski, K. J., Parker, D., Eaton, M. A. W., Millican, A. T., Millar, K., Beeley, N. R. A., and Boyce, B. A. (1990) Synthesis of C- and N-functionalized derivatives of 1,4,7-triazacyclononane-1,4,7-triyltriacetic acid (NOTA), 1,4,7,10-tetraazacyclododecane-1,4,7,10-tetrayltetraacetic acid (DOTA), and diethylenetriaminepentaacetic acid (DTPA): Bifunctional complexing agents for the derivatization of antibodies. *J. Chem. Soc. Perkin Trans.* **1**, 2567–2576.

Cramer, E. M., Beesley, J. E., Pulford, K. A. F., Breton-Gorius, J., and Mason, D. Y. (1989) Colocalization of elastase and myeloperoxidase in human blood and bone marrow neutrophils using a monoclonal antibody and immunogold. *Am. J. Pathol.* **134**, 1275–1284.

Crestfield, A. M., Stein, W. H., and Moore, S. (1963) *J. Biol. Chem.* **238**, 2413.

Cuatrecasaes, P. (1970) Protein purification by affinity chromatography. Derivatizations of agarose and polyacrylamide beads. *J. Biol. Chem.* **245**, 3059.

Cuatrecaseas, P. (1972) *Adv. Enzymol.* **36**, 29.

Cuatrecaseas, P., and Parikh, I. (1972) Adsorbents for affinity chromatography. Use of N-hydroxysuccinimide esters of agarose. *Biochemistry* **11**, 2291–2299.

Cuatrecasas, P., and Wilchek, M. (1968) Single-step purification of avidin from egg white by affinity chromatography of biocytin-Sepharose columns. *Biochem. Biophys. Res. Commun.* **33**, 235–246.

Cubie, H. A., and Norval, M. (1989) Detection of human papilloma viruses in paraffin wax sections with biotinylated synthetic oligonucleotide probes and immunogold staining. *J. Clin. Pathol.* **42**, 988–991.

Cumber, A. J., Forrester, B. M. J., Foxwell, W. C. J., Ross., and Thorpe, P. W. (1985) Preparation of antibody–toxin conjugates. *in* "Methods in Enzymology" (K. J. Widder and R. Green, eds.), Vol. 112, pp. 207–224. Academic Press, New York.

Czworkowski, J., Odom, O. W., and Hardesty, B. (1991) Study of the topology of messenger RNA bound to the 30S ribosomal subunit of *Escherichia coli*. *Biochemistry* **30**, 4821.

Daemen, T., Veninga, A., Dijkstra, J., and Scherphof, G. (1989) Differential effects of liposome-incorporation on liver macrophage activating potencies of rough lipopolysaccharide, lipid A, and muramyl dipeptide: Differences in susceptibility to lysosomal enzymes. *J. Immunol.* **142**, 2469–2474.

Damjanovich, S., and Kleppe, K. (1966) *Biochim. Biophys. Acta* **122**, 145.

Danscher, G., and Rytter-Nörgaard (1983) Light microscopic visualization of colloidal gold on resin-embedded tissue. *J. Histochem. Cytochem.* **31**, 1394–1398.

Darzynkiewicz, Z., and Crissman, H. A. (eds.) (1990) Flow cytometry. *Methods Cell Biol.*

Das, M., and Fox, C. F. (1979) Chemical cross-linking in biology. *Annu. Rev. Biophys. Bioeng.* **8**, 165–196.

Davidson, R. S., and Hilchenbach, M. M. (1990) The use of fluorescent probes in immunochemistry. *Photochem. Photobiol.* **52**, 431.

Davies, G. E., and Kaplan, J. G. (1972) Use of diimidoester cross-linking reagent to examine the subunit structure of rabbit muscle pyruvate kinase. *Can. J. Biochem.* **50**, 416–422.

Davies, G. E., and Palek, J. (1981) ^{125}I-labeling of platelet proteins with Bolton–Hunter reagent. *Anal. Biochem.* **115**, 383–387.

Davies, G. E., and Stark, G. R. (1970) Use of dimethyl suberimidate, a cross-linking reagent, in studying the subunit structure of oligomeric proteins. *Proc. Natl. Acad. Sci. U.S.A.* **66**, 651.

Davis, F. F., Van Es, T., and Palczuk, N. C. (1979) Nonimmunogenic polypeptides. U.S. Patent 4,179,337.

Debye, P. (1947) *J. Phys. Colloid Chem.* **51**, 18.

Delgado, C., Francis, G. F., and Fisher, D. (1992) The uses and properties of PEG-linked proteins. *Crit. Rev. Ther. Drug Carrier Syst.* **9**, 249–304.

Dell'Arciprete, L., Colombatti, M., Rappuoli, R., and Tridente, G. (1988) A C terminus cysteine of diphtheria toxin B chain involved in immunotoxin cell penetration and cytotoxicity. *J. Immunol.* **140**, 2466–2471.

Della-Penna, D., Christofferson, R. E., and Bennett, A. B. (1986) Biotinylated proteins as molecular weight standards on Western blots. *Anal. Biochem.* **152**, 329–332.

DeMar, J. C., Jr., Disher, R. M., and Wensel, T. G. (1992) HPLC analysis of protein-linked fatty acids using fluorescence detection of 4-(diazomethyl)-7-diethylaminocoumarin derivatives. *Biophys. J.* **61**, A81, Abstract 465.

De Mey, J., Moeremans, M., Geuens, G., Nuydens, R., and De Brabander, M. (1981) High resolution light and electron microscopic localization of tubulin with the IgS (immuno gold staining) method. *Cell Biol. Int. Rep.* **5**, 889.

Denney, J. B., and Blobel, G. (1984) ^{125}I-Labeled cross-linking reagent that is hydrophilic, photoactivatable, and cleavable through an azo linkage. *Proc. Natl. Acad. Sci. U.S.A.* **81**, 5286–5290.

DePont, J. J. H. H. M. (1979) Reversible inactivation of $(Na^+ + K^+)$-ATPase by use of a cleavable bifunctional reagent. *Biochim. Biophys. Acta* **567**, 247–256.

dePont, J. J., Schoot, B. M., and Bonting, S. L. (1980) Use of mon- and bifunctional group-specific reagents in the study of the renal Na^+-K^+-ATPase. *Int. J. Biochem.* **12**, 307–313.

Derksen, J. T. P., and Scherphof, G. L. (1985) An improved method for the covalent coupling of proteins to liposomes. *Biochim. Biophys. Acta* **814**, 151–155.

Dermer, O. C., and Ham, G. E. (1969) "Ethylenimine and Other Aziridines," pp. 327–333. Academic Press, New York.

de Rosario, R. B., Wahl, R. L., Brocchini, S. J., Lawton, R. G., and Smith, R. H. (1990) Sulfhydryl site-specific cross-linking and labeling of monoclonal antibodies by a fluorescent equilibrium transfer alkylation cross-link reagent. *Bioconjugate Chem.* **1**, 51–59.

Deryagin, B. V., and Landau, L. (1941) Theory of the stability of strongly charged lyophobic sols and of the adhesion of strongly charged particles in solutions of electrolytes. *Acta Physiochim. U.R.S.S.* **14**, 633–662.

De Waele, M., Renmans, W., Segers, E., De Valck, V., Jochmans, K., and van Camp, B. (1989) An immunogold-silver staining method for detection of cell surface antigens in cell smears. *J. Histochem. Cytochem.* **37**, 1855–1862.

Dewey, R. E., Timothy, D. H., and Levings III, C. S. (1987) A mitochondrial protein associated with cytoplasmic male sterility in the T cytoplasm of maize. *Proc. Natl. Acad. Sci. U.S.A.* **84**, 5374–5378.

Dewey, T. G. (ed.) (1991) "Biophysical and Biochemical Aspects of Fluorescence Spectroscopy." Plenum, New York.

Didenko, V. V. (1993) Biotinylation of DNA on membrane supports: A procedure for preparation and easy control of labeling of nonradioactive single-stranded nucleic acid probes. *Anal. Biochem.* **213**, 75–78.

Dintzis, R. Z., Okajima, M., Middleton, M. H., Greene, G., and Dintzis, H. M. (1989) The immunogenicity of soluble hapenated polymers is determined by molecular mass and hapten valence. *J. Immunol.* **143**, 1239–1244.

Dixon, H. B. F., and Perham, R. N. (1968) *Biochem. J.* **109**, 312–314.

Domen, P. L., and Hermanson, G. T. (1992) Cationized carriers for immunogen production. U.S. Patent No. 5,142,027.

Domen, P. L., Muckerheide, A., and Michael, J. G. (1987) Cationization of protein antigens III. Abrogation of oral tolerance. *J. Immunol.* **139**, 3195–3198.

Domen, P. L., Nevens, J. R., Mallia, A. K., Hermanson, G. T., and Klenk, D. C. (1990) Site-directed immobilization of proteins. *J. Chromatogr.* **510**, 293–302.

Donovan, J. A., and Jennings, M. L. (1986) N-Hydroxysulfosuccinimido active esters and the L-(+)-lactate transport protein in rabbit erythrocytes. *Biochemistry* **25**, 1538–1545.

Dottavio-Martin, D., and Ravel, J. M. (1978) Radiolabeling of proteins by reductive alkylation with [^{14}C]-formaldehyde and sodium cyanoborohydride. *Anal. Biochem.* **87**, 562.

Dower, S. K., DeLisi, C., Titus, J. A., and Segal, D. M. (1981) Mechanism of binding of multivalent immune complexes to Fc receptors. 1. Equilibrium binding. *Biochemistry* **20**, 6326–6334.

Drafler, F. L., and Marinetti, G. V. (1977) Synthesis of a photoaffinity probe for the β-adrenergic receptor. *Biochem. Biophys. Res. Commun.* **79**, 1.

Draper, D. E., and Gold, L. (1980) A method for linking fluorescent labels to polynucleotides: Application to studies of ribosome–ribonucleic acid interactions. *Biochemistry* **19**, 1774-1781.

Dreborg, S., and Akerblom, E. B. (1990) Immunotherapy with monomethoxypolyethylene glycol modified allergens. *Crit. Rev. Ther. Drug Carrier Syst.* **6**, 315–365.

Duband, J.-L., Nuckolls, G., Ishihara, A., Hasegawa, T., Yamada, K., Thiery, J. P., and Jacobson, K. (1988) Fibronectin receptor exhibits high lateral mobility in embryonic locomoting cells but is immobile in focal contacts and fibrillar streaks in stationary cells. *J. Cell Biol.* **107**, 1385–1396.

Duijndam, W. A. L., Wiegant, J., Van Duijn, P., and Haaijman, J. J. (1988) A simple method for labeling the carbohydrate moieties of antibodies with fluorochromes. *J. Immunol. Methods* **109**, 289–290.

Dunbar, B. S. (1987) "Two-Dimensional Electrophoresis and Immunological Techniques," pp. 229–335. Plenum, New York.

Duncan, R., and Kopecek, J. (1984) Soluble synthetic polymers as potential drug carriers. *Adv. Polym. Sci.* **57**, 53–101.

Duncan, R. J. S., Weston, P. D., and Wrigglesworth, R. (1983) A new reagent which may be used to introduce sulfhydryl groups into proteins, and its use in the preparation of conjugates for immunoassay. *Anal. Biochem.* **132**, 68–73.

Dunn, B. M., and Affinsen, C. B. (1974) Kinetics of Woodward's reagent K hydrolysis and reaction with staphylococcal nuclease. *J. Biol. Chem.* **249**, 3717.

Dunn, R. L., and Ottenbrite, R.M., (eds.) (1991) "Polymeric Drugs and Drug Delivery Systems." American Chemical Society, Washington, D. C.

Dunnick, J. K., McDougall, R., Aragon, S., Goris, M., and Kriss, J. (1975) *J. Nucl. Med.* **16**, 483.

Durand, R. E., and Olive, P. L. (1983) Flow cytometry techniques for studying cellular thiols. *Radiat. Res.* **95**, 456.

du Vigneaud, V., Melville, D. B., Gyorgy, P., and Rose, C. S. (1940) On the identity of vitamin H with biotin. *Science* **92**, 62–63.

Ebrahim, H., and Dakshinamurti, K. (1986) A fluorometric assay for biotinidase. *Anal. Biochem.* **154**, 282–286.

Ebrahim, H., and Dakshinamurti, K. (1987) Determination of biocytin. *Anal. Biochem.* **162**, 319–324.

Edelhoch, H., Katchalsk, E., Maybury, R. H., Hughes, W. L., Jr., and Edsall, J. T. (1953) Dimerization of serum mercaptalbumin in the presence of mercurials. I. Kinetic and equilibrium studies with mercuric salts. *J. Am. Chem. Soc.* **75**, 5058.

Edelman, G. M., Gall, W. E., Waxdal, M. J., and Konigsberg, W. H. (1968) The covalent structure of a human γG-immunoglobulin. I. Isolation and characterization of the whole molecules, the polypeptide chains, and the tryptic fragments. *Biochemistry* **7**, 1950–1958.

Edsall, J. T., Maybury, R. H., Simpson, R. B., and Straessle, R. (1954) Dimerization of serum mercaptalbumin in the presence of mercurials. II. Studies with a bifunctional organic mercurial. *J. Am. Chem. Soc.* **76**, 3131.

Edwards, D. C., Ross, W. C. J., Cumber, A. J., McIntosh, D., Smith, A., Thorpe, P. E., Brown, A., Williams, R. H., and Davies, A. J. S. (1982) A comparison of the *in vitro* and *in vivo* activities of conjugates of anti-mouse lymphocytes globulin and abrin. *Biochim. Biophys. Acta* **717**, 272.

Edwards, J. O., and Pearson, R. G. (1962) The factors determining nucleophilic reactivities. *J. Chem. Soc.* **84**, 26.

Edwards, R. J., Singleton, A. M., Boobis, A. R., and Davies, D. S. (1989) Cross-reaction of antibodies to coupling groups used in the production of anti-peptide antibodies. *J. Immunol. Methods* **117**, 215–220.

Eiklid, K., Olsnes, S., and Pihl, A. (1980) Entry of lethal doses of abrin, ricin, and modeccin into the cytosol of Hela cells. *Exp. Cell Res.* **126**, 321–326.

Eisen, H. N., Belman, S., and Carsten, M. E. (1953) The reaction of 2,4-dinitrobenzenesulfonic acid with free amino groups of proteins. *J. Am. Chem. Soc.* **75**, 4583.

Eldjarn, L., and Jellum, E. (1963) *Acta Chem. Scand.* **17**, 2610–2621.

Ellis, I. O., Bell, J., and Bancroft, J. D. (1988) An investigation of optimal gold particle size for immunohistological immunogold and immunogold-silver staining to be viewed by polarized incident light (EPI polarization) microscopy. *J. Histochem. Cytochem.* **36**, 121–124.

Ellman, G. L. (1958) *Arch. Biochem. Biophys.* **74**, 443.

Ellman, G. L. (1959) Tissue sulfhydryl groups. *Arch. Biochem. Biophys.* **82**, 70–77.

Englund, P. T., King, T. P., and Craig, L. C. (1968) Studies on ficin. I. Its isolation and characterization. *Biochemistry* **7**, 163–174.

Enoch, H. G., and Strittmatter, P. (1979) Formation and properties of 100-Å-diameter, single-bilayer phospholipid vesicles. *Proc. Natl. Acad. Sci. U.S.A.* **76**, 145–149.

Epps, D. E., *et al.* (1992) Spectral characterization of environment-sensitive adducts of interleukin 1b. *J. Biol Chem.* **267**, 3129.

Ernsting, N. P., Asimov, M., and Shaefer, F. P. (1982) The electronic origin of the π-π* absorption of amino coumarins studied in a supersonically cooled free jet. *Chem. Phys. Lett.* **91**, 231.

Eschrich, T. C., and Morgan, T. J. (1985) Dye laser radiation in the 370–760 nm region pumped by a xenon monofluoride excimer laser. *Appl. Opt.* **24**, 937.

Ewig, R. A. G., and Kohn, K. W. (1977) DNA–protein cross-linking and DNA interstrand cross-linking by haloethylnitrosoureas in L1210 cells. *Cancer Res.* **38**, 3197.

Fahien, L. A., Ruoho, A. E., and Kmiotek, E. (1978) A study of glutamate dehydrogenase•aminotransferase complexes with a bifunctional imidate. *J. Biol. Chem.* **253**, 5745–5751.

Falke, J. J., Dernburg, A. F., Sternberg, D. A., Zalkin, N., Milligan, D. L., and Koshland, D. E., Jr. (1988) Structure of a bacterial sensory receptor. *J. Biol. Chem.* **263**, 14850–14858.

Farmer, J. G., and Castaneda, M. (1991) An improved preparation and purification of oligonucleotide–alkaline phosphatase conjugates. *BioTechniques* **11**, 588–589.

Farr, A. G., and Nakane, P. K. (1981) Immunohistochemistry with enzyme labeled antibodies: A brief review. *J. Immunol. Methods* **47**, 129–144.

Farries, T. C., and Atkinson, J. P. (1989) Biosynthesis of properdin. *J. Immunol.* **142**, 842–847.

Fasold, H., Groschel-Stewart, U., and Turba, F. (1963) Azophenyl-dimaleimide als spaltbare peptidbrucken-bildende reagentien zwischen cysteinresten. *Biochem. Z.* 337, 425.

Faulk, W. P., and Taylor, G. M. (1971) An immunocolloid method for the electron microscope. *Immunochemistry* 8, 1081–1083.

Fearnley, C., and Speakman, J. B. (1950) Cross-linkage formation in keratin. *Nature (London).* 166, 743.

Fein, M. L., and Filachione, E. M. (1957) Tanning studies with aldehydes. *J. Am. Leather Chem. Assoc.* 52, 17.

Feinberg, A. P., and Vogelstein, B. (1983) A technique for radiolabeling DNA restriction endonuclease fragments to high specific activity. *Anal. Biochem.* 132, 6–13.

Feinberg, A. P., and Vogelstein, B. (1984) A technique for radiolabeling DNA restriction endonuclease fragments to high specific activity. (Addendum). *Anal. Biochem.* 137, 266–267.

Feller, A. C., Parwaresch, M. R., Wacker, H.-H., Radzun, H.-J., and Lennert, K. (1983) Combined immunohistochemical staining for surface IgD and T-lymphocyte subsets with monoclonal antibodies in human tonsils. *Histochem. J.* 15, 557–562.

Ferguson, B., and Yang, D. (1986) Localization of noncovalently bound ethidium in free and methionyl-tRNA synthetase bound tRNA(fMet) by singlet–singlet energy transfer. *Biochemistry* 25, 5298.

Finlay, T. H., Troll, V., Levy, M., Johnson, A. J., and Hodgins, L. T. (1978) New methods for the preparation of biospecific adsorbents and immobilized enzymes utilizing trichloro-*s*-triazine. *Anal. Biochem.* 87, 77–90.

Fizgerald, D., Morris, R. E., and Saelinger, C. B. (1980) Receptor-mediated internalization of *Pseudomonas* toxin by mouse fibroblasts. *Cell (Cambridge, Mass.)* 21, 867.

Fitzgerald, D. J., Willingham, M. C., and Pastan, I. (1988) *Pseudomonas* exotoxin–immunotoxin. *in* "Immunotoxins" (A. E. Frankel, ed.), p. 161. Kluwer, Boston.

Fok, K.-F., Ohga, K., Incefy, G. S., and Erickson, B. W. (1982) *Mol. Immunol.* 19, 1667.

Ford, D. J., Radin, R., and Pesce, A. J. (1978) Characterization of glutaraldehyde coupled alkaline phosphatase–antibody and lactoperoxidase–antibody conjugates. *Immunochemistry* 15, 237.

Forster, A. C., McInnes, J. L., Skingle, D. C., and Symons, R. H. (1985) Non-radioactive hybridization probes prepared by the chemical labeling of DNA and RNA with a novel reagent, photobiotin. *Nucleic Acid Res.* 13, 745–761.

Fraenkel-Conrat, H. (1959) Methods for investigating the essential groups for enzyme activity. *in* "Methods in Enzymology" (S. P. Colowick and N. O. Kaplan, eds.), Vol. 4, p. 247–269. Academic Press, New York.

Fraker, P. J., and Speck, J. C., Jr. (1978) Protein and cell membrane iodinations with a sparingly soluble chloroamide, 1,3,4,6-tetrachloro-3α,6α-diphenylglycouril. *Biochem. Biophys. Res. Commun.* 80, 849–857.

Freedberg, W. B., and Hardman, J. K. (1971) Structural and functional roles of the cysteine residues in the α-subunit of the *Escherichia coli* tryptophan synthetase. *J. Biol. Chem.* 246, 1439.

Freedman, M. H., Grossberg, A. L., and Pressman, D. (1968) The effects of complete modification of amino groups on the antibody activity of antihapten antibodies. Reversible inactivation with maleic anhydride. *Biochemistry* 7, 1941–1950.

Freytag, J. W., Lau, H. P., and Wadsley, J. J. (1984a) Affinity-column-mediated immunoenzymometric assays: Influence of affinity-column ligand and valency of antibody–enzyme conjugates. *Clin. Chem.* 30, 1494–1498.

Freytag, J. W., Dickinson, J. C., and Tseng, S. Y. (1984b) A highly sensitive affinity-column-mediated immunometric assay, as exemplified by digoxin. *Clin. Chem.* 30, 417–420.

Friden, P. M., Walus, L. R., Watson, P., Doctrow, S. R., Kozarich, J. W., Backman, C., Bergman, H., Hoffer, B., Bloom, F., and Granholm, A.-C. (1993) Blood–brain barrier penetration and *in vivo* activity of an NGF conjugate. *Science* 259, 373–378.

Fried, V. A., Ando, M. E., and Bell, A. J. (1985) Protein quantitation at the picomole level: An o-phthaldialdehyde pre-TSK column-derivatization assay. *Anal. Biochem.* **146**, 271–276.

Friede, M., Van Regenmortel, M. H. V., and Schuber, F. (1993) Lyophilized liposomes as shelf items for the preparation of immunogenic liposome–peptide conjugates. *Anal. Biochem.* **211**, 117–122.

Friedman, M. L., and Ball, W. J., Jr. (1989) Determination of monoclonal antibody-induced alterations in Na^+/K^+-ATPase conformations using fluorescein-labeled enzyme. *Biochim. Biophys. Acta* **995**, 42.

Friedrich, K., Woolley, P., and Steinhauser, K. G. (1988) Fluorimetric distance determination by resonance energy transfer. *Eur. J. Biophys.* **173**, 233.

Frytak, S., Creagan, E. T., Brown, M. L., Salk, D., and Nelp, W. (1993) A technetium labeled monoclonal antibody for imaging metastatic melanoma. *Am. J. Clin. Oncol.* **14**, 156–161.

Fuji, N., Akaji, K., Hayashi, Y., and Yajima, H. (1985) Studies on peptides. CXXV. 3-(3-p-methoxybenzylthiopropionyl)-thiazolidine-2-thione and its analogs as reagents for the introduction of the mercapto group into peptides and proteins. *Chem. Pharm. Bull.* **33**, 362–367.

Fujiwara, K., Matsumoto, N., Yagisawa, S., Tanimori, H., Kitagawa, T., Hirota, M., Hiratani, K., Fukushima, K., Tomonaga, A., Hara, K., and Yamamoto, K. (1988) Sandwich enzyme immunoassay of tumor-associated antigen sialosylated Lewis[x] using β-D-galactosidase coupled to a monoclonal antibody of IgM isotype. *J. Immunol. Methods* **112**, 77–83.

Gabizon, A., Meshorer, A., and Barenholz, Y. (1986) Comparative long-term study of the toxicities of free and liposome-associated doxorubicin in mice after intravenous administration. *J. Natl. Cancer Inst.* **77**, 459–469.

Gaffney, B. J., Willingham, G. L., and Schopp, R. S. (1983) Synthesis and membrane interactions of a spin-label bifunctional reagent. *Biochemistry* **22**, 881.

Gahmberg, C. G. (1978) Tritium labeling of cell-surface glycoproteins and glycolipids using galactose oxidase. *in* "Methods in Enzymology" (V. Ginsburg, ed.), Vol. 50, pp. 204–206. Academic Press, New York.

Gailit, J. (1993) Restoring free sulfhydryl groups in synthetic peptides. *Anal. Biochem.* **214**, 334–335.

Galardy, R. E., Craig, L. C., Jamieson, J. D., and Printz, M. P. (1974) Photoaffinity labeling of peptide hormone binding sites. *J. Biol. Chem.* **249**, 3510–3518.

Galardy, R. E. *et al.* (1978) Biologically active derivatives of angiotensin for labeling cellular receptors. *J. Med. Chem.* **21**, 1279.

Gaur, R. K. (1991) *Nucleoside Nucleotides* **10**, 895.

Gaur, R., Sharma, S., and Gupta, K. C. (1989) A simple method for the introduction of thiol group at 5'-termini of oligodeoxynucleotides. *Nucleic Acids Res.* **17**, 4404.

Gee, B., Warhol, M. J., and Roth, J. (1991) Use of an anti-horseradish peroxidase antibody gold complex in the ABC technique. *J. Histochem. Cytochem.* **39**, 863–870.

Gegg, C. V., and Etzler, M. E. (1993) Directional coupling of synthetic peptides to poly-L-lysine and applications to the ELISA. *Anal. Biochem.* **210**, 309–313.

Geiger, B., and Singer, S. J. (1980) Association of microtubules and intermediate filaments in chicken gizzard cells as detected by double immunofluorescence. *Proc. Natl. Acad. Sci. U.S.A.* **77**, 4769.

Geoghegan, K. F., and Stroh, J. G. (1992) Site-directed conjugation of nonpeptide groups to peptides and proteins via periodate oxidation of a 2-amino alcohol. Applications to modification at N-terminal serine. *Bioconjugate Chem.* **3**, 138–146.

Geoghegan, W. D. (1988) The effect of three variables on adsorption of rabbit IgG to colloidal gold. *J. Histochem. Cytochem.* **36**, 401–407.

Geoghegan, W. D., Ambegaonkar, N., and Calvanico, N. (1980) Passive gold agglutination: An alternative to passive hemagglutination. *J. Immunol. Methods* **34**, 11.

Germain, R. N. (1986) The ins and outs of antigen processing and presentation. *Nature (London)* **322**, 687–689.

Gershoni, J. M., Bayer, E. A., and Wilchek, M. (1985) Blot analysis of glycoconjugates: Enzyme–hydrazide—a novel reagent for the detection of aldehydes. *Anal. Biochem.* **146**, 59–63.

Ghebrehiwet, B., Bossone, S., Erdei, A., and Reid, K. B. M. (1988) Reversible biotinylation of Clq with a cleavable biotinyl derivative. Application in Clq receptor (ClqR) purification. *J. Immunol. Methods* **110**, 251–260.

Ghetie, V., Ghetie, M.-A., Uhr, J. W., and Vitetta, E. S. (1988) Large scale preparation of immunotoxins constructed with the Fab' fragment of IgGl murine monoclonal antibodies and chemically deglycosylated ricin A chain. *J. Immunol. Methods* **112**, 267–277.

Ghetie, V., Till, M. A., Ghetie, M.-A., Tucker, T., Porter, J., Patzer, E. J., Richardson, J. A., Uhr, J. W., and Vitetta, E. S. (1990) Preparation and characterization of conjugates of recombinant CD4 and deglycosylated ricin A chain using different crosslinkers. *Bioconjugate Chem.* **1**, 24–31.

Ghetie, V., Thorpe, P., Ghetie, M.-A., Knowles, P., Uhr, J. W., and Vitetta, E. S. (1991) The GLP large scale preparation of immunotoxins containing deglycosylated ricin A chain and a hindered disulfide bond. *J. Immunol. Methods* **142**, 223–230.

Ghetie, V., Swindel, E., Uhr, J. W., and Vitetta, E. S. (1993) Purification and properties of immunotoxins containing one vs. two deglycosylated ricin A chains. *J. Immunol. Methods* **166**, 117–122.

Ghosh, S. S., Kao, P. M., and Kwoh, D. Y. (1989) Synthesis of 5'-oligonucleotide hydrazide derivatives and their use in preparation of enzyme-nucleic acid hybridization probes. *Anal. Biochem.* **178**, 43–51.

Ghosh, S. S., Kao, P. M., McCue, A. W., and Chappelle, H. L. (1990) Use of maleimide-thiol coupling chemistry for efficient syntheses of oligonucleotide-enzyme conjugate hybridization probes. *Bioconjugate Chem.* **1**, 71–76.

Gilchrist, T. L., and Rees, C. W. (1969) "Carbenes, Nitrenes, and Arynes" (Studies in Modern Chemistry, p. 131. Nelson Publ., London.

Gill, D. M., Pappenheimer, A. M., Jr., Brown, R., and Kurnick, J. T. (1969) *J. Exp. Med.* **129**, 1.

Gilles, M. A., Hudson, A. Q., and Borders, C. L. (1990) Stability of water-soluble carbodiimides in aqueous solution. *Anal. Biochem.* **184**, 244–248.

Gillitzer, R., Berger, R., and Moll, H. (1990) A reliable method for simultaneous demonstration of two antigens using a novel combination of immunogold-silver staining and immunoenzymatic labeling. *J. Histochem. Cytochem.* **38**, 307–313.

Gitlin, G., Bayer, E. A., and Wilchek, M. (1987) Studies on the biotin-binding site of avidin. Lysine residues involved in the active site. *Biochem. J.* **242**, 923–926.

Gitlin, G., Bayer, E. A., and Wilchek, M. (1988) Studies on the biotin-binding site of avidin. Tryptophan residues involved in the active site. *Biochem. J.* **250**, 291–294.

Gitman, A. G., Kahane, L., and Loyter, A. (1985a) Use of virus-attached antibodies or insulin molecules to mediate fusion between Sendai virus envelopes and neuraminidase-treated cells. *Biochemistry* **24**, 2762–2768.

Gitman, A. G., Graessmann, A., and Loyter, A. (1985b) Targeting of loaded Sendai virus envelopes by covalently attached insulin molecules to virus receptor-depleted cells: Fusion-mediated microinjection of ricin A and simian 40 DNA. *Proc. Natl. Acad. Sci. U.S.A.* **82**, 7209–7313.

Glacy, S. (1983) Subcellular distribution of rhodamine-actin microinjected into living fibroblastic cells. *J. Cell Biol.* **97**, 1207.

Glazer, A. N. (1981) Photosynthetic accessory proteins with bilin prosthetic groups. *Biochem. Plants* **8**, 51–96.

Glazer, A. N. (1985) *Annu. Rev. Biophys. Biophys. Chem.* **14**, 47.

Glazer, A. N., and Hixson, C. S. (1977) Subunit structure and chromophore composition of rhodophytan phycoerythrins. *Porphyridium cruentum* B-phycoerythrin and b-phycoerythrin. *J. Biol. Chem.* **252**, 32–42.

Glazer, A. N., and Stryer, L. (1983) Fluorescent tandem phycobiliprotein conjugates: Emission wavelength shifting by energy transfer. *Biophys. J.* **43**, 383–386.

Glazer, A. N., Delange, R. J., and Sigman, D. S. (1975) *Lab. Tech. Biochem. Mol. Biol.* **1**, 205.

Goding, J. W. (1976) Conjugation of antibodies with fluorochromes: Modifications to the standard methods. *J. Immunol. Methods* **13**, 215–226.

Goding, J. W. (1986a) *in* "Monoclonal Antibodies: Principles and Practice," pp. 6–58. Academic Press, Orlando, Florida.

Goding, J. W. (1986b) *in* "Monoclonal Antibodies: Principles and Practice," pp. 35. Academic Press, Orlando, Florida.

Golds, E. E., and Braun, P. E. (1978) Protein associations and basic protein conformation in the myelin membrane. *J. Biol. Chem.* **253**, 8162–8170.

Goodfellow, V. S., Settineri, M., and Lawton, R. G. (1989) *p*-Nitrophenyl 3-diazopyruvate and diazopyruvamides, a new family of photoactivatable cross-linking bioprobes. *Biochemistry* **28**, 6346.

Goodlad, G. A. J. (1957) Cross-linking of collagen by sulfur- and nitrogen-mustards. *Biochim. Biophys. Acta* **25**, 202.

Goodson, R. J., and Katre, N. V. (1990) Site-directed pegylation of recombinant interleukin-2 at its glycosylation site. *Bio/Technology* **8**, 343–346.

Gorecki, M., and Patchornik, A. (1973) *Biochim. Biophys. Acta* **303**, 36.

Gorecki, M., and Patchornik, A. (1975) U.S. Patent No. 3,914,205.

Gorecki, M., Wilchek, M., and Patchornik, A. (1971) The conversion of 3-monoazotyrosine to 3-aminotyrosine in peptides and proteins. *Biochim. Biophys. Acta* **220**, 590–595.

Gorin, G., Martin, P. A., and Doughty, G. (1966) Kinetics of the reaction of N-ethylmaleimide with cysteine and some congeners. *Arch. Biochem. Biophys.* **115**, 593.

Gorman, J. J. (1984) Fluorescent labeling of cysteinyl residues to facilitate electrophoretic isolation of proteins suitable for amino-terminal sequence analysis. *Anal. Biochem.* **160**, 376.

Gorman, J. J., and Folk, J. E. (1980) Transglutaminase amine substrates for photochemical labeling and cleavable cross-linking of proteins. *J. Biol. Chem.* **255**, 1175.

Gorman, J. J., Corino, G. L., and Mitchell, S. J. (1987) Fluorescent labeling of cysteinyl residues. Application to extensive primary structure analysis of protein on a microscale. *Eur. J. Biochem.* **168**, 169–179.

Gotoh, Y., Tsukada, M., and Minoura, N. (1993) Chemical modification of silk fibroin with cyanuric chloride-activated poly(ethylene glycol: Analysis of reaction site by 1H-NMR spectroscopy and conformation of the conjugates. *Bioconjugate Chem.* **4**, 554–559.

Gounaris, A. D., and Perlman, G. E. (1967) *J. Biol. Chem.* **242**, 2739.

Goundalkar, A., Ghose, T., and Mezei, M. (1983) Covalent binding of antibodies to liposomes using a novel lipid derivative. *J. Pharm. Pharmacol.* **36**, 465–466.

Grabarek, Z., and Gergely, J. (1990) Zero-length cross-linking procedure with the use of active esters. *Anal. Biochem.* **185**, 131–135.

Grabowski, J., and Gantt, E. (1978) Photophysical properties of phycobiliproteins from phycobilisomes: Fluorescence lifetimes, quantum yields, and polarization spectra. *Photochem. Photobiol.* **28**, 39–45.

Granata, A. R., and Kitai, S. T. (1992) Intracellular analysis *in vivo* of different barosensitive bulbospinal neurons in the rat rostral ventrolateral medulla. *J. Neurosci.* **12**, 1–20.

Grassetti, D. R., and Murray, J. F. (1967) *Arch. Biochem. Biophys.* **119**, 41.

Gray, G. R. (1974) The direct coupling of oligosaccharides to proteins and derivatized gels. *Arch. Biochem. Biophys.* **163**, 426–428.

Gray, G. R. (1978) Antibodies to carbohydrates: Preparation of antigens by coupling carbohydrates to proteins by reductive amination with cyanoborohydride. *in* "Methods in Enzymology" (V. Ginsburg, ed.), Vol. 50, pp. 155–160. Academic Press, New York.

Grayeski, M. L., and DeVasto, J. K. (1987) Coumarin derivatizing agents for carboxylic acid detection using peroxyoxalate chemiluminescence with liquid chromatography. *Anal. Chem.* **59**, 1203.

Green, N. M. (1963) Stability at extremes of pH and dissociation into sub-units by guanidine hydrochloride. *Biochem. J.* **89**, 609–620.

Green, N. M. (1965) A spectrophotometric assay for avidin and biotin based on binding of dyes by avidin. *Biochem. J.* **94**, 23c–24c.

Green, N. M. (1975) Avidin. *Adv. Protein Chem.* **29**, 85–133.

Green, N. M., Konieczny, L., Toms, E. J., and Valentine, R. C. (1971) The use of bifunctional biotinyl compounds to determine the arrangement of subunits in avidin. *Biochem. J.* **125**, 781–984.

Greene, L. E. (1986) Cooperative binding of myosin subfragment one to regulated actin as measured by fluoresce changes of troponin 1 modified within different fluorophores. *J. Biol. Chem.* **261**, 1279.

Greenwood, F. C., Hunter, W. M., and Glover, J. S. (1963) The preparation of ^{131}I-labelled human growth hormone of high specific radioactivity. *Biochem. J.* **89**, 114.

Gregoriadis, G. (1984) "Liposome Technology," Vol. 3. CRC Press, Boca Raton, Florida.

Gregoriadis, G., and Senior, J. (1980) The phospholipid component of small unilamellar liposomes controls the rate of clearance of entrapped solutes from the circulation. *FEBS Lett.* **119**, 43–46.

Gregory, J. D. (1955) The stability of N-ethylmaleimide and its reaction with sulfhydryl groups. *J. Am. Chem. Soc.* **77**, 3922.

Gretch, D. R., Suter, M., and Stinski, M. F. (1987) The use of biotinylated monoclonal antibodies and streptavidin affinity chromatography to isolate herpesvirus hydrophobic proteins or glycoproteins. *Anal. Biochem.* **163**, 270–277.

Gros, O., Gros, P., Jansen, F. K., and Vidal, H. (1985) Biochemical aspects of immunotoxin preparation. *J. Immunol. Methods* **81**, 283.

Grossman, S. H., Pyle, J., and Steiner, R. J. (1981) Kinetic evidence for active monomers during the reassembly of denatured creatine kinase. *Biochemistry* **21**, 6122.

Guesdon, J.-L., Ternynck, T., and Avrameas, S. (1979) The use of avidin–biotin interaction in immunoenzymatic techniques. *J. Histochem. Cytochem.* **27**, 1131–1139.

Guire, P. (1976) Stepwise thermophotochemical cross-linking agents for enzyme stabilization and immobilization. *Fed. Proc.* **35**, 1632.

Gurd, F. R. N. (1967) Carboxymethylation. *in* "'Methods in Enzymology" (C. H. W. Hirs, ed.), Vol. 11, p. 532. Academic Press, New York.

Gutteridge, S., and Robb, D. A. (1973) *Biochem. Soc. Trans.* **1**, 519.

Habeeb, A. F. S. A. (1966) Determination of free amino groups in protein by trinitrobenzene sulfonic acid. *Anal. Biochem.* **14**, 328.

Habeeb, A. F. S. A., and Atassi, M. Z. (1970) Enzymatic and immunochemical properties of lysozyme. Evaluation of several amino group reversible blocking reagents. *Biochemistry* **9**, 4939–4944.

Habeeb, A. F. S. A., Cassidy, H. G., and Singer, S. J. (1958) *Biochim. Biophys. Acta* **29**, 587.

Habili, N., McInnes, J. K., and Symons, R. H. (1987) Non-radioactive photobiotin-labelled DNA probes for the routine diagnosis of barley yellow dwarf virus. *J. Virol. Methods* **16**, 225–237.

Hadi, U. A. M., Malcolme-Lawes, J., and Oldham, G. (1979) Rapid radiohalogenations of small molecules-II. Radiobromination of tyrosine, uracil, and cytosine. *Int. J. Appl. Radiat. Isot.* **30**, 709–712.

Hajdu, J., Dombradi, V., Bot, G., and Friedrich, P. (1979) Structural changes in glycogen phosphorylase as revealed by cross-linking with bifunctional diimidates: Phosphorylase b. *Biochemistry* **18**, 4037–4041.

Hall, L. D., and Yalpani, M. (1980) Synthesis of luminescent probe-sugar conjugates of either protected or unprotected sugars. *Carbohydr. Res.* **78**, C4.

Hamada, H., and Tsuro, T. (1987) Determination of membrane antigens by a covalent cross-linking method with monoclonal antibodies. *Anal. Biochem.* **160**, 483–488.

Hamaguchi, Y., Yoshitake, S., Ishikawa, E., Endo, Y., and Ohtaki, S. (1979) Improved procedure for the conjugation of rabbit IgG and Fab' antibodies with β-D-galactosidase from *Escherichia coli* using N,N'-o-phenylenedimaleimide. *J. Biochem. (Tokyo)* **85**, 1289–1300.

Hardy, P. M., Nicholls, A. C., and Rydon, H. N. (1969) The nature of glutaraldehyde in aqueous solution. *Chem. Commun.* **65**, 525.

Hardy, P. M., Nicholls, A. C., and Rydon, H. N. (1976) The nature of the cross-linking of proteins by glutaraldehyde. Interaction of glutaraldehyde with the amino-groups of 6-aminohexanoic acid and of β-N-acetyl-lysine. *J. Chem. Soc., Perkin Trans.* **1**, 958.

Harlow, E., and Lane, D. (1988a) "Antibodies: A Laboratory Manual," pp. 23–135. Cold Springs Harbor Laboratory, Cold Spring Harbor, New York.

Harlow, E., and Lane, D. (1988b) "Antibodies: A Laboratory Manual," pp. 7–22. Cold Spring Harbor Laboratory, Cold Spring Harbor, New York.

Harlow, E., and Lane, D. (1988c) "Antibodies: A Laboratory Manual," pp. 319–358. Cold Spring Harbor Laboratory, Cold Spring Harbor, New York.

Harrigan, P. R., Madden, T. D., and Cullis, P. R. (1990) *Chem. Phys. Lipids* **52**, 139.

Harris, J. M., Stuck, E. C., Case, M. G., Paley, M. S., Yalpani, M., van Alstine, J. M., and Brooks, D. E. (1984) Synthesis and characterization of PEG derivatives. *J. Polym. Sci. Polym. Chem. Ed.* **22**, 341–352.

Harrison, J. K., Lawton, R. G., and Gnegy, M. E. (1989) Development of a novel photoreactive calmodulin derivative: Cross-linking of purified adenylate cyclase from bovine brain. *Biochemistry* **28**, 6023.

Hartman, F. C., and Wold, F. (1966) *J. Am. Chem. Soc.* **88**, 3890–3891.

Hartman, F. C., and Wold, F. (1967) Cross-linking of bovine pancreatic ribonuclease A with dimethyl adipimidate. *Biochemistry* **6**, 2439–2448.

Hashida, S., and Ishikawa, E. (1985) Use of normal IgG and its fragments to lower the nonspecific binding of Fab'-enzyme conjugates in sandwich enzyme immunoassay. *Anal. Lett.* **18**(B9), 1143–1155.

Hashimoto, K., Loader, J. E., and Kinsky, S. C. (1986) Iodoacetylated and biotinylated liposomes: Effect of spacer length on sulfhydryl ligand binding and avidin precipitability. *Biochim. Biophys. Acta* **856**, 556–565.

Hashimoto, N., Takatsu, K., Masuho, Y., Kishida, K., Hara, T., and Hamaoka, T. (1984) Selective elimination of a B cell subset having acceptor site(s) for T cell-replacing factor (TRF) with biotinylated antibody to the acceptor site(s) and avidin–ricin A chain conjugate. *J. Immunol.* **132**, 129–135.

Hassell, J., and Hand, A. (1974) Tissue fixation with diimidoesters as an alternative to aldehydes. I. Comparison of cross-linking and ultrastructure obtained with dimethylsuberimidate and glutaraldehyde. *J. Histochem. Cytochem.* **22**, 223–239.

Hatakeyama, T., Kohzake, H., and Yamasaki, N. (1992) A microassay for proteases using succinylcasein as a substrate. *Anal. Biochem.* **204**, 181–184.

Haugaard, N., Cutler, J., and Ruggieri, M. R. (1981) Use of N-ethylmaleimide to prevent interference by sulfhydryl reagents with the glucose oxidase assay for glucose. *Anal. Biochem.* **116**, 341–343.

Haugland, R. P. (1991) Fluorescent labels. *in* "Biosensors with Fiberoptics" (D. L. Wise and L. B. Wingard, eds.), pp. 85–109. Humana Press, Totowa, New Jersey.

Hearn, M. T. W. (1987) 1,1′-Carbonyldiimidazole-mediated immobilization of enzymes and affinity ligands. *in* "Methods in Enzymology" (K. Mosbach, ed.), Vol. 135, pp. 102–117. Academic Press, Orlando, Florida.

Hearn, M. T. W., Bethell, G. S., Ayers, J. S., and Hancock, W. S. (1979) Application of 1,1′-carbonyldiimidazole-activated agarose for the purification of proteins. *J. Chromatogr.* **185**, 463–470.

Hearn, M. T. W., Smith, P. K., Mallia, A. K., and Hermanson, G. T. (1983) Preparative and analytical applications of CDI-mediated affinity chromatography. *in* "Affinity Chromatography and Biological Recognition," pp. 191–196. Academic Press, New York.

Hearn, S. A. (1987) Electron microscopic localization of chromogranin A in osmium-fixed neuroendocrine cells with a protein A–gold technique. *J. Histochem. Cytochem.* **35**, 795–801.

Heath, T. D., Edwards, D. C., and Ryman, B. E. (1976) The adjuvant properties of liposomes. *Biochem. Soc. Trans.* **4**, 129.

Heath, T. D., Macher, B. A., and Papahadjopoulos, D. (1981) Covalent attachment of immunoglobulins to liposomes via glycosphingolipids. *Biochim. Biophys. Acta* **640**, 66–81.

Heath, T. D., Montgomery, J. A., Piper, J. R., and Papahadjopoulos, D. (1983) Antibody-targeted liposomes: Increase in specific toxicity of methotrexate-γ-aspartate. *Proc. Natl. Acad. Sci. U.S.A.* **80**, 1377–1381.

Heath, T. D., Bragman, K. S., Matthay, K. K., Lopez-Straubinger, N. G., and Papahadjopoulos, D. (1984) Antibody-directed liposomes: The development of a cell-specific cytotoxic agent. *Biochem. Soc. Trans.* **12**, 340.

Hebert, G. A., Pittman, B., and Cherry, W. B. (1967) Factors affecting the degree of nonspecific staining given by fluorescent isothiocyanate labeled globulins. *J. Immunol.* **98**, 1204–1212.

Heilmann, H. D., and Holzner, M. (1981) The spatial organization of the active sites of the bifunctional oligomeric enzyme tryptophan synthetase: Cross-linking by a novel method. *Biochem. Biophys. Res. Commun.* **99**, 1146.

Heindel, N. D., Zhao, H., Leiby, J., VanDongen, J. M., Lacey, C. J., Lima, D. A., Shabsoug, B., and Buzby, J. H. (1990) Hydrazide pharmaceuticals as conjugates to polyaldehyde dextran: Syntheses, characterization, and stability. *Bioconjugate Chem.* **1**, 77–82.

Heindel, N. D., Zhao, H., Egolf, R. A., Chang, C.-H., Schray, K. J., Emrich, J. G., McLaughlin, J. P., and Woo, D. V. (1991) A novel heterobifunctional linker for formyl to thiol coupling. *Bioconjugate Chem.* **2**, 427–430.

Heindel, N. D., Kauffman, M. A., Akyea, E. K., Engel, S. A., Frey, M. F., Lacey, C. J., and Egolf, R. A. (1994) Carboxymethyldextran lactone: A preactivated polymer for amine conjugations. *Bioconjugate Chem.* **5**, 98–100.

Heinmark, R. L., Hershey, J. W. B., and Traut, R. R. (1976) Cross-linking of initiation factor IF2 to proteins L7/L12 in 70S ribosomes of *Escherichia coli. J. Biol. Chem.* **251**, 7779–7784.

Heitz, J. R., Anderson, C. D., and Anderson, B. M. (1968) Inactivation of yeast alcohol dehydrogenase by N-alkylmaleimides. *Arch. Biochem. Biophys.* **127**, 627.

Helenius, A., Kartenbech, J., Simons, K., and Fries, E. (1980) On the entry of Semliki forest virus into BHK-21 cells. *J. Cell Biol.* **84**, 404.

Helmeste, D. M., Hammonds, R. G., Jr., and Li, C. H. (1986) Preparation of [^{125}I-Tyr27, Leu5]β^h-endorphin and its use for cross-linking of opioid binding sites in human striatum and NG108-15 neuroblastoma-glioma cells. *Proc. Natl. Acad. Sci. U.S.A.* **83**, 4622–4625.

Hemmila, I. (1988) Lanthanides as probes for time-resolved fluorometric immunoassays. *Scand. J. Clin. Lab. Invest.* **48**, 389–400.

Herbener, G. H. (1989) Use of the protein A–gold immunocytochemical and enzyme–gold cytochemical techniques in studies of vitellogenesis. *Am. J. Anat.* **185**, 244–254.

Hermanson, G. T., Mallia, A. K., and Smith, P. K. (1992) *in* "Immobilized Affinity Ligand Techniques." Academic Press, San Diego.

Herriott, R. M. (1947) Reactions of native proteins with chemical reagents. *Adv. Protein Chem.* **3**, 169.

Hersckovits, T. (1988) Recent aspects of the subunit organization and dissociation of hemocyanins. *Comp. Biochem. Physiol.* **91B**, 597–611.

Herzberg, M. (1984) Molecular genetic probe, assay technique, and a kit using this molecular genetic probe. Eur. Patent Appl. 0128018.

Hillel, Z., and Wu, C. W. (1977) Subunit topography of RNA polymerase from *Escherichia coli*. A cross-linking study with bifunctional reagents. *Biochemistry* **16**, 3334–3342.

Hines, K. (1992) Pierce Chemical, unpublished observations.

Hiratsuka, T. (1987) Nucleotide-induced change in the interaction between the 20- and 26-kilodalton heavy-chain segments of myosin adenosine triphosphatase revealed by chemical cross-linking via the reactive thiol SH2. *Biochemistry* **26**, 3168.

Hiratsuka, T. (1988) Cross-linking of three heavy-chain domains of myosin adenosine triphosphatase with a trifunctional alkylating agent. *Biochemistry* **27**, 4110.

Hirsch, R. E., Zukin, R. S., and Nagel, R. L. (1986) Steady-state fluorescence emission from the fluorescent probe 5-iodoacetamido-fluorescein, bound to hemoglobin. *Biochem. Biophys. Res. Commun.* **138**, 4889.

Hnatowich, D. J. (1990) Antibody radiolabeling, problems and promises. *Nucl. Med. Biol.* **17**, 49–55.

Hnatowich, D. J., Layne, W. W., and Childs, R. L. (1982) The preparation and labeling of DTPA-coupled albumin. *Int. J. Appl. Radiat. Isot.* **33**, 327–332.

Hnatowich, D. J., Virzi, F., and Rusckowski, M. (1987) Investigations of avidin and biotin for imaging applications. *J. Nucl. Med.* **28**, 1294–1302.

Ho, R. J. Y., Rouse, B. T., and Huang, L. (1986) Target-sensitive immunoliposomes: Preparation and characterization. *Biochemistry* **25**, 5500.

Hoare, D. G., and Koshland, D. E. (1966) *J. Am. Chem. Soc.* **88**, 2057.

Hoare, D. G., and Koshland, D. E., Jr. (1967) *J. Biol. Chem.* **242**, 2447–2453.

Hochman, J. H., Shimizu, Y., DeMars, R., and Edidin, M. (1988) Specific associations of fluorescent B-2 microglobulin with cell surfaces. *J. Immunol.* **140**, 2322–2329.

Hoffman, W. L., and O'Shannessy, D. J. (1988) Site-specific immobilization of antibodies by their oligosaccharide moieties to new hydrazide derivatized solid supports. *J. Immunol. Methods* **112**, 113–120.

Hofmann, K., Finn, F. M., Friesen, H.-J., Diaconescu, C., and Zahn, H. (1977) Biotinylinsulins as potential tools for receptor studies. *Proc. Natl. Acad. Sci. U.S.A.* **74**, 2697–2700.

Hofmann, K., Wood, S. W., Brinton, C. C., Montibeller, J. A., and Finn, F. M. (1980) Iminobiotin affinity columns and their application to retrieval of streptavidin. *Proc. Natl. Acad. Sci. U.S.A.* **77**, 4666–4668.

Holgate, C., Jackson, P., Cowen, P., and Bird, C. (1983) ImmunoGold-silver staining: New method of immunostaining with enhanced sensitivity. *J. Histochem. Cytochem.* **31**, 938.

Holmberg, A., and Meurling, L. (1993) Preparation of sulfhydrylborane–dextran conjugates for boron neutron capture therapy. *Bioconjugate Chem.* **4**, 570–573.

Höltke, H.-J., Seibl, R., Burg, J., Mühlegger, K., and Kessler, C. (1990) Non-radioactive labeling and detection of nucleic acids: II. Optimization of the digoxigenin system. *Mol. Gen. Hoppe-Seyler* **371**, 929–938.

Honjo, J., Nishizuka, Y., Hayaishi, O., and Kato, I. (1968) Diphtheria toxin-dependent adenosine diphosphate ribosylation of aminoacyl transferase II and inhibition of protein synthesis. *J. Biol. Chem.* **243**, 3553–3555.

Hopman, A. H. N., Wiegant, J., Tesser, G. I., and Van Duijn, P. (1986) A nonradioactive *in situ* hybridization method based on mercurated nucleic acid probes and sulfhydryl–hapten ligands. *Nucleic Acids Res.* **14**, 6471–6488.

Hopp, T. P. (1984) *Mol. Immunol.* **21**, 13.

Hopwood, D. (1969) Comparison of the cross-linking abilities of glutaraldehyde, formaldehyde, and α-hydroxyadipaldehyde with bovine serum albumin and casein. *Histochemie* **17**, 151.

Hordern, J. S., Leonard, J. D., and Scraba, D. G. (1979) Structure of the mengo virion. *Virology* **97**, 131–140.

Horikawa, K., and Armstrong, W. E. (1988) A versatile means of intracellular labeling: Injection of biocytin and its detection with avidin conjugates. *J. Neurosci. Methods* **25**, 1–11.

Horisberger, M. (1979) Evaluation of colloidal gold as a cytochemical marker for transmission and scanning electron microscope. *Biol. Cell* **36**, 253–258.

Horisberger, M., and Clerc, M. F. (1985) Labeling of colloidal gold with protein A. *Histochemistry* **82**, 219.

Horisberger, M., and Rosset, J. (1977) Colloidal gold, a useful marker for transmission and scanning electron microscopy. *J. Histochem. Cytochem.* **25**, 295.

Horisberger, M., and Tacchini-Vonlanthen, M. (1983) Ultrastructural localization of Kunitz inhibitor on thin sections of *Glycine max* (soybean) cv. Maple Arrow by the gold method. *Histochemistry* **77**, 37–50.

Horisberger, M., Rosset, J., and Bauer, H. (1975) Colloidal gold granules as markers for cell surface receptors in the scanning electron microscope. *Experientia* **31**, 1147–1149.

Hornsey, V. S., Prowse, C. V., and Pepper, D. S. (1986) Reductive amination for solid-phase coupling of protein. A practical alternative to cyanogen bromide. *J. Immunol. Methods* **93**, 83–88.

Howard, A., de La Baume, S., Gioannini, T. L., Hiller, J. M., and Simon, E. J. (1985) Covalent labeling of opioid receptors with human β-endorphin. *J. Biol. Chem.* **260**, 10833–10839.

Huang, A., Huang, L., and Kennel, S. J. (1980) Monoclonal antibody covalently coupled with fatty acid. *J. Biol. Chem.* **255**, 8015–8018.

Huang, C. (1969) Studies of phospholipid vesicles. Formation and physical characteristics. *Biochemistry* **8**, 344.

Huang, K.-H., Fairclough, R. H., and Cantor, C. R. (1975) Singlet energy transfer studies of the arrangement of proteins in the 30S *Escherichia coli* ribosome. *J. Mol. Biol.* **97**, 443.

Hudson, E. N., and Weber, G. (1973) Synthesis and characterization of two fluorescent sulfhydryl reagents. *Biochemistry* **12**, 4154.

Hughes, W. L., and Straessle, R. (1950) Preparation and properties of serum and plasma proteins. XXIV. Iodination of human serum albumin. *J. Am. Chem. Soc.* **72**, 452–457.

Hunter, M. J., and Ludwig, M. L. (1962) The reaction of imidoesters with protein and related small molecules. *J. Am. Chem. Soc.* **84**, 3491.

Hurwitz, E., Maron, R., Arnon, R., Wilchek, M., and Sela, M. (1978) Daunomycin immunoglobulin conjugates, uptake and activity *in vitro*. *Eur. J. Cancer* **14**, 1213.

Hurwitz, E., Wilchek, M., and Phita, J. (1980) Soluble macromolecules as carriers for daunomycin. *J. Appl. Biochem.* **2**, 25.

Hurwitz, E., Arnon, R., Sahar, E., and Danon, Y. (1983a) A conjugate of adriamycin and monoclonal antibodies to Thy-1 antigen inhibits human neuroblastoma cells *in vitro*. *Ann. N.Y. Acad. Sci.* **417**, 125.

Hurwitz, E., Kashi, R., Burowsky, D., Arnon, R., and Haimovich, J. (1983b) Site-directed chemotherapy with a drug bound to antiidiotypic antibody to a lymphoma cell-surface IgM. *Int. J. Cancer* **31**, 745.

Hurwitz, E., Kashi, R., Arnon, R., Wilchek, M., and Sela, M. (1985) The covalent linking of two nucleotide analogues to antibodies. *J. Med. Chem.* **28**, 137.

Husain, S. S., and Lowe, G. (1968) Evidence for histidine in the active sites for ficin and stembromelain. *Biochem. J.* **110**, 53.

Huse, W. D., Sastry, L., Iverson, S. A., Kang, A. S., Alting-Mees, M., Burton, D. R., Benkovic, S. J., and Lerner, R. A. (1989) Generation of a large combinatorial library of the immunoglobulin repertoire in phage lambda. *Science* **246**, 1275–1281.

Hutchinson, F. J., Francis, S. E., Lyle, I. G., and Jones, M. N. (1989) The characterization of liposomes with covalently attached proteins. *Biochim. Biophys. Acta* **978**, 17–24.

Hynes, R. O. (1987) Integrins: A family of cell surface receptors. *Cell (Cambridge, Mass.)* **48**, 549–554.

Ikai, A., and Yanagita, Y. (1980) A cross-linking study of apo-low density lipoprotein. *J. Biochem. (Tokyo)* **88**, 1359–1364.

Imagawa, M., Yoshitake, S., Hamguchi, Y., Ishikawa, E., Niitsu, Y., Urushizaki, I., Kanazawa, R., Tachibana, S., Nakazawa, N., and Ogawa, H. (1982) Characteristics and evaluation of antibody–horseradish peroxidase conjugates prepared by using a maleimide compound, glutaraldehyde, and periodate. *J. Appl. Biochem.* **4**, 41–57.

Inada, Y., Takahashi, K., Yoshimoto, T., Ajima, A., Matsushima, A., and Saito, Y. (1986) Applications of polyethylene glycol-modified enzymes in biotechnological processes: Organic solvent-soluble enzymes. *Trends Biotechnol.* **4**, 190–194.

Ingalls, H. M., Goodloe-Holland, C. M., and Luna, E. J. (1986) Junctional plasma membrane domains isolated from aggregating *Dictyostelium discoideum* amebae. *Proc. Natl. Acad. Sci. U.S.A.* **83**, 4779–4783.

Inman, J. K. (1985) Functionalization of agarose beads via carboxymethylation and aminoethylamide formation. *in* "Affinity Chromatography—A Practical Approach" (P. D. G. Dean, W. S. Johnson, and F. A. Middle, eds.), pp. 53–59. IRL Press, Washington, D.C.

Inman, J. K., and Dintzis, H. M. (1969) *Biochemistry* **8**, 4074.

Innis, M. A., Gelfand, D. H., Sninsky, J. J., and White, T. J. (1990) "PCR Protocols. A Guide to Methods and Applications." Academic Press, New York.

Irvin, J. D. (1983) *Pharmacol. Ther.* **21**, 371–387.

Isaacs, B. S., Husten, E. J., Esmon, C. T., and Johnson, A. E. (1986) A domain of membrane-bound blood coagulation factor Va is located far from the phospholipid surface. A fluorescence energy transfer measurement. *Biochemistry* **25**, 4958–4969.

Ishi, Y., and Lehrer, S. S. (1986) Effects of the state of the succinimido-ring on the fluorescence and structural properties of pyrene maleimide-labeled $\alpha\alpha$-tropomyosin. *Biophys. J.* **50**, 75–80.

Ishikawa, E., Yamada, Y., and Yoshitake, S. (1981a) Enzyme labeling with N,N'-o-phenylenedimaleimide. *in* "Enzyme Immunoassay" (E. Ishikawa, T. Kawai, and K. Miyazi, eds.), pp. 67–80. Tokyo.

Ishikawa, E., Yamada, Y., Yoshitake, S., and Hamaguchi, Y. (1981b) A more stable maleimide, N-(4-carboxycyclohexylmethyl)maleimide, for enzyme labeling. *in* "Enzyme Immunoassay" (E. Ishikawa, T. Kawai, and K. Miyazi, eds.), pp. 90–105. Tokyo.

Ishikawa, E., Imagawa, M., Hashida, S., Yoshitake, S., Hamaguchi, Y., and Ueno, T. (1983a) Enzyme-labeling of antibodies. *J. Immunoassay* **4**, 209–327.

Ishikawa, E., Imagawa, M., and Hashida, S. (1983b) Ultra sensitive enzyme immunoassay using fluorogenic, luminogenic, radioactive and related substances and factors to limit the sensitivity. Proceedings 2nd International Symp. Immunoenzymatic Tech.

Ito, K., and Maruyama, J. (1983) Studies on stable diazoalkanes as potential fluorogenic reagents. I. 7-substituted 4-diazomethylcoumarins. *Chem. Pharm. Bull.* **31**, 3014.

Ito, K., and Sawanobori, J. (1982) 4-Diazomethyl-7-methoxycoumarin as a new type of stable aryldiazomethane reagent. *Synth. Commun.* **12**, 665.

Iwai, K., Fukuoka, S.-I., Fushiki, T., Kido, K., Sengoku, Y., and Semba, T. (1988) Preparation of a verifiable peptide–protein immunogen: direction-controlled conjugation of a synthetic fragment of the monitor peptide with myoglobin and application for sequence analysis. *Anal. Biochem.* **171**, 277–282.

Izzo, P. N. (1991) A note on the use of biocytin in anterograde tracing studies in the central nervous system: Application at both light and electron microscopic level. *J. Neurosci. Methods* **36**, 155–166.

Jablonski, E., Moomaw, E. W., Tullis, R. H., and Ruth, J. L. (1986) Preparation of oligodeoxynucleotide–alkaline phosphatase conjugates and their use as hybridization probes. *Nucleic Acids Res.* **14**, 6115–6129.

Jackson, P., Dockey, D. A., Lewis, F. A., and Wells, M. (1990) Application of 1-nm gold probes on paraffin wax sections for *in situ* hybridization histochemistry. *J. Clin. Pathol.* **43**, 810–812.

Jaffe, C. L., Lis, H., and Sharon, N. (1980) New cleavable photoreactive heterobifunctional cross-linking reagents for studying membrane organization. *Biochemistry* **19**, 4423.

Jagannath, C., and Sehgal, S. (1989) Enhancement of the antigen-binding capacity of incomplete IgG antibodies to *Brucella melitensis* through Fc region interactions with Staphylococcal protein A. *J. Immunol. Methods* **124**, 251–257.

Jansen, F. L., Blythman, H. E., Carriere, D., Casellas, P., Diaz, J., Gros, P., Hennequin, J. R., Paolucci, F., Pau, B., Poncelet, P., Richer, G., Salhi, S. L., Vidal, H., and Voisin, G. A. (1980) High specific cytotoxicity of antibody-toxin hybrid molecules (immunotoxins) for target cells. *Immunol. Lett.* **2**, 97.

Jansons, V. K., and Mallet, P. L. (1980) Targeted liposomes: A method for preparation and analysis. *Anal. Biochem.* **111**, 54–59.

Jayabaskaran, C., Davison, P. F., and Paulus, H. (1987) *Prep. Biochem.* **17**, 121.

Jeanloz, R. W. (1963) Mucopolysaccharides (acidic glycosaminoglycans) *in* "Comprehensive Biochemistry" (M. Florkin and E. Stotz, eds.), Vol. 3, pp. 266–267. Elsevier, New York.

Jeanson, A., Cloes, J. M., Bouchet, M., and Rentier, B. (1988) Preparation of reproducible alkaline phosphatase–antibody conjugates for enzyme immunoassay using a heterobifunctional linking agent. *Anal. Biochem.* **172**, 392.

Johansson, G. (1992) Affinity partitioning in PEG-containing two-phase systems. *in* "Poly(Ethylene Glycol) Chemistry: Biotechnical and Biomedical Applications" (J. M. Harris, ed.), pp. 73–84. Plenum, New York.

Johnson, J. D., Collins, J. H., and Potter, J. D. (1978) Dansylaziridine labeled troponin C. A fluorescent probe of calcium ion binding to the calcium ion-specific regulatory sites. *J. Biol. Chem.* **253**, 6451.

Joiris, E., Basin, B., and Thornback, J. A. (1991) A new method of labeling of monoclonal antibodies and their fragments with ^{99m}Tc. *Nucl. Med. Biol.* **18**, 353–356.

Jones, B. N., and Gilligan, J. P. (1983) o-Phthaldialdehyde precolumn derivatization and reversed phase high-performance liquid chromatography of polypeptide hydrolysates and physiological fluids. *J. Chromatogr.* **266**, 471–482.

Jones, G., Bergmark, W. R., and Jackson, W. R. (1984) Products of photodegradation for coumarin laser dyes. *Opt. Commun.* **50**, 320.

Jones, G., Jackson, W. R., Choi, C. Y., and Bergmark, W. R. (1985) Solvent effects on emission yields and lifetime for coumarin laser dyes. Requirements for the rotatory decay mechanism. *J. Phys. Chem.* **89**, 294.

Jones, O. T., Kunze, D. L., and Angelides, K. J. (1989) Localization and mobility of w-conotoxin-sensitive Ca^{+2} channels in hippocampal CA1 neurons. *Science* **244**, 1189.

Joshi, S., and Burrows, R. (1990) ATP synthase complex from bovine heart mitochondria. *J. Biol. Chem.* **265**, 14518–14525.

Jue, R., Lambert, J. M., Pierce, L. R., and Traut, R. R. (1978) Addition of sulfhydryl groups to *Escherichia coli* ribosomes by protein modification with 2-iminothiolane (methyl 4-mercaptobutyrimidate). *Biochemistry* **17**, 5399–5405.

Jeffrey, A. M., Zopf, D. A., and Ginsburg, V. (1975) *Biochem. Biophys. Res. Commun.* **62**, 608.

Jellum, E. (1964) *Acta Chem. Scand.* **18**, 1887–1895.

Jemmerson, R., and Agre, M. (1987) Monoclonal antibodies to different epitopes on a cell-surface enzyme, human placental alkaline phosphatase, effect different patterns of labeling with protein A–colloidal gold. *J. Histochem. Cytochem.* **35**, 1277–1284.

Jennings, M. L., and Nicknish, J. S. (1985) Localization of a site of intermolecular cross-linking in human red blood cell band 3 protein. *J. Biol. Chem.* **260**, 5472–5479.

Jentoft, N. (1990) Why are proteins O-glycosylated? *Trends Biochem. Sci.* **15**, 291–294.

Jeon, W. M., Lee, K. N., Birckbichler, P. J., Conway, E., and Patterson, M. K., Jr. (1989) Colorimetric assay for cellular transglutaminase. *Anal. Biochem.* **182**, 170–175.

Ji, I., and Ji, T. H. (1981) Both α and β subunits of human choriogonadotropin photoaffinity label the hormone receptor. *Proc. Natl. Acad. Sci. U.S.A.* **78**, 5465–5469.

Ji, I., Shin, J., and Ji, T. H. (1985) Radioiodination of a photoactivatable heterobifunctional reagent. *Anal. Biochem.* **151**, 348–349.

Ji, T. H. (1979) The application of chemical cross-linking for studies of cell membrane and the identification of surface reporters. *Biochim. Biophys. Acta* **559**, 39.

Ji, T. H. (1983) Bifunctional reagents. *Methods Enzymol.* **91**, 580.

Ji, T. H., and Ji, I. (1982) Macromolecular photoaffinity labeling with radioactive photoactivatable heterobifunctional reagents. *Anal. Biochem.* **121**, 286–289.

Jobbagy, A., and Jobbagy, G. M. (1972) *J. Immunol. Methods* **2**, 159.

Jobbagy, A., and Kiraly, K. (1966) Chemical characterization of fluorescein isothiocyanate–protein conjugates. *Biochim. Biophys. Acta* **124**, 166.

Jung, S. M., and Moroi, M. (1983) Cross-linking of platelet glycoprotein Ib by N-succinimidyl-(4-azidophenyldithio)propionate and 3,3'-dithiobis(sulfosuccinimidyl propionate). *Biochim. Biophys. Acta* **761**, 152–162.

Kagedal, L. (1989) *in* "Protein Purification: Principles, High Resolution Methods and Applications" (J.-C. Janson and L. Ryden, eds.), p. 227. VCH Publ., New York.

Kang, J., Tarscafalvi, A., Fujimoto, E., Shahrokh, Z., Shohet, S., and Ikemoto, N. (1991) Specific labeling of the foot protein moiety of the triad with a novel fluorescent probe: Application to the studies of conformational changes of the foot protein. *Biophys. J.* **59**, Tu-Pos 81, p. 249a (Abstract).

Kaplan, M. R., Calef, E., Bercovici, T., and Gitler, C. (1983) *Biochim. Biophys. Acta* **728**, 112.

Kareva, V. V., Dobrovol'sky, A. B., Baratova, L. A., Friedrich, P., and Gusev, N. B. (1986) Ca^{2+}-induced structural change in the Ca^{2+}/Mg^{2+} domain of troponin C detected by cross-linking. *Biochim. Biophys. Acta* **869**, 322.

Kasina, S., Rao, T. N., Srinivasan, A., Sanderson, J. A., Fitzner, J. N., Reno, J. M., Beaumier, P. L., and Fritzberg, A. R. (1991) Development and biologic evaluation of a kit for preformed chelate ^{99m}Tc. *J. Nucl. Med.* **32**, 1445–1451.

Kato, K., Hamaguchi, Y., Fukui, H., and Ishikawa, E. (1975a) Enzyme-linked immunoassay. I. Novel method for synthesis of the insulin-β-D-galactosidase conjugate and its applicability for insulin assay. *J. Biochem. (Tokyo)* **78**, 235.

Kato, K., Hamaguchi, Y., Fukui, H., and Ishikawa, E. (1975b) Enzyme-linked immunoassay. II. A simple method for synthesis of the rabbit antibody-β-D-galactosidase complex and its general applicability. *J. Biochem. (Tokyo)* **78**, 423.

Katz, M. J., Lasek, R. J., Osdoby, P., Whittaker, J. R., and Caplan, A. I. (1982) Bolton–Hunter reagent as a vital stain for developing systems. *Dev. Biol.* **90**, 419–429.

Kawamura, A., Jr. (ed.) (1977) "Fluorescent Antibody Techniques and Their Application." Univ. of Tokyo Press. Baltimore, Maryland.

Kay, C. M., and Edsall, J. T. (1956) Dimerization of mercaptalbumin in the presence of mercurials. III. Bovine mercaptalbumin in water and in concentrated urea solutions. *Arch. Biochem. Biophys.* **65**, 354.

Kay, E., Shannon, L. M., and Lew, J. Y. (1967) Peroxidase isozymes from horseradish roots. II. Catalytic properties. *J. Biol. Chem.* **242**, 2470.

Keana, J. F. W., and Cai, S. X. (1990) New reagents for photoaffinity labeling: Synthesis and photolysis of functionalized perfluorophenyl azides. *J. Org. Chem.* **55**, 3640.

Keen, J. H., Maxfield, F. R., Hardegree, M. C., and Habig, W. H. (1982) Receptor-mediated endocytosis of diphtheria toxin by cell in culture. *Proc. Natl. Acad. Sci. U.S.A.* **79**, 2912.

Keller, G. H., and Manak, M. M. (1989) "DNA Probes." Stockton, New York.

Keller, G. H., Cumming, C. U., Huang, D. P., Manak, M. M., and Ting, R. (1988) A chemical method for introducing haptens onto DNA probes. *Anal. Biochem.* **170**, 441–450.

Keller, G. H., Huang, D.-P., and Manak, M. M. (1989) Labeling of DNA probes with a photoactivatable hapten. *Anal. Biochem.* **177**, 392–395.

Kellogg, D. R., Michison, T. J., and Alberts, B. M. (1988) Behavior of microtubules and actin filaments in living *Drosophila* embryos. *Development (Cambridge, UK)* **103**, 675.

Kenny, J. W., Fanning, T. G., Lambert, J. M., and Traut, R. R. (1979) The subunit interface of the *Escherichia coli* ribosome. Cross-linking of 30S protein S9 to proteins of the 50S subunit. *J. Mol. Biol.* **135**, 151–170.

Kessler, C., Höltke, H.-J., Seibl, R., Burg, J., and Muhlegger, K. (1990) Nonradioactive labeling and detection of nucleic acids: I. A novel DNA labeling and detection system based on digoxigenin:antidigoxigenin ELISA principle (digoxigenin system). *Mol. Gen. Hoppe-Seyler* **371**, 917–927.

Khalfan, H., Abuknesha, R., Rand-Weaver, M., Price, R. G., and Robinson, D. (1986) Aminomethyl coumarin acetic acid: A new fluorescent labeling reagent for proteins. *Histochem. J.* **18**, 497–499.

Khan, A. M., and Wright, P. J. (1987) Detection of flavivirus RNA in infected cells using photobiotin-labelled hybridization probes. *J. Virol. Methods* **15**, 121–130.

Khanna, P. L., and Ullman, E. F. (1980) 4′,5′-Dimethoxy-6-carboxyfluorescein: A novel dipole–dipole coupled fluorescence energy transfer acceptor useful for fluorescence immunoassays. *Anal. Biochem.* **108**, 156.

Kiehm, D., and Ji, T. H. (1977) Photochemical cross-linking of cell membranes. *J. Biol. Chem.* **252**, 8524–8531.

Kim, C. G., and Sheffrey, M. (1990) Physical characterization of the affinity purified CCAAT transcription, α-CP1. *J. Biol. Chem.* **265**, 13362–13369.

Kim, E. E., and Wyckoff, H. W. (1991) Reaction mechanism of alkaline phosphatase based on crystal structures. Two metal ion catalysis. *J. Mol. Biol.* **218**, 449–464.

Kim, S., and Martin, G. M. (1981) Preparation of cell-size unilamellar liposomes with high captured volume and defined size distribution. *Biochim. Biophys. Acta* **646**, 1–9.

Kim, S., Tuker, M. S., Chi, E. Y., Sela, S., and Martin, G. M. (1983) Preparation of multivesicular liposomes. *Biochim. Biophys. Acta* **728**, 339–348.

Kimura, A., Orn, A., Holmquist, G., Wizzell, H., and Ersson, B. (1979) Unique lectin-binding characteristics of cytotoxic T-lymphocytes allowing their distribution from natural killer cells and "K" cells. *Eur. J. Immunol.* **9**, 575.

King, M. A., Louis, P. M., Hunter, B. E., and Walker, D. W. (1989) Biocytin: A versatile anterograde neuroanatomical tract-tracing alternative. *Brain Res.* **497**, 361–367.

King, P., Li, Y., and Kochoumian, L. (1978) Preparation of protein conjugates via intermolecular disulfide bond formation. *Biochemistry* **17**, 1499.

Kirley, T. L. (1989) Reduction and fluorescent labeling of cyst(e)ine-containing proteins for subsequent structural analysis. *Anal. Biochem.* **180**, 231–236.

Kitagawa, T., and Aikawa, T. (1976) Enzyme coupled immunoassay of insulin using a novel coupling reagent. *J. Biochem. (Tokyo)* **79**, 233–236.

Kitagawa, T., Fujitake, T., Taniyama, H., and Aikawa, T. (1978) Enzyme immunoassay of viomycin. *J. Biochem. (Tokyo)* **83**, 1493–1501.

Kitagawa, T., Kawasaki, T., and Munechika, H. (1982) *J. Biochem. (Tokyo)* **92**, 585–590.

Kiyama, H., Emson, P. C., and Tokyama, M. (1992) *In situ* hybridization histochemistry using alkaline phosphatase-labeled oligodeoxynucleotide probe. *in* "Methods in Molecular Biology, Volume 13: Protocols in Molecular Neurobiology" (A. Longstaff and P. Revest, eds.), pp. 167–179. Humana Press, Totowa, New Jersey.

Klapper, M. H., and Klotz, I. M. (1972) Acylation with dicarboxylic acid anhydrides. *in* "Methods in Enzymology" (C. H. W. Hirs and S. N. Timasheff, eds.), Vol. 25, pp. 531–552. Academic Press, New York.

Klibanov, A. L., Maruyama, K., Beckerleg, A. M., Torchilin, V. P., and Huang, L. (1991)

Activity of amphipathic poly(ethylene glycol) 5000 to prolong the circulation time of liposomes depends on the liposome size and is unfavorable for immunoliposome binding to target. *Biochim. Biophys. Acta* **1062**, 142–148.

Klotz, I. M. (1967) Succinylation. *in* "Methods in Enzymology" (C. H. W. Hirs, ed.), Vol. 11, p. 576. Academic Press, New York.

Klotz, I. M., and Heiney, R. E. (1962) Introduction of sulfhydryl groups into proteins using acetylmercaptosuccinic anhydride. *Arch. Biochem. Biophys.* **96**, 605.

Klotz, I. M., and Keresztes-Nagy, S. (1962) *Nature (London)* **195**, 900.

Knauf, M. J., Bell, D. P., Hirtzer, P., Luo, Z.-P., Young, J. D., and Katre, N. V. (1988) Relationship of effective molecular size to systemic clearance in rats of recombinant interleukin-2 chemically modified with water-soluble polymers. *J. Biol. Chem.* **263**, 15064–15070.

Knoller, S., Shpungin, S., and Pick, E. (1991) The membrane-associated component of the amphiphile-activated, cytosol-dependent superoxide-forming NADPH oxidase of macrophages is identical to cytochrome b559. *J. Biol. Chem.* **266**, 2795–2804.

Kobayashi, M., and Ichishima, E. (1991) Application of periodate oxidized glucans to biochemical reactions. *J. Carbohydro Chem.* **10**, 635–644.

Kohn, K. W., Spears, C. L., and Doty, P. (1966) Interstrand cross-linking of DNA by nitrogen mustard. *J. Mol. Biol.* **19**, 87.

Konigsberg, W. (1972) Reduction of disulfide bonds in proteins with dithiothreitol. *in* "Methods in Enzymology" (C. H. W. Hirs and S. N. Timaseff, eds.) Vol. 25, p. 185. Academic Press, New York.

Konishi, K., and Fujioka, M. (1987) Chemical modification of a functional arginine residue of rat liver glycine methyltransferase. *Biochemistry* **26**, 8496–8502.

Konno, K., and Morales, M. F. (1985) Exposure of actin thiols by the removal of tightly held calcium ions. *Proc. Natl. Sci. Acad. U.S.A.* **82**, 7904–7908.

Korn, A. H., Feairheller, S. H., and Filachione, E. M. (1972) Glutaraldehyde: Nature of the reagent. *J. Mol. Biol.* **65**, 525–529.

Kornblatt, J. A., and Lake, D. F. (1980) Cross-linking of cytochrome oxidase subunits with difluorodinitrobenzene. *Can. J. Biochem.* **58**, 219-224.

Kornfield, R., and Kornfield, S. (1985) Assembly of asparagine-linked oligosaccharides. *Annu. Rev. Biochem.* **54**, 631.

Kotite, N. J., Staros, J. V., and Cunningham, L. W. (1984) Interaction of specific platelet membrane proteins with collagen: Evidence from chemical cross-linking. *Biochemistry* **23**, 3099–3104.

Kovacic, P., and Hein, R. W. (1959) Cross-linking of polymers with dimaleimide. *J. Am. Chem. Soc.* **81**, 1187.

Kozulic, B., Barbaric, S., Ries, B., and Mildner, P. (1984) Study of the carbohydrate part of yeast acid phosphatase. *Biochem. Biophys. Rev. Commun.* **122**, 1083.

Kraehenbuhl, J. P., Galardy, R. E., and Jamieson, J. D. (1974) Preparation and characterization of an immunoelectron microscope tracer consisting of a heme-octapeptide coupled to Fab. *J. Exp. Med.* **139**, 208.

Kricka, L. J. (1992) "Nonisotopic DNA Probe Techniques." Academic Press, New York.

Krieg, U. C., Walter, P., and Johnson, A. E. (1986) Photocross-linking of the signal sequence of nascent preprolactin to the 54-kilodalton polypeptide of the signal recognition particle. *Proc. Natl. Acad. Sci. U.S.A.* **83**, 8604–8608.

Kronick, M. N. (1986) The use of phycobiliproteins as fluorescent labels in immunoassay. *J. Immunol. Methods* **92**, 1–13.

Kull, F. C., Jr., Jacobs, S., and Cuatrecasas, P. (1985) Cellular receptor for ^{125}I-labeled tumor necrosis factor: Specific binding, affinity labeling, and relationship to sensitivity. *Proc. Natl. Acad. Sci. U.S.A.* **82**, 5756–5760.

Kumar, A., and Malhotra, S. (1992) A simple method for introducing -SH group at 5' OH terminus of oligonucleotides. *Nucleosides Nucleotides* **11**, 1003–1007.

Kumar, A., Tchen, P., Roullet, F., and Cohen, J. (1988) Nonradioactive labeling of synthetic oligonucleotide probes with terminal deoxynucleotidyl transferase. *Anal. Biochem.* **169**, 376–382.

Kung, V. T., Maxim, P. E., Veltri, R. W., and Martin, F. J. (1985) Antibody-bearing liposomes improve agglutination of latex particles used in clinical diagnostic assays. *Biochim, Biophys. Acta* **839**, 105–109.

Kurzchalia, T. V., Wiedmann, M., Breter, H., Zimmermann, W., Bauschke, E., and Rapoport, T. A. (1988) tRNA-mediated labeling of proteins with biotin. A nonradioactive method for the detection of cell-free translation products. *Eur. J. Biochem.* **172**, 663–668.

Kuwata, K., Uebori, M., Yamada, K., and Yamazaki, Y. (1982) Liquid chromatographic determination of alkylthiols via derivatization with 5,5′-dithiobis(2-nitrobenzoic acid). *Anal. Chem.* **54**, 1082–1087.

Labbe, J. P., Mornet D., Roseau, G., and Kassab, R. (1982) Cross-linking of F-actin to skeletal muscle myosin subfragment 1 with bis(imido esters): Further evidence for the interaction of myosin-head heavy chain with an actin dimer. *Biochemistry* **21**, 6897–6902.

Laburthe, M., Breant, B., and Rouyer-Fessard, C. (1984) Molecular identification of receptors for vasoactive intestinal peptide in rat intestinal epithelium by covalent cross-linking. *Eur. J. Biochem.* **139**, 181–187.

Lacey, B., and Grant, W. N. (1987) Photobiotin as a sensitive probe for protein labeling. *Anal. Biochem.* **163**, 151–158.

Laemmli, U. K. (1970) Cleavage of structural proteins during the assembly of the head of the bacteriophage T4. *Nature (London)* **277**, 680–685.

Lakowicz, J. R. (ed.) (1991) "Topics in Fluorescence Spectroscopy," Vols. 1–3. Plenum, New York.

Lambert, J. M., Boileau, G., Cover, J. A., and Traut, R. R. (1983) Cross-links between ribosomal proteins of 30S subunits in 70S tight couples and in 30S subunits. *Biochemistry* **22**, 3913–3920.

Lambert, J. M., Senter, P. D., Yau-Young, A., Blattler, W. A., and Goldmacher, V.S. (1985) Purified immunotoxins that are reactive with human lymphoid cells: monoclonal antibodies conjugated to the ribosome-inactivating proteins gelonin and the pokeweed antiviral proteins. *J. Biol. Chem.* **260**, 12035–12041.

Lambert, J. M., Blattler, W. A., McIntyre, G. D., Golmacher, V. S., and Scott, C. F., Jr. (1988) Immunotoxins containing single chain ribosome-inactivating proteins, *in* "Immunotoxins" (A. E. Frankel, ed.), p. 175. Kluwer, Boston.

Langer, P. R., Waldrop, A. A., and Ward, D. C. (1981) Enzymatic synthesis of biotin-labeled polynucleotides: Novel nucleic acid affinity probes. *Proc. Natl. Acad. Sci. U.S.A.* **78**, 6633–6637.

Langone, J. J. (1980) Radioiodination by use of the Bolton–Hunter and related reagents. *in* "Methods in Enzymology" (H. Van Vunakis and J. J. Langone, eds.), Vol. 70, pp. 221–243. Academic Press, New York.

Langone, J. J. (1981) Radioiodination by use of the Bolton–Hunter and related reagents. *in* "Methods in Enzymology" (J. J. Langone and H. Van Vunakis, eds.), Vol. 73, pp. 113–127. Academic Press, New York.

Lanier, L. L., and Recktenwald, D. J. (1991) Multicolor immunofluorescence and flow cytometry. *Methods (San Diego)* **2**, 192.

Lanteigne, D., and Hnatowich, D. J. (1984) The labeling of DTPA-coupled proteins with ^{99m}Tc. *Int. J. Appl. Radiat. Isot.* **35**, 617–621.

Lapidot, Y., Rappoport, S., and Wolman, Y. (1967) *J. Lipid Res.* **8**, 142.

LaRochelle, W. J., and Froehner, S. C. (1986a) Determination of the tissue distributions and relative concentrations of the postsynaptic 43-kDa protein and the acetylcholine receptor in Torpedo. *J. Biol. Chem.* **261**, 5270–5274.

LaRochelle, W. J., and Froehner, S. C. (1986b) Immunochemical detection of proteins biotinylated on nitrocellulose replicas. *J. Immunol. Methods* **92**, 65–71.

Larsson, P.-O., and Mosbach, K. (1971) *Biotechnol. Bioeng.* **13**, 393.

Leary, J. J., Brigati, D. J., and Ward, D. C. (1981) Enzymatic synthesis of biotin-labelled nucleotides: Novel nucleic acid affinity probes. *Proc. Natl. Acad. Sci. U.S.A.* **80**, 4045–4049.

Leary, J. J., Waldrop, A. A., and Ward, D. C. (1983) Rapid and sensitive colorimetric method for visualizing biotin-labeled DNA probes hybridized to DNA or RNA immobilized on nitrocellulose: Bio-blots. *Proc. Natl. Acad. Sci. U.S.A.* **80**, 4045–4049.

Lee, A. C. J., Powell, J. E., Tregear, G. W., Niall, H. D., and Stevens, V. C. (1980) *Mol. Immunol.* **17**, 749.

Lee, D. S. C., and Griffiths, B. W. (1984) Comparative studies of Iodo-Bead and chloramine-T methods for the radioiodination of human alpha-fetoprotein. *J. Immunol. Methods* **74**, 181–189.

Lee, J. A., and Fortes, P. A. G. (1985) Labeling of the glycoprotein subunit of (Na,K)ATPase with fluorescent probes. *Biochemistry* **24**, 322–330.

Lee, K. N., Maxwell, M. D., Patterson, M. K., Jr., Birckbichler, P. J., and Conway, E. (1992) Identification of transglutaminase substrates in HT29 colon cancer cells: use of 5-(biotinamido)pentylamine as a transglutaminase-specific probe. *Biochim. Biophys. Acta* **1136**, 12–16.

Lee, K. Y., Birckbichler, P. J., and Patterson, M. K., Jr. (1988) Colorimetric assay of blood coagulation factor XIII in plasma. *Clin. Chem.* **34**, 906–910.

Lee, L., Kelly, R. E., Pastra-Landis, S. C., and Evans, D. R. (1985) Oligomeric structure of the multifunctional protein CAD that initiates pyrimidine biosynthesis in mammalian cells. *Proc. Natl. Acad. Sci. U.S.A.* **82**, 6802–6806.

Lee, W. T., and Conrad, D. H. (1984) The murine lymphocyte receptor for IgE. II. Characterization of the multivalent nature of the B lymphocyte receptor for IgE. *J. Exp. Med.* **159**, 1790–1795.

Lee, W. T., and Conrad, D. H. (1985) The murine lymphocyte receptor for IgE. III. Use of chemical cross-linking reagents to further characterize the B lymphocyte Fcε receptor. *J. Immunol.* **134**, 518–525.

Leffak, I. M. (1983) Decreased protein staining after chemical cross-linking. *Anal. Biochem.* **135**, 95–101.

Lennarz, W. J. (ed.) (1980) "The Biochemistry of Glycoproteins and Proteoglycans." Plenum, New York.

Lerner, R. A., Green, N., Alexander, H., Liu, F.-T., Sutcliffe, J. G., and Shinnick, T. M. (1981) Chemically synthesized peptides predicted from the nucleotide sequence of the hepatitis B virus genome elicit antibodies reactive with the native envelope protein of Dane particles. *Proc. Natl. Acad. Sci. U.S.A.* **78**, 3403–3407.

Leserman, L., and Machy, P. (1987) Ligand targeting of liposomes. *in* "Liposomes: From Biophysics to Therapeutics" (M. J. Ostro, ed.), pp. 157–194. Dekker, New York.

Lethias, C., Hartmann, D. J., Masmejean, M., Ravazzola, M., Sabbagh, I., Ville, G., Herbage, D., and Eloy, R. (1987) Ultrastructural immunolocalization of elastic fibers in rat blood vessels using the protein A–gold technique. *J. Histochem. Cytochem.* **35**, 15–21.

Levison, M. E., *et al.* (1969) *Experientia* **25**, 126–127.

Lewis, R. V., Roberts, M. F., Dennis, E. A., and Allison, W. S. (1977) Photoactivated heterobifunctional cross-linking reagents which demonstrate the aggregation state of phospholipase A2. *Biochemistry* **16**, 5650–5654.

Li, C. H. (1945) Iodination of tyrosine groups in serum albumin and pepsin. *J. Am. Chem. Soc.* **67**, 1065–1069.

Li, M., and Meares, C. F. (1993) Synthesis, metal chelate stability studies, and enzyme digestion of a peptide-liked DOTA derivatives and its corresponding radiolabeled immunoconjugates. *Bioconjugate Chem.* **4**, 275–283.

Liener, I. E., and Friedenson, B. (1970) Ficin. *in* "Methods in Enzymology" (G. E. Perlmann and L. Lorand, eds.), Vol. 19, pp. 261–273. Academic Press, New York.

Lindley, H. (1956) A new synthetic substrate for trypsin and its application to the determination of the amino acid sequence of proteins. *Nature (London)* **178**, 647.

Liu, F.-T., Zinnecker, M., Hamaoka, T., and Katz, D. H. (1979) New procedures for preparation and isolation of conjugates of proteins and a synthetic copolymer of D-amino acids and immunochemical characterization of such conjugates. *Biochemistry* **18**, 690–697.

Liu, S. C., Fairbanks, G., and Palek, J. (1977) Spontaneous reversible protein cross-linking in the human erythrocyte membrane. Temperature and pH dependence. *Biochemistry* **16**, 4066.

Liu, Y., and Wu, C. (1991) Radiolabeling monoclonal antibodies with metal chelates. *Pure Appl. Chem.* **63**, 427–463.

Lloyd, R. V., Jin, L., and Fields, K. (1990) Detection of chromogranins A and B in endocrine tissues with radioactive and biotinylated oligonucleotide probes. *Am. J. Surg. Pathol.* **14**, 35–43.

Loken, M. R., Keij, J. F., and Kelley, K. A. (1987) Comparison of helium–neon and dye lasers for the excitation of allophycocyanin. *Cytometry* **8**, 96.

Lomant, A. J., and Fairbanks, G. (1976) Chemical probes of extended biological structures: Synthesis and properties of the cleavable cross-linking reagent [^{35}S] dithiobis(succinimidyl propionate). *J. Mol. Biol.* **104**, 243–261.

Lonnerdal, B., and Keen, C. L. (1982) *J. Appl. Biochem.* **4**, 203.

Lord, J. M., Spooner, R. A., Hussain, K., and Roberts, L. M. (1988) Immunotoxins: Properties, applications, and current limitations. *Adv. Drug Delivery Rev.* **2**, 297.

Lotan, R., Debray, H., Cacan, M., Cacan, R., and Sharon, N. (1975) Labeling of soybean agglutinin by oxidation with sodium periodate followed by reduction with [^{3}H]borohydride. *J. Biol. Chem.* **250**, 1955–1957.

Louis, C. F., Saunders, M. J., and Holroyd, J. A. (1977) The cross-linking of rabbit skeletal muscle sarcoplasmic reticulum protein. *Biochim. Biophys. Acta* **493**, 78–92.

Lowe, C. R. (1979) Immobilized nucleotides and coenzymes for affinity chromatography. *Pure Appl. Chem.* **51**, 1429–1441.

Lowe, C. R., and Dean, P. D. G. (1971) Affinity chromatography of enzymes on insolubilized cofactors. *FEBS Lett.* **14**, 313–316.

Lowe, C. R., and Dean, P. D. G. (1974) "Affinity Chromatography," pp. 228–229. Wiley, New York.

Lowe, C. R., Harvey, M. J., Craven, D. B., and Dean, P. D. G. (1973) *Biochem. J.* **133**, 499.

Lu, R. C., and Wong, A. (1989) Glutamic acid-88 is close to SH-1 in the tertiary structure of myosin subfragment-1. *Biochemistry* **28**, 4826.

Luduena, R. F., Roach, M. C., Trcka, P. P., and Weintraub, S. (1982) Bioiodoacetyldithio-ethylamine: A reversible cross-linking reagent for protein sulfhydryl group. *Anal. Biochem.* **117**, 76.

Ludwig, F. R., and Jay, F. A. (1985) *Eur. J. Biochem.* **151**, 83–87.

Lundblad, R. (1991) "Chemical Reagents for Protein Modification." CRC Press, Boca Raton, Florida.

McBroom, C. R., Samanen, C. H., and Goldstein, I. J. (1976) Carbohydrate antigens: Coupling of carbohydrates to proteins by diazonium and phenylisothiocyanate reactions. *in* "Methods in Enzymology" (W. B. Jakoby, ed.), Vol. 2, p. 212. Academic Press, New York.

McCleary, B. V., and Matheson, N. K. (1986) Enzymic analysis of polysaccharide structure. *Adv. Carbohydr. Chem. Biochem.* **44**, 147–276.

McGown, L. B., and Warner, I. M. (1990) Molecular fluorescence, phosphorescence, and chemiluminescence spectroscopy. *Anal. Chem.* **190**, 255R.

McInnes, J. L., Dalton, S., Vize, P. D., and Robins, A. J. (1987) Non-radioactive photobiotin-labeled probes detect single copy genes and low abundance mRNA. *Bio/Technology* **5**, 269–272.

McKinney, R. M., Spillane, J. T., and Pearce, G. W. (1964) Factors affecting the rate of reaction of fluorescein isothiocyanate with serum proteins. *J. Immunol.* **93**, 232–242.

Mahan, D. E., Morrison, L., Watson, L., and Haugneland, L. S. (1987) Phase change enzyme immunoassay. *Anal. Biochem.* **162**, 163–170.

Malcolm, A. D. B., and Nicolas, J. L. (1984) Detecting a polynucleotide sequence and labelled polynucleotides useful in this method. WO Patent Appl. 8403520.

Male, D., Champion, B., and Cooke, A. (1987) "Advanced Immunology," (Section 8.1–8.8. Lippincott, Gower Medical Publ., London.

Mallia, A. K. (1992) Pierce Chemical, personal communications.

Manabe, Y., Tsubota, T., Haruta, Y., Okazaki, M., Haisa, S., Nakamura, K., and Kimura, I. (1983) Production of monoclonal antibody–bleomycin conjugate utilizing dextran T40 and the antigen-targeting cytotoxicity of the conjugate. *Biochem. Biophys. Res. Commun.* **115**, 1009.

Mandy, W. J., Rivers, M. M., and Nisonoff, A. (1961) Recombination of univalent subunits derived from rabbit antibody, *J. Biol. Chem.* **236**, 3221.

Mannik, M., and Downey, W. (1973) Studies on the conjugation of horseradish peroxidase to Fab fragments. *J. Immunol. Methods* **3**, 233.

Manske, J. M., Buchsbaum, D. J., and Vallera, D. A. (1989) The role of ricin B chain in the intracellular trafficking of anti-CD5 immunotoxins. *J. Immunol.* **142**, 1755–1766.

Marcholonis, J. J. (1969) Biochem. J. **113**, 299.

Marcus, S. L., and Balbinder, E. (1972) Use of affinity matrices in determining steric requirements for substrate binding: Binding of anthranilate 5-phosphoribosyl-pyrophosphate phosphoribosyltransferase from *Salmonella typhimurium* to Sepharose-anthranilate derivatives. *Anal. Biochem.* **48**, 448–459.

Marfey, S. P., and Tsai, K. H. (1975) Cross-linking of phospholipids in human erythrocyte membrane. *Biochem. Biophys. Res. Commun.* **65**, 31–38.

Markwell, M. A. K. (1982) A new solid-state reagent to iodinate proteins: Conditions for the efficient labeling of antiserum. *Anal. Biochem.* **125**, 427–432.

Markwell, M. A. K., and Fox, C. F. (1978) Surface-specific iodination of membrane proteins of viruses and eukaryotic cells using 1,3,4,6-tetrachloro-3α,6α-diphenylglycouril. *Biochemistry* **17**, 4807–4817.

Markwell, M. A. K., and Fox, C. F. (1980) *J. Virol.* **33**, 152–166.

Martin, F. J., and Papahadjopoulos, D. (1982) Irreversible coupling of immunoglobulin fragments to preformed vesicles. *J. Biol. Chem.* **257**, 286–288.

Martin, F. J., Hubbell, W., and Papahyadjopoulos, D. (1981) Immunospecific targeting of liposomes to cells: A novel and efficient method for covalent attachment of Fab′ fragments via disulfide bonds. *Biochemistry* **20**, 4229–4238.

Martin, F. J., Heath, T. D., and New, R. R. C. (1990) Covalent attachment of proteins to liposomes. *in* "Liposomes, a Practical Approach," pp. 163–182. IRL Press, New York.

Martinez-Ramon, A., Knecht, E., Rubio, V., and Grisolia, S. (1990) Levels of carbamoyl phosphate synthetase I in livers of young and old rats assessed by activity and immunoassays and by electron microscopic immunogold procedures. *J. Histochem. Cytochem.* **38**, 371–376.

Masamune, S., Palmer, M. A. J., Gamboni, R., Thompson, S., Davis, J. T., Williams, S. F., Peoples, O. P., Sinskey, A. J., and Walsh, C. T. (1989) Bio-Claisen condensation catalyzed by thiolase from *Zoogloea ramigera*. Active site cysteine residues. *Chemtracts: Org. Chem.* **2**, 247–251.

Massague, J., Guillette, B. J., Czech, M. P., Morgan, C. J., and Bradshaw, R. A. (1981) Identification of a nerve growth factor receptor protein in sympathetic ganglia membranes by affinity labeling. *J. Biol. Chem.* **256**, 9419–9424.

Masuho, Y., Hara, T., and Noguchi, T. (1979) Preparation of hybrid of fragment Fab′ of antibody and fragment A of diphtheria toxin and its cytotoxicity. *Biochem. Biophys. Res. Commun.* **90**, 320.

Masuho, Y., Kishida, K., Saito, M., Umemoto, N., and Hara, T. (1982) Importance of the antigen-binding valency and the nature of the cross-linking bond in ricin A-chain conjugates with antibody. *J. Biochem. (Tokyo)* **91**, 1583.

Matteucci, M. D., and Caruthers, M. H. (1980) The synthesis of oligodeoxypyrimidines on a polymer support. *Tetrahedron Lett.* **21**, 719–722.

Matthay, K. K., Heath, T. D., and Papahadjopoulos, D. (1984) Specific enhancement of drug delivery to AKR lymphoma by antibody-targeted small unilamellar vesicles. *Cancer Res.* **44**, 1880–1886.

Matthay, K. K., Heath, T. D., Badger, C. C., Bernstein, I. D., and Papahadjopoulos, D. (1986) Antibody-directed liposomes: Comparison of various ligands for association, endocytosis and drug delivery. *Cancer Res.* **46**, 4904.

Matthews, J. A., and Kricka, I. J. (1988) Analytical strategies for the use of DNA probes. *Anal. Biochem.* **169**, 1–25.

Mattiasson, B., and Nilsson, H. (1977) An enzyme immunoelectrode. *FEBS Lett.* **78**, 251.

Mayhew, E., Lazo, R., Vail, W. J., King, J., and Green, A. M. (1984) Characterization of liposomes prepared using a microemulsifier. *Biochim. Biophys. Acta* **775**, 169–174.

Mazaitis, J. K., Francis, B. E., Eckelman, W. C., Gibson, R. E., Reba, R. C., Barnes, J. W., Bentley, G. E., Grant, P. M., and O'Brien, H. A. (1981) No-carrier-added bromination of estrogens with chloramine-T and Na[77]. *Br. J. Labelled Compd. Radiopharm.* **18**, 1033–1038.

Means, G. E., and Feeney, R. E. (1971) "Chemical Modification of Proteins." p. 20. Holden-Day, San Francisco.

Meares, C. F. (1986) Chelating agents for the binding of metal ions to antibodies. *Nucl. Med. Biol.* **13**, 311–318.

Meige, J. B., and Wang, Y.-L. (1986) Reorganization of alpha-actin and vinculin induced by a phorboll ester in living cells. *J. Cell Biol.* **102**, 1430.

Meighen, E. A., Nicolim, M. Z., and Hustings, J. W. (1971) Hybridization of bacterial luciferase with a variant produced by chemical modification. *Biochemistry* **10**, 4062.

Meijer, E. W., Nijhuis, S., and Vroonhoven, F. C. B. M. (1988) Poly-1,2-azepines by the photopolymerization of phenyl azides. Precursors for conducting polymer films. *J. Am. Chem. Soc.* **110**, 7209–7210.

Melchers, F., and Messer, W. (1973) The activity of individual molecules of hybrid β-galactosidase reconstituted from the wild-type and an inactive-mutant enzyme. *Eur. J. Biochem.* **34**, 228.

Mentzer, W. C., Jr., and Lubin, B. H. (1979) The effect of cross-linking agents on red-cell shape. *Semin. Hematol.* **16**, 115-127.

Mentzer, W. C., Jr., Lewis, S., Pennathur-Das, R., Halpin, R., Cerrone, K. L., Lubin, B., and Kenyon, G. L. (1982) Formation of 5-carbomethoxyvaleramidine during hydrolysis of the protein cross-linking agent dimethyl adipimidate. *J. Protein Chem.* **1**, 141–155.

Metz, D. H., and Brown, G. L. (1969) The investigation of nucleic acid secondary structure by means of chemical modification with a carbodiimide reagent. I. The reaction between N-cyclohexyl-N'-β-(4-methylmorpholinium)ethyl carbodiimide and model nucleotides. *Biochemistry* **8**, 2312–2328.

Meyne, J. (1993) Chromosome mapping by fluorescent *in situ* hybridization, *in* "Methods in Nonradioactive Detection" (G. C. Howard, ed.), pp. 263–268. Appleton & Lange, Norwalk, Connecticut.

Mikkelsen, R. B., and Wallach, D. F. H. (1976) Photoactivated cross-linking of protein within the erythrocyte membrane core. *J. Biol. Chem.* **251**, 7413.

Millar, J. B., and Rozengur, E. (1990) Chronic desensitization to bombesin by progressive down-regulation of bombesin receptors in Swiss 3T3 cells. *J. Biol. Chem.* **265**, 12052–12058.

Miller, M. D., Hata, S., De Waal Malefyt, R., and Krangel, M. S. (1989) A novel polypeptide secreted by activated human T lymphocytes. *J. Immunol.* **143**, 2907–2916.

Miron, T., and Wilchek, M. (1993) A simplified method for the preparation of succinimidyl carbonate polyethylene glycol for coupling to proteins. *Bioconjugate Chem.* **4**, 568–569.

Miskimins, W. K., and Shimizu, N. (1979) Synthesis of cytotoxic insulin cross-linked to diphtheria toxin fragment A capable of recognizing insulin receptors. *Biochem. Biophys. Res. Commun.* **91**, 143.

Mittal, B., Sanger, J. M., and Sanger, J. W. (1987) Visualization of myosin in living cells. *J. Cell Biol.* **105**, 1753–1760.

Miyakawa, T., Takemoto, L. J., and Fox, C. F. (1978) *J. Supramol. Struct.* **8**, 303–310.

Moeremans, M., Daneels, G., Van Dijck, A., Langanger, G., and De Mey, J. (1984) Sensitive visualization of antigen–antibody reactions in dot and blot immuno overlay assays with the immunogold and immunogold/silver staining. *J. Immunol. Methods* **74**, 353–360.

Moi, M. K., Meares, C. F., McCall, M. J., Cole, W. C., and DeNardo, S. J. (1985) Copper chelates as probes of biological systems: Stable copper complexes with a macrocyclic bifunctional chelating agent. *Anal. Biochem.* **148**, 249–253.

Monsan, P., Puzo, G., and Mazarguil, H. (1975) *Biochimie* **57**, 1281.

Montesano, L., Cawley, D., and Herschman, H. R. (1982) Disuccinimidyl suberate cross-linked ricin does not inhibit cell-free protein synthesis. *Biochem. Biophys. Res. Commun.* **109**, 7–13.

Moore, J. E., and Ward, W. H. (1956) Cross-linking of bovine plasma albumin with wool keratin. *J. Am. Chem. Soc.* **78**, 2414.

Moreland, R. B., Smith, P. K., Fujimoto, E. K., and Dockter, M. E. (1982) Synthesis and characterization of N-(4-azidophenylthio)-phthalimide. *Anal. Biochem.* **121**, 321.

Morgan, C. J., and Stanley, E. R. (1984) Chemical crosslinking of the mononuclear phagocyte specific growth factor CSF-1 to its receptor at the cell surface. *Biochem. Biophys. Res. Commun.* **119**, 35–41.

Moroney, J. V., Warncke, K., and McCarthy, R. E. (1982) The distance between thiol groups in the gamma subunit of coupling factor 1 influences the protein permeability of thylakoid membranes. *J. Bioenerg. Biomembr.* **14**, 347.

Moroney, S. E., D'Alarcao, L. J., Goldmacher, V. S., Lambert, H. M., and Blattler, W. A. (1987) Modification of the binding site(s) of lectins by an affinity column carrying an activated galactose-terminated ligand. *Biochemistry* **26**, 8390.

Morris, R. E., and Saelinger, C. B. (1984) Visualization of intracellular trafficking: Use of biotinylated ligands in conjunction with avidin–gold colloids. *J. Histochem. Cytochem.* **32**, 124–128.

Morrison, L. E. (1992) Detection of energy transfer and fluorescence quenching. *in* "Nonisotopic DNA Probe Techniques" (L. J. Kricka, ed.), pp. 311–352. Academic Press, New York.

Morrison, M., and Bayse, G. S. (1970) Catalysis of iodination by lactoperoxidase. *Biochemistry* **9**, 2995–3000.

Mossberg, K., and Ericsson, M. (1990) Detection of doubly stained fluorescent specimens using confocal microscopy. *J. Microsc.* **158**, 215.

Motta-Hennessy, C., Eccles, S. A., Dean, C., and Coghlan, G. (1985) Preparation of ^{67}Ga-labeled human IgG and its Fab fragments using deferoxamine as chelating agent. *Eur. J. Nucl. Med.* **11**, 240–245.

Muckerheide, A., Apple, R. J., Pesce, A. J., and Michael, J. G. (1987a) Cationization of protein antigens. I. Alteration of immunogenic properties. *J. Immunol.* **138**, 833–837.

Muckerheide, A., Domen, P. L., and Michael, J. G. (1987b) Cationization of protein antigens. II. Alteration of regulatory properties. *J. Immunol.* **138**, 2800–2804.

Mudd, J. A., and Swanson, R. E. (1978) *Virology* **88**, 263–280.

Mukkala, V.-M., Mikola, H., and Hemmila, I. (1989) The synthesis and use of activated N-benzyl derivatives of diethylenetriaminetetraacetic acids: Alternative reagents for labeling of antibodies with metal ions. *Anal. Biochem.* **176**, 319–325.

Mumtaz, S., and Bachhawat, B. K. (1991) Conjugation of proteins and enzymes with hydrophilic polymers and their applications. *Indian J. Biochem. Biophys.* **28**, 346–351.

Murachi, T. (1976) Bromelain enzymes. *in* "Methods in Enzymology" (L. Lorand, ed.), Vol. 45, pp. 475–485. Academic Press, New York.

Muramoto, K., Kamiya, H., and Kawauchi, H. (1984) The application of fluorescein isothiocyanate and high performance liquid chromatography for the microsequencing of proteins and peptides. *Anal. Biochem.* **141**, 446.

Murayama, Y., Satoh, S., Oka, T., Imanishi, J., and Noishiki, Y. (1988) Reduction of the antigenicity and immunogenicity of xenografts by a new cross-linking reagent. *ASAIO Trans.* **34**, 546.

Murphy, F. R., Jorgensen, E. D., and Cantor, C. R. (1982) Kinetics of histone endocytosis in Chinese hamster cells. A flow cytofluorometric analysis. *J. Biol. Chem.* **257**, 1895.

Myers, D. E., Uckun, F. M., Swaim, S. E., and Vallera, D. A. (1989) The effects of aromatic and aliphatic maleimde cross-linkers on anti-CD5 ricin immunotoxins. *J. Immunol. Methods* **121**, 129–142.

Nagai, Y., *et al.* (1978) *Arch. Virol.* **58**, 15–28.

Nakajima, M., Ito, N., Nishi, K., Okamura, Y., and Hirota, T. (1988) Cytochemical localization of blood group substances in human salivary glands using lectin–gold complexes. *J. Histochem. Cytochem.* **36**, 337–348.

Nakamura, S., Kato, A., and Kobayashi, K. (1990) Novel bifunctional lysozyme–dextran conjugate that acts on both gram-negative and gram-positive bacteria. *Agric. Biol. Chem.* **54**, 3057–3059.

Nakane, P. K. (1975) Recent progress in the peroxidase-labeled antibody method. *Ann. N. Y. Acad. Sci.* **254**, 203.

Nakane, P. K., and Kawaoi, A. (1974) Peroxidase-labeled antibody. A new method of conjugation. *J. Histochem. Cytochem.* **22**, 1084–1091.

Nathan, A., Zalipsky, S., Ertel, S. I., Agathos, S. N., Yarmush, M. L., and Kohn, J. (1993) Copolymers of lysine and polyethylene glycol: A new family of functionalized drug carriers. *Bioconjugate Chem.* **4**, 54–62.

Newhall, J., Sawyer, W. D., and Haak, R. A. (1980) Cross-linking analysis of the outer membrane proteins of *Neisseria gonorrhoeae. Infect. Immun.* **28**, 785–791.

Ngo, T. T., Yam, C. F., Lenhoff, H. M., and Ivy, J. (1981) *p*-Azidophenylglyoxal: A heterobifunctional photoactivatable cross-linking reagent selective for arginyl residues. *J. Biol. Chem.* **256**, 11313–11318.

Nicolas, J.-C., Balaguer, P., Terouanne, B., Villebrun, M. A., and Boussioux, A.-M. (1992) Detection of glucose 6-phosphate dehydrogenase by bioluminescence, *in* "Nonisotopic DNA Probe Techniques" (L. J. Kricka, ed.), p. 207. Academic Press, New York.

Nicolson, G. L. (1978) Ultrastructural localization of lectin receptors. *in* "Advanced Techniques in Biological Electron Microscopy" (M. Koehler, ed.), p. 1. Springer-Verlag, New York.

Nielsen, M. H., Bastholm, L., Chatterjee, S., Koga, J., and Norrild, B. (1989) Simultaneous triple-immunogold staining of virus and host cell antigens with monoclonal antibodies of virus and host cell antigens in ultrathin cryosections. *Histochemistry* **92**, 89–93.

Nillson, K., and Mosbach, K. (1984) Immobilization of ligands with organic sulfonyl chlorides. *Methods Enzymol.* **104**, 56–69.

Niman, H. L., Thompson, A. M. H., Yu, A., Markman, M., Willems, J. J., Herwig, K. R., Habib, N. A., Wood, C. B., Houghten, R. A., and Lerner, R. A. (1985) Anti-peptide antibodies detect oncogene-related proteins in urine. *Proc. Natl. Acad. Sci. U.S.A.* **82**, 7924–7928.

Nithipatikom, K., and McGown, L. B. (1987) Homogeneous immunochemical technique for determination of human lactoferrin using excitation tranfer and phase-resolved fluorometry. *Anal. Chem.* **59**, 423.

Noguchi, A., Takahashi, T., Yamaguchi, T., Kitamura, K., Takakura, Y., Hashida, M., and Sezaki, H. (1992) Preparation and properties of the immunoconjugate composed of antihuman colon cancer monoclonal antibody and mitomycin C–dextran conjugate. *Bioconjugate Chem.* **3**, 132–137.

Norde, W. (1986) Adsorption of proteins from solution at the solid-liquid interface. *Adv. Colloid Interface Sci.* **25**, 267.

Novak-Hofer, I., and Siegenthaler, P. (1978) *Plant Physiol.* **62**, 368–372.

Novick, D., Orchansky, P., Revel, M., and Rubenstein, M. (1987) The human interferon-γ receptor. *J. Biol. Chem.* **262**, 8483–8487.

Nur, I., Reinhartz, A., Hyman, H. C., Razin, S., and Herzberg, M. (1989) Chemiprobe, a nonradioactive system for labeling nucleic acid. *Ann. Biol. Clin.* **47**, 601–606.

O'Keefe, E. T., Mordick, T., and Bell, J. E. (1980) Bovine galactosyltransferase: Interaction with α-lactalbumin and the role of α-lactalbumin in lactose synthase. *Biochemistry* **19**, 4962–4966.

O'Malley, J. J., and Weaver, J. L. (1972) Subunit structure of glucose oxidase from *Aspergillus niger*. *Biochemistry* **11**, 3527.

O'Shannessy, D. J., and Quarles, R. H. (1985) Specific conjugation reactions of the oligosaccharide moieties of immunoglobulins. *J. Appl. Biochem.* **7**, 347–355.

O'Shannessy, D. J., and Wilchek, M. (1990) Immobilization of glycoconjugates by their oligosaccharides: Use of hydrazido-derivatized matrices. *Anal. Biochem.* **191**, 1–8.

O'Shannessy, D. J., Doberson, M. J., and Quarles, R. H. (1984) A novel procedure for labeling immunoglobulins by conjugation to oligosaccharide moieties. *Immunol. Lett.* **8**, 273–277.

O'Shannessy, D. J., Voorstad, P. J., and Quarles, R. H. (1987) Quantitation of glycoproteins on electroblots using the biotin–streptavidin complex. *Anal. Biochem.* **163**, 204–209.

O'Sullivan, M., Gnemmi, E., Morris, D., Chieregatti, G., Simmonds, A., Simmons, M., Bridges, J., and Marks, V. (1979) Comparison of two methods of preparing enzyme–antibody conjugates: Application of these conjugates for enzyme immunoassay. *Anal. Biochem.* **100**, 100–108.

Obrist, R., Schmidli, J., and Obrecht, J. P. (1988) Chemotactic monoclonal antibody conjugates: A comparison of four different f-Met–peptide conjugates. *Biochem. Biophys. Res. Commun.* **155**, 1139–1144.

Odom, O. W., Jr., Robins, D. J., Lynch, J., Dottavio-Martin, D., Kramer, G., and Hardesty, B. (1980) Distances between 3′ ends of ribosomal ribonucleic acids reassembled into *Escherichia coli* ribosomes. *Biochemistry* **19**, 5947–5954.

Odom, O. W., Dabbs, E. R., Dionne, C., Muller, M., and Hardesty, B. (1984) The distance between S1, S21, and the 3′ end of 16S RNA in 30S ribosomal subunits. The effect of poly(uridylic acid) and 50S subunits on these distances. *Eur. J. Biochem.* **142**, 261.

Odom, O. W., Picking, W. D., and Hardesty, B. (1990) Movement of tRNA but not the nascent peptide during peptide bond formation on ribosomes. *Biochemistry* **29**, 10734–10744.

Oeltmann, T. N. (1985) Synthesis and *in vitro* activity of a hormone-diphtheria toxin fragment A hybrid. *Biochem. Biophys. Res. Commun.* **133**, 430.

Oeltmann, T. N., and Forbes, J. T. (1981) Inhibition of mouse spleen cell function by diphtheria toxin fragment A coupled to anti-mouse Thy-1.2 and by ricin A chain coupled to anti-mouse IgM. *Arch. Biochem. Biophys.* **209**, 362.

Oi, V. T., Glazer, A. N., and Stryer, L. (1982) Fluorescent phycobiliprotein conjugates for analyses of cells and molecules. *J. Cell Biol.* **93**, 981–986.

Olsnes, S. (1978) Binding, entry, and action of abrin, ricin, and modeccin. *in* "Transport of Macromolecules in Cellular Systems" (S. C. Silverstein, ed.), pp. 103–116. Dahlem Konferenzen, Berlin.

Olsnes, S., and Pihl, A. (1976) Abrin, ricin, and their associated agglutinins. *in* "The Specificity of Animal, Bacterial and Plant Toxins. Receptors and Recognition" (P. Cuatrecasas, ed.), Series B, Vol. 1, pp. 129–173. Chapman & Hall, London.

Olsnes, S., and Pihl, A. (1982a) Cytotoxic proteins with intracellular site of action: Mechanism of action and anti-cancer properties. *Cancer Surv.* **3**, 467–487.

Olsnes, S., and Pihl, A. (1982b) Chimeric toxins. *Pharmacol. Ther.* **15**, 355.

Olsnes, S., and Pihl, A. (1982c) Toxic lectins and related proteins. *in* "Molecular Action of Toxins and Viruses" (P. Cohen and S. van Heynigen, eds.), p. 51. Elsevier, New York.

Olson, F., Hunt, C. A., Szoka, F. C., Vail, W., Mayhew, E., and Paphadjopoulos, D. (1980) *Biochim. Biophys. Acta.* **601**, 559.

Ondetti, M. A., and Thomas, P. L. (1965) Synthesis of a peptide lactone related to vernamycin Bα. *J. Am. Chem. Soc.* **87**, 4373–4380.

Oparka, K. J., Murant, E. A., Wright, K. M., Prior, D. A. M., and Harris, N. (1991) The drug probenecid inhibits the vacuolar accumulation of fluorescent anions in onion epidermal cells. *J. Cell Sci.* **99**, 557–563.

Order, S. E. (1982) Monoclonal antibodies potential in radiation therapy and oncology. *Int. J. Radiat. Oncol. Biol. Phys.* **8**, 1193–1201.

Order, S. E. (1989) Therapeutic use of radioimmunoconjugates. *Antibody, Immunoconjugates, Radiopharm.* **2**, 235.

Orlandi, R., Gussow, D. H., Jones, P. T., and Winter, G. (1989) Cloning immunoglobulin variable domains for expression by the polymerase chain reaction. *Proc. Natl. Acad. Sci. U.S.A.* **86**, 3833–3837.

Ormerod, M. G. (ed.) (1990) "Flow Cytometry. A Practical Approach." IRL Press, New York.

Orr, G. A. (1981) The use of the 2-iminobiotin–avidin interaction for the selective retrieval of labeled plasma membrane components. *J. Biol. Chem.* **256**, 761–766.

Osborn, M., and Weber, K. (1982) Immunofluorescence and immunocytochemical procedures with affinity purified antibodies: Tubulin-containing structures. *Methods Cell Biol.* **24**, 97–132.

Otsuka, F. L., and Welch, M. J. (1987) Methods to label monoclonal antibodies for use in tumor imaging. *Nucl. Med. Biol.* **14**, 243–249.

Ozawa, H. (1967) Bridging reagent for protein. II. The reaction of *N,N'*-polymethylene-bis(iodoacetamide) with cysteine and rabbit muscle aldolase. *J. Biochem. (Tokyo)* **62**, 531.

Ozawa, S., Ueda, M., Ando, N., Abe, O., Minoshima, S., and Shimizu, N. (1989) Selective killing of squamous carcinoma cells by an immunotoxin that recognizes the EGF receptor. *Int. J. Cancer* **43**, 152.

Packman, L. C., and Perham, R. N. (1982) Quaternary structure of the pyruvate dehydrogenase multienzyme complex of *Bacillus stearothermophilus* studied by a new reversible cross-linking procedure with bis(imidoesters). *Biochemistry* **21**, 5171–5175.

Paganelli, G., Riva, P., Deleide, G., Clivio, A., Chiolerio, F., Scassellati, G. A., Malcovati, M., and Siccardi, A. G. (1988) *In vivo* labeling of biotinylated monoclonal antibodies by radioactive avidin: A strategy to increase tumor radiolocalization. *Int. J. Cancer* **2**, 121–125.

Palmer, J. L., and Nissonoff, A. (1963) *J. Biol. Chem.* **238**, 2393.

Park, L. S., Friend, D., Gillis, S., and Urdal, D. L. (1986) Characterization of the cell surface receptor for a multi-lineage colony-stimulating factor (CSF-2α). *J. Biol. Chem.* **261**, 205–210.

Park, Y. S., and Huang, L. (1992) Cryoprotective activity of synthetic glycophospholipids and their interactions with trehalose. *Biochim. Biophys. Acta* **1124**, 241–248.

Parker, D. J., and Allison, W. S. (1969) The mechanism of inactivation of glyceraldehyde 3-phosphate dehydrogenase by tetrathionate, *o*-iodosobenzoate, and iodine monochloride. *J. Biol. Chem.* **244**, 180–189.

Partis, M. D., Griffiths, D. G., Roberts, G. C., and Beechey, R. B. (1983) Cross-linking of protein by ω-maleimido alkanoyl *N*-hydroxysuccinimido esters. *J. Protein Chem.* **2**, 263–277.

Pascual, A., Casanova, J., and Samuels, H. H. (1982) Photoaffinity labeling of thyroid hormone nuclear receptors in intact cells. *J. Biol. Chem.* **257**, 9640–9647.

Pathy, L., and Smith, E. L. (1975) Reversible modification of arginine residues: Application to sequence studies by restriction of tryptic hydrolysis to lysine residues. *J. Biol. Chem.* **250**, 557.

Paul, R., and Anderson, G. W. (1960) N,N'-Carbonyldiimidazole, a new peptide forming reagent. *J. Am. Chem. Soc.* **82**, 4596–4600.

Paul, R., and Anderson, G. W. (1962) N,N'-Carbonyldiimidazole in peptide synthesis. III. A synthesis of isoleucine-5-angiotensin II amide-1. *J. Org. Chem.* **27**, 2094–2099.

Pawley, J., and Albrecht, R. (1988) Imaging colloidal gold labels in LVSEM. *Scan. Microsc.* **10**, 184–189.

Pearson, R. G., Sobel, H., and Songstad, J. (1968) Nucleophilic reactivity constants toward methyl iodide and trans-[Pt(py)$_2$Cl$_2$]. *J. Am. Chem. Soc.* **90**, 319–326.

Peeters, J. M., Hazendonk, T. G., Beuvery, E. C., and Tesser, G. I. (1989) Comparison of four bifunctional reagents for coupling peptides to proteins and the effect of the three moieties on the immunogenicity of the conjugates. *J. Immunol. Methods* **120**, 133–143.

Peng, L., Calton, G. J., and Burnett, J. W. (1987) Effect of borohydride reduction on antibodies. *Appl. Biochem. Biotechnol.* **14**, 91–99.

Pennathur-Das, R., Heath, R., Mentzer, W. C., and Lubin, B. (1982) Modification of hemoglobin s with dimethyl adipimidate. Contribution of individual reacted subunits to changes in properties. *Biochim. Biophys. Acta* **704**, 389–397.

Pepinsky, R. B. *et al.* (1980) *Virology* **102**, 205–210.

Perham, R. N., and Jones, G. M. T. (1967) *Eur. J. Biochem.* **2**, 84.

Perham, R. N., and Thomas, J. O. (1971) *J. Mol. Biol.* **62**, 415.

Peters, K., and Richards, F. M. (1977) *Annu. Rev. Biochem.* **46**, 523.

Petruzzelli, L., Herrer, R., Garcia-Arenas, R., and Rosen, R. M. (1985) Acquisition of insulin-dependent protein tyrosine kinase activity during *Drosophila* embryogenesis. *J. Biol. Chem.* **226**, 16072–16075.

Pfeuffer, E., Dreher, R.-M., and Pfeuffer, T. (1985) Catalytic unit of adenylate cyclase purification and identification by affinity cross-linking. *Proc. Natl. Acad. Sci. U.S.A.* **82**, 3086–3090.

Pick, U. (1981) Liposomes with a large trapping capacity prepared by freezing and thawing of sonicated phospholipid mixtures. *Arch. Biochem. Biophys.* **212**, 186.

Pidgeon, C., Hung, A. H., and Dittrich, K. (1986) *Pharm. Res.* **3**, 23.

Pihl, A., and Lange, R. (1962) *J. Biol. Chem.* **237**, 1356.

Pikuleva, I. A., and Turko, I. V. (1989) A new method of preparing hemin conjugate with rabbit IgG. *Bioorg. Khim.* **15**, 1480.

Pillai, V. N. R., and Mutter, M. (1980) New, easily removable polyethylene glycol supports for liquid phase method of peptide synthesis. *J. Org. Chem.* **45**, 5364–5367.

Pimm, M. V., Raiput, R. S., Frier, M., and Gribben, S. J. (1991) Anomalies in reduction-mediated technetium-99m labeling of monoclonal antibodies. *Eur. J. Nucl. Med.* **18**, 973–976.

Pirker, R., Fitzgerald, D. J. P., Hamilton, T., Ozols, R. F., Laird, W., Frankel, A. E., Willingham, M. C., and Pastan, I. (1986) Characterization of immunotoxins active against ovarian cancer cell lines. *J. Clin. Invest.* **76**, 1261.

Plank, L., and Ware, B. R. (1987) Acanthamoeba profiln binding to fluorescein-labelled actin. *Biophys. J.* **51**, 985.

Plant, A. L., Brizgys, M. V., Lacasio-Brown, L., and Durst, R. A. (1989) Generic liposome reagent for immunoassays. *Anal. Biochem.* **176**, 420–426.

Plapp, B. V., Raftery, M. A., and Cole, R. D. (1967) *J. Biol. Chem.* **242**, 265.

Ploem, J. S., and Tanke, H. J. (1987) "Introduction to Fluorescence Microscopy." Oxford Univ. Press, London.

Plotz, P. H., and Rifai, A. (1982) Stable, soluble, model immune complexes made with a versatile multivalent affinity-labeling antigen. *Biochemistry* **21**, 301.

Podhradsky, D., Drobnica, L., and Kristian, P. (1979) Reactions of cysteine, its derivatives, glutathione, coenzyme A, and dihydrolipoic acid with isothiocyanates. *Experientia* **35**, 154.

Politz, S. M., Noller, H. F., and McWhirter, P. D. (1981) *Biochemistry* **20**, 372–378.

Porath, J. (1976) General methods and coupling procedures. *in* "Methods in Enzymology" (W. B. Jakoby and M. Wilchek, eds.), Vol. 34, p. 13. Academic Press, New York.

Porath, J., and Belew, M. (1983) *in* "Affinity Chromatography and Biological Recognition" (I. M. Chaiken, M. Wilchek, and I. Parikh, eds.), pp. 173. Academic Press, San Diego.

Porath, J., and Olin, B. (1983) Immobilized metal ion affinity adsorption and immobilized metal ion affinity chromatography of biomaterials. Serum protein affinities for gel-immobilized iron and nickel ions. *Biochemistry* 22, 1621–1630.

Porath, J., Carlsson, J., Olsson, I., and Belfrage, G. (1975) *Nature (London)* 258, 598.

Porstmann, B., Porstmann, T., Nugel, E., and Evers, U. (1985) Which of the commonly used marker enzymes gives the best results in colorimetric and fluorimetric enzyme immunoassays: Horseradish peroxidase, alkaline phosphatase or β-galactosidase. *J. Immunol. Methods* 79, 27–37.

Posnett, D. N., McGrath, H., and Tam, J. P. (1988) A novel method for producing anti-peptide antibodies: Production of site-specific antibodies to the T-cell antigen receptor β-chain. *J. Biol. Chem.* 263, 1719–1725.

Pow, D. V., and Crook, D. K. (1993) Extremely high titre polyclonal antisera against small neurotransmitter molecules: Rapid production, characterization and use in light- and electron-microscopic immunocytochemistry. *J. Neuroscience Methods* 48, 51–63.

Pow, D. V., and Morris, J. F. (1991) Membrane routing during exocytosis and endocytosis in neuroendocrine neurons and endocrine cells: Use of colloidal gold particles and immunocytochemical discrimination of membrane compartments. *Cell Tissue Res.* 264, 299–316.

Powsner, E. R. (1994) Basic principles of radioactivity and its measurement. *in* "Tietz Textbook of Clinical Chemistry" (C. A. Burtis and E. R. Ashwood, eds.) pp. 256–282, Saunders, Philadelphia, Pennsylvania.

Preis, J. (ed.) (1980) "Carbohydrates: Structure and Function. Volume 3 of The Biochemistry of Plants: A Comprehensive Treatise" (P. K. Stumpf and E. E. Conn, eds.), Academic Press, New York.

Pressman, D., and Keighley, G. (1948) The zone of activity of antibodies as determined by the use of radioactive tracers; The zone of activity of nephritoxic antikidney serum. *J. Immunol.* 59, 141–146.

Prestayko, A. W., Baker, L. H., Crooke, S. T., Carter, S. K., and Schein, P. S. (1981) "Nitrosoureas. Current Status and New Developments," Chapter 4. Academic Press, New York.

Price, M. R., Sekowski, M., Hooi, D. S. W., Durrant, L. G., Hudecz, F., and Tendler, S. J. B. (1993) Measurement of antibody binding to antigenic peptides conjugated *in situ* to albumin-coated microtitre plates. *J. Immunol. Methods* 159, 277–281.

Pulliam, M. W., Boyd, L. F., Baylan, N. C., and Bradshaw, R. A. (1975) Specific binding of covalently cross-linked mouse nerve growth factor to responsive peripheral neurons. *Biochem. Biophys. Res. Commun.* 67, 1281–1289.

Raftery, M. A., and Cole, R. D. (1963) *Biochem. Biophys. Res. Commun.* 10, 467.

Raftery, M. A., and Cole, R. D. (1966) *J. Biol. Chem.* 241, 3457.

Ralph, R. K., Young, R. J., and Khorana, H. G. (1962) The labeling of phosphomonoester end groups in amino acid acceptor ribonucleic acids and its use in the determination of nucleotide sequences. *J. Am. Chem. Soc.* 84, 1490–1491.

Ranadive, G. N., Rosenzweig, H. S., Epperly, M. W., Seskey, T., and Bloomer, W. D. (1993) A new method of technetium-99m labeling of monoclonal antibodies through sugar residues. A study with TAG-72 specific CC-49 antibody. *Nucl. Med. Biol.* 20, 719–726.

Rao, A., Martin, P., Reithmeier, R. A. F., and Cantley, L. C. (1979) Location of the stilbenedisulfonate binding site of the human erythrocyte anion-exchange system by resonance energy transfer. *Biochemistry* 18, 4505–4516.

Rashidbaigi, A., Langer, J. A., Jung, V., Jones, C., Morse, R. G., Tischfield, J. A., Trill, J. J., Kung, H.-F., and Pestka, S. (1986) The gene for the human immune interferon receptor is located on chromosome 6. *Proc. Natl. Acad. Sci. U.S.A.* 83, 384–388.

Raso, V., and Basala, M. (1984) A highly cytotoxic human transferrin–ricin A chain conjugate used to select receptor-modified cells. *J. Biol. Chem.* **259**, 1143.

Raso, V., and Basala, M. (1985) Study of the transferrin receptor using a cytotoxic human transferrin–ricin A chain conjugate. *in* "Receptor-Mediated Targeting of Drugs" (G. Gregoriadis, ed.), Vol. 2, p. 73. Plenum, New York.

Rebek, J., and Feitler, D. (1974) Mechanism of the carbodiimide reaction. II. Peptide synthesis on the solid phase. *J. Am. Chem. Soc.* **96**, 1606–1607.

Reese, C. B. (1973) *in* "Protective Groups in Organic Chemistry" (McOmie, ed.), p. 95. Plenum, New York.

Reeves, J. P., and Dowben, R. M. (1969) Formation and properties of thin-walled phospholipid vesicles. *J. Cell Physiol.* **73**, 49.

Regoeczi, E. (1984) Methods of protein iodination. *in* "Iodine-labeled Plasma Proteins," Vol. 1, pp. 35–102. CRC Press, New York.

Reisfeld, A., Rothenberg, J. M., Bayer, E. A., and Wilchek, M. (1987) Nonradioactive hybridization probes prepared by the reaction of biotin hydrazide with DNA. *Biochem. Biophys. Res. Commun.* **142**, 519–526.

Reisfeld, R. A., Yang, H. M., Muller, B., Wargalla, U. C., Schrappe, M., and Wrasidlo, W. (1989) Promises, problems, and prospects of monoclonal antibody–drug conjugates for cancer therapy. *Antibody, Immunoconjugates, Radiopharm.* **2**, 217–224.

Reiter, Y., and Fishelson, Z. (1989) Targeting of complement to tumor cells by heteroconjugates composed of antibodies and of the complement component C_{3b}. *J. Immunol.* **142**, 2771.

Renn, O., and Meares, C. F. (1992) Large scale synthesis of the bifunctional chelating agent 2-*p*-nitrobenzyl-1,4,7,10-tetraazacyclododecane-*N,N',N''N'''*-tetraacetic acid and the determination of its enantiomeric purity by chiral chromatography. *Bioconjugate Chem.* **3**, 563–569.

Rhodes, B.A. (1991) Direct labeling of proteins with ^{99m}Tc. *Nucl. Med. Biol.* **18**, 667–676.

Richard, F. M., and Knowles, J. R. (1968) Glutaraldehyde as a protein cross-linking reagent. *J. Mol. Biol.* **37**, 231.

Riddles, P. W., Blakeley, R. L., and Zerner, B. (1979) Ellman's reagent: 5,5'-dithiobis(2-nitrobenzoic acid)—a reexamination. *Anal. Biochem.* **94**, 75–81.

Riehm, J. P., and Scheraga, H. A. (1965) Structural studies of ribonuclease. XVII. A reactive carboxyl group in ribonuclease. *Biochemistry* **4**, 772.

Rifai, A., and Wong, S. S. (1986) Preparation of phosphorylcholine-conjugated antigens. *J. Immunol. Methods* **94**, 25.

Rigby, P. W. J., Dieckmann, M., Rhodes, C., and Berg, P. (1977) Labeling deoxyribonucleic acid to high specific activity *in vitro* by nick translation with DNA polymerase I. *J. Mol. Biol.* **113**, 237–251.

Riordan, J. F., and Vallee, B. L. (1963) Acetylcarboxypeptidase. *Biochemistry* **2**, 1460.

Riordan, J. F., and Vallee, B. L. (1964) Succinylcarboxy peptidase. *Biochemistry* **3**, 1768.

Riordan, J. F., and Vallee, B. L. (1972) Diazonium salts as specific reagents and probes of protein conformation. *in* "Methods in Enzymology" (C. H. W. Hirs and S. N. Timasheff, eds.), Vol. 25, p. 521. Academic Press, New York.

Roffler, S. R., and Tseng, T.-L. (1994) Enhanced serum half-life and tumor localization of PEG-modified antibody-enzyme conjugates for targeted prodrug activation. "Antibody Engineering Conference." San Diego, California.

Roffman, E., Spiegel, Y., and Wilchek, M. (1980) Ferritin hydrazide, a novel conalent electron dense reagent for the ultrastructural localization of glycoconjugates. *Biochem. Biophys. Res. Commun.* **97**, 1192–1198.

Roitt, I. (1977) "Essential Immunology," p. 21. Blackwell, London.

Rosenberg, M. B., Hawrot, E., and Breakefield, X. O. (1986) Receptor binding activities of biotinylated derivatives of β-nerve growth factor. *J. Neurochem.* **46**, 641–648.

Ross, S. E., Carson, S. D., and Fink, L. M. (1986) Effects of detergents on avidin–biotin interaction. *BioTechniques* **4**, 350–354.

Ross, W. C. J. (1953) The chemistry of cytotoxic alkylating agents. *Adv. Cancer Res.* **1**, 397.

Roth, J. (1983) Application of lectin–gold complexes for electron microscopic localization of glycoconjugates on thin sections. *J. Histochem. Cytochem.* **31**, 987.

Roth, J., and Binder, M. (1978) Colloidal gold, ferritin, and peroxidase as markers for electron microscopic double labeling lectin techniques. *J. Histochem. Cytochem.* **26**, 163.

Roth, J., Taatjes, D. J., and Warhol, M. J. (1989) Prevention of non-specific interactions of gold-labeled reagents on tissue sections. *Histochemistry* **92**, 47–56.

Rothenberg, J. M., and Wilchek, M. (1988) *p*-Diazobenzoyl-biocytin: A new biotinylating reagent for DNA. *Nucleic Acids Res.* **16**, 7197–7198.

Rothfus, J. A., and Smith, E. L. (1963) *J. Biol. Chem.* **238**, 1402.

Rousseaux, J., Rousseaux-Prevost, R., and Bazin, H. (1983) Optimal conditions for the preparation of Fab and F(ab')$_2$ fragments from monoclonal IgG of different rat IgG subclasses. *J. Immunol. Methods* **64**, 141–146.

Roychoudhury, R., Tu, C.-P.D., and Wu, R. (1979) Influence of nucleotide sequence adjacent to duplex DNA termini on 3'-terminal labeling by terminal transferase. *Nucleic Acids Res.* **6**, 1323–1333.

Ruegg, U. T., and Rudingder, J. (1977) Reductive cleavage of cystine disulfides with tributylphosphine. *in* "Methods in Enzymology" (C. H. W. Hirs and S. N. Timasheff, eds), Vol. 47, p. 111. Academic Press, New York.

Ruiz-Carrillo, A., and Allfrey, V. G. (1973) A method for the purification of histone fraction F3 by affinity chromatography. *Arch. Biochem. Biophys.* **154**, 185–191.

Ruth, J. L. (1993) Direct attachment of enzymes to DNA probes. *in* "Methods in Nonradioactive Detection" (G. C. Howard, ed.), pp. 153–177. Appleton & Lange, Norwalk, Connecticut.

Saiki, R. K., Scharf, S., Faloona, F., Mullis, K. B., Horn, G. T., Erlich, H. A., and Arnheim, N. (1985) Enzymatic amplification of beta-globin genomic sequences and restriction site analysis for diagnosis of sickle cell anemia. *Science* **230**, 1350–1354.

Saiki, R. K., Gelfand, D. H., Stoffel, S., Scharf, S. J., Higuchi, R., Horn, G. T., Mullis, K. B., and Erlich, H. A. (1988) Primer-directed enzymatic amplification of DNA with a thermostable DNA polymerase. *Science* **239**, 487–491.

Sakamoto, H., Traincard, F., Vo-Quang, T., Ternynck, T., Guesdon, J. L., and Avrameas, S. (1987) 5-Bromodeoxyuridin *in vivo* labeling of M13 DNA, and its use as a nonradioactive probe for hybridization experiments. *Mol. Cell. Probes* **1**, 109–120.

Salmon, S. E. (1989) Monoclonal antibody immunoconjugates for cancer. *Antibody, Immunoconjugates, and Radiopharm.* **2**, 63–70.

Sanderson, C. J., and Wilson, D. V. (1971) *Immunology* **20**, 1061–1065.

Sandvig, K., and Olsnes, S. (1981) Rapid entry of nicked diphtheria toxin into cells at low pH. Characterization of the entry process and effects of low pH on the toxin molecule. *J. Biol. Chem.* **256**, 9068.

Sashidhar, R. B., Capoor, A. K., and Ramana, D. (1994) Quantitation of ε-amino group using amino acids as reference standards by trinitrobenzene sulfonic acid. *J. Immunol. Methods* **167**, 121–127.

Sastry, L., Alting-Mees, M., Huse, W. D., Short, J. M., Sorge, J. A., Hay, B. N., Janda, K. D., Benkovic, S. J., and Lerner, R. A. (1989) Cloning of the immunological repertoire in *Escherichia coli* for generation of monoclonal catalytic antibodies: Construction of a heavy chain variable region-specific cDNA library. *Proc. Natl. Acad. Sci. U.S.A.* **86**, 5728–5732.

Sato, S., and Nakao, M. (1981) Cross-linking of intact erythrocyte membrane with a newly synthesized cleavable bifunctional reagent. *J. Biochem. (Tokyo)* **90**, 1177.

Savage, D., Mattson, G., Desai, S., Nielander, G., Morgensen, S., and Conklin, E. (1992) "Avidin–Biotin Chemistry: A Handbook." Pierce Chemical Company, Rockford, Illinois.

Sawin, K. E., and Mitchison, T. J. (1991) Mitotic spindle assembly by two different pathways *in vitro*. *J. Cell Biol.* **112**, 925.

Sawyer, S. T., Krantz, S. B., and Luna, J. (1987) Identification of the receptor for erythropoietin by cross-linking to Friend virus-infected erythroid cells. *Proc. Natl. Acad. Sci. U.S.A.* **84**, 3690–3694.

Scherson, T., Kreis, T. E., Schlessinger, J., Littauer, U., Borisy, G. G., and Geiger, B. (1984) Dynamic interactions of fluorescently labeled microtubule-associated proteins in living cells. *J. Cell Biol.* **99**, 425–434.

Schewale, J. G., and Brew, K. (1982) Effects of Fe^{+3} binding on the microenvironments of individual amino groups in human serum transferrin as determined by different kinetic labeling. *J. Biol. Chem.* **257**, 9406.

Schimitschek, E. J., Trias, J. A., Hammond, P. R., and Atkins, R. L. (1974) Laser performance and stability of fluorinated coumarin dyes. *Opt. Commun.* **11**, 352.

Schlom, J. (1986) Basic principles and applications of monoclonal antibodies in the management of carcinomas. *Cancer Res.* **46**, 3225.

Schmer, G. (1972) *Hoppe-Seyler's Z. Physiol. Chem.* **353**, 810.

Schmitt, M., Painter, R. G., Jesaitis, A. J., Preissner, K., Sklar, L. A., and Cochrane, C. G. (1983) Photoaffinity labeling of the N-formyl peptide receptor binding site of intact human polymorphonuclear leukocytes. *J. Biol. Chem.* **258**, 649–654.

Schnapp, K. A., and Platz, M. S. (1993) A laser flash photolysis study of di-, tri- and tetrafluorinated phenylnitrenes; Implications for photoaffinity labeling. *Bioconjugate Chem.* **4**, 178–183.

Schnapp, K. A., Poe, R., Leyva, E., Soundararajan, N., and Platz, M. S. (1993) Exploratory photochemistry of fluorinated aryl azides. Implications for the design of photoaffinity labeling reagents. *Bioconjugate Chem.* **4**, 172–177.

Schneede, J., and Ueland, P. M. (1992) Formation in an aqueous matrix and properties and chromatographic behavior of 1-pyrenyldiazomethane derivatives of methylmalonic acid and other short-chain dicarboxylic acids. *Anal. Chem.* **64**, 315.

Schneider, C., Newman, R. A., Sutherland, D. R., Asser, U., and Greaves, M. F. (1982) A 1 step purification of membrane proteins using a high efficiency immuno matrix. *J. Biol. Chem.* **257**, 10766–10769.

Schroeder, W. A., Shelton, J. R., and Robberson, B. (1967) *Biochim. Biophys. Acta* **147**, 590.

Schwartz, B. A., and Gray, G. R. (1977) Proteins containing reductively aminated disaccharides. Synthesis and chemical characterization. *Arch. Biochem. Biophys.* **181**, 542–549.

Schwartz, W. E., Smith, P. K., and Royer, G. P. (1980) N-(β-iodoethyl)trifluoroacetamide: A new reagent for the aminoethylation of thiol groups in proteins. *Anal. Biochem.* **106**, 43–48.

Schwinghamer, E. A. (1980) A method for improved lysis of gram-negative bacteria. *FEMS Microbiol. Lett.* **7**, 157–162.

Scopes, R. (1982) "Protein Purification," p. 30. Springer-Verlag, New York.

Scott, C. J., Jr., Goldmacher, V. S., Lambert, J. M., Chari, R. V., Bolender, S., Gauthier, M. N., and Blattler, W. A. (1987) The antileukemic efficacy of an immunotoxin composed of a monoclonal anti-Thy-1 antibody disulfide linked to the ribosome-inactivating protein gelonin. *Cancer Immunol. Immunother.* **25**, 31.

Scouten, W. H., and Van der Tweel, W. (1984) Chromophoric sulfonyl chloride agarose for immobilizing bioligands. *Annal. N.Y. Acad. Sci.* **434**, 249.

Seela, F., and Waldeck, S. (1975) Agarose linked adenosine and guanosine-5'-monophosphate; a new general method for the coupling of ribonucleotides to polymers through their cis-diols. *Nucleic Acids Res.* **2**, 2343–2349.

Segal, D. M., and Hurwitz, E. (1976) Dimers and trimers of immunoglobulin G covalently cross-linked with a bivalent affinity label. *Biochemistry* **15**, 5253.

Sela, M., and Hurwitz, E. (1987) Conjugates of antibodies with cytotoxic drugs. *in* "Immunoconjugates: Antibody Conjugates in Radioimaging and Therapy of Cancer" (C.-W. Vogel, ed.), p. 189. Oxford Univ. Press, New York.

Seligsberger, L., and Sadlier, C. (1957) New developments in tanning with aldehydes. *J. Am. Leather Chem. Assoc.* **52**, 2.

Senozan, N. *et al.* (1981) Hemocyanin of the giant keyhole limpet, *Megathura crenulata. in*

"Invertebrate Oxygen Binding Proteins: Structure, Active Sites, and Function" (J. Lamy and J. Lamy, eds.), pp. 703–717. Dekker, New York.

Senter, P. D., Saulnier, M. G., Schreiber, G. J., Hirschberg, D. L., Brown, J. P., Hellstrom, I., and Hellstrom, K. E. (1988) Anti-tumor effects of antibody–alkaline phosphatase conjugates in combination with etoposide phosphate. *Proc. Natl. Acad. Sci. U.S.A.* **85**, 4842.

Sgro, J., Jacrot, B., and Chroboczek, J. (1986) Identification of regions of brome mosaic virus coat protein chemically cross-linked *in situ* to viral RNA. *Eur. J. Biochem.* **154**, 69–76.

Shabarova, Z. A. (1988) Chemical development in the design of oligonucleotide probes for binding to DNA and RNA. *Biochimie* **70**, 1323–1334.

Shabarova, Z. A., Ivanovskaya, M. G., and Isaguliants, M. G. (1983) DNA-like duplexes with repetitions: Efficient template-guided polycondensation of decadeoxyribonucleotide imidazolide. *FEBS Lett.* **154**, 288–292.

Shaltiel, S. (1967) Thiolysis of some dinitrophenyl derivatives of amino acids. *Biochem. Biophys. Res. Commun.* **29**, 178.

Shanahan, M. F., Wadzinski, B. E., Lowndes, J. M., and Ruoho, A. E. (1985) Photoaffinity labeling of the human erythrocyte monosaccharide transporter with an aryl azide derivative of D-glucose. *J. Biol. Chem.* **260**, 10897–10900.

Shannon, L. M., Kay, E., and Lew, J. Y. (1966) Peroxidase isozymes from horseradish roots. I Isolation and physical properties. *J. Biol. Chem.* **241**, 2166.

Shapiro, R., and Weisgras, J. M. (1970) *Biochem. Biophys. Res. Commum.* **40**, 839–843.

Shapiro, R., Braverman, B., Louis, J. B., and Servis, R. E. (1973) Nucleic acid reactivity and conformation. II. Reaction of cytosine and uracil with sodium bisulfite. *J. Biol. Chem.* **248**, 4060–4064.

Shapiro, R., DiFate, V., and Welcher, M. (1974) Deamination of cytosine derivatives by bisulfite. Mechanism of the reaction. *J. Am. Chem. Soc.* **96**, 906–912.

Sheehan, D. C., and Hrapchak, B. B. (1980) "Theory and Practice of Histotechnology," 2nd Ed. Mosby, St. Louis, Missouri.

Sheehan, J. C., and Hess, G. P. (1955) A new method of forming peptide bonds. *J. Am. Chem. Soc.* **77**, 1067–1068.

Sheehan, J. C., and Hlavka, J. J. (1956) The use of water-soluble and basic carbodiimides in peptide synthesis. *J. Org. Chem.* **21**, 439–441.

Sheehan, J. C., Cruickshank, P. A., and Boshart, G. L. (1961) A convenient synthesis of water-soluble carbodiimides. *J. Org. Chem.* **26**, 2525–2528.

Sheehan, J. C., Preston, J., and Cruickshank, P. A. (1965) A rapid synthesis of oligonucleotide derivatives without isolation of intermediates. *J. Am. Chem. Soc.* **87**, 2492–2493.

Shek, P. N., and Heath, T. D. (1983) Immune response mediated by liposome-associated protein antigens. III. Immunogenicity of bovine serum albumin covalently coupled to vesicle surface. *Immunology* **50**, 101.

Shek, P. N., and Sabiston, B. H. (1982a) Immune response mediated by liposome-associated protein antigens. I. Potentiation of the plaque-forming cell response. *Immunology* **45**, 349.

Shek, P. N., and Sabiston, B. H. (1982b) Immune response mediated by liposome-associated protein antigens. II. Comparison of the effectiveness of vesicle-entrapped and surface-associated antigens in immunopotentiation. *Immunology* **47**, 627.

Shephard, E. G., DeBeer, F. C., von Holt, C., and Hapgood, J. P. (1988) The use of sulfosuccinimidyl-2-(*p*-azidosalicylamido)-1,3'-dithiopropionate as a cross-linking reagent to identify cell surface receptors. *Anal. Biochem.* **168**, 306–313.

Sherry, A. D., Brown III, R. D., Geraldes, C. F. C., Koeng, S. H., and Kuan, K.-T. (1989) Synthesis and characterization of the gadolinium (3+) complex of DOTA-propylamide: A model DOTA–protein conjugate. *Inorg. Chem.* **28**, 620–622.

Shetty, J. K., and Kinsella, J. E. (1980) Ready separation of proteins from nucleoprotein complexes by reversible modification of lysine residues. *Biochem. J.* **191**, 269–272.

Shetty, K. J., and Rao, M. S. N. (1978) Effect of succinylation on the oligomeric structure of arachin. *Int. J. Pept. Protein Res.* **11**, 305.

Shiao, D. D. F., Lumry, R., and Rejender, S. (1972) Modification of protein properties by change in charge. Succinylated chymotrypsinogen. *Eur. J. Biochem.* **29**, 377.

Shih, L. B., Goldenberg, D. M., Xuan, H, Lu, H., Sharkey, R. M., and Hall, T. C. (1991) Anthracycline immunoconjugates prepared by a site-specific linkage via an aminodextran intermediate carrier. *Cancer Res.* **51**, 4192–4198.

Shimisu, N., Nickimins, W. K., and Shimizu, Y. (1980) A cytotoxic epidermal growth factor cross-linked to diphtheria toxin A-fragment. *FEBS Lett.* **118**, 274.

Shimkus, M., Levy, J., and Herman, T. (1985) A chemically cleavable biotinylated nucleotide: Usefulness in the recovery of protein-DNA complexes from avidin affinity columns. *Proc. Natl. Acad. Sci. U.S.A.* **82**, 2593–2597.

Shimomura, S., and Fukui, T. (1978) Characterization of the pyridoxal phosphate site in glycogen phosphorylase b from rabbit muscle. *Biochemistry* **17**, 5359.

Shivdasani, R. A., and Thomas, D. W. (1988) Molecular associations of IA antigens after T-B cell interactions. *J. Immunol.* **141**, 1252–1260.

Sia, C. L., and Horecker, B. L. (1968) Biochem. Biophys. Res. Commun. **31**, 731.

Siezen, R. J., Bindels, J. G., and Hoenders, H. J. (1980) The quaternary structure of bovine a-crystallin. Chemical cross-linking with bifunctional imido esters. *Eur. J. Biochem.* **107**, 243–249.

Silman, H. I., Albu-Weissenberg, M., and Katchalski, E. (1966) *Biopolymers* **4**, 441.

Silvius, J. R., Leventis, R., Brown, P. M., and Zuchermann, M. (1987) Novel fluorescent phospholipids for assays of lipid mixing between membranes. *Biochemistry* **26**, 4279–4287.

Simon, J. R., and Taylor, D. L. (1988) Preparation of a fluorescent analog: Acetamidofluoresceinyl labeled dictyostelium discoideum a-actin. *in* "Methods in Enzymology" (R. B. Vallee, ed.), Vol. 134, p. 47. Academic Press, San Diego.

Simon, S. R., and Konigsberg, W. H. (1966) Chemical modification of hemoglobins: A study of conformation restraint by internal bridging. *Proc. Natl. Acad. Sci. U.S.A.* **56**, 749.

Singer, S. J., Fothergill, J. E., and Shainoff, J. R. (1960) A general method for the isolation of antibodies. *J. Am. Chem. Soc.* **82**, 565.

Singh, V., Sairam, M. R., Bhargavi, G. N., and Akhras, R. G. (1989) Hormonotoxins: Preparation and characterization of ovine luteinizing hormone–gelonin conjugate. *J. Biol. Chem.* **264**, 3089–3095.

Singh, V., Mavila, A. K., and Kar, S. K. (1993) Comparison of the cytotoxic effect of hormonotoxins prepared with the use of heterobifunctional cross-linking agents N-succinimidyl 3-(2-pyridyldithio)propionate and N-succinimidyl 6-[3-(2-pyridyldithio)propionamido]-hexanoate. *Bioconjugate Chem.* **4**, 473–482.

Singhal, A., and Gupta, C. M. (1986) Antibody-mediated targeting of liposomes to red cells *in vivo*. *FEBS Lett.* **201**, 321.

Sipe, D. M., Jesurum, A., and Murphy, R. F. (1991) Absence of Na^+, K^+-ATPase regulation of endosomal acidification in K562 erythroleukemia cells. *J. Biol. Chem.* **266**, 3469.

Sippel, T. O. (1981) New fluorochromes for thiols: Maleimide and iodoacetamide derivatives of 3-phenylcoumarin fluorophore. *J. Histochem. Cytochem.* **29**, 314.

Sivakoff, S. I., and Janes, C. J. (1988) Automated high performance gel-filtration chromatography (HPGFC) processing of avidin coupled β-galactosidase. *Biochromatography* **3**, 62–68.

Skold, S.-E. (1983) Chemical crosslinking of elongation factor G to the 23S RNA in 70S ribosomes from *Escherichia coli*. *Nucleic Acids Res.* **11**, 4923.

Slaughter, T. F., Achyuthan, K. E., Lai, T.-S., and Greenberg, C. S. (1992) A microtiter plate transglutaminase assay utilizing 5-(biotinamido)pentylamine as substrate. *Anal. Biochem.* **205**, 1–6.

Slinkin, M. A., Klibanov, A. L., Khaw, B. A., and Torchilin, V. P. (1990) Succinylated polylysine as a possible link between an antibody molecule and deferoxamine. *Bioconjugate Chem.* **1**, 291–295.

Slinkin, M. A., Klibanov, A. L., and Torchilin, V. P. (1991) Terminal-modified polylysine-based chelating polymers: Highly efficient coupling to antibody with minimal loss in immunoreactivity. *Bioconjugate Chem.* **2**, 342–348.

Slot, J. W., and Geuze, H. J. (1984) Gold markers for single and double immunolabeling of ultrathin cryosections. *in* "Immunolabeling for Electron Microscopy" (J. M. Polak and I. M. Varndess, eds.), p. 139. Elsevier, New York.

Smith, D. F., Zopf, D. A., and Ginsburg, V. (1978) Carbohydrate antigens: Coupling of oligosaccharide phenethylamine-isothiocyanate derivatives to bovine serum albumin. *in* "Methods in Enzymology" (V. Ginsburg, ed.), Vol. 50, pp. 169–171. Academic Press, New York.

Smith, L. M., Fung, S., Hunkapiller, M. W., Hunkapiller, T. J., and Hood, L. E. (1985) The synthesis of oligonucleotides containing an aliphatic amino group at the 5′ terminus: Synthesis of fluorescent DNA primers for use in DNA sequence analysis. *Nucleic Acids Res.* **13**, 2399–2412.

Smith, P. K., Krohn, R. I., Hermanson, G. T., Mallia, A. K., Gartner, F. H., Provenzano, M. D., Fujimoto, E. K., Goeke, N. M., Olson, B. J., and Klenk, D. C. (1985) Measurement of protein using bicinchoninic acid. *Anal. Biochem.* **150**, 76–85.

Smith, R. A. G., and Knowles, J. R. (1973) Aryldiazirines. Potential reagents for photolabeling of biological receptor sites. *J. Am. Chem. Soc.* **95**, 5072–5073.

Smith, R. J., Capaldi, R. A., Muchmore, D., and Dahlquist, F. (1978) Cross-linking of ubiquinone cytochrome c reductase (complex III) with periodate-cleavable bifunctional reagents. *Biochemistry* **17**, 3719–3723.

Smyth, D. G. (1967) *J. Biol. Chem.* **242**, 1592.

Smyth, D. G., Nagamatsu, A., and Fruton, J. S. (1960) Reactions of N-ethylmaleimide. *J. Am. Chem. Soc.* **82**, 4600.

Smyth, D. G., Blumenfeld, O. O., and Konigsberg, W. (1964) Reaction of N-ethylmaleimide with peptides and amino acids. *Biochem. J.* **91**, 589.

Snyder, B., and Hammes, G. G. (1984) Structural mapping of chloroplast coupling factor. *Biochemistry* **23**, 5787–5795.

Snyder, B., and Hammes, G. G. (1985) Structural organization of chloroplast coupling factor. *Biochemistry* **24**, 2324–2331.

Sokolovsky, M., Riordan, J. F., and Vallee, B. L. (1967) Conversion of 3-nitrotyrosine to 3-aminotyrosine in peptides and proteins. *Biochem. Biophys. Res. Commun.* **27**, 20.

Sorensen (1993) Method for isolation and purification of enzyme-antibody conjugates. U.S. Patent No. 5,266,686.

Sorenson, P., Farber, N. M., and Krystal, G. (1986) Identification of the interleukin-3 receptor using an iodinatable, cleavable, photoreactive cross-linking agent. *J. Biol. Chem.* **261**, 9094–9097.

Soundararajan, N., Liu, S. H., Soundararajan, S., and Platz, M. S. (1993) Synthesis and binding of new polyfluorinated aryl azides to a-chymotrypsin. New reagents for photoaffinity labeling. *Bioconjugate Chem.* **4**, 256–261.

Souza, E. D., Ginsberg, M. H., Lam, S., and Plow, E. F. (1988) Chemical cross-linking of arginyl-glycyl-aspartic acid peptides to an adhesion receptor on platelets. *J. Biol. Chem.* **263**, 3943–3951.

Spiegal, S., Skutelsky, E., Bayer, E. A., and Wilchek, M. (1982) *Biochim. Biophys. Acta* **687**, 27.

Spiegel, S., Wilchek, M., and Fishman, P. H. (1983) Fluorescent labeling of cell surface glycoconjugates with Lucifer Yellow CH. *Biochem. Biophys. Res. Commun.* **112**, 872–877.

Spiegel, S., Yamada, K. M., Hom, B. E., Moss, J., and Fishman, P. H. (1985) Fluorescent gangliosides as probes for the retention and organization of fibronectin by ganglioside-deficient mouse cells. *J. Cell Biol.* **100**, 721–726.

Srinivasachar, K., and Neville, D. M., Jr. (1989) New protein cross-linking reagents that are cleaved by mild acid. *Biochemistry* **28**, 2501.

Stahlberg, T., Markela, E., Mikola, H., Mottram, P., and Hemmila, I. (1993) Europium and samarium in time-resolved fluoroimmunoassays. *American Laboratory* Dec., 15–20.

Staros, J. V. (1982) *N*-Hydroxysulfosuccinimide active esters: Bis(*N*-hydroxysulfosuccinimide) esters of two dicarboxylic acids are hydrophilic, membrane impermeant, protein cross-linkers. *Biochemistry* **21**, 3950–3955.

Staros, J. V. (1988) Membrane-impermeant cross-linking reagents: Probes of the structure and dynamics of membrane proteins. *Acc. Chem. Res.* **21**, 435–441.

Staros, J. V., and Kakkad, B. P. (1983) Cross-linking and chymotryptic digestion of the extra-cytoplasmic domain of the anion exchange channel in intact human erythrocytes. *J. Membr. Biol.* **74**, 247–254.

Staros, J. V., Bayley, H., Standring, D. N., and Knowles, J. R. (1978) Reduction of aryl azides by thiols: Implication for the use of photoaffinity reagents. *Biochem. Biophys. Res. Commun.* **80**, 568.

Staros, J. V., Wright, R. W., and Swingle, D. M. (1986) Enhancement by *N*-hydroxysulfosuc-cinimide of water-soluble carbodiimide-mediated coupling reactions. *Anal. Biochem.* **156**, 220–222.

Staros, J. V., Lee, W. T., and Conrad, D. H. (1987) Membrane-impermeant cross-linking reagents: Application to the study of the cell surface receptor for IgE. *in* "Methods in Enzymology" (G. Di Sabato, ed.), Vol. 150, pp. 503–512. Academic Press, Orlando, Florida.

Steer, C. J., and Ashwell, G. (1986) Hepatic membrane receptors for glycoproteins. *Prog. Liver Dis.* **8**, 99–123.

Sternberger, L. A. (1986) "Immunocytochemistry." Wiley, New York.

Stewart, J. M., and Young, J. D. (1984) "Solid Phase Peptide Synthesis," 2nd Ed., p. 31. Pierce Chemical Company, Rockford, Illinois.

Stewart, W. W. (1978) Functional connections between cells as revealed by dye-coupling with a highly fluorescent naphthalimide tracer. *Cell (Cambridge, Mass.)* **14**, 741.

Stewart, W. W. (1981a) Lucifer dyes—highly fluorescent dyes for biological tracing. *Nature (London)* **292**, 17.

Stewart, W. W. (1981b) Synthesis of 3,6-disulfonate 4-aminonaphthalimides. *J. Am. Chem. Soc.* **103**, 7615.

Stickel, S. K., and Wang, Y.-L. (1988) Synthetic peptide GRGDS induces dissociation of alpha-actin and vinculin from the sites of focal contacts. *J. Cell Biol.* **107**, 1231.

Stirpe, F., Olsnes, S., and Pihl, A. (1980) Gelonin, a new inhibitor of protein synthesis, nontoxic to intact cells. Isolation, characterization, and preparation of cytotoxic complexes with concanavalin A. *J. Biol. Chem.* **255**, 6947–6953.

Straubinger, R. M., Lopez, N. G., Debs, R. J., Hong, K., and Papahajopoulos, D. (1988) Liposome-based therapy of human ovarian cancer: Parameters determining potency of nega-tively charged and antibody-targeted liposomes. *Cancer Res.* **48**, 5237–5245.

Strickland, E., Hardin, E. K., Shannon, L. M., and Horwitz, J. (1968) Peroxidase isoenzymes from horseradish roots. III Circular dichroism of isoenzymes and apoisoenzymes. *J. Biol. Chem.* **243**, 3560.

Strottmann, J. M., Robinson, J. B., Jr., and Stellwagen, E. (1983) Advantages of preelectrophore-tic conjugation of polypeptides with fluorescent dyes. *Anal. Biochem.* **132**, 334–337.

Stuchbury, T., Shipton, M., Norris, R., Malthouse, J. P. G., and Brocklehurst, K. (1975) Report-er groups delivery system with both absolute and selective specificity for thiol groups and an improved fluorescent probe containing the 7-nitrobenzo-2-oxa-1,3-diazole moiety. *Bio-chem. J.* **151**, 417–432.

Stump, R. F., Pfeiffer, J. R., Schneebeck, M. C., Seagrave, J. C., and Oliver, J. M. (1989) Mapping gold-labeled receptors on cell surfaces by backscattered electron imaging and digital image analysis: Studies of the IgE receptor on mast cells. *Am. J. Anat.* **185**, 128–141.

Subramanian, R., and Meares, C. F. (1991) Bifunctional chelating agents for radiometal-labeled monoclonal antibodies. *in* "Cancer Imaging with Radiolabeled Antibodies" (D. M. Goldenberg, ed.), pp. 183–199. Kluwer, Boston.

Sulkowski, E. (1985) *Trends Biotechnol.* 3, 1.

Sun, J.-S., Francois, J.-C., Lavery, R., Saison-Behmoaras, T., Montenay-Garestier, T., Thuong, N. T., and Helene, C. (1988) Sequence-targeted cleavage of nucleic acids by oligo-a--thymidylate-phenanthroline conjugates: Parallel and antiparallel double helices are formed with DNA and RNA, respectively. *Biochemistry* 27, 6039–6045.

Sun, T. T., Bollen, A., Kahan, L., and Traut, R. R. (1974) Topography of ribosomal proteins of the *Escherichia coli* 30S subunit as studied with the reversible cross-linking reagent methyl 4-mercaptobutyrimidate. *Biochemistry* 13, 2334–2340.

Sundberg, L., and Porath, J. (1974) Preparation of adsorbents for biospecific affinity chromatography. I. Attachment of group containing ligands to insoluble polymers by means of bufunctional oxiranes. *J. Chromatogr.* 90, 87–98.

Suter, M., and Butler, J. E. (1986) The immunochemistry of sandwich ELISAs II. A novel system prevents the denaturation of capture antibodies. *Immunol. Lett,* 13, 313–316.

Sutoh, K., Yamamoto, K., and Wakabayashi, T. (1984) Electron microscopic visualization of the SH$_1$ thiol of myosin by the use of an avidin–biotin system. *J. Mol. Biol.* 178, 323–339.

Swanson, J., Burke, E., and Silverstein, S. C. (1987) Tubular lysosomes accompany stimulated pinocytosis in macrophages. *J. Cell Biol.* 104, 1217.

Swanson, S. J., Lin, B.-F., Mullenix, M. C., and Mortensen, R. F. (1991) A synthetic peptide corresponding to the phosphorylcholine (PC)-binding region of human C-reactive protein possesses the TEPC-15 myeloma PC-idiotype. *J. Immunol.* 146, 1596–1601.

Sweeley, C. C., and Nunez, H. A. (1985) Structural analysis of glycoconjugates by mass spectrometry and nuclear magnetic resonance spectroscopy. *Annu. Rev. Biochem.* 54, 765–801.

Sykaluk, L. (1994) unpublished data, Pierce Chemical.

Szabo, J., Kruger, S. R., and Beall, G. N. (1989) Detection of cells producing anti-idiotypic antibody to thyroid stimulating hormone-reactive antibodies. *Immunol. Invest.* 18, 879–884.

Szoka, F. C., Olson, F., Hunt, C. A., Vail, W., Mayhew, E., and Paphadjopoulos, D. (1980) Preparation of unilamellar liposomes of intermediate size (0.1–0.2 μm) by a combination of reverse phase evaporation and extrusion through polycarbonate membranes. *Biochim. Biophys. Acta* 601, 559–571.

Tadayoni, B. M., Friden, P. M., Walus, L. R., and Musso, G. F. (1993) Synthesis *in vitro* kinetics, and *in vivo* studies on protein conjugates of AZT: Evaluation as a transport system to increase brain delivery. *Bioconjugate Chem.* 4, 139–145.

Tager, H. S. (1976) Coupling of peptides to albumin with difluorodinitrobenzene. *Anal. Biochem.* 71, 367–375.

Takadate, A., Irikura, M., Suehiro, T., Fujino, H., and Goya, S. (1985) New labeling reagents for alcohols in fluorescence high-performance liquid chromatography. *Chem. Pharm. Bull.* 33, 1164–1169.

Takahashi, K. (1968) The reaction of phenylglyoxal with arginine residues in proteins. *J. Biol. Chem.* 243, 6171–6179.

Talsma, H., van Steenberg, M. J., and Crommelin, D. J. A. (1991) *Int. J. Pharm.* 77, 119–126.

Tam, J. P. (1988) Synthetic peptide vaccine design: Synthesis and properties of a high-density multiple antigenic peptide system. *Proc. Natl. Acad. Sci. U.S.A.* 85, 5409–5413.

Tanaka, E. M., and Kirschner, M. W. (1991) Microtubule behavior in the growth cones of living neurons during axon elongation. *J. Cell Biol.* 115, 345.

Tanford, C., and Hauenstein, J. D. (1956) Hydrogen ion equilibria of ribonuclease. *J. Am. Chem. Soc.* 78, 5287–5291.

Tao, T., Lamkin, M., and Scheiner, C. (1984) Studies on the proximity relationships between thin filament proteins using benzophenone-4-maleimide as a site-specific photoreactive crosslinker. *Biophys. J.* 45, 261.

Tarentino, A. L., Phelan, A. W., and Plummer, T. H., Jr. (1993) 2-Iminothiolane: A reagent for the introduction of sulfhydryl groups into oligosaccharides derived from asparagine-linked glycans. *Glycobiology* **3**, 279–285.

Tarvers, R. C., Noyes, C. M., Roberts, H. R., and Lundblad, R. L. (1982) Influence of metal ions on prothrombin self-association. *J. Biol. Chem.* **257**, 10708–10714.

Tawney, P. O., Snyder, R. H., Conger, R. P., Leibbrand, K. A., Stiteler, C. H., and Williams, A. R. (1961) Maleimide and derivatives. II. Maleimide and N-methylmaleimide. *J. Org. Chem.* **26**, 15.

Taylor, K. E., and Wu, Y. C. (1980) A thiolation reagent for cell surface carbohydrate. *Biochem. Int.* **1**, 353.

Teale, J. M., and Kearney, J. R. (1986) Clonotypic analysis of the fetal B cell repertoire: Evidence for an early and predominant expression of idiotypes associated with the VH 36-60 family. *J. Mol. Cell. Immunol.* **2**, 283–292.

Thakur, M. L., and DeFulvio, J. D. (1991) Technetium-99m labeled monoclonal antibodies for immunoscintigraphy. *J. Immunol. Methods* **137**, 217–224.

Theis, F. V., and Freeland, M. R. (1940) *Arch. Surg.* **40**, 190.

Thelen, P., and Deuticke, B. (1988) Chemo-mechanical leak formation in human erythrocytes upon exposure to a water-soluble carbodiimide followed by very mild shear stress. II. Chemical modifications involved. *Biochim. Biophys. Acta* **944**, 297–307.

Theorell, H. (1951) The iron-containing enzymes. B. Catalases and peroxidases. "Hydroperoxidases." *in* "The Enzymes" (J. B. Sumner and K. Myrback, eds.), Vol. 2, Part 1, p. 397. Academic Press, New York.

Therien, H.-M., and Shahum, E. (1989) Importance of physical association between antigen and liposomes in liposome adjuvanticity. *Immunol. Lett.* **22**, 253–258.

Thevenin, B., Shahrokh, Z., Williard, R., Fujimoto, E., Ikemoto, N., and Shohet, S. (1991) A novel reagent for functionally-directed site-specific fluorescent labeling of proteins. *Biophys. J.* **59**, Tu-Pos 476, p. 358a (Abstract).

Thorpe, P. E., Brown, A. N. F., Ross, W. C. J., Cumber, A. J., Detre, S. I., Edwards, D. C., Davies, A. J. S., and Stirpe, F. (1981) Cytotoxicity acquired by conjugation of an anti-Thy 1.1 monoclonal antibody and the ribosome-inactivating protein, gelonin. *Eur. J. Biochem.* **116**, 447–454.

Thorpe, P. E., Mason, D. W., Brown, A. N. F., Simmonds, S. J., Ross, W. C. J., Cumber, A. J., and Forrester, J. A. (1982) Selective killing of malignant cells in a leukaemic rat bone marrow using an antibody–ricin conjugate. *Nature (London)* **297**, 594.

Thorpe, P. E., Ross, W. C. J., Brown, A. N. F., Myers, C. D., Cumber, A. F., and Foxwell, B. M. J. (1984) Blockade of the galactose-binding sites of ricin by its linkage to antibody specific cytotoxic effects of the conjugate. *Eur. J. Biochem.* **140**, 63.

Thorpe, P. E., Wallace, P. M., Knowles, P. P., Relf, M. G., Brown, A. N., Watson, G. L., Knyba, R. E., Wawrzynczak, E. J., and Blakey, D. C. (1987) New coupling agents for the synthesis of immunotoxins containing a hindered disulfide bond with improved stability *in vivo*. *Cancer Res.* **47**, 5924–5931.

Tinglu, G., Ghosh, A., and Ghosh, B. K. (1984) Subcellular localization of alkaline phosphatase in *Bacillus licheniformis* 749/C by immunoelectron microscopy with colloidal gold. *J. Bacteriol.* **159**, 668.

Titus, J. A., Haugland, R. P., Sharrow, D. M., and Segal, J. (1982) Texas Red, a hydrophilic, red-emitting fluorophore for use with fluorescein in dual parameter flow microfluorometric and fluorescence microscopic studies. *J. Immunol. Methods* **50**, 193–204.

Tjerneld, F. (1992) Aqueous two-phase partitioning on an industrial scale. *in* "Poly(Ethylene Glycol) Chemistry: Biotechnical and Biomedical Applications" (J. M. Harris, ed.), pp. 85–102. Plenum, New York.

Tokuyasu, K. T. (1983) Present state of immunocryoultramicrotomy. *J. Histochem. Cytochem.* **31**, 164.

Torchilin, V. P., Klibanov, A. L., Slinkin, M. A., Danilov, S. M., Levitsky, D. O., and Khow, B. A. (1989) Antibody-linked chelating polymers for immunoimaging *in vivo*. *J. Controlled Release* **11**, 297–303.

Torchilin, V. P., Trubetskoy, V. S., Narula, J., Khaw, B. A., Klibanov, A. L., and Slinkin, M. A. (1993) Chelating polymer modified monoclonal antibodies for radioimmunodiagnostics and radioimmunotherapy. *J. Controlled Release* **24**, 111–118.

Trail, P. A., Willner, D., Lasch, S. J., Henderson, A. J., Hofstead, S., Casazza, A. M., Firestone, R. A., Hellstrom, I., and Hellstrom, K. E. (1993) Cure of xenografted human carcinomas by BR96-doxorubicin immunoconjugates. *Science* **261**, 212–215.

Traincard, F., Ternynck, T., Danchin, A., and Avrameas, S. (1983) An immunoenzymic procedure for the demonstration of nucleic acid molecular hybridization. *Ann. Immunol.* **134**, 339–405.

Traut, R. R., Bollen, A., Sun, R. R., Hershey, J. W. B., Sundberg, J., and Pierce, L. R. (1973) *Biochemistry* **12**, 3266.

Traut, R. R., Casiano, C., and Zecherle, N. (1989) Cross-linking of protein subunits and ligands by the introduction of disulfide bonds. *in* "Protein Function—A Practical Approach" (T. E. Creighton, ed.), pp. 101–133. IRL Press at Oxford Univ., Oxford.

Trubetskoy, V. S., Narula, J., Khaw, B. A., and Torchilin, V. P. (1993) Chemically optimized antimyosin Fab conjugates with chelating polymers: Importance of the nature of the protein-polymer single site covalent bond for biodistribution and infarction localization. *Bioconjugate Chem.* **4**, 251–255.

Truneh, A., Machy, P., and Horan, P. K. (1987) Antibody-bearing liposomes as multicolor immunofluorescent markers for flow cytometry and imaging. *J. Immunol. Methods* **100**, 59–71.

Tsien, R. Y., and Waggoner, A. (1990) Fluorophores for confocal microscopy: Photophysics and photochemistry. *in* "Handbook of Biological Confocal Microscopy" (J. B. Pawley, ed.), p. 169. Plenum, New York.

Tsomides, T. J., Walker, B. D., and Eisen, H. N. (1991) An optimal viral peptide recognized by $CD8^+T$ cells binds very tightly to the restricting class I major histocompatibility complex protein on intact cells but not to the purified class I protein. *Proc. Natl. Acad. Sci. U.S.A.* **88**, 11276–11280.

Tsudo, M., Kozak, R. W., Goldman, C. K., and Waldmann, T. A. (1987) Demonstration of a non-Tac peptide that binds interleukin 2: A potential participant in a multichain interleukin 2 receptor complex. *Proc. Natl. Acad. Sci. U.S.A.* **83**, 9694–9698.

Tsukamoto, Y., and Wakil, S. J. (1988) Isolation and mapping of the β-hydroxyacyl dehydratase activity of chicken liver fatty acid synthase. *J. Biol. Chem.* **263**, 16225–16229.

Tu, C.-P. C., and Cohen, S. (1980) 3′-End labeling of DNA with $[\alpha\text{-}^{32}P]$cordycepin-5′-triphosphate. *Gene* **10**, 177–183.

Tussen, P., and Kurstak, E. (1984) Highly efficient and simple methods for the preparation of peroxidase and active peroxidase–antibody conjugates for enzyme immunoassays. *Anal. Biochem.* **136**, 451–457.

Uchino, O., Mizunami, T., Maida, M., and Miyazoe, Y. (1979) Efficient dye lasers pumped by an XeCl excimer laser. *Appl. Phys.* **19**, 35.

Uchiumi, T., Terao, K., and Ogata, K. (1980) Identification of neighboring protein pairs in rat liver 60S ribosomal subunits cross-linked with dimethyl suberimidate or dimethyl 3,3′-dithiobispropionimidate. *J. Biochem. (Tokyo)* **88**, 1033–1044.

Uraki, Z., Terminiello, L., Bier, M., and Nord, F. F. (1957) *Arch. Biochem. Biophys.* **69**, 644.

Urdea, M. S., Warner, B. D., Running, J. A., Stempien, M., Clyne, J., and Horn, T. (1988) A comparison of non-radioactive hybridization assay methods using fluorescent, chemiluminescent and enzyme-labeled synthetic oligodeoxyribonucleotide probes. *Nucleic Acids Res.* **16**, 4937–4956.

Uto, I., Ishimatsu, T., Hirayama, H., Ueda, S., Tsuruta, J., and Kambara, T. (1991) Determination of urinary Tamm–Horsfall protein by ELISA using a maleimide method for enzyme–antibody conjugation. *J. Immunol. Methods* **138**, 87–94.

Uyeda, K. (1969) *Biochemistry* **8**, 2366.

van Dalen, J. P. R., and Haaijman, J. J. (1974) Determination of the molar absorbance coefficient of bound tetramethyl rhodamine isothiocyanate relative to fluorescein isothiocyanate. *J. Immunol. Methods* **5**, 103–106.

Vandelen, R. L., Arcuri, K. E., and Napier, M. A. (1985) Identification of a receptor for atrial natriuretic factor in rabbit aorta membranes by affinity cross-linking. *J. Biol. Chem.* **260**, 10889–10892.

van den Brink, W., van der Loos, C., Volkers, H., Lauwen, R., van den Berg, F., Houthoff, H.-J., and Das, P. K. (1990) Combined β-galactosidase and immunogold/silver staining for immunohistochemistry and DNA *in situ* hybridization. *J. Histochem. Cytochem.* **38**, 325–329.

van der Horst, G. T. J., Mancini, G. M. S., Brossmer, R., Rose, U., and Verheijen, F. W. (1990) Photoaffinity labeling of a bacterial sialidase with an aryl azide derivative of sialic acid. *J. Biol. Chem.* **265**, 10801–10804.

van Dongen, J. J. M., Hooijkaas, H., Comans-Bitter, W. M., Benne, K., van Os, T. M., and de Josselin de Jong, J. (1985) Triple immunological staining with colloidal gold, fluorescein, and rhodamine as labels. *J. Immunol. Methods* **80**, 1.

Vanin, E. F., and Ji, T. H. (1981) Synthesis and application of cleavable photoactivatable heterobifunctional reagents. *Biochemistry* **20**, 6754–6760.

Van Lenten, L., and Ashwell, G. (1971) Studies on the chemical and enzymatic modification of glycoproteins. A general method for the tritiation of sialic acid-containing glycoproteins. *J. Biol. Chem.* **246**, 1889–1894.

Van Regenmortal, M. H. V., Briand, J. P., Muller, S., and Plaue, S. (1988) Synthetic polypeptides as antigens. *Lab. Tech. Biochem. Mol. Biol.* **19**, 121–125.

Verwey, E. J. W., and Overbeek, J. T. G. (1948) "Theory of the Stability of Lyophobic Colloids." Elsevier, New York.

Vigers, G. P. A., Coue, J. R., and Mcintosh, J. (1988) Fluorescent microtubules break up under illumination. *J. Cell Biol.* **107**, 1011.

Vilja, P. (1991) One- and two-step non-competitive avidin–biotin immunoassays for monomeric and heterodimeric antigen. *J. Immunol. Methods* **136**, 77.

Vincent, J. P., Lazdunski, M., and Delaage, M. (1970) Use of tetranitromethane as a nitration reagent. Reaction of phenol sidechains in bovine and porcine trypsinogens and trypsins. *Eur. J. Biochem.* **12**, 250.

Viscidi, R. P., Connelly, C. J., and Yolken, R. H. (1986) Novel chemical method for the preparation of nucleic acids for nonisotopic hybridization. *J. Clin. Microbiol.* **23**, 311–317.

Vitetta, E. S., and Thorpe, P. E. (1985) Immunotoxins containing ricin A or B chains with modified carbohydrate residues act synergistically in killing neoplastic B cells *in vitro*. *Cancer Drug Delivery* **2**, 191.

Vithayathil, P. J., and Richards, F. M. (1960) *J. Biol. Chem.* **235**, 2343.

Vliegenthart, J. F. G., Dorland, L., and van Halbeek, H. (1983) High-resolution, 1H-nuclear magnetic resonance spectroscopy as a tool in the structural analysis of carbohydrates related to glycoproteins. *Adv. Carbohydr. Chem. Biochem.* **41**, 209–374.

Vogel, C.-W. (1987) Antibody conjugate without inherent toxicity: The targeting of cobra venom factor and other biological response modifiers. *in* "Immunoconjugates: Antibody Conjugates in Radioimaging and Therapy of Cancer" (C.-W. Vogel, ed.), p. 170. Oxford Univ. Press, New York.

Vogel, C.-W., and Muller-Eberhard, H. J. (1984) Cobra venom factor: Improved method for purification and biochemical characterization. *J. Immunol. Methods* **73**, 203.

von der Haar, F., Schlimme, E., and Gauss, D. H. (1971) *in* "Proceedings of Nucleic acids Research" (G. L. Cantoni and D. R. Davies, eds.), Vol. 2, pp. 643–664. Harper & Row, New York.

Wade, D. P., Knight, B. L., and Soutar, A. K. (1985) Detection of the low-density lipoprotein receptor with biotin-low-density lipoprotein. A rapid new method for ligand blotting. *Biochem. J.* **229**, 785–790.

Waggoner, A. S. (1990) Fluorescent probes for cytometry. *in* "Flow Cytometry and Sorting" (M. R. Melamed, T. Lindmo, and M. L. Mendelsohn, eds.), 2nd ed., pp. 209–225. Wiley-Liss, New York.

Waldmann, T. A. (1991) Monoclonal antibodies in diagnosis and therapy. *Science* **252**, 1657.

Wallenfels, K., and Weil, R. (1972) *in* "The Enzymes," (P. D. Boyer, ed.), 3rd Ed., Vol. 7, p. 617. Academic Press, New York.

Walling, C., and Gibian, M. J. (1965) Hydrogen abstraction reactions by the triplet states of ketones. *J. Am. Chem. Soc.* **87**, 3361.

Wang, D., Wilson, G., and Moore, S. (1976) Preparation of cross-linked dimers of pancreatic ribonuclease. *Biochemistry* **15**, 660–665.

Wang, K. (1974) Ph.D. Thesis, Yale University, New Haven, Connecticut.

Wang, K., and Richards, F. (1974) An approach to nearest neighbor analysis of membrane proteins. Application to the human erythrocyte membrane of a method employing cleavable cross-linkages. *J. Biol. Chem.* **249**, 8005–8018.

Wang, K., and Richards, F. (1975) Reaction of dimethyl-3,3′-dithiobispropionimidate with intact human erythrocytes. Cross-linking of membrane proteins and of hemoglobin. *J. Biol. Chem.* **250**, 6622–6626.

Wang, S. S., and Carpenter, F. H. (1968) Kinetic studies at high pH of the trypsin-catalyzed hydrolysis of N$^\alpha$-benzoyl derivatives of L-arginamide, L-lysinamide, and S-2-aminoethyl-L-cysteinamide and related compounds. *J. Biol. Chem.* **243**, 3702–3710.

Wang, Y.-L. (1985) Exchange of actin subunits at the leading edge of living fibroblasts: Possible role of treadmilling. *J. Cell Biol.* **101**, 597.

Waterman, M. R., Yanaoka, K., Chuang, A. H., and Cottam, G. L. (1975) Anti-sickling nature of dimethyl adipimidate. *Biochem. Biophys. Res. Commun.* **63**, 580–587.

Watson, J. V. (ed.) (1991) "Introduction to Flow Cytometry." Univ. Press.

Waugh, S. M., DiBella, E. E., and Pilch, P. F. (1989) Isolation of a proteolitically derived domain of the insulin receptor containing the major site of cross-linking/binding. *Biochemistry* **28**, 3448–3455.

Wedekind, F., Baer-Pontzen, K., Bala-Mohan, S., Choli, D., Zahn, H., and Brandenburg, D. (1989) Hormone binding site of the insulin receptor: Analysis using photoaffinity-mediated avidin complexing. *Biol. Chem. Hoppe-Seyler* **370**, 251–258.

Wells, J. A., Knoeber, C., Sheldon, M. C., Werber, M. M., and Yount, R. G. (1980) Cross-linking of myosin subfragment 1. Nucleotide-enhanced modification by a variety of bifunctional reagents. *J. Biol. Chem.* **255**, 11135.

Weltman, J. K., Hohnson, S.-A., Langevin, J., and Riester, E. F. (1983) N-Succinimidyl(4-iodoacetyl)aminobenzoate: A new heterobifunctional cross-linker. *BioTechniques* **1**, 148–152.

Wessels, B. W., and Rogus, R. D. (1984) Radionuclide selection and model absorbed dose calculations for radiolabeled tumor associated antibodies. *Med. Phys.* **11**, 638–645.

Wessendorf, M. W. (1990) Characterization and use of multi-color fluorescence microscopic techniques. "Handbook of Chemical Neuroanatomy," Vol. 8, Chapter 1.

Weston, P. D., Devries, J. A., and Wrigglesworth, R. (1980) *Biochim. Biophys. Acta* **612**, 40–49.

Wheat, T., Shelton, J. A., Gonzales-Prevatt, V., and Goldberg, E. (1985) *Mol. Immunol.* **22**, 1195.

Whitaker, J. E., Haugland, R. P., Moore, P. L., Hewitt, P. C., Reese, M., and Haugland, R. P. (1991) Cascade blue derivatives: Water soluble, reactive, blue emission dyes evaluated as fluorescent labels and tracers. *Anal. Biochem.* **198**, 119–130.

Wieder, K. J., Palczuk, N. C., van Es, T., and Davis, F. F. (1979) Some properties of polyethylene glycol: Phenylalanine ammonia–lyase adducts. *J. Biol. Chem.* **254**, 12579–12587.

Wiels, J., Junqua, S., Dujardin, P., Le Pecq, J. B., and Tursz, T. (1984) Properties of immunotoxins against a glycolipid antigen associated with Burkitt's lymphoma. *Cancer Res.* **44**, 129.

Wilbur, D. S. (1992) Radiohalogenation of proteins: An overview of radionuclides, labeling methods, and reagents for conjugate labeling. *Bioconjugate Chem.* **3**, 433–470.

Wilchek, M., and Bayer, E. A. (1987) Labeling glycoconjugates with hydrazide reagents. *in* "Methods in Enzymology" (V. Ginsburg, ed.), Vol. 138, pp. 429–442. Academic Press, Orlando Florida.

Wilchek, M., and Bayer, E. A. (1988) The avidin–biotin complex in bioanalytical applications. *Anal. Biochem.* **171**, 1–32.

Wilchek, M., and Givol, D. (1977) Affinity cross-linking of heavy and light chains. *in* "Methods in Enzymology" (W. B. Jakoby and M. Wilchek, eds.), Vol. 46, p. 501. Academic Press, New York.

Wilchek, M., Spiegel, S., and Spiegel, Y. (1980) Fluorescent reagents for the labeling of glycoconjugates in solution and on cell surfaces. *Biochem. Biophys. Res. Commun.* **92**, 1215.

Wilchek, M., Ben-Hur, H., and Bayer, E. A. (1986) *p*-Diazobenzoyl biocytin—a new biotinylating reagent for the labeling of tyrosines and histidines in proteins. *Biochem. Biophys. Res. Commun.* **138**, 872–879.

Wiley, D. C., Skehel, J. J., and Waterfield, M. (1977) *Virology* **79**, 446–448.

Williams, A., and Ibrahim, I. A. (1981) A mechanism involving cyclic tautomers for the reaction with nucleophiles of the water-soluble peptide coupling reagent 1-ethyl-3-(3-dimethylaminopropyl) carbodiimide (EDC). *J. Am. Chem. Soc.* **103**, 7090–7095.

Williams, D. C., Huff, G. F., and Seitz, W. R. (1976) Glucose oxidase chemiluminescence measurement of glucose in urine compared with the hexokinase method. *Clin. Chem.* **22**, 372.

Willingham, G. L., and Gaffney, B. J. (1983) Reactions of spin-label cross-linking reagents with red blood cell proteins. *Biochemistry* **22**, 892.

Willner, D., Trail, P. A., Hofstead, S. J., King, H. D., Lasch, S. J., Braslawsky, G. R., Greenfield, R. S., Kaneko, T., and Firestone, R. A. (1993) (6-Maleimidocaproyl)hydrazone of doxorubicin—a new derivative for the preparation of immunoconjugates of doxorubicin. *Bioconjugate Chem.* **4**, 521–527.

Wilson, M. B., and Nakane, P. K. (1978) Recent developments in the periodate method of conjugating horseradish peroxidase (HRPO) to antibodies. *In* "Immunofluorescence and Related Staining Techniques" (W. Knapp, K. Holuber, and G. Wick, eds.), pp. 215–224. Elsevier/North-Holland Biomedical Press, Amsterdam.

Wirth, P., Souppe, J., Tritsch, D., and Biellmann, J. F. (1991) Chemical modification of horseradish peroxidase with ethanal-MePEG: Solubility in organic solvents, activity and properties. *Bioorg. Chem.* **19**, 133–142.

Wojchowski, D. M., and Sytkowski, A. J. (1986) Hybridoma production by simplified avidin-mediated electrofusion. *J. Immunol. Methods* **90**, 173–177.

Wold, F. (1961) Reaction of bovine serum albumin with the bifunctional reagent p,p'-difluoro-m,m'-dinitrodiphenylsulfone. *J. Biol. Chem.* **236**, 106.

Wold, F. (1972) Bifunctional reagents. *in* "Methods in Enzymology" (C. H. W. Hirs and S. N. Timasheff, eds.), Vol. 25, p. 623. Academic Press, New York.

Wollenweber, H.-W., and Morrison, D. C. (1985) Synthesis and biochemical characterization of a photoactivatable, iodinatable, cleavable bacterial lipopolysaccharide derivative. *J. Biol. Chem.* **260**, 15068–15074.

Wood, C. L., and O'Dorisio, M. S. (1985) Covalent cross-linking of vasoactive intestinal polypeptide to its receptors on intact human lymphoblasts. *J. Biol. Chem.* **260**, 1243–1247.

Woodward, R. B., and Olofson, R. A. (1961) The reaction of isoxazolium salts with bases. *J. Am. Chem. Soc.* **83**, 1010.

Woodward, R. B., Olofson, R. A., and Mayer, H. (1961) A new synthesis of peptides. *J. Am. Chem. Soc.* **83**, 1007–1009.

Wright, B. S., Tyler, G. A., O'Brien, R., Coporale, L. H., and Rosenblatt, M. (1987) Immunoprecipitation of the parathyroid hormone receptor. *Proc. Natl. Acad. Sci. U.S.A.* **84**, 26–30.

Wu, C.-W., Yarbrough, L. R., and Wu, F. Y.-H. (1976) N-(1-pyrene)maleimide: A fluorescent cross-linking reagent. *Biochemistry* **15**, 2863–2867.

Yamada, H., Imoto, T., Fujita, K., Okazaki, K., and Motomura, M. (1981) Selective modification of aspartic acid-101 in lysozyme by carbodiimide reaction. *Biochemistry* **20**, 4836–4842.

Yamamoto, K., Sekine, T., and Sutoh, K. (1984) Spatial relationship between SH_1 and the actin binding site on myosin subfragment-1 surface. *FEBS Lett.* **176**, 75–78.

Yamasaki, R. B., Shimer, D. A., and Feeney, R. E. (1981) Colorimetric determination of arginine residues in proteins by p-nitrophenylglyoxal. *Anal. Biochem.* **111**, 220.

Yan, M., Cai, S. X., Wybourne, M. N., and Keana, J. F. W. (1994) N-Hydroxysuccinimide ester functionalized perfluorophenyl azides as novel photoactivatable heterobifunctional cross-linking reagents. The covalent immobilization of biomolecules to polymer surfaces. *Bioconjugate Chem.* **5**, 151–157.

Yasuda, T., Dancey, G. F., and Kinsky, S. C. (1977) Immunogenicity of liposomal model membranes in mice: Dependence on phospholipid composition. *Proc. Natl. Acad. Sci. U.S.A.* **74**, 1234–1236.

Yem, A. W., *et al.* (1992) Site-specific chemical modification of interleukin 1b by Acrylodan at cysteine 8 and lysine 103. *J. Biol. Chem.* **267**, 3122.

Yeung, C. W. T., Moule, M. L., and Yip, C. C. (1980) Photoaffinity labeling of insulin receptor with an insulin analogue selectively modified at the amino terminal of the B chain. *Biochemistry* **19**, 2196–2203.

Yokota, S. (1988) Effect of particle size on labeling density for catalase in protein A–gold immunocytochemistry. *J. Histochem. Cytochem.* **36**, 107–109.

Yoshitake, S., Yamada, Y., Ishikawa, E., and Masseyeff, R. (1979) Conjugation of glucose oxidase from *Aspergillus niger* and rabbit antibodies using N-hydroxysuccinimide ester of N-(4-carboxycyclohexylmethyl)maleimide. *Eur. J. Biochem.* **101**, 395–399.

Yoshitake, S., Imagawa, M., Ishikawa, E., Niitsu, Y., Urushizaki, I., Nishiura, M., Kanazawa, R., Kurosaki, H., Tachibana, S., Nakazawa, N., and Ogawa, H. (1982a) Mild and efficient conjugation of rabbit Fab' and horseradish peroxidase using a maleimide compound and its use for enzyme immunoassay. *J. Biochem. (Tokyo)* **92**, 1413–1424.

Yoshitake, S., Imagawa, M., and Ishikawa, E. (1982b) Efficient preparation of rabbit Fab'-horseradish peroxidase conjugates using maleimide compounds and its use for enzyme immunoassay. *Anal. Lett.* **15**(B2), 147–160.

Youle, R. J., and Nevelle, D. M., Jr. (1980) Anti-Thy 1.2 monoclonal antibody linked to ricin is a potent cell-type-specific toxin. *Proc. Natl. Acad. Sci. U.S.A.* **77**, 5483–5486.

Young, J. L. (1979) The effect of dimethyl 3,3'-dithiobispropionimidate on the adenylate cyclase activity of bovine corpus luteum. *FEBS Lett.* **104**, 294–296.

Yu, R., and Schweinberger, F. (1979) *Z. Pflanzenphsiol.* **94**, 135–142.

Zahn, H., and Lumper, L. (1968) Specificity of bifunctional sulfhydryl reagents and synthesis of a defined dimer of bovine serum albumin. *Hoppe-Seyler's Z. Physiol. Chem.* **349**, 485.

Zahn, H., and Meinhoffer, J. (1958) Reactions of 1,5-difluoro-2,4-dinitrobenzene with insulin. *Makromol. Chem.* **26**, 153.

Zalipsky, S., and Lee, C. (1992) Use of functionalized poly(ethylene glycol)s for modification of polypeptides. *in* "Poly(Ethylene Glycol) Chemistry: Biotechnical and Biomedical Applications" (J. M. Harris, ed.), pp. 347–370. Plenum, New York.

Zalipsky, S., Seltzer, R., and Nho, K. (1991) Succinimidyl carbonates of polyethylene glycol: Useful reactive polymers for preparation of protein conjugates. *in* "Polymeric Drugs and Drug Delivery Systems" (R. L. Dunn and R. M. Ottenbrite, eds.), pp. 91–100. American Chemical Society, Washington, D.C.

Zalipsky, S., Seltzer, R., and Menon-Rudolph, S. (1992) Evaluation of a new reagent for covalent attachment of polyethylene glycol to proteins. *Biotechnol. Appl. Biochem.* **15**, 100–114.

Zanocco, J., Krohn, R., Sykaluk, L., and Olson, B. (1993) unpublished observations. Pierce Chemical.

Zara, J. J., *et al.* (1991) A carbohydrate-directed heterobifunctional cross-linking reagent for the synthesis of immunoconjugates. *Anal. Biochem.* **194**, 156–162.

Zarling, D. A., Watson, A., and Bach, F. H. (1980) Mapping of lymphocyte surface polypeptide antigens by chemical cross-linking with BSOCOES. *J. Immunol.* **124**, 913–920.

Zarling, D. A., Miskimen, J. A., Fan, D. P., Fujimoto, E. K., and Smith, P. K. (1982) Association of Sendai virion envelope and a mouse surface membrane polypeptide on newly infected cells: Lack of association with H-2K/D or alteration of viral immunogenicity. *J. Immunol.* **128**, 251–257.

Zecherle, G. N. (1990) Doctoral Dissertation. University of California at Davis.

Zeheb, R., Chang, V., and Orr, G. A. (1983) An analytical method for the selective retrieval of iminobiotin-derivatized plasma membrane proteins. *Anal. Biochem.* **129**, 156–161.

Zhao, H., and Heindel, N. D. (1991) Determination of degree of substitution of formyl groups in polyaldehyde dextran. *Pharm. Res.* **8**, 400–402.

Zlateva, T. P., Krysteva, M., Balajthy, Z., and Elodi, P. (1988) Properties of chymotrypsin bound covalently to dextran. *Acta Biochim. Biophys. Hung.* **23**, 225–230.

Zola, H., Neoh, S. H., Bantzioris, B. X., Webster, J., and Loughman, M. S. (1990) Detection by immunofluorescence of surface molecules present in low copy numbers. *J. Immunol. Methods* **135**, 247–255.

Zopf, D. A., Smith, D. F., Drzeniek, Z., Tsai, C.-M., and Ginsburg, V. (1978a) Affinity purification of antibodies using oligosaccharide–phenethylamine derivatives coupled to Sepharose. *in* "Methods in Enzymology" (V. Ginsburg, ed.), Vol. 50, pp. 171–175. Academic Press, New York.

Zopf, D. A., Tsai, C.-M., and Ginsburg, V. (1978b) Carbohydrate antigens: Coupling of oligosaccharide–phenethylamine derivatives to edestin by diazotization and characterization of antibody specificity by radioimmunoassay. *in* "Methods in Enzymology" (V. Ginsburg, ed.), Vol. 50, pp. 163–169. Academic Press, New York.

Zsigmondy, R. (1905) "Zur Erkenntnis der Kolloide." Jena, Germany.

Suppliers of Reagents and Devices for Bioconjugate Applications

Aldrich
PO Box 355
Milwaukee, WI 53201
General organic chemicals

Amersham
2636 South Clearbrook
  Drive
Arlington Heights, IL 60005
*Radiolabeled molecules,
  radioisotopes*

Amicon
24 Cherry Hill Drive
Danvers, MA 01923
*Microconcentrators,
  membranes*

Avanti Polar Lipids
5001 A Whitling Drive
Pelham, AL 35214
Lipids and lipid derivatives

Bio-Rad
3300 Regatta Boulevard
Richmond, CA 94804
Electrophoresis equipment

**Biotechnology Development
  Corporation**
Medicontrol Corporation
44 Mechanic Street
Newton, MA 02164
Microfluidizers

**Boehringer Mannheim
  Biochemicals**
PO Box 50414
Indianapolis, IN 46250
*Molecular biology reagents
  and enzymes*

**Branson Ultrasonics
  Corporation**
Eagle Road
Danbury, CT 06810
Sonicators

Brinkmann Instruments
Catiague Road
Westbury, NY 11590
Homogenizers

Calbiochem
PO Box 12087
San Diego, CA 92112
*Protein modification
  reagents*

E-Y Labs
107 North Amphlett
  Boulevard
San Mateo, CA 94401
Colloidal gold reagents

Genzyme
1 Kendall Square
Cambridge, MA 02139
Enzymes, lipids

Hoefer Scientific
PO Box 77387
San Francisco, CA 94107
*Electrophoresis and blotting
  equipment*

**Janssen Life Sciences
  Products**
40 Kingsbridge Road
Piscataway, NJ 08854
Colloidal gold reagents

J. T. Baker
222 Red School Lane
Phillipsburg, NJ 08865
*Organic solvents, buffer
  salts*

Kodak
343 State Street
Rochester, NY 14652
*Chemicals, fluorescent
  probes*

Liposome Technology, Inc.
1050 Hamilton Court
Menlo Park, CA 94025
*Liposome technology for
  therapeutic applications*

Matreya
500 Tressler Street
Pleasant Gap, PA 16823
Lipids

Molecular Probes
PO Box 22010
Eugene, OR 97402
*Fluorescent probes, protein
  modification reagents*

Nanoprobes, Inc.
25 East Loop Road, Suite
  124
Stony Brook, NY 11790
Colloidal gold derivatives

New England Nuclear
El du Pont de Nemours &
  Co.
NEN Products
331 Treble Cove Road
North Billerica, MA 01862
Radiochemicals

Pharmacia LKB
800 Centennial Avenue
Piscataway, NJ 00854
Chromatography supports,
electrophoresis equipment

Pierce Chemical Company
PO Box 117
Rockford, IL 61105
Cross-linking agents, protein
modification reagents,
fluorescent probes,
biotin/avidin systems,
chromatography supports,
kits for production of
various conjugates

Polysciences
400 Valley Road
Warrington, PA 18976
Polymers, reactive
monomers

ProChem, Inc.
826 Roosevelt Road
Rockford, IL 61109
Cross-linkers, protein
modification reagents

Promega
2800 South Fish Hatchery
Road
Madison, WI 53711
Molecular biology enzymes
and labeling kits

Sigma
PO Box 14508
St. Louis, MO 63178
General biochemicals

Waters
34 Maple Street
Milford, MA 01757
Automated chromatography
systems

Worthington Biochemical
Corp.
Freehold, NJ 07728
Enzymes

Index